Metals Handbook® Ninth Edition

Volume 12
Fractography

Bright-field image

Dark-field image

Secondary electron image

Backscattered electron image

Comparison of light microscope (top row) and scanning electron microscope (bottom row) fractographs showing the intergranular fracture appearance of an experimental nickel-base precipitation-hardenable alloy rising-load test specimen that was tested in pure water at 95 °C (200 °F). All shown at 50×. Courtesy of G.F. Vander Voort and J.W. Bowman, Carpenter Technology Corporation. Additional comparisons of fractographs obtained by light microscopy and scanning electron microscopy can be found in the article "Visual Examination and Light Microscopy" in this Volume.

Metals Handbook® Ninth Edition
Volume 12
Fractography

Prepared under the direction of
the ASM Handbook Committee

Kathleen Mills, Manager of Editorial Operations
Joseph R. Davis, Senior Technical Editor
James D. Destefani, Technical Editor
Deborah A. Dieterich, Production Editor
Heather J. Frissell, Editorial Supervisor
George M. Crankovic, Assistant Editor
Diane M. Jenkins, Word Processing Specialist

Robert L. Stedfeld, Director of Reference Publications

Donald F. Baxter Jr., Consulting Editor

Editorial Assistance
Robert T. Kiepura
Bonnie R. Sanders

METALS PARK, OHIO 44073

Copyright © 1987
by
ASM INTERNATIONAL
All rights reserved

No part of this book may be reproduced, stored in a retrieval system, or transmitted, in any form or by any means, electronic, mechanical, photocopying, recording, or otherwise, without the written permission of the copyright owner.

First printing, March 1987

Metals Handbook is a collective effort involving thousands of technical specialists. It brings together in one book a wealth of information from world-wide sources to help scientists, engineers, and technicians solve current and long-range problems.

Great care is taken in the compilation and production of this volume, but it should be made clear that no warranties, express or implied, are given in connection with the accuracy or completeness of this publication, and no responsibility can be taken for any claims that may arise.

Nothing contained in the Metals Handbook shall be construed as a grant of any right of manufacture, sale, use, or reproduction, in connection with any method, process, apparatus, product, composition, or system, whether or not covered by letters patent, copyright, or trademark, and nothing contained in the Metals Handbook shall be construed as a defense against any alleged infringement of letters patent, copyright, or trademark, or as a defense against liability for such infringement.

Comments, criticisms, and suggestions are invited, and should be forwarded to ASM International.

Library of Congress Cataloging in Publication Data

ASM International

Metals handbook.

Includes bibliographies and indexes.
Contents: v. 1. Properties and selection—[etc.]—
v. 9. Metallography and Microstructures—[etc.]—
v. 12. Fractography.

1. Metals—Handbooks, manuals, etc. I. ASM
International. Handbook Committee.
TA459.M43   1978   669   78-14934
ISBN 0-87170-007-7 (v. 1)
SAN 204-7586

Printed in the United States of America

Foreword

Volume 12 of the 9th Edition of *Metals Handbook* is the culmination of 43 years of commitment on the part of ASM to the science of fracture studies. It was at the 26th Annual Convention of the Society in October of 1944 that the term "fractography" was first introduced by Carl A. Zapffe, the foremost advocate and practitioner of early microfractography. Since then, the usefulness and importance of this tool have gained wide recognition.

This Handbook encompasses every significant element of the discipline of fractography. Such depth and scope of coverage is achieved through a collection of definitive articles on all aspects of fractographic technique and interpretation. In addition, an Atlas of Fractographs containing 1343 illustrations is included. The product of several years of careful planning and preparation, the Atlas supplements the general articles and provides Handbook readers with an extensive compilation of fractographs that are useful when trying to recognize and interpret fracture phenomena of industrial alloys and engineered materials.

The successful completion of this project is a tribute to the collective talents and hard work of the authors, reviewers, contributors of fractographs, and editorial staff. Special thanks are also due to the ASM Handbook Committee, whose members are responsible for the overall planning of each volume in the Handbook series. To all these men and women, we express our sincere gratitude.

Raymond F. Decker
President,
ASM International

Edward L. Langer
Managing Director,
ASM International

**The Ninth Edition of Metals Handbook
is dedicated to the memory of
TAYLOR LYMAN, A.B. (Eng.), S.M., Ph.D.
(1917–1973)
Editor, Metals Handbook, 1945–1973**

Preface

The subject of fractography was first addressed in a *Metals Handbook* volume in 1974. Volume 9 of the 8th Edition, *Fractography and Atlas of Fractographs*, provided systematic and comprehensive treatment of what was at that time a relatively new body of knowledge derived from examination and interpretation of features observed on the fracture surfaces of metals. The 8th Edition volume also documented the resurgence of engineering and scientific interest in fracture studies, which was due largely to the development and widespread use of the transmission electron microscope and the scanning electron microscope during the 1960s and early '70s.

During the past 10 to 15 years, the science of fractography has continued to mature. With improved methods for specimen preparation, advances in photographic techniques and equipment, the continued refinement and increasing utility of the scanning electron microscope, and the introduction of quantitative fractography, a wealth of new information regarding the basic mechanisms of fracture and the response of materials to various environments has been introduced. This new volume presents in-depth coverage of the latest developments in fracture studies.

Like its 8th Edition predecessor, this Handbook is divided into two major sections. The first consists of nine articles that present over 600 photographic illustrations of fracture surfaces and related microstructural features. The introductory article provides an overview of the history of fractography and discusses the development and application of the electron microscope for fracture evaluation. The next article, "Modes of Fracture," describes the basic fracture modes as well as some of the mechanisms involved in the fracture process, discusses how the environment affects material behavior and fracture appearance, and lists material defects where fracture can initiate. Of particular interest in this article is the section "Effect of Environment on Fatigue Fracture," which reviews the effects of gaseous environments, liquid environments, vacuum, temperature, and loading on fracture morphology.

The following two articles contribute primarily to an understanding of proper techniques associated with fracture analysis. Care, handling, and cleaning of fractures, procedures for sectioning a fracture and opening secondary cracks, and the effect of nondestructive inspection on subsequent evaluation are reviewed in "Preparation and Preservation of Fracture Specimens." "Photography of Fractured Parts and Fracture Surfaces" provides extensive coverage of proper photographic techniques for examination of fracture surfaces by light microscopy, with the emphasis on photomacrography.

The value of fractography as a diagnostic tool in failure analyses involving fractures can be appreciated when reading "Visual Examination and Light Microscopy." Information on the application and limitations of the light microscope for fracture studies is presented. A unique feature of this article is the numerous comparisons of fractographs obtained by light microscopy with those obtained by scanning electron microscopy.

The next article describes the design and operation of the scanning electron microscope and reviews the application of the instrument to fractography. The large depth of field, the wide range of magnifications available, the simple nondestructive specimen preparation, and the three-dimensional appearance of SEM fractographs all contribute to the role of the scanning electron microscope as the principal tool for fracture studies.

Although the transmission electron microscope is used far less today for fracture work, it remains a valuable tool for specific applications involving fractures. These applications are discussed in the article "Transmission Electron Microscopy," along with the various techniques for replicating and shadowing a fracture surface. A point-by-point comparison of TEM and SEM fractographs is also included.

Quantitative geometrical methods to characterize the nonplanar surfaces encountered in fractures are reviewed in the articles "Quantitative Fractography" and "Fractal Analysis of Fracture Surfaces." Experimental techniques (such as stereoscopic imaging and photogrammetric methods), analytical procedures, and applications of quantitative fractography are examined.

An Atlas of Fractographs constitutes the second half of the Handbook. The 270-page Atlas, which incorporates 31 different alloy and engineered material categories, contains 1343 illustrations, of which 1088 are SEM, TEM, or light microscope fractographs. The remainder are photographs, macrographs, micrographs, elemental dot patterns produced by scanning Auger electron spectroscopy or energy-dispersive x-ray analysis, and line drawings that serve primarily to augment the information in the fractographs. The introduction to the Atlas describes its organization and presentation. The introduction also includes three tables that delineate the distribution of the 1343 figures with respect to type of illustration, cause of fracture, and material category.

The Editors

Officers and Trustees of ASM International

Raymond F. Decker
President and Trustee
Universal Science Partners, Inc.

William G. Wood
Vice President and Trustee
Materials Technology

John W. Pridgeon
Immediate Past President and Trustee
John Pridgeon Consulting Company

Frank J. Waldeck
Treasurer
Lindberg Corporation

Trustees

Stephen M. Copley
University of Southern California

Herbert S. Kalish
Adamas Carbide Corporation

William P. Koster
Metcut Research Associates, Inc.

Robert E. Luetje
Kolene Corporation

Gunvant N. Maniar
Carpenter Technology Corporation

Larry A. Morris
Falconbridge Limited

Richard K. Pitler
Allegheny Ludlum Steel Corporation

C. Sheldon Roberts
Consultant
Materials and Processes

Klaus M. Zwilsky
National Materials Advisory Board
National Academy of Sciences

Edward L. Langer
Managing Director

Members of the ASM Handbook Committee (1986-1987)

Dennis D. Huffman
(Chairman 1986-; Member 1983-)
The Timken Company

Roger J. Austin (1984-)
Materials Engineering Consultant

Peter Beardmore (1986-)
Ford Motor Company

Deane I. Biehler (1984-)
Caterpillar Tractor Company

Robert D. Caligiuri (1986-)
SRI International

Richard S. Cremisio (1986-)
Rescorp International Inc.

Thomas A. Freitag (1985-)
The Aerospace Corporation

Charles David Himmelblau (1985-)
Lockheed Missiles & Space
Company, Inc.

John D. Hubbard (1984-)
HinderTec, Inc.

L.E. Roy Meade (1986-)
Lockheed-Georgia Company

Merrill I. Minges (1986-)
Air Force Wright Aeronautical
Laboratories

David V. Neff (1986-)
Metaullics Systems

David LeRoy Olson (1982-)
Colorado School of Mines

Paul E. Rempes (1986-)
Champion Spark Plug Company

Ronald J. Ries (1983-)
The Timken Company

E. Scala (1986-)
Cortland Cable Company, Inc.

David A. Thomas (1986-)
Lehigh University

Peter A. Tomblin (1985-)
De Havilland Aircraft of Canada Ltd.

Leonard A. Weston (1982-)
Lehigh Testing Laboratories, Inc.

Previous Chairmen of the ASM Handbook Committee

R.S. Archer
(1940-1942) (Member, 1937-1942)

L.B. Case
(1931-1933) (Member, 1927-1933)

T.D. Cooper
(1984-1986) (Member, 1981-1986)

E.O. Dixon
(1952-1954) (Member, 1947-1955)

R.L. Dowdell
(1938-1939) (Member, 1935-1939)

J.P. Gill
(1937) (Member, 1934-1937)

J.D. Graham
(1966-1968) (Member, 1961-1970)

J.F. Harper
(1923-1926) (Member, 1923-1926)

C.H. Herty, Jr.
(1934-1936) (Member, 1930-1936)

J.B. Johnson
(1948-1951) (Member, 1944-1951)

L.J. Korb
(1983) (Member, 1978-1983)

R.W.E. Leiter
(1962-1963) (Member, 1955-1958, 1960-1964)

G.V. Luerssen
(1943-1947) (Member, 1942-1947)

G.N. Maniar
(1979-1980) (Member, 1974-1980)

J.L. McCall
(1982) (Member, 1977-1982)

W.J. Merten
(1927-1930) (Member, 1923-1933)

N.E. Promisel
(1955-1961) (Member, 1954-1963)

G.J. Shubat
(1973-1975) (Member, 1966-1975)

W.A. Stadtler
(1969-1972) (Member, 1962-1972)

R. Ward
(1976-1978) (Member, 1972-1978)

M.G.H. Wells
(1981) (Member, 1976-1981)

D.J. Wright
(1964-1965) (Member, 1959-1967)

Policy on Units of Measure

By a resolution of its Board of Trustees, ASM International has adopted the practice of publishing data in both metric and customary U.S. units of measure. In preparing this Handbook, the editors have attempted to present data primarily in metric units based on Système International d'Unités (SI), with secondary mention of the corresponding values in customary U.S. units. The decision to use SI as the primary system of units was based on the aforementioned resolution of the Board of Trustees, the widespread use of metric units throughout the world, and the expectation that the use of metric units in the United States will increase substantially during the anticipated lifetime of this Handbook.

For the most part, numerical engineering data in the text and in tables are presented in SI-based units with the customary U.S. equivalents in parentheses (text) or adjoining columns (tables). For example, pressure, stress, and strength are shown both in SI units, which are pascals (Pa) with a suitable prefix, and in customary U.S. units, which are pounds per square inch (psi). To save space, large values of psi have been converted to kips per square inch (ksi), where 1 kip = 1000 lb. Some strictly scientific data are presented in SI units only. For example, fatigue crack growth rates have been given only in millimeters per cycle (mm/cycle).

To clarify some illustrations that depict machine parts described in the text, only one set of dimensions is presented on artwork. References in the accompanying text to dimensions in the illustrations are presented in both SI-based and customary U.S. units.

On graphs and charts, grids correspond to SI-based units, which appear along the left and bottom edges; where appropriate, corresponding customary U.S. units appear along the top and right edges. In some instances, only customary U.S. units have been used.

Data pertaining to a specification published by a specification-writing group may be given in only the units used in that specification or in dual units, depending on the nature of the data. For example, the typical yield strength of aluminum sheet made to a specification written in customary U.S. units would be presented in dual units, but the thickness specified in that specification might be presented only in inches.

Data obtained according to standardized test methods for which the standard recommends a particular system of units are presented in the units of that system. Wherever feasible, equivalent units are also presented.

Conversions and rounding have been done in accordance with ASTM Standard E 380, with careful attention to the number of significant digits in the original data. For example, an annealing temperature of 1570 °F contains three significant digits. In this instance, the equivalent temperature would be given as 855 °C; the exact conversion to 854.44 °C would not be appropriate. For an invariant physical phenomenon that occurs at a precise temperature (such as the melting of pure silver), it would be appropriate to report the temperature as 961.93 °C or 1763.5 °F. In many instances (especially in tables and data compilations), temperature values in °C and °F are alternatives rather than conversions.

The policy on units of measure in this Handbook contains several exceptions to strict conformance to ASTM E 380; in each instance, the exception has been made to improve the clarity of the Handbook. The most notable exception is the use of $MPa\sqrt{m}$ rather than $MN \cdot m^{-3/2}$ or $MPa \cdot m^{0.5}$ as the SI unit of measure for fracture toughness. Other examples of such exceptions are the use of "L" rather than "l" as the abbreviation for liter and the use of g/cm^3 rather than kg/m^3 as the unit of measure for density (mass per unit volume).

SI practice requires that only one virgule (diagonal) appear in units formed by combination of several basic units. Therefore, all of the units preceding the virgule are in the numerator and all units following the virgule are in the denominator of the expression; no parentheses are required to prevent ambiguity.

Authors and Reviewers

D.L. Bagnoli
  Mobil Research & Development Corporation
Kingshuk Banerji
  Georgia Institute of Technology
Bruce Boardman
  Deere & Company
R.D. Bucheit
  Battelle Columbus Laboratories
H. Burghard
  Southwest Research Institute
Theodore M. Clarke
  J.I. Case Company
E. Philip Dahlberg
  Metallurgical Consultants, Inc.
Barbra L. Gabriel
  Packer Engineering Associates, Inc.
J. Gurland
  Brown University
R.W. Hertzberg
  Lehigh University
Jan Hinsch
  E. Leitz, Inc.
Brian H. Kaye
  Laurentian University
Victor Kerlins
  McDonnell Douglas Astronautics Company
Campbell Laird
  University of Pennsylvania
Robert McCoy
  Youngstown State University
W.C. McCrone
  McCrone Research Institute
C.R. Morin
  Packer Engineering Associates, Inc.
Alex J. Morris
  Olin Corporation
J.C. Murza
  The Timken Company
D.E. Passoja
  Technical Consultant
R.M. Pelloux
  Massachusetts Institute of Technology
Austin Phillips
  Technical Consultant
Robert O. Ritchie
  University of California at Berkeley
Cyril Stanley Smith
  Technical Consultant
Ervin E. Underwood
  Georgia Institute of Technology
George F. Vander Voort
  Carpenter Technology Corporation
George R. Yoder
  Naval Research Laboratory
F.G. Yost
  Sandia National Laboratory
Richard D. Zipp
  J.I. Case Company

Contributors of Fractographs

R. Abrams
  Howmedica, Division of Pfizer Hospital Products Group, Inc.
C. Alstetter
  University of Illinois
C.-A. Baer
  California Polytechnic State University
R.K. Bhargava
  Xtek Inc.
H. Birnbaum
  University of Illinois
R.W. Bohl
  University of Illinois
W.L. Bradley
  Texas A&M University
E.V. Bravenec
  Anderson & Associates, Inc.
C.R. Brooks
  University of Tennessee
N. Brown
  University of Pennsylvania
C. Bryant
  De Havilland Aircraft Company of Canada Ltd.
D.A. Canonico
  C-E Power Systems Combustion Engineering Inc.
G.R. Caskey, Jr.
  Atomic Energy Division DuPont Company
S.-H. Chen
  Norton Christensen
A. Choudhury
  University of Tennessee
L. Clements
  San Jose State University
R.H. Dauskardt
  University of California
D.R. Diercks
  Argonne National Laboratory
S.L. Draper
  NASA Lewis Research Center
D.J. Duquette
  Rensselaer Polytechnic Institute
L.M. Eldoky
  University of Kansas
Z. Flanders
  Packer Engineering Associates Inc.
L. Fritzmeir
  Columbia University
M. Garshasb
  Syracuse University
D. Gaydosh
  NASA Lewis Research Center
E.P. George
  University of Pennsylvania
R. Goco
  California Polytechnic State University
G.M. Goodrich
  Taussig Associates Inc.
R.J. Gray
  Consultant
J.E. Hanafee
  Lawrence Livermore National Laboratory
S. Harding
  University of Texas
C.E. Hartbower
  Consultant
H.H. Honnegger
  California Polytechnic State University
G. Hopple
  Lockheed Missiles & Space Company, Inc.
T.E. Howson
  Columbia University
D. Huang
  Fuxin Mining Institute People's Republic of China
T.J. Hughel
  General Motors Research Laboratories
N.S. Jacobson
  NASA Lewis Research Center
W.L. Jensen
  Lockheed-Georgia Company
A. Johnson
  University of Louisville
J.R. Kattus
  Associated Metallurgical Consultants Inc.
J.R. Keiser
  Oak Ridge National Laboratory
C. Kim
  Naval Research Laboratory
H.W. Leavenworth, Jr.
  U.S. Bureau of Mines
P.R. Lee
  United Technologies
I. Le May
  Metallurgical Consulting Services Ltd.

R. Liu
University of Illinois
X. Lu
University of Pennsylvania
S.B. Luyckx
University of the Witwatersrand
South Africa
J.H. Maker
Associated Spring, Barnes Group Inc.
K. Marden
California Polytechnic State
University
H. Margolin
Polytechnic Institute of New York
D. Matejczyk
Columbia University
A.J. McEvily
University of Connecticut
C.J. McMahon, Jr.
University of Pennsylvania
E.A. Metzbower
Naval Research Laboratory
R.V. Miner
NASA Lewis Research Center
A.S. Moet
Case Western Reserve University
D.W. Moon
Naval Research Laboratory
M.J. Morgan
University of Pennsylvania
J.M. Morris
U.S. Department of Transportation
V.C. Nardonne
Columbia University
N. Narita
University of Illinois

F. Neub
University of Toronto
J.E. Nolan
Westinghouse Hanford Company
T. O'Donnell
California Institute of Technology
J. Okuno
California Institute of Technology
A.R. Olsen
Oak Ridge National Laboratory
D.W. Petrasek
NASA Lewis Research Center
D.P. Pope
University of Pennsylvania
B. Pourlaidian
University of Kansas
N. Pugh
University of Illinois
R.E. Ricker
University of Notre Dame
J.M. Rigsbee
University of Illinois
R.O. Ritchie
University of California at Berkeley
D. Roche
California Polytechnic State University
R. Ruiz
California Institute of Technology
J.A. Ruppen
University of Connecticut
E.A. Schwarzkopf
Columbia University
R.J. Schwinghamer
NASA Marshall Space Flight Center
H.R. Shetty
Zimmer Inc.

A. Shumka
California Institute of Technology
J.L. Smialek
NASA Lewis Research Center
H.J. Snyder
Snyder Technical Laboratory
S.W. Stafford
University of Texas
J. Stefani
Columbia University
J.E. Stulga
Columbia University
F.W. Tatar
Factory Mutual Research Corporation
J.K. Tien
Columbia University
P. Tung
California Institute of Technology
R.V. Vijayaraghavan
Polytechnic Institute of New York
R.C. Voigt
University of Kansas
R.W. Vook
Syracuse University
P.W. Walling
Metcut Research Associates, Inc.
D.C. Wei
Kelsey-Hayes Company
A.D. Wilson
Lukens Steel Company
F.J. Worzala
University of Wisconsin
D.J. Wulpi
Consultant
R.D. Zipp
J.I. Case Company

Contents

History of Fractography 1
Modes of Fracture 12
Preparation and Preservation of Fracture Specimens 72
Photography of Fractured Parts and Fracture Surfaces 78
Visual Examination and Light Microscopy 91
Scanning Electron Microscopy 166
Transmission Electron Microscopy 179
Quantitative Fractography 193
Fractal Analysis of Fracture Surfaces 211
Atlas of Fractographs 216
    Pure Irons 219
    Gray Irons 225
    Ductile Irons 227
    Malleable Irons/White Irons 238
    Low-Carbon Steels 240
    Medium-Carbon Steels 253
    High-Carbon Steels 277
    AISI/SAE Alloy Steels 291
    ASTM/ASME Alloy Steels 345
    Austenitic Stainless Steels 351
    Martensitic Stainless Steels 366
    Precipitation-Hardening Stainless Steels 370
    Tool Steels 375
    Maraging Steels 383
    Iron-Base Superalloys 388
    Nickel-Base Superalloys 389
    Nickel Alloys 396
    Cobalt Alloys 398
    Copper Alloys 399
    Cast Aluminum Alloys 405
    Wrought Aluminum Alloys 414
    P/M Aluminum Alloys 440
    Titanium Alloys 441
    Miscellaneous Metals and Alloys 456
    Metal-Matrix Composites 465
    Cemented Carbides 470
    Ceramics 471
    Concrete and Asphalt 472
    Resin-Matrix Composites 474
    Polymers 479
    Electronic Materials 481
Metric Conversion Guide 489
Abbreviations and Symbols 492
Index 495

History of Fractography

FRACTOGRAPHY is the term coined by Carl A. Zapffe in 1944 following his discovery of a means for overcoming the difficulty of bringing the lens of a microscope sufficiently near the jagged surface of a fracture to disclose its details within individual grains (Ref 1). The purpose of fractography is to analyze the fracture features and to attempt to relate the topography of the fracture surface to the causes and/or basic mechanisms of fracture (Ref 2).

Etymologically, the word fractography is similar in origin to the word metallography; fracto stems from the Latin *fractus*, meaning fracture, and graphy derives from the Greek term *grapho*, meaning descriptive treatment. Alternate terms used to describe the study of fracture surfaces include fractology, which was proposed in 1951 (Ref 3). Further diversification brought such terms as macrofractography and microfractography for distinguishing the visual and low magnification ($\leq 25 \times$) from the microscopic, and optical fractography and electron fractography for distinguishing between studies conducted using the light (optical) microscope and electron microscope.

This article will review the historical development of fractography, from the early studies of fracture appearance dating back to the sixteenth century to the current state-of-the-art work in electron fractography and quantitative fractography. Additional information can be obtained from the cited references and from subsequent articles in this Volume.

Fracture Studies Before the Twentieth Century

Valuable information has long been known to exist in the fracture surfaces of metals, and through the years various approaches have been implemented to obtain and interpret this information (Ref 4). According to metallurgical historian Cyril Stanley Smith, fracture surfaces have been analyzed to some degree since the beginning of the Bronze Age (Ref 5). Early metalsmiths and artisans most likely observed specific fracture characteristics of metal tools and weapons and related them to variables in smelting or melting procedures.

Sixteenth to Eighteenth Centuries. The first specific written description of the use of fracture appearance to gage the quality of a metallurgical process was by Vannoccio Biringuccio in *De La Pirotechnia*, published in 1540 (Ref 6). He described the use of fracture appearance as a means of quality assurance for both ferrous and nonferrous (tin and copper-tin bronzes) alloys.

Another early authority was Lazarus Ercker, who discussed fracture tests in a 1574 publication (Ref 7). The quality of copper, for example, was determined by examining the fracture surface of an ingot that had been notched and then broken by a transverse blow. Brass was similarly tested. A gray fracture surface was found to be associated with subsequent cracking during working; this gray surface was the result of the use of a special variety of calamine, which caused lead contamination of the ingot. Brittle fractures of silver were traced to lead and tin contamination.

In 1627, Louis Savot described in greater detail the use of the fracture test as a method of quality control of copper-tin-bismuth cast bells (Ref 8). He recorded observations of grain size in fracture control samples as a guide for composition adjustments to resist impact fracture when the bells were struck. In the same year, Mathurin Jousse described a method of selecting high-quality grades of iron and steel, based on the appearance of fracture samples (Ref 9).

One of the most significant early contributions to the study of metal fractures was by de Réaumur (Ref 10), who published a book in 1722 that contained engravings illustrating both the macroscopic and microscopic appearance of fracture surfaces of iron and steel (although the microscope was invented *circa* 1600, at the time of de Réaumur it was necessary to sketch what one saw and then transfer the sketch to metal, wood, or stone by engraving). In this classical work, de Réaumur listed and illustrated seven classes of fracture appearance in iron and steel. These are described below and shown in Fig. 1:

- *Type I fracture*: Large, irregularly arranged, mirrorlike facets, indicating inferior metal (Fig. 1a and b)
- *Type II fracture*: More regular distribution and smaller facets, indicating a slightly improved metal (Fig. 1c to e)
- *Type III fracture*: Interposed areas of fibrous metal between facets (Fig. 1f to h)
- *Type IV fracture*: Fibrous metal, with very few reflecting facets (Fig. 1j)
- *Type V fracture*: Framelike area surrounding an entirely fibrous center (Fig. 1k and m)
- *Type VI fracture*: An unusual type, with a few small facets in a fibrous background (Fig 1n, p, and q)
- *Type VII fracture*: Characterized by a woody appearance (Fig. 1r)

A second plate from de Réaumur's book (Fig. 2) concerns the use of fracture surfaces in appraising the completeness of conversion of iron to steel by the then current process of cementation (carburization). In his meticulous reproduction of detail, he included phenomena still bothersome to metallurgists today, such as blistering, burning (overheating), brittle fracture, and woody fracture. Descriptions of the fractures characteristic of the various stages of conversion are given with Fig. 2. In summary, they are:

- Woody fractures characteristic of iron (Fig. 2a to c)
- Fractures characteristic of partly converted metal (Fig. 2d, f, and j)
- Fractures characteristic of steel (Fig. 2e and g)

Figure 2(h) shows a fracture that is typical of an iron that will convert easily to steel; Fig. 2(j), a fracture typical of an iron that will not convert to steel.

A third plate from de Réaumur (Fig. 3) displays his fracture studies at high magnifications and his fracture test. The minute platelets shown in Fig. 3(f), which may have been pearlite or cementite in some form, were recorded a full century and a half before the founding of metallography.

Encouraged by his studies of the changes of fracture surfaces accompanying the conversion of iron into steel, de Réaumur published in 1724 studies of cast metal fractures, including plates illustrating fractured antimony and lead ingots (Ref 11). The results of these studies are reviewed by Smith in Ref 5.

Information on the nature of fracture of copper-zinc alloys was published in 1725 by Geoffroy (Ref 12). In his studies, Geoffroy investigated the influence of the copper-to-zinc ratio on the appearance of the fracture surface and on grain size.

Somewhat later (1750), Gellert described the fracture characteristics of metals and semimetals (Ref 13), mentioning the use of a fracture test for distinguishing among steel, wrought iron, and cast iron. The test was also used to appraise the effects of carburization and heat

2 / History of Fractography

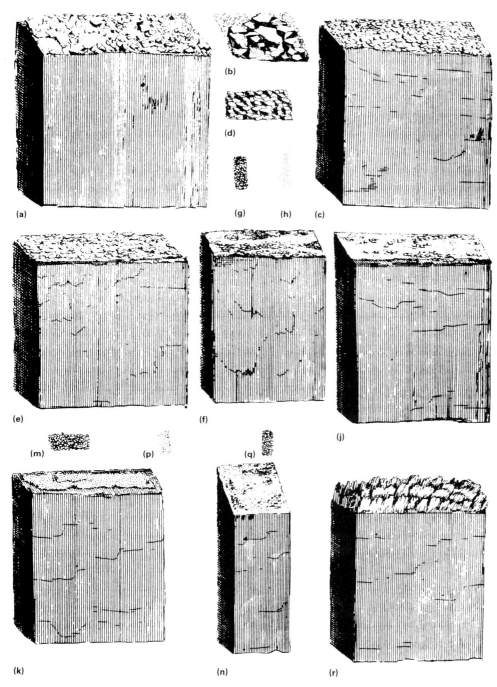

Fig. 1 Sketches from R.A.F. de Réaumur (Ref 10) depicting seven categories of fracture appearance in iron and steel. (a) Type I fracture; large mirrorlike facets. (b) Same as (a), but as viewed with a hand lens. (c) Type II fracture; smaller facets and more regular distribution. (d) Same as (c), but as viewed with a hand lens. (e) Another type II fracture, but improved in regularity of distribution and reduced facet size compared with (c). (f) Type III fracture; advantageous occurrence of interposed areas of fibrous metal between facets. (g) Detail of facets in (f). (h) Detail of fibrous metal in (f). (j) Type IV fracture; fibrous with very few reflecting facets. (k) Type V fracture; framelike area surrounding an entirely fibrous center. (m) Same as (k), but as viewed with a hand lens; a type VI fracture would look like this, except for finer grain size. (n) Type VI fracture; an unusual type, with tiny facets in a fibrous background. (p) Detail of fibrous area in (n). (q) Detail of small facets in (n). (r) Type VII fracture; woody appearance

treatment. Further, Gellert discussed causes of embrittlement of metals as disclosed by inspection of fracture surfaces.

The German physicist and chemist Karl Franz Achard, in carrying out his studies on the properties of alloys, also realized the importance of the appearance of fracture (Ref 14). Archard noted the appearance of the broken surfaces of nearly all of the 896 alloys he tested (Ref 5). This number represented virtually every possible combination of all metals known at that time.

Nineteenth Century. With the development of metallography as a metallurgical tool, interest in the further development of fracture studies waned. An important exception to this was Mallet (Ref 15), who published a paper in 1856 that related fracture details in cannon barrels to the mode of solidification, referring to planes of weakness resulting from sharp angles in the contours of the barrels. This may have been the first example of failure analysis and the first recognition of the deleterious effects of stress concentration in design. At the same time, the U.S. Army Ordnance Corps implemented fracture evaluation with mechanical testing for the study of ruptured cannon barrels (Ref 16).

In 1858, Tunner published a list categorizing fracture characteristics, citing such conditions as hot shortness, overheating, and various types of tears (Ref 17). In 1862, Kirkaldy correlated the change in fracture appearance from fibrous to crystalline with specimen configuration, heat treatment, and strain rate (Ref 18). He reported that crystalline fractures were at 90° to the tensile axis, whereas fibrous fractures were irregular and at angles other than 90°.

The doctoral dissertation of E.F. Dürre in 1868 contains an excellent description of the many different textures and details to be seen in the fracture of cast irons as well as a summary of the literature of the time (Ref 19). Dürre advocated the use of low magnification to study the fracture of castings, but considered the high-magnification microscope impractical for this purpose (Ref 5).

Two papers on steel, published by the Russian metallurgist D.K. Tschernoff, contributed significantly to fracture studies. The first, published in 1868, discussed fracture grain size in relation to heat treatment and carbon content (Ref 20). In a later paper, Tschernoff described the fracture of large-grain steel and, for the first time in history, accurately illustrated the true shape of metal grains (Ref 21).

John Percy, a prolific author on metallurgical subjects, described by 1875 six general types of fracture patterns (Ref 22):

- *Crystalline*, with facets as in zinc, antimony, bismuth, and spiegeleisen
- *Granular*, with smaller facets, as in pig iron
- *Fibrous*, a general criterion of quality
- *Silky*, a finer variety of fibrous, such as in copper
- *Columnar*, typical of high-temperature fracture
- *Vitreous*, or glasslike

Adolf Martens (for whom martensite is named) undertook studies of metal structure by examining newly fractured surfaces and polished-and-etched sections, both under the microscope. He published his first findings in Germany in 1878 (Ref 23, 24). His illustrations were hand engravings that reproduced meticu-

History of Fractography / 3

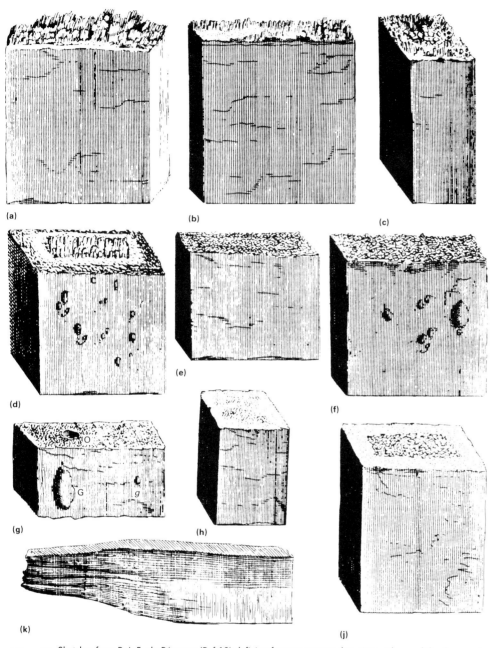

Fig. 2 Sketches from R.A.F. de Réaumur (Ref 10) defining fracture aspects that give evidence of the degree of conversion of iron to steel by the cementation process. (a) Woody fracture, but without the distinctly clustered appearance of the fracture in Fig. 1(r). (b) Woody fracture combined with minutely granular areas. (c) Fracture exhibiting a combination of brittle facets, woody texture, and minutely granular areas. These fractures are typical of iron. (d) Fracture in iron bar partly converted to steel, the outer minutely granular zone giving way to an inner framework of brittle facets, which in turn surround the woody center. (e) Fracture in steel produced from iron by cementation, showing a mass of small facets throughout the fracture, those in the center being somewhat larger. (f) Fracture in iron bar converted to steel only from *a-a* to *b-b* and remaining as iron from *a-a* to *c-c* because of overheating in the furnace. (g) Fracture in steel; the lusterless, rough facets resulted from holding the specimen too long in the furnace. (h) Fracture in a type of iron that always produces very small facets when converted to steel. (j) A type II fracture in iron (see Fig. 1), indicating that the iron will fail to convert to steel in the center and will also produce an inferior fracture frame. (k) Fracture in forged steel showing folding at left end, which could later cause cracking during heat treatment. In (d), (f), and (g), small blisters are indicated by the letter *g*, large blisters by the letter G, and porosity by the letter O.

lous pencil drawings in some figures and photomicrographs in others.

A plate from a later article (1887) by Martens is shown in part in Fig. 4. This plate consists of fractographs produced by photography and then printed by a photogravure process. The fractures shown in Fig. 4 illustrate features that Martens termed Bruchlinien (fracture lines), which today would be called radial marks. The description of this fracture form by Martens predates reports by other investigators who are usually credited for first treatment of this fracture feature.

In the field of macrofractography, Martens observed the fracture surfaces obtained in tension, torsion, bending, and fatigue. In describing the topography of these surfaces, he differentiated between coarse radial shear elements and fine radial marks. He recognized that sharper radial marks occurred in fine-grain material and that all radial marks diverge from the fracture origin; that is, they point backward to the origin. More detailed information on the contributions of Martens to metallurgy can be found in Ref 5.

Another important paper on fracture was written by Johann Augustus Brinell (the inventor of the Brinell hardness test) in 1885 (Ref 26). Brinell discussed the influence of heat treating and the resulting change in the state of carbon on the appearance of steel fractures. Henry Marion Howe critically analyzed and subsequently praised Brinell's findings, stating that the latter's work represented "the most important fracture studies ever made" (Ref 27).

Classic cup-and-cone tensile fracture was described by B. Kirsh before 1889 (Ref 28). He postulated a concept of crack propagation in tensile specimens that is retained today. He theorized that the crack origin was at the tensile axis in the necked region, that the origin grew concentrically in a transverse direction to produce the "bottom of the cup," and that the sides of the cone were formed by a maximum shear stress at final separation.

However, because of an overriding interest in metallography, many noted authorities in metallurgy, such as French scientist and metallographer Floris Osmond, dismissed microfractography as leading "to nothing either correct or useful." Microfractography thus became a forgotten art until well into the twentieth century, with nothing of the earlier techniques and findings being taught or acknowledged in the universities.

Development of Microfractography

Most of the microscopical studies of metals in the early 1900s were limited to examinations of polished specimens. In the 1930s, a number of investigators recognized that the properties of steels could be correlated with the macroscopic coarseness or fineness of the fracture surface. For example, Arpi developed a set of standard fracture tests (the Jernkontoret fracture tests) that was believed to cover the entire grain size range (Ref 29, 30). Similarly, Shepherd developed a set of standards for evaluating grain size in hardened tool and die steels (Ref 31-33). His method remains in limited use.

However, it was not until the work of Zapffe and his co-workers (Ref 1, 3, 34-50) in the decade 1940 to 1950 that significant, detailed studies of the microscopic elements of fractures

4 / History of Fractography

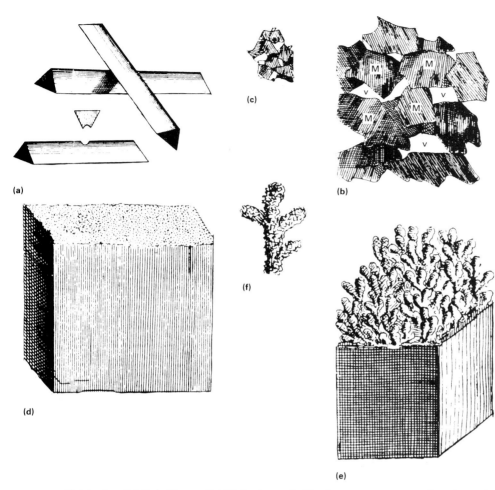

Fig. 3 Sketches from R.A.F. de Réaumur (Ref 10) showing notched-bar fracture test specimen, enlarged view of grains, and details of fracture in cast iron. (a) A notched-bar fracture test, which allows two identical steel bars to be broken with a single blow. (b) Grain "... prodigiously magnified [i.e., of the order of 50×] showing the voids V and molecules M ..." (probably crystallites) of which the grain is composed. (c) One of the individual molecules of (b) showing the units of which it is composed. (d) Fracture in gray iron, strongly resembling one in steel, except that the surface is brownish and the grains are coarser. (e) Detail of (d) at high magnification showing the fracture to comprise "... an infinity of branches ..." Here de Réaumur's magnification clearly exceeded 100×, although some of the enlargement may have been contributed to drawing what he saw. (f) Detail of an individual branch (dendrite) showing structure of minute platelets (possibly pearlite or cementite) placed one on another

were brought to the attention of the scientific community. Zapffe's work on application of the light microscope to fractography was regarded by many as definitive (Ref 51).

Although bothered by the relatively small depth of focus of the light microscope, Zapffe and his co-workers were able to orient the facets of a fracture relative to the axis of the microscope so that examinations could be made at relatively high magnifications. (Zapffe routinely took fractographs with magnifications as high as 1500 to 2000×.) Most of Zapffe's work was done on brittle fractures in ingot iron and steels (notably welded ship plate, Ref 48), bismuth (Ref 35), zinc (Ref 37), antimony (Ref 42), molybdenum (Ref 44), and tungsten (Ref 45), from which he described in considerable detail the appearance and crystallography of cleavage facets. Figures 5(a) and (b) are examples of Zapffe's early work.

In the case of cleavage, Zapffe made a detailed study of the relationship among crystallographic orientation, structure, and the characteristics of the fracture surface, particularly in iron-silicon (Ref 43) and iron-chromium alloys (Ref 38, 43, and 49). Zapffe and Clogg also described the modifications to the appearance of the fracture surface when a second phase is present (Ref 1). In the case of tungsten, Zapffe and Landgraf observed, depending on the composition, pure cleavage fracture or a mixed fracture mode consisting of cleavage and intergranular fracture (Ref 45). They succeeded in photographing the intergranular zones at high magnification, the features of which are analogous to those observed in electron fractography (Ref 51).

Metallic cleavage has subsequently been the subject of a large number of optical fractography studies. Two papers were of particular importance. One was by Tipper and Sullivan (Ref 52) on the relationship between cleavage and mechanical twinning in iron-silicon alloys. The other was by Klier (Ref 53), who in 1951 used x-ray diffraction in addition to the light microscope in his work on cleavage in ferrite. In a written discussion following Klier's paper, Zapffe wrote, "Dr. Klier's photographs are splendid from both a photographic and a technical standpoint. He has in addition brought the important tool of X-ray diffraction to bear upon the problem, also the electron microscope."

According to Henry and Plateau (Ref 51), Zapffe and his colleagues were also the first to observe striations on fatigue fracture surfaces (Ref 50). In describing the striations observed in an aluminum alloy 75S-T6 (equivalent to present-day 7075) Zapffe wrote, "This fine lamellar structure seems clearly to be a fatigue phenomenon, suggesting a stage of minute structural ordering advanced beyond the grosser platy structure, and apparently favored by an increasing number of stress cycles. The lamellae are approximately parallel to the platy markings; and both sets of markings lie approximately in the bending plane perpendicular to the stress motion, as one would expect for a structural rearrangement due to this type of flexion." Figure 6 shows one of the fractographs included in Zapffe and Worden's 1951 paper "Fractographic Registrations of Fatigue" (Ref 50).

Although the above-mentioned studies with the light microscope were of tremendous value, it must be pointed out that they were mainly limited to cases in which the fractures consisted of relatively large flat facets, ideal subjects for optical fractography. Consequently, detailed studies of ductile fracture morphologies were not made possible until the advent of electron fractography. For additional information on the applications and limitations of the light microscope for fracture studies, see the article "Visual Examination and Light Microscopy" in this Volume.

Electron Fractography

The development of both the transmission electron microscope and scanning electron microscope and their widespread use beginning in the 1960s provided vast amounts of new information regarding the micromechanisms of fracture processes and made fractography an indispensable tool in failure analysis. Among the advances in fracture studies using electron fractography are (Ref 4):

- The micromechanism of ductile fracture, that is, the initiation, growth, and coalescence of microvoids, has been confirmed, and correlations between void (dimple) size/shape and stress state and material cleanliness have been developed
- New models to explain the mechanisms of fatigue fracture have evolved, and correla-

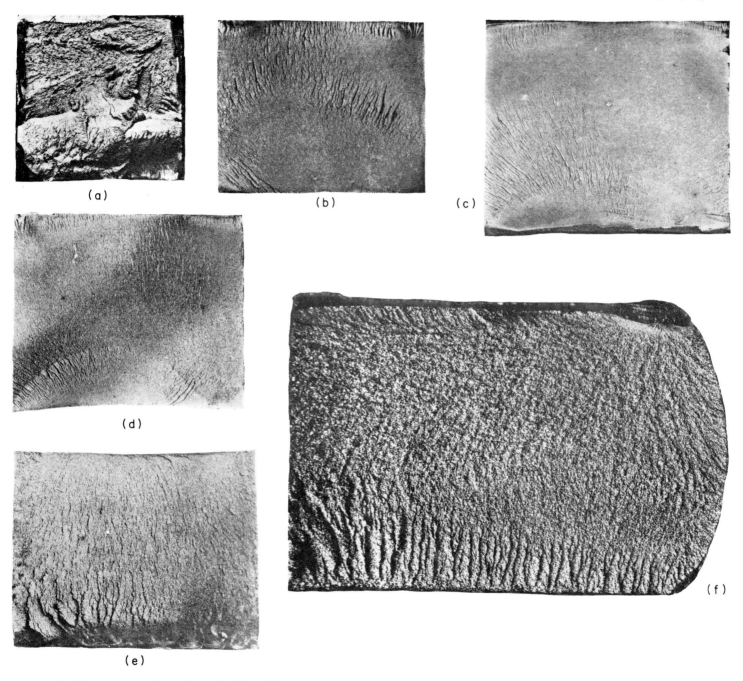

Fig. 4 Steel fractures recorded by A. Martens (Ref 25, p 237, Plate X). Martens called attention to the radial fracture marks in these fractographs, terming them Bruchlinien (fracture lines). (a) Ingot steel; tensile strength, 765 MPa (111 ksi). (b) through (e) Tool steels from Böhler Bros. of Vienna and others: (b) extra hard; (c) very hard, special; (d) moderately hard; (e) ductile. (f) Chisel steel with fracture lines. (a) through (e) Actual size. (f) 6×.

tions between fatigue striations, load cycles, striation spacing, and loading conditions have been developed. Conclusive experimental evidence regarding initiation mechanisms of fatigue fracture has also been acquired from electron fractography studies

In brittle fracture, explanations have been offered for the cleavage patterns that occur on fracture surfaces, and the form of the patterns has been successfully used to determine fracture direction and initiation points

Historically, as will be described below, the transmission and scanning electron microscopes were both demonstrated in an experimental form in Germany between 1930 and 1940. However, the transmission electron microscope received priority in development. As a result, electron fractography studies were first carried out using the transmission electron microscope.

The Transmission Electron Microscope

Historical Development. The origins of the transmission electron microscope can be traced back to developments in electron optics during the 1920s and 1930s (Ref 54-57, 58). In 1926, after 15 years of intermittent work on the

Fig. 5 Cleavage fractures in room-temperature impact specimens examined by C.A. Zapffe. (a) Cast polycrystalline antimony (99.83Sb-0.04S-0.035As-0.035Pb-0.015Fe-0.01Cu) (b) Vacuum-arc-cast high-oxygen molybdenum

Fig. 6 Fatigue striations observed by Zapffe (Ref 50) in an aluminum alloy specimen tested in completely reversed bending, at a maximum stress of 172 MPa (25 ksi) at room temperature, to failure at 336 × 10^3 cycles

trajectory of electrons in magnetic fields, Busch published a paper in which he demonstrated that magnetic or electric fields possessing axial symmetry act as lenses for electrons or other charged particles (Ref 58).

In 1932, Ruska developed the magnetic lens and published the first account of a magnetic electron microscope (Ref 59). In the same year, Brüche and Johannson produced images of an emitting (heated) oxide cathode with an electron microscope system (Ref 56, 57). In 1934, Ruska described an improved instrument built specifically for achieving high resolution (Ref 60). There is some debate as to who developed the first electron microscope with a resolving power greater than that obtainable with the light microscope. Cosslett (Ref 57) credited Ruska (Ref 61), while Hillier (Ref 55) stated that Driest and Müller (Ref 61) adapted Ruska's 1934 microscope and were the first to achieve resolutions exceeding those of the conventional microscope.

The first practical instrument for general laboratory use was described by von Borries and Ruska in 1938 (Ref 62). In its early form, this instrument was capable of resolutions of 10 nm (Ref 55, 56). Meanwhile, Prebus and Hillier, working independently in Toronto, developed a magnetic electron microscope of equal capability (Ref 63). Within 5 years, commercial instruments were being produced by a number of manufacturers, and by 1950, transmission electron microscopes with resolutions of 2 to 1 nm were widely available.

Application to Fractography. Transmission electron microscopes were first used to study fractures of metals in the 1950s and this method of fracture examination remained the most extensively used until the late 1960s. Although the limitations of the light microscope, such as its limited depth of field and magnification range, were eliminated by the use of the transmission electron microscope, several new problems were created (Ref 2). The most significant of these were (1) the problems introduced by the necessity of preparing a replica (primarily the time and effort required to make good replicas and the possibilities of misinterpretation because of the introduction of artifacts into the images as a result of the replication process) and (2) the difficulties of interpretation (because the images produced were considerably different in appearance from those obtained optically).

With the commercial development of and subsequent improvements in the scanning electron microscope in the mid-1960s, the role of the transmission electron microscope and replicas changed dramatically. Today, direct replication is used in fractography for a few special problems, such as examining the surface of a large component without cutting it or examining fine striations produced by fatigue crack propagation. Nonetheless, from a historical viewpoint, fracture studies of replicated surfaces using the transmission electron microscope represent an important contribution to modern fractography.

It should be noted, however, that the transmission electron microscope remains a vital tool in the field of analytical electron microscopy and enables the simultaneous examination of microstructural features through high-resolution imaging and the acquisition of chemical and crystallographic information from submicron regions of the specimen. The principles, instrumentation, and applications of the analytical transmission electron microscope are extensively reviewed in Ref 64.

Specimens for transmission electron microscopy must be reasonably transparent to electrons, must have sufficient local variations

History of Fractography / 7

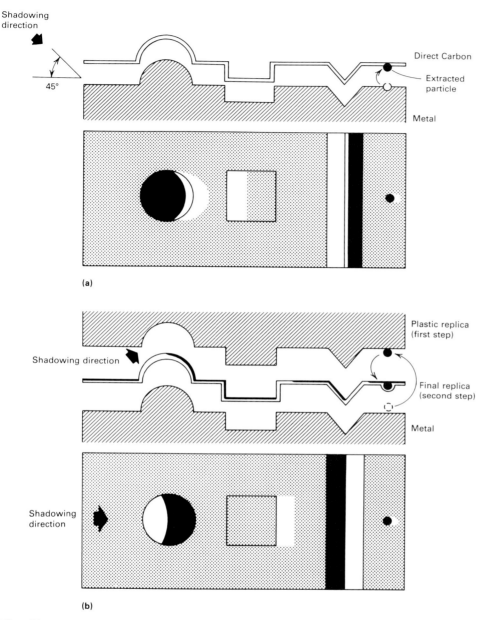

Fig. 7 Schematic of the two types of replicas. (a) One-step replica. (b) Two-step replica

in thickness, density, or both to provide adequate contrast in the image, and must be small enough to fit within the specimen-holder chamber of the transmission electron microscope. Transparency to electrons is provided by plastic or carbon replicas of the fracture surface. Fractures are usually too rough to permit electrolytic thinning.

The development of carbon replica techniques opened the way to significant progress in microfractography. As early as 1953, Robert *et al.* proposed the preparation of thin carbon films by the evaporation of graphite onto a glass plate coated with glycerine (Ref 51, 65). In 1954, Bradley prepared replicas by the volatization of carbon under vacuum onto first-stage plastic replicas (Ref 51, 66). Two years later, it was the method of direct evaporation of carbon onto the specimen developed by Smith and Nutting that, when adapted to the study of fractures, gave the best results (Ref 51, 67). A brief review of commonly used replicating techniques follows.

One-step replicas are the most accurate of the replicating techniques (Fig. 7a). The carbon film is directly evaporated onto the fracture surface and released by dissolving the base metal. The shadowing angle is usually not critical, because of the roughness of the surface (shadowing is discussed below). The continuity of the carbon film is ensured by the surface diffusion of the evaporated carbon. Direct carbon replicas can be extracted electrolytically or chemically.

In two-step replicas, the details of the fracture surface are transferred to a plastic mold, which is easy and convenient to dissolve in order to release the final carbon replica (Fig. 7b). The plastic mold can be obtained by applying successive layers of a varnish or Formvar or by simply pressing a softened piece of cellulose acetate to the fracture surface.

Preparation of the two-step replicas includes metal shadowing to enhance the contrast. The shadowing angle should coincide with the macroscopic direction of crack propagation to facilitate the orientation of the replica in the electron microscope. The shadowing angle and the direction of the carbon film is not critical. If it is available, rotary shadowing is recommended. Two-step replicas are chemically extracted, normally using acetone.

Shadowing. To increase the contrast and to give the replica a three-dimensional effect, a process known as shadowing is used. Shadowing is an operation by which a heavy metal is deposited at an oblique angle to the surface by evaporating it from an incandescent filament or an arc in a vacuum chamber. The vaporized metal atoms travel in essentially straight, parallel lines from the filament to strike the surface at an oblique angle. Upon contact, the metal condenses where it strikes, and certain favorably oriented surface features will receive a thicker metal deposit than others.

In the direct carbon method, the fracture surface itself is shadowed. In the two-step plastic-carbon technique, the plastic replica is usually shadowed before carbon deposition. Whether the direct carbon or the two-stage technique is used, it is recommended that the replica or fracture surface be oriented, if possible, such that the shadow direction relates to the macroscopic fracture direction.

Excellent reviews on replicating and shadowing techniques and methods for replica extraction are available in Ref 51, 54, and 68. Additional information on the use of replicas can be found in the article "Transmission Electron Microscopy" in this Volume.

Important Literature. Although a complete survey of the published work on the use of the transmisson electron microscope in fractography is beyond the scope of this article, the outstanding contributions of Crussard, Plateau, and Henry (Ref 51, 69-72), Beachem (Ref 73-91), and Pelloux (Ref 91-95) merit special mention. These sources provide hundreds of excellent fractographs of ductile and brittle fracture modes that were obtained from replicas. In addition, an extensive bibliography on electron fractography, which covers the late 1950s to mid-1960s, is available in Ref 68.

The Scanning Electron Microscope

Historical Development. The development of the scanning electron microscope can be traced back to the work of Knoll in 1935 during his studies of secondary electron emission from surfaces (Ref 96, 97). In 1938, a scanning electron microscope suitable for transparent specimens was built by von Ardenne

(Ref 98, 99). In 1942, Zworykin, Hillier, and Snyder gave an account of a scanning electron microscope that was more closely related to present-day instruments (Ref 100). This microscope was hindered by and later abandoned because of its unsatisfactory signal-to-noise ratio.

During the 1950s, developmental work progressed rapidly and concurrently in France and England. In the decade 1940 to 1950, a scanning electron microscope was constructed in France by Léauté and Brachet (Ref 96). The theory of the scanning electron microscope was proposed by Brachet in 1946 (Ref 101), who predicted that a resolution of 10 nm should be attainable if a noise-free electron detector could be used. Later, French workers under the direction of Bernard and Davoine (Ref 102, 103) built improved instruments and studied mechanically strained metals by the secondary electron method (Ref 104).

In 1948 at the Cambridge University Engineering Department in England, C.W. Oatley became interested in the scanning electron microscope, and a series of Ph.D. projects was initiated that resulted in the most significant contributions to the modern scanning electron microscope. An excellent review of the work conducted at Cambridge from 1948 to 1968 was provided by W.C. Nixon (Ref 105), who co-supervised research on the scanning electron microscope with Oatley beginning in 1959. With the advent of improved electron detectors developed by Everhart and Thornley in 1960 (Ref 106) and the improved instruments made by Crewe in 1963, which utilized field-emission electron guns (Ref 107), the scanning electron microscope had reached the point at which a commercial version seemed viable. The first commercial scanning electron microscope (the Stereoscan) was announced by Steward and Snelling in 1965 (Ref 108).

Since the development of the Stereoscan, significant changes have taken place in these instruments, including improvements in resolution, dependability, and ease of operation, as well as reductions in size. The cost of the instrument in constant dollars has fallen dramatically, and today it is quite common to have a scanning electron microscope in a laboratory where fracture studies and failure analyses are performed. The modern scanning electron microscope provides two outstanding improvements over the light microscope: it extends the resolution limits so that picture magnification can be increased from 1000 to 2000× (maximum useful magnification for the light microscope) up to 30 000 to 60 000×, and it improves the depth-of-field resolution from 100 to 200 nm for the light microscope to 4 to 5 nm (Ref 109).

Application to Fractography. The first paper to discuss the use of the scanning electron microscope for the study of fracture surfaces was published in 1959 by Tipper, Dagg, and Wells (Ref 110). Cleavage fractures in α-iron specimens were shown. Two years later, Laird and Smith used the scanning electron microscope to show that fatigue striations occur at the beginning of fracture in a high stress failure; this was not apparent when using optical fractography (Ref 111). Soon afterward, McGrath *et al.* used the scanning electron microscope to study fracture surfaces of copper tested in fatigue and 24S-T aluminum alloy (equivalent to present-day 2024) tested in fatigue and impact (Ref 112). The fractographs in this report approached the quality of those published today.

However, because of the slow commercial development of the instrument and the popularity of the transmission electron microscope and associated fracture replication techniques, the potential of the scanning electron microscope for fracture studies was not realized until the early 1970s (Ref 2, 113, 114). Today, fractography is one of the most popular applications of the scanning electron microscope. The large depth of focus, the possibility of changing magnification over a wide range, very simple nondestructive specimen preparation with direct inspection, and the three-dimensional appearance of scanning electron microscope fractographs make the instrument a vital and essential tool for fracture research.

Additional information on the scanning electron microscope and its application to fractography can be found in the article "Scanning Electron Microscopy" in this Volume. An extensive review of the principles, instrumentation, and applications of the scanning electron microscope is available in Ref 109.

Important Literature. Over the past 15 years, hundreds of papers have been published featuring scanning electron microscope fractographs. Of particular note are several Handbooks and Atlases, which illustrate the utility of the instrument for fracture studies.

In August of 1974, the American Society for Metals published Volume 9, *Fractography and Atlas of Fractographs*, of the 8th Edition of *Metals Handbook*. This was the first extensive collection of scanning electron microscope fractographs ever published.

From 15 October 1973 to 15 June 1975, engineers at McDonnell Douglas Astronautics Company prepared the *SEM/TEM Fractography Handbook*, which was subsequently published in December of 1975 (Ref 115). Unique to this Volume were the numerous comparisons of scanning electron fractographs with transmission electron fractographs obtained from replicas.

From 1969 to 1972, funded research performed at IIT Research Institute under the direction of Om Johari resulted in the *IITRI Fracture Handbook*, published in January of 1979 (Ref 116). Hundreds of fractographs of ferrous materials, aluminum-base alloys, nickel-base alloys, and titanium-base alloys were shown.

In 1981, Engel and Klingele published *An Atlas of Metal Damage* (Ref 117). This Atlas illustrates fracture surfaces as well as surfaces damaged due to wear, chemical attack, melting of metals or glasses, or high-temperature gases.

Quantitative Fractography (Ref 118)

The availability of the scanning electron microscope opened up new avenues toward the understanding of fracture surfaces in three dimensions and the subsequent interest in quantitative fractography. The goal of quantitative fractography is to express the features and important characteristics of a fracture surface in terms of the true surface areas, lengths, sizes, numbers, shapes, orientations, and locations, as well as distributions of these quantities. With an enhanced capability for quantifying the various features of a fracture, engineers can perform better failure analyses, can better determine the relationship of the fracture mode to the microstructure, and can develop new materials and evaluate their response to mechanical, chemical, and thermal environments.

Detailed descriptions of the historical development of quantitative fractography and associated quantification techniques can be found in the articles "Quantitative Fractography" and "Fractal Analysis of Fracture Surfaces" in this Volume. Supplementary information can be found in the article "Scanning Electron Microscopy."

ACKNOWLEDGMENT

ASM wishes to express its appreciation to the following individuals for their assistance in compiling the historical data used in this article: G.F. Vander Voort, Carpenter Technology Corporation; C.S. Smith, Massachusetts Institute of Technology; R.O. Ritchie, University of California at Berkeley; C. Laird, University of Pennsylvania; J. Gurland, Brown University; R.T. Kiepura, American Society for Metals.

REFERENCES

1. C.A. Zapffe and M. Clogg, Jr., *Fractography—A New Tool for Metallurgical Research*, Preprint 36, American Society for Metals, 1944; later published in *Trans. ASM*, Vol 34, 1945, p 71-107
2. J.L. McCall, "Failure Analysis by Scanning Electron Microscopy," MCIC Report, Metals and Ceramics Information Center, Dec 1972
3. C.A. Zapffe and C.O. Worden, Temperature and Stress Rate Affect Fractology of Ferrite Stainless, *Iron Age*, Vol 167 (No. 26), 1951, p 65-69
4. J.L. McCall, Electron Fractography—Tools and Techniques, in *Electron Fractography*, STP 436, American Society for Testing and Materials, 1968, p 3-16
5. C.S. Smith, *A History of Metallography*, The University of Chicago Press, 1960, p 97-127
6. V. Biringuccio, *De La Pirotechnia*, 1540; see translation by M.T. Gnudi and C.S. Smith, American Institute of Mining and Metallurgical Engineers, 1942

7. L. Ercker, *Beschreibung Allerfürnemisten Mineralischen Ertzt und Berckwercksarten*, 1st ed., G. Schwartz, 1574; see translation of 2nd ed. by A.G. Sisco and C.S. Smith, *Lazarus Ercker's Treatise on Ores and Assaying*, University of Chicago Press, 1951; see also E.V. Armstrong and H.S. Lukens, Lazarus Ercker and His "Probierbuch"; Sir John Pettus and His "Fleta Minor," *J. Chem. Educ.*, Vol 16, 1939, p 553-562
8. L. Savot, *Discours sur les Médailles Antiques*, 1627; see C.S. Smith, *A History of Metallography*, University of Chicago Press, 1960, p 99
9. M. Jousee, *Fidelle Ouverture de l'art de Serrurier*, La Fleche, 1627; see translation by C.S. Smith and A.G. Sisco, *Technol. Culture*, Vol 2 (No. 2), 1961, p 131-145
10. R.A.F. de Réaumur, *L'Art de Convertir le Fer Forgé en Acier, et L'Art d'Adoucir le Fer Fondu*, (The Art of Converting Wrought Iron to Steel and the Art of Softening Cast Iron), Michel Brunet, 1722; see translation by A.G. Sisco, *Réaumur's Memoirs on Steel and Iron*, University of Chicago Press, 1956
11. R.A.F. de Réaumur, De l'Arrangement que Prennent les Parties des Matières Métalliques et Minerales, Lorsqu'après Avoir été Mises en Fusion, Elles Viennent à se Figer, *Mém. Acad. Sci.*, 1724, p 307-316
12. C.F. Geoffroy, Observations sur un Métal que Résulte de L'alliage du Cuivre & du Zinc, *Mém. Acad. Sci.*, 1725, p 57-66
13. C.E. Gellert, *Anfangsgründe der Metallurgischen Chemie* (Elements of Metallurgical Chemistry), J. Wendler, 1750; see translation by J. Seiferth, *Metallurgic Chemistry*, T. Bechet, 1776
14. K.F. Achard, *Recherches sur les Propriétés des Alliages Métalliques*, 1788
15. R. Mallet, *Physical Conditions Involved in the Construction of Artillery*, 1856
16. Officers of Ordnance Dept., U.S. Army, "Reports on the Strength and Other Properties of Metals for Cannons," H. Baird, 1856
17. P. Tunner, *Das Eisenhüttenwesen im Schweden*, Engelhardt, 1858
18. D. Kirkaldy, *Results of an Experimental Inquiry into the Tensile Strength and Other Properties of Various Kinds of Wrought Iron and Steel*, 1862
19. E.F. Dürre, *Über die Constitution des Roheisens und der Werth seiner Physikalischen Eigenschaften*, 1868; preliminary version in *Berg- und Hüttenmannischen Zeitung* (1865 and 1868) and in *Zeitschrift für das Berg-Hütten- und Salienen-wesen*, Vol 16, 1868, p 70-131, 271-301
20. D.K. Tschernoff, Kriticheskii Obzor Statei gg. Lavrova y Kalakutzkago o Stali v Stalnikh Orudiakh' i Sobstvennie ego Izsledovanie po Etomuje Predmetu (Critical Review of Articles by Messrs. Lavrov and Kalakutzkii on Steel and Steel Ordnance, with Original Investigations on the Same Subject), in *Zapiski Russkago Tekhnicheskago Obshestva*, 1868, p 399-440; see English translation by W. Anderson of Tschernoff's original contribution (p 423-440 only), On the Manufacture of Steel and the Mode of Working It, *Proc. Instn. Mech. Engrs.*, 1880, p 286-307; see also French translation, *Rev. Univ. Mines*, Vol 7, 1880, p 129
21. D.K. Tschernoff, Izsledovanie, Otnosiashchiasia po Struktury Litikh Stalnykh Bolvanok (Investigations on the Structure of Cast Steel Ingots), in *Zapiski Imperatorskago Russkago Tekhnicheskago Obshestva*, 1879, p 1-24; see English translation by W. Anderson, *Proc. Instn. Mech. Engrs.*, 1880, p 152-183
22. J. Percy, *Metallurgy*, Vol 1 to 4, John Murray, 1861-1880
23. A. Martens, Über die Mikroskopische Untersuchung des Eisens, *Z. Deut. Ing.*, Vol 22, 1878, p 11-18
24. A. Martens, Zur Mikrostruktur des Spiegeleisens, *Z. Deut. Ing.*, Vol 22, 1878, p 205-274, 481-488
25. A. Martens, Ueber das Kleingefüge des Schmiedbaren Eisens, *Stahl Eisen*, Vol 7, 1887, p 235-242
26. J.A. Brinell, Über die Texturveränderungen des Stahls bei Erhitzung und bei Abkühlung, *Stahl Eisen*, Vol 11, 1885, p 611-620
27. H.M. Howe, *The Metallography of Steel and Cast Iron*, McGraw-Hill, 1916, p 527, and Table 29 on p 534-535. (Translation and condensation, with slight amendments, of original diagram from Ref 19, p 611)
28. B. Kirsh, Beiträge zum Studium des Fliessens, *Mitt Hlg*, 1887, p 67; 1888, p 37; 1889, p 9; see also G.C. Hennings, *Adolf Martens, Handbook of Testing Materials*, Part I, John Wiley & Sons, 1899, p 103, 105
29. R. Arpi, Report on Investigations Concerning the Fracture Test and the Swedish Standard Scale, *Jernkontorets Ann.*, Vol 86, 1931, p 75-95 (in Swedish); see abstract in *Stahl Eisen*, Vol 51, 1931, p 1483-1484
30. R. Arpi, The Fracture Test as Used for Tool Steel in Sweden, *Metallugia*, Vol 11, 1935, p 123
31. B.F. Shepherd, The P-F. Characteristics of Steel, *Trans. ASM*, Vol 22, 1934, p 979-1016
32. B.F. Shepherd, Carburization of Steel, *Trans. ASST*, Vol 4, Aug 1923, p 171-196
33. B.F. Shepherd, A Few Notes on the Shimer Hardening Process, *Trans. ASST*, Vol 5, May 1924, p 485-490
34. C.A. Zapffe and G.A. Moore, A Micrographic Study of the Cleavage of Hydrogenized Ferrite, *Trans. AIME*, Vol 154, 1943, p 335-359
35. C.A. Zapffe, Fractographic Structures in Bismuth, *Met. Prog.*, Vol 50, Aug 1946, p 283-286
36. C.A. Zapffe, "Neumann Bands and Planar-Pressure Theory of Hydrogen Embrittlement," Iron and Steel Institute, Aug 1946
37. C.A. Zapffe, Fractographic Structures in Zinc, *Met. Prog.*, Vol 51, March 1947, p 428-431
38. C.A. Zapffe, Étude Fractographique des Alliages Fer-Chrome, *Rev. Met. (Paris)*, Vol 44 (No. 3 and 4), 1947, p 91-96; see also C.A. Zapffe and F.K. Landgraf, Tearline Patterns in Ferrochromium, *J. Appl. Phys.*, Vol 21 (No. 11), 1950, p 1197-1198
39. C.A. Zapffe, C.O. Worden, and F.K. Landgraf, Cleavage Patterns Disclose "Toughness" of Metals, *Science*, Vol 108 (No. 2808), 1948, p 440-441
40. C.A. Zapffe, F.K. Landgraf, and C.O. Worden, Transgranular Cleavage Facets in Cast Molybdenum, *Met. Prog.*, Vol 54 (No. 3), 1948, p 328-331
41. C.A. Zapffe, F.K. Landgraf, and C.O. Worden, History of Crystal Growth Revealed by Fractography, *Science*, Vol 107 (No. 2778), 1948, p 320-321
42. C.A. Zapffe, Fractographic Structures in Antimony, *Met. Prog.*, Vol 53, March 1948, p 377-381
43. C.A. Zapffe, F.K. Landgraf, and C.O. Worden, Fractography: The Study of Fractures at High Magnification, *Iron Age*, Vol 161, April 1948, p 76-82
44. C.A. Zapffe, F.K. Landgraf, and C.O. Worden, Fractographic Study of Cast Molybdenum, *Trans. AIME*, Vol 180, 1949, p 616-636
45. C.A. Zapffe and F.K. Landgraf, Fractographic Examination of Tungsten, *Trans. ASM*, Vol 41, 1949, p 396-418
46. C.A. Zapffe and C.O. Worden, Deformation Phenomena on a Cleavage Facet of Iron, *Met. Prog.*, Vol 55, March 1949, p 640-641
47. C.A. Zapffe and C.O. Worden, Fractographic Study of Deformation and Cleavage in Ingot Iron, Preprint 31, American Society for Metals, 1949; later published in *Trans. ASM*, Vol 42, 1950, p 577-602; discussion, p 602-603
48. C.A. Zapffe, F.K. Landgraf, and C.O. Worden, Fractographic Examination of Ship Plate, *Weld. J.*, Vol 28, March 1949, p 126s-135s
49. C.A. Zapffe and F.K. Landgraf, Tearline Patterns in Ferrochromium, *J. Appl. Phys.*, Vol 21, Nov 1950, p 1197-1198
50. C.A. Zapffe and C.O. Worden, Fractographic Registrations of Fatigue, Preprint 32, American Society for Metals, 1950;

later published in *Trans. ASM*, Vol 43, 1951, p 958-969; discussion, p 969
51. G. Henry and J. Plateau, *La Microfractographie*, Institute de Recherches de la Sidérurgie Francaise [1966]; see translation by B. Thomas with Preface by C. Crussard, *Éditions Métaux* [1967]
52. C.F. Tipper and A.M. Sullivan, Fracturing of Silicon-Ferrite Crystals, *Trans. ASM*, Vol 43, 1951, p 906-928; discussion, p 929-934
53. E.P. Klier, A Study of Cleavage Surfaces in Ferrite, *Trans. ASM*, Vol 43, 1951, p 935-953; discussion, p 953-957
54. The Transmission Electron Microscope and Its Application to Fractography, in *Fractography and Atlas of Fractographs*, Vol 9, 8th ed., *Metals Handbook*, American Society for Metals, 1974, p 54-63
55. V.K. Zworykin and J. Hillier, *Microscopy: Electron, Medical Physics*, Vol II, 1950, p 511-529
56. J. Hiller, Electron Microscope, in *Encyclopedia Britannica*, 1960
57. V.E. Cosslett, *Practical Electron Microscopy*, Academic Press, 1951, p 41-46
58. H. Busch, Calculation of Trajectory of Cathode Rays in Electromagnetic Fields of Axial Symmetry, *Ann. d. Physik*, Vol 81, 1926, p
59. E. Ruska and M. Knoll, The Electron Microscope, *Ztschr. f. Physik*, Vol 78, 1932, p 318
60. E. Ruska, Advance in Building and Performing of Magnetic Electron Microscope, *Ztschr. f. Physik*, Vol 87, 1934, p 580
61. E. Driest and H.O. Müller, Electron Micrographs of Chitin Substances, *Ztschr. f. wissensch. Mikr.*, Vol 52, 1935, p 53
62. B. von Borries and E. Ruska, Development and Present Efficiency of the Electron Microscope, *Wissensch. Veröfent. Siemens-Werke*, Vol 17, 1938, p 99
63. A. Prebus and J. Hillier, Construction of Magnetic Electron Microscope of High Resolving Power, *Can. J. Res.*, Vol A17, 1939, p 49
64. A.D. Romig, Jr. et al., Analytical Transmission Electron Microscopy, in *Materials Characterization*, Vol 10, 9th ed., *Metals Handbook*, American Society for Metals, 1986, p 429-489
65. L. Robert, J. Bussot, and J. Buzon, First International Congress for Electron Microscopy, *Rev. d'Optique*, 1953, p 528
66. D.E. Bradley, *Br. J. Appl. Phys.*, Vol 5, 1954, p 96
67. E. Smith and J. Nutting, *Br. J. Appl. Phys.*, Vol 7, 1956, p 214
68. A. Phillips, V. Kerlins, R.A. Rawe, and B.V. Whiteson, Ed., *Electron Fractography Handbook*, sponsored by Air Force Materials Laboratory, Air Force Wright Aeronautical Laboratories, Air Force Systems Command, published by Metals and Ceramics Information Center, Battelle Columbus Laboratories, March 1968 (limited quantities), June 1976 (unlimited distribution)
69. C. Crussard, R. Borione, J. Plateau, Y. Morillon, and F. Maratray, A Study of Impact Tests and the Mechanism of Brittle Fracture, *J. Iron Steel Inst.*, Vol 183, June 1956, p 146
70. C. Crussard and R. Tamhankar, High Temperature Deformation of Steels: A Study of Equicohesion, Activation Energies and Structural Modifications, *Trans. AIME*, Vol 212, 1958, p 718
71. C. Crussard, J. Plateau, R. Tamhankar, and D. Lajeunesse, *A Comparison of Ductile and Fatigue Fractures*, (Swampscott Conference, 1959), John Wiley & Sons, 1959
72. J. Plateau, G. Henry, and C. Crussard, Quelque Nouvelles Applications de la Microfractographie, *Rev. Metall.*, Vol 54 (No. 3), 1957
73. C.D. Beachem, "Characterizing Fractures by Electron Fractography, Part IV, The Slow Growth and Rapid Propagation of a Crack Through a Notched Type 410 Stainless Steel Wire Specimen," NRL Memorandum Report 1297, Naval Research Laboratory, April 1962
74. C.D. Beachem, "Gases in Steel (Electron Fractographic Examination of Ductile Rupture Tearing in a Notched Wire Specimen)," NRL Problem Report, Naval Research Laboratory, Aug 1962
75. C.D. Beachem, "An Electron Fractographic Study of the Mechanism of Ductile Rupture in Metals," NRL Report 5871, Naval Research Laboratory, 31 Dec 1962
76. C.D. Beachem, "Effect of Test Temperature Upon the Topography of Fracture Surfaces of AMS 6434 Sheet Steel Specimens," NRL Memorandum Report 1293, Naval Research Laboratory, March 1962
77. C.D. Beachem, "Characterizing Fractures by Electron Fractography, Part V, Several Fracture Modes and Failure Conditions in Four Steel Specimens," NRL Memorandum Report 1352, Naval Research Laboratory, Aug 1962
78. C.D. Beachem, Electron Fractographic Studies of Mechanical Fracture Processes in Metals, *J. Basic Eng. (Trans. ASME)*, 1964
79. C.D. Beachem and D.A. Meyn, "Illustrated Glossary of Fractographic Terms," NRL Memorandum Report 1547, Naval Research Laboratory, June 1964
80. C.D. Beachem, B.F. Brown, and A.J. Edwards, "Characterizing Fractures by Electron Fractography, Part XII, Illustrated Glossary, Section 1, Quasi-Cleavage," NRL Memorandum Report 1432, Naval Research Laboratory, June 1963
81. C.D. Beachem, An Electron Fractographic Study of the Influence of Plastic Strain Conditions Upon Ductile Rupture Processes in Metals, *Trans. ASM*, Vol 56 (No. 3), Sept 1963, p 318
82. C.D. Beachem, "The Interpretation of Electron Microscope Fractographs," NRL Report 6360, Naval Research Laboratory, 21 Jan 1966
83. C.D. Beachem, Electron Fractographic Studies of Mechanical Processes in Metals, *Trans. ASME*, Series D, Vol 87, 1965
84. C.D. Beachem and D.A. Meyn, "Fracture by Microscope Plastic Deformation Process," Paper 41, presented at the Seventieth Annual Meeting of ASTM, Boston, MA, American Society for Testing and Materials, 25-30 June 1967
85. C.D. Beachem, "The Formation of Cleavage Tongues in Iron," NRL Report, Naval Research Laboratory, Feb 1966
86. C.D. Beachem, "The Interpretation of Electron Microscope Fractographs," NRL Report No. 6360, Naval Research Laboratory, 21 Jan 1966
87. C.D. Beachem, Microscopic Fatigue Fracture Surface Features in 2024-T3 Aluminum and the Influence of Crack Propagation Angle Upon Their Formation, *Trans. ASM*, Vol 60, 1967, p 324
88. C.D. Beachem, "Origin of Tire Tracks," NRL Progress Report, Naval Research Laboratory, May 1966
89. C.D. Beachem, "The Usefulness of Fractography," Paper 46, presented at the Seventieth Annual Meeting of ASTM, Boston, MA, American Society for Testing and Materials, 25-30 June, 1967
90. C.D. Beachem, "The Crystallographic of Herringbone Fractures in an Iron-Chromium-Aluminum Alloy," NRL Report, Naval Research Laboratory, Jan 1967
91. C.D. Beachem and R.M.N. Pelloux, Electron Fractography—A Tool for the Study of Micromechanisms of Fracturing Processes, in *Fracture Toughness Testing and Its Applications*, STP 381, American Society for Testing and Materials, 1964, p 210-245
92. R.M.N. Pelloux, "Influence of Constituent Particles on Fatigue Crack Propagation in Aluminum Alloys," DI-82-0297, Boeing Scientific Research Laboratories, Sept 1963
93. R.M.N. Pelloux, "The Analysis of Fracture Surfaces by Electron Microscopy," DI-82-0169-RI, Boeing Scientific Research Laboratories, Dec 1963
94. R.M.N. Pelloux and J.C. McMillan, The Analysis of Fracture Surfaces by Electron Microscopy, in *Proceedings of the First International Conference of Fracture*, Vol 2, Japanese Society for Strength and Fracture of Materials, 1966
95. R.M.N. Pelloux, The Analysis of Fracture Surfaces by Electron Microscopy, *Met. Eng. Quart.*, Nov 1965, p 26-37

96. O.C. Wells, A. Boyde, E. Lifshin, and A. Rezanowich, *Scanning Electron Microscopy*, McGraw-Hill, 1974
97. M. Knoll, Static Potential and Secondary Emission of Bodies Under Electron Irradiation, *Z. Tech. Phys.*, Vol 11, 1935, p 467-475 (in German)
98. M. von Ardenne, The Scanning Electron Microscope: Practical Construction, *Z. Tech. Phys.*, Vol 19, 1938, p 407-416 (in German)
99. M. von Ardenne, The Scanning Electron Microscope: Theoretical Fundamentals, *Z. Tech. Phys.*, Vol 109, 1938, p 553-572
100. V.K. Zworykin, J. Hillier, and R.L. Snyder, A Scanning Electron Microscope, *ASTM Bull.*, No. 117, 1942, p 14-23
101. C. Brachet, Note on the Resolution of the Scanning Electron Microscope, *Bull. L'Assoc. Tech. Mar. et Aero.*, No. 45, 1946, p 369-378 (in French)
102. R. Bernard and F. Davione, The Scanning Electron Microscope, *Ann. L'Univ. de Lyon, Ser. 3, Sci., B*, Vol 10, 1957, p 78-86 (in French)
103. F. Davoine, Scanning Electron Microscopy, in *Proceedings of the Fourth International Conference on Electron Microscopy*, Springer Verlag, 1960, p 273-276 (in French)
104. F. Davoine, "Secondary Electron Emission of Metals Under Mechanical Strain," Ph.D. Dissertation, L'Universelle de Lyon, 1957 (in French)
105. W.C. Nixon, "Twenty Years of Scanning Electron Microscopy, 1948-1968, In the Engineering Department, Cambridge University, England," Paper presented at The Scanning Electron Microscope—the Instrument and Its Applications symposium, Chicago, IL, IIT Research Institute, April 1968
106. T.E. Everhart and R.F.M. Thornley, Wide-Band Detector for Micro-Microampere Low-Energy Electron Currents, *J. Sci. Instrum.*, Vol 37, 1960, p 246-248
107. A.V. Crewe, A New Kind of Scanning Microscope, *J. Microsc.*, Vol 2, 1963, p 369-371
108. A.D.G. Stewart and M.A. Snelling, "A New Scanning Electron Microscope," in Titlebach, 1965, p 55-56
109. J.D. Verhoeven, Scanning Electron Microscopy, in *Materials Characterization*, Vol 10, 9th ed., *Metals Handbook*, 1986, p 490-515
110. C.F. Tipper, D.I. Dagg, and O.C. Wells, Surface Fracture Markings on Alpha Iron Crystals, *J. Iron Steel Inst.*, Vol 193, Oct 1959, p 133-141
111. C. Laird and G.C. Smith, Crack Propagation in High Stress Fatigue, *Philos. Mag.*, Vol 7, 1962, p 847-857
112. J.T. McGrath, J.G. Buchanan, and R.C.A. Thurston, A Study of Fatigue and Impact Features With the Scanning Electron Microscope, *J. Inst. Met.*, Vol 91, 1962, p 34-39
113. T. Inoue, S. Matsuda, Y. Okamura, and K. Aoki, The Fracture of a Low Carbon Tempered Martensite, *Trans. Jpn. Inst. Met.*, Vol 11, Jan 1970, p 36-43
114. A. Rukwied and D.B. Ballard, Scanning Electron Microscope Fractography of Continuously Cast High Purity Copper After High Temperature Creep, *Met. Trans.*, Vol 3, Nov 1972, p 2999-3008
115. G.F. Pittinato, V. Kerlins, A. Phillips, and M.A. Russo, Ed., *SEM/TEM Fractography Handbook*, sponsored by Air Force Materials Laboratory, Air Force Wright Aeronautical Laboratories, Air Force Systems Command, Wright-Patterson Air Force Base, published by Metals and Ceramics Information Center, Battelle Columbus Laboratories, Dec 1975
116. S. Bhattacharya, V.E. Johnson, S. Agarwal, and M.A.H. Howes, Ed., *IITRI Fracture Handbook—Failure Analysis of Metallic Materials by Scanning Electron Microscopy*, Metals Research Division, IIT Research Institute, Jan 1979
117. L. Engel and H. Klingele, *An Atlas of Metal Damage*, S. Murray, Trans., Prentice Hall, 1981
118. E.E. Underwood, Quantitative Fractography, in *Applied Metallography*, G.F. Vander Voort, Ed., Van Nostrand Reinhold, 1986, to be published

Modes of Fracture

Victor Kerlins, McDonnell Douglas Astronautics Company
Austin Phillips, Metallurgical Consultant

METALS FAIL in many different ways and for different reasons. Determining the cause of failure is vital in preventing a recurrence. One of the most important sources of information relating to the cause of failure is the fracture surface itself. A fracture surface is a detailed record of the failure history of the part. It contains evidence of loading history, environmental effects, and material quality. The principal technique used to analyze this evidence is electron fractography. Fundamental to the application of this technique is an understanding of how metals fracture and how the environment affects the fracture process.

This article is divided into three major sections. The section "Fracture Modes" describes the basic fracture modes as well as some of the mechanisms involved in the fracture process. The section "Effect of Environment" discusses how the environment affects metal behavior and fracture appearance. The final section, "Discontinuities Leading to Fracture," discusses material flaws where fracture can initiate.

Fracture Modes

Fracture in engineering alloys can occur by a transgranular (through the grains) or an intergranular (along the grain boundaries) fracture path. However, regardless of the fracture path, there are essentially only four principal fracture modes: dimple rupture, cleavage, fatigue, and decohesive rupture. Each of these modes has a characteristic fracture surface appearance and a mechanism or mechanisms by which the fracture propagates.

In this section, the fracture surface characteristics and some of the mechanisms associated with the fracture modes will be presented and illustrated. Most of the mechanisms proposed to explain the various fracture modes are often based on dislocation interactions, involving complex slip and crystallographic relationships. The discussion of mechanisms in this section will not include detailed dislocation models or complex mathematical treatments, but will present the mechanisms in more general terms in order to impart a practical understanding as well as an ability to identify the basic fracture modes correctly.

Dimple Rupture

When overload is the principal cause of fracture, most common structural alloys fail by a process known as microvoid coalescence. The microvoids nucleate at regions of localized strain discontinuity, such as that associated with second-phase particles, inclusions, grain boundaries, and dislocation pile-ups. As the strain in the material increases, the microvoids grow, coalesce, and eventually form a continuous fracture surface (Fig. 1). This type of fracture exhibits numerous cuplike depressions that are the direct result of the microvoid coalescence. The cuplike depressions are referred to as dimples, and the fracture mode is known as dimple rupture.

The size of the dimples on a fracture surface is governed by the number and distribution of microvoids that are nucleated. When the nucleation sites are few and widely spaced, the microvoids grow to a large size before coalescing and the result is a fracture surface that contains large dimples. Small dimples are formed when numerous nucleating sites are activated and adjacent microvoids join (coalesce) before they have an opportunity to grow to a larger size. Extremely small dimples are often found in oxide dispersion strengthened materials.

The distribution of the microvoid nucleation sites can significantly influence the fracture surface appearance. In some alloys, the nonuniform distribution of nucleating particles and the nucleation and growth of isolated microvoids early in the loading cycle produce a fracture surface that exhibits various dimple sizes (Fig. 2). When microvoids nucleate at the grain boundaries (Fig. 3), intergranular dimple rupture results.

Dimple shape is governed by the state of stress within the material as the microvoids form and coalesce. Fracture under conditions of uniaxial tensile load (Fig. 1a) results in the formation of essentially equiaxed dimples bounded by a lip or rim (Fig. 3 and 4a). Depending on the microstructure and plasticity of the material, the dimples can exhibit a very deep, conical shape (Fig. 4a) or can be quite shallow (Fig. 4b). The formation of shallow dimples may involve the joining of microvoids by shear along slip bands (Ref 1).

Fracture surfaces that result from tear (Mode I) or shear (Modes II and III) loading conditions (Fig. 5) exhibit elongated dimples (Ref 2, 3). The characteristics of an elongated dimple are that it is, as the name implies, elongated (one axis of the dimple is longer than the other) and that one end of the dimple is open; that is, the dimple is not completely surrounded by a rim. In the case of a tear fracture (Fig. 6a), the elongated dimples on both fracture faces are oriented in the same direction, and the closed ends point to the fracture origin. This characteristic of the tear dimples can be used to establish the fracture propagation direction (Ref 4) in thin sheet that ruptures by a full-slant fracture (by combined Modes I and III), which consists entirely of a shear lip and exhibits no macroscopic fracture direction indicators, such as chevron marks. A shear fracture, however, exhibits elongated dimples that point in opposite directions on mating fracture faces (Fig. 6b). Examples of typical elongated dimples are shown in Fig. 7.

It should be noted that the illustrations representing equiaxed and elongated dimple formation and orientation were deliberately kept simple in order to convey the basic concepts of the effect on dimple shape and orientation of loading or plastic-flow directions in the immediate vicinity where the voids form, such as at the crack tip. In reality, matching dimples on mating fracture faces are seldom of the same size or seldom show equivalent angular correspondence. Because actual fractures rarely occur by pure tension or shear, the various combinations of loading Modes I, II, and III, as well as the constant change in orientation of the local plane of fracture as the crack propagates, result in asymmetrical straining of the mating fracture surfaces.

Figure 8 shows the effect of such asymmetry on dimple size. The surface (B) that is strained after fracture exhibits longer dimples than its mating half (A). When fracture occurs by a combination of Modes I and II, examination of the dimples on mating fracture surfaces can reveal the local fracture direction (Ref 5). As illustrated in Fig. 8, the fracture plane containing the longer dimples faces the region from which the crack propagated, while the mating fracture plane containing the shorter dimples faces away from the region. With the different

Modes of Fracture / 13

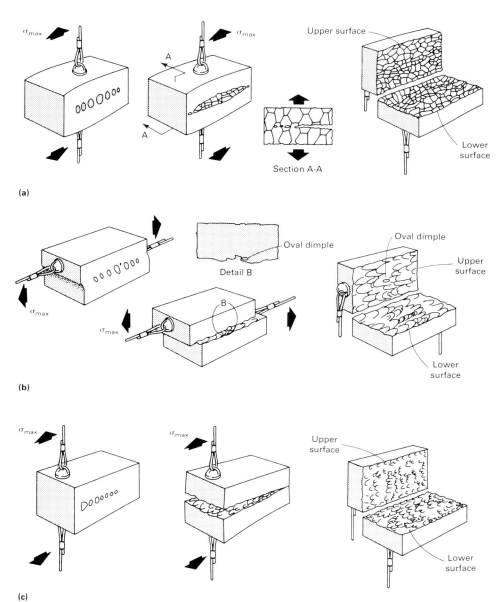

Fig. 1 Influence of direction of maximum stress (σ_{max}) on the shape of dimples formed by microvoid coalescence. (a) In tension, equiaxed dimples are formed on both fracture surfaces. (b) In shear, elongated dimples point in opposite directions on matching fracture surfaces. (c) In tensile tearing, elongated dimples point toward fracture origin on matching fracture surfaces.

combinations of Modes I, II, and III, there could be as many as 14 variations of dimple shape and orientation on mating fracture surfaces (Ref 5).

Metals that undergo considerable plastic deformation and develop large dimples frequently contain deformation markings on the dimple walls. These markings occur when slip-planes at the surface of the dimples are favorably oriented to the major stress direction. The continual straining of the free surfaces of the dimples as the microvoids enlarge produces slip-plane displacement at the surface of the dimple (Ref 6), as shown in Fig. 9. When first formed, the slip traces are sharp, well defined, and form an interwoven pattern that is generally referred to as serpentine glide (Fig. 10). As the slip process proceeds, the initial sharp slip traces become smooth, resulting in a surface structure that is sometimes referred to as ripples.

Oval-shaped dimples are occasionally observed on the walls of large elongated dimples. An oval dimple is formed when a smaller subsurface void intersects the wall of a larger void (dimple). The formation of oval dimples is shown schematically in Fig. 1(b) and 6(b).

Cleavage

Cleavage is a low-energy fracture that propagates along well-defined low-index crystallographic planes known as cleavage planes. Theoretically, a cleavage fracture should have perfectly matching faces and should be completely flat and featureless. However, engineering alloys are polycrystalline and contain grain and subgrain boundaries, inclusions, dislocations, and other imperfections that affect a propagating cleavage fracture so that true, featureless cleavage is seldom observed. These imperfections and changes in crystal lattice orientation, such as possible mismatch of the low-index planes across grain or subgrain boundaries, produce distinct cleavage fracture surface features, such as cleavage steps, river patterns, feather markings, chevron (herringbone) patterns, and tongues (Ref 7).

As shown schematically in Fig. 11, cleavage fractures frequently initiate on many parallel cleavage planes. As the fracture advances, however, the number of active planes decreases by a joining process that forms progressively higher cleavage steps. This network of cleavage steps is known as a river pattern. Because the branches of the river pattern join in the direction of crack propagation, these markings can be used to establish the local fracture direction.

A tilt boundary exists when principal cleavage planes form a small angle with respect to one another as a result of a slight rotation about a common axis parallel to the intersection (Fig. 11a). In the case of a tilt boundary, the cleavage fracture path is virtually uninterrupted, and the cleavage planes and steps propagate across the boundary. However, when the principal cleavage planes are rotated about an axis perpendicular to the boundary, a twist boundary results (Fig. 11b). Because of the significant misalignment of cleavage planes at the boundary, the propagating fracture reinitiates at the boundary as a series of parallel cleavage fractures connected by small (low) cleavage steps. As the fracture propagates away from the boundary, the numerous cleavage planes join, resulting in fewer individual cleavage planes and higher steps. Thus, when viewing a cleavage fracture that propagates across a twist boundary, the cleavage steps do not cross but initiate new steps at the boundary (Fig. 11b). Most boundaries, rather than being simple tilt or twist, are a combination of both types and are referred to as tilt-twist boundaries. Cleavage fractures exhibiting twist and tilt boundaries are shown in Fig. 12(a) and 13, respectively.

Feather markings are a fan-shaped array of very fine cleavage steps on a large cleavage facet (Fig. 14a). The apex of the fan points back to the fracture origin. Large cleavage steps are shown in Fig. 14(b).

Tongues are occasionally observed on cleavage fractures (Fig. 12b). They are formed when a cleavage fracture deviates from the cleavage plane and propagates a short distance along a twin orientation (Ref 8).

Wallner lines (Fig. 15) constitute a distinct cleavage pattern that is sometimes observed on fractured surfaces of brittle nonmetallic materials or on brittle inclusions or intermetallic compounds. This structure consists of two sets

14 / Modes of Fracture

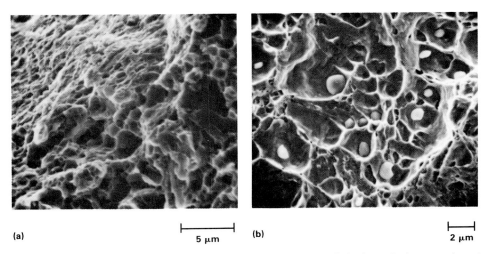

Fig. 2 Examples of the dimple rupture mode of fracture. (a) Large and small dimples on the fracture surface of a martempered type 234 tool steel saw disk. The extremely small dimples at top left are nucleated by numerous, closely spaced particles. (D.-W. Huang, Fuxin Mining Institute, and C.R. Brooks, University of Tennessee). (b) Large and small sulfide inclusions in steel that serve as void-nucleating sites. (R.D. Buchheit, Battelle Columbus Laboratories)

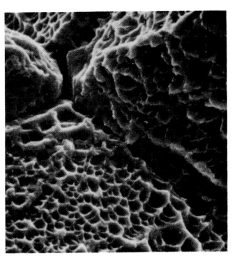

Fig. 3 Intergranular dimple rupture in a steel specimen resulting from microvoid coalescence at grain boundaries.

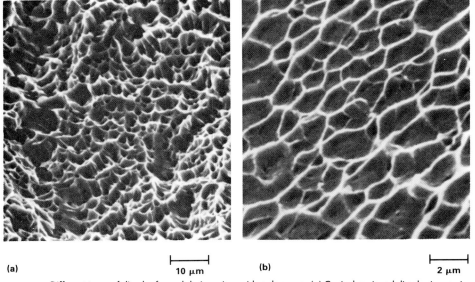

Fig. 4 Different types of dimples formed during microvoid coalescence. (a) Conical equiaxed dimples in a spring steel specimen. (b) Shallow dimples in a maraging steel specimen

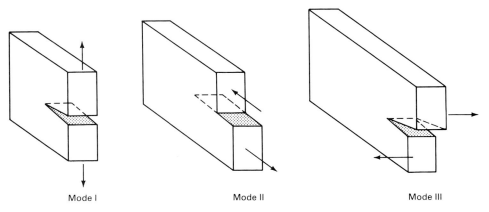

Fig. 5 Fracture loading modes. Arrows show loading direction and relative motion of mating fracture surfaces.

of parallel cleavage steps that often intersect to produce a crisscross pattern. Wallner lines result from the interaction of a simultaneously propagating crack front and an elastic shock wave in the material (Ref 9).

Fatigue

A fracture that is the result of repetitive or cyclic loading is known as a fatigue fracture. A fatigue fracture generally occurs in three stages: it initiates during Stage I, propagates for most of its length during Stage II, and proceeds to catastrophic fracture during Stage III.

Fatigue crack initiation and growth during Stage I occurs principally by slip-plane cracking due to repetitive reversals of the active slip systems in the metal (Ref 10-14). Crack growth is strongly influenced by microstructure and mean stress (Ref 15), and as much as 90% of the fatigue life may be consumed in initiating a viable fatigue crack (Ref 16). The crack tends to follow crystallographic planes, but changes direction at discontinuities, such as grain boundaries. At large plastic-strain amplitudes, fatigue cracks may initiate at grain boundaries (Ref 14). A typical Stage I fatigue fracture is shown in Fig. 16. Stage I fatigue fracture surfaces are faceted, often resemble cleavage, and do not exhibit fatigue striations. Stage I fatigue is normally observed on high-cycle low-stress fractures and is frequently absent in low-cycle high-stress fatigue.

The largest portion of a fatigue fracture consists of Stage II crack growth, which generally occurs by transgranular fracture and is more influenced by the magnitude of the alternating stress than by the mean stress or microstructure (Ref 15, 17, 18). Fatigue fractures generated during Stage II fatigue usually exhibit crack-arrest marks known as fatigue striations (Fig. 17 to 22), which are a visual record

Modes of Fracture / 15

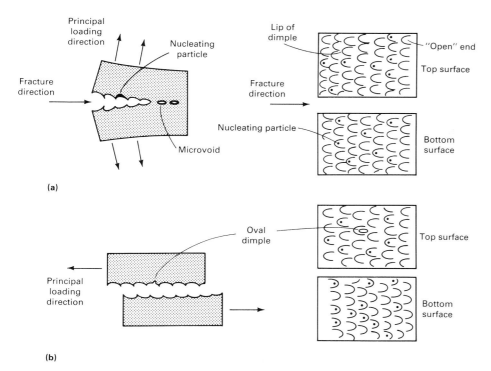

Fig. 6 Formation of elongated dimples under tear and shear loading conditions. (a) Tear fracture. (b) Shear fracture

of the position of the fatigue crack front during crack propagation through the material.

There are basically two models that have been proposed to explain Stage II striation-forming fatigue propagation. One is based on plastic blunting at the crack tip (Ref 11). This model cannot account for the absence of striations when a metal is fatigue tested in vacuum and does not adequately predict the peak-to-peak and valley-to-valley matching of corresponding features on mating halves of the fracture (Ref 8, 19-23).

The other model, which is based on slip at the crack tip, accounts for conditions where slip may not occur precisely at the crack tip due to the presence of lattice or microstructural imperfections (Ref 19-21). This model (Fig. 23) is more successful in explaining the mechanism by which Stage II fatigue cracks propagate. The concentration of stress at a fatigue crack results in plastic deformation (slip) being confined to a small region at the tip of the crack while the remainder of the material is subjected to elastic strain. As shown in Fig. 23(a), the crack opens on the rising-tension portion of the load cycle by slip on alternating slip planes. As slip proceeds, the crack tip blunts, but is resharpened by partial slip reversal during the declining-load portion of the fatigue cycle. This results in a compressive stress at the crack tip due to the relaxation of the residual elastic tensile stresses induced in the uncracked portion of the material during the rising load cycle (Fig. 23b). The closing crack does not reweld, because the new slip surfaces created during the crack-opening displacement are instantly oxidized (Ref 24), which makes complete slip reversal unlikely.

The essential absence of striations on fatigue fracture surfaces of metals tested in vacuum tends to support the assumption that oxidation reduces slip reversal during crack closure, which results in the formation of striations (Ref 19, 25, 26). The lack of oxidation in hard vacuum promotes a more complete slip reversal (Ref 27), which results in a smooth and relatively featureless fatigue fracture surface. Some fracture surfaces containing widely spaced fatigue striations exhibit slip traces on the leading edges of the striation and relatively smooth trailing edges, as predicted by the model (Fig. 23). Not all fatigue striations, however, exhibit distinct slip traces, as suggested by Fig. 23, which is a simplified representation of the fatigue process.

As shown schematically in Fig. 24, the profile of the fatigue fracture can also vary, depending on the material and state of stress. Materials that exhibit fairly well-developed striations display a sawtooth-type profile (Fig. 24a) with valley-to-valley or groove-to-groove matching (Ref 23, 28). Low compressive stresses at the crack tip favor the sawtooth profile; however, high compressive stresses promote the groove-type fatigue profile, as shown in Fig. 24(c) (Ref 23, 28). Jagged, poorly formed, distorted, and unevenly spaced striations (Fig. 24b), sometimes termed quasi-striations (Ref 23), show no symmetrical matching profiles. Even distinct sawtooth and groove-type fatigue surfaces may not show symmetrical matching. The local microscopic plane of a fatigue crack often deviates from the normal to the principal stress. Consequently, one of the fracture surfaces will be deformed more by repetitive cyclic slip than its matching counterpart (Ref 29) (for an analogy, see Fig. 8). Thus, one fracture surface may show well-developed striations, while its counterpart exhibits shallow, poorly formed striations.

Under normal conditions, each striation is the result of one load cycle and marks the position of the fatigue crack front at the time the striation was formed. However, when there is a sudden decrease in the applied load, the crack can temporarily stop propagating, and no striations are formed. The crack resumes propagation only after a certain number of cycles are applied at the lower stress (Ref 4, 23, 30). This phenomenon of crack arrest is believed to be due to the presence of a residual compressive-stress field within the crack tip plastic zone produced after the last high-stress fatigue cycle (Ref 23, 30).

Fatigue crack propagation and therefore striation spacing can be affected by a number of variables, such as loading conditions, strength of the material, microstructure, and the environment, for example, temperature and the presence of corrosive or embrittling gases and fluids. Considering only the loading conditions—which would include the mean stress, the alternating stress, and the cyclic frequency—the magnitude of the alternating stress ($\sigma_{max} - \sigma_{min}$) has the greatest effect on striation spacing. Increasing the magnitude of the alternating stress produces an increase in the striation spacing (Fig. 25a). While rising, the mean stress can also increase the striation spacing; this increase is not as great as one for a numerically equivalent increase in the alternating stress. Within reasonable limits, the cyclic frequency has the least effect on striation spacing. In some cases, fatigue striation spacing can change significantly over a very short distance (Fig. 25b). This is due in part to changes in local stress conditions as the crack propagates on an inclined surface.

For a Stage II fatigue crack propagating under conditions of reasonably constant cyclic loading frequency and advancing within the nominal range of 10^{-5} to 10^{-3} mm/cycle*, the crack growth rate, da/dN, can be expressed as a function of the stress intensity factor K (Ref 15, 31, 32):

$$\frac{da}{dN} = C(\Delta K)^m \qquad \text{(Eq 1)}$$

where a is the distance of fatigue crack advance, N is the number of cycles applied to advance the distance a, m and C are constants, and $\Delta K = K_{max} - K_{min}$ is the difference between the maximum and minimum stress intensity factor for each fatigue load cycle. The

---
*All fatigue crack growth rates in this article are given in millimeters per cycle (mm/cycle). To convert to inches per cycle (in./cycle), multiply by 0.03937. See also the Metric Conversion Guide in this Volume.

16 / Modes of Fracture

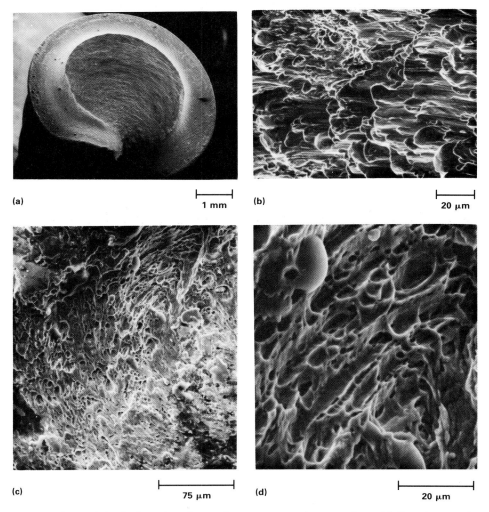

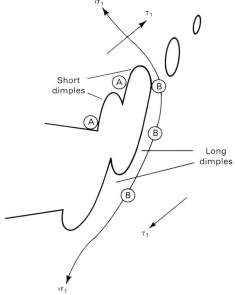

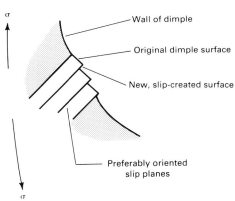

Fig. 7 Elongated dimples formed on shear and torsion specimen fracture surfaces. (a) Shear fracture of a commercially pure titanium screw. Macrofractograph shows spiral-textured surface of shear-off screw. Typical deformation lines are fanning out on the thread. (b) Higher-magnification view of (a) shows uniformly distributed elongated shear dimples. (O.E.M. Pohler, Institut Straumann AG). (c) Elongated dimples on the surface of a fractured single-strand copper wire that failed in torsion. (d) Higher-magnification view of the elongated dimples shown in (c). (R.D. Lujan, Sandia National Laboratories)

Fig. 9 Slip step formation resulting in serpentine glide and ripples on a dimple wall

stress intensity factor, K, describes the stress condition at a crack and is a function of the applied stress and a crack shape factor, generally expressed as a ratio of the crack depth to length.

When a fatigue striation is produced on each loading cycle, da/dN represents the striation spacing. Equation 1 does not adequately describe Stage I or Stage III fatigue crack growth rates; it tends to overestimate Stage I and often underestimates Stage III growth rates (Ref 15).

Stage III is the terminal propagation phase of a fatigue crack in which the striation-forming mode is progressively displaced by the static fracture modes, such as dimple rupture or cleavage. The rate of crack growth increases during Stage III until the fatigue crack becomes unstable and the part fails. Because the crack propagation is increasingly dominated by the static fracture modes, Stage III fatigue is sensitive to both microstructure and mean stress (Ref 17, 18).

Characteristics of Fractures With Fatigue Striations. During Stage II fatigue, the crack often propagates on multiple plateaus that are at different elevations with respect to one another (Fig. 26). A plateau that has a concave surface curvature exhibits a convex contour on the mating fracture face (Ref 29). The plateaus are joined either by tear ridges or walls that contain fatigue striations (Fig. 19 and 20a). Fatigue striations often bow out in the direction of crack propagation and generally tend to align perpendicular to the principal (macroscopic) crack propagation direction. However, variations in local stresses and microstructure can change the orientation of the plane of fracture and alter the direction of striation alignment (Fig. 27).

Large second-phase particles and inclusions in a metal can change the local crack growth rate and resulting fatigue striation spacing. When a fatigue crack approaches such a particle, it is briefly retarded if the particle remains intact or is accelerated if the particle cleaves (Fig. 18). In both cases, however, the crack growth rate is changed only in the immediate vicinity of the particle and therefore does not significantly affect the total crack growth rate. However, for low-cycle (high-stress) fatigue, the relatively large plastic zone at the crack tip can cause cleavage and matrix separation at the particles at a significant distance ahead of the advancing fatigue crack. The cleaved or matrix-separated particles, in effect, behave as cracks or voids that promote a tear or shear fracture between themselves and the fatigue crack, thus significantly advancing the crack front (Ref 33,

Modes of Fracture / 17

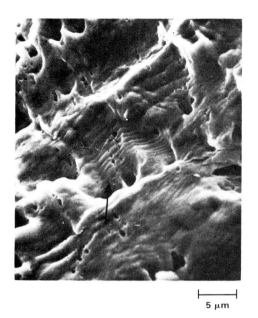

Fig. 10 Serpentine glide formation (arrow) in oxygen-free high-conductivity copper specimen

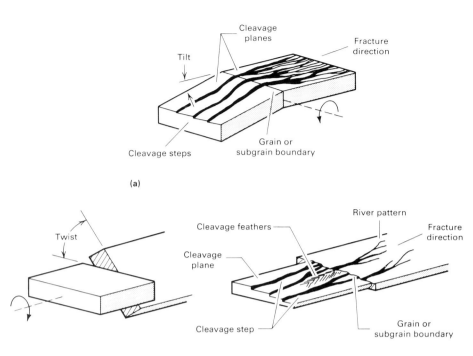

Fig. 11 Schematic of cleavage fracture formation showing the effect of subgrain and grain boundaries. (a) Tilt boundary. (b) Twist boundary

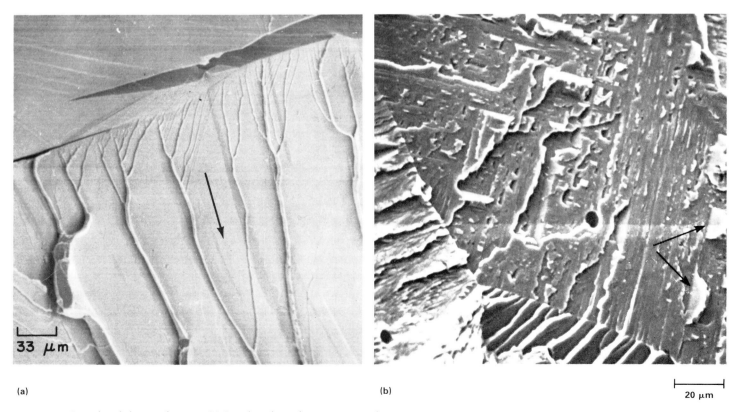

Fig. 12 Examples of cleavage fractures. (a) Twist boundary, cleavage steps, and river patterns in an Fe-0.01C-0.24Mn-0.02Si alloy that was fractured by impact. (b) Tongues (arrows) on the surface of a 30% Cr steel weld metal that fractured by cleavage

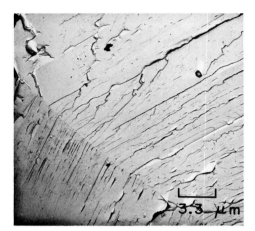

Fig. 13 Cleavage fracture in Armco iron showing a tilt boundary, cleavage steps, and river patterns. TEM p-c replica

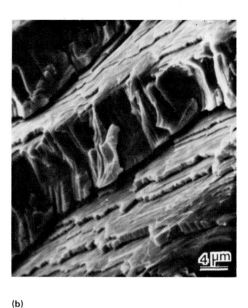

Fig. 14 Examples of cleavage fractures. (a) Feather pattern on a single grain of a chromium steel weld metal that failed by cleavage. (b) Cleavage steps in a Cu-25 at.% Au alloy that failed by transgranular stress-corrosion cracking. (B.D. Lichter, Vanderbilt University)

34). Relatively small, individual particles have no significant effect on striation spacing (Fig. 17b).

The distinct, periodic markings sometimes observed on fatigue fracture surfaces are known as tire tracks, because they often resemble the tracks left by the tread pattern of a tire (Fig. 28). These rows of parallel markings are the result of a particle or a protrusion on one fatigue fracture surface being successively impressed into the surface of the mating half of the fracture during the closing portion of the fatigue cycle (Ref 23, 29, 34). Tire tracks are more common for the tension-compression than the tension-tension type of fatigue loading (Ref 23). The direction of the tire tracks and the change in spacing of the indentations within the track can indicate the type of displacement that occurred during the fracturing process, such as lateral movement from shear or torsional loading. The presence of tire tracks on a fracture surface that exhibits no fatigue striations may indicate that the fracture occurred by low-cycle (high-stress) fatigue (Ref 35).

Decohesive Rupture

A fracture is referred to as decohesive rupture when it exhibits little or no bulk plastic deformation and does not occur by dimple rupture, cleavage, or fatigue. This type of fracture is generally the result of a reactive environment or a unique microstructure and is associated almost exclusively with rupture along grain boundaries. Grain boundaries contain the lowest melting point constituents of an alloy system. They are also easy paths for diffusion and sites for the segregation of such elements as hydrogen, sulfur, phosphorus, antimony, arsenic, and carbon; the halide ions, such as chlorides; as well as the routes of penetration by the low melting point metals, such as gallium, mercury, cadmium, and tin. The presence of these constituents at the boundaries can significantly reduce the cohesive strength of the material at the boundaries and promote decohesive rupture (Fig. 29).

Decohesive rupture is not the result of one unique fracture process, but can be caused by several different mechanisms. The decohesive processes involving the weakening of the atomic bonds (Ref 36), the reduction in surface energy required for localized deformation (Ref 37-39), molecular gas pressure (Ref 40), the rupture of protective films (Ref 41, 42), and anodic dissolution at active sites (Ref 43) are associated with hydrogen embrittlement and stress-corrosion cracking (SCC). Decohesive rupture resulting from creep fracture mechanisms is discussed at the end of this section.

The fracture of weak grain-boundary films (such as those resulting from grain-boundary penetration by low melting point metals), the rupture of melted and resolidified grain-boundary constituents (as in overheated aluminum alloys), or the separation of melted material in the boundaries (Ref 44) before it solidifies (as in the cracking at the heat-affected zones, HAZs, of welds, a condition known as hot cracking) can produce a decohesive rupture. Figures 30 to 32 show examples of decohesive rupture. A decohesive rupture resulting from hydrogen embrittlement is shown in Fig. 30. Figure 31 shows a decohesive rupture in a precipitation-hardenable stainless steel due to SCC. A fracture along a low-strength grain-boundary film resulting from the diffusion of liquid mercury is shown in Fig. 32. More detailed information on hydrogen embrittlement, SCC, and liquid-metal embrittlement can be found later in this article in the section "Effect of Environment." When a decohesive rupture occurs along flattened, elongated grains that form nearly uninterrupted planes through the material, as in severely extruded alloys and along the parting planes of some forgings, a relatively smooth, featureless fracture results (Fig. 33).

Creep rupture is a time-dependent failure that results when a metal is subjected to stress for extended periods at elevated temperatures that are usually in the range of 40 to 70% of the absolute melting temperature of the metal. With few exceptions (Ref 45-49), creep ruptures

Fig. 15 Wallner lines (arrow) on the surface of a fractured WC-Co specimen. TEM formvar replica. Etched with 5% HCl. (S.B. Luyckx, University of the Witwatersrand)

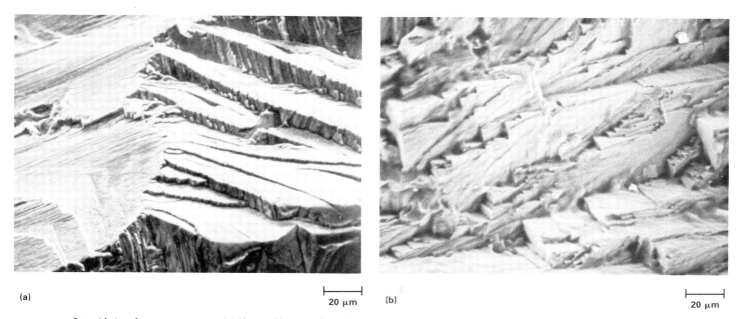

Fig. 16 Stage I fatigue fracture appearance. (a) Cleavagelike, crystallographically oriented Stage I fatigue fracture in a cast Ni-14Cr-4.5Mo-1Ti-6Al-1.5Fe-2.0(Nb + Ta) alloy. (b) Stair-step fracture surface indicative of Stage I fatigue fracture in a cast ASTM F75 cobalt-base alloy. SEM. (R. Abrams, Howmedica, Div. Pfizer Hospital Products Group Inc.)

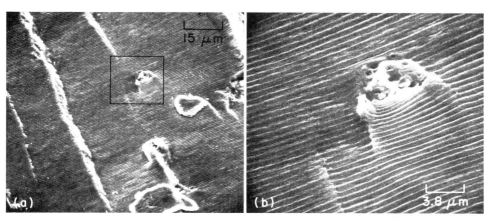

Fig. 17 Uniformly distributed fatigue striations in an aluminum 2024-T3 alloy. (a) Tear ridge and inclusion (outlined by rectangle). (b) Higher-magnification view of the region outlined by the rectangle in (a) showing the continuity of the fracture path through and around the inclusion. Compare with Fig. 18.

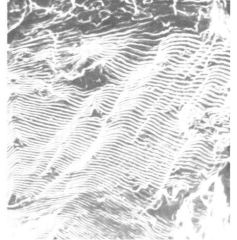

Fig. 18 Local variations in striation spacing in a Ni-0.04C-21Cr-0.6Mn-2.5Ti-0.7Al alloy that was tested under rotating-bending conditions. Compare with Fig. 17(b).

exhibit an intergranular fracture surface. Transgranular creep ruptures, which generally result from high applied stresses (high strain rates), fail by a void-forming process similar to that of microvoid coalescence in dimple rupture (Ref 45-47). Because transgranular creep ruptures show no decohesive character, they will not be considered for further discussion. Intergranular creep ruptures, which occur when metal is subjected to low stresses (often well below the yield point) and to low strain rates, exhibit decohesive rupture and will be discussed in more detail.

Creep can be divided into three general stages: primary, secondary, and tertiary creep. The fracture initiates during primary creep, propagates during secondary or steady-state creep, and becomes unstable, resulting in failure, during tertiary or terminal creep. From a practical standpoint of the service life of a structure, the initiation and steady-state propagation of creep ruptures are of primary importance, and most efforts have been directed toward understanding the fracture mechanisms involved in these two stages of creep.

As shown schematically in Fig. 34, intergranular creep ruptures occur by either of two fracture processes: triple-point cracking or grain-boundary cavitation (Ref 50-63). The strain rate and temperature determine which fracture process dominates. Relatively high strain rates and intermediate temperatures promote the formation of wedge cracks (Fig. 34a). Grain-boundary sliding as a result of an applied tensile stress can produce sufficient stress concentration at grain-boundary triple points to initiate and propagate wedge cracks (Ref 50-52, 55, 56, 58-61). Cracks can also nucleate in the grain boundary at locations other than the triple point by the interaction of primary and secondary slip steps with a sliding grain boundary (Ref 61). Any environment that lowers grain-boundary cohesion also promotes cracking (Ref 59). As sliding proceeds, grain-boundary

20 / Modes of Fracture

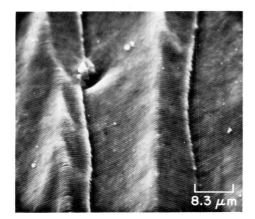

Fig. 19 Fatigue striations in a 2024-T3 aluminum alloy joined by tear ridges

cracks propagate and join to form intergranular decohesive fracture (Fig. 35a and b).

At high temperatures and low strain rates, grain-boundary sliding favors cavity formation (Fig. 34b). The grain-boundary cavities resulting from creep should not be confused with microvoids formed in dimple rupture. The two are fundamentally different; the cavities are principally the result of a diffusion-controlled process, while microvoids are the result of complex slip. Even at low strain rates, a sliding grain boundary can nucleate cavities at irregularities, such as second-phase inclusion particles (Ref 54, 57, 63, 64). The nucleation is believed to be a strain-controlled process (Ref 63, 64), while the growth of the cavities can be described by a diffusion growth model (Ref 65-67) and by a power-law growth relationship (Ref 68, 69). Irrespective of the growth model,

as deformation continues, the cavities join to form an intergranular fracture. Even though the fracture resulting from cavitation creep exhibits less sharply defined intergranular facets (Fig. 35c), it would be considered a decohesive rupture.

Instead of propagating by a cracking or a cavity-forming process, a creep rupture could occur by a combination of both. There may be no clear distinction between wedge cracks and cavities (Ref 70-72). The wedge cracks could be the result of the linkage of cavities at triple points.

The various models proposed to describe the creep process are mathematically complex and were not discussed in detail. Comprehensive reviews of the models are available in Ref 59, 63, 73, and 74.

Unique Fractures

Some fractures, such as quasi-cleavage and flutes, exhibit a unique appearance but cannot be readily placed within any of the principal fracture modes. Because they can occur in common engineering alloys under certain failure conditions, these fractures will be briefly discussed.

Quasi-cleavage fracture is a localized, often isolated feature on a fracture surface that exhibits characteristics of both cleavage and plastic deformation (Fig. 36 and 37). The term quasi-cleavage does not accurately describe the fracture, because it implies that the fracture resembles, but is not, cleavage. The term was coined because, although the central facets of a quasi-cleavage fracture strongly resembled cleavage (Ref 75), their identity as cleavage planes was not established until well after the term had gained widespread acceptance (Ref 76-83). In steels, the cleavage facets of quasi-cleavage fracture occur on the $\{100\}$, $\{110\}$, and possibly the $\{112\}$ planes. The term quasi-cleavage can be used to describe the distinct fracture appearance if one is aware that quasi-cleavage does not represent a separate fracture mode.

A quasi-cleavage fracture initiates at the central cleavage facets; as the crack radiates, the cleavage facets blend into areas of dimple rupture, and the cleavage steps become tear ridges. Quasi-cleavage has been observed in steels, including quench-and-temper hardenable, precipitation-hardenable, and austenitic stainless steels; titanium alloys; nickel alloys; and even aluminum alloys. Conditions that impede plastic deformation promote quasi-cleavage fracture—for example, the presence of a triaxial state of stress (as adjacent to the root of a notch), material embrittlement (as by hydrogen or stress corrosion), or when a steel is subjected to high strain rates (such as impact loading) within the ductile-to-brittle transition range.

Flutes. Fractography has acquired a number of colorful and descriptive terms, such as dimple rupture, serpentine glide, ripples, tongues, tire tracks, and factory roof, which describes a ridge-to-valley fatigue fracture topography resulting from Mode III antiplane shear loading (Ref 84). The term flutes should also be included in this collection. Flutes exhibit elongated grooves or voids (Fig. 38 and 39) that connect widely spaced cleavage planes (Ref 85-90). The fracture process is known as fluting. The term flutes was apparently chosen because the fractures often resemble the long, parallel grooves on architectural columns or the pleats in drapes.

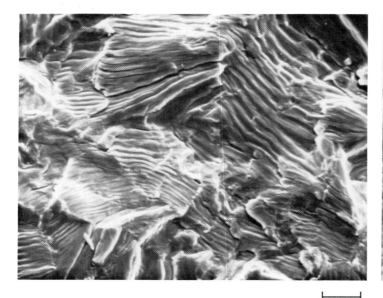

Fig. 20 Fatigue striations on adjoining walls on the fracture surface of a commercially pure titanium specimen. (O.E.M. Pohler, Institut Straumann AG)

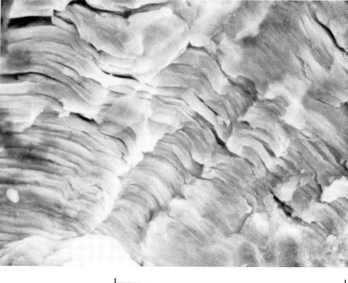

Fig. 21 Fatigue striations on the fracture surface of a tantalum heat-exchanger tube. The rough surface appearance is due to secondary cracking caused by high-cycle low-amplitude fatigue. (M.E. Blum, FMC Corporation)

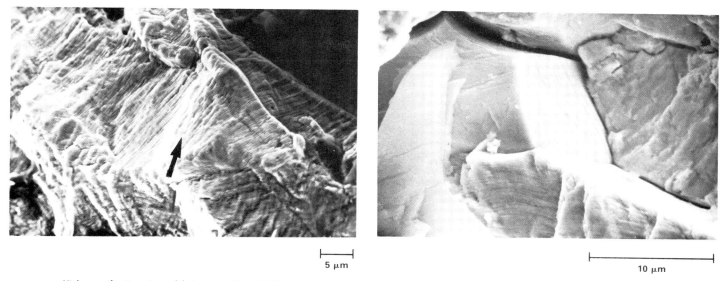

Fig. 22 High-magnification views of fatigue striations. (a) Striations (arrow) on the fracture surface of an austenitic stainless steel. (C.R. Brooks and A. Choudhury, University of Tennessee). (b) Fatigue striations on the facets of tantalum grains in the heat-affected zone of a weldment. (M.E. Blum, FMC Corporation)

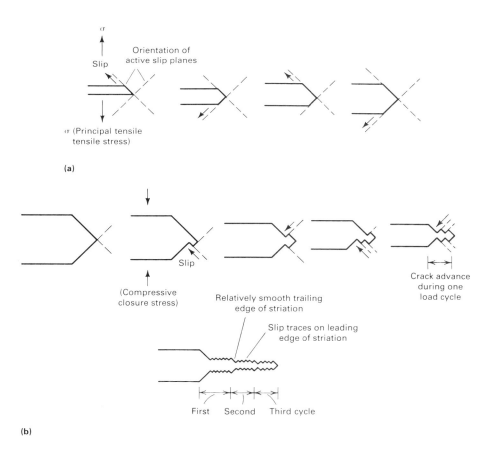

Fig. 23 Mechanism of fatigue crack propagation by alternate slip at the crack tip. Sketches are simplified to clarify the basic concepts. (a) Crack opening and crack tip blunting by slip on alternate slip planes with increasing tensile stress. (b) Crack closure and crack tip resharpening by partial slip reversal on alternate slip planes with increasing compressive stress

Although flutes are not elongated dimples, they are the result of a plastic deformation process. Flutes are the ruptured halves of tubular voids believed to be formed by a planar intersecting slip mechanism (Ref 85, 88, 89) and have matching tear ridges on opposite fracture faces. The tear ridges join in the direction of fracture propagation, forming an arrangement that resembles cleavage river patterns (Ref 89). Although fluting has been observed primarily in hexagonal close-packed (hcp) metal systems, such as titanium and zirconium alloys, evidence of fluting has also been reported on a hydrogen-embrittled type 316 austenitic stainless steel (Ref 90). Titanium alloys having a relatively high oxygen or aluminum content (α-stabilizers) that are fractured at cryogenic temperatures or fail by SCC may exhibit fluting (Ref 89).

Tearing Topography Surface

A tentative fracture mode called tearing topography surface (TTS) has been identified and described (Ref 91). The TTS fracture occurs in a variety of alloy systems, including steels, aluminum, titanium, and nickel alloys, and under a variety of fracture conditions, such as overload, hydrogen embrittlement (Ref 92), and fatigue. Examples of TTS fractures are shown in Fig. 40 to 42.

Although the precise nucleation and propagation mechanism for TTS fracture has not been identified, the fracture appears to be the result of a microplastic tearing process that operates on a very small (submicron) scale (Ref 91). The TTS fractures do not exhibit as much plastic deformation as dimple rupture, although they are often observed in combination with dimples (Fig. 40 and 41). The fractures are generally characterized by relatively smooth, often flat, areas or facets that usually contain thin tear ridges. Tearing topography surface fractures may be due to closely spaced microvoid nucleation and limited growth before coalescence, resulting in extemely shallow dimples. However, this hypothesis does not appear to be probable, because TTS is often observed along with well-developed dimples in alloys having relatively uniform carbide dispersions, such as HY-130 steel, and because TTS

22 / Modes of Fracture

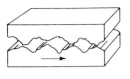 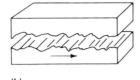 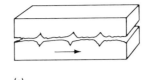

Fig. 24 Sawtooth and groove-type fatigue fracture profiles. Arrows show crack propagation direction. (a) Distinct sawtooth profile (aluminum alloy). (b) Poorly formed sawtooth profile (steel). (c) Groove-type profile (aluminum alloy). Source: Ref 23

is observed under varying stress states. A detailed discussion of the TTS fracture mode is available in Ref 91.

Effect of Environment

The environment, which refers to all external conditions acting on the material before or during fracture, can significantly affect the fracture propagation rate and the fracture appearance. This section will present some of the principal effects of such environments as hydrogen, corrosive media, low-melting metals, state of stress, strain rate, and temperature. Where applicable, the effect of the environment on the fracture appearance will be illustrated.

Effect of Environment on Dimple Rupture

The Effect of Hydrogen. When certain body-centered cubic (bcc) and hcp metals or alloys of such elements as iron, nickel, titanium, vanadium, tantalum, niobium, zirconium, and hafnium are exposed to hydrogen, they are susceptible to a type of failure known as hydrogen embrittlement. Although the face-centered cubic (fcc) metals and alloys are generally considered to have good resistance to hydrogen embrittlement, it has been shown that the 300 series austenitic stainless steels (Ref 95-98) and certain 2000 and 7000 series high-strength aluminum alloys are also embrittled by hydrogen (Ref 99-107). Although the result of hydrogen embrittlement is generally perceived to be a catastrophic fracture that occurs well below the ultimate strength of the material and exhibits no ductility, the effects of hydrogen can be quite varied. They can range from a slight decrease in the percent reduction of area at fracture to premature rupture that exhibits no ductility (plastic deformation) and occurs at a relatively low applied stress.

The source of hydrogen may be a processing operation, such as plating (Fig. 30) or acid cleaning, or the hydrogen may be acquired from the environment in which the part operates. If hydrogen absorption is suspected, prompt heating at an elevated temperature (usually about 200 °C, or 400 °F) will often restore the original properties of the material.

The effect of hydrogen is strongly influenced by such variables as the strength level of the alloy, the microstructure, the amount of hydrogen absorbed (or adsorbed), the magnitude of the applied stress, the presence of a triaxial state of stress, the amount of prior cold work, and the degree of segregation of such contaminant elements as phosphorus, sulfur, nitrogen, tin, or antimony at the grain boundaries. In general, an increase in strength, higher absorption of hydrogen, an increase in the applied stress, the presence of a triaxial stress state, extensive prior cold working, and an increase in the concentration of contaminant elements at the grain boundaries all serve to intensify the embrittling effect of hydrogen. However, for an alloy exhibiting a specific strength level and microstructure, there is a stress intensity, K_I, below which, for all practical purposes, hydrogen embrittlement cracking does not occur. This threshold crack tip stress intensity factor is determined experimentally and is designated as K_{th}.

A number of theories have been advanced to explain the phenomenon of hydrogen embrittlement. These include the exertion of an internal gas pressure at inclusions, grain boundaries, surfaces of cracks, dislocations, or internal voids (Ref 40, 108, 109); the reduction in atomic and free-surface cohesive strength (Ref 110-116); the attachment of hydrogen to dislocations, resulting in easier dislocation breakaway from the pinning effects of carbon and nitrogen (Ref 38, 112, 117-122); enhanced nucleation of dislocations (Ref 112, 123); enhanced nucleation and growth of microvoids (Ref 109, 110, 113, 116, 122, 124-126); enhanced shear and decrease of strain for the onset of shear instability (Ref 112, 127, 128);

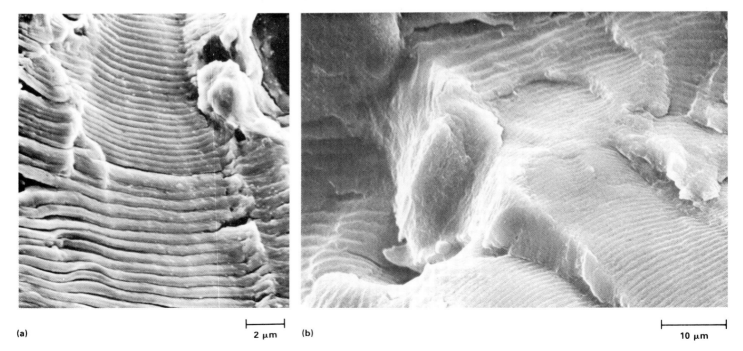

Fig. 25 Variations in fatigue striation spacing. (a) Spectrum-loaded fatigue fracture in a 7475-T7651 aluminum alloy test coupon showing an increase in striation spacing due to higher alternating stress. (b) Local variation in fatigue striation spacing in a spectrum-loaded 7050-T7651 aluminum alloy extrusion. (D. Brown, Douglas Aircraft Company)

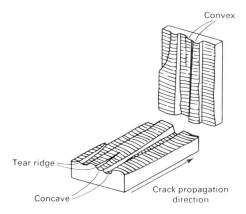

Fig. 26 Schematic illustrating fatigue striations on plateaus

Fig. 27 Striations on two joining, independent fatigue crack fronts on a fracture surface of aluminum alloy 6061-T6. The two arrows indicate direction of local crack propagation. TEM p-c replica

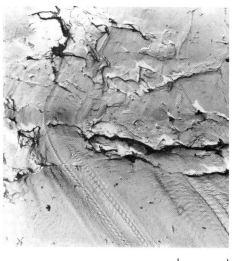

Fig. 28 Tire tracks on the fatigue fracture surface of a quenched-and-tempered AISI 4140 steel. TEM replica. (I. Le May, Metallurgical Consulting Services Ltd.)

the formation of methane gas bubbles at grain boundaries (Ref 129, 130); and, especially for titanium alloys, the repeated formation and rupture of the brittle hydride phase at the crack tip (Ref 131-137). Probably no one mechanism is applicable to all metals, and several mechanisms may operate simultaneously to embrittle a material. Whatever the mechanism, the end result is an adverse effect on the mechanical properties of the material.

If the effect of hydrogen is subtle, such as when there is a slight decrease in the reduction of area at fracture as a result of a tensile test, there is no perceivable change in the dimple rupture fracture appearance. However, the dimples become more numerous but are more shallow at a greater loss in ductility (Fig. 43).

Hydrogen Embrittlement of Steels. At low strain rates or when embrittlement is more severe, the fracture mode in steels can change from dimple rupture to quasi-cleavage, cleavage, or intergranular decohesion. These changes in fracture mode or appearance may not occur over the entire fracture surface and are usually more evident in the region of the fracture origin. Figure 44 shows an example of a hydrogen-embrittled AISI 4340 steel that exhibits quasi-cleavage.

When an annealed type 301 austenitic stainless steel is embrittled by hydrogen, the fracture occurs by cleavage (Fig. 45a). An example in which the mode of fracture changed to intergranular decohesion in a hydrogen-embrittled AISI 4130 steel is shown in Fig. 45b.

When a hydrogen embrittlement fracture propagates along grain boundaries, the presence of such contaminant elements as sulfur, phosphorus, nickel, tin, and antimony at the boundaries can greatly enhance the effect of hydrogen (Ref 111, 139). For example, the segregation of contaminant elements at the grain boundaries enhances the hydrogen embrittlement of high-strength low-alloy steels tempered above 500 °C (930 °F) (Ref 92). The presence of sulfur at grain boundaries promotes hydrogen embrittlement of nickel, and for equivalent concentrations, the effect of sulfur is nearly 15 times greater than that of phosphorus (Ref 140).

Hydrogen Embrittlement of Titanium. Although titanium and its alloys have a far greater tolerance for hydrogen than high-strength steels, titanium alloys are embrittled by hydrogen. The degree and the nature of the embrittlement is strongly influenced by the alloy, the microstructure, and whether the hydrogen is present in the lattice before testing or is introduced during the test. For example, a Ti-8Al-1Mo-1V alloy that was annealed at 1050 °C (1920 °F), cooled to 850 °C (1560 °F), and water quenched to produce a coarse Widmanstätten structure exhibited cracking along the α-β interfaces when tested in 1-atm hydrogen gas at room temperature (Ref 137). The fracture surface, which exhibited crack-arrest markings, is shown in Fig. 46(a). The arrest markings are believed to be due to the discontinuous crack propagation as a result of the repeated rupture of titanium hydride phase at the crack tip (Ref 137). Also, Fig. 46(b) shows a hydrogen embrittlement fracture in a Ti-5Al-2.5Sn alloy containing 90 ppm H that was β processed at 1065 °C (1950 °F) and aged for 8 h at 950 °C (1740 °F). The fracture occurred by cleavage.

Cleavage was also the mode of fracture for a Ti-6Al-4V alloy having a microstructure consisting of a continuous, equiaxed α phase with a fine, dispersed β phase at the α grain boundaries embrittled by exposure to hydrogen gas at a pressure of 1 atm (Fig. 47a). However, when the same Ti-6Al-4V alloy having a microstructure consisting of a medium, equiaxed α phase with a continuous β network was embrittled by 1-atm hydrogen gas, the fracture occurred by intergranular decohesion along the α-β boundaries (Fig. 47b and c).

Hydrogen Embrittlement of Aluminum. There is conclusive evidence (Ref 99-107) that some aluminum alloys, such as 2124, 7050, 7075, and even 5083 (Ref 143), are embrittled by hydrogen and that the embrittlement is apparently due to some of the mechanisms already discussed, namely enhanced slip and trapping of hydrogen at precipitates within grain boundaries. The embrittlement in alumi-

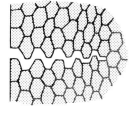
(a)

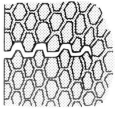
(b)

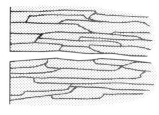
(c)

Fig. 29 Schematic illustrating decohesive rupture along grain boundaries. (a) Decohesion along grain boundaries of equiaxed grains. (b) Decohesion through a weak grain-boundary phase. (c) Decohesion along grain boundaries of elongated grains

24 / Modes of Fracture

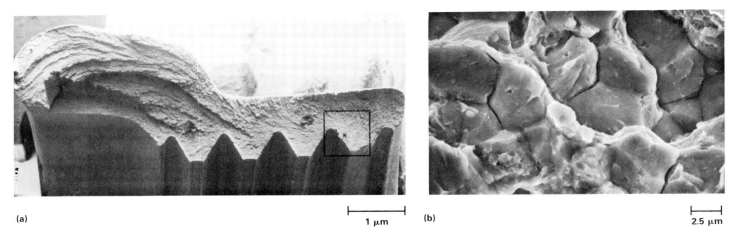

Fig. 30 Decohesive rupture in an AISI 8740 steel nut due to hydrogen embrittlement. Failure was due to inadequate baking following cadmium plating; thus, hydrogen, which was picked up during the plating process, was not released. (a) Macrograph of fracture surface. (b) Higher-magnification view of the boxed area in (a) showing typical intergranular fracture. (W.L. Jensen, Lockheed Georgia Company)

num alloys depends on such variables as the microstructure, strain rate, and temperature. In general, underaged microstructures are more susceptible to hydrogen embrittlement than the peak or overaged structures. For the 7050 aluminum alloy, a low (0.01%) copper content renders all microstructures more susceptible to embrittlement than those of normal (2.1%) copper content (Ref 106). Also, hydrogen embrittlement in aluminum alloys is more likely to occur at lower strain rates and at lower temperatures.

The effect of hydrogen on the fracture appearance in aluminum alloys can vary from no significant change in an embrittled 2124 alloy (Ref 99) to a dramatic change from the normal dimple rupture to a combination of cleavagelike transgranular fracture and intergranular decohesion in the high-strength 7050 (Ref 106) and 7075 (Ref 105) aluminum alloys. Figure 48 shows an example of a fracture in a hydrogen-embrittled (as measured by a 21% decrease in the reduction of area at fracture) 2124-UT (underaged temper: aged 4 h at 190 °C, or 375 °F) aluminum alloy. It can be seen that there is little difference in fracture appearance between the nonembrittled and embrittled specimens. However, when a low-copper (0.01%) 7050 in the peak-aged condition (aged 24 h at 120 °C, or 245 °F) is hydrogen embrittled, a cleavagelike transgranular fracture results (Fig. 49a). This same alloy in the underaged condition (aged 10 h at 100 °C, or 212 °F) fails by a combination of intergranular decohesion and cleavagelike fracture (Fig. 49b).

The Effect of a Corrosive Environment. When a metal is exposed to a corrosive environment while under stress, SCC, which is a form of delayed failure, can occur. Corrosive environments include moist air; distilled and tap water; seawater; gaseous ammonia and ammonia in solutions; solutions containing chlorides or nitrides; basic, acidic, and organic solutions; and molten salts. The susceptibility of a material to SCC depends on such variables as strength, microstructure, magnitude of the applied stress, grain orientation (longitudinal or short transverse) with respect to the principal applied stress, and the nature of the corrosive environment. Similar to the K_{th} in hydrogen

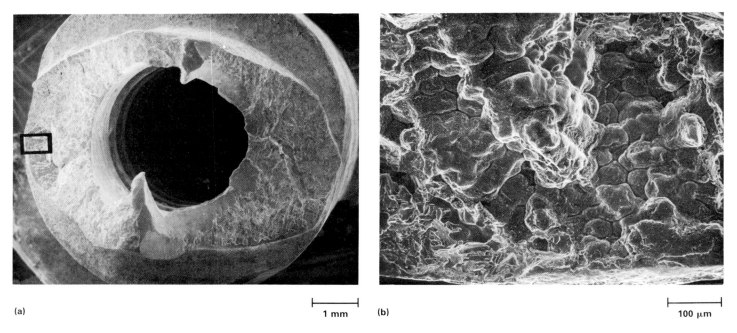

Fig. 31 17-4 PH stainless steel main landing-gear deflection yoke that failed because of intergranular SCC. (a) Macrograph of fracture surface. (b) Higher-magnification view of the boxed area in (a) showing area of intergranular attack. (W.L. Jensen, Lockheed Georgia Company)

embrittlement, there is also a threshold crack tip stress intensity factor, K_{ISCC}, below which a normally susceptible material at a certain strength, microstructure, and testing environment does not initiate or propagate stress-corrosion cracks. Stress-corrosion cracks normally initiate and propagate by tensile stress; however, compressive-stress SCC has been observed in a 7075-T6 aluminum alloy and a type 304 austenitic stainless steel (Ref 144).

Stress-corrosion cracking is a complex phenomenon, and the basic fracture mechanisms are still not completely understood. Although such processes as dealloying (Ref 145-148) in brass and anodic dissolution (Ref 149, 150, 151) in other alloy systems are important SCC mechanisms, it is apparent that the principal SCC mechanism in steels, titanium, and aluminum alloys is hydrogen embrittlement (Ref 38, 100, 107, 137, 143, 152-166). In these alloys, SCC occurs when the hydrogen generated as a result of corrosion diffuses into and embrittles the material. In these cases, SCC is used to describe the test or failure environment, rather than a unique fracture mechanism.

Mechanisms of SCC. The basic processes that lead to SCC, especially in environments containing water, involve a series of events that begin with the rupture of a passive surface film (usually an oxide), followed by metal dissolution, which results in the formation of a pit or crevice where a crack eventually initiates and propagates. When the passive film formed during exposure to the environment is ruptured by chemical attack or mechanical action (creep-strain), a clean, unoxidized metal surface is exposed. As a result of an electrochemical potential difference between the new exposed metal surface and the passive film, a small electrical current is generated between the anodic metal and the cathodic film. The relatively small area of the new metal surface compared to the large surface area of the surrounding passive film results in an unfavorable anode-to-cathode ratio. This causes a high local current density and induces high metal dissolution (anodic dissolution) at the anode as the new metal protects the adjacent film from corrosion; that is, the metal surface acts as a sacrificial anode in a galvanic couple.

If the exposed metal surface can form a new passive film (repassivate) faster than the new metal surface is created by film rupture, the corrosion attack will stop. However, if the repassivation process is suppressed, as in the presence of chlorides, or if the repassivated film is continuously ruptured by strain, as when the material creeps under stress, the localized corrosion attack proceeds (Ref 167-172). The result is the formation and progressive enlargement of a pit or crevice and an increase in the concentration of hydrogen ions and an accompanying decrease in the pH of the solution within the pit.

The hydrogen ions result from a chemical reaction between the exposed metal and the water within the cavity. The subsequent

Fig. 32 Fracture surface of a Monel specimen that failed in liquid mercury. The fracture is predominantly intergranular with some transgranular contribution. (C.E. Price, Oklahoma State University)

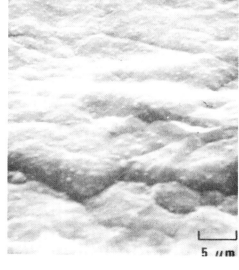

Fig. 33 Stress-corrosion fracture that occurred by decohesion along the parting plane of an aluminum alloy forging

reduction of the hydrogen ions by the acquisition of electrons from the environment results in the formation of hydrogen gas and the diffusion of hydrogen into the metal. This absorption of hydrogen produces localized cracking due to a hydrogen embrittlement mechanism (Ref 173, 174). Because the metal exposed at the crack tip as the crack propagates by virtue of hydrogen embrittlement and the applied stress is anodic to the oxidized sides of the crack and the adjacent surface of the material, the electrochemical attack continues, as does the evolution and absorption of hydrogen. The triaxial state of stress and the stress concentration at the crack tip enhance hydrogen embrittlement and provide a driving force for crack propagation.

In materials that are insensitive to hydrogen embrittlement, SCC can proceed by the anodic dissolution process with no assistance from hydrogen (Ref 149, 155, 161). Alloys are not homogeneous, and when differences in chemical composition or variations in internal strain occur, electrochemical potential differences arise between various areas within the microstructure. For example, the grain boundaries are usually anodic to the material within the grains and are therefore subject to preferential

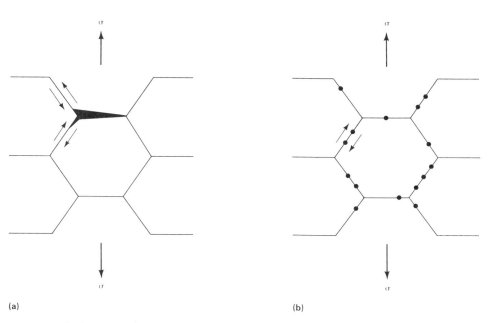

Fig. 34 Triple-point cracking (a) and cavitation (b) in intergranular creep rupture. Small arrows indicate grain-boundary sliding.

26 / Modes of Fracture

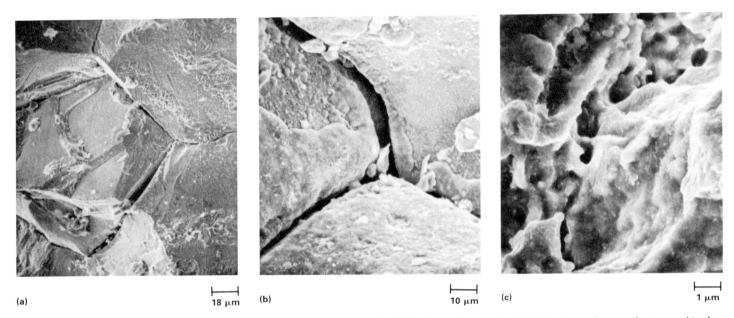

Fig. 35 Examples of intergranular creep fractures. (a) Wedge cracking in Inconel 625. (b) Wedge cracking in Incoloy 800. (c) Intergranular creep fracture resulting from grain-boundary cavitation in PE-16. Source: Ref 59

anodic dissolution when exposed to a corrosive environment. Inclusions and precipitates can exhibit potential differences with respect to the surrounding matrix, as can plastically deformed (strained) and undeformed regions within a material. These anode-cathode couplings can initiate and propagate dissolution cracks or fissures without regard to hydrogen.

Although other mechanisms may operate (Ref 175-178), including the adsorption of unspecified damaging species (Ref 177) and the occurrence of a strain-induced martensitic transformation (Ref 178), dezincification or dealloying (Ref 145-148) appears to be the principal SCC mechanism in brass (copper-zinc and copper-zinc-tin alloys). Dezincification is the preferential dissolution or loss of zinc at the fracture interface during SCC, which can result in the corrosion products having a higher concentration of zinc than the adjacent alloy. This dynamic loss of zinc near the crack aids in propagating the stress-corrosion fracture.

Some controversy remains regarding the precise mechanics of dezincification. One mechanism assumed that both zinc and copper are dissolved and that the copper is subsequently redeposited, while the other process involves the diffusion of zinc from the alloy, resulting in a higher concentration of copper in the depleted zone (Ref 179). However, there is evidence that both processes may operate (Ref 180).

Like hydrogen embrittlement, SCC can change the mode of fracture from dimple rupture to intergranular decohesion or cleavage,

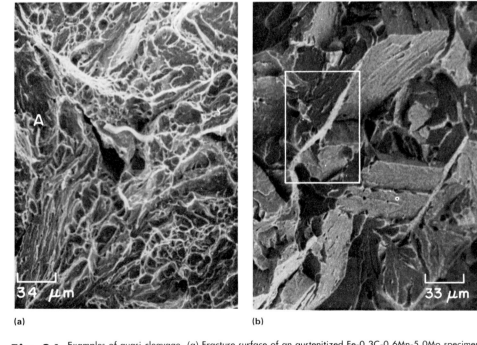

Fig. 36 Examples of quasi-cleavage. (a) Fracture surface of an austenitized Fe-0.3C-0.6Mn-5.0Mo specimen exhibiting large quasi-cleavage facets, such as at A; elsewhere, the surface contains rather large dimples. (b) Charpy impact fracture in an Fe-0.18C-3.85Mo steel. Many quasi-cleavage facets are visible. The rectangle outlines a tear ridge.

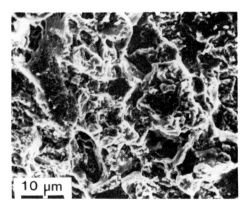

Fig. 37 Small and poorly defined quasi-cleavage facets connected by shallow dimples on the surface of a type 234 tool steel. (D.-E. Huang, Fuxin Mining Institute, and C.R. Brooks, University of Tennessee)

Modes of Fracture / 27

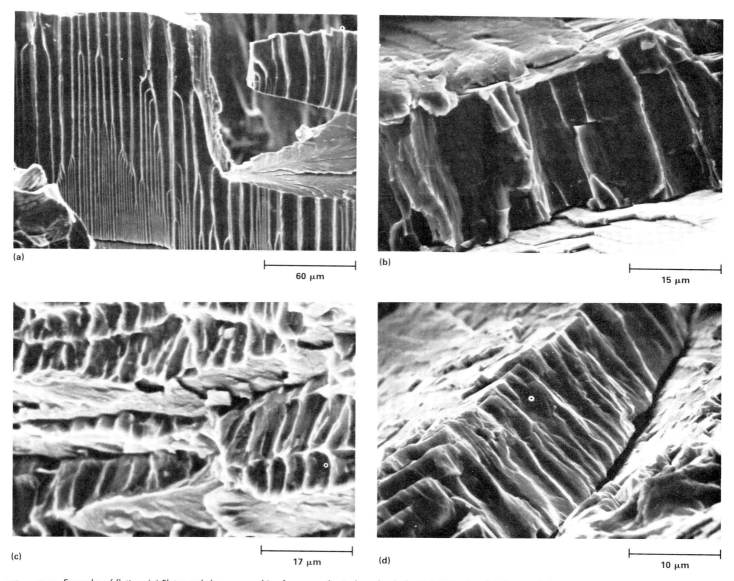

Fig. 38 Examples of fluting. (a) Flutes and cleavage resulting from a mechanical overload of a Ti-0.35O alloy. (b) Flutes and cleavage resulting from SCC at β-annealed Ti-8Al-1Mo-1V alloy in methanol. (c) Flutes and cleavage in β-annealed Ti-8Al-1Mo-1V resulting from sustained-load cracking in vacuum. (d) Flutes occurring near the notch on the fracture surface of mill-annealed Ti-8Al-1Mo-1V resulting from corrosion fatigue in saltwater. Source: Ref 89

although quasi-cleavage has also been observed. The change in fracture mode is generally confined to that portion of the fracture that propagated by SCC, but it may extend to portions of the rapid fracture if a hydrogen embrittlement mechanism is involved.

Stress-corrosion fractures that result from hydrogen embrittlement closely resemble those fractures; however, stress-corrosion cracks usually exhibit more secondary cracking, pitting, and corrosion products. Of course, pitting and corrosion products could be present on a clean hydrogen embrittlement fracture exposed to a corrosive environment.

SCC of Steels. Examples of known stress-corrosion fractures are shown in Fig. 50 to 56. Steels, including the stainless grades, stress corrode in such environments as water, seawater, chloride- and nitrate-containing solutions, and acidic as well as basic solutions, such as those containing sodium hydroxide or hydrogen sulfide. Stress-corrosion fractures in high-strength quench-and-temper hardenable or precipitation-hardenable steels occur primarily by intergranular decohesion, although some transgranular fracture may also be present.

Figure 50 shows a stress-corrosion fracture in an HY-180 quench-and-temper hardenable steel tested in aqueous 3.5% sodium chloride. The stress-corrosion fracture was believed to have occurred predominantly by hydrogen embrittlement (Ref 154). Increasing the stress intensity coefficient, K_I, resulted in a decreased tendency for intergranular decohesion; however, the oppposite was true for a cold-worked type 316 austenitic stainless steel tested in boiling aqueous magnesium chloride (Ref 181). It was shown that increasing K_I or increasing the negative electrochemical potential resulted in an increased tendency toward intergranular decohesion (Fig. 51). When the 300 type stainless steels are sensitized—a condition that results in the precipitation of chromium carbides at the grain boundaries, causing depletion of chromium in the adjacent material in the grains— the steel becomes susceptible to SCC, which occurs principally along grain boundaries.

Figure 52 shows the effect of the electrochemical potential, E, on the fracture path in a cold-worked AISI C-1018 low-carbon steel that stress corroded in a hot sodium hydroxide solution. At an electrochemical potential of $E = -0.76$ V_{SHE}, the fracture path is predominantly intergranular; at a freely corroding potential of $E = -1.00$ V_{SHE}, the fracture path is transgranular (Ref 182).

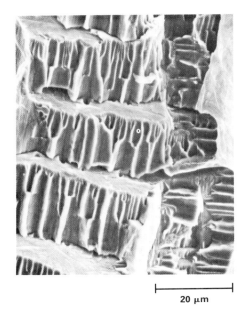

Fig. 39 Flutes and cleavage resulting from SCC of β-annealed Ti-8Al-1Mo-1V in methanol. Source: Ref 89

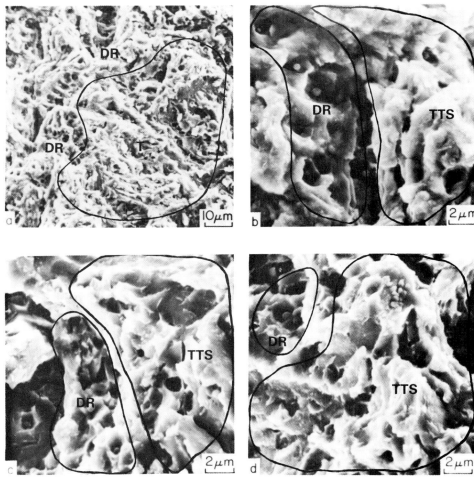

Fig. 40 Appearance of TTS fracture in bainitic HY-130 steel. (a) Areas of complex tearing (T) and dimple rupture (DR). (b) Detail from upper left corner of (a) showing particle-nucleated dimples (DR) and region of TTS. SEM fractographs in (c) and (d) show additional examples of TTS fractures. Source: Ref 91, 93

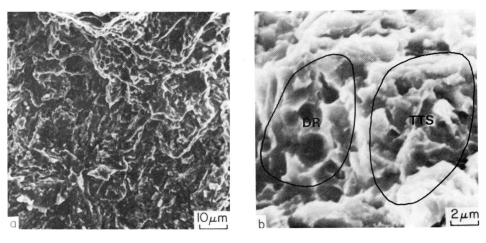

Fig. 41 Appearance of TTS fracture. (a) An essentially 100% pearlitic eutectoid steel (similar to AISI 1080) where fracture propagates across pearlite colonies. (b) Fractograph showing dimple rupture (DR) and TTS fracture in a quenched-and-tempered (martensitic) HY-130 steel. Source: Ref 91, 93

SCC of Aluminum. Aluminum alloys, especially the 2000 and 7000 series, that have been aged to the high-strength T6 temper or are in an underaged condition are susceptible to SCC in such environments as moist air, water, and solutions containing chlorides. The sensitivity to SCC depends strongly on the grain orientation with respect to the principal stress, the short-transverse direction being the most susceptible to cracking. Figure 53 shows examples of stress-corrosion fractures in a 7075-T6 (maximum tensile strength: 586 MPa, or 85 ksi) aluminum alloy that was tested in water. The fracture occurred primarily by intergranular decohesion.

SCC of brass in the presence of ammonia and moist air has long been recognized. The term season cracking was used to describe the SCC of brass that appeared to coincide with the moist weather in the spring and fall. Environments containing nitrates, sulfates, chlorides, ammonia gas and solutions, and alkaline solutions are known to stress corrode brass. Even distilled water and water containing as little as $5 \times 10^{-3}\%$ sulfur dioxide have been shown to attack brass (Ref 176, 178). Depending on the arsenic content of the Cu-30Zn brass, SCC in distilled water occurs either by intergranular decohesion or by a combination of cleavage and intergranular decohesion (Fig. 54). When brass containing 0.032% As is stress corroded in water containing minute amounts of sulfur dioxide, it exhibits a unique transgranular fracture containing relatively uniformly spaced, parallel markings (Fig. 55). These distinct periodic marks apparently represent the stepwise propagation of the stress-corrosion fracture.

SCC of titanium alloys has been observed in such environments as distilled water, seawater, aqueous 3.5% sodium chloride, chlorinated organic solvents, methanol, red fuming nitric acid, and molten salts. Susceptiblity depends on such variables as the microstructure (Ref 183-185), the amount of internal hydrogen (Ref 186), the state of stress (Ref 187, 188), and strength level (Ref 188). In general, microstructures consisting of large-grain α phase or containing substantial amounts of α phase in relation to β phase, high levels of

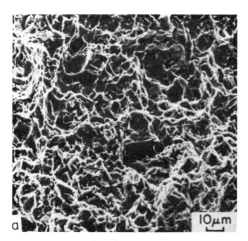
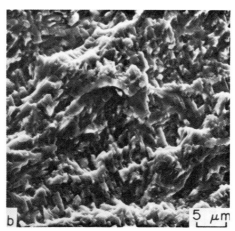

Fig. 42 Examples of TTS fracture in Ti-6Al-4V α-β alloys. (a) Solution treated and aged microstructure consisting of about 10-μm diam primary α particles in a matrix of about 70 vol% of fine Widmanstätten α and β. The microstructural constituents are not evident on the fracture surface as verified by the plateau-etching technique (Ref 91, 94). (b) Fractograph of a β-quenched Ti-6Al-4V alloy consisting of a fine Widmanstätten martensitic microstructure. The tearing portions of the fracture surface exhibit TTS.

internal hydrogen, the presence of a triaxial state of stress, and high yield strengths all promote the susceptibility of an alloy to SCC. If hydrogen is present in the corrosive environment, SCC will probably occur by a hydrogen embrittlement mechanism. Depending on the environment, alloy, and heat treatment (microstructure), mild stress-corrosion attack can exhibit a fracture that cannot be readily distinguished from normal overload, while more severe attack results in cleavage or quasi-cleavage fracture.

Figure 56 shows a stress-corrosion fracture in an annealed Ti-8Al-1Mo-1V alloy that was tested in aqueous 3.5% sodium chloride. The stress-corrosion fractures in titanium alloys exhibit both cleavage (along with fluting) and quasi-cleavage.

Corrosion products are a natural by-product of corrosion, particularly on most steels and aluminum alloys. They not only obscure fracture detail but also cause permanent damage, because a portion of the fracture surface is chemically attacked in forming the corrosion products. Therefore, removing the corrosion products will not restore a fracture to its original condition. However, if the corrosion damage is moderate, enough surface detail remains to identify the mode of fracture.

Depending on the alloy and the environment, corrosion products can appear as powdery residue, amorphous films, or crystalline deposits. Corrosion products may exhibit cleavage fracture and secondary cracking. Care must be exercised in determining whether these fractures are part of the corrosion product or the base alloy. Some of the corrosion products observed on an austenitic stainless steel and a niobium alloy are shown in Fig. 57 and 58, respectively. Detailed information on the cleaning of fracture surfaces is available in the article "Preparation and Preservation of Fracture Specimens" in this Volume.

Effect of Exposure to Low-Melting Metals. When metals such as certain steels, titanium alloys, nickel-copper alloys, and aluminum alloys are stressed while in contact with low-melting metals, including lead, tin, cadmium, lithium, indium, gallium, and mercury, they may be embrittled and fracture at a stress below the yield strength of the alloy. If the embrittling metal is in a liquid state during exposure, the failure is referred to as liquid-metal embrittlement (LME); when the metal is solid, it is known as solid-metal embrittlement (SME). Both failure processes are sometimes called stress alloying.

Temperature has a significant effect on the rate of embrittlement. For a specific embrittling metal species, the higher the temperature, the more rapid the attack. In addition, LME is a faster process than SME. In fact, under certain conditions, LME can occur with dramatic speed. For liquid indium embrittlement of steel, the time to failure appears to be limited primarily by the diffusion-controlled period required to form a small propagating crack (Ref 189). Once the crack begins to propagate, failure can occur in a fraction of a second. For example, when an AISI 4140 steel that was heat treated to an ultimate tensile strength of 1500 MPa (218 ksi) was tested at an applied stress of 1109 MPa (161 ksi) (the approximate proportional limit of the material) while in contact with liquid indium at a temperature of 158 °C (316 °F) (indium melts at 156 °C, or 313 °F), crack formation required about 511 s. The crack then propagated and fractured the 5.84-mm (0.23-in.) diam electropolished round bar specimen in only 0.1 s (Ref 189). In contrast, at 154 °C (309 °F), when the steel was in contact with solid indium, crack nucleation required 4.07×10^3 s (1.13 h), and failure required an additional 2.41×10^3 s (0.67 h) (Ref 189).

Although gallium and mercury rapidly embrittle aluminum alloys, all cases of LME and, especially, SME do not occur in such short time spans. The embrittlement of steels and titanium alloys by solid cadmium can occur over months of exposure; however, when long time spans are involved, the generation of hydrogen by the anodic dissolution of cadmium in a service environment can result in a hydrogen embrittlement assisted fracture. The magnitude of the applied stress, the strain rate, the amount of prior cold work, the grain size, and the grain-boundary composition can also influence the rate of embrittlement. In general, higher applied stresses and lower strain rates promote embrittlement (Ref 112), while an increase in the amount of cold work reduces embrittlement (Ref 189). The reduction in embrittlement from cold work is believed to be due to the increase in the dislocation density within grains providing a large number of additional diffusion paths to dilute the concentration of embrittling atoms at grain boundaries.

Smaller grain size should reduce embrittlement because of reduced stress concentration at grain-boundary dislocation pile-ups (Ref 189); however, in the embrittlement of Monel 400 by mercury, maximum embrittlement is observed at an approximate grain size of 250 μm (average grain diameter), and the embrittlement decreases for both the smaller and the larger grain sizes (Ref 112). The decrease in embrittlement at the smaller grain sizes was attributed to a difficulty in crack initiation, and for the larger grain sizes, the effect was due to enhanced plasticity (Ref 112). An example of a Monel specimen embrittled by liquid mercury is shown in Fig. 32.

When fracture occurs by intergranular decohesion, the presence of such elements as lead, tin, phosphorus, and arsenic at grain boundaries can affect the embrittlement mechanism. The segregation of tin and lead at grain boundaries of steel can make it more susceptible to embrittlement by liquid lead, while a similar grain-boundary enrichment by phosphorus and arsenic reduces it (Ref 190). Grain-boundary segregation of phosphorus has also been shown to reduce the embrittlement of nickel-copper alloys, such as Monel 400, by mercury (Ref 191, 192). It has been suggested that the beneficial effects of phosphorus are due to a modification in the grain-boundary composition that results in improved atomic packing at the boundary (Ref 192).

The mechanisms proposed to explain the low-melting metal embrittlement process are often similar to those suggested for hydrogen embrittlement. Some of the mechanisms assume a reduction in the cohesive strength and enhancement of shear as a result of adsorption of the embrittling metal atoms (Ref 112, 114, 189, 193). It has also been suggested that the diffusion of a low melting point metal into the alloy results in enhanced dislocation nucleation at the crack tip (Ref 123, 127, 194). A modified theory for crack initiation is based on stress and dislocation-assisted diffusion of the embrittling metal along dislocation networks and grain

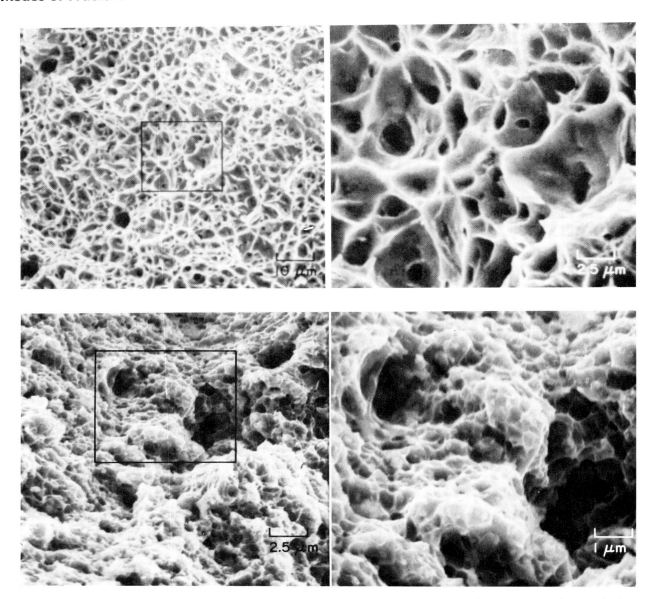

Fig. 43 Effect of hydrogen on fracture appearance in 13-8 PH stainless steel with a tensile strength of 1634 MPa (237 ksi). Top row: SEM fractographs of a specimen not embrittled by hydrogen. Bottom row: SEM fractographs of a specimen charged with hydrogen by plating without subsequent baking.

boundaries (Ref 189). The diffused atoms lower the crack resistance and make slip more difficult; when a critical concentration of the embrittling species has accumulated in the penetration zone, a crack initiates. The mechanism for the extremely rapid crack propagation for LME is not well understood.

Diffusion processes are far too slow to transport the embrittling liquid metal to the rapidly advancing crack front. For embrittlement by liquid indium, it has been proposed that the transport occurs by a bulk liquid flow mechanism (Ref 189, 195); for the SME mode, the crack propagation is sustained by a much slower surface self-diffusion of the embrittling metal to the crack tip (Ref 189).

Examples of low-melting metal embrittlement fractures are shown in Fig. 32 and 59 to 61. Figure 59 shows fractures in AISI 4140 steel resulting from testing in argon and in liquid lead. Figure 60 shows the embrittlement of a 7075-T6 aluminum alloy by mercury, and Fig. 61 shows the embrittlement of AISI 4140 steel by liquid cadmium. The articles "Liquid-Metal Embrittlement" and "Embrittlement by Solid-Metal Environments" in Volume 11 of the 9th edition of *Metals Handbook* provide additional information on the effect of exposure to low melting point metals.

Effect of State of Stress. This section will briefly discuss some effects of the direction of the principal stress as well as the state of stress, that is, uniaxial or triaxial, on the fracture modes of various metal systems. This section will not, however, present any mathematical fracture mechanics relationships describing the state of stress or strain in a material. The effects of stress will be discussed in general terms only.

The effect of the direction of the applied stress has been presented in the section "Dimple Rupture" in this article. Briefly, the direction of the principal stress affects the dimple shape. Stresses acting parallel to the plane of fracture (shear stresses) result in elongated dimples, while a principal stress acting normal to the plane of fracture results in primarily equiaxed dimples. Because the local fracture planes often deviate from the macroscopic plane and because the fracture is usually the result of the combined effects of tensile and shear stresses, it generally exhibits a variety of dimple shapes and orientations.

The state of stress affects the ability of a material to deform. A change from a uniaxial to biaxial to triaxial state of stress decreases the ability of a material to deform in response to the

Modes of Fracture / 31

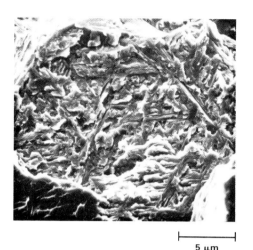

Fig. 44 Quasi-cleavage fracture in a hydrogen-embrittled AISI 4340 steel heat treated to an ultimate tensile strength of 2082 MPa (302 ksi). Source: Ref 138

applied stresses. As a result, metals sensitive to such changes in the state of stress exhibit a decrease in elongation or reduction of area at fracture and in extreme cases may exhibit a change in the fracture mode.

The fcc metals, such as the aluminum alloys and austenitic stainless steels, and the hcp metals, such as the titanium and zirconium alloys, are generally unaffected by the state of stress. Although there can be a change in the nature of the dimples under biaxial or triaxial stresses, namely a reduction in dimple size and depth (Ref 196, 197), fcc and hcp metal systems usually do not exhibit a change in the mode of fracture. However, the bcc metals, such as most iron-base alloys and refractory metals, can exhibit not only smaller and shallower dimples but also a change in the fracture mode in response to the restriction on plastic deformation. This response depends on such variables as the strength level, microstructure, and the intensity of the triaxial stress. When a change in the fracture mode does occur as a result of a triaxial state of stress, such as that present near the root of a sharp notch, the mode of rupture can change from the normal dimple rupture to quasi-cleavage or intergranular decohesion (Ref 198). These changes in fracture mode are most evident in the general region of the fracture origin and may not be present over the entire fracture surface.

Figure 62 shows the effect of a biaxial state of stress on dimples in a basal-textured Ti-6Al-4V alloy. Under a biaxial state of stress, the size and the depth of the dimples decreased. For a pearlitic AISI 4130 steel (Ref 198) and a PH 13-8 precipitation-hardenable stainless steel, a triaxial state of stress resulting from the presence of a notch with a stress concentration factor of at least $K_t = 2.5$ can change the fracture mode from dimple rupture to quasi-cleavage (Fig. 63). When a high-strength AISI 4340 steel is subjected to a triaxial stress, the mode of fracture can change from dimple rupture to intergranular decohesion.

Effect of Strain Rate. The strain rate is a variable that can range from the very low rates observed in creep to the extremely high strain rates recorded during impact or shock loading by explosive or electromagnetic impulse.

Very low strain rates (about 10^{-9} to 10^{-7} s^{-1}) can result in creep rupture, with the accompanying changes in fracture mode that have been presented in the section "Creep Rupture" in this article.

At moderately high strain rates (about 10^2 s^{-1}), such as experienced during Charpy impact testing, the effect of strain rate is generally similar to the effect of the state of stress, namely that the bcc metals are more affected by the strain rate than the fcc or the hcp metals. Because essentially all strain rate tests at these moderate strain rates are Charpy impact tests that use a notched specimen, the effect of strain rate is enhanced by the presence of the notch, especially in steels when they are tested below the transition temperature.

A moderately high strain rate either alters the size and depth of the dimples or changes the mode of fracture from dimple rupture to quasi-cleavage or intergranular decohesion. For example, when an AISI 5140 H steel that was tempered at 500 °C (930 °F) was tested at Charpy impact rates, it exhibited a decrease in the width of the stretched zone adjacent to the precrack and an increase in the amount of intergranular decohesion facets (Fig. 64). The same steel tempered at 600 °C (1110 °F) showed no significant effect of the Charpy impact test (Ref 199).

At very high strain rates, such as those observed during certain metal-shearing operations, high-velocity (100 to 3600 m/s, or 330 to 11 800 ft/s) projectile impacts or explosive rupture, materials exhibit a highly localized deformation known as adiabatic* shear (Ref

---
*Adiabatic process is a thermodynamic concept where no heat is gained or lost to the environment.

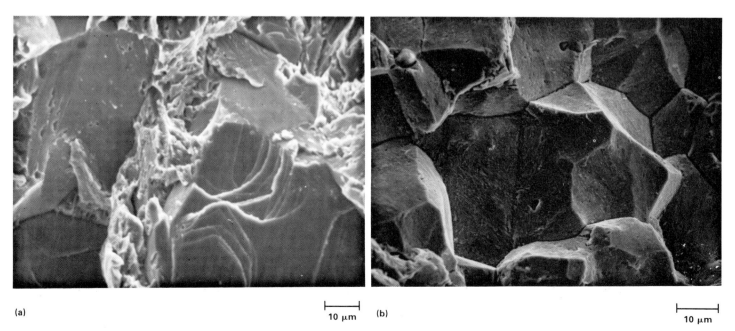

Fig. 45 Examples of hydrogen-embrittled steels. (a) Cleavage fracture in a hydrogen-embrittled annealed type 301 austenitic stainless steel. Source: Ref 98. (b) Intergranular decohesive fracture in an AISI 4130 steel heat treated to an ultimate tensile strength of 1281 MPa (186 ksi) and stressed at 980 MPa (142 ksi) while being charged with hydrogen. Source: Ref 111

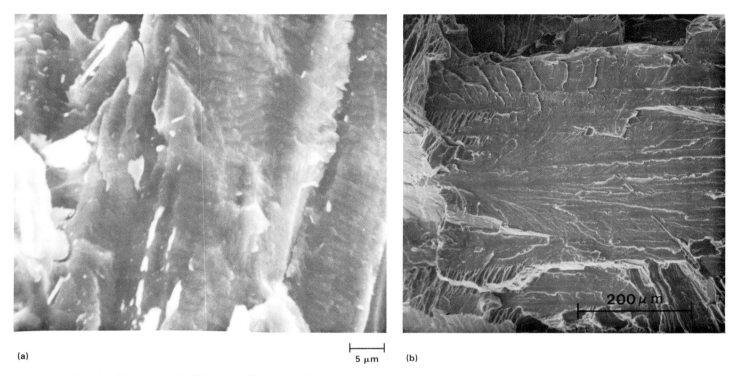

Fig. 46 Examples of hydrogen-embrittled titanium alloys. (a) Hydrogen embrittlement fracture in a Ti-8Al-1Mo-1V alloy in gaseous hydrogen. Note crack-arrest marks. Source: Ref 137. (b) Cleavage fracture in hydrogen-embrittled Ti-5Al-2.5Sn alloy containing 90 ppm H. Source: Ref 141

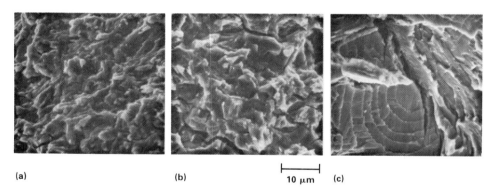

Fig. 47 Influence of heat treatment and resulting microstructure on the fracture appearance of a hydrogen-embrittled Ti-6A-4V alloy. Specimens tested in gaseous hydrogen at a pressure of 1 atm. (a) Transgranular fracture in a specimen heat treated at 705 °C (1300 °F) for 2 h, then air cooled. (b) Intergranular decohesion along α-β boundaries in a specimen heat treated at 955 °C (1750 °F) for 40 min, then stabilized. (c) Coarse acicular structure resulting from heating specimen at 1040 °C (1900 °F) for 40 min, followed by stabilizing. The relatively flat areas of the terraced structure are the prior-β grain boundaries. See text for a discussion of the microstructures of these specimens. Source: Ref 142

200-208). In adiabatic shear, the bulk of the plastic deformation of the material is concentrated in narrow bands within the relatively undeformed matrix (Fig. 65 to 67). Adiabatic shear has been observed in a variety of materials, including steels, aluminum and titanium alloys, and brass.

These shear bands are believed to occur along slip planes (Ref 201, 202), and it has been estimated that under certain conditions, such as from the explosive-driven projectile impact of a steel target, the local strain rate within the adiabatic shear bands in the steel can reach 9×10^5 s^{-1} and the total strain in the band can be as high as 532% (Ref 204). An estimated $3 \times 10^6 \cdot$s^{-1} strain rate has been reported for shear bands in a 2014-T6 aluminum alloy block impacted by a gun-fired (up to 900 m/s, or 2950 ft/s) steel projectile (Ref 205).

The extremely high strain rates within the adiabatic shear bands result in a rapid increase in temperature as a large portion of the energy of deformation is converted to heat. It has been estimated that the temperature can go high enough to melt the material within the bands (Ref 205, 206). The heated material also cools very rapidly by being quenched by the large mass of the cool, surrounding matrix material; therefore, in quench-and-temper hardenable steels, the material within the bands can contain transformed untempered martensite. This transformed zone is shown schematically in Fig. 65(b).

The hardness in the transformed bands is sometimes higher than can be obtained by conventional heat treating of the steel. This increase in hardness has been attributed to the additive effects of lattice hardening due to supersaturation by carbon on quenching and the extremely fine grain size within the band (Ref 203). However, for an AISI 1060 carbon steel, the hardness of the untempered martensite bands was no higher than that which could be obtained by conventional heat treating (Ref 206). In both cases, the hardness of the adiabatic shear bands was independent of the initial hardness of the steel. For a 7039 aluminum alloy, however, the hardness of the shear bands was dependent on the hardness of the base material. The adiabatic shear bands in an 80-HV material exhibited an average peak hardness of about 100 HV, while those in a 150-HV material had an average peak hardness of about 215 HV (Ref 208). For the Ti-6Al-4V STA alloy shown in Fig. 66, there was no significant difference in hardness between the shear bands and the matrix. In materials that do not exhibit a phase transformation, or if the temperature generated during deformation is not high enough for the transformation to occur, the final hardness of

Modes of Fracture / 33

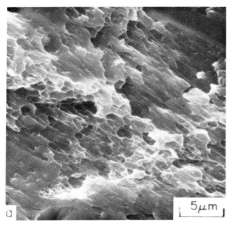 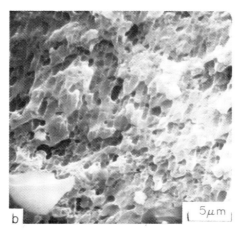

Fig. 48 Hydrogen-embrittled 2124-UT aluminum alloy that shows no significant change in the fracture appearance. (a) Not embrittled. (b) Hydrogen embrittled. Source: Ref 99

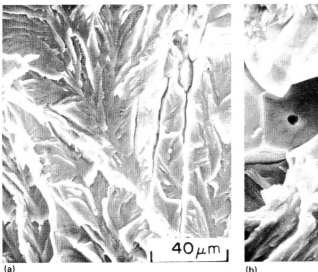 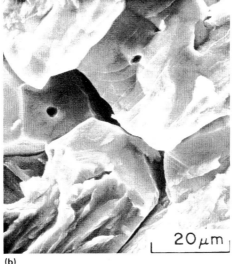

Fig. 49 Effect of heat treatment on the fracture appearance of a hydrogen-embrittled low-copper 7050 aluminum alloy. (a) Transgranular cleavagelike fracture in a peak-aged specimen. (b) Combined intergranular decohesion and transgranular cleavagelike fracture in an underaged specimen. Source: Ref 106

the adiabatic shear band is the net result of the competing effects of the increase in hardness due to the large deformation and the softening due to the increase in temperature.

The width of the adiabatic shear bands depends on the hardness (strength) of the material (Ref 206, 208). Generally, the harder the material, the narrower the shear bands. In a 7039 aluminum alloy aged to a hardness of 80 HV, the average band width resulting from projectile impact was 90 μm, while in a 150-HV material, the band width was only 20 μm (Ref 208). The average width of the shear bands observed in a Ti-6Al-4V STA alloy (average hardness, 375 HV_{1kg}) was 3 to 6 μm.

When an adiabatic shear band cracks or separates during deformation, the fractured surfaces often exhibit a distinct topography referred to as knobbly structure (Ref 205-208). The name is derived from the surface appearance, which resembles a mass of knoblike structures. The knobbly structure, which has been observed in 2014-T6 and 7039-T6 aluminum alloys, as well as in an AISI 4340 steel (Fig. 67) and AISI 1060 carbon steel, is believed to be the result of melting within the shear bands (Ref 205, 206). Although the cracked surfaces of adiabatic shear bands can exhibit a unique appearance, adiabatic shear failure is easiest to identify by metallographic, rather than fractographic, examination.

Effect of Temperature. Depending on the material, the test temperature can have a significant effect on the fracture appearance and in many cases can result in a change in the fracture mode. However, for materials that exhibit a phase change or are subject to a precipitation reaction at a specific temperature, it is often difficult to separate the effect on the fracture due to the change in temperature from that due to the solid-state reactions. In general, slip, and thus plastic deformation, is more difficult at low temperatures, and materials show reduced ductility and an increased tendency for more brittle behavior than at high temperatures.

A convenient means of displaying the fracture behavior of a specific material is a fracture map. When sufficient fracture mode data are available for an alloy, areas of known fracture mode can be outlined on a phase diagram or can be plotted as a function of such variables as the test temperature and strain rate (Fig. 68). Similar maps can also be constructed for low-temperature fracture behavior.

Effect of Low Temperature. Similar to the effect of the state of stress, low temperatures affect the bcc metals far more than the fcc or hcp metal systems (see the section "Effect of the State of Stress" in this article). Although lower temperatures can result in a decrease in the size and depth of dimples in fcc and hcp metals, bcc metals often exhibit a change in the fracture mode, which generally occurs as a change from dimple rupture or intergranular fracture to cleavage. For example, a fully pearlitic AISI 1080 carbon steel tested at 125 °C (255 °F) showed a fracture that consisted entirely of dimple rupture; at room temperature, only 30% of the fracture was dimple rupture, with 70% exhibiting cleavage. At −125 °C (−195 °F), the amount of cleavage fracture increased to 99% (Ref 210). This transition in fracture mode is illustrated in Fig. 69.

Charpy impact testing of an AISI 1042 carbon steel whose microstructure consisted of slightly tempered martensite (660 HV) as well as one containing a tempered martensite (335 HV) microstructure at 100 °C (212 °F) and at −196 °C (−320 °F) produced results essentially identical to those observed for the AISI 1080 steel. In both conditions, the fracture mode changed from dimple rupture at 100 °C (212 °F) to cleavage at −196 °C (−320 °F), as shown in Fig. 70. Similar changes in the fracture mode, including a change to quasi-cleavage, can be observed for other quench-and-temper and precipitation-hardenable steels.

A unique effect of temperature was observed in a 0.39C-2.05Si-0.005P-0.005S low-carbon steel that was tempered 1 h at 550 °C (1020 °F) to a hardness of 30 HRC and Charpy impact tested at room temperature and at −85 °C (−120 °F) (Fig. 71). In this case, the fracture exhibited intergranular decohesion at room temperature and changed to a combination of intergranular decohesion and cleavage at −85 °C (−120 °F). This behavior was attributed to the intrinsic reduction in matrix toughness by the silicon in the alloy, because when nickel is substituted for the silicon the matrix toughness

34 / Modes of Fracture

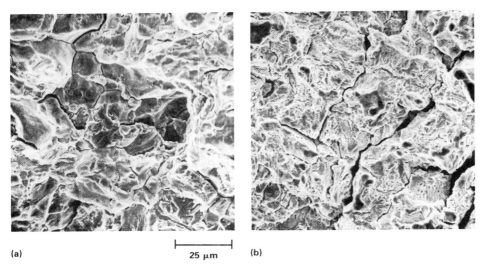

Fig. 50 Stress-corrosion fractures in HY-180 steel with an ultimate strength of 1450 MPa (210 ksi). The steel was tested in aqueous 3.5% sodium chloride at an electrochemical potential of $E = -0.36$ to -0.82 V_{SHE} (SHE, standard hydrogen electrode). Intergranular decohesion is more pronounced at lower values of stress intensity, $K_I = 57$ MPa$\sqrt{m}$ (52 ksi$\sqrt{in.}$) (a), than at higher values, $K_I = 66$ MPa$\sqrt{m}$ (60 ksi$\sqrt{in.}$) (b). Source: Ref 154

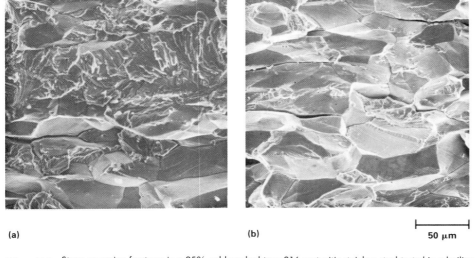

Fig. 51 Stress-corrosion fractures in a 25% cold-worked type 316 austenitic stainless steel tested in a boiling (154 °C, or 309 °F) aqueous 44.7% magnesium chloride solution. At low (14 MPa$\sqrt{m}$, or 12.5 ksi $\sqrt{in.}$) K_I values, the fracture exhibits a combination of cleavage and intergranular decohesion (a). At higher (33 MPa$\sqrt{m}$, or 30 ksi$\sqrt{in.}$) values of K_I the principal mode of fracture is intergranular decohesion (b). Source: Ref 181

is increased and no cleavage is observed (Ref 211).

The temperature at which a sudden decrease in the Charpy impact energy occurs is known as the ductile-to-brittle transition temperature for that specific alloy and strength level. Charpy impact is a severe test because the stress concentration effect of the notch, the triaxial state of stress adjacent to the notch, and the high strain rate due to the impact loading combine to add to the reduction in ductility resulting from the decrease in the testing temperature. Although temperature has a strong effect on the fracture process, a Charpy impact test actually measures the response of a material to the combined effect of temperature and strain rate.

The effects of high temperature on fracture are more complex because solid-state reactions, such as phase changes and precipitation, are more likely to occur, and these changes affect bcc as well as fcc and hcp alloys. As shown in Fig. 72, the size of the dimples generally increases with temperature (Ref 209, 212, 213). The dimples on transgranular fractures and those on intergranular facets in a 0.3C-1Cr-1.25Mo-0.25V-0.7Mn-0.04P steel that was heat treated to an ultimate strength of 880 MPa (128 ksi) show an increase in size when tested at temperatures ranging from room temperature to 600 °C (1110 °F) (Ref 213).

Figure 73 shows the effect of temperature on the fracture mode of an ultralow-carbon steel. The steel, which normally fractures by dimple rupture at room temperature, fractured by intergranular decohesion when tensile tested at a strain rate of 2.3×10^{-2} s^{-1} at 950 °C (1740 °F). The change in fracture mode was due to the precipitation of critical submicron-size MnS precipitates at the grain boundaries. This embrittlement can be eliminated by aging at 1200 °C (2190 °F), which coarsens the MnS precipitates (Ref 209).

A similar effect was observed for Inconel X-750 nickel-base alloy that was heat treated by a standard double-aging process and tested at a nominal strain rate of 3×10^{-5} s^{-1} at room temperature and at 816 °C (1500 °F). The fracture path was intergranular at room temperature and at 816 °C (1500 °F), except that the room-temperature fracture exhibited dimples on the intergranular facets and those resulting from fracture at 816 °C (1500 °F) did not (Fig. 74). The fracture at room temperature exhibited intergranular dimple rupture because the material adjacent to the grain boundaries is weaker due to the depletion of coarse γ' precipitates. The absence of dimples at 816 °C (1500 °F) was the result of intense dislocation activity along the grain boundaries, producing decohesion at $M_{23}C_6$ carbide/matrix interfaces within the boundaries (Ref 214).

A distinct change in fracture appearance was also noted during elevated-temperature tensile testing of Haynes 556, which had the following composition:

Element	Composition, %
Iron	28.2
Chromium	21.5
Nickel	22.2
Cobalt	19.0
Tungsten	2.9
Molybdenum	2.9
Tantalum	0.8
Manganese	1.4
Silicon	0.5
Copper	0.1
Nitrogen	0.1

Three specimens were tested at a strain rate of approximately 1 s^{-1} at increasing temperatures. At 1015 °C (1860 °F), specimen 1 underwent a 72% reduction of area and fractured by dimple rupture (Fig. 75a). At 1253 °C (2287 °F), specimen 2 exhibited an 8% reduction of area. Fracture occurred intergranularly by grain-boundary decohesion (Fig. 75b). Specimen 3, which was tested at 1333 °C (2431 °F), fractured because of local eutectic melting of TaC + austenite, with 0% reduction of area (Fig. 75c).

A final example of the effect of temperature on the fracture process is the behavior of a titanium alloy when tested at room temperature and at 800 °C (1470 °F). At room temperature, the fracture in a Ti-6Al-2Nb-1Ta-0.8Mo alloy

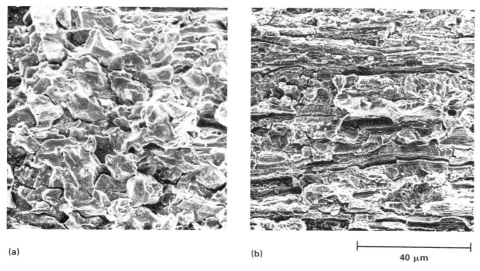

Fig. 52 Effect of electrochemical potential on the stress-corrosion fracture path in a cold-worked AISI C-1018 low-carbon steel with a 0.2% offset yield strength of 63 MPa (9 ksi). The steel was tested in a 92-°C (198-°F) aqueous 33% sodium hydroxide solution. At a potential of $E = -0.76\,V_{SHE}$, the fracture propagates along grain boundaries (a) by a metal dissolution process; however, at a freely corroding potential of $E = -1.00\,V_{SHE}$, the fracture path is transgranular and occurs by a combination of hydrogen embrittlement and metal dissolution (b). Source: Ref 182

Fig. 53 Stress-corrosion fractures from two different areas in a 7075-T6 aluminum alloy specimen exposed to water at ambient temperature. The fracture exhibits intergranular decohesion, although some dimple rupture is present near center of fracture in (a).

Modes of Fracture / 35

Figures 77 to 79 show the effects of high-temperature air exposure on the overload fracture surfaces of two titanium alloys and a steel. The progressive deterioration of the fracture surface of an annealed Ti-6Al-2Sn-4Zr-6Mo alloy exposed for various times at 700 °C (1290 °F) in air is illustrated in Fig. 77. As seen in Fig. 77(b), the oxide formed after only a 3-min exposure already obscured the fine ridges of the smaller dimples. An example of an extremely severe oxidation attack is shown in Fig. 78. The oxide cover is so complete that it is not possible to identify the fracture mode. A similar result was observed for a 300M high-strength steel fracture exposed for only 5 min at 700 °C (1290 °F) (Fig. 79). The relatively short exposure formed an oxide film that completely covered the fracture surface and rendered even the most prominent features unrecognizable.

Effect of Environment on Fatigue

Of the different fracture modes, fatigue is the most sensitive to environment. Because fatigue in service occurs in a variety of environments, it is important to understand the effects of these environments on the fracture process. Environments can include reactive gases, corrosive liquids, vacuum, the way the load is applied, and the temperature at which the part is cycled.

Just as environments vary, the effects of the environments also vary, ranging from large increases in the fatigue crack propagation rates in embrittling or corrosive environments to substantial decreases in vacuum and at low temperatures. Because fatigue is basically a slip process, any environment that affects slip also affects the rate at which a fatigue crack propagates. In general, conditions that promote easy slip, such as elevated temperatures, or interfere with slip reversal (oxidation), enhance fatigue crack propagation and increase the fatigue striation spacing. Environments that suppress slip, such as low temperatures, or enhance slip reversal by retarding or preventing oxidation (vacuum) of newly formed slip surfaces decrease crack propagation rates, which decreases the fatigue striation spacing and in extreme cases obliterates striations.

In some embrittling or corrosive environments, however, the fatigue crack propagation rate can be affected not only by interfering with the basic slip process but also by affecting the material ahead of the crack front. This can result in the formation of brittle striations and in the introduction of quasi-cleavage, cleavage, or intergranular decohesion fracture modes. When a fatigue crack advances by one of these other fracture modes, the fracture segments technically are not fatigue fractures. However, because the occurrence of these fracture modes is a natural consequence of a fatigue crack propagating under the influence of a specific environment, these mixed fracture modes can be considered as valid a part of a fatigue fracture as fatigue striations.

(heat treated to produce a basket-weave structure consisting of Widmanstätten α + β + grain-boundary α with equiaxed prior-β grains) that was tested at an approximate strain rate of $3.3 \times 10^{-4}\,s^{-1}$ occurred predominantly by transgranular dimple rupture. At 800 °C (1470 °F), however, the alloy exhibited low ductility, which is associated with an intergranular dimple rupture (Fig. 76). The low ductility can be explained by void formation and coalescence along prior-β grain boundaries because of strain localization in the α phase within the grain boundaries (Ref 215).

As has been shown, testing temperature can significantly affect fracture appearance. However, in addition to temperature, such factors as the strain rate and solid-state reactions must be considered when evaluating the effect of temperature on the fracture process.

Effect of Oxidation. A natural consequence of high-temperature exposure is oxidation. Engineering alloys exposed to elevated temperatures in the presence of an oxidizing medium, such as oxygen (air), form oxides. The degree of oxidation depends on the material, the temperature, and the time at temperature. Oxidation, which consumes a part of the fracture surface in forming the oxide, can also obscure significant fracture detail.

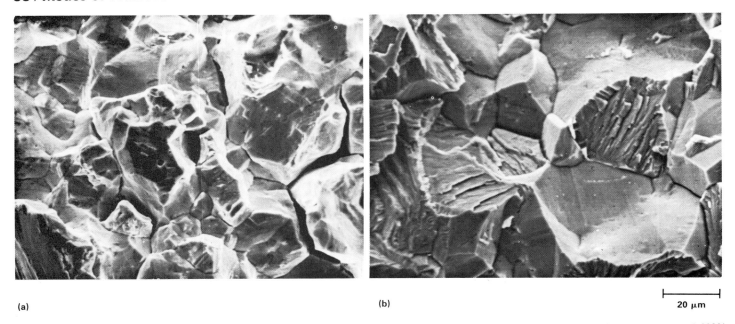

Fig. 54 Stress-corrosion fractures in a Cu-30Zn brass tested in distilled water at a potential of $E = 0$ V_{SCE} (SCE, saturated calomel electrode). Brass containing 0.002% As fails by predominantly intergranular decohesion (a), and one with 0.032% As fails by a combination of cleavage and intergranular decohesion (b). Source: Ref 176

Fig. 55 Stress-corrosion fracture in a Cu-30Zn brass with 0.032% As tested in water containing 5×10^{-3}% sulfur dioxide at a potential of $E = 0.05$ V_{SCE}. The periodic marks are believed to be the result of a stepwise mode of crack propagation. Source: Ref 176

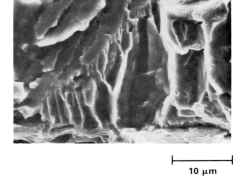

Fig. 56 Stress-corrosion fracture in an annealed Ti-8Al-1Mo-1V alloy tested in aqueous 3.5% sodium chloride. The fracture surface exhibits cleavage and fluting. Source: Ref 89

The effect of environment on fatigue fracture is divided into five principal categories:

- Effect of gaseous environments
- Effect of liquid environments
- Effect of vacuum
- Effect of temperature
- Effect of loading

The effect of these environments on fatigue fracture will be discussed, and where applicable, the effects will be illustrated with fractographs.

Effect of Gaseous Environments. If an environment such as air or hydrogen gas affects a material under static load conditions, it will generally affect the rate at which fatigue cracks propagate in that material, especially at low fatigue crack growth rates* when the environment has the greatest opportunity to exert its influence. The effect of hydrogen on steels of various compositions and strength levels is to increase the fatigue crack growth rate (Ref 216-222). When compared to dry air or an inert gas atmosphere at equivalent cyclic load conditions, hydrogen in steels accelerates the Stage II crack growth rate, often by a factor of ten or more (Ref 216-218, 221) and promotes the onset of Stage III and premature fracture (Ref

---

*The fatigue crack growth rate is expressed as da/dN, where a is the distance the fatigue crack advances during the application of N number of load cycles. When a fatigue striation is formed on each load cycle, the fatigue crack growth rate will about equal the striation spacing.

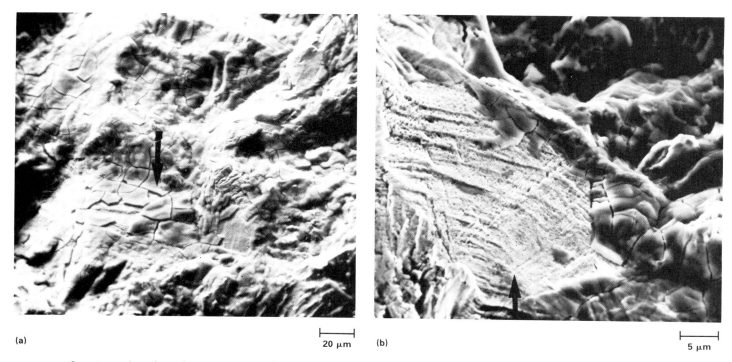

Fig. 57 Corrosion products observed on an austenitic stainless steel hip implant device. (a) View of the fracture surface showing a mud crack pattern (arrow) that obscures fracture details. (b) Surface after cleaning in acetone in an ultrasonic cleaner. Arrow points to region exhibiting striations and pitting. (C.R. Brooks and A. Choudhury, University of Tennessee)

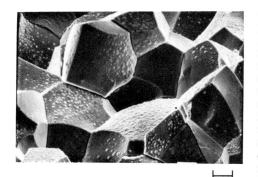

Fig. 58 Corrosion products on the intergranular fracture surface of an Nb-106 alloy. These corrosion products, which are residues from acid cleaning, contributed to failure by SCC. (L. Kashar, Scanning Electron Analysis Laboratories, Inc.)

218, 223, 224). Depending on the degree of embrittlement, the effect of hydrogen on the fracture appearance can range from one that is barely perceptible to one that exhibits brittle striations; however, the more common effect is the addition of quasi-cleavage, cleavage, or intergranular decohesion to the fracture modes visible on the fatigue fracture surfaces. The basic embrittlement mechanisms responsible for these changes are essentially the same as those discussed in the section "Effect of Environment on Dimple Rupture" in this article.

Effect of Gases on Steels. Fatigue testing of an API-5LX, grade X42 (0.26C-0.82Mn-0.014 Si-0.02Cu), 511-MPa (74-ksi) ultimate strength pipeline steel in hydrogen resulted in up to a 300-fold increase in the fatigue crack growth rate as compared to the crack growth rate in nitrogen gas (Ref 218). The fatigue tests were conducted at room temperature in dry nitrogen and dry hydrogen atmospheres (both at a pressure of 6.9 MPa, or 1 ksi), using a stress intensity range of ΔK = 6 to 20 MPa$\sqrt{m}$ (5.5 to 18 ksi$\sqrt{in.}$), a load ratio of R = 0.1 to 0.8, and a cyclic frequency of 1 to 5 Hz. An example of the effect of hydrogen on the fracture appearance is illustrated in Fig. 80. The fatigue fracture in hydrogen showed more bands of intergranular decohesion, which was associated with the ferrite in the microstructure, and fewer regions of a serrated fracture than the specimens tested in nitrogen. The mechanism responsible for the serrated fracture was not established; however, more than one mechanism may be involved. Compared to nitrogen, the serrated fracture in hydrogen exhibited little deformation, and at high values of ΔK, the serrated areas resembled cleavage or quasi-cleavage (Ref 218).

Although testing of the grade X42 pipeline steel in hydrogen increased the fatigue crack growth rate by a factor of nearly 300 over that in nitrogen gas, prechared ASTM A533B class 2 (0.22C-1.27Mn-0.46Mo-0.68Ni-0.15Cr-0.18Si) 790-MPa (115-ksi) ultimate tensile strength commercial pressure vessel steel showed only a maximum fivefold increase in the fatigue crack growth rate as compared to uncharged specimens (Ref 223). Lightly charged (240 h at 550 °C, or 1020 °F, in 17.2 MPa, or 2.5 ksi, hydrogen gas) and severely charged (1000 h at 550 °C, or 1020 °F, in 13.8 MPa, or 2 ksi, hydrogen gas) specimens were both fatigue tested in air at room temperature with a stress intensity range of ΔK = 7 to 50 MPa$\sqrt{m}$ (65.5 ksi$\sqrt{in.}$), a load ratio of R = 0.05 to 0.75, and a cyclic frequency of 50 Hz.

Compared to uncharged material, the lightly charged specimens were found to show only a moderate increase in the fatigue crack growth rate and only at crack propagation rates of less than 10^{-6} mm/cycle. There was no significant change in fracture appearance. The severely charged material showed a large decrease in mechanical properties: the 0.2% yield strength decreased from 660 to 242 MPa (96 to 35 ksi), the ultimate tensile strength from 790 to 315 MPa (115 to 46 ksi), the percent elongation from 22.4 to 9%, and the percent reduction of area from 73 to 5%. However, there was only a slight increase in the fatigue crack growth rate at growth rates of less than 10^{-6} and greater than 10^{-5} mm/cycle, although the fatigue fracture surfaces showed evidence of substantial hydrogen attack in the form of cavitated intergranular fracture (Fig. 81). The cavities on the intergranular facets were due to the formation of methane gas bubbles or cavities at the grain boundaries.

The reaction of hydrogen with the carbon in the steel results in methane gas production and depletion of carbon from the matrix (decarburization). The surprisingly small effect that the severe hydrogen damage had on the fatigue crack growth rate was due to the competing effects of two opposing processes. The accel-

38 / Modes of Fracture

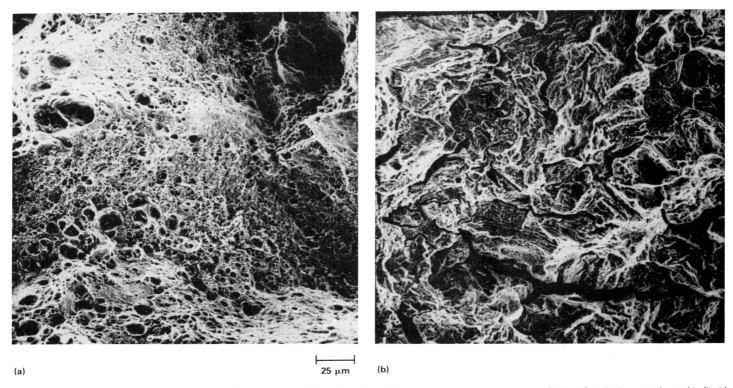

Fig. 59 Influence of lead on the fracture morphology of an AISI 4340 steel. (a) Ductile failure after testing in argon at 370 °C (700 °F). (b) Same steel tested in liquid lead at 370 °C (700 °F) showing brittle intergranular fracture

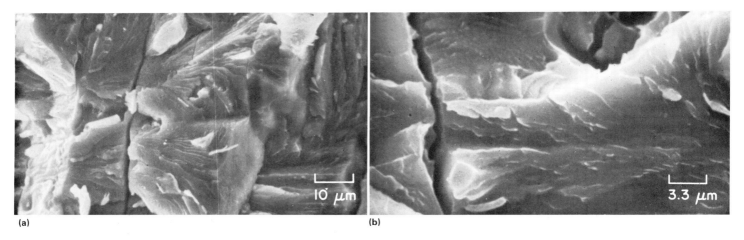

Fig. 60 Cleavage fracture resulting from exposure of a 7075-T6 aluminum alloy to mercury vapor during a slow-bend fracture toughness test. Both (a) and (b), which is at a higher magnification, clearly show cleavage facets and secondary cracking.

erating effect of the hydrogen damage was nearly offset by the decarburization softening-induced enhancement of crack closure (the premature contact between crack walls during the declining-load portion of the fatigue cycle), which lowers the crack tip stress intensity (Ref 223). Another factor that contributed to a lowering of the crack propagation rate was the tortuous crack path resulting from grain-boundary cavitation.

A unique effect was observed in an iron-nickel-chromium-molybdenum-vanadium (0.24C-3.51Ni-1.64Cr-0.39Mo-0.11V-0.28Mn-0.01Si) 882-MPa (128-ksi) ultimate tensile strength rotor steel that was fatigue tested in air and hydrogen. The fatigue tests were conducted at 93 °C (200 °F) in 30 to 40% relative humidity air and in 448-kPa (65-psi) dry hydrogen gas at a stress intensity range of $\Delta K = 3$ to 30 MPa$\sqrt{m}$ (2.5 to 27 ksi$\sqrt{in.}$), a load ratio of $R = 0.1$ to 0.8, and a cyclic frequency of 120 Hz. From near threshold ($\Delta K \cong 3$ MPa$\sqrt{m}$, or 2.5 ksi$\sqrt{in.}$) fatigue conditions, the fatigue crack growth rate in hydrogen was found to be lower than in air (Ref 220). As the ΔK increased, the crack growth rates in hydrogen increased at a rate faster than those in air, and at about $\Delta K = 12$ MPa$\sqrt{m}$ (11 ksi$\sqrt{in.}$), the fatigue crack growth rate in hydrogen began to exceed the crack growth rate in air and kept exceeding it at the higher levels of ΔK. Although the precise mechanism for this unusual behavior is not known, it was postulated that at the lower values of ΔK water from the moist air could be adsorbed and dissociated at the crack tip to supply atomic hydrogen to the steel faster than could be supplied by the dissociation of the hydrogen gas. However, at ΔK levels exceeding 12 MPa$\sqrt{m}$ (11 ksi$\sqrt{in.}$), the dissociation

Modes of Fracture / 39

Fig. 61 Intergranular fracture surface of an AISI 4140 low-alloy steel nut that failed because of embrittlement by liquid cadmium

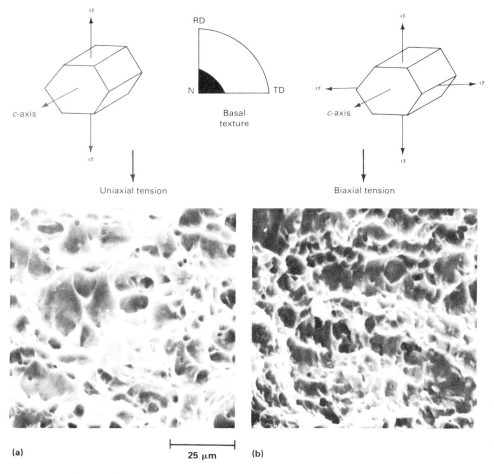

Fig. 62 Effect of balanced biaxial tension on dimple rupture in a hot-rolled basal-textured Ti-6Al-4V alloy. The dimples on the biaxially fractured specimen (b) are smaller and more shallow when compared to the uniaxially fractured specimen (a). Source: Ref 196

of gaseous hydrogen was faster and more efficient in embrittling the steel than the dissociation of water.

Another interesting observation in this material was the appearance of the fracture surface. Although the fatigue fractures in air were transgranular at all levels of ΔK, the fractures in hydrogen exhibited transgranular and intergranular fracture, with the intergranular fracture being a function of ΔK. The fatigue fractures in hydrogen were transgranular in the near-threshold region, the amount of intergranular fracture increased until it peaked at 30% at $\Delta K \cong 11$ MPa$\sqrt{m}$ (10 ksi$\sqrt{in.}$) at $R = 0.1$, then gradually decreased again to near zero at the highest levels of ΔK (Ref 220). There was no correlation between the amount of intergranular fracture and the rate at which the fatigue cracks propagated.

The general effect of hydrogen on austenitic grades of stainless steels follows the trends observed in most heat-treatable steels. For example, when compared to air, the fatigue crack growth rate in hydrogen gas for an annealed austenitic type 302 stainless steel showed a threefold increase, and a type 301 stainless steel exhibited as much as a tenfold increase in the fatigue crack growth rate when tested at mean stress of 66 MPa (9.6 ksi) and a stress intensity factor range of $\Delta K = 60$ MPa$\sqrt{m}$ (54.5 ksi$\sqrt{in.}$) (Ref 217). The fatigue tests were conducted at 25 °C (75 °F) in laboratory air and in dry hydrogen gas, which was at slightly above atmospheric pressure, with a stress intensity range of about $\Delta K = 50$ to 70 MPa$\sqrt{m}$ (45.5 to 64 ksi$\sqrt{in.}$), a mean stress of 66 and 72 MPa (9.6 and 10.4 ksi), a load ratio of $R = 0.05$, and a cyclic frequency of 0.6 Hz. Although the actual test conditions in this investigation covered a much broader range of ΔK and mean stresses, narrower testing parameters were selected for this discussion in order to overlap the air and hydrogen data.

The fracture surfaces of annealed type 301 and 302 austenitic stainless steels are shown in Fig. 82. The type 302 stainless steel exhibited fatigue striations in all atmospheres, but the type 301 stainless steel formed few striations under any fatigue condition. In the presence of hydrogen gas, however, the type 301 stainless steel fatigue fractures exhibited a significant number of relatively flat, cleavagelike facets and smaller, parallel plateaus of facets, each at a different elevation, forming a stepped structure. The difference in the response between the two stainless steels was due to the presence of unstable austenite in the type 301 stainless steel, which could transform to martensite as a result of strain at the crack tip. Also, cold-worked (19 to 40% reduction in thickness) material of both stainless steels was more susceptible to the effects of hydrogen than the annealed material. This was also due to the presence of strain-induced martensite in the cold-worked stainless steels, which showed no fatigue striations under any testing condition (Ref 217).

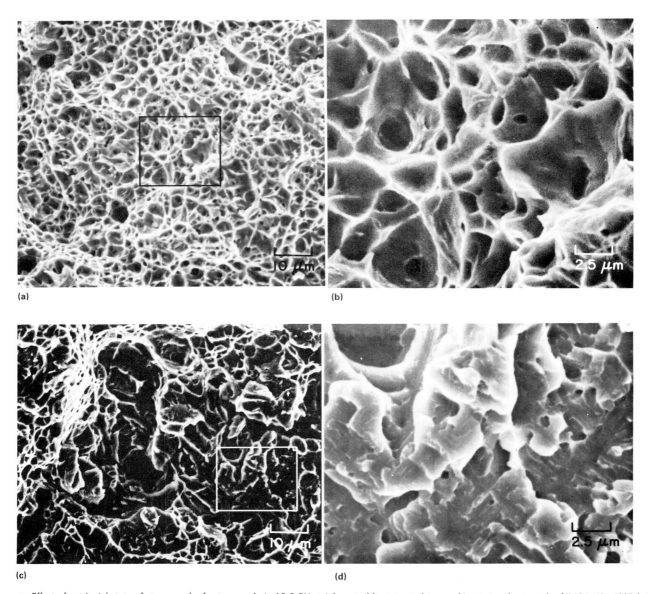

Fig. 63 Effect of a triaxial state of stress on the fracture mode in 13-8 PH stainless steel heat treated to an ultimate tensile strength of 1634 MPa (237 ksi). (a) and (b) Equiaxed dimples on the fracture surface of an unnotched specimen. (c) and (d) The quasi-cleavage fracture appearance of a notched specimen

Effect of Gases on Nonferrous Alloys. Steels are not the only alloys affected by gaseous environments; some aluminum alloys are also susceptible. When a 2219-T851 aluminum alloy was fatigue tested at room temperature within a stress intensity range of $\Delta K = 10$ to 24 MPa$\sqrt{\text{m}}$ (9 to 22 ksi$\sqrt{\text{in.}}$), a load ratio of $R = 0.05$, and a cyclic frequency of the 5 Hz in 0.2-torr water vapor and 760-torr (1-atm) dry argon gas (at a cyclic test frequency of 20 Hz), the fatigue crack growth rate in the water vapor was three times greater than that in the argon gas (Ref 225). The increase in the crack propagation rate was attributed to hydrogen embrittlement and was essentially equivalent to the fatigue rates observed when the 2219-T851 aluminum alloy is tested in 40 to 60% humidity air, distilled water, and even a 3.5% sodium chloride solution (Ref 225, 226). A significant change in the fatigue fracture appearance (Fig. 83) accompanies the increase in the fatigue crack growth rate.

The nickel-base superalloys are not immune to the effects of reactive atmospheres. Although hydrogen showed no effect, gases containing hydrogen sulfide and sulfur dioxide increased the crack propagation rate substantially in an Inconel alloy 718 that had received a modified heat treatment that improved the notch rupture properties of the material (Ref 227).

The fatigue tests were conducted at 650 °C (1200 °F) at slightly above atmospheric pressure in dry hydrogen gas, dry helium gas, helium with 0.5% hydrogen sulfide, helium with 5% sulfur dioxide, air, and air with 0.5 and 5% sulfur dioxide (other gas atmospheres were also tested). The approximate stress intensity range was $\Delta K = 40$ to 70 MPa$\sqrt{\text{m}}$ (36.5 to 63.5 ksi$\sqrt{\text{in.}}$), the load ratio $R = 0.1$, and the cyclic frequency 0.1 Hz. At a stress intensity range of $\Delta K = 40$ MPa$\sqrt{\text{m}}$ (36.5 ksi$\sqrt{\text{in.}}$), the fatigue crack growth rate in air was five times greater than that in dry helium gas, and at $\Delta K = 60$ MPa$\sqrt{\text{m}}$ (54.5 ksi$\sqrt{\text{in.}}$), the crack growth rate in air containing 5% sulfur dixoide was about four times greater than that in air.

The greatest increase, however, was observed when testing at a stress intensity range of $\Delta K = 40$ MPa$\sqrt{\text{m}}$ (36.5 ksi$\sqrt{\text{in.}}$) in atmospheres containing helium with 0.5% hydrogen sulfide and helium with 5% sulfur dioxide, which increased the fatigue crack growth rate by almost a factor of 30 over the rate in dry helium gas. Except for the fatigue fractures produced in air that contained 0.5% sulfur dioxide, the fracture surfaces produced in

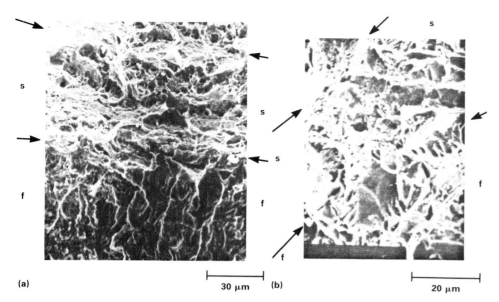

Fig. 64 Effect of Charpy impact strain rate on the fracture appearance of an AISI 5140 H steel tempered at 500 °C (930 °F) and tested at room temperature. (a) Fatigue-precracked specimen tested at a strain velocity of 5×10^{-2} mm/s (2×10^{-3} in./s). (b) Fatigue-precracked specimen tested at a strain velocity of 5400 mm/s (17.7 ft/s). The more rapid strain rate results in a reduction in the width of the stretched zone adjacent to the crack and the presence of some intergranular decohesion with ductile tearing on the facets. f, fatigue crack; s, stretched zone. Source: Ref 199

sulfur-containing atmospheres were covered with thick, adherent scales that obscured the fracture detail. The fracture surfaces of specimens tested in hydrogen gas exhibited distinct fatigue striations whose spacing approximated the fatigue crack growth rate. The fatigue fractures in air consisted of intergranular decohesion among the principally transgranular fracture, which contained fatigue striations with spacings approximating the fatigue crack growth rate. Although there was no significant difference in the fatigue propagation rates between the air and air containing 0.5% sulfur dioxide, the fractures produced in the latter atmosphere showed no fatigue striations and consisted of intergranular decohesion and flat, transgranular facets. All atmospheres that increased the fatigue crack growth rate tended to promote intergranular fracture, and atmospheres containing hydrogen sulfide were more aggressive than those containing sulfur dioxide. Atmospheres that produced a significant increase in the fatigue crack growth rate often showed no appreciable corrosion attack on unstressed material.

The increase in the fatigue crack growth rate in air was probably due to the diffusion of oxygen along the grain boundaries and to the addition of oxide growth stresses at the crack tip. The mechanism of sulfur attack and acceleration of the fatigue crack growth rate could involve the formation of low-melting (eutectic) liquids at grain boundaries, or it could be due to the formation and subsequent fracture of brittle sulfur compounds (Ref 227).

A similar accelerating effect of sulfur-containing atmospheres was observed when an annealed Incoloy 800 was fatigue tested in air, dry helium gas, helium with 0.35% hydrogen sulfide, and helium with 0.75 to 9.3% sulfur dioxide (all helium-containing gases were at slightly above atmospheric pressure) in the range of 316 to 650 °C (600 to 1200 °F) (Ref 228). The fatigue tests were conducted in the approximate stress intensity range of $\Delta K = 28$ to 44 MPa$\sqrt{\text{m}}$ (25.5 to 40 ksi$\sqrt{\text{in.}}$), at load ratios of $R = 0.1$ to 0.33, and at a cyclic frequency of 0.1 Hz.

At equivalent concentrations, the atmospheres containing hydrogen sulfide were found to be a far more aggressive environment than those containing sulfur dioxide, although above 400 °C (750 °F) all of the sulfur-containing atmospheres, when compared to helium gas, produced a four- to fivefold increase in the fatigue crack growth rate. The atmospheres containing sulfur dioxide were apparently less aggressive because of the inhibiting effect of oxygen, which is produced by the reaction of sulfur dioxide with the alloy. The mechanism for the general increase in the fatigue crack growth rate in sulfur-containing atmospheres is believed to be sulfur enrichment of crack tip region (Ref 228), and this seems to be supported by fractographic evidence.

In the 316- to 427-°C (600- to 800-°F) temperature range (the fracture surfaces were obscured at higher temperatures), the fatigue fracture surfaces produced in sulfur-containing atmospheres exhibited ridges and fatigue striations (Fig. 84). The ridges and stretched zones were probably formed by a localized, rapid advance of the crack front. This effect is believed to be due to the periodic local arrest of the crack by the formation of sulfide particles in the sulfur enrichment zone at the crack tip.

Modes of Fracture / 41

Although the crack front as a whole is accelerating, the crack may slow down locally. When this occurs, sulfide particles have time to form in the enrichment zone. The sulfide particles (Fig. 85) appear to blunt and stop the growth of the crack, and when it finally propagates past the particle, it does so by sudden fracture, which results in a ridge and a stretched zone and locally accelerated fatigue growth, as indicated by the increase in the striation spacing immediately adjacent to the ridge-type structures. Unlike Inconel 718, no significant intergranular fracture was observed, and testing in air had no appreciable effect on the fatigue crack growth rate.

Effect of Liquid Environments. Corrosive liquid environments, such as water, brines, organic fluids, basic or acid media, and molten salts, can affect the rate at which fatigue cracks propagate and the fracture appearance. Fatigue that occurs in environments that are corrosive to the material is referred to as corrosion fatigue. An overview of the basic mechanisms involved in fractures due to corrosive attack is provided in the section ''Stress Corrosion Cracking'' in this article. In general, any environment that promotes the initiation of fatigue cracks, such as pitting, enhanced corrosion, or oxidation; that allows an embrittling species to enter the material; that promotes strain-hardening relief (which lowers the fracture stress) at the crack tip; or that interferes with crack tip slip reversal will accelerate the corrosion fatigue crack growth rate and decrease the fatigue life of the material.

Mechanism of Corrosion Fatigue. In order to understand the mechanism of corrosion fatigue crack initiation, the process has been approached from a purely theoretical standpoint by using basic corrosion kinetics and fracture mechanics (Ref 229). The theoretical analyses have been applied to three specific corrosion conditions: general, pitting, and passive corrosion. The basic mathematics describing the kinetics and fracture mechanics involved for each of the three conditions is complex; however, the analyses yielded the following conclusions.

First, under conditions of general corrosion, corrosion fatigue initiation is controlled by the corrosion rate and the applied alternating stress range. General corrosion is defined as fairly uniform attack that results in loss of material and a reduction in the load-bearing cross section.

Second, in pitting corrosion, in which the corrosion attack is localized, corrosion fatigue initiation is controlled by the rate at which pits nucleate and grow to a critical depth. The critical depth of a pit is a function of the applied alternating stress range. Under pitting conditions, no corrosion fatigue limit exists.

Third, for passive corrosion, the corrosion fatigue crack initiates at a site where unprotected new metal exposed by slip steps penetrating the passive surface film is dissolved during repassivation. As these local slip emer-

42 / Modes of Fracture

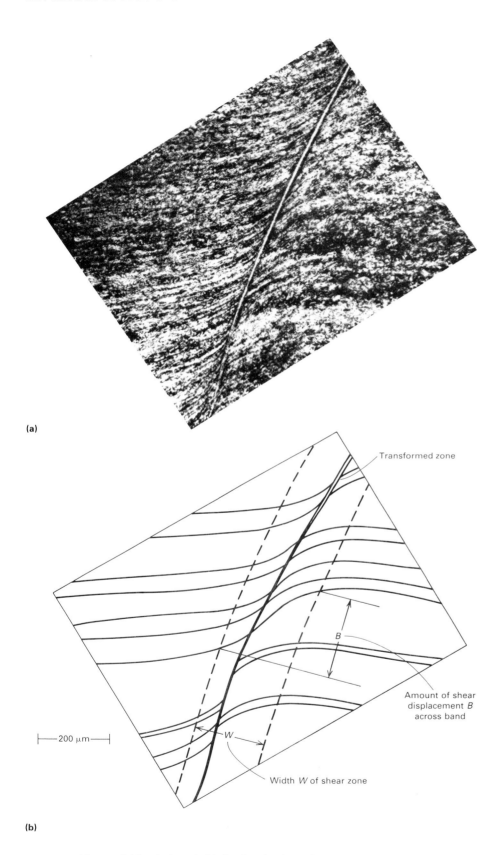

Fig. 65 Micrograph (a) and schematic (b) of a shear band in a plate of rolled medium carbon steel produced by ballistic impact showing the transformed zone and the zone of strain localization. (D.A. Shockey, SRI International)

gence and repassivation dissolution cycles proceed, a small, sharp notch can form along the slip band. A corrosion fatigue crack initiates when the notch reaches a critical depth determined by the applied cyclic stress range. For passive corrosion, therefore, the corrosion fatigue initiation is controlled by the repassivation kinetics and the critical notch depth. It was also concluded that under conditions of passive corrosion, there is a critical repassivation current density below which corrosion fatigue crack initiation cannot occur and crack initiation is controlled by air fatigue behavior (Ref 229).

In some cases, cracks propagating under a static load in a corrosive environment, such as a type 304 austenitic stainless steel tested in a boiling magnesium chloride solution, exhibit periodic, parallel crack-arrest marks on the fracture surface that strongly resemble brittle fatigue striations (Ref 230-232). The crack can proceed by very short bursts of cleavage fracture with about a 1-μm spacing between the crack-arrest lines (Ref 232). This type of crack propagation is believed to be due to localized embrittlement by hydrogen at the crack interface (Ref 230, 231). Also, when tested under similar stress intensity range, ΔK, conditions, cleavage fracture due to corrosion attack and cleavagelike corrosion fatigue cracks have been shown to exhibit similar crack growth rates and fracture appearances (Ref 232).

In view of the similarities, a unified theory, developed using the principles of corrosion kinetics and fracture mechanics, has been proposed linking SCC and corrosion fatigue (Ref 169). The theory states that passive film rupture is a critical event in both fracture processes. Film rupture causes anodic dissolution of metal as a result of galvanic coupling between the unprotected bare metal surface at the rupture site and the adjacent passive oxidized suface film.

In SCC, the initial film rupture occurs as a result of strain from the applied load, and when repassivation is retarded or arrested by some critical ion, such as chloride, that is present in the environment, the anodic dissolution is maintained and the crack propagates. During corrosion fatigue, the strain due to the cyclic loading continuously ruptures the protective film and mechanically prevents it from repassivating. The anodic dissolution at the rupture sites relieves strain hardening at surface slip bands during crack initiation and promotes cleavage on prismatic lattice planes and/or decohesion along grain boundaries in the plane strain region adjacent to the crack tip during crack propagation. The mechanism for the strain-hardening relief and subsequent fracture is the generation and the movement of divacancies. There is evidence from numerous alloy systems and corrosive environments that the mechanism for SCC and corrosion fatigue involves the relief of strain hardening at the crack tip by corrosion-generated divacancies (Ref 169).

Modes of Fracture / 43

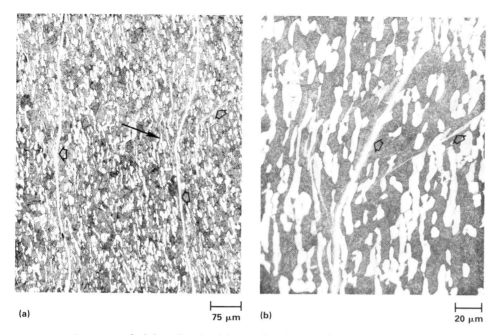

Fig. 66 Appearance of adiabatic shear bands in an explosively ruptured Ti-6Al-4V STA alloy rocket motor. The material exhibits multiple, often intersecting, shear bands (open arrows). Slender arrow points to portion of shear band shown in more detail in (b). Note that the intense deformation has obliterated the α-β microstructure within the band. The 1.9-mm (0.075-in.) sheet thickness direction is left to right. (V. Kerlins, McDonnell Douglas Astronautics Company)

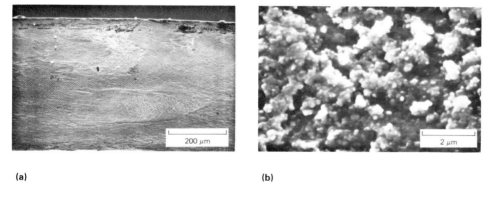

Fig. 67 Low-magnification (a) and higher-magnification (b) views of a failure surface produced in vacuum-arc remelted AISI 4340 steel (40 HRC) by dynamically shearing in a split Hopkinson torsion bar at a nominal shear strain rate of 6000 s^{-1}. The knobbly fracture surface suggests that local melting occurred. (J.H. Giovanola, SRI International)

The theory also proposes that because the crack propagation rate is often increased by a lower cyclic frequency and longer hold times (dwell at maximum load), elevated-temperature fatigue in vacuum and lower temperature corrosion fatigue may also occur by the same mechanism. At elevated temperatures, thermally generated vacancies can cause dislocation rearrangement; in corrosion fatigue, corrosion-induced divacancies may perform the same function (Ref 169).

Corrosion Fatigue of Martensitic Stainless Steels. The drastic effect that pitting corrosion has on the fatigue life of a material is illustrated by the corrosion fatigue testing of an SUS 410J1 stainless steel (Japanese equivalent of type 410 stainless steel) that was tempered at 600 and 750 °C (1110 and 1380 °F) to an ultimate tensile strength of 930 and 700 MPa (135 and 102 ksi), respectively (Ref 233). The fatigue testing was performed at 25 °C (75 °F) in air and in a 3% sodium chloride solution (pH 7) with a stress intensity range of ΔK = 3 to 20 MPa$\sqrt{m}$ (2.5 to 18 ksi$\sqrt{in.}$), a load ratio of R = 0, and a cyclic frequency of 60 Hz for the fatigue life tests and a frequency of 30 Hz for the crack propagation rate tests.

Fatigue life tests for the material tempered at 600 °C (1110 °F) showed that at a constant 10^7 cycles to failure fatigue life the 500-MPa (72.5-ksi) fatigue strength observed for specimens tested in air was reduced to about 100 MPa (14.5 ksi) in the brine. At a constant 500-MPa (72.5-ksi) fatigue strength, the 10^7 cycles to failure air endurance limit was reduced to 10^5 cycles in the salt solution. This drastic reduction in fatigue life was due to early corrosion fatigue crack initiation from corrosion pits. However, crack growth rate tests indicated that at ΔK values of less than about 18 MPa$\sqrt{m}$ (16.5 ksi$\sqrt{in.}$) the corrosion fatigue crack growth rate in brine was slightly lower than that in air. This anomaly was believed to be due to the relatively high (30 Hz) testing frequency, which resulted in fluid not being able to escape from the region of the crack tip during the decreasing portion of the cyclic loading. The wedging action of the trapped fluid reduced the effective stress intensity at the crack tip and thus reduced the crack growth rate.

Both tempers showed an increase in the amount of intergranular fracture when tested in the salt solution; however, the amount of intergranular fracture varied with stress intensity. The 750-°C (1380-°F) tempered material showed a maximum of 30 to 60% intergranular fracture (versus about 12 to 26% in air) at a maximum stress intensity of K_{max} = 15 MPa$\sqrt{m}$ (13.5 ksi$\sqrt{in.}$). A similar effect was observed for a nickel-chromium-molybdenum-vanadium steel tested in moist air (see Ref 220 and the section "Effect of Gaseous Environments" in this article) and for a type 410 stainless steel that was corrosion fatigue tested in various environments (Ref 234). Analysis of a large number of corrosion fatigue tests indicated that there was little correlation between the amount of intergranular fracture and the crack propagation rate in this stainless steel (Ref 234).

Corrosion Fatigue of Aluminum Alloys. A decrease in the fatigue strength was observed for a 7075-T6 high-strength aluminum alloy tested at room temperature in dry air and in a 3% sodium chloride solution at a constant mean stress of 207 MPa (30 ksi), a load ratio of 0 < R < 0.7, and a cyclic frequency of 30 Hz (Ref 235). At a constant 3 × 10^6 cycles to failure, the fatigue strength in air was 115 MPa (17 ksi) (cyclic stress: $\Delta\sigma$ = ±115 MPa, or 17 ksi) and 50 MPa (7.25 ksi) in the saline solution. This difference in fatigue strength was reflected in the fracture appearance (Fig. 86).

The fatigue fracture in air initiated along grain boundaries, but changed immediately to a cleavagelike fracture with river patterns reminiscent of Stage I fatigue (Fig. 86a). In the saline solution, the crack exhibited substantial intergranular decohesion at the fracture origin and for some distance from the origin before the crack size had increased (ΔK increased) sufficiently to result in the crack changing to a principally cleavagelike fracture with distinct river patterns (Fig. 86b). The cleavagelike transgranular fracture in air exhibited finer, feathery river pattern lines, and the fracture extended over a number of grains before changing direction. In the 3% sodium chloride solution, the river pattern lines were generally

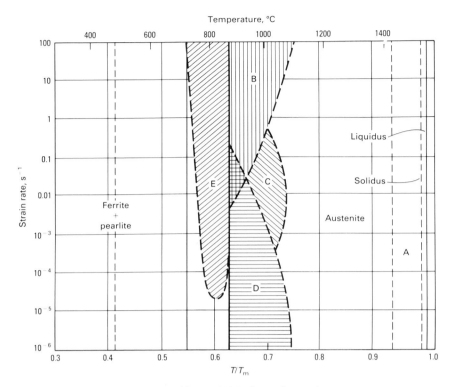

Fig. 68 Possible fracture zones mapped for a 0.2% C plain carbon steel in strain rate temperature space. T, testing temperature; T_m, melting temperature. The zones, which are shaded in the diagram, are as follows: A, subsolidus intergranular fracture due to segregation of sulfur and phosphorus; B, high strain rate intergranular fracture associated with MnS; C, ductile intergranular fracture—may or may not be preceded by B or D; D, low strain rate intergranular fracture; E, two-phase mixture with fracture at second-phase particles in the weaker preferentially strained ferrite. Source: Ref 209

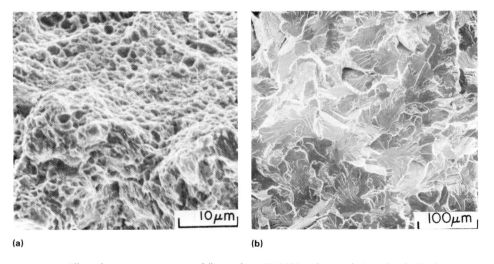

Fig. 69 Effect of test temperature on a fully pearlitic AISI 1080 carbon steel. Smooth cylindrical specimens tensile tested at a strain rate of 3.3×10^{-4} s^{-1}. Specimens tested at 125 °C (255 °F) show fractures consisting entirely of dimple rupture (a), while at −125 °C (−195 °F), the fractures exhibit 99% cleavage (b). The size of the cleavage facets approximates the prior-austenite grain size. Source: Ref 210

coarser and more distinct, and a set of lines was confined to a single grain; a new set of lines initiated at the grain boundary as the fracture entered the adjacent grain. In the fatigue striation forming portions of the fatigue cracks, the striations in air were ductile, while the striations that formed in the saline solution had a brittle character. Although the reduction in the fatigue life of the specimens tested in the 3% sodium chloride solution was reflected in the fracture appearance, the mechanism responsible for the decrease was hydrogen embrittlement. The embrittlement attack was enhanced by the cyclic mechanical rupture of the passive film and the prevention of repassivation by the presence of the chloride ion (Ref 235).

Corrosion Fatigue of Titanium Alloys. Chloride-containing liquid environments are damaging not only to iron and aluminum alloys but also to titanium and its alloys. This was shown by fatigue tests conducted using a basal-textured IMI 155 (British designation for commercially pure titanium containing 0.34% oxygen alloy at room temperature in laboratory air and in a 3.5% sodium chloride solution (Ref 236). The fatigue tests were conducted within a stress intensity range of $\Delta K = 5$ to 25 MPa$\sqrt{m}$ (4.5 to 23 ksi$\sqrt{in.}$), a load ratio of 0.35, and a cyclic frequency of 130 Hz.

The fatigue tests showed that at a stress intensity range of $\Delta K = 8$ MPa$\sqrt{m}$ (7.5 ksi$\sqrt{in.}$) the crack propagation rate in the saline solution was up to 10 times greater than that in air. The fracture surfaces of specimens tested in the saline solution contained more cleavage facets than those tested in air. Because cleavage fractures within individual grains can propagate up to ten times faster than the other fracture modes, the greater number of cleavage facets on the specimens fatigued in brine could account for the greater corrosion fatigue crack growth rate. Although not conclusively established, the cleavage fracture probably resulted from hydrogen embrittlement by hydrogen picked up at the crack tip from the dissociation of water at the new metal surfaces formed during the fatigue crack cyclic advance.

An examination of the air and brine fracture surfaces indicated that the fatigue cracks advanced by three distinct fracture types: cleavage, furrows, and striations (Ref 236). The air fatigue fractures exhibited all three fracture types, while brine corrosion fatigue fractures showed only cleavage and furrow fractures. Cleavage fractures and the fatigue striations resulting from a duplex slip process are two fracture modes that are fairly well understood and do not require further elaboration.

Furrow-type fatigue fractures (Fig. 87) are of interest because they occurred in both environments and were the dominant fracture type in air above a ΔK of 10 MPa$\sqrt{m}$ (9 ksi$\sqrt{in.}$). As the name suggests, furrows are grooves or trenches in the fracture surface. The furrows, which were about 1 μm wide and 1 μm deep and exhibited a separation of about 5 to 10 μm between the grooves, were always oriented parallel to the crack growth direction. The fracture between the furrows exhibited very fine lines that made an angle of 30° with the furrows. From electron channeling patterns and metallography, it was concluded that the furrow fracture was the result of shear on secondary slip systems in the titanium metal (Ref 236). Furrow-type fracture apparently occurs when the fatigue crack approaches a grain that is not favorably oriented for either cleavage or a striation mode of fracture.

Corrosion Fatigue of Austenitic Stainless Steels. Type 304 austenitic stainless steel and

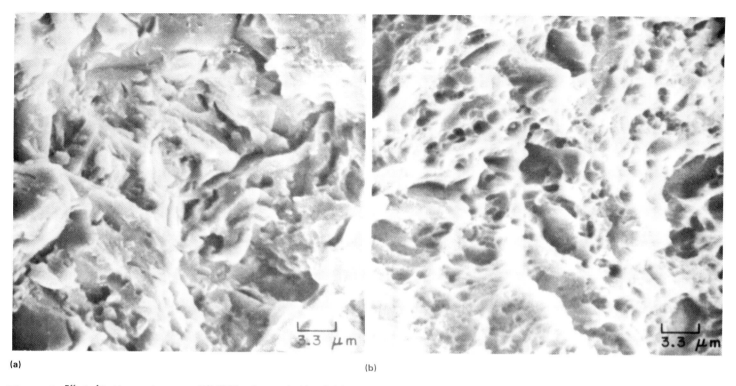

Fig. 70 Effect of test temperature on an AISI 1042 carbon steel with a slightly tempered martensitic (660 HV) microstructure that was Charpy impact tested at −196 and 100 °C (−320 and 212 °F). The fracture at −196 °C (−320 °F) consists entirely of cleavage (a), and at 100 °C (212 °F), it is dimple rupture (b).

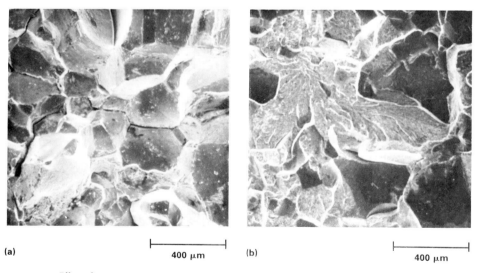

Fig. 71 Effect of test temperature on a 0.39C-2.05Si-0.005P-0.005S steel that was heat treated to a hardness of 30 HRC and Charpy impact tested at room temperature at −85 °C (−120 °F). The fracture at room temperature occurs by intergranular fracture (a) and by a combination of intergranular fracture and cleavage (b) at −85 °C (−120 °F). Source: Ref 211

nickel-base Inconel 600 are often used as heat-exchanger tubing materials in nuclear power plants, where thermal cycling and vibration can subject the alloys to fatigue in the presence of a caustic environment. To evaluate the effect of a caustic environment, an annealed type 304 stainless steel and a solution-annealed Inconel 600 were fatigue tested in 140-°C (285-°F) air and in a boiling 140-°C (285-°F) 17.5 M (46 wt%) sodium hydroxide solution (Ref 237, 238). Round bar type 304 stainless steel specimens were fatigue tested at a mean stress of 248 MPa (36 ksi) (the 0.2% offset yield strength of the material was 228 MPa, or 33 ksi), a cyclic stress range of about 25 to 200 MPa (3 to 29 ksi), a cyclic frequency of 10 Hz, and with and without a protective anodic potential. The anodic potential had no effect on the fatigue properties. No fatigue endurance limit was observed in the boiling caustic solution, and the air fatigue life of greater than 10^6 cycles to failure at a cyclic stress of 172 MPa (25 ksi) was reduced to 4×10^4 cycles (Ref 237). However, when an annealed type 304 stainless steel was fatigue tested under identical conditions with a precracked specimen instead of a round bar specimen, no significant effect of the boiling sodium hydroxide solution on the corrosion fatigue crack growth rate was observed (Ref 238).

The results of these two tests are not contradictory, but indicated that because the fatigue life of smooth, round bar specimens is strongly influenced by fatigue initiation the boiling caustic solution apparently assisted in crack initiation by interfering with the repassivation of the anodic film (Ref 237). The lack of a significant effect of the boiling sodium hydroxide solution on fatigue crack propagation was confirmed by examining the fracture surfaces, which showed that the air and caustic fractures had a very similar appearance.

Corrosion Fatigue of Nickel-Base Alloys. The round bar solution-annealed Inconel 600 specimens were fatigue tested at a mean stress of 248 MPa (36 ksi) (the 0.2% offset yield strength of the alloy was 225 MPa, or 33 ksi), a cyclic stress range of 140 to 220 MPa (20 to 32 ksi), a cyclic frequency of 0.1 to 1 Hz, and both with and without an anodic passivation

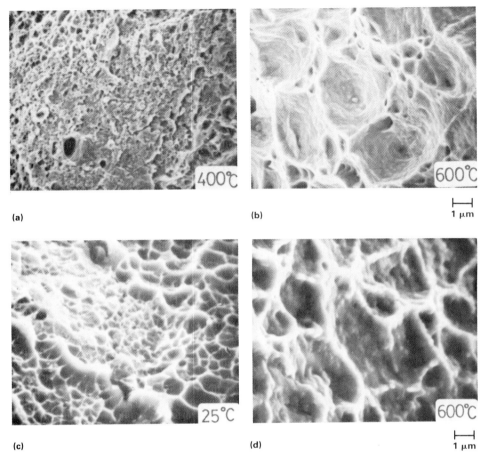

Fig. 72 Effect of temperature on dimple size in a 0.3C-1Cr-1.25Mo-0.25V-0.7Mn-0.04P steel that was heat treated to an ultimate strength level of 880 MPa (128 ksi). (a) and (b) Dimples on transgranular facets. (c) and (d) Dimples on intergranular facets. Note that dimple size increased with temperature. Source: Ref 213

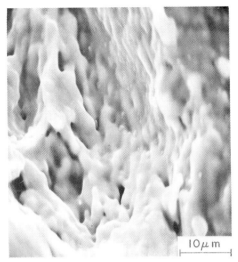

Fig. 73 Effect of temperature on the fracture of an ultralow-carbon steel. The 0.05C -0.82Mn-0.28Si steel containing 180 ppm S was annealed for 5 min at 1425 °C (2620 °F), cooled to 950 °C (1740 °F), and held for 3 min before tensile testing at a strain rate of 2.3×10^{-4} s^{-1}. The fracture, which occurs by dimple rupture when tested at room temperature, exhibits intergranular decohesion at 950 °C (1740 °F). Source: Ref 209

potential (Ref 238). Fatigue testing in boiling sodium hydroxide solution under conditions identical to those used to test the round bar type 304 stainless steel produced completely opposite results. Instead of a substantial decrease in the fatigue life observed for the stainless steel, the fatigue life of the Inconel 600 was actually higher in the boiling caustic solution, particularly at the lower cyclic stress levels, than in air (Ref 238). For example, when fatigued at a 180-MPa (26-ksi) cyclic stress and a frequency of 1 Hz, the cycles to failure were greater than 2×10^6 in the boiling sodium hydroxide solution versus 6×10^5 in air, and anodic passivation had a positive effect.

Because the air and the sodium hydroxide fatigue fracture surfaces exhibited identical fracture appearances (slip plane cracking during Stage I and well-defined striations in Stage II), it was concluded that for Inconel 600 the slip at the crack tip was controlled more by local strain and strain rates than the caustic environment. The fracture appearance and the greater increase in the fatigue life at the lower cyclic stresses indicated that, as in the case of the type 304 stainless steel, the boiling sodium hydroxide principally affected the process by which fatigue cracks initiate. Therefore, the increase in fatigue life was believed to be primarily due to a delay in crack initiation by the caustic solution dissolving or blunting the incipient cracks, although some crack tip blunting during fatigue crack propagation may also occur (Ref 238).

Effect of Vacuum. The general effect of vacuum is to retard the rate at which fatigue cracks propagate. Because a hard vacuum (usually 10^{-6} torr or better) essentially excludes such aggressive environments as water (moisture) and oxygen that are present in laboratory air, embrittling reactions that accelerate crack growth in air are eliminated, and the oxidation of newly formed slip surfaces at the crack tip is significantly reduced. This permits more complete slip reversal during the unloading portion of the fatigue cycle, resulting in a decrease or even arrest in the propagation of the crack. Crack advance can also be retarded by the partial rewelding of the fracture surfaces at the crack tip. Whatever the retardation mechanism, the fracture surfaces of fatigue cracks propagating in vacuum usually exhibit poorly formed fatigue striations or no striations at all (Ref 19, 25, 239-242).

Effect of Vacuum on Aluminum Alloys. Compared to fatigue testing in air, some of the greatest improvements in the fatigue crack propagation resistance in vacuum have been exhibited by a 7475 aluminum alloy in an underaged condition (aged 6 h at 120 °C, or 250 °F) with a 0.2% offset yield strength of 451 MPa (65 ksi) and an average grain size of 80 μm (expressed as the diameter of a sphere having a volume that is approximately equivalent to that of an average grain) (Ref 239). The fatigue tests were conducted at approximately 22 °C (72 °F) in air with a relative humidity of 50% and a vacuum at a pressure of 10^{-6} torr, using a stress intensity range of $\Delta K = 3$ to 15 MPa$\sqrt{m}$ (2.5 to 13.5 ksi$\sqrt{in.}$), a load ratio of $R = 0.1$, and a cyclic frequency of 30 Hz.

At a stress intensity range of $\Delta K = 10$ MPa$\sqrt{m}$ (9 ksi$\sqrt{in.}$), the fatigue crack growth rate in vacuum was almost 1000 times lower than that in air (Ref 239). Despite the large difference in the fatigue crack propagation rates, the fracture surfaces exhibited fairly similar appearances consisting of facets on crystallographic planes and transgranular fracture with some fatigue striations, although the vacuum fractures had a somewhat rougher topography with secondary cracking. It was concluded that the large decrease in the fatigue crack growth rate in vacuum was due to a number of factors, including enhanced slip reversibility, irregular fracture path with crack branching, mixed Mode I and II crack loading, the absence of embrittlement at the crack tip, and the possibility of rewelding of the fracture surfaces on unloading (Ref 239).

Modes of Fracture / 47

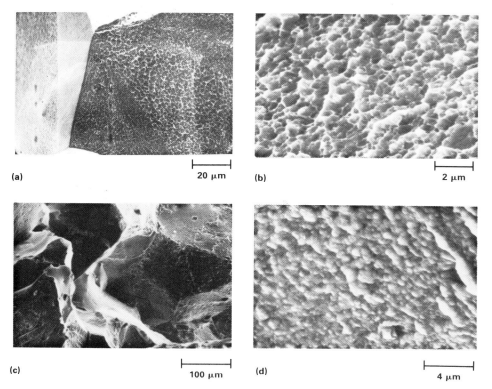

Fig. 74 Effect of temperature on double-aged Inconel X-750 that was tested at a nominal strain rate of 3 × 10^{-5} s^{-1}. (a) and (b) The fracture at room temperature occurs by intergranular dimple rupture. Note the evidence of dimple rupture network on the intergranular walls. (c) and (d) At 816 °C (1500 °F), the fracture shows intergranular decohesion with no dimple rupture. However, the intergranular facets are roughened by the presence of $M_{23}C_6$ carbides. Source: Ref 214.

tests were conducted at stress amplitudes of 670 to 950 MPa (97 to 138 ksi), a load ratio of $R = -1$, and a cyclic frequency of 80 Hz. The fatigue crack propagation tests were performed at a stress intensity range of about $\Delta K = 7$ to 15 MPa$\sqrt{m}$ (6.5 to 13.5 ksi$\sqrt{in.}$), a load ratio of $R = 0.2$, and a cyclic frequency of 10 Hz. With the loading perpendicular to the basal planes in the alloy, the fatigue strength in vacuum was 875 MPa (127 ksi) versus 690 MPa (100 ksi) in air; at a constant stress amplitude of 875 MPa (127 ksi), more than 10^7 cycles were required to fail the material in vacuum, although only about 10^4 cycles were required to fail the alloy in air. The fatigue crack propagation rate exhibited a similar trend in vacuum. At a stress intensity range of $\Delta K = 10$ MPa$\sqrt{m}$ (9 ksi$\sqrt{in.}$), the fatigue crack growth rate in vacuum decreased by about a factor of six compared to the rate in air.

In fatigue life tests, any delay in fatigue crack initiation increases the fatigue life. Therefore, the orientation of the (0002) basal lattice planes with respect to the loading direction is important. In vacuum, when loading normal to the basal planes, fatigue crack initiation is delayed because of the high yield stress in the direction normal to the basal planes and because of the impeded slip reversibility on the active slip planes, which results in smaller slip steps at the specimen surface and more difficult fatigue crack nucleation. Once a crack is nucleated, it propagates more slowly because of a more ductile, less crystallographically controlled fracture. When loading normal to the basal planes in moist air, which is considered a corrosive medium, the fatigue life does not benefit from the increased yield strength and

Effect of Vacuum on Titanium Alloys. A substantial increase in the fatigue life in vacuum and a corresponding decrease in the fatigue crack growth rate were observed for a basal-transverse textured, solution treated and aged Ti-6Al-4V alloy tested at room temperature in air (50% relative humidity) and in vacuum (10^{-6} torr) (Ref 240). The fatigue life

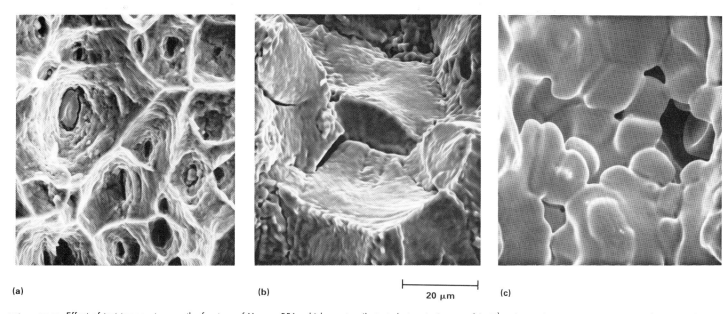

Fig. 75 Effect of test temperature on the fracture of Haynes 556, which was tensile tested at a strain rate of 1 s^{-1} at increasing temperature. (a) Dimple rupture fracture at 1015 °C (1860 °F). At the bottom of many of the dimples are TaC inclusions, which initiated microvoid coalescence. (b) Intergranular decohesion at 1523 °C (2287 °F). Secondary intergranular cracks are also visible. (c) Local eutectic melting of TaC + austenite at 1333 °C (2431 °F). (J.J. Stephens, M.J. Cieslak, R.J. Lujan, Sandia National Laboratories)

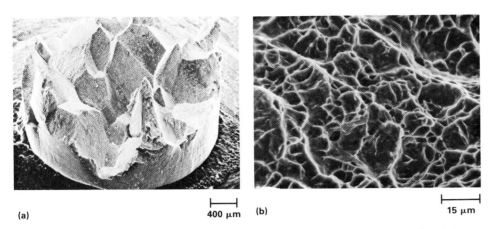

Fig. 76 Intergranular dimple rupture in a Ti-6Al-2Nb-1Ta-0.8Ta alloy tested at 800 °F (1470 °F). The fracture path in this alloy, tensile tested at a strain rate of about 3.3×10^{-4} s^{-1}, changes at 800 °C (1470 °F) from a predominantly transgranular dimple rupture at room temperature (not shown) to intergranular dimple rupture. Source: Ref 215

impeded slip reversal, because loading normal to the basal planes renders the material highly susceptible to hydrogen embrittlement attack, which results in cleavage along the (0002) planes. This negates the beneficial characteristics of the textured structure observed in vacuum and results in a reduction in the fatigue life and an increase in the fatigue crack growth rate (Ref 240).

Another Ti-6Al-4V alloy, except in the recrystallized annealed condition with a 0.2% offset yield strength of 862 MPa (125 ksi), showed a two- to threefold decrease in the fatigue crack growth rate in vacuum (10^{-5} torr) compared to humid air (50% relative humidity) when tested at a stress intensity range of ΔK = 10 MPa$\sqrt{m}$ (9 ksi$\sqrt{in.}$) at room temperature (Ref 241). The alloy was fatigue tested within a stress intensity range of about ΔK = 7 to 15 MPa$\sqrt{m}$ (6.5 to 13.5 ksi$\sqrt{in.}$) at a load ratio of R = 0.2 to 0.3 and a cyclic frequency range of 1 to 5 Hz.

By directly observing the propagation of fatigue cracks in the vacuum environment of the scanning electron microscope, it was discovered that at low values of ΔK the fatigue crack did not advance on each application of load. The specimen had to be cycled a considerable number of times to accumulate a critical amount of strain damage at the crack tip before crack advance occurred (Ref 241). Examination of the fracture surfaces showed no obvious evidence of crack arrest. The air fracture contained fatigue striations that were easy to identify, but the vacuum fractures were relatively featureless, with fatigue striations being difficult to discern.

The objective of this investigation was to use such concepts as the crack-opening displacement and the state of strain at the tip of the crack to verify the ability of a crack tip geometric model to predict the spacing of resulting fatigue striations. Therefore, the differences between vacuum and air crack growth rates were examined from a standpoint of fracture mechanics and grain size rather than crack rewelding or ease of slip reversal. Despite the complex environmental effects on the material, the fatigue striation spacing could be fairly successfully predicted by using microstructural features as well as measured and calculated properties, such as the crack-opening displacement, ΔK, strain, slip-line length, displacement per slip line, number of slip lines leaving the crack tip, and the cyclic work-hardening coefficient (Ref 241, 243).

Effect of Vacuum on Astroloy. The nickel-base superalloy Astroloy also exhibits a lower fatigue crack growth rate in vacuum (5×10^{-6} torr) than in air. Fatigue testing of Astroloy, aged at 845 °C (1555 °F) and having an average grain size of 5 μm, within a stress-intensity range of ΔK = 20 to 50 MPa$\sqrt{m}$ (18 to 45.5 ksi$\sqrt{in.}$), a load ratio of R = 0.05, and using a triangular wave form with a cyclic frequency of 0.33 Hz at 650 °C (1200 °F) reduced the fatigue crack growth rate in vacuum by a factor of three at ΔK = 50 MPa$\sqrt{m}$ (45.5 ksi$\sqrt{in.}$) and by a factor of ten at ΔK = 20 MPa$\sqrt{m}$ (18 ksi$\sqrt{in.}$) (Ref 244). Along with the reduction in the crack growth rate, the fracture path changed from predominantly intergranular in air to transgranular in vacuum. Figure 88 shows the typical fracture appearance when tested in air and vacuum. The decrease in the fatigue crack growth rate in vacuum was believed to be due to the change in the fracture path; in air, the loss of oxide-forming elements at the grain boundaries during high-temperature oxidation attack weakened or embrittled the boundaries, resulting in easier intergranular crack propagation (Ref 244).

Effect of Vacuum on Type 316 Stainless Steel. A different effect was observed when a 20% cold-worked type 316 stainless steel was fatigue tested in air and vacuum (better than 10^{-6} torr) at a temperature of 593 °C (1100 °F), a stress intensity range of ΔK = 28 to 80 MPa$\sqrt{m}$ (25.5 to 73 ksi$\sqrt{in.}$), a load ratio of R = 0.05, and a cyclic frequency of 0.17 Hz. In this stainless steel, the fatigue fractures in air were predominantly transgranular with distinct fatigue striations; the fractures in vacuum exhibited a more intergranular appearance with little evidence of striations (Fig. 89). No specific discussion on the fracture process in air or vacuum was presented because this work was principally concerned with the effect of temperature and hold times on the fatigue of the stainless steel.

Effect of Vacuum on Inconel X-750. A material that was essentially unaffected by vacuum exposure at 25 °C (75 °F) but showed a five- to tenfold decrease in the fatigue crack growth rate at 650 °C (1200 °F) is Inconel X-750 (Ref 245). The annealed Inconel X-750 was fatigue tested in air and vacuum (10^{-5} torr) at a stress intensity range of ΔK = 20 to 50 MPa$\sqrt{m}$ (18 to 45.5 ksi$\sqrt{in.}$) with a load ratio of R = 0.05 and a triangular wave form having a cyclic frequency of 10 Hz.

The largest decrease in the fatigue crack growth rate in vacuum at 650 °C (1200 °F) occurred at the lower ΔK testing levels. Figure 90 shows the air and vacuum fatigue fractures produced at a stress intensity range of ΔK = 20 MPa$\sqrt{m}$ (18 ksi$\sqrt{in.}$). The fractures in both environments were predominantly transgranular. The air fatigue fractures exhibited a more crystallographic, faceted appearance than those in vacuum. The vacuum fractures contained periodic markings that were believed to be due to slip offsets resulting from stress gradients produced by crack front jumps (Ref 245, 246). The large decrease in the fatigue crack growth rate in vacuum at 650 °C (1200 °F) was attributed to the absence of oxygen, which can diffuse along slip bands to embrittle the material ahead of the crack tip and increase the rate at which a fatigue crack propagates. No reason was given for the identical fatigue propagation rates in air and vacuum at 25 °C (75 °F), although the high oxidation and corrosion resistance of this alloy at ambient temperatures rendered it immune to any adverse effects of air.

Effect of Vacuum on Commercially Pure Titanium. A reversal in the general effect of a reduction in the fatigue crack growth rate in vacuum was observed in an IMI 155 (British designation) commercially pure titanium containing 0.34% O (Ref 236). The fatigue tests were conducted in laboratory air and vacuum (2×10^{-6} torr) at room temperature at a stress intensity range of ΔK = 11 to 20 MPa$\sqrt{m}$ (10 to 18 ksi$\sqrt{in.}$), a load ratio of R = 0.35, and a cyclic frequency of 130 Hz. Over the entire stress intensity range, the fatigue crack growth rate in vacuum was about three times greater than that in air. The difference in the crack propagation rates was attributed to the nature of the crack paths in air and vacuum. In air, the cleavage fractures are associated with crack branching and irregular crack fronts, which tend to result in lower fatigue crack growth rates as compared to vacuum fractures, which exhibited less branching of the

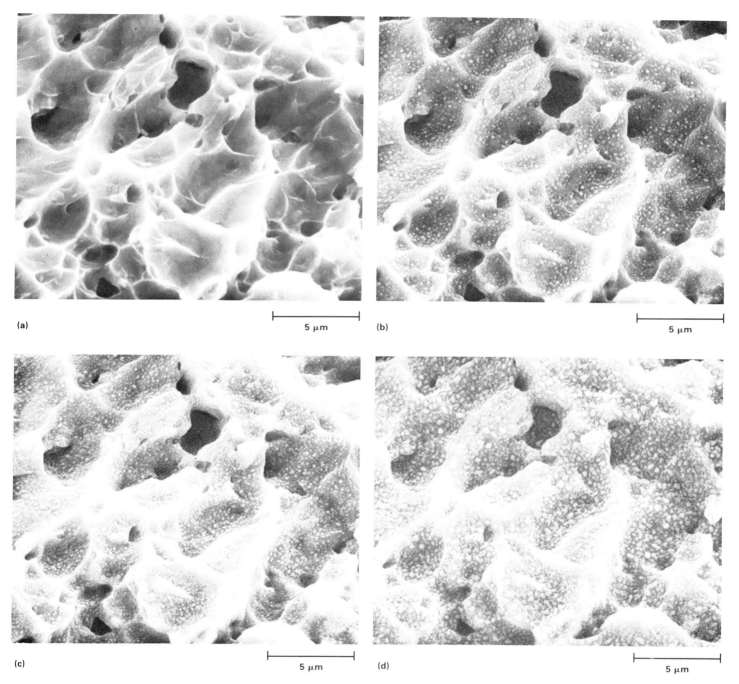

Fig. 77 Effect of a 700-°C (1290-°F) air exposure on an annealed Ti-6Al-2Sn-4Zr-6Mo alloy dimple rupture overload fracture. (a) As fractured. (b) The identical area as in (a) except exposed for 3 min. (c) 10 min. (d) 30 min. As time at temperature increases, the fracture surface becomes progressively more obscured by the oxide. (V. Kerlins, McDonnell Douglas Astronautics Company)

cleavage facets and more uniform crack fronts (Ref 236).

Effect of Temperature. As an environment, temperature is unique because it can substantially alter the basic mechanical properties of materials and can affect the activity of other environments that depend on oxidation or diffusion reactions to exert their influence. The principal mechanical properties or material characteristics that affect fatigue are the yield strength (more precisely, the elastic limit), the elastic modulus (E), and the cyclic work-hardening rate, which elevates the yield strength. As the yield strength, elastic limit, and cyclic work-hardening rate decrease, slip and plastic deformation can occur more readily, resulting in easier fatigue crack initiation and more rapid propagation. These are the primary reasons the fatigue properties are generally degraded at elevated and enhanced at cryogenic temperatures. Because the diffusion and oxidation rates increase with temperature, the effect of elevated temperatures is to raise the effectiveness of environments that influence fatigue properties by oxidation or by the diffusion of an embrittling species. The opposite is true at cyrogenic temperatures.

Because there is a large variation in the effect of temperature on materials, the effect of temperature on fracture appearance also varies. However, at constant testing conditions, the striation spacing generally increases, the striations become less distinct, and, depending on

50 / Modes of Fracture

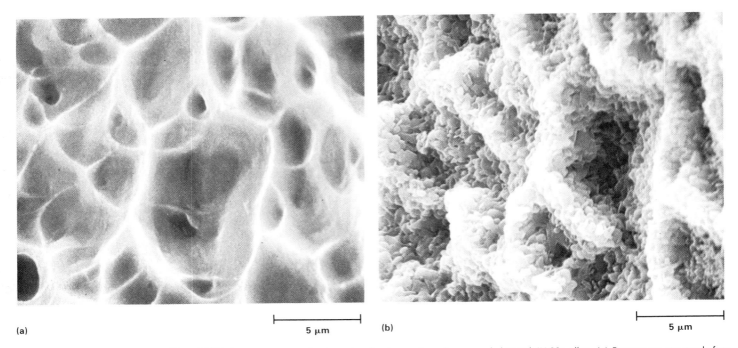

Fig. 78 Effect of a 15-min 800-°C (1470-°F) air exposure on a dimple rupture fracture surface of an annealed Ti-6Al-6V-2Sn alloy. (a) Fracture appearance before exposure. (b) The identical fracture surface after exposure. The oxide buildup is so great that it is impossible to identify the fracture mode. (V. Kerlins, McDonnell Douglas Astronautics Company)

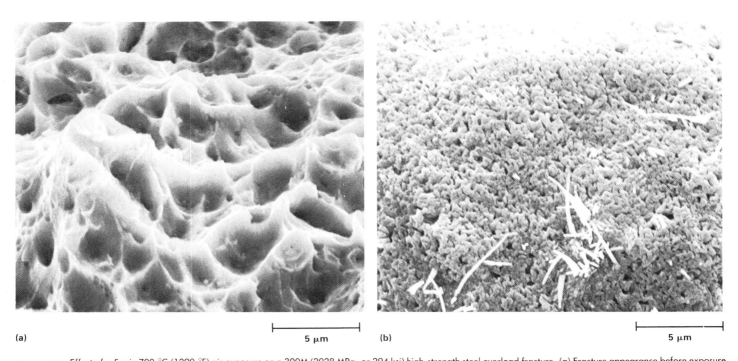

Fig. 79 Effect of a 5-min 700-°C (1290-°F) air exposure on a 300M (2028 MPa, or 294 ksi) high-strength steel overload fracture. (a) Fracture appearance before exposure. (b) The same area after exposure. The entire fracture, which exhibited dimple rupture, was covered by an oxide that obscured all fracture detail. Some areas on the oxidized surface contained needlelike whiskers, which are probably an iron oxide. (V. Kerlins, McDonnell Douglas Astronautics Company)

the environment, the fracture surfaces become more oxidized as the temperature increases. The effect of temperature on fatigue in several different materials is illustrated in the following examples.

Effect of Temperature on Austenitic Stainless Steels. Increasing the temperature from 25 to 593 °C (75 to 1100 °F) for an annealed type 316 austenitic stainless steel increased the fatigue crack growth rate by a factor of 3.5 when tested in vacuum (better than 10^{-6} torr) at a stress intensity range of $\Delta K = 34$ to 60 MPa$\sqrt{m}$ (31 to 54.5 ksi$\sqrt{in.}$), a load ratio of $R = 0.05$, and a cyclic frequency of 0.17 Hz (Ref 242). The increase in temperature had no significant effect

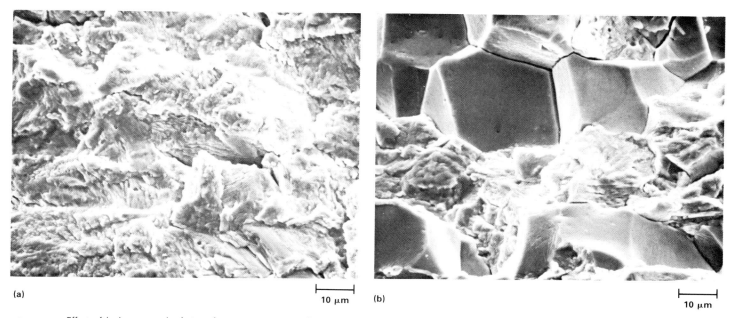

Fig. 80 Effect of hydrogen on the fatigue fracture appearance of a grade X42 pipeline steel. (a) Tested in dry nitrogen gas at a stress intensity range of $\Delta K = 20$ MPa$\sqrt{m}$ (18 ksi$\sqrt{in.}$), a load ratio of $R = 0.1$, and a cyclic frequency of 5 Hz. (b) Tested in dry hydrogen gas at a stress intensity range of $\Delta K = 10$ MPa$\sqrt{m}$ (9 ksi$\sqrt{in.}$), a load ratio of $R = 0.1$, and a cyclic frequency of 1 Hz. Both room-temperature fatigue tests resulted in a fatigue crack growth rate of approximately 2×10^{-4} mm/cycle. The fatigue fracture in nitrogen exhibited principally a serrated transgranular fracture, along with occasional bands of intergranular decohesion (not shown); however, the fracture in hydrogen exhibited fewer regions of serrated fracture and more bands of intergranular decohesion. Source: Ref 218

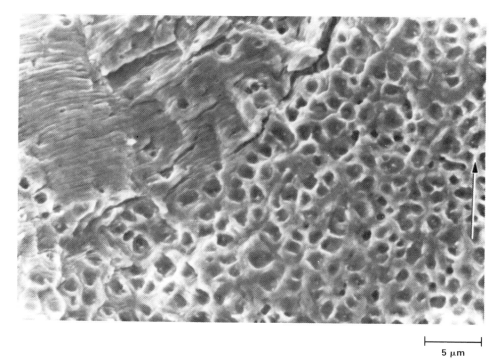

Fig. 81 Effect of severe hydrogen attack on the fatigue fracture appearance of an ASTM 533B pressure vessel steel. The severely charged material was tested at room temperature at a stress intensity range of $\Delta K = 20$ MPa$\sqrt{m}$ (18 ksi$\sqrt{in.}$), a load ratio of $R = 0.1$, and a cyclic frequency of 50 Hz. The fracture exhibited fatigue striations (upper left) as well as cavitated intergranular facets resulting from methane gas bubble formation at grain boundaries. Arrow indicates crack growth direction. Source: Ref 223

on the fracture appearance; both fractures, especially at the lower ΔK values, were transgranular and contained crystallographic facets (Fig. 91). The absence of fatigue striations was probably the result of testing in a vacuum environment. The increase in the fatigue crack growth rate was attributed to the decrease in the elastic modulus at elevated temperature.

The way in which slight differences in composition can have a significant effect on how materials respond to changes in testing temperature is illustrated by annealed type 304, 316, 321, and 347 austenitic stainless steels. These stainless steels have a basically similar composition, except that type 321 and 347 stainless steels are stabilized by the addition of small amounts (less than 1%) of titanium and niobium, respectively, to prevent the depletion of chromium from material adjacent to the grain boundaries. Chromium depletion can occur when chromium carbides precipitate at elevated temperatures.

The steels were fatigue tested in air at 420 to 800 °C (790 to 1470 °F) by using a triangular wave form, a zero mean strain, a total strain range of 1%, and strain rates of 6.7×10^{-3} s^{-1} and 6.7×10^{-5} s^{-1} (Ref 247). For materials having an ASTM grain size greater than 3, testing at a strain rate of 6.7×10^{-3} s^{-1} showed a continuously decreasing (by a factor of five to seven times) fatigue life with increasing temperature for all four stainless steels. However, when testing at a strain rate of 6.7×10^{-5} s^{-1}, the fatigue life of the type 304 and 316 stainless steels decreased by a factor of three up to a temperature of 600 °C (1110 °F) and remained essentially unchanged to 800 °C (1470 °F). The type 321 and 347 stainless steels exhibited a continued decrease, by up to a factor of 15, in the fatigue life with increasing temperature in the range of 420 to 800 °C (790 to 1470 °F). The reason for this discrepancy in

52 / Modes of Fracture

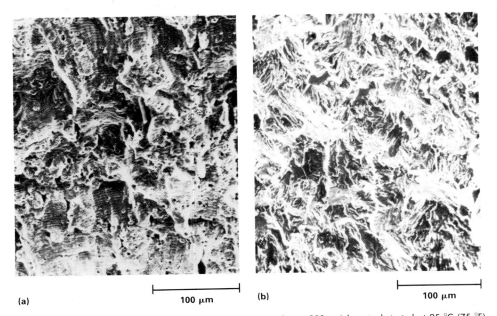

Fig. 82 Fatigue fracture surfaces of annealed type 301 and type 302 stainless steels tested at 25 °C (75 °F) in 1 atm hydrogen gas. The type 302 stainless steel (a) showed well-developed fatigue striations. The type 301 stainless steel (b) showed a more brittle-appearing fracture surface with few striations but containing flat, cleavagelike facets and small, parallel facets at different elevations joined by tear ridges. Source: Ref 217

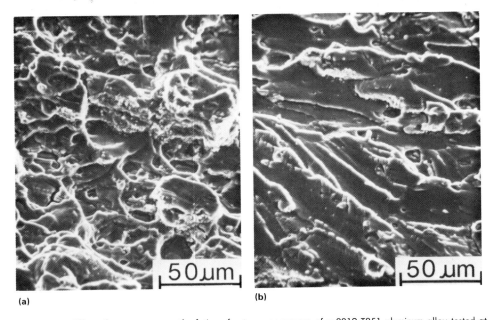

Fig. 83 Effect of water vapor on the fatigue fracture appearance of a 2219-T851 aluminum alloy tested at room temperature. (a) 760-torr dry argon, $\Delta K = 16.5$ MPa$\sqrt{\text{m}}$ (15 ksi$\sqrt{\text{in.}}$), $R = 0.05$, and $f = 20$ Hz. (b) 0.2-torr water vapor, $\Delta K = 16.5$ MPa$\sqrt{\text{m}}$ (15 ksi$\sqrt{\text{in.}}$), $R = 0.05$, and $f = 5$ Hz. The magnifications are too low to resolve the fatigue striations clearly, but the general change in the fracture morphology is apparent. The crack propagation direction is from left to right. Source: Ref 225

behavior became apparent when the dislocation structures of one of the stainless steels from each group were examined in the transmission electron microscope.

The examination revealed that the dislocation substructure for the type 316 stainless steel was cell-type up to 600 °C (1110 °F) and changed to one of subboundary-type above 600 °C (1110 °F); for type 321 stainless steel, the dislocation substructure remained cell-type even above 600 °C (1110 °F). This indicated that the type 316 stainless steel had experienced recovery from cyclic strain hardening but that the type 321 stainless steel had not. If recovery occurs, the material experiences less cyclic work hardening, and both transgranular slip and intergranular sliding can occur at the low strain rates, resulting in a mixed transgranular-intergranular fracture path and a smaller decrease in the fatigue life at elevated temperatures. This is because transgranular fracture generally exhibits a lower fatigue crack growth rate. If recovery does not occur, the material within the grains is cyclically strain hardened, decreasing the amount of transgranular fracture as intergranular sliding becomes the easier fracture process, with a resulting continued increase in the fatigue crack growth rate and a corresponding decrease in the fatigue life (Ref 247).

Effect of Temperature on High-Temperature Materials. Alloys such as Inconel X-750 and Incoloy 800 exhibit varying degrees of increase in the fatigue crack growth rate with increasing temperature. Annealed Inconel X-750 that was fatigue tested in air at a stress intensity range of $\Delta K = 20$ to 30 MPa$\sqrt{\text{m}}$ (18 to 27.5 ksi$\sqrt{\text{in.}}$), a load ratio of $R = 0.05$, a triangular wave form, and a cyclic frequency of 10 Hz showed about a tenfold increase in the fatigue crack growth rate at $\Delta K = 20$ MPa$\sqrt{\text{m}}$ (18 ksi$\sqrt{\text{in.}}$) when the temperature was increased from 25 to 650 °C (75 to 1200 °F); a sixfold increase was seen at $\Delta K = 30$ MPa$\sqrt{\text{m}}$ (27.5 ksi$\sqrt{\text{in.}}$) (Ref 245). Although the fracture appearance of fatigue specimens tested at 25 °C (75 °F) was not discussed, the fractures at elevated temperatures had a crystallographic character, along with relatively flat, narrow plateaus with fine fatigue striations at the 20-MPa$\sqrt{\text{m}}$ (18-ksi$\sqrt{\text{in.}}$) stress intensity range and fractures that consisted primarily of ductile fatigue striations at the 30-MPa$\sqrt{\text{m}}$ (27.5-ksi$\sqrt{\text{in.}}$) stress intensity range. The principal increase in the fatigue crack growth rate was due to transgranular oxidation-assisted cracking as a result of oxygen diffusion along slip bands to the material ahead of the crack tip (Ref 245).

Annealed Incoloy 800 that was fatigue tested in air and dry helium gas using a stress intensity range of $\Delta K = 33$ MPa$\sqrt{\text{m}}$ (30 ksi$\sqrt{\text{in.}}$), a load ratio of $R = 0.10$, and a cyclic frequency of 0.1 Hz showed a slight 1.5-time increase in the fatigue crack growth rate as a result of raising the test temperature from 316 to 650 °C (600 to 1200 °F) (Ref 228). No discussion of fracture appearance or reason for the modest increase in the fatigue crack growth rate was given, because the primary objective of this investigation was to determine the effect of sulfur-containing atmospheres on the fatigue crack growth rates in Incoloy 800 at elevated temperatures.

The effect of cryogenic temperature is to decrease the fatigue crack growth rate. This phenomenon has been observed in a number of different materials, including iron alloys (Ref 248), copper (Ref 249), and aluminum and its alloys (Ref 250, 251). For example, annealed 1100 commercially pure aluminum and an annealed and cold-rolled 99.99% pure aluminum that were fatigue tested at a stress intensity range of $\Delta K = 2$ to 12 MPa$\sqrt{\text{m}}$ (1.8 to 11 ksi$\sqrt{\text{in.}}$),

Modes of Fracture / 53

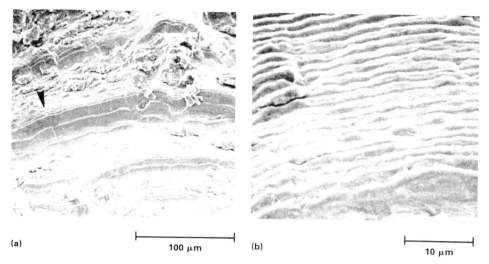

Fig. 84 Typical ridges and fatigue striations on the fracture surface of an annealed Incoloy 800 specimen tested in a sulfidizing atmosphere in the 316- to 427-°C (600- to 800-°F) temperature range. (b) Higher-magnification fractograph of the area indicated by arrow in (a). Note the width of the smooth stretched areas at the ridges. Source: Ref 228

a load ratio of $R = 0.05$, and a cyclic frequency of 30 Hz showed a substantial decrease in the fatigue crack growth rate when cooled from 25 to -196 °C (75 to -320 °F) (temperature of liquid nitrogen) (Ref 252).

The fatigue crack growth rate for the 1100 aluminum decreased by a factor of 20 at -196 °C (-320 °F). For the 99.99% pure aluminum in both the annealed and cold-rolled condition, the fatigue crack growth rate decreased by a factor of 80. The large decreases in the crack growth rate at cryogenic temperature were attributed to the substantial increase in the cyclic yield stress and to the increase in the rate of work hardening, resulting in high plastic work per unit area of crack surface produced. The high cyclic yield stresses and high plastic work were associated with a relatively homogeneous distribution of dislocations that did not form cell structures (which result in softening) because of difficult dislocation climb and cross-slip at cryogenic temperatures (Ref 252). The fracture surfaces were not examined.

Effect of Loading. The way in which a cyclic load is applied can have as great an influence on the fatigue crack growth rate as some of the environments already discussed. The loading variables that affect the fatigue crack growth rate include such parameters as the stress intensity factor range (ΔK), the mean stress intensity factor (K_{mean}), the load or stress ratio (R), the cyclic loading frequency (f), and the wave form. The effect on the fatigue crack growth rate of the principal independent fatigue parameters ΔK, frequency, and wave form will be evaluated in such environments as vacuum, dry helium, argon, nitrogen, and moist air. The inert environments (vacuum and the dry gases) were selected because they allow for a more accurate evaluation of the effect that a change in the value of a fatigue parameter has on the fatigue crack growth rate without the additional influence of an external environment. The moist air was included because it affords an opportunity to assess the crack behavior in a realistic service environment.

When a change occurs in the magnitude (numerical value) of a fatigue parameter such as ΔK, it affects the fatigue crack growth rate as well as the fracture appearance. An increase in the fatigue crack growth rate for a crack propagating by a striation-forming mechanism in an inert environment results in an increase in the fatigue striation spacing. If the increase in the fatigue crack growth rate is large enough to produce Stage III fatigue propagation, overload fracture modes, such as dimple rupture, will intermix with the fatigue striations, which exhibit increasing spacings.

However, for a fatigue crack propagating in a relatively corrosive environment such as moist air, especially at low fatigue crack propagation rates when the environment can exert its greatest influence (see the section "Effect of Gaseous Environments" in this article), a change in the crack growth rate is usually the result of the combined effects of the change in the magnitude of the fatigue parameter as well as the corrosive environment. When an aggressive environment contributes to the increase in the fatigue crack growth rate, it can affect the fracture appearance by changing the ductile fatigue striations to brittle striations and by adding quasi-cleavage, cleavage, or intergranular decohesion fracture among the fatigue-striated areas. Although quasi-cleavage, cleavage, and intergranular decohesion are technically not fatigue fractures, they will be

 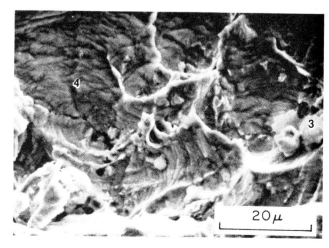

Fig. 85 Discrete sulfide particles on fracture surfaces of an Incoloy 800 tested in a sulfidizing atmosphere. Energy-dispersive x-ray analysis indicated that sulfur was present at particles 1 and 3 and that no sulfur was detected on surfaces 2 and 4. The sulfur-containing particles are believed to be responsible for locally arresting the fatigue crack. Source: Ref 228

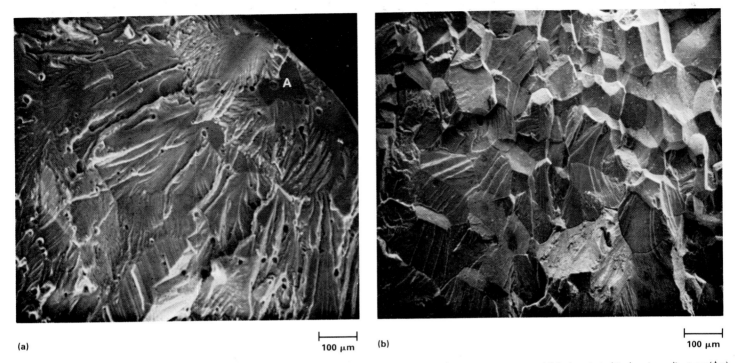

Fig. 86 Effect of environment on the fatigue fracture appearance in a 7075-T6 aluminum alloy tested at room temperature. (a) Fatigue tested in dry air; cyclic stress ($\Delta\sigma$) = ±110 MPa (16 ksi), cycles to failure (N_f) ≅ 6 × 10⁵. (b) Corrosion fatigue tested in a 3% sodium chloride solution, $\Delta\sigma$ = ±68 MPa (9.9 ksi), N_f ≅ 6 × 10⁵. Note that the fatigue fracture in air initiated at a grain boundary (A) but propagated by cleavagelike fracture, with river patterns emanating from the origin. The corrosion fatigue fracture in the salt solution initiated and propagated some distance by intergranular decohesion. However, as the crack depth increased (ΔK increased), the fracture changed to a primarily transgranular, cleavagelike fracture with river patterns. Source: Ref 235

Fig. 87 Furrow-type fatigue fracture in a commercially pure titanium alloy IMI 155 tested at room temperature in laboratory air. ΔK = 16 MPa$\sqrt{\text{m}}$ (14.5 ksi$\sqrt{\text{in.}}$), da/dN = 10⁻⁸ m/cycle. Source: Ref 236

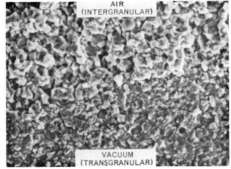

Fig. 88 Typical fatigue fracture appearance of Astroloy tested in air and vacuum at 650 °C (1200 °F). In air (upper portion of fractograph) the fracture exhibits a predominantly intergranular character; in vacuum (lower half of fractograph) the fracture is transgranular. Source: Ref 244

considered part of a normal fatigue fracture surface because they occur as a natural consequence of a fatigue crack propagating in a corrosive environment.

At higher fatigue crack growth rates, the crack is relatively unaffected by a corrosive environment, the crack growth rates begin to approach those observed in the inert environments, and the changes in the rate and fracture appearance are again due more to changes in the magnitude of the fatigue parameter and less to the environment.

The fatigue parameters ΔK, K_{mean}, and R are based on load or stress. The cyclic frequency and the wave form are independent variables that are unrelated to load or the other fatigue parameters. Of the load-based parameters, ΔK is the most significant and the most widely used to evaluate fatigue crack behavior because it is the only parameter that can successfully describe the fatigue crack growth rate. For this reason and because K_{mean} and R are dependent on ΔK (see the section "Effect of the Stress Intensity Range, ΔK" in this article), ΔK will be the only load-based fatigue parameter whose effect on the fatigue crack growth rate will be presented in more detail below.

From a technical standpoint, cyclic frequency (the rate at which the fatigue load is applied) and the wave form (the shape of the load versus time curve) are two independent variables. They are unrelated because the same frequency can exhibit different wave forms, such as sinusoidal or triangular. However, from a practical standpoint, the frequency and the wave form are the only two fatigue parameters that are not load related and are similar enough to be discussed in the same section.

In addition to discussing ΔK, frequency, and wave form, the effect of these parameters on the fatigue crack growth rate and the fracture appearance will be illustrated by using the results of fatigue tests conducted on different materials under different testing conditions. Each example will contain such data as a description of the material, the testing conditions, and the results of the tests.

Effect of the Stress Intensity Range, ΔK. The stress intensity factor, K, is a fracture mechanics parameter that expresses the stress condition in the material adjacent to the tip of a

Modes of Fracture / 55

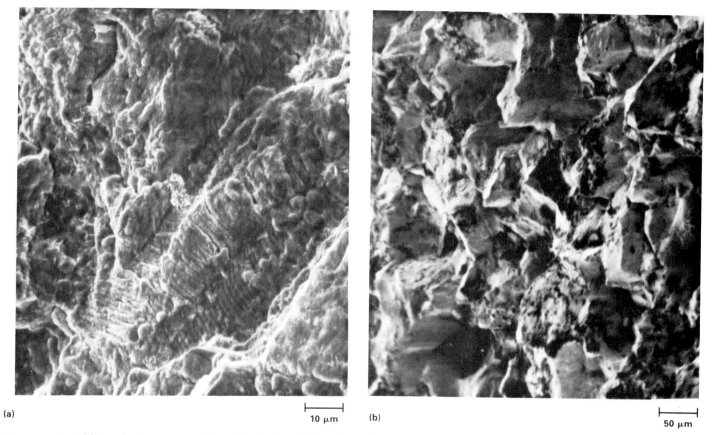

Fig. 89 Typical fatigue fracture appearance in a cold-worked type 316 stainless steel tested at 593 °C (1100 °F) in air and vacuum. (a) The fracture in air was primarily transgranular with distinct fatigue striations. (b) The fracture in vacuum was predominantly intergranular with no distinct fatigue striations. Source: Ref 242

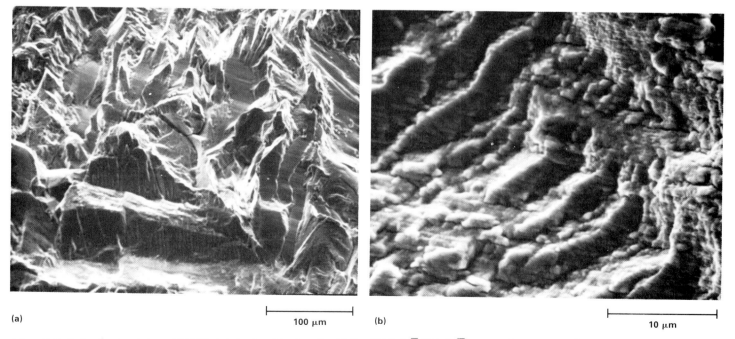

Fig. 90 Fatigue fractures in Inconel X-750 tested at a stress intensity ratio of $\Delta K = 20$ MPa$\sqrt{m}$ (18 ksi$\sqrt{in.}$) in air and vacuum at 650 °C (1200 °F). The crack propagation direction is from bottom to top. The fracture in air (a) exhibited a faceted, crystallographic appearance; the vacuum fracture (b) was less faceted and contained distinct, periodic crack front markings. Source: Ref 245

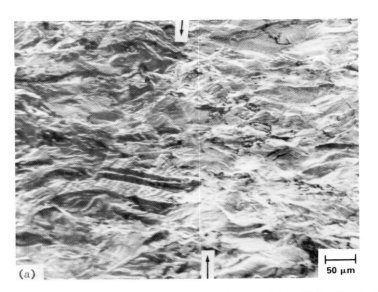 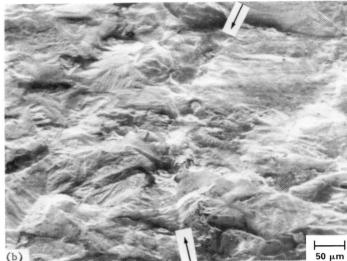

Fig. 91 Typical fatigue fracture appearance of an annealed type 316 stainless steel tested in vacuum at 25 °C (75 °F) (a) and 593 °C (1100 °F) (b). The crack propagation direction is from left to right. Small arrows point to the interface between the precracked region and the propagating fatigue crack. Both fractures were transgranular and contained crystallographically oriented facets. Source: Ref 242

crack and is a function of the applied load and the crack shape factor, which for a partial thickness crack is expressed as a function of the depth and length of the crack (Ref 15, 31, 32). A numerically larger K represents a more severe state of stress.

In fatigue, ΔK represents the range of the cyclic stress intensity factor; however, it is generally referred to as the stress intensity range. The stress intensity range, ΔK, is defined as:

$$\Delta K = K_{max} - K_{min} \qquad \text{(Eq 2)}$$

where K_{max} is the maximum stress intensity factor, and K_{min} is the minimum stress intensity factor. The values K_{max} and K_{min} are calculated by using the maximum and minimum stress (load) associated with each fatigue load cycle. For an actual fatigue test using an edge-cracked specimen, ΔK can be calculated directly from (Ref 244):

$$\Delta K = \frac{\Delta P}{BW} \gamma \frac{a}{w} \qquad \text{(Eq 3)}$$

where ΔP is load amplitude ($P_{max} - P_{min}$), B is specimen thickness, W is specimen width (distance from point where load is applied and the uncracked edge of specimen), a is crack length, and γ (a/W) is the K calibration function.

The fatigue crack growth rate always increases with the value of ΔK because a numerically larger ΔK represents a greater mechanical driving force to propagate the crack. The relationship between the fatigue crack growth rate, da/dN and ΔK can be expressed as (Ref 31, 32):

$$\frac{da}{dN} = C(\Delta K)^m \qquad \text{(Eq 4)}$$

where a is the distance of crack growth (advance), N is the number of cycles applied to advance the crack a distance (a), da/dN is the fatigue crack growth rate, and C and m are constants. If the fatigue crack propagates exclusively by a striation-forming mechanism, da/dN represents the average striation spacing, which increases with the value of ΔK.

The fatigue parameters K_{mean} and R are defined as:

$$K_{mean} = \frac{(K_{min} + K_{max})}{2} \qquad \text{(Eq 5)}$$

and

$$R = \frac{K_{min}}{K_{max}} \qquad \text{(Eq 6)}$$

By using Eq 2, 5, and 6, it can be shown that K_{mean} and R are related to ΔK as follows:

$$K_{mean} = K_{min} + \frac{\Delta K}{2} \qquad \text{(Eq 7)}$$

and

$$R = 1 - \frac{\Delta K}{K_{max}} \qquad \text{(Eq 8)}$$

Because K_{mean} and R are both directly related to ΔK, by determining the effect of ΔK on the fatigue crack growth rate, the effect of K_{mean} and R can be deduced from Eq 7 and 8.

In the examples given later in this section, the fatigue crack growth rate in a material is sometimes given at low and high numerical values of ΔK for moist air as well as the inert environments. As discussed previously, at low fatigue crack growth rates (low values of ΔK), the corrosive effects of moist air can result in an increase in the fatigue crack growth rate above that in the inert environments, but at high fatigue crack growth rates (high values of ΔK), the moist air environment has little effect and all environments exhibit similar fatigue growth rates. For these reasons, the span of the fatigue crack growth rates between the low and high values of ΔK is generally smaller for the fatigue cracks propagating in the moist air than in inert environments.

The effect of ΔK on the fatigue crack growth rate, da/dN, in various materials is summarized below. Unless specified, the cyclic wave form is sinusoidal and the gas pressures are at or slightly above atmospheric.

Example 1. An API-5LX Grade X42 carbon-manganese pipeline steel with an ultimate tensile strength of 490 to 511 MPa (71 to 74 ksi) was tested under the following conditions: $R = 0.1, f = 5$ Hz, dry nitrogen gas at a pressure of 69 atm, at room temperature (Ref 218). The results are listed below:

| ΔK | | da/dN, |
MPa$\sqrt{m}$	ksi$\sqrt{in.}$	mm/cycle
7	6.5	1×10^{-6}
40	36.5	6×10^{-4}

The fatigue crack growth rate at $\Delta K = 40$ MPa$\sqrt{m}$ (36.5 ksi$\sqrt{in.}$) was 600 times greater than at 7 MPa$\sqrt{m}$ (6.5 ksi$\sqrt{in.}$).

Example 2. A nickel-chromium-molybdenum-vanadium rotor steel (Fe-0.24C-0.28Mn-3.51Ni-1.64Cr-0.11V-0.39Mo) with an ultimate tensile strength of 882 MPa (128 ksi) was tested under the following conditions: $R = 0.5$, $f = 120$ Hz, air (30 to 40% relative humidity), at room temperature (Ref 220). The results are listed below:

ΔK		da/dN,
MPa$\sqrt{\text{m}}$	ksi$\sqrt{\text{in.}}$	mm/cycle
4	3.5	6×10^{-7}
20	18	8×10^{-5}

The fatigue crack growth rate at $\Delta K = 20$ MPa$\sqrt{\text{m}}$ (18 ksi$\sqrt{\text{in.}}$) was about 130 times greater than at 4 MPa$\sqrt{\text{m}}$ (3.5 ksi$\sqrt{\text{in.}}$).

Example 3. An AISI 4130 steel with an ultimate tensile strength of 1330 MPa (193 ksi) and a hardness of 43 HRC was tested under the following conditions: $R = 0.1, f = 1$ to 50 Hz, moist air (relative humidity not specified), at room temperature (Ref 251). The results are listed below:

ΔK		da/dN,
MPa$\sqrt{\text{m}}$	ksi$\sqrt{\text{in.}}$	mm/cycle
10	9	7×10^{-6}
40	36.5	2×10^{-4}

The fatigue crack growth rate at $\Delta K = 40$ MPa$\sqrt{\text{m}}$ (36.5 ksi$\sqrt{\text{in.}}$) was about 30 times greater than at 10 MPa$\sqrt{\text{m}}$ (9 ksi$\sqrt{\text{in.}}$).

Example 4. An AISI 4340 steel with an ultimate tensile strength of 2082 MPa (302 ksi) was tested under the following conditions: $R = 0.1$, $f = 20$ Hz, dry argon gas, at room temperature (Ref 253). The results are listed below:

ΔK		da/dN,
MPa$\sqrt{\text{m}}$	ksi$\sqrt{\text{in.}}$	mm/cycle
15	13.5	7×10^{-5}
30	27.5	1.2×10^{-4}

The fatigue crack growth rate at $\Delta K = 30$ MPa$\sqrt{\text{m}}$ (27.5 ksi$\sqrt{\text{in.}}$) was less than twice as great as that at 15 MPa$\sqrt{\text{m}}$ (13.5 ksi$\sqrt{\text{in.}}$).

Example 5. An SUS 410J1 martensitic stainless steel (Japanese equivalent of type 410 stainless steel), which was tempered at 600 °C (1110 °F) for 3 h, was tested under the following conditions: $R = 0$, $f = 30$ Hz, laboratory air (relative humidity not specified), at room temperature (Ref 233). The results are listed below:

ΔK		da/dN,
MPa$\sqrt{\text{m}}$	ksi$\sqrt{\text{in.}}$	mm/cycle
4	3.5	2×10^{-7}
15	13.5	2×10^{-5}

The fatigue crack growth rate at $\Delta K = 15$ MPa$\sqrt{\text{m}}$ (13.5 ksi$\sqrt{\text{in.}}$) was 100 times greater than at 4 MPa$\sqrt{\text{m}}$ (3.5 ksi$\sqrt{\text{in.}}$).

Example 6. An annealed type 316 stainless steel was tested under the following conditions: $R = 0.05$, $f = 0.17$ Hz, vacuum (better than 10^{-6} torr), at 25 °C (75 °F) (Ref 242). The results are listed below:

ΔK		da/dN,
MPa$\sqrt{\text{m}}$	ksi$\sqrt{\text{in.}}$	mm/cycle
35	32	2×10^{-4}
70	63.5	7×10^{-3}

The fatigue crack growth rate at $\Delta K = 70$ MPa$\sqrt{\text{m}}$ (63.5 ksi$\sqrt{\text{in.}}$) was 35 times greater than at 35 MPa$\sqrt{\text{m}}$ (32 ksi$\sqrt{\text{in.}}$).

Example 7. Annealed type 301 and 302 stainless steel specimens were tested under the following conditions: $R = 0.05$, $f = 0.6$ Hz, dry argon gas, at room temperature (Ref 217). The results are listed below:

Specimen	ΔK		da/dN,
	MPa$\sqrt{\text{m}}$	ksi$\sqrt{\text{in.}}$	mm/cycle
Type 301	50	45.5	4×10^{-4}
	100	90	3×10^{-3}
Type 302	50	45.5	6×10^{-4}
	100	90	5×10^{-3}

The fatigue crack growth rate at $\Delta K = 100$ MPa$\sqrt{\text{m}}$ (90 ksi$\sqrt{\text{in.}}$) was about eight times greater than at 50 MPa$\sqrt{\text{m}}$ (45.5 ksi$\sqrt{\text{in.}}$) for both stainless steels.

Example 8. A 7475 aluminum alloy, aged 6 h at 120 °C (250 °F), that had an average grain size of 18 μm was tested under the following conditions: $R = 0.1$, $f = 30$ Hz, air (50% relative humidity) and vacuum (10^{-6} torr), at room temperature (Ref 239). The results are listed below:

Environment	ΔK		da/dN,
	MPa$\sqrt{\text{m}}$	ksi$\sqrt{\text{in.}}$	mm/cycle
Air	4	3.5	4×10^{-6}
	12	11	4×10^{-4}
Vacuum	4	3.5	1×10^{-7}
	12	11	2×10^{-5}

The fatigue crack growth rate at $\Delta K = 12$ MPa$\sqrt{\text{m}}$ (11 ksi$\sqrt{\text{in.}}$) was 100 times greater in air and 200 times greater in vacuum than at 4 MPa$\sqrt{\text{m}}$ (3.5 ksi$\sqrt{\text{in.}}$).

Example 9. A mill-annealed Ti-6Al-6V-2Sn alloy with a 0.2% offset yield strength of 1100 MPa (160 ksi) and an ultimate tensile strength of 1170 MPa (170 ksi) was tested under the following conditions: $R = 0.1$, $f = 10$ Hz, air (relative humidity not specified), at room temperature (Ref 254). The results are listed below:

ΔK		da/dN,
MPa$\sqrt{\text{m}}$	ksi$\sqrt{\text{in.}}$	mm/cycle
10	9	1×10^{-5}
40	36.5	1×10^{-2}

The fatigue crack growth rate at $\Delta K = 40$ MPa$\sqrt{\text{m}}$ (36.5 ksi$\sqrt{\text{in.}}$) was 1000 times greater than at 10 MPa$\sqrt{\text{m}}$ (9 ksi$\sqrt{\text{in.}}$).

Example 10. A recrystallized annealed Ti-6Al-4V alloy was tested under the following conditions: $R = 0.2$ to 0.3, $f = 1$ to 5 Hz, air (50% relative humidity) and vacuum (10^{-5} torr), at room temperature (Ref 244). The results are listed below:

Environment	ΔK		da/dN,
	MPa$\sqrt{\text{m}}$	ksi$\sqrt{\text{in.}}$	mm/cycle
Air	7	6.5	1×10^{-8}
	15	13.5	7×10^{-8}
Vacuum	7	6.5	3×10^{-9}
	15	13.5	3.5×10^{-8}

The fatigue crack growth rate at $\Delta K = 15$ MPa$\sqrt{\text{m}}$ (13.5 ksi$\sqrt{\text{in.}}$) was 7 times greater in air and 12 times greater in vacuum than at 7 MPa$\sqrt{\text{m}}$ (6.5 ksi$\sqrt{\text{in.}}$).

Example 11. Annealed IMI 155 (British designation for commercially pure titanium with 0.34% O) specimens were tested under the following conditions: $R = 0.35$, $f = 130$ Hz, air, (relative humidity not specified) and vacuum (2×10^{-6} torr), at room temperature (Ref 236). The results are listed below:

Environment	ΔK		da/dN,
	MPa$\sqrt{\text{m}}$	ksi$\sqrt{\text{in.}}$	mm/cycle
Air(a)	11	10	3×10^{-6}
	21	19	2×10^{-4}
Vacuum	11	10	3×10^{-6}
	21	19	2×10^{-4}

(a) There was considerable scatter in the data from four individual tests. The data shown represent the approximate average fatigue crack growth rate from the four tests. The air fatigue rates were smaller than those in vacuum because of irregular fatigue crack fronts and crack branching observed on air fatigue fractures.

The fatigue crack growth rate at $\Delta K = 21$ MPa$\sqrt{\text{m}}$ (19 ksi$\sqrt{\text{in.}}$) was 70 times greater in air and 7 times greater in vacuum than at 11 MPa$\sqrt{\text{m}}$ (10 ksi$\sqrt{\text{in.}}$).

Example 12. An annealed Inconel X-750 alloy was tested under the following conditions: $R = 0.05, f = 10$ Hz, triangular wave form, air (relative humidity not specified), at 25 and 650 °C (75 and 1200 °F) (Ref 245). The results are listed below:

Temperature		ΔK		da/dN,
°C	°F	MPa$\sqrt{\text{m}}$	ksi$\sqrt{\text{in.}}$	mm/cycle
25	75	18	16.5	9×10^{-6}
		50	45.5	1×10^{-3}
650	1200	18	16.5	1×10^{-4}
		50	45.5	3×10^{-3}

The fatigue crack growth rate at $\Delta K = 50$ MPa$\sqrt{\text{m}}$ (45.5 ksi$\sqrt{\text{in.}}$) was 110 times greater at 25 °C (75 °F) and 30 times greater at 650 °C (1200 °F) than at 18 MPa$\sqrt{\text{m}}$ (16.5 ksi$\sqrt{\text{in.}}$). The effect of ΔK on the fracture appearance of Inconel X-750 tested at 650 °C (1200 °F) in air can be seen by comparing Fig. 90(a) ($\Delta K = 20$ MPa$\sqrt{\text{m}}$, or 18 ksi$\sqrt{\text{in.}}$) with Fig. 92 ($\Delta K = 35$ MPa$\sqrt{\text{m}}$, or 32 ksi$\sqrt{\text{in.}}$).

Example 13. A nickel-base superalloy (Astroloy), aged at 845 °C (1555 °F), was tested under the following conditions: $R = 0.05, f = $

Fig. 92 Fatigue fracture appearance of Inconel X-750 tested in air at 650 °C (1200 °F) at $\Delta K = 35$ MPa$\sqrt{m}$ (32 ksi$\sqrt{in.}$). The fracture path is transgranular with ductile fatigue striations. The crack propagation direction is indicated by arrow. Compare with the fracture appearance of the same alloy shown in Fig. 90(a), which was tested at $\Delta K = 20$ MPa$\sqrt{m}$ (18 ksi$\sqrt{in.}$). Source: Ref 245

0.33 Hz, triangular wave form, air and vacuum (better than 5×10^{-6} torr), at 650 °C (1200 °F) (Ref 244). The results are listed below:

Environment	ΔK MPa$\sqrt{m}$	ksi$\sqrt{in.}$	da/dN, mm/cycle
Air	20	18	5×10^{-4}
	50	45.5	5×10^{-3}
Vacuum	20	18	5×10^{-5}
	50	45.5	2×10^{-3}

The fatigue crack growth rate at $\Delta K = 50$ MPa$\sqrt{m}$ (45.5 ksi$\sqrt{in.}$) was 10 times greater in air and 40 times greater in vacuum than at 20 MPa$\sqrt{m}$ (18 ksi$\sqrt{in.}$).

Example 14. An Inconel 718 alloy, which was heat treated at 925 °C (1700 °F) for 10 h, air cooled, then aged at 730 °C (1345 °F) for 48 h, was tested under the following conditions: $R = 0.1$, $f = 0.1$ Hz, air and dry helium gas, at 650 °C (1200 °F) (Ref 227). The results are listed below:

Environment	ΔK MPa$\sqrt{m}$	ksi$\sqrt{in.}$	da/dN, mm/cycle
Air	40	36.5	2×10^{-3}
	60	54.5	1×10^{-2}
Helium	40	36.5	4×10^{-4}
	60	54.5	4×10^{-3}

The fatigue crack growth rate at $\Delta K = 60$ MPa$\sqrt{m}$ (54.5 ksi$\sqrt{in.}$) was five times greater in air and ten times greater in helium than at 40 MPa$\sqrt{m}$ (36.5 ksi$\sqrt{in.}$).

Example 15. An annealed Incoloy 800 alloy was tested under the following conditions: $R = 0.1$, $f = 0.1$ Hz, air and dry helium gas, at 427 °C (800 °F) (Ref 228). The results are listed below:

ΔK MPa$\sqrt{m}$	ksi$\sqrt{in.}$	da/dN, mm/cycle
28	25.5	3×10^{-4}
44	40	2×10^{-3}

In both air and helium, the fatigue crack growth rate at $\Delta K = 44$ MPa$\sqrt{m}$ (40 ksi$\sqrt{in.}$) was seven times greater than at 28 MPa$\sqrt{m}$ (25.5 ksi$\sqrt{in.}$).

Effect of Frequency and Wave Form. Frequency, expressed as hertz (Hz = cycles/s), is the cyclic rate at which a fatigue load is applied. Strain rate, expressed as $\dot{\epsilon}$ (s^{-1}), is a form of frequency that indicates the rate at which a material is strained.

In general, the effect of testing frequency is similar to the effect of ΔK; that is, if a material is susceptible to environmental attack, the ef-

(a)

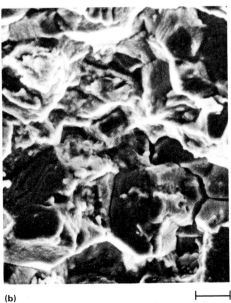
(b)

(c)

$\vdash\!\!-\!\!\dashv$ 4 μm

Fig. 93 Effect of frequency on the fracture appearance of an IN-718 nickel-base superalloy that was creep fatigue tested at 650 °C (1200 °F): $R = 0.1$, $K_{max} = 40$ MPa$\sqrt{m}$ (36.5 ksi$\sqrt{in.}$). (a) Striation formation at 10 Hz. (b) Mixture of intergranular and transgranular cracking at 0.5 Hz. (c) Fully intergranular cracking at 0.001 Hz

Table 1 Effect of strain rate on the fatigue life of type 316 and 321 stainless steels

Alloy	Strain rate ($\dot{\epsilon}$ = s^{-1})	600 °C (1110 °F) Fatigue life (N_f = cycles to failure)	Fatigue Life, 700 °C (1290 °F) (N_f = cycles to failure)
Type 316(a) (ASTM grain size = 2)	6.7×10^{-3}	2×10^3	1.3×10^3
	6.7×10^{-5}	6×10^2	7×10^2
Type 321(a) (ASTM grain size = 1)	6.7×10^{-3}	2×10^3	1×10^3
	6.7×10^{-5}	7×10^2	2×10^2

(a) Although the data are not shown, the behavior of type 304 stainless steel parallels that of type 316 stainless steel, and the behavior of type 347 stainless steel parallels that of type 321 stainless steel.

fect of the environment is greatest at the lower frequencies when the longer cycle times allow the environment to affect the material ahead of the crack tip. This environmental effect is generally manifested as an increase in the fatigue crack growth rate. The environmental contribution to the fatigue crack growth rate and its effect on the fracture appearance are similar to that observed at low ΔK values, namely the addition of nonfatigue fracture modes to the fatigue fracture, as shown in Fig. 93 (see also the section "Effect of Stress Intensity Range, ΔK" in this article). When the environment is not a factor, frequency usually has little effect on the fatigue crack growth rate.

The wave form, which is the shape of the load or strain versus time curve, can affect the fatigue crack growth rate and the fracture appearance. As in the case of ΔK and frequency, any time the environment can affect the material, there is a change, generally an increase, in the fatigue crack growth rate. The wave form affords this opportunity by imposing low ramp rates (ramp is the slope of the increasing and decreasing segments of the cyclic load curve) and long dwell or hold times (periods when the cyclic loading is stopped at any position along the load curve), especially at maximum tensile loads.

In addition to the usual fatigue crack growth rate accelerating effects of an aggressive environment, the addition of creep effects can complicate the fracture process, especially at elevated temperatures and long dwell times. The effect of creep is unique because it occurs even in inert environments. When the wave form does affect the fatigue crack growth rate, the effect is often reflected in the fracture appearance, which can exhibit a change in the fatigue striation spacing or character and show the addition of such fractures as quasi-cleavage, cleavage, intergranular decohesion, areas of grain-boundary sliding, or cavitated fracture surfaces—fractures usually associated with embrittling environments and creep rupture (Ref 242, 247, 253, 255-257). There is no significant difference in the effect on the fatigue crack growth rate between sinusoidal and triangular wave forms.

The following examples illustrate the effect of frequency and wave form on fatigue properties. Examples 16 to 20 discuss the effect of frequency; examples 21 to 27, the effect of wave form.

Example 16. An AISI 4340 steel with an ultimate tensile strength of 2028 MPa (294 ksi) was tested under the following conditions: ΔK = 25 MPa$\sqrt{m}$ (23 ksi$\sqrt{in.}$), R = 0.1, moist air (water vapor pressure: 4.4 torr), at room temperature (Ref 253). The results are listed below:

Frequency, Hz	da/dN, mm/cycle
10	1×10^{-4}
1	3×10^{-4}
0.1	3×10^{-3}

The fatigue crack growth rate at 0.1 Hz was 30 times greater than at 10 Hz. At a frequency of 10 Hz, the fracture was primarily transgranular, typical of a normal fatigue fracture. As the frequency decreased to 1 Hz and lower, the fracture path gradually changed from transgranular to one consisting primarily of intergranular decohesion along the prior-austenite grain boundaries. The increase in the fatigue crack growth rate and the change in the fracture appearance were attributed to the embrittling effect of the hydrogen liberated during the dissociation of the water vapor at the crack tip.

Example 17. An AISI 4130 steel with an 0.2% offset yield strength of 1330 MPa (193 ksi) and a hardness of 43 HRC was tested under the following conditions: ΔK = 10 to 40 MPa$\sqrt{m}$ (9 to 36.5 ksi$\sqrt{in.}$), R = 0.1, moist air (relative humidity not specified), at room temperature (Ref 251). The fatigue crack growth rate did not change significantly when tested at a frequency range of 1 to 50 Hz. The fracture exhibited a transgranular path and showed no evidence of embrittlement. At these test frequencies, there apparently was insufficient moisture in the air to permit hydrogen embrittlement to occur, even at 1 Hz.

Example 18. Annealed type 304, 316, 321, and 347 stainless steels were tested under the following conditions: total strain range = 1%, mean strain = 0, triangular wave form, air, at 600 and 700 °C (1110 and 1290 °F) (Ref 247). The results are listed in Table 1. At 600 °C (1110 °F), the type 304/316 stainless steels and the type 321/347 stainless steels showed an essentially equal threefold decrease in the fatigue life when the strain rate was changed from a relatively high rate (6.7×10^{-3} s^{-1}) to a low rate (6.7×10^{-5} s^{-1}). At 700 °C (1290 °F), however, the fatigue life for the type 304/316 stainless steels decreased by a factor of two when the strain rate changed from a high to a low rate; for the type 321/347 stainless steels, the fatigue life decreased by a factor of five. The decrease in the fatigue life at 600 °C (1100 °F) with decreasing strain rate for all of the stainless steels was associated with a change in the fracture appearance from one of principally transgranular with fatigue striations to one that was increasingly intergranular.

At 700 °C (1290 °F), all the stainless steels exhibited a predominantly transgranular fracture at the high strain rate. At the low strain rate, the type 304/316 stainless steels still showed a mostly transgranular fracture with discernible fatigue striations, along with some intergranular fracture. However, the type 321/347 stainless steels exhibited a completely intergranular fracture. The difference in the high-temperature (700 °C, or 1290 °F) response between the two groups of stainless steels was attributed to the inability of the type 321/347 stainless steels to recover from the cyclic strain hardening to the same degree that the type 304/316 stainless steels could. With less recovery, the material within the grains is strengthened by cyclic strain hardening, making grain-boundary sliding an easier fracture process than transgranular fatigue. If recovery occurs, the material within the grains is strengthened less, and although some intergranular fracture still takes place, transgranular fatigue, which exhibits a lower fatigue crack growth rate than intergranular sliding, dominates (Ref 247).

Example 19. An annealed Inconel X-750 alloy was tested under the following conditions: ΔK = 35 MPa$\sqrt{m}$ (32 ksi$\sqrt{in.}$), R = 0.05, triangular wave form, air and vacuum (better than 10^{-5} torr), at 650 °C (1200 °F) (Ref 245). The results are listed below:

Environment	Frequency, Hz	da/dN, mm/cycle
Air	10	5×10^{-4}
	0.1	1.5×10^{-3}
	0.01	2×10^{-2}
Vacuum	10	2×10^{-4}
	0.1	8×10^{-4}
	0.01	2×10^{-3}

In air, the fatigue crack growth rate at 0.01 Hz was 40 times greater than at 10 Hz. In vacuum, the fatigue crack growth rate at 0.01 Hz was only ten times greater than at 10 Hz.

In air, the substantial increase in the fatigue crack growth rate with decreasing frequency was believed to result from the combined effects of creep and enhanced crack growth due to transgranular and intergranular oxidation by diffusion of oxygen in the air. Therefore, the fracture mode changed from a principally fatigue-dominated transgranular fracture with fatigue striations at the high frequencies (1 and

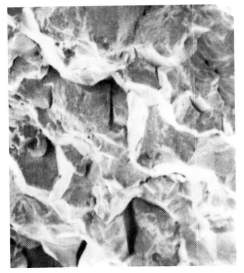

Fig. 94 Fracture appearance of a cold-worked type 316 stainless steel fatigue tested in vacuum at 593 °C (1100 °F) under a 1-min dwell time. Note the more intergranular nature of this fracture when compared to Fig. 89(b), which shows the fracture appearance of the same alloy tested under the same conditions, except at zero dwell time. Source: Ref 242

Table 2 Effect of dwell time at maximum tensile strain on the fatigue life of type 316 and 321 stainless steels

Alloy	Dwell time, min	Fatigue life (N_f), at 600 °C (1110 °F)	Fatigue life (N_f), at 700 °C (1290 °F)
Type 316 (ASTM grain size = 2)	0	2×10^3	1.3×10^3
	30	3×10^2	2×10^2
Type 321 (ASTM grain size = 1)	0	2×10^3	1×10^3
	30	2×10^2	1×10^2

10 Hz) to a primarily creep-dominated intergranular fracture at the low frequencies (0.1 and 0.01 Hz) (a similar change in fracture appearance in an IN-718 nickel-base alloy is illustrated in Fig. 93). In vacuum, the increase in the fatigue crack growth rate with decreasing frequency was also due to a transition from a transgranular fatigue-dominated fracture to a primarily creep-dominated intergranular fracture. The smaller increase in the fatigue crack growth rate in vacuum may have been due to the absence of the additional embrittling effect of oxygen (Ref 245).

Example 20. An annealed Inconel 600 alloy was tested under the following conditions: ΔK = 30 to 50 MPa$\sqrt{\text{m}}$ (27.5 to 45.5 ksi$\sqrt{\text{in.}}$), R = 0.05, air (relative humidity not specified), at room temperature (Ref 258). Frequency had no significant effect on the fatigue crack growth rate when tested in the range of 1 to 10 Hz. This is because room-temperature air is essentially an inert environment for Inconel 600 and because room temperature is too low for any significant creep to occur.

Example 21. A 20% cold-worked type 316 stainless steel was tested under the following conditions: ΔK = 30 MPa$\sqrt{\text{m}}$ (27.5 ksi$\sqrt{\text{in.}}$), R = 0.05, f = 0.17 Hz, with and without a 1-min dwell time at K_{max}, vacuum (better than 10^{-6} torr), at 593 °C (1100 °F) (Ref 242). The results are listed below:

Dwell time, min	da/dN mm/cycle
0	4×10^{-4}
1	4×10^{-2}

The fatigue crack growth rate with a 1-min dwell time was 100 times greater than when no dwell time was present. The large increase in the fatigue crack growth rate was due to a creep interaction with the fatigue fracture. Consequently, the fracture acquired a more intergranular character (Fig. 94).

Example 22. Annealed type 304, 316, 321, and 347 stainless steels were tested under the following conditions: total strain range = 1%, mean strain = 0, triangular wave form, ramp strain rate = 6.7×10^{-3} s^{-1}, with and without a 30-min dwell time at maximum tensile strain, air, at 600 and 700 °C (1110 and 1290 °F) (Ref 247). The results are given in Table 2. At 600 and 700 °C (1110 and 1290 °F), the type 316 stainless steel showed about a 6.5-time decrease in the fatigue life, N_f, because of the imposition of a 30-min dwell period at maximum tensile strain. The type 321 stainless steel exhibited a tenfold decrease. Although the data are not shown, the behavior of type 304 stainless steel parallels that of type 316 stainless steel, and the data for the type 347 stainless steel are similar to those of the type 321 stainless steel.

Imposing a 30-min dwell time at the maximum tensile strain had an effect similar to that of decreasing the strain rate from 6.7×10^{-3} to 6.7×10^{-5} s^{-1}, except that the magnitude of the decrease in the fatigue life was greater for the dwell-time tests, and the decrease was greater for the type 321/347 stainless steels than for the type 304/316 stainless steels. This larger decrease in the fatigue life was reflected in the fracture appearance. All of the steels showed an intergranular appearance with cavitated intergranular facets reminiscent of creep rupture; however, the type 304/316 stainless steels exhibited more of a mixed intergranular fracture and transgranular fatigue with fatigue striations, while the type 321/347 stainless steel fractures were predominantly intergranular.

Example 23. An IMI-685 titanium alloy (Ti-6Al-5Zr-0.5Mo-0.25Si and with 40 ppm H) with a microstructure consisting of aligned α within prior-β grains was tested under the following conditions: ΔK = 22 to 40 MPa$\sqrt{\text{m}}$ (20 to 365 ksi$\sqrt{\text{in.}}$), R = 0.05, f = 1 Hz, 5-min dwell time at maximum tensile load, air (relative humidity not specified), at room temperature (Ref 257). The tests were conducted by continuous fatigue cycling until the crack had grown to a length of 5 mm (0.2 in.), the 5-min dwell was then applied, and the normal cycling was again resumed after the 5-min dwell.

Because of appreciable scatter in the dwell test data, no specific numerical results will be given; however, the 5-min dwell at maximum tensile stress produced at least an order of magnitude increase in the fatigue crack growth rate. This increase was accompanied by changes in the fracture appearance. These changes to the predominantly fatigue-striated fracture included a large increase over a short distance immediately after the dwell period, changes in the local fracture direction (local cracks often propagated at right angles to the macroscopic fracture direction), and the introduction of quasi-cleavage and cleavage fractures (Fig. 95). Although the nondwell fractures also exhibited some quasi-cleavage and cleavage regions, the dwell fractures contained more of these features. Both the local increase in the striation spacing and the changes in the local fracture direction resulted from the linkup of the fatigue crack with the dwell-induced nonfatigue cracked areas ahead of the main crack front. The increase in the fatigue crack growth rate and the changes in the fracture mode were due to the combined action of creep and hydrogen embrittlement ahead of the crack tip that occurred during the dwell period (Ref 257).

Example 24. An IMI-685 titanium alloy (Ti-6Al-5Zr-0.5Mo-0.25Si and with 140 ppm H), which was β annealed and stress-relieved, and a Ti-5Al-2.5Sn alloy with 90 ppm H, which was β annealed and aged, were tested under the following conditions: load-controlled, maximum load = 95% of 0.2% offset yield strength, R = 0.3, f = 0.33 Hz, triangular wave form with and without a 5-min dwell at maximum load, air (relative humidity not specified), at room temperature (Ref 141).

Both alloys exhibited a drastic decrease in the fatigue life, N_f, as a result of the 5-min dwell at 95% of the 0.2% offset yield strength. The fatigue life of the IMI-685 alloy decreased from a no-dwell N_f of over 30 000 to 173 cycles, while the fatigue life of the Ti-5Al-2.5Sn alloy decreased from 12 300 to 103 cycles. The dwell fractures exhibited extensive areas of cleavage. The large decrease in the fatigue lives was attributed to hydrogen embrittlement, which produced cleavage in the α

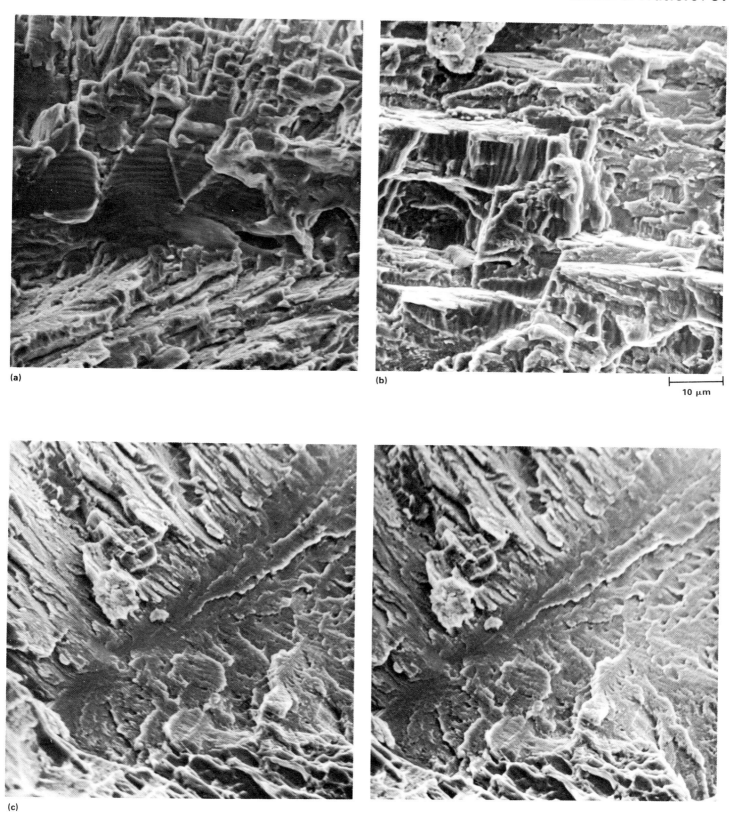

Fig. 95 Changes in the principally fatigue-striated fracture appearance of a Ti-6Al-5Zr-0.5Mo-0.25Si alloy as a result of a 5-min dwell at maximum tensile stress. The changes included a local increase in the striation spacing after the dwell period (a); a change in the local fracture direction, as indicated by the orientation of the fatigue striations (central portion of photograph), which are normal to the macroscopic fracture propagation direction (b); and the introduction of cleavage fracture shown in the stereo pair in (c). The macroscopic crack growth direction is vertical. Source: Ref 257

62 / Modes of Fracture

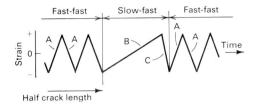

Fig. 96 Fatigue test strain wave form. A, strain rate = 50×10^{-5} s^{-1} (fast-fast cycles); B, strain rate = 1.4×10^{-5} s^{-1} (the slow-rising strain ramp of the slow-fast cycle); C, strain rate = 28×10^{-5} s^{-1} (the fast-declining strain ramp of the slow-fast cycle). Source: Ref 255

phase. There was no evidence of a creep contribution to the fracture process (Ref 141).

Example 25. A Ti-6Al-4V alloy containing 44 ppm H and a Ti-6Al-2Sn-4Zr-6Mo alloy containing 68 ppm H, which were annealed and basal textured, were tested to determine the influence of dwell time at maximum tensile stress on the fatigue crack growth rates of the alloys. The Ti-6Al-4V alloy was tested under the following conditions: three ΔK ranges: 12 to 14 MPa$\sqrt{\text{m}}$ (11 to 12.5 ksi$\sqrt{\text{in.}}$), 21 to 25 MPa$\sqrt{\text{m}}$ (19 to 23 ksi$\sqrt{\text{in.}}$), and 37 to 41 MPa$\sqrt{\text{m}}$ (33.5 to 37 ksi$\sqrt{\text{in.}}$); $R = 0.01$; $f = 0.3$ and 6 Hz; with and without a 45-min dwell at maximum tensile stress. The Ti-6Al-2Sn-4Zr-6Mo alloy was tested as follows: ΔK = approximately 23 to 28 MPa$\sqrt{\text{m}}$ (21 to 25.5 ksi$\sqrt{\text{in.}}$) and $\Delta K = 38.5$ MPa$\sqrt{\text{m}}$ (35 ksi$\sqrt{\text{in.}}$), $R = 0.01$, $f = 0.3$, with and without a 45-min dwell at maximum tensile stress for the $\Delta K = 23$ to 28 MPa$\sqrt{\text{m}}$ (21 to 25.5 ksi$\sqrt{\text{in.}}$) tests, with and without a 10-min dwell for the $\Delta K = 38.5$ MPa$\sqrt{\text{m}}$ (35 ksi$\sqrt{\text{in.}}$) test. All of the fatigue tests were conducted using displacement-controlled constant stress intensity (K), in air (relative humidity not specified), at room temperature.

The imposition of a 45-min dwell period at maximum tensile stress resulted in a very slight increase in the fatigue crack growth rate for the Ti-6Al-4V alloy when tested at the 37- to 41-MPa$\sqrt{\text{m}}$ (33.5- to 37-ksi$\sqrt{\text{in.}}$) ΔK range. There was no significant increase in the fatigue crack growth rate at the lower values of ΔK. The Ti-6Al-2Sn-4Zr-6Mo alloy showed a nominal two- to threefold increase in the total fatigue crack growth rate as a result of the 10-min dwell at $\Delta K = 38.5$ MPa$\sqrt{\text{m}}$ (35 ksi$\sqrt{\text{in.}}$), but there was little effect on the fatigue crack growth rate at the lower values of ΔK. The small changes in the fatigue crack growth rate for both alloys were due to the crack advance by cleavage during the dwell periods. The cleavage fracture was the result of hydrogen embrittlement (Ref 259).

Example 26. An annealed SUS 304 stainless steel (SUS 304 is the Japanese equivalent of AISI type 304 stainless steel) with ultimate tensile strengths of 648 MPa (94 ksi) and 356 MPa (52 ksi) at room temperature and 600 °C (1110 °F), respectively, was tested to determine the effect of the type of cyclic load on the

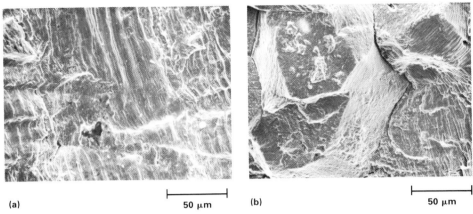

Fig. 97 Effect of one slow-fast cycle on the fracture appearance of an SUS 304 stainless steel that was fatigue tested in air at 600 °C (1110 °F). (a) Increase in fatigue striation spacing after the introduction of the slow-fast cycle at a half crack length = 0.97 mm (0.038 in.). (b) Intergranular fracture after the application of one slow-fast cycle at a half crack length of 1.0 mm (0.039 in.). Crack growth direction in both figures is right to left. Source: Ref 255

Type	Abbreviation	Wave form
Fast-fast	FF	
Slow-slow	SS	
Truncated wave with tensile and compressive dwell (hold)	TW-TC	
Slow-fast	SF	
Truncated wave with tensile dwell	TW-T	
Fast-slow	FS	
Truncated wave with compressive dwell	TW-C	

Fig. 98 Strain wave forms used in fatigue testing. Source: Ref 256

fatigue crack growth rate (Ref 255). Through-thickness cracks in thin-wall cylindrical test specimens were grown to half crack lengths ranging from 0.55 to 1.23 mm (0.02 to 0.048 in.) using fast-fast strain cycling. At the predetermined half length, one slow-fast cycle was introduced, and the fast-fast strain cycling was resumed. The fatigue test wave form is illustrated in Fig. 96. The wave form was triangular, fully reversed, with a total strain range of 0.8 to 1.5%. The fast-fast cycles were applied at a strain rate of $\pm 50 \times 10^{-5}$ s^{-1}, and the slow-fast cycle consisted of a 1.4×10^{-5} s^{-1} rising strain ramp and 28×10^{-5} s^{-1} declining strain ramp. All tests were conducted in air at a temperature of 600 °C (1110 °F).

The introduction of one slow-fast cycle always increased the fatigue crack growth rate, sometimes by up to a factor of two. When the fast-fast cycling was resumed after the slow-fast cycle, the crack propagated by a fatigue striation forming mechanism, in which case there was a local increase in the striation spacing, or by intergranular fracture (Fig. 97). Whether the crack propagated by a striation-forming mechanism or by intergranular fracture was determined by the ratio $\Delta J_c/\Delta J_f$, where ΔJ_c = range of the creep J-integral and ΔJ_f = range of the cyclic J-integral.* The values of ΔJ_c and ΔJ_f can be determined from plots of load versus crack-opening displacement at half crack length for the slow-fast cycle, ΔJ_c, and the fast-fast cycles, ΔJ_f (Ref 255). When the value of $\Delta J_c/\Delta J_f$ is less than 0.1, the crack propagates by a striation-forming mechanism; if the value exceeds 0.1, the crack propagates by intergranular fracture when fast-fast cycling is resumed. In general, fatigue striations were formed when half crack lengths were less than about 0.85 mm (0.033 in.) and the total strain range was less than 1%. Intergranular fractures occurred at the longer half crack lengths and the higher total strain ranges.

From a material standpoint, the factors that determined whether the cracking occurred by

---

*The basic J-integral is a fracture mechanics parameter, and in the elastic case, the J-integral is related to the strain energy release rate and is a function of ΔK (the range of the stress intensity factor, K) and E (elastic modulus).

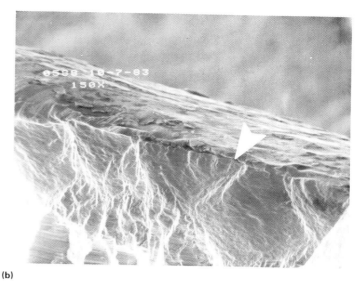

Fig. 99 Fractures in AISI 5160 wire springs that originated at seams. (a) Longitudinal fracture originating at a seam. (b) Fracture origin at a very shallow seam. the arrow indicates the base of the seam. (J.H. Maker, Associated Spring)

striation fatigue or intergranular fracture were the degree of creep damage and the size of the damaged zone produced during the slow-fast cycle. If the creep-damaged zone ahead of the crack tip was smaller than the strain-deformed region resulting from the fast-fast cycling before the slow-fast cycle, the crack propagated by striation-forming fatigue. However, if the creep-damaged zone was larger than that of the fast-fast cycling and if wedge cracks were produced at the grain boundaries adjacent to the crack tip, intergranular fracture resulted. Regardless of the cracking mechanism, the fatigue crack growth rate is accelerated until the fatigue crack escapes the creep-damaged zone (Ref 255).

Example 27. A fully heat treated IN-738 nickel-base superalloy was tested to determine the effect of wave form on fatigue life (Ref 256). Seven strain wave forms (Fig. 98) were used to conduct fatigue life tests at a total strain range of 0.4 to 1.2%. The strain rates for the fast-fast and the slow-slow wave forms were 10^{-2} and 10^{-5} s^{-1}, respectively. For the slow-fast and fast-slow sawtooth wave forms, the slow strain rate was 10^{-5} s^{-1} and the fast rate was 10^{-2} s^{-1}. The ramp rate for the truncated wave forms was 10^{-2} s^{-1}, and the dwell times at maximum tensile or compressive strains varied from 400 to 700 s. All tests were performed in air at a temperature of 850 °C (1560 °F).

The fatigue life to crack initiation, defined as the number of cycles required to initiate a 0.5-mm (0.02-in.) long crack, was lower for the SS, TW-TC, SF, and TW-T wave forms than for the FF wave form (see Fig. 98 for explanation of abbreviations). The FS and TW-C types were also lower than FF, but generally not by as much as the SS, TW-TC, SF, and TW-T types.

Crack initiation and propagation for the FF wave form were transgranular; the SS, TW-TC, SF, and TW-T forms showed intergranular crack initiation but a mixed propagation. For the FS and TW-C wave forms, fatigue crack initiation was either intergranular or transgranular; the determining factor was the total strain range. The higher range favored intergranular initiation and mixed propagation, while the lower ranges promoted transgranular initiation and propagation.

In addition to the wave forms influencing the fracture path, the dwell and ramp rates were also shown to affect the mean stress. It was discovered that a dwell on one side of the strain wave developed a mean stress on the opposite side of the wave. For example, a dwell at maximum tensile strain (maximum tensile stress) developed a compressive mean stress in the fatigue load cycle. This shift in the mean stress was apparently due to creep during dwell at the maximum tensile (or compressive) strain so that returning the specimen to the maximum compressive (or tensile) strain resulted in a shift in the initial zero-mean stress to the compressive (or tensile) side of the fatigue curve. Although not as large as in the dwell wave forms, mean stresses sometimes developed in wave forms having unequal ramp rates, with the mean stress occurring on the side of the faster ramp rate.

The one clear trend in the fatigue life tests showed that the FF wave forms exhibited the longest fatigue lives. Because of scatter in the data, no clear trends were established for the other wave forms, although a few individual specimens had longer fatigue lives than some of the shorter-life FF specimens. Because FF loading always produced the optimum (low fatigue crack growth rate) transgranular crack path, the occasional longer fatigue lives of individual non-FF wave form specimens were attributed to such factors as the presence of a compressive mean stress, multiple crack propagation, and crack branching, all of which would contribute to an increase in the fatigue life (Ref 256).

Discontinuities Leading to Fracture

Fracture of a stressed part is often caused by the presence of an internal or a surface discontinuity. The manner in which these types of discontinuities cause fracture and affect the features of fracture surfaces will be described and fractographically illustrated in this section.

Discontinuities such as laps, seams, cold shuts, previous cracks, porosity, inclusions, segregation, and unfavorable grain flow in forgings often serve as nuclei for fatigue fractures or stress-corrosion fractures because they increase both local stresses and reactions to detrimental environments. Large discontinuities may reduce the strength of a part to such an extent that it will fracture under a single application of load. However, a discontinuity should not be singled out as the sole cause of fracture without considering other possible causes or contributing factors. Thorough failure analysis may show that the fracture would have occurred even if the discontinuity had not been present.

Fractures that originate at, or pass through, significant metallurgical discontinuities usually show a change in texture, surface contour, or coloration near the discontinuity. Examination of a suspect area at several different magnifications and under several different lighting conditions will often help to determine whether a significant discontinuity is present and may provide information about its size and type. Varying the angle of incident light during examination with a low-power stereomicro-

64 / Modes of Fracture

Fig. 100 Laps formed during thread rolling of a 300M steel stud. (a) Light fractograph showing laps (arrows). (b) SEM fractograph giving detail of a lap. (c) SEM fractograph showing heavily oxidized surfaces of a lifted lap; the oxidation indicates that the lap was present before heat treatment of the stud. Arrow at right points to area shown in fractograph (d), and arrow at left points to area shown in fractograph (e). (d) and (e) SEM fractographs showing oxidized surfaces of the lifted lap in fractograph (c). See Fig. 101 for views of the stress-corrosion crack initiated by the laps.

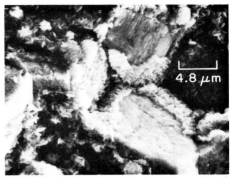

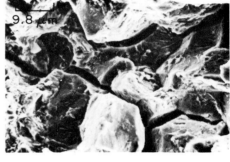

Fig. 101 SEM views of the corrosion products (a) and the intergranular fracture and secondary grain-boundary cracks (b) that were the result of the laps shown in Fig. 100

scope may be especially helpful. Segregation or unfavorable grain flow sometimes contributes to fracture without showing evidence that can be detected by direct visual examination. Even when visual indications of a metallurgical discontinuity are present, corroborating evidence should be obtained from other sources, such as examination of metallographic sections through the suspect area or study of local variations in chemical composition by electron microprobe analysis or Auger electron spectroscopy (AES).

Even though cracks usually originate at discontinuities, the type of discontinuity does not necessarily determine fracture mechanism. For example, fracture from a gross discontinuity, such as a rolling lap, can occur by any of the common fracture mechanisms. In general, discontinuities act as fracture initiation sites and cause fracture initiation to occur earlier, or at lower loads, than it would in material free from discontinuities. Additional information on material defects that contribute to fracture/failure is available in the Atlas of Fractographs in this Volume and in Volume 11 of the 9th Edition of *Metals Handbook*.

Laps, Seams, and Cold Shuts. An observer familiar with the characteristics of various types of fractures in the material under examination can usually find indications of a discontinuity if one was present at the fracture origin. A flat area that, when viewed without magnification, appears black or dull gray and does not exhibit the normal characteristics of fracture indicates the presence of a lap, a seam, or a cold shut. Such an area may appear to have resulted from the peeling apart of two metal surfaces that were in intimate contact but not strongly bonded together. A lap, a seam, or a cold shut is fairly easy to identify under a low-power stereomicroscope because the area of any of these discontinuities is distinctly different in texture and color from the rest of the fracture surface.

Failures in valve springs that originated at a seam are shown in Fig. 99. The failure shown in Fig. 99(a) began at the seam that extended more than 0.05 mm (0.002 in.) below the spring wire surface. The fatigue fracture front progressed downward from several origins. Each one of these fronts produces a crack that is triangular in outline and is without fine detail due to sliding of the opposing surfaces during the later stages of fracture. This occurs when the fracture plane changes to an angle with the wire axis in response to the torsional strain. These surfaces are visible in the lower part of Fig. 99(a).

The failure shown in Fig. 99(b) has many of the characteristics of that shown in Fig. 99(a), except that the seam is scarcely deeper than the folding of the surface that results from shot

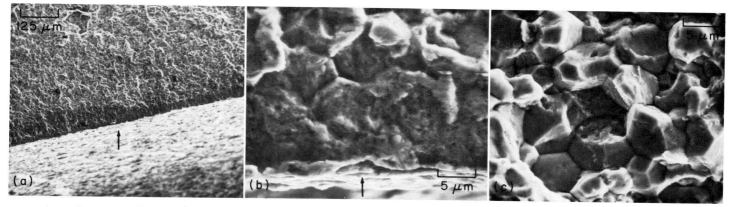

Fig. 102 Fracture caused by a portion of a pre-existing intergranular stress-corrosion crack that was not removed in reworking. The part was made of AISI 4340 steel that was heat treated to a tensile strength of 1790 to 1930 MPa (260 to 280 ksi). (a) and (b) Remains of an old crack along the edge of the surface of the part (arrows); note dark zone in (a) and extensively corroded separated grain facets in (b). (c) Clean intergranular portion of crack surface that formed at the time of final fracture

peening. Observation of it requires close examination of the central portion of the fractograph. This spring operated at a very high net stress and failed at less than 10^6 cycles.

The fractographs in Fig. 100 show laps that had been rolled into the thread roots of a 300M high-strength steel stud during thread rolling. The laps served as origins of a stress-corrosion crack that partially severed the stud. Both surfaces of a lifted lap (Fig. 100c) were heavily oxidized (Fig. 100d and e), indicating that the lap was formed before the stud was heat treated (to produce a tensile strength of 1930 to 2070 MPa, or 280 to 300 ksi). The stress-corrosion crack near the origin is shown in Fig. 101.

Cracks. The cause and size of a pre-existing crack are of primary importance in fracture mechanics, as well as in failure analysis, because of their relationship to the critical crack length for unstable crack growth. Figure 102 shows a fracture in a highly stressed AISI 4340 steel part. A narrow zone of corroded intergranular fracture at the surface of the part is adjoined by a zone of uncorroded intergranular fracture, which is in turn adjoined by a dimpled region. The part had been reworked to remove

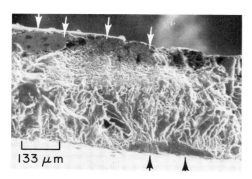

Fig. 103 Fracture in a weld in commercially pure titanium showing incomplete fusion. Unfused regions, on both surfaces (arrows), served as nuclei of fatigue cracks that developed later under cyclic loading.

general corrosion products shortly before fracture. It was concluded that the rework failed to remove about 0.1 mm (0.004 in.) of a pre-existing stress-corrosion crack, which continued to grow after the part was returned to service.

The heat treat cracks that most commonly contribute to service fractures are the transformation stress cracks and quenching cracks that occur in steel. When a heat treat crack is broken open, the surface of the crack usually has an intercrystalline or intergranular texture. If a crack has been open to an external surface of the part (so that air or other gases could penetrate the crack), it usually has been blackened by oxidation during subsequent tempering treatments or otherwise discolored by exposure to processing or service environments.

Heat treatment in the temperature range of 205 to 540 °C (400 to 1000 °F) may produce temper colors (various shades of straw, blue, or brown) on the surface of a crack that is open to an external surface. The appearance of temper colors is affected by the composition of the steel, the time and temperature of exposure, the furnace atmosphere, and the environment subsequent to the heat treatment that produced the temper color. Additional information on heat treat cracks and the appearance of temper colors can be found in the article "Visual Examination and Light Microscopy" in this Volume.

Incomplete fusion or inadequate weld penetration can produce a material discontinuity similar to a crack. Subsequent loading can cause the discontinuity to grow, as in Fig. 103, which shows a fracture in a weld in commercially pure titanium that broke by fatigue from crack nuclei, on both surfaces, that resulted from incomplete fusion during welding.

Inclusions. Discontinuities in the form of inclusions, such as oxides, sulfides, and silicates, can initiate fatigue fractures in parts subjected to cyclic loading (see, for example, Fig. 579 to 583 and Fig. 588 to 598 in the Atlas of Fractographs in this Volume, which illustrate the effect of inclusions on the fatigue crack propagation in ASTM A533B steel). In addition, such inclusions have been identified as initiation sites of ductile fractures in aluminum alloys and steels. At relatively low strains, microvoids form at inclusions, either by fracture of the inclusion or by decohesion of the matrix/inclusion interface.

The very large inclusion shown in Fig. 104 was found in the fracture surface of a case-hardened AISI 9310 steel forging that broke in service. X-ray analysis of the inclusion led to the deduction that it was a fragment of the firebrick lining of the pouring ladle.

Figure 105 shows a large inclusion in a fracture surface of a cast aluminum alloy A357-T6 blade of a small, high-speed air turbine and two views of the fracture-surface features around this inclusion.

Figure 106 shows the fracture features associated with inclusions in AISI 4340 steel with a tensile strength of 1790 to 1930 MPa (260 to 280 ksi). Entrapped flux in a brazed joint can effectively reduce the strength of the brazement and also can create a long-term corrosion problem. A 6061 aluminum alloy attachment bracket was dip brazed to an actuator of the same alloy in a flux consisting of a mixture of sodium, potassium and lithium halides, then heat treated to the T6 temper after brazing. The flux inclusion, shown in Fig. 107, reduced the cross section of the joint, and an overload fracture occurred in the Al-12Si brazing alloy.

Stringers are elongated nonmetallic inclusions, or metallic or nonmetallic constituents, oriented in the direction of working. Nonmetallic stringers usually form from deoxidation products or slag, but may also result from the intentional addition of elements such as sulfur to enhance machinability. Figure 108 shows unidentified stringers on the fracture surface of an AISI 4340 steel forging. Overload cracking occurred during straightening after the forging had been heat treated to a tensile strength of 1380 to 1520 MPa (200 to 220 ksi).

Porosity is the name applied to a condition of fine holes or pores in a metal. It is most common in castings and welds, but residual

66 / Modes of Fracture

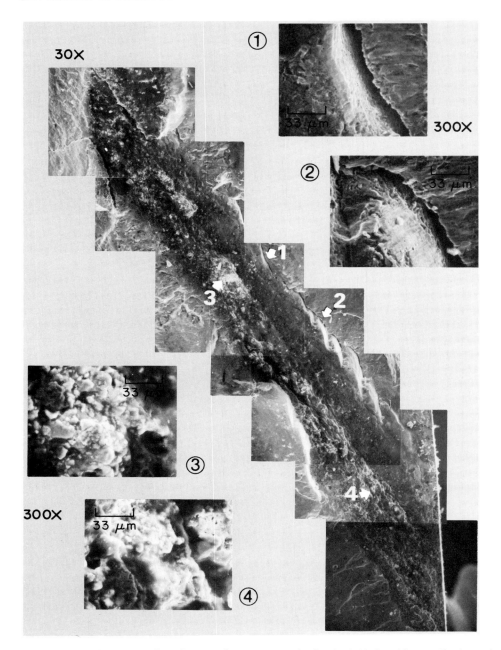

Fig. 104 Inclusion in a surface of a service fracture in a case-hardened AISI 9310 steel forging. The diagonal view is a composite of several fractographs showing a very large inclusion, which was a fragment of the pouring-ladle firebrick lining. Fractographs 1 to 4 are higher-magnification views of areas indicated by arrows 1 to 4, respectively, in the diagonal view.

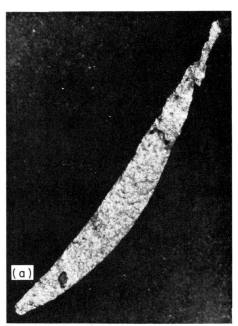
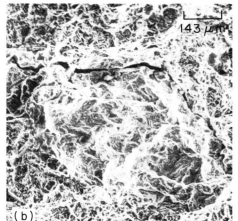
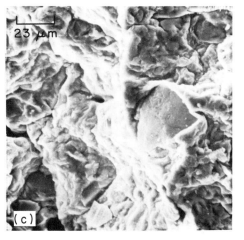

Fig. 105 Fracture surface of a cast aluminum alloy A357-T6 air-turbine blade. (a) Overall view of the fracture surface showing a large inclusion (dark) near the tip of the blade. Approximately 0.4X. (b) and (c) Decohesion at the interfaces between the inclusion and the aluminum matrix

porosity from the cast ingot sometimes still persists in forgings. In fractures that occur through regions of excessive porosity, numerous small depressions or voids (sometimes appearing as round-bottom pits) or areas with a dendritic appearance can be observed. At low magnification, fractures through regions of excessive porosity may appear dirty or sooty because of the large number of small voids, which look like black spots.

Figure 109 shows random porosity (with pores surrounded by dimples) in a fracture of a cast aluminum alloy A357 air-turbine blade. Fracture was caused by overload from an impact.

Figure 110 shows a shrinkage void intersected by the fracture surface of a cast aluminum alloy A357-T6 gear housing. The dendrite nodules in the void indicate that the cavity was caused by unfavorable directional solidification during casting. The fracture was caused by overload.

Figure 111 shows SEM fractographs of the surface of a fatigue fracture in a resistance spot

Modes of Fracture / 67

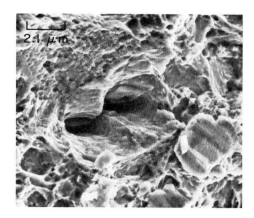

Fig. 106 Fracture surface of an AISI 4340 steel that was heat treated to a tensile strength of 1790 to 1930 MPa (260 to 280 ksi) showing deep dimples containing the inclusions that initiated them. (J. Kilpatrick, Delta Air Lines)

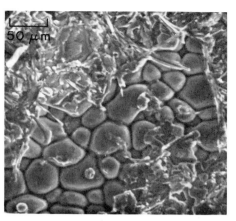

Fig. 107 Halide-flux inclusion (rounded granules) in the joint between an actuator and an attachment bracket of aluminum alloy 6061 that were joined by dip brazing using an Al-12Si brazing alloy

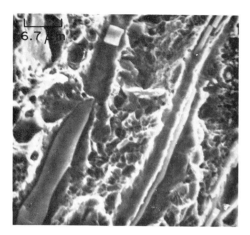

Fig. 108 Stringers on the surface of a fracture that occurred during straightening of an AISI 4340 steel forging that had been heat treated to a tensile strength of 1380 to 1520 MPa (200 to 220 ksi). Stringers are visible as parallel features inclined about 30° to right of vertical. They were not identified as to composition, but may be accidentally entrapped slag that was elongated in the major direction of flow during forging.

Fig. 109 Porosity in a fracture of a cast aluminum alloy A357 blade from a small air turbine. The blade fractured by overload from an impact to its outer edge.

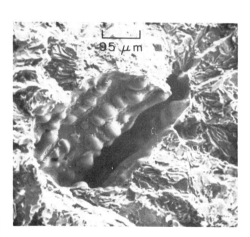

Fig. 110 Shrinkage void with dendrite nodules on a fracture surface of a cast aluminum alloy A357-T6 gear housing that broke by overload

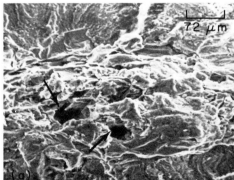
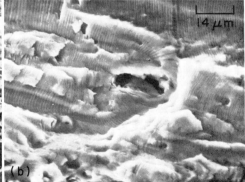

Fig. 111 Fatigue fracture surface of a resistance spot weld that broke during bond testing of an aluminum alloy 7075-T6 specimen. (a) Note voids (arrows) caused by molten-metal shrinkage in the weld nugget. (b) Both fatigue striations and shrinkage voids are evident, which indicates that the fracture path favored the porous areas.

weld that broke during bond testing of an aluminum alloy 7075-T6 specimen. The voids in the weld nugget are apparent in both fractographs. The fatigue fracture path appears to favor the voids.

Segregation. The portion of a fracture in a region of segregation may appear either more brittle or more ductile than the portion in the surrounding regions. Differences in fracture texture may be slight and therefore difficult to evaluate. Fractographic evidence of segregation should always be confirmed by comparing the microstructure and chemical composition of the material in the suspect region with those in other locations in the same part.

Unfavorable Grain Flow. Grain flow in an unfavorable direction may be indicated by a woody fracture in some materials and by a flat, delaminated appearance in others. An area of woody fracture is indicated in region B in Fig. 112, which shows a fatigue fracture in a forged AISI 4340 steel aircraft landing-gear axle. The fatigue fracture occurred in an area where the resistance of the material to fatigue cracking was low because the fluctuating loads were applied nearly perpendicular to the direction of grain flow. In high-strength aluminum alloy extrusions and hot-rolled products, tension loads are occasionally applied perpendicular to

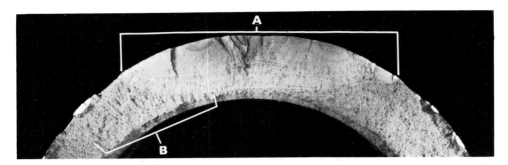

Fig. 112 Fatigue fracture through a region of unfavorable grain flow and large inclusions in an aircraft landing-gear axle forged from AISI 4340 steel. Several fatigue crack origins were found in region A. Numerous small discontinuous marks nearly perpendicular to the direction of crack propagation were determined by metallographic examination to be related to forging flow lines and large, elongated sulfide inclusions. Woody appearance outside fatigue zone, region B, also suggests unfavorable grain flow. Light fractograph. 1.8X

the flow direction, which may cause splitting along flow lines. This somewhat resembles the pulling apart of laminated material. Splitting may also appear as secondary cracks perpendicular to the primary fracture when these materials are broken by bending loads.

ACKNOWLEDGMENT

The valuable technical discussions and advice provided by Richard A. Rawe, McDonnell Douglas Astronautics Company, during the preparation of this article is sincerely appreciated.

REFERENCES

1. I. Le May, *Metallography in Failure Analysis*, J.L. McCall and P.M. French, Ed., American Society for Metals, 1977
2. F.P. McClintock and G.R. Irwin, in *Fracture Toughness Testing and Its Applications*, STP 381, American Society for Testing and Materials, 1965, p 84-113
3. P.C. Paris and G.C. Sih, in *Fracture Toughness Testing and Its Applications*, STP 381, American Society for Testing and Materials, 1965, p 30-81
4. B.V. Whiteson, A. Phillips, V. Kerlins, and R.A. Rawe, Ed., in *Electron Fractography*, STP 436, American Society for Testing and Materials, 1968, p 151-178
5. C.D. Beachem, *Metall. Trans. A*, Vol 6A, 1975, p 377-383
6. C.D. Beachem and D.A. Meyn, in *Electron Fractography*, STP 436, American Society for Testing and Materials, 1968, p 59
7. J. Friedel, in *Fracture*, Proceedings of the Swampscott Conference, MIT Press, 1959, p 498
8. C.D. Beachem, *Liebowitz Fracture I*, Academic Press, 1968, p 243-349
9. H. Schardin, in *Fracture*, Proceedings of the Swampscott Conference, MIT Press, 1959, p 297
10. J.P.E. Forsyth, *Acta Metall.*, Vol 11, July 1963, p 703
11. C. Laird and G.C. Smith, *Philos. Mag.*, Vol 7, 1962, p 847
12. W.A. Wood and A.K. Head, *J. Inst. Met.*, Vol 79, 1950, p 89
13. W.A. Wood, in *Fracture*, Proceedings of the Swampscott Conference, MIT Press, 1959, p 412
14. C.V. Cooper and M.E. Fine, *Metall. Trans. A*, Vol 16A (No. 4), 1985, p 641-649
15. R.O. Ritchie, in *Environment-Sensitive Fracture of Engineering Materials*, Z.A. Foroulis, Ed., The Metallurgical Society, 1979, p 538-564
16. O. Buck, W.L. Morris, and M.R. James, *Fracture and Failure: Analyses, Mechanisms and Applications*, P.P. Tung, S.P. Agrawal, A. Kumar, and M. Katcher, Ed., Proceedings of the ASM Fracture and Failure Sessions at the 1980 Western Metal and Tool Exposition and Conference, Los Angeles, CA, American Society for Metals, 1981
17. C.E. Richards and T.C. Lindley, *Eng. Fract. Mech.*, Vol 4, 1972, p 951
18. R.O. Ritchie and J.F. Knott, *Mater. Sci. Eng.*, Vol 14, 1974, p 7
19. R.M.N. Pelloux, *Trans. ASM*, Vol 62, 1969, p 281-285
20. D. Broek and G.O. Bowles, *Int. J. Fract. Mech.*, Vol 6, 1970, p 321-322
21. P. Neumann, *Acta Metall.*, Vol 22, 1974, p 1155-1178
22. R.M.N. Pelloux, in *Fracture*, Chapman and Hall, 1969, p 731
23. R. Koterazawa, M. Mori, T. Matsni, and D. Shimo, *J. Eng. Mater. Technol., (Trans. ASME)*, Vol 95 (No. 4), 1973, p 202
24. F.E. Fujita, *Acta Metall.*, Vol 6, 1958, p 543
25. D.A. Meyn, *Trans. ASM*, Vol 61 (No. 1), 1968, p 42
26. D.L. Davidson and J. Lankford, *Metall. Trans. A*, Vol 15A, 1984, p 1931-1940
27. R.D. Carter, E.W. Lee, E.A. Starke, Jr., and C.J. Beevers, *Metall. Trans. A*, Vol 15A, 1984, p 555-563
28. J.C. McMillan and R.M.N. Pelloux, *Eng. Fract. Mech.*, Vol 2, 1970, p 81-84
29. C.D. Beachem, *Trans. ASM*, Vol 60 (No. 3), 1967, p 325
30. R.W. Hertzberg, Fatigue Fracture Surface Appearance, in *Fatigue Crack Propagation*, STP 415, American Society for Testing and Materials, 1967, p 205
31. P.C. Paris and F. Erdogan, *J. Basic Eng., (Trans. ASME)*, D, Vol 85, 1963, p 528
32. H.H. Johnson and P.C. Paris, *Eng. Fract. Mech.*, Vol 1, 1968, p 3
33. R.M.N. Pelloux, *Trans. ASM*, Vol 57, 1964, p 511
34. D. Broek, in *Fracture*, Chapman and Hall, 1969, p 754
35. D. Broek, Report NLR TR 72029U (AD-917038), National Aerospace Laboratory, 1972
36. A.R. Troiano, *Trans. ASM*, Vol 52, 1960, p 54
37. N.J. Petch, *Philos. Mag.*, Vol 1, 1956, p 331
38. C.D. Beachem, *Metall. Trans. A*, Vol 3A, 1972, p 437
39. J.A. Clum, *Scr. Metall.*, Vol 9, 1975, p 51
40. C.A. Zapffe and C.E. Sims, *Trans. AIME*, Vol 145, 1941, p 225
41. A.J. Forty, *Physical Metallurgy of Stress Corrosion Cracking*, Interscience, 1959, p 99
42. H.L. Logan, *J. Res. Natl. Bur. Stand.*, Vol 48, 1952, p 99
43. T.P. Hoar and J.G. Hines, *Stress Corrosion Cracking and Embrittlement*, John Wiley & Sons, 1956, p 107
44. I. Yamauchi and F. Weinberg, *Metall. Trans. A*, Vol 14A, 1983, p 939-946
45. K.E. Puttick, *Philos. Mag.*, Vol 4, 1959, p 964-969
46. F.A. McClintock, *J. Appl. Mech. (Trans. ASME)*, Vol 35, 1968, p 363-371
47. A.S. Argon, J. Im, and R. Safoglu, *Metall. Trans. A*, Vol 6A, 1975, p 825-837
48. M.F. Ashby, C. Gandhi, and D.M.R. Taplin, *Acta Metall.*, Vol 27, 1979, p 699-729
49. T. Veerasooriya and J.P. Strizak, Report ONRL/TM-7255, Oak Ridge National Laboratory, 1980
50. I. Servi and N.J. Grant, *Trans. AIME*, Vol 191, 1951, p 909-922
51. J.N. Greenwood, D.R. Miller, and J.W. Suiter, *Acta Metall.*, Vol 2, 1954, p 250-258
52. R.W. Baluffi and L.L. Seigle, *Acta Metall.*, Vol 3, 1965, p 170-177
53. A.J. Perry, *J. Mater. Sci.*, Vol 9, 1974, p 1016-1039
54. D.A. Miller and R. Pilkington, *Metall. Trans. A*, Vol 9A, 1978, p 489-494
55. R. Raj and M.F. Ashby, *Acta Metall.*, Vol 23, 1975, p 653-666
56. J.A. Williams, *Acta Metall.*, Vol 15, 1967, p 1119-1124, 1559-1562
57. C.C. Law and M.J. Blackburn, *Metall. Trans. A*, Vol 11A (No. 3), 1980, p 495-507

58. D.S. Wilkinson, K. Abiko, N. Thyagarajan, and D.P. Pope, *Metall. Trans. A*, Vol 11A (No. 11), 1980, p 1827-1836
59. K. Sadananda and P. Shahinian, *Met. Sci. J.*, Vol 15, 1981, p 425-432
60. J.L. Bassani, *Creep and Fracture of Engineering Materials and Structures*, B. Wilshire and D.R. Owen, Ed., Pineridge Press, 1981, p 329-344
61. T. Watanabe, *Metall. Trans. A*, Vol 14A (No. 4), 1983, p 531-545
62. I-Wei Chen, *Metall. Trans. A*, Vol 14A (No. 11), 1983, p 2289-2293
63. M.H. Yoo and H. Trinkaus, *Metall. Trans. A*, Vol 14A (No. 4), 1983, p 547-561
64. S.H. Goods and L.M. Brown, *Acta Metall.*, Vol 27, 1979, p 1-15
65. D. Hull and D.E. Rimmer, *Philos. Mag.*, Vol 4, 1959, p 673-687
66. R. Raj, H.M. Shih, and H.H. Johnson, *Scr. Metall.*, Vol 11, 1977, p 839-842
67. R.L. Coble, *J. Appl. Phys.*, Vol 34, 1963, p 1679
68. F.C. Monkman and N.J. Grant, *Proc. ASTM*, Vol 56, 1956, p 593-605
69. R. Raj, *Acta Metall.*, Vol 26, 1978, p 341-349
70. A.N. Stroh, *Adv. Phys.*, Vol 6, 1957, p 418
71. J.O. Stiegler, K. Farrell, B.T.M. Loh, and H.E. McCoy, *Trans. ASM*, Vol 60, 1967, p 494-503
72. I-Wei Chen and A.S. Argon, *Acta Metall.*, Vol 29, 1981, p 1321-1333
73. D.A. Miller and T.G. Langdon, *Metall. Trans. A*, Vol 10A (No. 11), 1979, p 1635-1641
74. K. Sadananda and P. Shahinian, *Metall. Trans. A*, Vol 14A (No. 7), 1983, p 1467-1480
75. C.D. Beachem, B.F. Brown, and A.J. Edwards, Memorandum Report 1432, Naval Research Laboratory, 1963
76. T. Inoue, S. Matsuda, Y. Okamura, and K. Aoki, *Trans. Jpn. Inst. Met.*, Vol 11, 1970, p 36
77. I.M. Bernstein, *Metall. Trans. A*, Vol 1A, 1970, p 3143
78. C.D. Beachem, *Metall. Trans. A*, Vol 4A, 1973, p 1999
79. Y. Kikuta, T. Araki, and T. Kuroda, in *Fractography in Failure Analysis*, STP 645, B.M. Strauss and W.M. Cullen, Jr., Ed., American Society for Testing and Materials, 1978, p 107
80. F. Nakasoto and I.M. Bernstein, *Metall. Trans. A*, Vol 9A, 1978, p 1317
81. Y. Kikuta and T. Araki, in *Hydrogen Effects in Metals*, I.M. Bernstein and A.W. Thompson, Ed., The Metallurgical Society, 1981, p 309
82. Y.H. Kim and J.W. Morris, Jr., *Metall. Trans. A*, Vol 14A, 1983, p 1883-1888
83. A.R. Rosenfield, D.K. Shetty, and A.J. Skidmore, *Metall. Trans. A*, Vol 14A, 1983, p 1934-1937
84. R.O. Ritchie, F.A. McClintock, H. Nayeb-Hashemi, and M.A. Ritter, *Metall. Trans. A*, Vol 13A, 1982, p 101
85. I. Aitchison and B. Cox, *Corrosion*, Vol 28, 1972, p 83
86. J. Spurrier and J.C. Scully, *Corrosion*, Vol 28, 1972, p 453
87. D.B. Knorr and R.M. Pelloux, *Metall. Trans. A*, Vol 13A, 1975, p 73
88. R.J.H. Wanhill, *Corrosion*, Vol 32, 1976, p 163
89. D.A. Meyn and E.J. Brooks, in *Fractography and Material Science*, STP 733, L.N. Gilbertson and R.D. Zipp, Ed., American Society for Testing and Materials, 1981, p 5-31
90. H. Hänninen and T. Hakkarainen, *Metall. Trans. A*, Vol 10A, 1979, p 1196-1199
91. A.W. Thompson and J.C. Chesnutt, *Metall. Trans. A*, Vol 10A, 1979, p 1193
92. M.F. Stevens and I.M. Bernstein, *Metall. Trans. A*, Vol 16A, 1985, p 1879
93. C. Chen, A.W. Thompson, and I.M. Bernstein, OROC. 5th Bolton Landing Conference, Claitor's, Baton Rouge, LA
94. J.C. Chesnutt and R.A. Spurling, *Metall. Trans. A*, Vol 8A, 1977, p 216
95. M.B. Whiteman and A.R. Troiano, *Corrosion*, Vol 21, 1965, p 53
96. M.L. Holtzworth, *Corrosion*, Vol 25, 1969, p 107
97. R. Langenborg, *J. Iron Steel Inst.*, Vol 207, 1967, p 363
98. S. Singh and C. Altstetter, *Metall. Trans. A*, Vol 13A, 1982, p 1799
99. D.A. Hardwick, M. Taheri, A.W. Thompson, and I.M. Bernstein, *Metall. Trans. A*, 1982, Vol 13A, p 235
100. R.J. Gest and A.R. Troiano, *Corrosion*, Vol 30, 1974, p 274
101. D.G. Chakrapani and E.N. Pugh, *Metall. Trans. A*, Vol 7A, 1976, p 173
102. J. Albrecht, B.J. McTiernan, I.M. Bernstein, and A.W. Thompson, *Scr. Metall.*, Vol 11, 1977, p 393
103. M. Taheri, J. Albrecht, I.M. Bernstein, and A.W. Thompson, *Scr. Metall.*, Vol 13, 1975, p 871
104. J. Albrecht, A.W. Thompson, and I.M. Bernstein, *Metall. Trans. A*, Vol 10A, 1979, p 1759
105. J. Albrecht, I.M. Bernstein, and A.W. Thompson, *Metall. Trans. A*, Vol 13A, 1982, p 811
106. D.A. Hardwick, A.W. Thompson, and I.M. Bernstein, *Metall. Trans. A*, Vol 14A, 1983, p 2517
107. L. Christodoulou and H.M. Flower, *Acta Metall.*, Vol 28, 1980, p 481
108. T.D. Lee, T. Goldberg, and J.P. Hirth, *Fracture 1977*, Vol 2, Proceedings of the 4th International Conference on Fracture, Waterloo, Canada, 1977, p 243
109. R. Garber, I.M. Bernstein, and A.W. Thompson, *Scr. Metall.*, Vol 10, 1976, p 341
110. N.J. Petch and P. Stables, *Nature*, Vol 169, 1952, p 842
111. B.D. Craig, *Metall. Trans. A*, Vol 13A, 1982, p 907
112. C.E. Price and R.S. Fredell, *Metall. Trans. A*, Vol 17A, 1986, p 889
113. A.W. Thompson, *Mater. Sci. Eng.*, Vol 14, 1974, p 253
114. N.S. Stoloff and T.L. Johnson, *Acta Metall.*, Vol 11, 1963, p 251
115. J.P. Hirth and H.H. Johnson, *Corrosion*, Vol 32, 1976, p 3
116. R.A. Oriani and P.H. Josephic, *Metall. Trans. A*, Vol 11A, 1980, p 1809
117. H.C. Rogers, *Acta Metall.*, Vol 4, 1956, p 114
118. L.C. Weiner and M. Gensamer, *Acta Metall.*, Vol 5, 1957, p 692
119. K. Takita, M. Niikura, and K. Sakamoto, *Scr. Metall.*, Vol 7, 1973, p 989
120. K. Takita and K. Sakomoto, *Scr. Metall.*, Vol 10, 1976, p 399
121. A. Cracknell and N.J. Petch, *Acta Metall.*, Vol 3, 1955, p 200
122. H. Cialone and R.J. Asaro, *Metall. Trans. A*, Vol 12A, 1981, p 1373
123. S.P. Lynch, *Scr. Metall.*, Vol 13, 1979, p 1051
124. T. Goldberg, T.D. Lee, and J.P. Hirth, *Metall. Trans. A*, Vol 10A, 1979, p 199
125. H. Cialone and R.J. Asaro, *Metall. Trans. A*, Vol 10A, 1979, p 367
126. R.A. Oriani and P.H. Josephic, *Acta Metall.*, Vol 27, 1979, p 997
127. S.P. Lynch, *Acta Metall.*, Vol 32, 1984, p 79
128. S.C. Chang and J.P. Hirth, *Metall. Trans. A*, Vol 16A, 1985, p 1417
129. T.J. Hakkarainen, J. Wanagel, and C.Y. Li, *Metall. Trans. A*, Vol 11A, 1980, p 2035
130. M.J. Yacaman, T.A. Parthasarathy, and J.P. Hirth, *Metall. Trans. A*, Vol 15A, 1984, p 1485
131. J.P. Blackledge, *Metal Hydrides*, Academic Press, 1968, p 2
132. D.A. Meyn, *Metall. Trans. A*, Vol 3A, 1972, p 2302
133. G.H. Nelson, D.P. Williams, and J.C. Stein, *Metall. Trans. A*, Vol 3, 1972, p 469
134. N.E. Paton and R.A. Spurling, *Metall. Trans. A*, Vol 7A, 1976, p 1769
135. H.G. Nelson, *Metall. Trans. A*, Vol 7A, 1976, p 621
136. K.P. Peterson, J.C. Schwanebeck, and W.W. Gerberich, *Metall. Trans. A*, Vol 9A, 1978, p 1169
137. G.H. Koch, A.J. Bursle, R. Liu, and E.N. Pugh, *Metall. Trans. A*, Vol 12A, 1981, p 1833
138. M. Gao, M. Lu, and R.P. Wei, *Metall. Trans. A*, Vol 15A, 1984, p 735
139. S.M. Bruemmer, R.H. Jones, M.T. Thomas, and D.R Baer, *Metall. Trans. A*, Vol 14A, 1983, p 223
140. R.H. Jones, S.M. Bruemmer, M.T.

Thomas, and D.R. Baer, in *Effect of Hydrogen on Behavior of Metals*, I.M. Bernstein and A.W. Thompson, Ed., The Metallurgical Society, 1980, p 369
141. J.E. Hack and G.R. Leverant, *Metall. Trans. A*, Vol 13A, 1982, p 1729
142. H.G. Nelson, D.P. Williams, and J.E. Stein, in *Hydrogen Damage*, C.D. Beachem, Ed., American Society for Metals, 1977, p 274
143. J.R. Pickens, J.R. Gordon, and J.A.S. Green, *Metall. Trans. A*, Vol 14A, 1983, p 925
144. W.Y. Chu, J. Yao, and C.M. Hsiao, *Corrosion*, Vol 40, 1984, p 302
145. A. Parthasarathi and N.W. Polan, *Metall. Trans. A*, Vol 13A, 1982, p 2027
146. H. Leidheiser, Jr. and R. Kissinger, *Corrosion*, Vol 28, 1972, p 218
147. G.T. Burstein and R.C Newmann, *Corrosion*, Vol 36, 1980, p 225
148. N.W. Polan, J.M. Popplewell, and M.J. Pryor, *J. Electrochem. Soc.*, Vol 126, 1979, p 1299
149. W.Y. Chu, C.M. Hsiao, and J.W. Wang, *Metall. Trans. A*, Vol 16A, 1985, p 1663
150. A. Kawashima, A.K. Agrawal, and R.W. Staehle, *J. Electrochem. Soc.*, Vol 124, 1977, p 1822
151. J.C. Scully, *Corros. Sci.*, Vol 20, 1980, p 997
152. N.A. Nielsen, in *Hydrogen Damage*, C.D. Beachem, Ed., American Society for Metals, 1977, p 219
153. G.W. Simmons, P.S. Pao, and R.P. Wei, *Metall. Trans. A*, Vol 9A, 1978, p 1147
154. D. Tromans, *Metall. Trans. A*, Vol 12A, 1981, p 1445
155. S.R. Bala and D. Tromans, *Metall. Trans. A*, Vol 11A, 1980, p 1161
156. S.R. Bala and D. Tromans, *Metall. Trans. A*, Vol 11A, 1980, p 1187
157. A.W. Thompson, in *Environment-Sensitive Fracture of Engineering Materials*, Z.A. Foroulis, Ed., The Metallurgical Society, 1979, p 379
158. A.W. Thompson and I.M. Bernstein, *Advances in Corrosion Science and Technology*, Vol 7, Plenum Press, 1980, p 53
159. W.Y. Chu, S.Q. Li, C.M. Hsiao, and S.Y. Yu, *Corrosion*, Vol 37, 1981, p 514
160. V. Provenzano, K. Törrönen, D. Sturm, and W.H. Cullen, in *Fractography and Material Science*, STP 733, L.N. Gilbertson and R.D. Zipp, Ed., American Society for Testing and Materials, 1981, p 70
161. J.A.S. Green and H.W. Hayden, in *Hydrogen in Metals*, I.M. Bernstein and A.W. Thompson, Ed., American Society for Metals, 1974, p 575
162. R.K. Viswanadham, T.S. Sun, and J.A.S. Green, *Metall. Trans. A*, Vol 11A, 1980, p 85
163. L. Montgrain and P.R. Swann, in *Hydrogen in Metals*, I.M. Bernstein and A.W. Thompson, Ed., American Society for Metals, 1974, p 575
164. G.M. Scamans and A.S. Rehal, *J. Mater. Sci.*, Vol 14, 1979, p 2459
165. G.M. Scamans, *J. Mater. Sci.*, Vol 13, 1978, p 27
166. R.M. Latanision, O.H. Gastine, and C.R. Compeau, in *Environment-Sensitive Fracture of Engineering Materials*, Z.A. Foroulis, Ed., The Metallurgical Society, 1979, p 48
167. D.A. Vermilyea, *Stress Corrosion Cracking and Hydrogen Embrittlement of Iron-Base Alloys*, R.W. Staehle, Ed., National Association of Corrosion Engineers, 1977, p 208
168. F.P. Ford, *Corrosion Process*, R.N. Parkins, Ed., Applied Science, 1982, p 271
169. D.A. Jones, *Metall. Trans. A*, Vol 16A, 1985, p 1133
170. F.P. Vaccaro, R.F. Hehemann, and A.R. Troiano, *Corrosion*, Vol 36, 1980, p 530
171. N. Nielsen, *J. Mater.*, Vol 5, 1970, p 794
172. J.C. Scully, *Corros. Sci.*, Vol 7, 1967, p 197
173. A.A. Seys, M.F. Brabers, and A.A. Van Haute, *Corrosion*, Vol 30, 1974, p 47
174. P.R. Rhodes, *Corrosion*, Vol 25, 1969, p 462
175. J.A. Beavers and N.E. Pugh, *Metall. Trans. A*, Vol 11A, 1980, p 809
176. R.N. Parkins, C.M. Rangel, and J. Yu, *Metall. Trans. A*, Vol 16A, 1985, p 1671
177. M.B. Hintz, L.J. Nettleton, and L.A. Heldt, *Metall. Trans. A*, Vol 16A, 1985, p 971
178. S.S. Birley and D. Thomas, *Metall. Trans. A*, Vol 12A, 1981, p 1215
179. H.W. Pickering and P.J. Byrne, *J. Electrochem. Soc.*, Vol 116, 1969, p 1492
180. R.H. Hiedersbach, Jr. and E.D. Verink, *Metall. Trans. A*, Vol 3A, 1972, p 397
181. A.J. Russell and D. Tromans, *Metall. Trans. A*, Vol 10A, 1979, p 1229
182. D. Singbeil and D. Tromans, *Metall. Trans. A*, Vol 13A, 1982, p 1091
183. R.E. Curtis and S.F. Spurr, *Trans. ASM*, Vol 61, 1968, p 115
184. R.E. Curtis, R.R. Boyer, and J.C. Williams, *Trans. ASM*, Vol 62, 1969, p 457
185. W.F. Czyrklis and M. Levy, in *Environment-Sensitive Fracture of Engineering Materials*, Z.A. Foroulis, Ed., The Metallurgical Society, 1979, p 303
186. D.A. Meyn, *Metall. Trans. A*, Vol 5A, 1974, p 2405
187. D.N. Williams, *Mater. Sci. Eng.*, Vol 24, 1976, p 53
188. H.J. Rack and J.W. Munford, in *Environment-Sensitive Fracture of Engineering Materials*, Z.A. Foroulis, Ed., The Metallurgical Society, 1979, p 284
189. P. Gordon and H.H. An, *Metall. Trans. A*, Vol 13A, 1982, p 457
189. S. Dinda and W.R. Warke, *Mater. Sci. Eng.*, Vol 24, 1976, p 199
189. L.P. Costas, *Corrosion*, Vol 31, 1975, p 91
192. A.W. Funkebusch, L.A. Heldt, and D.F. Stein, *Metall. Trans. A*, Vol 13A, 1982, p 611
193. A.R.C. Westwood and M.H. Kamdar, *Philos. Mag.*, Vol 8, 1963, p 787
194. S.P. Lynch, in *Hydrogen Effects in Metals*, I.M. Bernstein and A.W. Thompson, Ed., The Metallurgical Society, 1981, p 863
195. P. Gordon, *Metall. Trans. A*, Vol 9A, 1978, p 267
196. K.S. Chan and D.A. Koss, *Metall. Trans. A*, Vol 14A, 1983, p 1343
197. G.F. Pittinato, V. Kerlins, A. Phillips, and M.A. Russo, *SEM/TEM Fractography Handbook*, MCIC-HB-06, Metals and Ceramics Information Center, 1975, p 214, 606
198. A. Phillips, V. Kerlins, R.A. Rawe, and B.V. Whiteson, *Electron Fractography Handbook*, MCIC-HB-08, Metals and Ceramics Information Center, 1976, p 3-8
199. J.R. Klepaczko and A. Solecki, *Metall. Trans. A*, Vol 15A, 1984, p 901
200. C. Zener and J. Holloman, *J. Appl. Phys.*, Vol 15, 1944, p 22
201. E. Manin, E. Beckman, and S.A. Finnegan, in *Metallurgical Effects at High Strain Rates*, R.W. Rohde, Ed., Plenum Press, 1973, p 531
202. S.M. Doraivelu, V. Gopinathan, and V.C. Venkatesh, in *Shock Waves and High-Strain-Rate Phenomena in Metals*, M.A. Meyers and L.E. Murr, Ed., Plenum Press, 1981, p 75
203. H.C. Rogers and C.V. Shastry, in *Shock Waves and High-Strain-Rate Phenomena in Metals*, M.A. Meyers and L.E. Murr, Ed., Plenum Press, 1981, p 285
204. G.L. Moss, in *Shock Waves and High-Strain-Rate Phenomena in Metals*, M.A. Meyers and L.E. Murr, Ed., Plenum Press, 1981, p 299
205. T.A.C. Stock and K.R.L. Thompson, *Metall. Trans. A*, Vol 1A, 1970, p 219
206. J.F. Velez and G.W. Powell, *Wear*, Vol 66, 1981, p 367
207. A.L. Wingrove, *Metall. Trans. A*, Vol 4A, 1973, p 1829
208. P.W. Leech, *Metall. Trans. A*, Vol 16A, 1985, p 1900
209. P.J. Wray, *Metall. Trans. A*, Vol 15A, 1984, p 2059
210. J.J. Lewandowsky and A.W. Thompson, *Metall. Trans. A*, Vol 17A, 1986, p 461
211. H. Kwon and C.H. Kim, *Metall. Trans. A*, Vol 17A, 1986, p 1173
212. A. Phillips, V. Kerlins, R.A. Rawe, and B.V. Whiteson, *Electron Fractography Handbook*, MCIC-HB-08, Metals and Ceramics Information Center, 1976, p 3-5
213. T. Takasugi and D.P. Pope, *Metall. Trans. A*, Vol 13A, 1982, p 1471

214. W.J. Mills, *Metall. Trans. A*, Vol 11A, 1980, p 1039
215. D.M. Bowden and E.A. Starke, Jr., *Metall. Trans. A*, Vol 15A, 1984, p 1687
216. H.G. Nelson, in *Effect of Hydrogen on Behavior of Materials*, A.W. Thompson and I.M. Bernstein, Ed., The Metallurgical Society, 1976, p 603
217. G. Schuster and C. Altstetter, *Metall. Trans. A*, Vol 14A, 1983, p 2085
218. H.J. Cialone and J.H. Holbrook, *Metall. Trans. A*, Vol 16A, 1985, p 115
219. R.J. Walter and W.T. Chandler, in *Effect of Hydrogen on Behavior of Materials*, A.W. Thompson and I.M. Bernstein, Ed., The Metallurgical Society, 1976, p 273
220. P.K. Liaw, S.J. Hudak, Jr., and J.K. Donald, *Metall. Trans. A*, Vol 13A, 1982, p 1633
221. W.G. Clark, Jr., in *Hydrogen in Metals*, I.M. Bernstein and A.W. Thompson, Ed., American Society for Metals, 1974, p 149
222. J.D. Frandsen and H.L. Marcus, *Metall. Trans. A*, Vol 8A, 1977, p 265
223. R.D. Pendse and R.O. Ritchie, *Metall. Trans. A*, Vol 16A, 1985, p 1491
224. S.C. Chang and J.P. Hirth, *Metall. Trans. A*, Vol 16A, 1985, p 1417
225. R.P. Wei, P.S. Pao, R.G. Hart, T.W. Weir, and G.W. Simmons, *Metall. Trans. A*, Vol 11A, 1980, p 151
226. R.P. Wei, N.E. Fennelli, K.D. Unangst, and T.T. Shih, AFOSR Final Report IFSM-78-88 (Air Force Office of Scientific Research), Lehigh University, 1978
227. S. Floreen and R.H. Kane, *Metall. Trans. A*, Vol 10A, 1979, p 1745
228. S. Floreen and R.H. Kane, *Metall. Trans. A*, Vol 13A, 1982, p 145
229. M. Müller, *Metall. Trans. A*, Vol 13A, 1982, p 649
230. D. Eliezer, D.G. Chakrapani, C.J. Altstetter, and E.N. Pugh, in *Hydrogen-Induced Slow Crack Growth in Austenitic Stainless Steels*, P. Azou, Ed., Second International Congress on Hydrogen in Metals (Paris), Pergamon Press, 1977
231. L.H. Keys, A.J. Bursle, H.R. Kemp, and K.R.L. Thompson, in *Hydrogen-Induced Slow Crack Growth in Austenitic Stainless Steels*, P. Azou, Ed., Second International Congress on Hydrogen in Metals (Paris), Pergamon Press, 1977
232. L.H. Keys, A.J. Bursle, K.R.L. Thompson, I.A. Ward, and P.J. Flower, in *Environment-Sensitive Fracture of Engineering Materials*, Z.A. Foroulis, Ed., The Metallurgical Society, 1979, p 614
233. H. Ishii, Y. Sakakibara, and R. Ebara, *Metall. Trans. A*, Vol 13A, 1982, p 1521
234. I.L.W. Wilson and B.W. Roberts, in *Environment-Sensitive Fracture of Engineering Materials*, Z.A. Foroulis, Ed., The Metallurgical Society, 1979, p 595
235. E.F. Smith III and D.J. Duquette, *Metall. Trans. A*, Vol 17A, 1986, p 339
236. C.M. Ward-Close and C.J. Beevers, *Metall. Trans. A*, Vol 11A, 1980, p 1007
237. A. Boateng, J.A. Begley, and R.W. Staehle, *Metall. Trans. A*, Vol 10A, 1979, p 1157
238. A. Boateng, J.A. Begley, and R.W. Staehle, *Metall. Trans. A*, Vol 14A, 1983, p 67
239. R.D. Carter, E.W. Lee, E.A. Starke, Jr., and C.J. Beevers, *Metall. Trans. A*, Vol 15A, 1984, p 555
240. M. Peters, A. Gysler, and G. Lütjering, *Metall. Trans. A*, Vol 15A, 1984, p 1597
241. D.L. Davidson and J. Lankford, *Metall. Trans. A*, Vol 15A, 1984, p 1931
242. K. Sadananda and P. Shahinian, *Metall. Trans. A*, Vol 11A, 1980, p 267
243. D.L. Davidson, *Acta Metall.*, Vol 32, 1984, p 707
244. J. Gayda and R.V. Miner, *Metall. Trans. A*, Vol 14A, 1983, p 2301
245. F. Gabrielli and R.M. Pelloux, *Metall. Trans. A*, Vol 13A, 1982, p 1083
246. W.J. Mills and L.A. James, *Fatigue Eng. Mater. Struct.*, Vol 3, 1980, p 159
247. K. Yamaguchi and K. Kanazawa, *Metall. Trans. A*, Vol 11A, 1980, p 1691
248. L.H. Burck and J. Weertman, *Metall. Trans. A*, Vol 7A, 1976, p 257
249. H. Ishii and J. Weertman, *Metall. Trans. A*, Vol 2A, 1971, p 3441
250. R.P. Wei, *Int. J. Fract. Mech.*, Vol 14, 1968, p 159
251. R.P. Gangloff, *Metall. Trans. A*, Vol 16A, 1985, p 953
252. P.K. Liaw and E. Fine, *Metall. Trans. A*, Vol 12A, 1981, p 1927
253. P.S. Pao, W. Wei, and R.P. Wei, in *Environment-Sensitive Fracture of Engineering Materials*, Z.A. Foroulis, Ed., The Metallurgical Society, 1979, p 565
254. D.B. Dawson, *Metall. Trans. A*, Vol 12A, 1981, p 791
255. M. Okazaki, I. Hattori, and T. Koizumi, *Metall. Trans. A*, Vol 15A, 1984, p 1731
256. M.Y. Nazmy, *Metall. Trans. A*, Vol 14, 1983, p 449
257. W.J. Evans and G.R. Gostelow, *Metall. Trans. A*, Vol 10A, 1979, p 1837
258. G.S. Was, H.H. Tischner, R.M. Latanision, and R.M. Pelloux, *Metall. Trans. A*, Vol 12A, 1981, p 1409
259. A.W. Sommer and D. Eylon, *Metall. Trans. A*, Vol 14A, 1983, p 2179

Preparation and Preservation of Fracture Specimens

Richard D. Zipp, J.I. Case Company
E. Philip Dahlberg, Metallurgical Consultants, Inc.

FRACTURE SURFACES are fragile and subject to mechanical and environmental damage that can destroy microstructural features. Consequently, fracture specimens must be carefully handled during all stages of analysis. This article will discuss the importance of care and handling of fractures and what to look for during the preliminary visual examination, fracture-cleaning techniques, procedures for sectioning a fracture and opening secondary cracks, and the effect of nondestructive inspection on subsequent evaluation.

Care and Handling of Fractures (Ref 1)

Fracture interpretation is a function of the fracture surface condition. Because the fracture surface contains a wealth of information, it is important to understand the types of damage that can obscure or obliterate fracture features and obstruct interpretation. These types of damage are usually classified as chemical and mechanical damage. Chemical or mechanical damage of the fracture surface can occur during or after the fracture event. If damage occurs during the fracture event, very little can usually be done to minimize it. However, proper handling and care of fractures can minimize damage that can occur after the fracture (Ref 2-4).

Chemical damage of the fracture surface that occurs during the fracture event is the result of environmental conditions. If the environment adjacent to an advancing crack front is corrosive to the base metal, the resultant fracture surface in contact with the environment will be chemically damaged. Cracking due to such phenomena as stress-corrosion cracking (SCC), liquid-metal embrittlement (LME), and corrosion fatigue produces corroded fracture surfaces because of the nature of the cracking process.

Mechanical damage of the fracture surface that occurs during the fracture event usually results from loading conditions. If the loading condition is such that the mating fracture surfaces contact each other, the surfaces will be mechanically damaged. Crack closure during fatigue cracking is an example of a condition that creates mechanical damage during the fracture event.

Chemical damage of the fracture surface that occurs after the fracture event is the result of environmental conditions present after the fracture. Any environment that is aggressive to the base metal will cause the fracture surface to be chemically damaged. Humid air is considered to be aggressive to most iron-base alloys and will cause oxidation to occur on steel fracture surfaces in a brief period of time. Touching a fracture surface with the fingers will introduce moisture and salts that may chemically attack the fracture surface.

Mechanical damage of the fracture surface that occurs after the fracture event usually results from handling or transporting of the fracture. It is easy to damage a fracture surface while opening primary cracks, sectioning the fracture from the total part, and transporting the fracture. Other common ways of introducing mechanical damage include fitting the two fracture halves together or picking at the fracture with a sharp instrument. Careful handling and transporting of the fracture are necessary to keep damage to a minimum.

Once mechanical damage occurs on the fracture surface, nothing can be done to remove its obliterating effect on the original fracture morphology. Corrosive attack, such as high-temperature oxidation, often precludes successful surface restoration. However, if chemical damage occurs and if it is not too severe, cleaning techniques can be implemented that will remove the oxidized or corroded surface layer and will restore the fracture surface to a state representative of its original condition.

Preliminary Visual Examination

The entire fracture surface should be visually inspected to identify the location of the fracture-initiating site or sites and to isolate the areas in the region of crack initiation that will be most fruitful for further microanalysis. The origin often contains the clue to the cause of fracture, and both low- and high-magnification analyses are critical to accurate failure analysis. Where the size of the failed part permits, visual examination should be conducted with a low-magnification wide-field stereomicroscope having an oblique source of illumination.

In addition to locating the failure origin, visual analysis is necessary to reveal stress concentrations, material imperfections, the presence of surface coatings, case-hardened regions, welds, and other structural details that contribute to cracking. The general level of stress, the relative ductility of the material, and the type of loading (torsion, shear, bending, and so on) can often be determined from visual analysis.

Finally, a careful macroexamination is necessary to characterize the condition of the fracture surface so that the subsequent microexamination strategy can be determined. Macroexamination can be used to identify areas of heavy burnishing in which opposite halves of the fracture have rubbed together and to identify regions covered with corrosion products. The regions least affected by this kind of damage should be selected for microanalysis. When stable crack growth has continued for an extended period, the region nearest the fast fracture is often the least damaged because it is the newest crack area. Corrodents often do not penetrate to the crack tip, and this region remains relatively clean.

The visual macroanalysis will often reveal secondary cracks that have propagated only partially through a cracked member. These part-through cracks can be opened in the laboratory and are often in much better condition than the main fracture. Areas for sectioning can be identified for subsequent metallography, chemical analysis, and mechanical-property determinations. Additional information on visual examination is available in the article "Visual

Examination and Light Microscopy" in this Volume.

Preservation Techniques (Ref 1)

Unless a fracture is evaluated immediately after it is produced, it should be preserved as soon as possible to prevent attack from the environment. The best way to preserve a fracture is to dry it with a gentle stream of dry compressed air, then store it in a desiccator, a vacuum storage vessel, or a sealed plastic bag containing a desiccant. However, such isolation of the fracture is often not practical. Therefore, corrosion-preventive surface coatings must be used to inhibit oxidation and corrosion of the fracture surface. The primary disadvantage of using these surface coatings is that fracture surface debris, which often provides clues to the cause of fracture, may be displaced during removal of the coating. However, it is still possible to recover the surface debris from the solvent used to remove these surface coatings by filtering the spent solvent and capturing the residue.

The main requirements for a surface coating are as follows:

- It should not react chemically with the base metal
- It should prevent chemical attack of the fracture from the environment
- It must be completely and easily removable without damaging the fracture

Fractures in the field may be coated with fresh oil or axle grease if the coating does not contain substances that might attack the base metal. Clear acrylic lacquers or plastic coatings are sometimes sprayed on the fracture surfaces. These clear sprays are transparent to the fracture surface and can be removed with organic solvents. However, on rough fracture surfaces, it can be difficult to achieve complete coverage and to remove the coating completely.

Another type of plastic coating that has been successfully used to protect most fracture surfaces is cellulose acetate replicating tape. The tape is softened in acetone and applied to the fracture surface with finger pressure. As the tape dries, it adheres tightly to the fracture surface. The main advantage of using replicating tape is that it is available in various thicknesses. Rough fracture surfaces can be coated with relatively thick replicating tape to ensure complete coverage. The principal limitation of using replicating tape is that on rough fracture surfaces it is difficult to remove the tape completely.

Solvent-cutback petroleum-base compounds have been used by Boardman *et al.* to protect fracture surfaces and can be easily removed with organic solvents (Ref 5). In this study, seven rust-inhibiting compounds were selected for screening as fracture surface coating materials. These inhibitor compounds were applied to fresh steel fracture surfaces and exposed to 100% relative humidity at 38 °C (100 °F) for 14 days. The coatings were removed by ultrasonic cleaning with the appropriate solvent, and the fracture surfaces were visually evaluated. Only the Tectyl 506 compound protected the fractures from rusting during the screening tests. Therefore, further studies were conducted with a scanning electron microscope to ensure that the Tectyl 506 compound would inhibit oxidation of the fracture surface and could be completely removed on the microscopic level without damaging the fracture surface.

Initially, steel Charpy samples and nodular iron samples were fractured in the laboratory by single-impact overload and fatigue, respectively. Representative fracture areas were photographed in the scanning electron microscope at various magnifications in the as-fractured condition. The fracture surfaces were then coated with Tectyl 506, exposed to 100% relative humidity at 38 °C (100 °F) for 14 days, and cleaned before scanning electron microscopy (SEM) evaluation by ultrasonically removing the coating in a naphtha solution. Figure 1 shows a comparison of identical fracture areas in the steel at increasing magnifications in the as-fractured

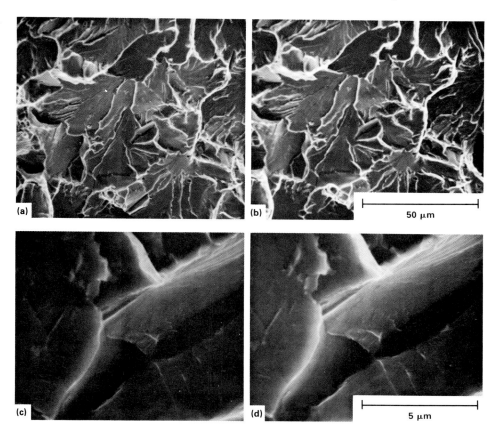

Fig. 1 Comparison of identical fracture areas of steel Charpy specimens at increasing magnifications. (a) and (c) show the as-fractured surface; (b) and (d) show the same fracture surface after coating with Tectyl 506, exposing to 100% relative humidity for 14 days, and cleaning with naphtha.

condition and after coating, exposing, and cleaning. These fractographs show that the solvent-cutback petroleum-base compound prevented chemical attack of the fracture surface from the environment and that the compound was completely removed in the appropriate solvent. It is interesting to note that Tectyl 506 is a rust-inhibiting compound that is commonly used to rustproof automobiles.

Fracture-Cleaning Techniques (Ref 1)

Fracture surfaces exposed to various environments generally contain unwanted surface debris, corrosion or oxidation products, and accumulated artifacts that must be removed before meaningful fractography can be performed. Before any cleaning procedures begin, the fracture surface should be surveyed with a low-power stereo binocular microscope, and the results should be documented with appropriate sketches or photographs. Low-power microscope viewing will also establish the severity of the cleaning problem and should also be used to monitor the effectiveness of each subsequent cleaning step. It is important to emphasize that the debris and deposits on the fracture surface can contain information that is vital to understanding the cause of fracture. Examples are fractures that initiate from such phenomena as SCC, LME, and corrosion fatigue. Often, knowing the nature of the surface debris and deposits, even when not essential to the fracture analysis, will be useful in determining the optimum cleaning technique.

The most common techniques for cleaning fracture surfaces, in order of increasing aggressiveness, are:

- Dry air blast or soft organic-fiber brush cleaning
- Replica stripping
- Organic-solvent cleaning
- Water-base detergent cleaning
- Cathodic cleaning
- Chemical-etch cleaning

The mildest, least aggressive cleaning procedure should be tried first, and as previously mentioned, the results should be monitored with a stereo binocular microscope. If residue is still left on the fracture surface, more aggressive cleaning procedures should be implemented in order of increasing aggressiveness.

Air Blast or Brush Cleaning. Loosely adhering particles and debris can be removed from the fracture surface with either a dry air blast or a soft organic-fiber brush. The dry air blast also dries the fracture surface. Only a soft organic-fiber brush, such as an artist's brush, should be used on the fracture surface because a hard-fiber brush or a metal wire brush will mechanically damage the fine details.

The replica-stripping cleaning technique is very similar to that described in the section "Preservation Techniques" in this article. However, instead of leaving the replica on the fracture surface to protect it from the environment, it is stripped off of the fracture surface, removing debris and deposits. Successive replicas are stripped until all the surface contaminants are removed. Figure 2 shows successive replicas stripped from a rusted steel fracture surface and demonstrates that the first replicas stripped from the fracture surface contain the most contaminants and that the last replicas stripped contain the least. Capturing these contaminants on the plastic replicas, relative to their position on the fracture surface, can be a distinct advantage. The replicas can be retained, and the embedded contaminants can be chemically analyzed if the nature of these deposits is deemed important.

The one disadvantage of using plastic replicas to clean a fracture surface is that on rough surfaces it is very difficult to remove the replicating material completely. However, if the fracture surface is ultrasonically cleaned in acetone after each successive replica is stripped from the fracture surface, removal of the residual replicating material is possible. Ultrasonic cleaning in acetone or the appropriate solvent should be mandatory when using the replica-stripping cleaning technique.

Organic solvents, such as xylene, naphtha, toluene, freon TF, ketones, and alcohols,

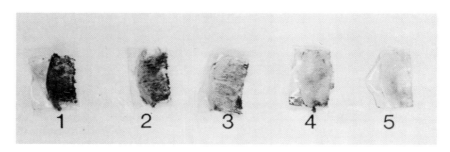

Fig. 2 Successive replicas (numbered 1 to 5) stripped from a rusted steel fracture surface. Note that the first replica stripped contains the most surface contaminants, while the last replica stripped is the cleanest. Actual size

Fig. 3 Fracture toughness specimen that has been intentionally corroded in a 5% salt steam chamber for 6 h. (a) Before ultrasonic cleaning in a heated Alconox solution for 30 min. (b) After ultrasonic cleaning

are primarily used to remove grease, oil, protective surface coatings, and crack-detecting fluids from the fracture surface. It is important to avoid use of the chlorinated organic solvents, such as trichloroethylene and carbontetrachloride, because most of them have carcinogenic properties. The sample to be cleaned is usually soaked in the appropriate organic solvent for an extended period of time, immersed in a solvent bath where jets from a pump introduce fresh solvent to the fracture surface, or placed in a beaker containing the solvent and ultrasonically cleaned for a few minutes.

The ultrasonic cleaning method is probably the most popular of the three methods mentioned above, and the ultrasonic agitation will also remove any particles that adhere lightly to the fracture surface. However, if some of these particles are inclusions that are significant for fracture interpretation, the location of these inclusions relative to the fracture surface and the chemical composition of these inclusions should be investigated before their removal by ultrasonic cleaning.

Water-base detergent cleaning assisted by ultrasonic agitation is effective in removing debris and deposits from the fracture surface and, if proper solution concentrations and times are used, does not damage the surface. A particular detergent that has proved effective in cleaning ferrous and aluminum materials is Alconox. The cleaning solution is prepared by dissolving 15 g of Alconox powder in a beaker containing 350 mL of water. The beaker is placed in an ultrasonic cleaner preheated to about 95 °C (205 °F). The fracture is then immersed in the solution for about 30 min, rinsed in water then alcohol, and air dried.

Figure 3(a) shows the condition of a laboratory-tested fracture toughness sample (AISI 1085 heat-treated steel) after it was intentionally corroded in a 5% salt steam spray chamber for 6 h. Figure 3(b) shows the condition of this sample after cleaning in a heated Alconox solution for 30 min. The fatigue precrack region is the smoother fracture segment located to the right of the rougher single-overload region. Figures 4(a) and (b) show identical views of an area in the fatigue precrack region before and after ultrasonic cleaning in a heated Alconox solution. Only corrosion products are visible, and the underlying fracture morphology is completely obscured in Fig. 4(a). Figure 4(b) shows that the water-base detergent cleaning has removed the corrosion products on the fracture surface. The sharp edges on the fracture features indicate that cleaning has not damaged the surface, as evidenced by the fine and shallow fatigue striations clearly visible in Fig. 4(b).

The effect of prolonged ultrasonic cleaning in the Alconox solution is demonstrated in Fig. 5(a) and (b), which show identical views of an area in the fatigue precrack region after cleaning for 30 min and 3.5 h, respectively. Figure 5(b) reveals that the prolonged exposure has not only chemically etched the fracture surface but has also dislodged the originally embedded inclusions. Any surface corrosion products not completely removed within the first 30 min of water-base detergent cleaning are difficult to remove by further cleaning; therefore, prolonged cleaning provides no additional benefits.

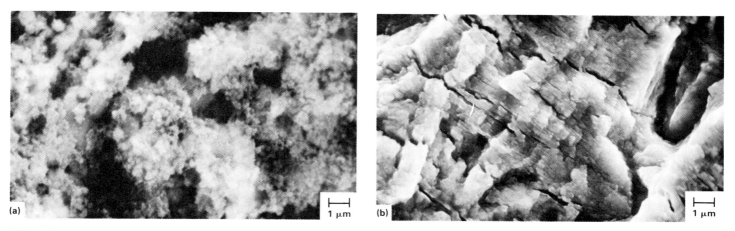

Fig. 4 Fatigue precrack region shown in Fig. 3. (a) Before ultrasonic cleaning in a heated Alconox solution for 30 min. (b) After ultrasonic cleaning

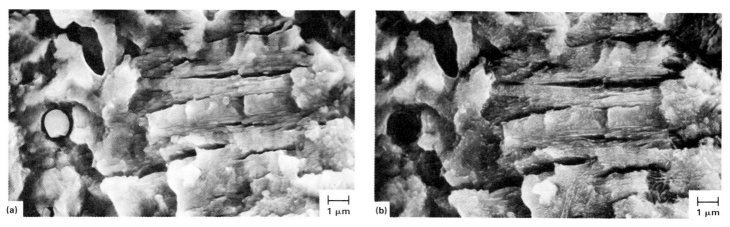

Fig. 5 Effect of increasing the ultrasonic cleaning time in a heated Alconox solution. (a) 30 min. (b) 3.5 h. Note the dislodging of the inclusion (left side of fractograph) and chemical etching of the fracture surface.

Cathodic cleaning is an electrolytic process in which the sample to be cleaned is made the cathode, and hydrogen bubbles generated at the sample cause primarily mechanical removal of surface debris and deposits. An inert anode, such as carbon or platinum, is normally used to avoid contamination by plating upon the cathode. During cathodic cleaning, it is common practice to vibrate the electrolyte ultrasonically or to rotate the specimen (cathode) with a small motor. The electrolytes commonly used to clean ferrous fractures are sodium cyanide (Ref 6, 7), sodium carbonate, sodium hydroxide solutions, and inhibited sulfuric acid (Ref 8). Because cathodic cleaning occurs primarily by the mechanical removal of deposits due to hydrogen liberation, the fracture surface should not be chemically damaged after elimination of the deposits.

The use of cathodic cleaning to remove rust from steel fracture surfaces has been successfully demonstrated (Ref 9). In this study, AISI 1085 heat-treated steel and EX16 carburized steel fractures were exposed to a 100% humidity environment at 65 °C (150 °F) for 3 days. A commercially available sodium cyanide electrolyte, ultrasonically agitated, was used in conjunction with a platinum anode for cleaning. A 1-min cathodic cleaning cycle was applied to the rusted fractures, and the effectiveness of the cleaning technique without altering the fracture morphology was demonstrated.

Figure 6 shows a comparison of an as-fractured surface with a corroded and cathodically stable ductile cracking region in a quenched-and-tempered 1085 carbon steel. The relatively low magnification (1000×) shows that the dimpled topography characteristic of ductile tearing was unchanged as a result of the corrosion and cathodic cleaning. High magnification (5000×) shows that the perimeters of the small interconnecting dimples were corroded away.

Chemical Etching. If the above techniques are attempted and prove ineffective, the chemical-etch cleaning technique, which involves treating the surface with mild acids or alkaline solutions, should be implemented. This technique should be used only as a last resort because it involves possible chemical attack of the fracture surface. In chemical-etch cleaning, the specimen is placed in a beaker containing the cleaning solution and is vibrated ultrasonically. It is sometimes necessary to heat the cleaning solution. Acetic acid, phosphoric acid, sodium hydroxide, ammonium citrate, ammonium oxalate solutions, and commercial solutions have been used to clean ferrous alloys (Ref 8). Titanium alloys are best cleaned with nitric acid (Ref 4). Oxide coatings can be removed from aluminum alloys by using a warmed solution containing 70 mL of orthophosphoric acid (85%), 32 g of chromic acid, and 130 mL of water (Ref 10). However, it has also been recommended that fracture surfaces of aluminum alloys be cleaned only with organic solvents (Ref 4).

Especially effective for chemical-etch cleaning are acids combined with organic corrosion inhibitors (Ref 11, 12). These inhibited acid solutions limit the chemical attack to the surface contaminants while protecting the base metal. For ferrous fractures, immersion of the samples for a few minutes in a 6 N hydrochloric acid solution containing 2 g/L of hexamethylene

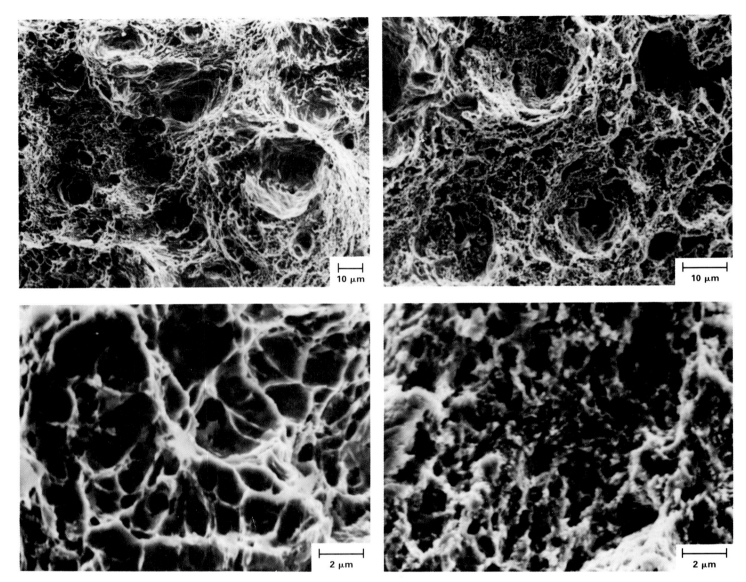

Fig. 6 Comparison of stable ductile crack growth areas from quenched-and-tempered 1085 carbon steel at increasing magnifications. The fractographs on the left show the as-fractured surface; those on the right show the fracture surface after corrosion exposure and cathodic cleaning.

tetramine has been recommended (Ref 6). Ferrous and nonferrous service fractures have been successfully cleaned by using the following inhibited acid solution: 3 mL of hydrochloric acid (1.19 specific gravity), 4 mL of 2-butyne-1,4-diol (35% aqueous solution), and 50 mL of deionized water (Ref 13). This study demonstrated the effectiveness of the cleaning solution in removing contaminants from the fracture surfaces of a low-carbon steel pipe and a Monel Alloy 400 expansion joint without damaging the underlying metal. Various fracture morphologies were not affected by the inhibited acid treatment when the cleaning time was appropriate to remove contaminants from these service fractures.

Sectioning a Fracture

It is often necessary to remove the portion containing a fracture from the total part, because the total part is to be repaired, or to reduce the specimen to a convenient size. Many of the examination tools—for example, the scanning electron microscope and the electron microprobe analyzer—have specimen chambers that limit specimen size. Records, either drawings or photographs, should be maintained to show the locations of the cuts made during sectioning.

All cutting should be done such that fracture faces and their adjacent areas are not damaged or altered in any way; this includes keeping the fracture surface dry whenever possible. For large parts, the common method of specimen removal is flame cutting. Cutting must be done at a sufficient distance from the fracture so that the microstructure of the metal underlying the fracture surface is not altered by the heat of the flame and so that none of the molten metal from flame cutting is deposited on the fracture surface.

Saw cutting and abrasive cutoff wheel cutting can be used for a wide range of part sizes. Dry cutting is preferable because coolants may corrode the fracture or may wash away foreign matter from the fracture. A coolant may be required, however, if a dry cut cannot be made at a sufficient distance from the fracture to avoid heat damage to the fracture region. In such cases, the fracture surface should be solvent cleaned and dried immediately after cutting.

Some of the coating procedures mentioned above may be useful during cutting and sectioning. For example, the fracture can be protected during flame cutting by taping a cloth over it and can be protected during sawing by spraying or coating it with a lacquer or a rust-preventive compound.

Opening Secondary Cracks

When the primary fracture has been damaged or corroded to a degree that obscures information, it is desirable to open any secondary cracks to expose their fracture surfaces for examination and study. These cracks may provide more information than the primary fracture. If rather tightly closed, they may have been protected from corrosive conditions, and if they have existed for less time than the primary fracture, they may have corroded less. Also, primary cracks that have not propagated to total fracture may have to be opened.

In opening these types of cracks for examination, care must be exercised to prevent damage, primarily mechanical, to the fracture surface. This can usually be accomplished if opening is done such that the two faces of the fracture are moved in opposite directions, normal to the fracture plane. A saw cut can usually be made from the back of the fractured part to a point near the tip of the crack, using extreme care to avoid actually reaching the crack tip. This saw cut will reduce the amount of solid metal that must be broken. Final breaking of the specimen can be done by:

- Clamping the two sides of the fractured part in a tensile-testing machine, if the shape permits, and pulling
- Placing the specimen in a vise and bending one half away from the other half by striking it with a hammer in a way that will avoid damaging the crack surfaces
- Gripping the halves of the fracture in pliers or vise grips and bending or pulling them apart

It is desirable to be able to distinguish between a fracture surface produced during opening of a primary or secondary crack. This can be accomplished by ensuring that a different fracture mechanism is active in making the new break; for example, the opening can be performed at a very low temperature. During low-temperature fracture, care should be taken to avoid condensation of water, because this could corrode the fracture surface.

Crack separations and crack lengths should be measured before opening. The amount of strain that occurred in a specimen can often be determined by measuring the separation between the adjacent halves of a fracture. This should be done before preparation for opening a secondary crack has begun. The lengths of cracks may also be important for analyses of fatigue fractures or for fracture mechanics considerations.

Effect of Nondestructive Inspection

Many of the so-called nondestructive inspection methods are not entirely nondestructive. The liquid penetrants used for crack detection may corrode fractures in some metals, and they will deposit foreign compounds on the fracture surfaces; corrosion and the depositing of foreign compounds could lead to misinterpretation of the nature of the fracture. The surface of a part that contains, or is suspected to contain, a crack is often cleaned for more critical examination, and rather strong acids that can find their way into a tight crack are frequently used. Many detections of chlorine on a fracture surface of steel, for example, which were presumed to prove that the fracture mechanism was SCC, have later been found to have been derived from the hydrochloric acid used to clean the part.

Even magnetic-particle inspection, which is often used to locate cracks in ferrous parts may affect subsequent examination. For example, the arcing that may occur across tight cracks can affect fracture surfaces. Magnetized parts that are to be examined by SEM will require demagnetization if scanning is to be done at magnifications above about 500×.

REFERENCES

1. R.D. Zipp, Preservation and Cleaning of Fractures for Fractography, *Scan. Elec. Microsc.*, No. 1, 1979, p 355-362
2. A. Phillips *et al.*, *Electron Fractography Handbook* MCIC-HB-08, Metals and Ceramics Information Center, Battelle Columbus Laboratories, June 1976, p 4-5
3. W.R. Warke *et al.*, Techniques for Electron Microscope Fractography, in *Electron Fractography*, STP 436, American Society for Testing and Materials, 1968, p 212-230
4. J.A. Fellows *et al.*, *Fractography and Atlas of Fractographs*, Vol 9, 8th ed., *Metals Handbook*, American Society for Metals, 1974, p 9-10
5. B.E. Boardman *et al.*, "A Coating for the Preservation of Fracture Surfaces," Paper 750967, presented at SAE Automobile Engineering Meeting, Detroit, MI, Society of Automotive Engineers, 13-17 Oct 1975
6. H. DeLeiris *et al.*, Techniques of De-Rusting Fractures of Steel Parts in Preparation for Electronic Micro-Fractography, *Mem. Sci. Rev. de Met.*, Vol 63, May 1966, p 463-472
7. P.M Yuzawich and C.W. Hughes, An Improved Technique for Removal of Oxide Scale From Fractured Surfaces of Ferrous Materials, *Pract. Metallogr.*, Vol 15, 1978, p 184-195
8. B.B. Knapp, Preparation & Cleaning of Specimens, in *The Corrosion Handbook*, John Wiley & Sons, 1948, p 1077-1083
9. E.P. Dahlberg and R.D. Zipp, Preservation and Cleaning of Fractures for Fractography—Update, *Scan. Elec. Microsc.*, No. 1, 1981, p 423-429
10. G.F. Pittinato *et al.*, *SEM/TEM Fractography Handbook*, MCIC-HB-06, Metals and Ceramics Information Center, Battelle Columbus Laboratories, Dec 1975, p 4-5
11. C.R. Brooks and C.D. Lundin, Rust Removal from Steel Fractures—Effect on Fractographic Evaluation, *Microstruc. Sci.*, Vol 3, 1975, p 21-23
12. G.G. Elibredge and J.C. Warner, Inhibitors and Passivators, in *The Corrosion Handbook*, John Wiley & Sons, 1948, p 905-916
13. E.P. Dahlberg, Techniques for Cleaning Service Failures in Preparation for Scanning Electron Microscope and Microprobe Analysis, *Scan. Elec. Microsc.*, 1974, p 911-918

Photography of Fractured Parts and Fracture Surfaces

Theodore M. Clarke, J.I. Case Company

PHOTOGRAPHY plays an important role in recording the features of a fracture. Most optical fractographs will be at magnifications of 1× and greater. Depth-of-field inadequacies usually limit the maximum magnification to about 50×. Optical fractographs at 1 to 50× are called photomacrographs when they are made with a single-lens system and not a compound microscope.

This article will discuss the preparation of photomacrographs of fracture surfaces. The details to be recorded arise from differences in topography, reflectivity, and sometimes color. Recording these details requires proper illumination of the fractures, adequate recording equipment, and knowledge of how to use the equipment.

The space allotted to photography in a metallurgical laboratory should be determined by the type of work under consideration. In general, the area should have a reasonably high ceiling, should be free of excessive vibrations, and, most important, should be capable of being darkened. Total darkness is not required, but when exposures are being made, the area should be dark enough to avoid extraneous light or reflections.

Visual Examination

The first step, before attempting to photograph a fracture surface, is to examine the specimen thoroughly in the as-received condition to determine which features are most important, which aspects are extraneous (such as dirt or postfracture mechanical damage), and whether special treatment of the surface will be required. This examination should begin with the unaided eye and proceed to higher-magnification examination of key features with a stereomicroscope. A hand-held magnifier may be necessary if the part is not easily positioned under a stereomicroscope or if this equipment is not available. A means of providing illumination at varying angles of incidence will be necessary. Fiber optic light sources, stereomicroscope illuminators, or even a penlight can provide suitable illumination.

Observations made in these surveys should be recorded for future review and determination of the probable causes of fracture. A list should be made of the features deserving photography, including the magnification that will probably be required.

The next step involves general photography of the entire fractured part and of the broken pieces to record their size and condition and to show how the fracture is related to the components of the part. This should be followed by careful examination of the fracture by studying its image on the ground-glass back of the camera or through the viewfinder. The examination should begin with the use of direct lighting and should proceed by using various angles of oblique lighting to assess how the fracture characteristics can be best delineated and emphasized. It should also assist in determining which areas of the fracture are of primary interest and which magnifications will be necessary (for a given picture size) to enhance the fine details. Once this evaluation is complete, the fracture should be photographed, recording what each photograph shows and how it relates to the other photographs.

Setups for Photography of Fractured Parts

Parts and units that are too large to bring into the laboratory must be photographed at the site of fracture. White or gray paper, or large cards or cloths, can be used to separate the object being photographed from a confusing background, or the angle at which the picture is taken can be selected to eliminate any confusing background.

If possible, it is preferable to bring the parts or assemblies into the laboratory for photographing. For larger parts, background paper, which is available in assorted colors and in rolls of various widths, can be used to provide a neutral background. To maintain a uniform background, the bottom portion of the roll should be cut off as it becomes dirty.

Modeling clay, blocks of wood, laboratory jacks, or blocks of foam plastic can be used to support the part. The support should be concealed, either by keeping it entirely under the part or by covering exposed areas with paper.

Use of middle-tone or black background papers will modify the apparent tonal relationships. Also, a black background will give clear areas on the negative, which can be marked with india ink for identification of the specimen. The identification will appear in white on the photographic prints. However, if the photograph is to be reproduced by offset lithography, large solid-black areas may not reproduce well.

View Cameras

For laboratories doing frequent analyses of fractures, a camera with a 4- × 5-in. (100- × 125-mm) Polaroid back is almost a necessity, because it provides immediate availability of prints with suitable sizes for most applications. Inadequacies in illumination or exposure can be quickly detected and corrected. Specialized view camera systems for photomacrography are commonly used for fractography (Fig. 1). These cameras are mounted on very rigid stands needed for higher-magnification photomacrography. These photomacrography systems lack the tilts and swings of conventional view cameras. The tilts are generally not needed for fractography, because a viewing direction perpendicular to the fracture is normally used for fractographs. The conventional view camera can be mounted on a rigid stand for high-magnification fractography and on a tripod for oblique photography of parts. The tilts and swings can be useful in the latter application.

35-mm Single-Lens-Reflex Cameras

The 35-mm single-lens-reflex (SLR) camera can provide, at much lower capital investment, the same image-recording capability as the view camera, but with the disadvantage that usable prints are not immediately available. The Polaroid Instant 35-mm Slide system reduces this disadvantage by providing a rapid means of assessing image quality. Polaroid prints can be quickly made from these slides. A

Photography of Fractured Parts and Fracture Surfaces

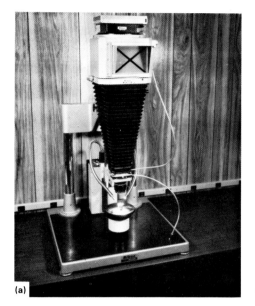

Fig. 1 Specialized view camera systems for photomacrography. (a) System with a bifurcated fiber optic source. Courtesy of Nikon, Inc. (b) View camera system. Courtesy of Polaroid Corporation

35-mm SLR system is almost a necessity if slides are needed for projection. Prints, in color or black and white, can be economically obtained from 35-mm slides or negatives. A 50-mm focal length macro lens and an extension ring can provide satisfactory print magnifications up to 8×. Bellows and special macro lenses are readily available for a print, or final magnification, up to 50×. The light-metering systems of modern 35-mm SLR cameras, especially those with spot metering, provide a distinct advantage in that the need to calculate and set the exposure is eliminated. However, long exposures that exceed the reciprocity range of the film require compensation for this effect with exposure increases (and use of color-compensating filters if color film is used).

Microscope Systems

Stereomicroscopes equipped for photography are available from many microscope suppliers (Fig. 2). They are well suited for magnifications above about 3× on a 4- × 5-in. (100- × 125-mm) format. A macroscope designed especially for photomacrography is shown in Fig. 2(b). The binocular viewing in this instrument is not stereo, because it uses a single, central optical path for photomacrography over a wide magnification range—3.2 to 64× in the 4- × 5-in. (100- × 125-mm) format.

Lenses

Lens Eq 1 to 4 show the relationships among the object-to-lens distance (object distance, S_o), the lens-to-film-plane distance (image distance, S_i), lens focal length, f, and camera magnification, M_c:

$$\frac{1}{S_o} + \frac{1}{S_i} = \frac{1}{f} \quad \text{(Eq 1)}$$

$$S_o = \frac{(M_c + 1)}{M_c} f \quad \text{(Eq 2)}$$

$$M_c = \frac{S_i}{S_o} \quad \text{(Eq 3)}$$

$$S_i = (M_c + 1) f \quad \text{(Eq 4)}$$

At a camera magnification of 1×, the object and image distances both equal $2f$. As the magnification exceeds 1×, the lens-to-film distance (bellows length) begins to exceed the lens-to-object distance and approaches a limiting value of $M_c f$. Similarly, at higher magnification, the lens-to-object distance approaches a limiting value of f.

To maintain a reasonable bellows length, S_i, it is necessary for the lens focal length to decrease for increasing magnification. The focal-length range for view camera lenses designed for photomacrography is about 16 to 160 mm. The lenses are designed for optimum image quality within the limited magnification ranges specified by the manufacturers. These ranges can be extended, especially at lower magnifications, when smaller apertures are used to increase depth of field while still providing good-quality images.

In general, the longest focal length lens that will conveniently give the desired magnification should be used. This will permit the greatest working distance between the lens and specimen. For images having equal magnification, the depth of field will be essentially the same regardless of the focal length of the lens used. However, because of the difference in the distance of the lens from the object being photographed, there may be some difference in perspective.

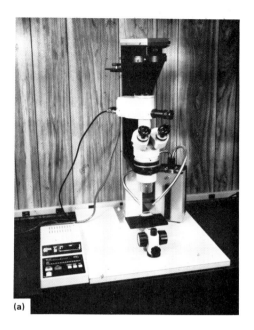

Fig. 2 Microscope systems equipped for photography. (a) Stereomicroscope equipped with an 80- × 105-mm (3¼- × 4¼-in.) camera back mounted on a rigid stand. Courtesy of Nikon, Inc., and Frank Fryer Company. (b) Macroscope mounted on a rigid stand. Courtesy of Wild Heerbrugg

Focusing

A magnifier should be used when focusing the image on the ground glass of a view camera. Viewfinder magnifiers are available for use with 35-mm SLR cameras. A clear spot in the center of the ground glass crossed with fiduciary marks can aid in focusing dim images. Screens of this type are available for 35-mm SLR cameras.

The plane of focus will be rendered with the highest resolution in the final image. The re-

Determination of Magnification

Camera magnification is easily determined by keeping the lens-to-film-plane distance constant and by imaging a millimeter scale. The image width, seen through a 35-mm camera viewfinder, is usually about 95 to 97% of the 35-mm film format. For view camera work, the image of the scale can be directly compared with a scale placed against the ground-glass screen or a scale ruled on the screen. Many photomacrography systems will have a bellows rail that is calibrated in camera magnification for various lenses. An actual check by viewing the image of a scale is the best assurance of correct magnification determination. The camera magnification is the ratio of the image size to the object size.

Selection of Lens Aperture

The microscopist knows from Abbe's criterion that the useful magnification of a light (optical) microscope ranges between 500 and 1000× numerical aperture.* This criterion also applies to photomacrographs, but is impractical to use because the lens-to-subject distance, which is variable, can be significantly greater than the focal length. Another concern is what aperture value to use to obtain an optimum balance of depth of field and image quality. The optimum-aperture problem has been solved analytically and the results verified experimentally (Ref 1). This analysis assumes that the resolution of the image will be diffraction limited and that points above and below the object focal plane will be blurred by the combined effects of diffraction and geometric blur. The method used to combine the blurs is described in Ref 2 and 3. Equations 5, 6, and 7 arise from the analysis:

$$\text{Optimum } f/\text{number} = \frac{220}{M_e (1 + M_c)} \quad \text{(Eq 5)}$$

$$\begin{array}{l}\text{Depth of field for} \\ \text{high resolution} \\ \text{(in mm using} \\ \text{optimum aperture)}\end{array} = \frac{70}{(M_e \times M_c)^2} \quad \text{(Eq 6)}$$

(for 6 line-pairs/mm print resolution)

$$\begin{array}{l}\text{Depth of field for} \\ \text{useful magnification} \\ \text{(in mm using} \\ \text{optimum aperture)}\end{array} = \frac{260}{(M_e \times M_c)^2} \quad \text{(Eq 7)}$$

(for 3 line-pairs/mm print resolution)

where M_e and M_c are the enlarging magnification and camera magnification, respectively. The total magnification, M_t, is the product of

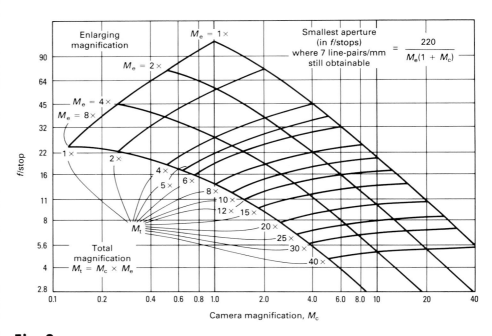

Fig. 3 Recommended f/numbers for optimum depth of field with high resolution. See Table 1. Source: Ref 4

*The numerical aperture, NA, is the measure of the light-collecting ability of an objective lens in a light microscope. It is defined as $NA = n \sin \alpha$ where n is the minimum refraction index of the medium (air or oil) between the specimen and the lens, and α is the half-angle of the most oblique light rays that enter the front lens of the objective. See the article "Optical Microscopy" in Volume 9 of the 9th Edition of Metals Handbook for additional information.

Table 1 Recommended f/numbers for optimum depth of field with high resolution
Exceeding these f/numbers by a full f/stop will seriously degrade image quality. See also Fig. 3.

Final image magnification, ×	Polaroid or contact print f/number	Polaroid or contact print Camera magnification, ×	4- × 6-in. 3.5- × 5-in. prints from 35 mm f/number	4- × 6-in. 3.5- × 5-in. prints from 35 mm Camera magnification, ×	8- × 10-in. 8- × 12-in. prints from 35 mm f/number	8- × 10-in. 8- × 12-in. prints from 35 mm Camera magnification, ×	Depth of field for resolution (6 line-pairs/mm), mm	Depth of field for resolution (3 line-pairs/mm), mm
1	110	1	44	0.25	24	0.12	70	260
2	73	2	37	0.50	22	0.25	17	65
4	44	4	27	1	18	0.50	4.5	16
8	24	8	18	2	14	1	1.1	4
10	20	10	16	2.5	12	1.2	0.70	2.6
15	14	15	12	3.7	10	1.9	0.31	1.1
20	10	20	9	5	8	2.5	0.17	0.65
30	7	30	6.5	7.5	5.8	3.7	0.08	0.29
40	5.4	40	5	10	4.6	5	0.044	0.16
Enlargement magnification in printing		1×		4×		8× High-resolution film required with very high resolution enlarging	Film and enlarging resolution losses will slightly decrease depth of field from the 6 line-pairs/mm depth values.	

f/number = 220/(final magnification + enlarging magnification). Depth of field for 6 line-pairs/mm resolution = 70/(final magnification)². Final magnification = camera magnification × enlarging magnification. Source: Ref 1

Table 2 Exposure factor to f/number conversion table for macro Luminar lenses

Manufactured by C. Zeiss, Inc.

Focal length, mm	Exposure factor						
	1	2	4	8	15	30	60
16	f/2.5	3.5	5	7	9.6	...	...
25	f/3.5	4.9	7	9.8	13.6	...	...
40	f/4.5	6.3	9	12.6	17.5	24.7	...
63	f/4.5	6.3	9	12.6	17.5	24.7	...
100	f/6.3	8.8	12.6	17.6	24.6	34.6	48.8

f/number = maximum lens aperture in f/numbers × $\sqrt{\text{exposure factor}}$
Source: Ref 1 and 5

Table 3 Aperture number to f/number conversion table for Macro-Nikkor lenses

Manufactured by Nikon Inc.

Focal length, mm	Aperture number, n						
	1	2	3	4	5	6	7
19	f/2.8	4	5.6	7.9	11.2	15.8	...
35	f/4.5	6.4	9	12.7	18	25	...
65	f/4.5	6.4	9	12.7	18	25	...
120	f/6.3	8.9	12.6	17.8	25.2	35.6	50

f/number = maximum lens aperture in f/numbers × $2^{(n-1)/2}$

the enlarging magnification and camera magnification.

The optimum-aperture concept, at higher magnification, is equivalent to 500× numerical aperture. Equations 5, 6, and 7 are plotted in more useful form in Table 1 and Fig. 3. The region of critical interest on the fracture should lie in the object plane of focus, with the fracture oriented to make optimum use of the zone of useful magnification, which is divided evenly above and below the object plane of focus. The focusing is usually done at maximum lens aperture because of the dim viewfinder image when the lens is stopped down to the optimum aperture for recording. The high illuminating intensity of fiber optic light sources often permits focusing with the lens stopped down.

Some macro lenses are calibrated by exposure factors relative to their maximum aperture, which is given a value of 1 and also listed as an f/number on the lens. Table 2 and its corresponding equation will aid in determining f/numbers for use with the optimum-aperture concept and Luminar lenses. Some lenses are calibrated in a sequence of numbers, typically 1 through 6. Each step indicates a one f/stop or factor of two exposure increase from the next smaller number. The maximum aperture, or smallest f/number, is given a value of 1 and is given as an f/number by the manufacturer. Table 3 and its corresponding equation will be useful for such calibrated lenses.

A common problem with the longer focal length view camera lenses, when used for Polaroid or contact prints of less than 3× magnification, is that these macro lenses do not have the apertures between f/45 and f/90 needed to use the optimum-aperture concept. This problem can be solved by using a lower camera magnification and by enlarging to the final magnification. However, this solution is inconvenient when an immediate print is desired.

Many of the stereomicroscopes equipped for photography, as well as the macroscope shown in Fig. 2(b), have adjustable diaphragms to control depth of field. The aperture that provides an optimum combination of resolution and depth of field is determined by observation through the eyepieces. Visual observation should be done at the same magnification as the final magnification of the photograph. This will be the same as the camera magnification if the camera image will not be subsequently enlarged.

Scanning Light Photomacrography

The seemingly inescapable consequence of the depth of field being inversely proportional to the square of the total magnification often results in a lack of sufficient depth of field to record all the desired details on a fracture. This depth-of-field limitation can be overcome by using scanning light photomacrography. This method should be ideal for fractography. Viewing the inside of cavities or deep valleys, for which this method is unsuitable, is generally not required for fractography.

Scanning light photomacrography was invented and patented in 1968 (Ref 6). Figure 4 compares fractographs obtained by conventional photography and scanning light photomacrography. The entire fracture surface is in sharp focus when scanning light photomacrography is used. This could not be accomplished in any other way except in a scanning electron microscope, which cannot record the colors of the object. Figure 5 shows a commercially available system that uses scanning light photomacrography.

Scanning light photomacrography functions by scanning the object, at a slow and constant velocity, through a thin sheet of light that illuminates a zone on the object that is thinner than the depth of field of the recording lens. The optical axis is normal to the sheet of light formed by light sources that image a thin slit source of light on the object. The convergence and divergence of the illuminating beams limit the object width that will fall within the depth of field of the lens. This convergence and divergence thickening of the beam can be reduced by using smaller apertures in the illuminating lenses. However, this becomes counterproductive at higher magnifications when larger apertures are required for the very thin light source to be imaged within the depth of field of the photography lens. At low magnifications, the depth of field is deep enough that the magnification may not be constant within this zone. The illuminating beam thickness must be restricted to avoid blurred images due to variable lateral magnification. The images recorded by scanning light photomacrography are isometric views and lack the visual cues regarding what is near and far in the image. Stereo images have been shown to overcome this minor problem (Ref 7).

Light Sources

Versatile lighting is required for fractography. Such light sources as stereomicroscope illuminators and fiber optic sources are very useful for illuminating small objects. If the fiber optic source can be positioned close to the object, more uniform illumination will be obtained without a focusing lens on the end of the fiber bundle. The most uniform illumination can be obtained with a Nicholas illuminator, which uses a double-condenser system. The construction and use of such an illuminator are discussed in Ref 8. Stereomicroscope illuminators of the Nicholas type are also commercially available. Broad light sources, such as linear and circular fluorescent-tube fixtures, are also often needed. Point sources are sometimes necessary for maximum contrast. Light from more than one source will often be needed to illuminate the fracture and surrounding surfaces properly. Photofloods or quartz halogen light sources, as used in fiber optic lighting systems, are the most suitable for color photography using daylight film with an 80A color-correcting filter or 3200 K balanced film and no filter (the article "Color Metallography" in Volume 9 of the 9th Edition of *Metals Handbook* includes a color filter nomograph to aid photographers in determining which color-correcting filters are required to match the film with the light source).

Electronic-flash lighting is rarely used for fractography. An exception might be field work with a 35-mm camera and with the flash on a

82 / Photography of Fractured Parts and Fracture Surfaces

Fig. 4 Photographs of a fractured valve stem. (a) Conventional photography. (b) Scanning light photography. Both 10×. Courtesy of Irvine Optical Corporation

Fig. 5 View camera with a scanning light photomacrography system. Courtesy of Irvine Optical Corporation

flexible cable to permit oblique lighting. This is most easily done with a camera that controls the flash duration by off-the-film-plane metering during the exposures. The low cost of 35-mm photography makes it practical to try many exposures with different lighting angles.

Lighting Techniques

Basic lighting for photographing fracture surfaces and small parts is shown in Fig. 6. Three other illumination setups are shown in Fig. 7 to 9. Shadowless lighting of a surface can be obtained with a white background by supporting the specimen on a piece of glass above the white background paper. For basic lighting, one spotlight is suggested. A piece of tracing paper placed between the light and the specimen will diffuse the light. The light is then raised or lowered and the beam adjusted from flood to spot to obtain the exact quality of lighting desired. Fracture texture is best illuminated when the light is placed at a relatively low angle. A spotlight beam without a diffuser or a bare bulb with a small filament will give high-contrast lighting. The advantages of this technique for revealing texture are illustrated in Fig. 10 for three different illumination angles. The angle of the incident light is much more important than its intensity.

Figure 11 gives further examples of the effect of different methods of illumination on the appearance of a fracture surface. For this par-

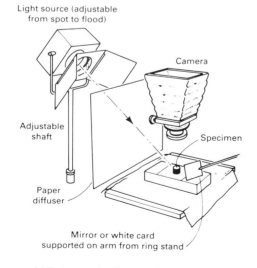
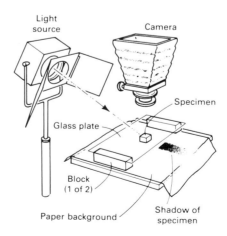
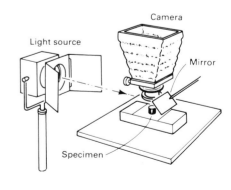

(a) Basic setup for photography of fracture specimens

(b) Shadowless background for photography of fracture specimens

(c) Horizontal illumination for close-up photography of fracture specimens

Fig. 6 Basic setups and lighting used to photograph fracture specimens and small parts. (a) General arrangement of camera, light source with diffuser, and specimen. Size and angle of the beam of light should be adjusted to give the best display of texture. A reflecting mirror or white card can be used to fill in shadows. (b) Placing the specimen on a raised glass plate will throw the shadow beyond the view range of the camera and provide a shadowless background. (c) If the magnification desired requires the close approach of the camera to the specimen, a horizontal light beam reflected by a small mirror positioned near the camera lens will provide the proper angle of illumination.

Photography of Fractured Parts and Fracture Surfaces / 83

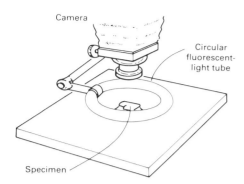

Fig. 7 Ring illumination by a circular fluorescent-light tube. This setup provides a soft, low-contrast, uniform illumination for 360° and generates a minimum of heat.

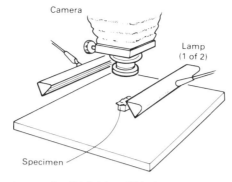

Fig. 8 Parallel lighting. With this setup, oblique lighting comes from two directions. It may be desirable to make one fluorescent lamp the principal source and to lighten the shadows it casts by positioning the other lamp farther from the specimen.

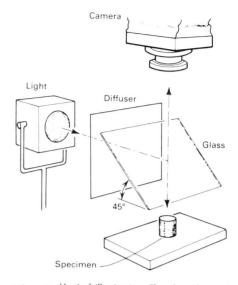

Fig. 9 Vertical illumination. The glass plate below the lens is set at a 45° angle. It reflects the light beam to the specimen surface. A portion of this light is reflected from the specimen back through the glass plate to the camera lens. This results in two losses of light: a portion of the incident beam passes through the plate, and a portion of the reflected light is returned toward the source. About one-quarter of the light (or even less, depending on the reflectance of the glass) reaches the lens.

ticular surface, ring illumination (Fig. 11a) is poorest and vertical illumination (Fig 11d) provides the best overall detail, primarily because of the greater reflectivity of the fatigue zone.

Vertical illumination is achieved by placing a glass plate below the lens at a 45° angle to reflect the incident light to the specimen and then back through the glass plate to the lens (Fig. 9). The effect is to convert light from a side source to illumination in the direction of the lens. This technique is especially useful for photographing the bottoms of cavities. Room lights should be dimmed to avoid extraneous reflections.

More complex versions of vertical illumination use a half-reflecting mirror and an antireflection coating, which greatly increase lighting efficiency. A condenser lens, about twice the field size, is used between a small light source and the 45° mirror. Uniform high-contrast lighting requires imaging the light source into the aperture of the macro lens. The source size should be sufficient to just fill the macro lens aperture. This method is described and illustrated in ASTM E 883 (Ref 9) and in Ref 10.

Another comparison of illumination techniques is provided in Fig. 12, which shows the difference in fracture surface appearance produced by a circular flashtube around the lens, by oblique photoflood illumination, by parallel oblique fluorescent illumination, and by 360° oblique illumination by circular fluorescent tubes.

Figure 13 shows a further illustration of the effects of different lighting. In Fig. 13(a), for which vertical illumination was used, the fatigue zone below the carburized case is clearly shown. In Fig. 13(b), for which oblique spotlighting with crossed polarization was used, the fatigue zone is not shown at all, but a large silicate inclusion at the fatigue crack origin is highlighted.

Figure 14 shows the importance of the position of the light source. The fracture was illuminated by a circular fluorescent-light tube for both photographs. For Fig. 14(a), the incident light had the proper obliquity to reveal all details of the fracture surface. For Fig. 14(b), the light was moved up 25 mm (1 in.), which changed the degree of obliquity, with consequent loss of detail.

Fracture surfaces usually have numerous planes that will reflect small light sources as a series of catchlights. The catchlights outside the

Fig. 10 Light fractographs showing effect of angle of incident light in delineating texture of a fracture surface of an unnotched laboratory tensile-test specimen of 4340 steel. The microstructure was tempered martensite. As the angle of illumination approaches 90°, the surface becomes featureless. The precise optimum angle of obliquity depends on the nature of the fracture markings. 7×

depth of field appear in the image as blur circles and have a tendency to degrade the quality of the image. The use of a controlled amount of diffusion minimizes the catchlight problem and decreases contrast in lighting, making it possible to obtain detail in some of the shadow areas. If a spotlight and tracing paper are used, the desired degree of diffusion can be obtained by adjusting the distance between the tracing paper and the part being photographed (the nearer the part, the lower the contrast) and by adjusting the light from spot source to flood source. Similar lighting can be obtained using other sources, such as small fluorescent lights, but these generally do not have the degree of control obtainable with spotlights.

When photographing fracture surfaces, the specimen should be rotated under the lens, and the effects of the lighting should be observed on the ground glass of the camera in order to reveal the most detail in the areas of interest. In general, the light should be across the fracture grain, not parallel to it. Dark shadows on the side of the specimen away from the light, or behind shear lips, can be filled in with light reflected from a mirror or a white card (Fig. 6a). It is preferable to use reflected light, such as from a white card, to fill in shadows rather than a second light, which will only cast more shadows. The light from the card is softer and is easily controlled by adjusting the distance of the card from the specimen. Too strong a fill light may overpower the texture lighting. In lighting specimens, keep in mind that the angle of reflection equals the angle of incidence.

Lights should not shine on lens surfaces. This can lead to ghost images and lens flare. Where possible, use a lens hood or keep the light off of the lens with a black-paper baffle. With coated lenses and antihalation films, the use of white backgrounds in overall shots generally causes no difficulty. On the other hand, when a lens with a very short focal length is used without a baffle to obtain high magnification, it must be brought close to the specimen during photography, and lens flare can become a problem if the specimen is placed on white paper (Fig. 15a). This can usually be corrected by using black paper under the specimen, as in Fig. 15(b). If the desired magnification can be achieved, use of a longer focal length lens can avoid the problem of proximity of the camera to the specimen. At magnifications of 15 to 20×, it may be difficult to light the fracture surface at the correct angle to delineate the details effectively. Fiber optic light sources are most effective in this situation.

Lighting of Etched Sections. Etched sections are best photographed with vertical illumination, which is perpendicular at the center of the field. Attempts to illuminate carburized or hardened cases obliquely usually result in tone reversal, with dark areas appearing white and white areas appearing dark. Steel weld specimens etched in ammonium persulfate solution can sometimes be photographed with an

Fig. 11 Effect of illumination technique on appearance of a fracture surface of a laboratory fatigue test specimen of 1039 steel quenched and tempered to a hardness of 42 HRC. (a) Photographed with fluorescent-tube ring illumination; note low contrast and overexposure. (b) Photographed with oblique light from a point source lamp. General fracture area exhibits more contrast than (a), but fatigue zone is not well defined. (c) Photographed with illumination from a mirror at camera lens. Quality is about the same as in (b). (d) Photographed with vertical illumination. For this particular fracture surface, vertical illumination gave the best contrast and detail of all features. Light fractographs. All 2×

oblique spotlight, but vertical lighting usually provides more information. The photomacrographs shown in Fig. 16 compare oblique and vertical lighting for revealing the structure of a weld; Fig. 17 shows a photograph of a properly illuminated weld.

Lighting of Highly Reflective Parts. Because of their high reflectivity, such parts as bearings, plated surfaces, and highly polished or ground objects require special lighting techniques. It is recommended that they be entirely encircled by a cone or cylinder of tracing paper to diffuse incident light (Fig. 18). This tent is translucent and can be illuminated by several lights from the sides. Because they obscure details, sprays and powders should not be used to reduce reflections.

Lighting of Deeply Etched Specimens to Show Grain Flow. Deeply etched specimens are photographed with a vertical setup and a single spotlight at a low angle (Fig. 19). The specimen can be supported on a glass plate over a piece of black paper to provide a black background. The light should be shielded such that it does not illuminate the background paper. The direction of incident light must be across the grain flow to show the texture properly (Fig. 20). Greater contrast is obtained if all room lights are off.

Photography with Ultraviolet Illumination. Black-and-white photography of fluorescent nondestructive-testing indications requires an ultraviolet illumination source, with a yellow (K2) filter placed over the lens to prevent the ultraviolet light from directly affecting the film. The camera is focused on the area to be photographed, using white light. All room lights are then turned off, and the exposure is

Photography of Fractured Parts and Fracture Surfaces / 85

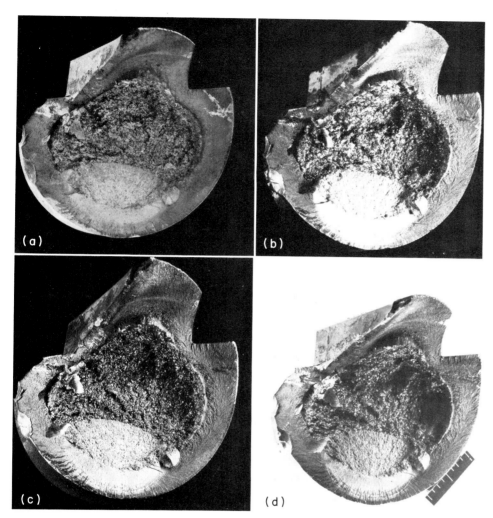

Fig. 12 Effect of illumination technique on appearance of a surface of a fatigue fracture in induction-hardened 15B28 steel. The fatigue crack originated at the hardened zone/core interface. (a) Illumination provided by a circular flash tube around the lens, resulting in a flat appearance. (b) Oblique illumination provided by photoflood lamp, highlighting the fatigue zone strongly. (c) Parallel oblique illumination provided by fluorescent tubes, resulting in less highlighting. (d) Ring lighting: 360° oblique illumination by circular fluorescent tubes, which gives good detail except in the central shadows. Light fractographs. All 0.5×

is desired. Cut film is available for 4- × 5-in. (100- × 125-mm) cameras in both color and black and white. This is best used after a test print is made on Polaroid film. Any of the fine-grain black-and-white or color films used for general photography can be used for 35-mm photomacrography. A comprehensive treatment of film selection, light filtration, processing, and printing is provided in Ref 8 and in ASTM E 883 (Ref 9).

Use of Light Meters

Correct exposure yields a print that shows adequate detail in both the light and dark regions without the need for custom printing involving dodging and burning. This requires proper exposure and an image brightness that does not vary by more than a factor of 2^5 or 5 stops.

Because fractures can be highly reflective or very dark, an incident-light meter is of little use for fractography. A reflected-light meter reading may be accurate only when the meter is placed along the optical axis. The most suitable location for a light meter reading is at the film plane. A spot-type meter may be useful to determine the variation in image brightness and to arrive at an average reading. Low image brightness may require that the meter reading be made with the lens at maximum aperture, or minimum f/number value. The factor used to multiply the exposure indicated for the larger aperture is given by:

$$\text{Exposure factor} = \frac{(f/\text{number})^2}{(\text{minimum } f/\text{number})^2} \quad \text{(Eq 8)}$$

The exposure factor can also be determined by counting the number of stops between the setting at which the meter reading is made and the setting for the photograph. The exposure increase factor is 2^n where n is the difference in number of stops. The final exposure value should be compared with the reciprocity characteristics of the film. Long exposures, typically 1 s or more, usually require multiplying by an additional factor to arrive at the final exposure value. When the meter readings are taken at the film plane, the bellows factor need not be considered. If the exposure is based on the incident-light meter reading, the effective aperture value is given by:

$$\text{Effective } f/\text{number} = \text{actual } f/\text{number } (1 + M_c) \quad \text{(Eq 9)}$$

Equation 9 indicates that the exposure must be increased with increasing magnification or that the illumination intensity must increase with increasing magnification. This is not the case, because at higher magnification the lens must be opened to a smaller f/number to maintain a constant standard of image resolution. If the lens aperture is chosen using the optimum-aperture concept, the effective aperture will be independent of the camera magnification and will be defined by:

made using only the ultraviolet source. Long exposures are generally required. Typically, an exposure of 2 min at f/11 may be required on ISO 400 film. This can be decreased to 15 to 20 s by using ISO 3000 Polaroid film. If only one light is used (as is customary), it should be moved during the exposure to obtain equal illumination. The motion is usually both circular and back and forth over distances of 150 to 300 mm (6 to 12 in.). The effect is reproducible because the long exposure time makes the motion noncritical. Background papers treated with brighteners that fluoresce under ultraviolet light should be avoided.

Film

Polaroid film is most commonly used with view camera systems. Polaroid Type 52 black-and-white film (400 ISO speed) is commonly used when no negative is needed for subsequent prints. The new Polaroid Type 53 and 553 should be even more useful than Type 52, because the new film print requires no coating and has a higher speed of 800 ISO. The original print can always be photographed with Polaroid Types 52 and 53 or, when a negative is needed, with Polaroid Type 55 P-N.

Any copies made photographically should be made at the aperture providing the best resolution at 1×. The setting will typically be f/22. Original exposures using Polaroid Type 55 P-N (50 ISO film speed) will be eight times longer than with Type 52, but will produce a high-quality negative. For an optimum-quality negative, the exposure giving the best-quality Polaroid print should be doubled. The 55 P-N negative material has finer grain and higher resolution than the 125 ISO speed black-and-white film commonly used in 35-mm cameras.

Polaroid Type 59 and Type 559 can be used for color prints when the self-processing feature

Effective f/number = $\dfrac{220}{M_c}$ (Eq 10)

where:

$$\text{Actual } f/\text{number} = \dfrac{220}{(M_c + 1) M_c}$$

Test Exposures

Exposures obtained with metering systems are subject to inaccuracies caused by differences in the brightness of the subject and the background, the color response of the meter compared to that of the film, corrections for bellows extension, shutter inaccuracies, and reciprocity failure (the nonproportional response of emulsions to light during very short and very long exposures). These variations can be bypassed by making a test exposure. Occasionally, such as for ultraviolet or infrared photography, the exposure must be determined mainly by experience and test exposures. Polaroid film is very helpful for this purpose. Using black-and-white film, typical corrections for reciprocity failure are:

Meter reading, s	Multiply reading by
1–2	1.4
2–6	2.0
6–16	2.8
16–35	4.0
35–70	5.6

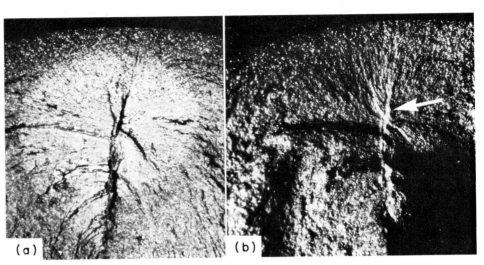

Fig. 13 Effect of illumination technique on appearance of a fracture surface of a laboratory fatigue test specimen of carburized-and-hardened 8620 steel. (a) Floodlight plus glass plate to give vertical illumination. Fatigue zone is well defined. (b) Oblique spotlight with crossed polarization. Fatigue zone is not shown, but a large silicate inclusion (arrow) is highlighted at the crack origin. Light fractographs. Both 1.5×

Assuming that a relatively long exposure is to be used, a step exposure should be produced by starting the exposure and then inserting the dark slide about 18 mm (0.7 in.) at intervals of 1, 2, 4, 8, 16, and 32 s with the shutter on the bulb setting. The film should then be processed and the most nearly correct exposure noted. If exposures are short, the shutter settings can be used. Exposure can be estimated from the image brightness, based on previous experience. After an exposure is made using that setting, the results can be compared with a previously prepared set of prints illustrating correct exposure, plus the results of one, two, and three f/stops over and under that exposure (Fig. 21). Corrections in exposure should be

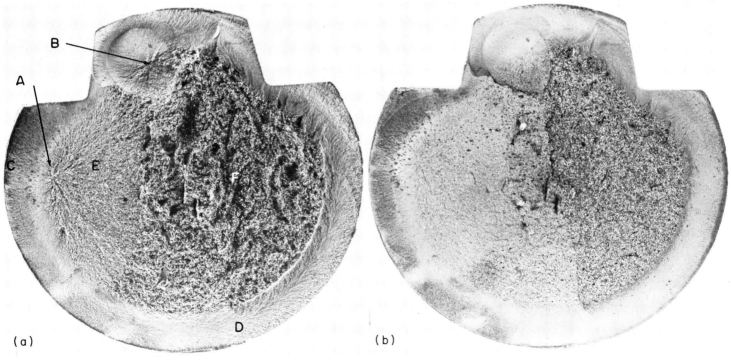

Fig. 14 Change in fracture detail that resulted from a shift in the position of the light source. Illumination was by a circular fluorescent-light tube. The details in the surface are well revealed in (a) but are missing in (b), for which the light was moved up 25 mm (1 in.). These fractographs are of the surface of a fatigue fracture in a 1541 steel axle that was induction hardened to 50 HRC and tested in rotating bending. The better-detailed fractograph at left shows that two fatigue cracks (one at A and the other at B) originated at the interface between the hardened zone and the soft core. The crack that began at A progressed around the hardened zone by fatigue in area C and by fast fracture in area D, before meeting the crack that began at B. The fatigue crack from A also penetrated the soft core, at area E. Final fast fracture occurred in the soft core, at area F. Light fractographs. Both 1.15×

Fig. 15 Surface of a fracture in an exhaust-valve stem of 21-4N (SAE EV 8) steel photographed with a 40-mm lens and with the specimen resting on (a) white paper and (b) black paper. In (a), lens flare and overexposure from the white background obscured some of the detail. A clear picture resulted when the specimen was placed on black paper (b), which eliminated the lens flare. Light fractographs. Both 10×

made and a second trial picture taken. If the film used in the test does not have the same ISO rating as the film to be used for the final picture, the exposure must be changed to account for the difference in film speeds. Alternatively, rather than change aperture or shutter settings, a neutral-density filter can be used. For example, if an ND4 filter with 400 ISO Polaroid film is used to make the test, it can be removed for exposures on 100 ISO cut film.

Stereo Images

The stereomicroscope is a valuable tool for fractography because of the depth perception it provides. The same information seen through a stereomicroscope can be recorded with stereo pairs. Loss of depth of field limits the useful magnification for photography to about 50×. Three stereo-pair fractographs of cleavage fractures in hydrogen-embrittled ingot iron were published by Zapffe in 1943 (Ref 11). Although at relatively low power (10×), these fractographs show the contours in three dimensions. Additional information on the pioneering work of Zapffe in the field of fractography can be found in the article "History of Fractography" in this Volume.

Most observers require a stereo-pair viewer to perceive the three-dimensional effect. The infrequent use of stereo-pair light (optical) fractography compared to scanning electron microscopy (SEM) stereo-pair usage may be due to the availability of the required equipment and the knowledge of its use rather than the need for a special viewer. A tilting stage, used for taking stereo-pair photomacrographs, is a standard feature on scanning electron microscopes. Additional information on stereo-pair SEM fractographs is available in the article "Scanning Electron Microscopy" in this Volume.

Figure 2(a) shows the only stereomicroscope currently in production designed for the rapid recording of stereo pairs. Figure 22 shows light stereo-pair fractographs obtained with the instrument shown in Fig. 2(a).

Earlier stereomicroscopes were sometimes available with optional cameras for recording stereo pairs from the two eyepieces, or optical paths. These earlier systems lack the modern exposure-metering system and adjustable dual irises available for the stereomicroscope shown in Fig. 2(a).

Stereo-pair photographs can be made with a standard photomacrography system by using a tilting stage. Oblique illumination may cause problems with this method if the incidence

Fig. 16 Effect of oblique and vertical lighting on structures revealed in a macroetched section through a multiple-pass weld in 1008 steel strip. (a) Oblique lighting reveals the grain structure of the steel base metal but no detail of the weld beads. (b) Vertical lighting shows the weld beads and the rim of the steel, but the grain structure is less distinct. Photomacrographs. Etched in ammonium persulfate. Both 4×

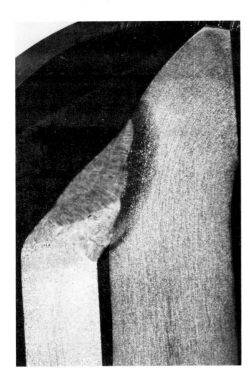

Fig. 17 A properly illuminated section through a weld. Vertical lighting was used to reveal flow lines in the base metal, the heat-affected zone, and the successive weld beads. Base metal was 1018 steel; component, an axle housing. Photomacrograph. Etched in ammonium persulfate. 4×

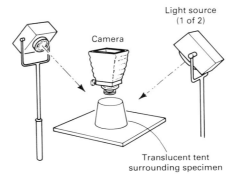

Fig. 18 Tent lighting used in photographing highly reflective parts

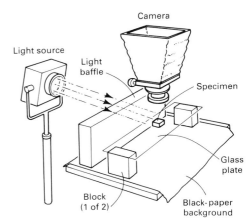

Fig. 19 Lighting for deeply etched specimens to show grain flow and inclusions

angle is not kept constant for the stereo-pair photographs. Constant illumination angles can be obtained by attaching fiber optic light sources to the tilting stage. The current market is so limited that tilting stages must usually be custom made for photomacrography.

There are two different types of tilting stages. One type operates like a standard stage used in a scanning electron microscope. A key detail that is in critical focus and along the vertical centerline, or y-axis, of the field is located. The sample is tilted about 6° on the tilt axis, which is parallel to the y-axis, and the x- and z-settings are adjusted until the key point is again in sharp focus and on the vertical centerline (the y-setting and the bellows distance must be kept constant). The image is recorded, and the process is repeated with the same amount of tilt in the other direction.

The other type of stage is better when there is no easily remembered detail to focus on. With this type of stage, the z-position of the key region on the specimen is raised or lowered relative to the tilt axis until there is no motion in the x-direction for points in critical focus along the vertical centerline, or y-axis, as the sample is rocked on the tilt axis. The camera system must be aligned so that the tilt axis is on the y-axis, which forms the vertical centerline of the field of view. The direction a point, in sharp focus along the vertical centerline, moves during tilting relative to the tilt rotation indicates whether the focus is above or below the tilt axis. It is important to use the longest focal length lens that provides suitable magnification. This precaution is necessary to minimize the magnification variation within the depth of field. This variation can make it difficult to obtain a good stereo image.

Depth of field increases greatly as the magnification of the final image drops below 2×. Taking advantage of the increased depth of field with low-magnification stereo photography requires abandoning the convergent-lens axes method as pictorial photography conditions are approached. Pictorial stereo pairs are made with the lens axes parallel and shifted by about 65 mm (2.6 in.). Extensive analysis and experimentation has been done within the transition region of final image magnification between about 0.2 to 5× (Ref 12). Both the convergent-axes and the parallel-axes methods can be used in this magnification range, but with a reduced convergence angle or parallel axes spacing. With the parallel-axes method, there will be some loss of field because only the overlapping portions of the two views can be used for the stereo pairs. In this situation, it is recommended that a basic rule be followed: the distance between the lens centers should be one-thirtieth of the distance from the lens to the nearest point on the object (Ref 12). This rule is claimed to be applicable for both the parallel-axes and convergent-axes methods. In addition, a plain background should be used. An alternative to using a tilting stage is to mount the camera on a tilting bar, with the camera system prefocused on the tilt axis and aligned so that the tilt axis bisects the field of view.

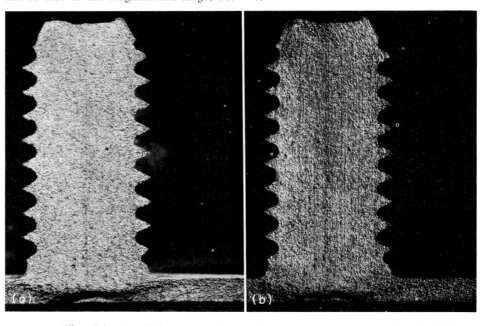

Fig. 20 Effect of direction of lighting in revealing grain flow in a deeply etched section through a 1010 steel stud that was attached to another part by resistance welding. Direction of grain flow was vertical. (a) With illumination from the top, parallel with the grain flow, little detail is shown. (b) With illumination from the side, across the grain flow, flow lines are revealed. Photomacrographs. Etched in ammonium persulfate. Both 3.5×

Photography of Fractured Parts and Fracture Surfaces / 89

(a) Decrease exposure to 1/8X

(b) Decrease exposure to 1/4X

(c) Decrease exposure to 1/2X

Correct exposure, 1X

Fig. 21 Exposure guide for Polaroid photographs of fracture surfaces. The photograph at left was made with the correct exposure. If a trial photograph happens to resemble one of those in the top row, the film was overexposed and the next exposure should be decreased to one-eighth (a), one-fourth (b), or one-half (c) of the trial exposure. This can be done either by stopping down the lens from the f/stop setting for the trial exposure (three stops to decrease to 1/8 of the trial exposure, two stops for 1/4 of the trial exposure, and one for 1/2 of the trial exposure) or by reducing the exposure time to 1/8, 1/4, or 1/2 of that used for the trial exposure. If a trial photograph resembles one of those in the bottom row, the film was underexposed and the next exposure should be two times (d), four times (e), or eight times (f) the trial exposure. These exposure increases can be made by appropriate adjustments of the f/stop or the exposure time. Exposure must be further adjusted if film of a different speed is to be used for the final photograph.

(d) Increase exposure by 2X

(e) Increase exposure by 4X

(f) Increase exposure by 8X

Auxiliary Equipment

A sturdy tripod is essential for general photography. The tripod should have a head that will permit the camera to be rotated, tilted upward or downward, and pivoted a full 90° sideways. An elevating center pole is also helpful. The use of a tripod will permit exposures of 1 s or longer when necessary with smaller apertures, which allow an increase in the depth of field; it will improve composition by putting the camera in a fixed, predetermined position; and it will avoid blurring due to camera movement. A good tripod should make the use of flash unnecessary. A cable release is useful for avoiding any camera motion during exposure.

The items listed below can be useful for photography of fracture specimens:

- Blocks of wood of various sizes for propping specimens in various positions
- Laboratory jacks for propping or supporting specimens
- Blocks of rigid foam plastic for propping specimens; the blocks can be cut to the desired shape and size
- Moldable materials, such as modeling clay, silicone putty, or jeweler's wax, for supporting small parts. A leveling device can be used to press the specimen into the material to produce a parallel face. Where possible, the material should be placed under the specimen or should be covered so it will not be visible in the photograph
- Laboratory ring stands with clamps for holding parts or reflectors
- Mirrors, of several sizes, to be used as reflectors to fill in shadow areas, to view the lens setting when the camera is in an awkward position, and to show both sides of the specimen in one picture, such as when documenting strain gage locations
- White cards, of several sizes, to be used as reflectors
- Draftsman's tracing paper or film for diffusion of light
- Masking tape for temporary support of paper backgrounds and diffusers and for holding offsize filters on the camera
- Glass sheets, 25- × 75-mm (1- × 3-in.) microscope slides, 50- × 50-mm (2- × 2-in.) or 80- × 100-mm (3¼- × 4-in.) cover glasses, and 100- × 125-mm (4- × 5-in.) or 125- × 175-mm (5- × 7-in.) or larger glass plates, for use as vertical illuminators or for supporting parts when shadowless backgrounds are desired. Clear acrylic plastic sheets can also be used to support parts
- Magnets for supporting or securing parts
- Scales or rule, of various sizes, to place in photographs to show degree of magnification or reduction

REFERENCES

1. T.M. Clarke, Method for Calculating Relative Apertures for Optimizing Diffraction-Limited Depth of Field in Photomacrog-

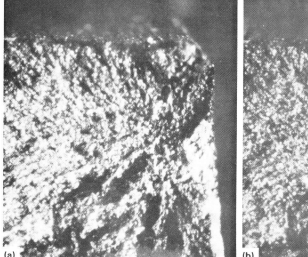

Fig. 22 Stereo-pair photographs of a slow-bend single-overload fracture of 1085 steel austempered to an upper-bainite microstructure. Note the thin shear lips and the elevation difference of the cleavage facets. Both 13×

raphy, *Microscope*, Vol 32 (No. 4), 1984, p 219-258
2. H.L. Gibson, "Close-Up Photography and Photomacrography," Publication No. N-12, Kodak
3. H.L. Gibson, "Photomacrography: Mathematical Analysis of Magnification and Depth of Detail," Publication No. N-15, Kodak
4. J. Gerakaris, Letter to the Editor, *Microscope*, Vol 34 (No. 1), 1986, p 85-91
5. J. Gerakaris, Letter to the Editor, *Microscope*, Vol 33 (No. 2), 1985, p 149-153
6. D.M. McLachlan, Jr., U.S. Patent 3,398,634, 27 Aug 1968
7. J. Gerakaris, A Second Look at Scanning Light Photomacrography, *Microscope*, Vol 34 (No. 1), 1986, p 1-8
8. R.P. Loveland, *Photomicrography—A Comprehensive Treatise*, Vol 1 and 2, Wiley-Interscience, 1970
9. "Standard Practice for Metallographic Photomicrography," E 883, *Annual Book of ASTM Standards*, American Society for Testing and Materials
10. T.M. Clarke, "Vertical Incident Illumination for Photomacrography of Metallographic Specimens," Paper presented at Inter/Micro—86, Chicago, IL, 21 July 1986, to be published in *Microscope*
11. C.A. Zapffe and G.A. Moore, Micrographic Study of Cleavage of Hydrogenized Ferrite, *Trans. AIME*, Vol 154, 1943, p 335-359
12. J.G. Ferwerda, *The World of 3-D—A Practical Guide to Stereo Photography*, Nederlandse Vereniging voor Stereofotografie, Reel 3-D Enterprises

Visual Examination and Light Microscopy

George F. Vander Voort, Carpenter Technology Corporation

THE VISUAL EXAMINATION of fractures is deeply rooted in the history of metals production and usage, as discussed in the article "History of Fractography" in this Volume. This important subject, referred to as macrofractography, or the examination of fracture surfaces with the unaided human eye or at low magnifications (≤50), is the cornerstone of failure analysis. In addition, a number of quality control procedures rely on visual fracture examinations. For failure analysis, visual inspection is performed to gain an overall understanding of the fracture, to determine the fracture sequence, to locate the fracture origin or origins, and to detect any macroscopic features relevant to fracture initiation or propagation. For quality control purposes, the fracture features are correlated to processing variables. In this article, examples of visual fracture examination will be given to illustrate the procedure as it applies to failure analysis and quality determination.

Although the light (optical) microscope can be used to examine fracture surfaces, most fracture examinations at magnifications above 50× (microfractography) are conducted with the scanning electron microscope, as described in the article "Scanning Electron Microscopy" in this Volume. However, if a scanning electron microscope is not available, light microscopy can be applied and usually provides satisfactory results.

Regardless of the equipment available to the analyst, it is still very useful to examine the fracture profile on a section perpendicular to the fracture origin. In this way, the origin of the fracture can be examined to determine if important microstructural abnormalities are present that either caused or contributed to fracture initiation. It is also possible to determine if the fracture path at the initiation site is transgranular or intergranular and to determine if the fracture path is specific to any phase or constituent present. Although such specimens can be examined by scanning electron microscopy (SEM), light microscopy is more efficient for such work, and certain information, such as the color or polarization response of constituents, can be assessed only by light microscopy. Interesting features can be marked with a scribe, microhardness indents, or a felt-tip pen and then examined by SEM and other procedures, such as energy-dispersive x-ray analysis, as required.

The techniques and procedures for the visual and light microscopic examination of fracture surfaces will be described and illustrated in this article. Results will also be compared and contrasted with those produced by electron metallographic methods, primarily SEM. Visual, light microscopic, and electron microscopic methods are complementary; each has particular advantages and disadvantages. Optimum results are obtained when the appropriate techniques are systematically applied.

In addition to examination of the gross fracture face, it is often useful to examine secondary cracks, when present. In many cases, the secondary cracks exhibit less damage than the gross fracture, which may often be damaged by rubbing, handling, postfracture events (such as repair attempts), or fire or corrosion. These fine cracks can be opened by careful sectioning and breaking, or they can be examined on polished cross sections. Also, it is often informative to examine the microstructure ahead of the secondary cracks, or adjacent to the primary fracture, to detect cracking in constituents or microvoids, either pre-existing or produced by the deformation associated with crack formation.

In this article, details will also be presented concerning the characteristic macro- and microscopic features associated with different fracture mechanisms. These features are detected and used to characterize the nature of both crack initiation and propagation. Examples will be presented, using a variety of techniques, to illustrate the procedure for classifying fractures.

Techniques

All fracture examinations should begin with visual inspection, perhaps aided by the use of a simple hand lens (up to 10×). In failure analyses, it may be necessary to examine the entire component or structure to find the broken sections, to determine the origin of the failure, and to separate the fractures according to the time sequence of failure, that is, which fractures existed before the event versus which ones occurred during the event. This article will assume that such work has already been accomplished and will concentrate on fracture examination and interpretation. Other related topics specific to failure analyses are discussed in Volume 11 of the 9th Edition of *Metals Handbook*.

Macroscopic Examination

Locating the fracture origin is a primary goal of fractography and is vital to successful failure analyses. The fracture markings formed during the event are like a road map that the analyst uses to evaluate the fracture. Fracture initiation and propagation produce certain characteristic marks on the fracture face, such as river marks, radial lines, chevrons, or beach marks, that indicate the direction of crack growth. The analyst traces these features backward to find the origin or origins. The appearance of these marks on the fracture face is a function of the type of loading, for example, tension, shear, bending, fatigue, or torsion; the nature of the stress system; its magnitude and orientation; the presence of stress concentrators; environmental factors; and material factors. Examples of these fracture patterns are illustrated in the section "Interpretation of Fractures" in this article and in the article "Modes of Fracture" in this Volume.

After a service failure has occurred or after a crack has been observed in a component, certain steps must be taken to ensure that the fracture features are not obliterated (Ref 1). In some cases, the fracture face is destroyed in the incident; alternatively, postfracture events can occur that drastically alter the fracture face and material condition. Such problems can often make a conclusive fracture interpretation and failure analysis difficult, if not impossible. In some instances, however, satisfactory results can be obtained (see the article "Failures of Locomotive Axles" in Volume 11 of the 9th Edition of *Metals Handbook*).

The fractured sections must be protected from further damage after the incident. It is often necessary to section the failed component or structure, sometimes at the failure site, so that it can be studied more extensively in the laboratory. Sectioning must be carried out in such a manner that the fracture and adjacent material is not altered. If burning is used, it should be conducted well away from the failure. Similarly, band saw cutting or abrasive wheel cutting must be conducted well away from the fracture. It is generally necessary to protect the fracture during such work. This subject is treated at length in the article "Preparation and Preservation of Fracture Specimens" in this Volume.

Macroscopic examination is the first step in fracture interpretation. In most cases, the origin must be determined in order to obtain conclusive results. Careful macroscopic examination should always precede any microscopic examination. Macroscopic examination will generally permit determination of the manner of loading, the relative level of applied stress, the mechanisms involved, the sequence of events, and the relative ductility or brittleness of the material. Other details can be revealed by gross fracture examination—for example, the presence of hardened cases; apparent grain size or variations of grain size; material imperfections, such as segregation, gross inclusions, or hydrogen flakes; and fabrication or machining imperfections that influenced failure.

The usual sequence for the examination of fractured components is as follows (Ref 1):

- Visually survey the entire component to obtain an overall understanding of the component and the significance of the fractured area
- Classify the fracture from a macroscopic viewpoint as ductile, brittle, fatigue, torsion, and so forth
- Determine the origin of failure by tracing the fracture back to its starting point or points
- Based on the observed fracture features, estimate the manner of loading (tension, compression, shear, bending, and so on), the relative stress level (high, medium, or low), and the stress orientation.
- Examine areas selected by macroscopic examination at higher magnifications by light microscopy, SEM, or replica transmission electron microscopy (TEM) to determine the fracture mode, to confirm the fracture mechanism (observation of cleavage facets, ductile dimples, fatigue striations, and so on), and to detect features at the fracture origin
- Examine metallographic cross sections containing the origin to detect any microstructural features that promoted or caused fracture initiation, and determine if crack propagation favors any microstructural constituent

Visual macroscopic examination is the most efficient procedure for fracture evaluation. This should be followed by stereoscopic examination of the fracture features using magnifications up to about 50×. Before the fracture is sectioned, relevant details should be recorded photographically (see the article "Photography of Fractured Parts and Fracture Surfaces" in this Volume). Sketches are also very useful. Dimensions should be recorded before cutting begins.

During this work, it is imperative that the fracture face be handled carefully. Excessive handling and touching of the fracture face should be avoided, and fractures should not be remated. Field fractures often exhibit debris on the fracture face that may be relevant to the diagnosis, particularly for corrosion-related failures. This contamination should not be removed without serious consideration. The debris should be analyzed *in situ* or after removal. Cleaning procedures, which are also discussed in the article "Preparation and Preservation of Fracture Specimens" in this Volume, should be performed only after the relevance of such debris has been determined.

If damage to the gross fracture face is extensive, it is often helpful to open secondary cracks for examination. The damage to secondary cracks is usually less than that of the main fracture. It is important to remember that the crack mechanism for propagation may be different from that at the initiation site. Damage done to the main fracture may prohibit successful microfractography, but the gross macroscopic features may still be visible and amenable to interpretation.

Mechanical damage to the fracture surface may occur during crack propagation, for exam-

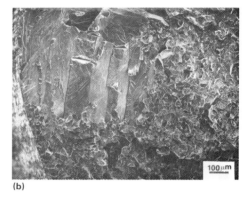

Fig. 1 Comparison of dark-field light microscope fractograph (a) and an SEM secondary electron image (b) of the same area in an iron-chromium-aluminum alloy. Both 50×

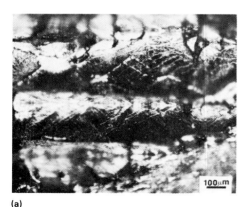

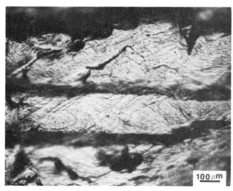

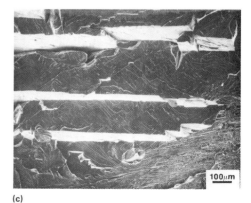

Fig. 2 Comparison of light microscope and SEM fractographs of the same area of an iron-chromium-aluminum alloy. (a) Bright-field light fractograph. (b) Dark-field light fractograph. (c) SEM secondary electron image. All 50×

ple, because of crack closure during fatigue crack growth. Although such damage may severely impair macro- and microscopic fractography, it does provide information about the manner of stressing.

Light Fractography

Several early researchers attempted to examine fracture surfaces with the light microscope, but the limitations of these early instruments precluded successful application. Little work of this type was performed until C.A. Zapffe demonstrated the usefulness of such examinations in the 1940s (Ref 2-5). Before this, fractures were examined only at high magnifications with the light microscope by using plated sections normal to the fracture surface. Although this technique remains very useful, Zapffe recognized that direct observation of the fracture surface would offer advantages and that the restricted depth of field of the light microscope would limit direct examination to brittle flat fractures (Ref 2). Consequently, Zapffe initially defined the new field of fractography as the micrographic study of cleavage facets on fractured metal specimens (Ref 2). Several years later, he generalized the definition of fractography, defining it as the study of detail on fracture surfaces (Ref 5).

Zapffe's technique consisted of obtaining a coarse brittle fracture, using low-power observation to find suitable cleavage facets, using a special sample holder to orient the facet perpendicular to the optical axis, and focusing at the desired examination magnification. Zapffe and his co-workers developed this method to a fine art. However, the inherently small depth of field, which decreases markedly with increasing magnification and numerical aperture, limited the technique. Despite the interesting results shown by Zapffe, relatively few optical fractographs have been published by others, and microfractography did not gain general acceptance until the development of TEM replication methods in the 1950s. With the commercial introduction of the scanning electron microscope in 1965, the field of microfractography gained popular applicability.

Although microfractography by SEM is a much simpler technique that produces equivalent images with far greater depth of field, the use of the light microscope in fractography should not be discarded. Lack of access to a scanning electron microscope should not prevent the innovative metallographer from pursuing fractography beyond macroscopic techniques. Direct optical examination of the fracture and examination of the fracture profile with cross sections will often provide the required information.

Zapffe's work involved bright-field illumination, which is perfectly suitable for examination of cleavage facets perpendicular to the optical axis. However, if a special device is not available for orienting cleavage facets perpendicular to the optical axis (few metallographers have

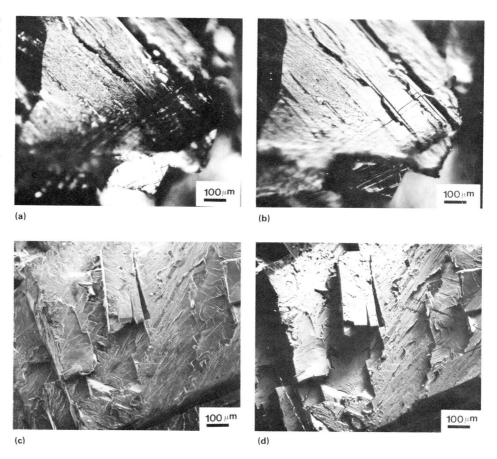

Fig. 3 Comparison of light microscope (a and b) and SEM (c and d) fractographs of cleavage facets in a coarse-grain Fe-2.5Si alloy broken at −195 °C (−320 °F). (a) Bright-field illumination. (b) Dark-field illumination. (c) Secondary electron image. (d) Everhart-Thornley backscattered electron image. All 60×

such devices), good results can still be obtained. For example, the fracture can be searched at low magnification for suitably oriented facets, or the fracture can be tilted slightly. In many cases, dark-field illumination can provide better images than bright-field illumination. The dark-field objective gathers the light that is scattered at angles away from the optical axis and often provides better image contrast with less glare. In some of the examples to be shown using both bright-field and dark-field illumination, certain features are best visible in one mode or the other. The photographs, however, cannot record the details that the observer can see as the focus point is moved up and down.

Figures 1 and 2 show two areas observed on the fracture of an iron-chromium-aluminum alloy. Figure 1 shows the same area using dark-field light microscopy and by SEM using secondary electrons. The dark-field image reveals several parallel steplike flat features surrounded by fine cleavage facets. The bright-field image was very poor and is not shown. The SEM image shows the same area more clearly because of the greater depth of field. The SEM negative was printed upside down so that the details are oriented in the same manner as those of the light micrograph. Figure 2 shows bright-field and dark-field light fractographs of similar steplike cleavage fractures in this iron-chromium-aluminum sample and the same area using secondary electron imaging. It is apparent that the dark-field image is superior to the bright-field image, but neither reveals the stepped nature as well as the SEM image, although the observer could see this effect by focusing in and out.

Figures 3 to 5 show coarse cleavage facets on a fractured Fe-2.5Si sample that was broken after cooling in liquid nitrogen. The SEM views are not of the same areas as the light fractographs. Figure 3 shows bright-field and dark-field images of a coarse facet. Certain areas are best observed on one or the other image because of the different light collection procedures. The two SEM views show a different area observed using secondary electrons and by backscattered electrons with the Everhart-Thornley detector. The latter procedure, because of its sensitivity to the specimen-detector orientation, provides greater height contrast compared to secondary electron images. Features that are not in a direct line with the detector are dark or are in shadow. Comparison of the secondary electron and Everhart-Thorn-

94 / Visual Examination and Light Microscopy

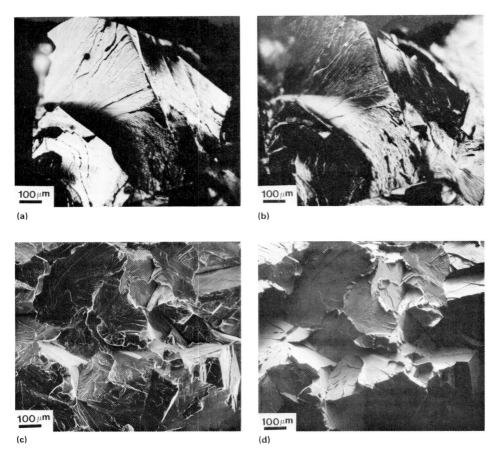

Fig. 4 Comparison of light microscope (a and b) and SEM (c and d) fractographs of cleavage facets in a coarse-grain Fe-2.5Si alloy impact specimen broken at −195 °C (−320 °F). (a) Bright-field illumination. (b) Dark-field illumination. (c) Secondary electron image. (d) Everhart-Thornley backscattered electron image. All 60×

ley backscattered electron images shows that the latter often reveals detail that is not as obvious in the secondary electron image. The secondary electron image is fully illuminated. Therefore, again, a combined presentation of both images reveals additional information.

Figure 4 shows similar bright-field and dark-field light fractographs of cleavage facets in the Fe-2.5Si alloy as well as secondary electron and Everhart-Thornley backscattered electron images of another area. The images shown in Fig. 4 are similar to those shown in Fig. 3. Figure 5 shows higher-magnification bright-field light fractographs of coarse cleavage facets in the Fe-2.5Si specimen and SEM secondary electron images at the same magnifications. All four fractographs are of different areas.

Figure 6 shows the interface between the fatigue precrack and test fracture of an X-750 nickel-base superalloy subsize Charpy rising-load test specimen after testing in pure water at 95 °C (200 °F). The interface region is shown by bright-field and dark-field light microscopy (same areas) and by secondary electron and Everhart-Thornley backscattered electron images (different location, same areas). At the magnification used, evidence of fatigue striations in the precracked region is barely visible in the light microscope images compared to the SEM images. The test fracture region is intergranular, but this is not obvious in the light fractographs. Figure 7 shows bright-field and dark-field fractographs of the fatigue-precracked region, in which the striations are more easily observed than in Fig. 6. Figure 8 shows secondary electron SEM fractographs of the striations in the precrack region at the same magnifications as in Fig. 7. Figure 9 shows bright-field and dark-field light fractographs of the intergranular region. The intergranular nature of this zone is more obvious in Fig. 9 than in Fig. 6. Figure 9 also shows corresponding secondary electron and Everhart-Thornley backscattered electron fractographs of the intergranular test fracture.

For comparison, Fig. 10 shows a similar X-750 rising-load specimen tested in air in which the test fracture is ductile. Figure 10 shows bright-field and dark-field light fractographs of the interface between the fatigue precrack and the ductile test fracture. A secondary electron fractograph is also included for comparison. Figure 11 shows high-magnification bright-field and dark-field light fractographs and a secondary electron fractograph of the ductile region of the test fracture at the same magnification. Microvoid coalescence (dimples) can be observed in all of the fractographs, but the limited depth of field of the light fractographs is obvious.

These examples demonstrate that the microscopic aspects of fractures can be assessed with the light microscope. Although the examination is easier and the results are better with SEM, light microscopy results are adequate in many cases. Such examination is easiest to accomplish when the fracture is relatively flat. For rougher, more irregular surfaces, SEM is far superior.

As a further note on the use of the light microscope to examine fractures, Fig. 12 shows a cleavage fracture in a low-carbon martensitic steel examined by using three direct and three replica procedures. Figure 12(a) shows an example of the examination of the fracture profile after nickel plating the surface. The flat, angular nature of the fracture surface is apparent. Figures 12(b) and (d) show a light microscope direct view of the cleavage surface and a light microscope view of a replica. Figures 12(c) and (e) show a direct SEM view of the fracture and a view of a replica of the fracture by SEM, respectively. Lastly, Fig. 12(f) shows a TEM replica of the fracture, but at a much higher magnification. The transmission electron microscope cannot be used at magnifications below about 2500×.

Replicas for Light Microscopy

In some situations, primarily in failure analysis, the fracture face cannot be sectioned, generally for legal reasons, so that it can fit within the chamber of the scanning electron microscope. In such cases, the fractographer can use replication procedures with examination by light microscopy, SEM, or TEM. The replication procedures for light microscopy are similar to those traditionally used for TEM fractography (Ref 6-8).

In general, the 0.25-mm (0.01-in.) thick cellulose acetate tape used for light microscopy is thicker than that used for TEM. The tape is moistened on one side with acetone, and this side is then pressed onto the fracture surface and held tightly in place, without motion, for 1 to 2 min. When thoroughly dry, the tape is carefully stripped from the fracture surface. An alternate procedure consists of, first, preparing a viscous solution of cellulose acetate tape dissolved in acetone and applying a thin coating to the fracture. Then, a piece of cellulose acetate tape is placed on top of this layer, pressed into the fracture, and held in place for 1 to 2 min. After drying, it is stripped from the fracture.

The stripped tape is a negative replica of the fracture and can be viewed as stripped from the fracture, and it can be photographed to record macroscopic fracture features (Fig. 13). This tape is a permanent record of the fracture for

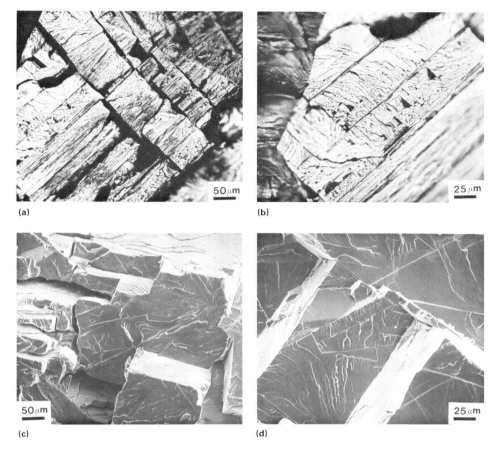

Fig. 5 Comparison of light microscope (a and b) and SEM (c and d) fractographs of cleavage facets in a coarse-grain Fe-2.5Si alloy impact specimen broken at −195 °C (−320 °F). (a) Bright-field illumination. (b) Dark-field illumination. (c) Secondary electron image. (d) Secondary electron image. (a) and (c) 120×. (b) and (d) 240×

future examination even if the fracture is sectioned.

Additional contrast can be obtained by shadowing the replica with either carbon or a heavy metal, such as chromium, molybdenum, gold, or gold-palladium, as is normally done in TEM fractography. Some fractographers also coat the back side of the replica with a reflective metal, such as aluminum, for reflected-light examination, or they tape the replica to a mirror surface. With an inverted microscope, some fractographers place the replica over the stage plate and then place a polished, unetched specimen against the tape to hold it flat and to reflect the light. Others prefer to examine the tape with transmitted light, but not all metallographers have access to a microscope with transmitted-light capability. Figure 14 illustrates low-power examination procedures for examination of replicas by light microscopy. Figure 14(a) shows the replica photographed with oblique illumination from a point source lamp, and Fig. 14(b) shows the same area using transmitted light. Carbon was then vapor deposited onto the replica, and it was photographed again using oblique light from a point source (Fig. 14c). Figure 14(c) exhibits the best overall contrast and sharpest detail. Figure 12(d) illustrates examination of a fracture replica by light microscopy using high magnification. This replica was shadowed with gold-palladium.

Fracture Profile Sections

Despite the progress made in direct examination of fracture surfaces, examination of sections perpendicular to the fracture, particularly those containing the initiation site, is a very powerful tool of the fractographer and is virtually indispensable to the failure analyst. If the origin of the fracture can be found, the failure analyst must examine the origin site by using metallographic cross sections. This is the only practical method for characterizing the microstructure at the origin and for assessing the role that the microstructure may have had in causing or promoting the fracture.

The safest procedure is to cut the sample to one side of the origin, but only after all prior nondestructive examinations have been completed. Cutting must be done in such a manner that damage is not produced. A water-cooled abrasive cutoff machine or a low-speed diamond saw is typically used. For optimum edge retention, it is recommended that the surface be plated, generally with electroless nickel, although some metals cannot be plated in this manner (Ref 9) and require other plating procedures. Figure 15 demonstrates the excellent edge retention that can be obtained with electroless nickel. A number of other edge retention procedures can also be used (Ref 9).

Examination of fracture profiles yields considerable information about the fracture mode and mechanism and about the influence of microstructure on crack initiation and propagation. This is accomplished by examining partially fractured (Ref 10-15) or completely fractured (Ref 16-26) specimens. Quantitative fractography makes extensive use of fracture profiles (additional information is available in the article "Quantitative Fractography" in this Volume). One interesting approach defines a fracture path preference index to describe the probability of a particular microstructural constituent being associated with a particular fracture mode to assess the relationship between fracture characteristics and microstructure (Ref 24).

In general, it is easier to assess the relationship between crack path and microstructure by using secondary cracks because both sides of the fracture can be examined. On a completely broken specimen, only one side can be examined; this makes the analysis more difficult.

Although most light micrographs are taken with bright-field illumination, the analyst should also try other illumination modes. Fractures should be initially examined on cross sections in the unetched condition and then should be examined after etching. Naturally, a high-quality polish is required, and the effort extended in achieving a high-quality surface is always rewarded with improved results and ease of correct interpretation. Errors in interpretation are made when specimens are not properly prepared.

As-polished samples should be examined first with bright-field illumination and then with dark-field illumination, differential interference contrast (DIC) or oblique light, and polarized light, if the specimen will respond to such illumination. Dark-field illumination is very useful and is highly suited to the examination of cracks and voids. Photography in dark field is more difficult, but not impossible if an automatic exposure device is available. Oblique light and DIC are very useful for revealing topographic (relief) effects. For example, Fig. 16 shows a fatigue crack in an aluminum alloy viewed with bright-field illumination and DIC where the specimen was not etched. Although the second-phase precipitates can be seen in both views, they are more clearly revealed with DIC (compare these views with the bright-field etched micrograph of this specimen shown in Fig. 71).

Another example of the examination of fracture profiles is shown in Fig. 17, which illustrates an impact fracture in an austenitic weld containing σ phase. This sample is in the as-polished condition and is shown examined with bright-field, DIC, and dark-field illumina-

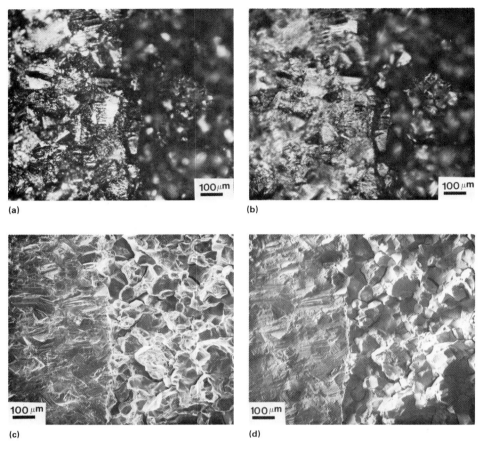

Fig. 6 Comparison of light microscope (a and b) and SEM (c and d) fractographs of the interface between the fatigue-precracked region and the test fracture in an X-750 nickel-base superalloy rising-load test specimen. The test was performed in pure water at 95 °C (200 °F). Note the intergranular nature of the fracture. (a) Bright-field illumination. (b) Dark-field illumination. (c) Secondary electron image. (d) Everhart-Thornley backscattered electron image. All 60×

tion. It is clear that full use of the light microscope can provide a better description of the relationship of the crack path to the microstructure.

Examination of properly polished specimens without etching often presents a clearer picture of the extent of fracture because etched microstructural detail does not obscure the crack detail. Etching presents other dark linear features, such as grain boundaries, that may be confused with the crack details. Therefore, it is always advisable to examine the specimens unetched first. Also, inclusions and other hard precipitates are more visible in unetched than in etched specimens. After careful examination of the as-polished specimen, the sample should be etched and the examination procedure repeated. Examples of fracture profile sections in the etched condition will be shown later in this article.

Taper Sections

Taper sections are often used to study fractures (Ref 9, 27-31). In this method, the specimen is sectioned at a slight angle to the fracture surface. Polishing of this plane produces a magnified view of the structure perpendicular to the fracture edge. The magnification factor is defined by the cosecant of the sectioning angle; an angle of 5° 43′ gives a tenfold magnification.

Etching Fractures

Zapffe (Ref 2) and others (Ref 32-37) have etched fracture surfaces in order to gain additional information. In general, etching is used to reveal the microstructure associated with the fracture surface (Ref 2, 36) or to produce etch pit attack to reveal the dislocation density and the crystallographic orientation of the fracture surfaces. Figure 18 shows an example of fracture surface etching of a cleavage fracture in a carbon steel sample. Although the etched fractures can be examined by light microscopy, SEM is simpler and produces better results.

Deep-Field Microscopy

The deep-field microscope provides greater depth of field for optical examination and photography of fractures (Ref 38-40). Its theoretical depth of field is 6 mm at 38× and 600 μm at 250× (Ref 39). The instrument uses a very thin beam of light to illuminate the specimen. The light beam is at a constant distance from the objective of the microscope and is at the focal plane. During photographic exposure, the specimen is moved at a constant rate up through the light beam. Only the illuminated portions of the specimen are recorded photographically, and all of the illuminated portions are in focus; therefore, the resultant photograph is in focus. The use of the deep-field microscope and the problems encountered in obtaining good fractographs are discussed in Ref 39.

Interpretation of Fractures

The study of fractures has been approached in several ways. One procedure is to categorize fractures on the basis of macro- or microscopic features, that is, by macro- or microfractography. The fracture path may be classified as transgranular or intergranular. Another approach is to classify all fractures as either ductile or brittle, with all others, such as fatigue, being special cases of one or the other. In general, all fractures can be grouped into four categories: ductile, brittle, fatigue, or creep. After these broad groupings, the fractures can be further classified on the basis of environmental influences, stress situations, or embrittlement mechanisms. In this section, the macro- and microscopic characteristics of fractures produced by the more common mechanisms will be described and illustrated, with emphasis on visual and light microscopy examination. Detailed information on this subject is also available in the article "Modes of Fracture" in this Volume.

Ductile Fractures

Ductile fractures have not received the same attention as other fracture mechanisms because their occurrence results from overloading under predictable conditions. From the standpoint of failure analysis, ductile failures are relatively uncommon, because their prevention through proper design is reasonably straightforward. Ductile failures, however, are of considerable interest in metal-forming operations and in quality control studies, such as materials evaluation.

Ductile failures occur through a mechanism known as microvoid coalescence (Ref 41-43). The microvoids are nucleated at any discontinuity where a strain discontinuity exists—for example, grain or subgrain boundaries, second-phase particles, and inclusions. These microvoids, shown in Fig. 10 and 11, are referred to as dimples. Because of the roughness of the dimples, they are best observed with the scanning electron microscope. On a typical ductile fracture, fine precipitates, generally inclusions, can usually be observed in nearly half of the dimples (there are two halves of the fracture; therefore, on any fracture face, only half of the dimples, or fewer, will contain precipitates). Ductile fracture is sometimes referred to as dimple rupture; such is the case in the article "Modes of Fracture" in this Volume.

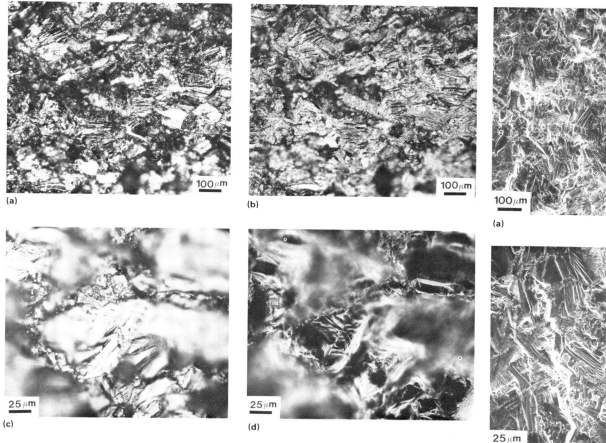

Fig. 7 Light microscope fractographs of the fatigue-precracked region of an alloy X-750 rising-load test specimen. (a) Bright-field image. (b) Dark-field image. (c) Bright-field image. (d) Dark-field image. (a) and (b) 60×. (c) and (d) 240×.

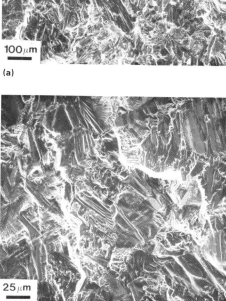

Fig. 8 Secondary electron images of the fatigue-precracked region of an alloy X-750 test specimen. (a) 65×. (b) 260×

Ductile fractures exhibit certain characteristic microscopic features:

- A relatively large amount of plastic deformation precedes the fracture
- Shear lips are usually observed at the fracture termination areas
- The fracture surface may appear to be fibrous or may have a matte or silky texture, depending on the material
- The cross section at the fracture is usually reduced by necking
- Crack growth is slow

The macroscopic appearance of a ductile fracture is shown by the controlled laboratory fracture of a spherical steel pressure vessel measuring 187 mm (7 3/8 in.) outside diameter and 3.2 mm (1/8 in.) in wall thickness that was made from quenched-and-tempered AISI 1030 aluminum-killed fine-grain steel with an impact transition temperature below −45 °C (−50 °F). At room temperature, ductile rupture occurred when the vessel was pressurized to 59 MPa (8500 psig) (Fig. 19). Figure 20 shows an identical vessel pressurized to failure at −45 °C (−50 °F) at 62 MPa (9000 psig). Greater pressure was required because the strength of the steel was greater at −45 °C (−50 °F) than at room temperature. Both failures are ductile. This vessel was designed to contain gas at 31 MPa (4475 psig). Therefore, the failures occurred by overloading.

From a microscopic viewpoint, a ductile fracture exhibits microvoid coalescence and transgranular fracture. The dimple orientation will vary with stress orientation. Dimple shapes and orientations on mating fractures as influenced by the manner of load application are summarized in Ref 44. Dimple shape is a function of stress orientation, with equiaxed dimples being formed under uniaxial tensile loading conditions. Shear stresses develop elongated, parabolically shaped dimples that point in opposite directions on the mating fracture surfaces. If tearing occurs, the elongated dimples point in the same direction on the mating fracture surfaces.

The number of dimples per unit area on the fracture surface depends on the number of nucleation sites and the plasticity of the material. If many nucleation sites are present, void growth is limited, because the dimples will intersect and link up. This produces many small, shallow dimples. If very few nucleation sites are present, there will be a small number of large dimples on the fracture surface.

The interface between two ductile phases can act as a nucleation site, but it is much more common for microvoids to form at interfaces between the matrix phase and inclusions or hard precipitates. Cementite in steel, either spheroidal or lamellar, can act as a nucleation site. For example, ductile fractures were studied in an iron-chromium alloy, and large dimples were observed at sulfide inclusions that subsequently merged with much finer voids at chromium carbides (Ref 45).

Ductile fractures in very pure low-strength metals can occur by shear or necking, and they can rupture without any evidence of microvoid formation. Such fractures can also be observed in sheet specimens in which triaxial stress development is negligible. Reference 46 discusses the problem of designing an experiment to prove conclusively whether or not ductile fracture can occur by void formation in the absence of hard particles. Results of the most carefully controlled experiments indicate that particles are required for void formation. Studies of high-purity metals have shown that ductile fracture occurs by rupture without void formation.

Markings can be observed within dimples because considerable plastic deformation is oc-

98 / Visual Examination and Light Microscopy

Fig. 9 Comparison of light microscope (a and b) and SEM (c and d) fractographs of the test fracture in an alloy X-750 rising-load test specimen. Test was performed in pure water at 95 °C (200 °F). Note the intergranular appearance of the fracture. (a) Bright-field image. (b) Dark-field image. (c) Secondary electron image. (d) Everhart-Thornley backscattered electron image. All 60×

curring. Such marks have been referred to as serpentine glide, ripples, or stretching.

Although light microscopy has rather limited value for the examination of ductile dimples (as shown in Fig. 10 and 11), dimples can be easily observed by using metallographic cross sections. Figure 21 compares the appearance of ductile and brittle fractures of a quenched-and-tempered low-alloy steel using nickel-plated cross sections. The ductile fracture consists of a series of connected curved surfaces, and it is easy to envision how this surface would appear if viewed directly. Also, near the fracture surface, small spherical voids, particularly at inclusions, are often seen that have opened during the fracture process. These should not be interpreted as pre-existing voids. The brittle fracture, by comparison, is much more angular, and crack intrusions into the matrix on cleavage planes can be observed (Fig. 21). The problems associated with the measurement of dimples are discussed in the article "Quantitative Fractography" in this Volume.

Tensile-Test Fractures

Tensile specimens are tested under conditions that favor ductile fracture, that is, room temperature, low strain rate, and a dry environment. Nevertheless, the relative amount of ductility exhibited by tensile specimens varies considerably. Although the percent elongation and percent reduction of area (% RA) are not intrinsic mechanical properties like the yield and tensile strength, they do provide useful comparative data.

The ideal tensile fracture is the classic cup-and-cone type (Fig. 22a). This type of fracture occurs in highly ductile materials. Pronounced necking is evident in Fig. 22(a). In comparison, in the brittle tensile fracture shown in Fig. 22(b), no necking has occurred, and the percent elongation and % RA values are nearly zero. In this type of tensile fracture, the yield and tensile strengths are essentially identical.

In one investigation, the classic cup-and-cone tensile fracture was studied by optical examination and by etch pitting (Ref 47). Shear-type fracture was found to be present all across the specimen, even in the flat central portion of the fracture. The approximate magnitude and distributions of longitudinal, radial, and circumferential stresses were determined in unfractured tensile bars strained at various amounts. The triaxial stresses in the necked section develop the highest shear stress at the center, not at the surface of the specimen. Shear

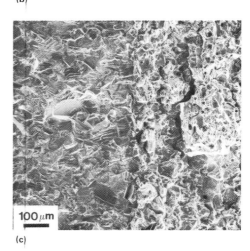

Fig. 10 Comparison of light microscope (a and b) and SEM (c) images of the interface between the fatigue-precrack area (left) and the test fracture region (right) of an alloy X-750 rising-load test specimen broken in air. The test fracture is ductile. (a) Bright-field image. (b) Dark-field image. (c) Secondary electron image. All 68×

fracture in cup-and-cone fractures begins at the interior of the specimen (Fig. 23) and progresses to the surface.

When a smooth tensile specimen is tested at room temperature, plastic deformation is ini-

Visual Examination and Light Microscopy / 99

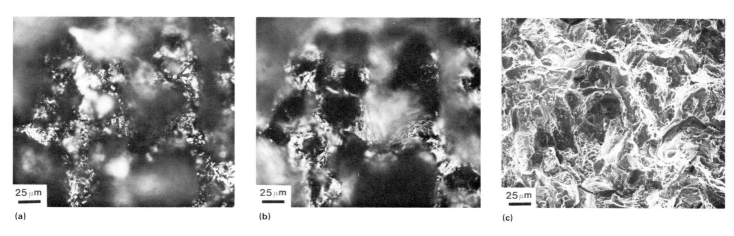

Fig. 11 Comparison of light microscope (a and b) and SEM (c) images of a ductile fracture in an alloy X-750 rising-load test specimen broken in air. (a) Bright-field image. (b) Dark-field image. (c) Secondary electron image. All 240×

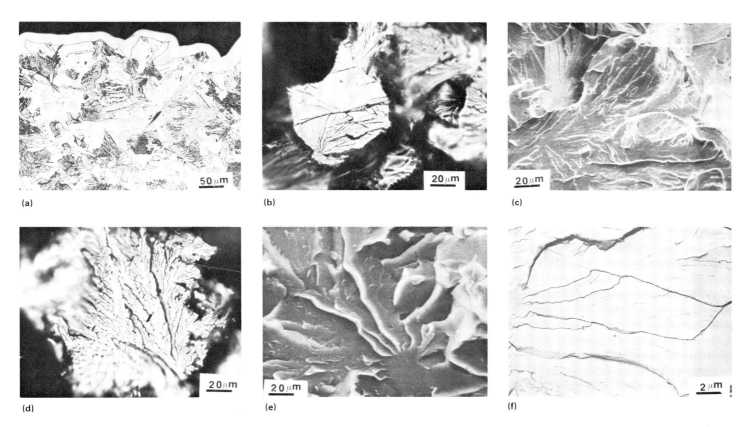

Fig. 12 Examples of three direct (a to c) and three replication procedures (d to f) for examination of a cleavage fracture in a low-carbon martensitic steel. (a) Light microscope cross section with nickel plating at top. (b) Direct light fractograph. (c) Direct SEM fractograph. (d) Light fractograph of replica. (e) SEM fractograph of replica. (f) TEM fractograph of replica

Fig. 13 Transparent tape replica of a fracture surface. See the article "Transmission Electron Microscopy" in this Volume for more information on replication techniques.

tially characterized by uniform elongation. At this stage, voids begin to form randomly at large inclusions and precipitates. With further deformation, plastic instability arises, producing localized deformation, or necking, and a shift from a uniaxial to a triaxial stress state. This results in void nucleation, growth, and coalescence at the center of the necked region, forming a central crack. Continued deformation is concentrated at the crack tips; this produces bands of high shear strain. These bands are oriented at an angle of 50 to 60° from the transverse plane of the test specimen. Sheets of voids are nucleated in the shear bands, and these voids grow and link up, producing a serrated pattern as the central crack expands radially. The cup-and-cone walls are formed when the crack grows to such an extent that the void sheets propagate in one large step to the surface.

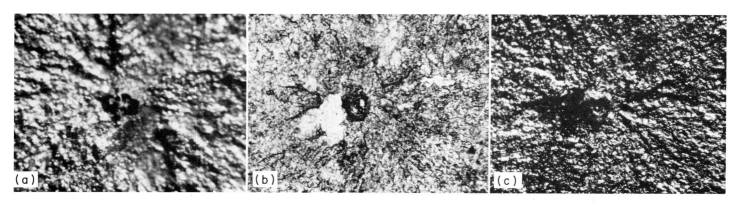

Fig. 14 Comparison of replica fractographs of a fatigue fracture in an induction-hardened 15B28 steel shaft. Fracture was initiated at the large inclusion in the center of the views during rotating bending. (a) Oblique illumination from a point source lamp. (b) Same area as (a), photographed using transmitted light. (c) Replica shadowed with a vapor-deposited coating and photographed using oblique illumination from a point source lamp. All 30×

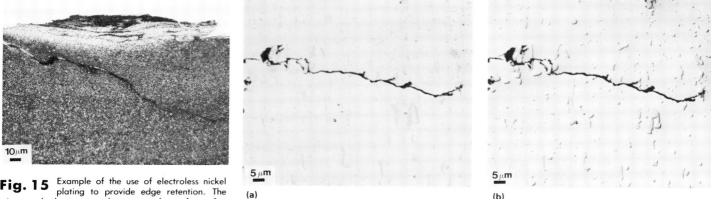

Fig. 15 Example of the use of electroless nickel plating to provide edge retention. The micrograph shows wear damage at the surface of a forged alloy steel Medart roll. Etched with 2% nital. 285×

Fig. 16 Comparison of bright-field (a) and Nomarski DIC (b) illumination for examination of a fatigue crack in an as-polished aluminum alloy. See also Fig. 71. Both 600×

Fig. 17 Comparison of bright-field (a), DIC (b), and dark-field (c), illumination for viewing a partially fractured (by impact) specimen of AISI type 312 weld metal containing substantial σ phase. All 240×

Tensile fractures of oxygen-free high-conductivity (OFHC) copper were studied by deforming the specimens in tension until necking and then halting the test (Ref 10). After radiographic examination, the specimens were sectioned and examined. This work also showed that ductile tensile fractures begin by void formation, with the voids linking up to form a central crack. The fracture can be completed either by continued void formation resulting in a cup-and-cone fracture or by an alternate slip method producing a double-cup fracture. Cup-and-cone fractures are observed in ductile iron-base alloys, brass, and Duralumin; double-cup fractures are seen in face-centered cubic (fcc) metals, such as copper, nickel, aluminum, gold, and silver.

Three types of tensile fractures have been observed in tests of fcc metals: chisel-point fractures, double-cup fractures, and cup-and-cone fractures (Ref 48, 49). The fracture mode changes from chisel-point to double-cup to cup-and-cone as the precipitate density and alloy content increase.

One investigation studied the influence of particle density and spacing and solute content on tensile fracture of aluminum alloys (Ref 50). As the particle density increases, there is a large initial decrease in ductility, followed by a gradual loss in ductility with further increases

Visual Examination and Light Microscopy / 101

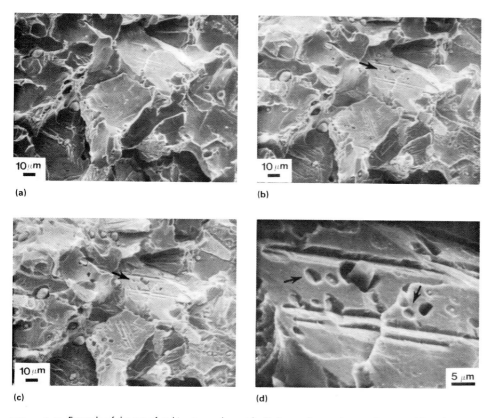

Fig. 18 Example of the use of etching to produce etch pits (arrows) on a cleavage fracture. (a) As-fractured. 320×. (b) Etched 60 s with nital. 320×. (c) Etched 360 s with nital. 320×. (d) Etched 360 s with nital. 1280×

in particle density. Other studies demonstrated that tensile ductility improves as the volume fraction of second-phase particles decreases (Ref 51, 52). Particle size and mean interparticle spacing also influence ductility, but the volume fraction is of greater importance. Dimple size has been shown to increase with increasing particle size or increasing interparticle spacing, but dimple density increases with increasing particle density (Ref 53). Tensile ductility in wrought materials is higher in longitudinally oriented specimens (fracture perpendicular to the fiber axis) than in transversely oriented specimens (fracture parallel to the fiber axis) because of the elongation of inclusions, and some precipitates, during hot working.

The tensile test is widely used for quality control and material acceptance purposes, and its value is well known to metallurgists and mechanical engineers. Tensile ductility measurements, although qualitative, are strongly influenced by microstructure. Transversely oriented tensile specimens have been widely used to assess material quality by evaluation of the transverse reduction of area (RAT). Numerous studies have demonstrated the structural sensitivity of RAT values (Ref 54-60).

The macroscopic appearance of tensile fractures is a result of the relative ductility or brittleness of the material being tested. Consequently, interpretation of macroscopic tensile fracture features is an important skill for the metallurgist. In addition to the nature of the material, other factors can influence the mac-

Fig. 19 Macrograph showing ductile overload fracture of a high-pressure steel vessel tested at room temperature. See also Fig. 20.

Fig. 20 Macrograph showing ductile overload fracture of a high-pressure steel vessel tested at −45 °C (−50 °F). See also Fig. 19.

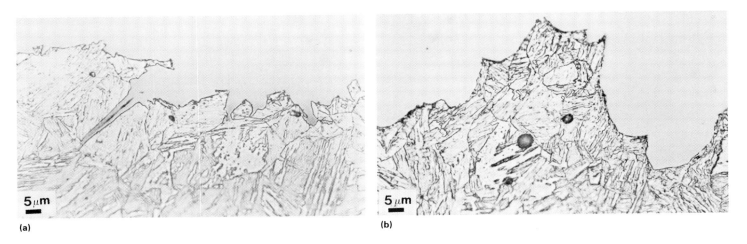

Fig. 21 Brittle (a) and ductile (b) crack paths in a fractured quenched-and-tempered low-alloy steel. Both etched with 2% nital. 800×

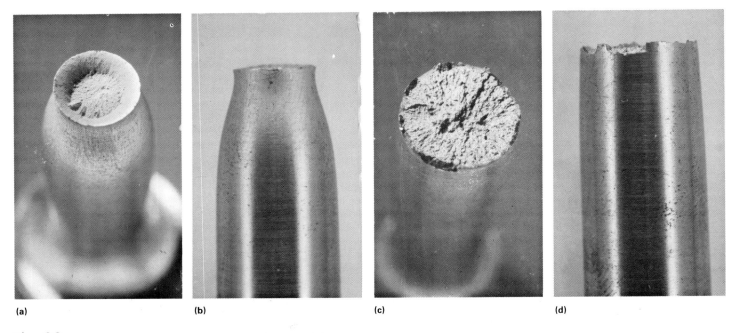

Fig. 22 Macroscopic appearance of ductile (a) and brittle (b) tensile fractures

roscopic tensile fracture appearance—for example, the size and shape of the test specimen and the product form from which it came, the test temperature and environment, and the manner of loading.

The classic cup-and-cone tensile fracture exhibits three zones: the inner flat fibrous zone where the fracture begins, an intermediate radial zone, and the outer shear-lip zone where the fracture terminates. Figure 22(a) shows each of these zones; the flat brittle fracture shown in Fig. 22(b) has no shear-lip zone.

The fibrous zone is a region of slow crack growth at the fracture origin that is usually at or very close to the tensile axis. The fibrous zone has either a random fibrous appearance or may exhibit a series of fine circumferential ridges; the latter is illustrated in Fig. 24. These ridges are normal to the direction of crack propagation from the origin to the surface of the specimen. The presence of such ridges indicates stable, subcritical crack growth that requires high energy (Ref 61). The fracture origin is usually located at or near the center of the fracture and on the tensile axis; it can often be observed to initiate at a hard second-phase constituent, such as an inclusion or a cluster of inclusions. Inspection of tensile fractures at low magnification—for example, with a stereomicroscope—will frequently reveal the initiating microstructural constituents (Ref 62).

The radial zone results when the crack growth rate becomes rapid or unstable. These marks trace the crack growth direction from either the edge of the fibrous zone or from the origin itself. In the latter case, it is easy to trace the radial marks backward to the origin. The radial marks may be fine or coarse, depending on the material being tested and the test temperature. The radial marks on tensile specimens of high-strength tempered martensite steels are usually rather fine. Tempering of such samples to lower strengths results in coarser radial marks. Low tensile-test temperatures result in finer radial marks than those produced with room-temperature tests. In a study of AISI 4340 steel, for example, fine radial marks were not produced by a shear mechanism but by quasi-cleavage, intergranular fracture, or both (Ref 61). Coarse radial marks on steel specimens are due to

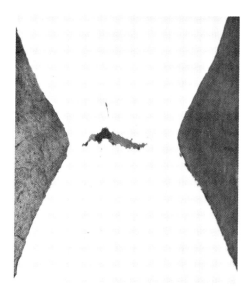

Fig. 23 Initiation of fracture in a tensile-test specimen. Note that the fracture initiated at the center of the specimen. 4.75×

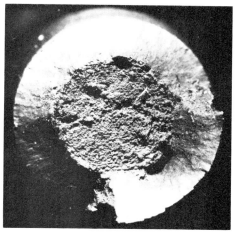

Fig. 24 Tensile fracture of a 4340 steel specimen tested at 120 °C (250 °F). The fracture contains a fibrous zone and a shear-lip zone. The steel microstructure consisted of tempered martensite; hardness was 46 HRC. The fracture started at the center of the fibrous zone, which shows circumferential ridges. The outer ring is the shear-lip zone. About 11×

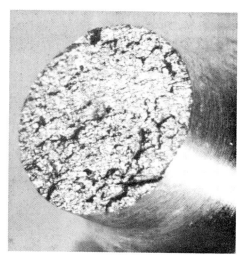

Fig. 25 Macrograph showing a granular brittle fracture in a cast iron tensile bar. Note the large cleavage facets. 2×

shear, and longitudinally oriented splits can be observed along the ridges or peaks.

Radial marks on tensile fractures are usually straight, but a special form of tensile fracture exhibiting coarse curved radial marks (the star or rosette pattern) can also occur, as discussed below. If the origin of the tensile fracture is off axis and if the fibrous zone is very small or absent, some curvature of the radial marks will be observed.

The appearance of the radial marks is partly the result of the ductility of the material. When tensile ductility is low, radial marks are fine with little relief. If a material is quite brittle with a coarse grain size, the amount of tensile ductility is extremely low, and the crack path will follow the planes of weakness in directions associated with each grain. Thus, cracking will be by cleavage, will be intergranular, or a combination of these. Figure 25 illustrates such a fracture in a cast iron tensile bar.

The outer shear-lip zone is a smooth, annular area adjacent to the free surface of the specimen. The size of the shear-lip zone depends on the stress state and the properties of the material tested. Changing the diameter of the test specimen will change the stress state and alter the nature of the shear lip. In many cases, the shear-lip width will be the same, but the percentage of the fracture it covers will change. However, exceptions to this have been noted (Ref 63).

An example of the influence of changes in the section size, the presence of notches, and the manner of loading on the appearance of fractures of round tensile specimens is provided by tests of 25-mm (1-in.) diam, 114-mm (4½ in.) long ASTM A490 high-strength bolts (Fig. 26 and 27). Bolt 1 was a full-size bolt with a portion of the shank turned to a diameter just smaller than the thread root diameter. Bolt 4

Fig. 26 Side views of four types of ASTM A490 high-strength steel bolt tensile specimens. See also Fig. 27. Left to right: bolts 1, 4, 6, and 7

had a major portion of the shank turned down to a diameter of 9 mm (0.357 in.). Bolt 6 had a major portion of the shank turned down to 13 mm (0.505 in.) in diameter and was then notched (60°) to 9 mm (0.357 in.) in diameter in the center of the turned section. Bolt 7 was a full-size bolt tested with a 10° wedge under the head. Bolts 1, 4, and 6 were axially loaded. Figure 26 shows side views of the four bolts after testing, and Fig. 27 shows their fracture surfaces. The fractures for bolts 1 and 4 exhibit the rosette star-type pattern, which is more fully developed in bolt 1. Bolt 6, which is notched,

exhibits a flat brittle fracture with a small split; bolt 7 has a slanted brittle fracture and a large split.

The presence of voids, such as microshrinkage cavities, can alter the fracture appearance. Figure 28 shows a tensile fracture from a carbon steel casting with a slant fracture due to the voids present.

Because the shape and size of the tensile specimen influence the stress state, fracture zones will be different for square or rectangular sections compared to those with the round cross sections discussed previously. For an un-

Fig. 27 Macrographs of fracture surfaces of ASTM A490 high-strength bolt tensile specimens shown in Fig. 26. Top left to right: bolts 1 and 4; bottom left to right: bolts 6 and 7

notched, rectangular test specimen, for example, the fibrous zone may be elliptical in shape, with the major axis parallel to the longer side of the rectangle. Figure 29 shows a schematic of such a test specimen, as well as two actual fracture faces. The radial zone of the test fracture is substantially altered by the shape of the specimen, particularly for the example in Fig. 29(c). As shown by the schematic illustrations in Fig. 30, as the section thickness decreases, the radial zone is suppressed in favor of a larger shear-lip zone (Ref 72). For very thin specimens (plane-stress conditions), there is no radial zone.

Tensile fractures of specimens machined with a transverse or short-transverse orientation from materials containing aligned second-phase constituents—for example, sulfide inclusions, slag stringers (wrought iron), or segregation—often exhibit a woody fracture appearance (Ref 13). In such fractures, the aligned second phase controls fracture initiation and propagation, and ductility is usually low or nonexistent.

Another unique macroscopic tensile fracture appearance is the star or rosette fracture, which exhibits a central fibrous zone, an intermediate region of radial shear, and an outer circumferential shear-lip zone (Ref 64-71). The nature and size of these zones can be altered by heat treatment, tensile size, and test temperature. Figure 27 shows two examples of this type of tensile fracture.

Rosette star-type tensile fractures are observed only in tensile bars taken parallel to the hot-working direction of round bar stock (Ref 65). The radial zone is the zone most characteristic of such fractures, and it exhibits longitudinally oriented cracks. The surfaces of these cracks exhibit quasi-cleavage, which is formed before final rupture. Rosette, fractures have frequently been observed in temper-embrittled steels (Ref 64), but are also seen in nonembrittled steels.

Certain tensile strength ranges appear to favor rosette, fracture formation in specimens machined from round bars. To illustrate this effect, Fig. 31 shows tensile fractures and test data for seven tensile specimens of heat-treated AISI 4142 alloy steel machined from a 28.6-mm (1⅛-in.) diam bar. Specimens 1 and 2 were oil quenched and tempered at 205 and 315 °C (400 and 600 °F) and exhibit classic cup-and-cone fractures. Specimens 3 to 7, which were tempered at 455, 510, 565, and 675 °C (850, 950, 1050, and 1250 °F), respectively, all exhibit rosette star-type tensile fractures; specimens 4 and 7 exhibit the best examples of such fractures. Specimens 3 to 7 exhibit longitudinal splitting. Shear lips are well developed on specimen 3, but are poorly developed on specimens 4 and 7 and are essentially absent on specimens 5 and 6. The testing temperature can also influence formation of the rosette star-type fracture. The texture produced when round bars are hot rolled appears to be an important criterion for the formation of such fractures (Ref 65).

Splitting also has been observed in ordinary cup-and-cone tensile fractures of specimens machined from plates (Ref 72). As with the rosette star-type tensile fracture, cup-and-cone fractures have been observed in quenched-and-tempered (205 to 650 °C, or 400 to 1200 °F) alloy steels. One study showed that the occurrence of splitting and the tensile fracture appearance varied with test temperature (Ref 72). Tests at 65 °C (150 °F) produced the cup-and-cone fracture, but lower test temperatures resulted in one or more longitudinally oriented splits. The splits were perpendicular to the plate surface and ran in the hot-working direction. The crack surfaces exhibited quasi-cleavage. When the same material was rolled to produce round bars, the tensile fracture exhibited rosette star-type fractures. Therefore, the split, layered cup-and-cone fracture was concluded to be a two-dimensional (plate) analog of the rosette, star fracture observed in heat-treated tensiles from round bars.

Splitting of fracture surfaces of both tensile specimens and Charpy V-notch impact specimens has been frequently observed in specimens machined from controlled-rolled plate (Ref 73-85) and line pipe steels (Ref 86-90). The splits, which are also referred to as delaminations, are fissures that propagate in the hot-rolling direction.

Visual Examination and Light Microscopy / 105

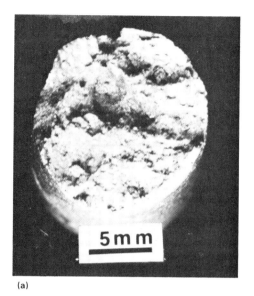

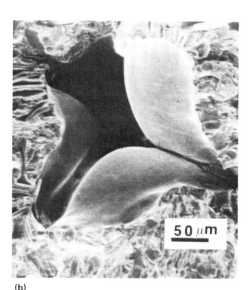

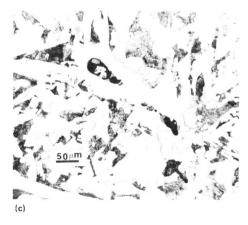

Fig. 28 Three views of an unusual tensile fracture from a carbon steel casting. (a) Macrograph of the fracture surface. (b) SEM view of voids on the fracture surface. (c) Light micrograph showing shrinkage cavities. (c) Etched with 2% nital

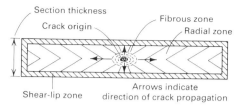

Fig. 29 Appearance of fracture surfaces in rectangular steel tensile specimens. (a) Schematic of tensile fracture features in a rectangular specimen. (b) Light fractograph with fracture features conforming to those of the schematic. (c) Light fractograph of a fracture similar to (b) but having a much narrower shear-lip zone. Source: Ref 63

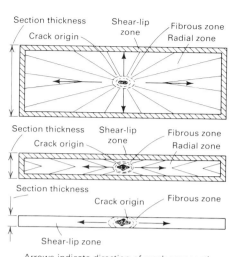

Fig. 30 Effect of section thickness on the fracture surface markings of rectangular tensile specimens. Schematics show the change in size of the radial zone of specimens of progressively decreasing section thickness. The thinnest of the three examples has a small fibrous zone surrounding the origin and a shear-lip zone, but no radial zone. Source: Ref 63

Various factors have been suggested as causes for these delaminations (Ref 82):

- Elongated ferrite grains produced by low finishing temperatures that promote grain-boundary decohesion
- Residual stress concentration
- Grain-boundary segregation
- Grain-boundary carbides
- Ferrite/pearlite banding
- Nonmetallic inclusions
- Cleavage on ⟨100⟩ planes
- Mechanical fibering
- Duplex ferrite grain size
- Increased amounts of deformed ferrite
- Prior-austenite grain boundaries

In a study of a relatively pure Fe-1Mn alloy that was essentially free of carbides and inclusions, delaminations were observed along grain boundaries; splitting occurred when the ferrite grains were deformed beyond a certain degree by controlled rolling (Ref 83). Examples of splitting in 12.7-mm² (0.02-in.²) tensile specimens from this study are shown in Fig. 32, which illustrates splitting on longitudinal and transverse tensile specimens where the hot-rolling finishing temperatures were 315 and 150 °C (600 and 300 °F). Splitting was observed in transverse and longitudinal specimens when the plates were finish rolled at 480 °C (895 °F) and below and at 370 °C (700 °F) and below, respectively. The splits were in the hot-rolling direction, and the frequency of splitting increased as the finishing temperature was lowered below the temperatures mentioned previously.

This study revealed that splitting followed the ferrite grain boundaries (Ref 83). The aspect ratio of the deformed ferrite grains was found to be related to the occurrence of splitting. Material that was susceptible to splitting continued to exhibit splitting after annealing until the ferrite grains were almost completely recrystallized.

As a final note on tensile fractures, numerous studies have used metallographic cross sections to assess the influence of second-phase constituents on fracture initiation and tensile properties. Many studies have shown void formation at the interface between hard constituents (carbides, intermetallics, and inclusions) and the matrix (Ref 20, 91-95); other studies have demonstrated quantitative relationships between inclusion parameters and tensile ductility (Ref 51, 96-102). The use of light microscopy has been of great importance in such studies.

Brittle Fractures

Brittle fractures can occur in body-centered cubic (bcc) and hexagonal close-packed (hcp) metals but not in fcc metals (except in certain specific cases). Brittle fractures are promoted by low service temperatures, high strain rates, the presence of stress concentrators, and certain environmental conditions. The ductile-to-brittle transition over a range of temperatures is a well-known characteristic behavior of steels

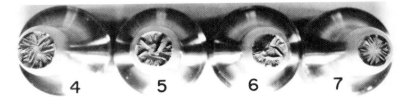

Specimen	Temper °C	Temper °F	Tensile strength MPa	Tensile strength ksi	0.2% yield strength MPa	0.2% yield strength ksi	Elongation, %	%RA
1	205	400	1970	285	1690	245	10	39
2	315	600	1730	251	1550	225	10	43
3	455	850	1410	204	1310	190	12.5	47
4	510	950	1250	181	1170	169	15	54
5	565	1050	1130	164	1030	150	16	58
6	620	1150	945	137	850	123	20	63
7	675	1250	770	112	670	97	24.5	66

Fig. 31 Macrographs of quenched-and-tempered AISI 4142 steel tensile specimens showing splitting parallel to the hot-working axis in specimens tempered at 455 °C (850 °F) or higher

and is influenced by such factors as strain rate, stress state, composition, microstructure, grain size, and specimen size. Macroscopic examination and light microscopy, as well as electron metallographic procedures, have played an important role in gaining an understanding of brittle fracture, and these analysis techniques are basic failure analysis tools.

Most metals, except fcc metals, exhibit a temperature-dependent brittleness behavior that has been studied by using a wide variety of impact-type tests. The Charpy V-notch impact test has had the greatest overall usage, and macroscopic examination of the fracture surfaces is used to assess the percentages of ductile and brittle fracture on the specimens as a function of test temperature. Figure 33 shows fractures of six Charpy V-notch impact specimens of a low-carbon steel tested between −18 and 95 °C (0 and 200 °F), along with the test data (absorbed energy, lateral expansion, and percent ductile, or fibrous, fracture). Figure 33 also shows SEM views of the fractures resulting from testing at −18 and 95 °C (0 and 200 °F) that illustrate cleavage and microvoid coalescence, respectively.

Another example of the macroscopic appearance of Charpy V-notch impact specimens is given in Fig. 34, which shows four specimens of heat-treated AISI 4340 tested between −196 and 40 °C (−321 and 104 °F), as well as plots of the test data (Ref 63). The test specimen at −80 °C (−112 °F), which is near the ductile-to-brittle transition temperature, shows a well-defined ductile zone surrounding an inner brittle zone. Such clear delineation between the ductile and brittle zones is not always obtained, as was the case with the samples shown in Fig. 33.

Splitting has been observed on Charpy V-notch specimens, as well as on tensile specimens, as discussed previously. One investigation (Ref 83) has shown that the orientation of the splits is always parallel to the rolling plane of plate in the longitudinal direction (Fig. 35). Therefore, splitting is always associated with planes of weakness in the rolling direction.

Figure 36 shows Charpy V-notch absorbed-energy curves for Fe-1Mn steel that was finish rolled at temperatures from 960 to 316 °C (1760 to 600 °F) from the study discussed in Ref 83. As the finishing temperature decreased, the absorbed-energy transition temperature (temperature for a certain level of absorbed energy, for example, 20 or 34 J, or 15 or 25 ft · lb) decreased, but the upper shelf energy also decreased. Figure 37 shows the fracture appearance of Charpy V-notch specimens used to produce the curves in Fig. 36. This demonstrates that the occurrence of splitting increased as the finishing temperature decreased. In general, lowering the test temperature for a given finishing temperature increased the number of splits unless the test temperature was low enough to produce a completely brittle cleavage fracture, for example, the −73-°C (−100-°F) samples of the plates finish rolled at 707 and 538 °C (1305 and 1000 °F).

Figure 38 shows the microstructural appearances of the splitting in two Charpy V-notch specimens in plate that was finish rolled at 315 °C (600 °F). The regions between the splits resemble a cup-and-cone tensile fracture.

Although brittle fractures are characterized by a lack of gross deformation, there is always some minor degree of plastic deformation preceding crack initiation and during crack growth. However, the amounts are quite low, and no macroscopically detectable deformation occurs.

Light microscopy has been used to study the initiation of brittle fracture. In polycrystalline bcc metals, grain boundaries can block the motion of slip or twinning, resulting in high tensile stresses and crack nucleation. If plastic deformation is not able to relax these tensile stresses, unstable crack growth results. The presence of hard second-phase particles (carbides, intermetallics, and nonmetallics) at the grain boundaries will facilitate crack nucleation. Because these particles are brittle, they inhibit relaxation at the tip of blocked slip bands, thus reducing the energy required for crack nucleation.

Another important feature is the relationship between the cleavage planes in these particles and in the neighboring grains. If the cleavage planes in the grain and in the particle are favorably oriented, little or no energy will be expended when the crack crosses the particle/matrix interface. If the cleavage planes are misoriented, considerable energy is required, and crack nucleation is more difficult. For a given stress level, the probability for crack nucleation increases with the number of grain-boundary particles. Particle shape and size are also very important, as is the degree of segregation of the particles to the grain boundaries. The strength of the interface and the presence of pre-existing voids at the particles also influence crack nucleation.

An interesting study of cleavage crack initiation in polycrystalline iron was conducted in which two vacuum-melted ferritic irons with low carbon contents (0.035 and 0.005%) were tested using tensile specimens broken between room temperature and −196 °C (−321 °F) (Ref 103, 104). For a given test temperature, microcracks were more frequent in the higher-carbon heat, and nearly all of the microcracks originated at cracked carbides. The microcracks were most often arrested at pre-existing twins or grain boundaries, but were also arrested by the initiation of twins and slip bands at the advancing crack tip.

Grain size also influences both initiation and propagation of brittle fractures. The quantitative description of this relationship is known as the Hall-Petch equation (Ref 105, 106):

$$\sigma_c = \sigma_0 + kd^{-1/2} \qquad \text{(Eq 1)}$$

where σ_c is the cleavage strength, d is the grain diameter, and σ_0 and k are constants. For single-phase metals, a finer grain size produces higher strength. Many studies of a very wide range of single-phase specimens (tensile) have demonstrated the validity of this relationship using several experimentally determined strengths (yield point, yield stress, or flow stress at certain values of strain). Equations similar to Eq 1 have been used to relate grain size to other properties, such as toughness, fatigue limit, creep rate, hardness, and fracture strength in stress-corrosion cracking (SCC).

Macroscopically, brittle fractures are characterized by the following:

- Little or no visible plastic deformation precedes the fracture
- The fracture is generally flat and perpendicular to the surface of the component
- The fracture may appear granular or crystalline and is often highly reflective to light. Facets may also be observed, particularly in coarse-grain steels
- Herringbone (chevron) patterns may be present
- Cracks grow rapidly, often accompanied by a loud noise

Figure 39 shows an example of a well-developed chevron fracture pattern in a railroad rail. This type of brittle fracture pattern is frequently observed in low-strength steels, such as structural steels. The origin of the chevron fracture is easy to find because the apexes of the chevrons point back to the origin. The origin is not on the sections shown in Fig. 39. Such fractures can propagate for some distance before being arrested. The wavy nature of the fracture edge is evident in Fig. 39. In general, the more highly developed the chevron pattern, the greater the edge waviness. Such patterns are not observed on more brittle materials.

The chevron fracture results from a discontinuous growth pattern. It proceeds first by initiation or initiations, followed by union of the initiation centers to form the fracture surface. One investigation of chevron-type brittle fractures, demonstrated that they are caused by discontinuous regions of cleavage fracture joined by regions of shear and that the chevron features are the ridges between the cleavage and shear zones (Ref 107). When such fractures occur, they tend to extend into the metal and produce cracks below the main fracture surface, particularly in the crack growth direction. The spreading of the crack toward the surfaces is not always equal, and this leads to asymmetrical chevron patterns. The front of the propagating crack tends to become curved, and traces of the curved front can often be observed.

The growing crack front in chevron fractures can be considered a parabolic envelope enclosing a number of individually initiated cracks that spread radially (Ref 108). Figure 40 shows a model for chevron formation and the analogy

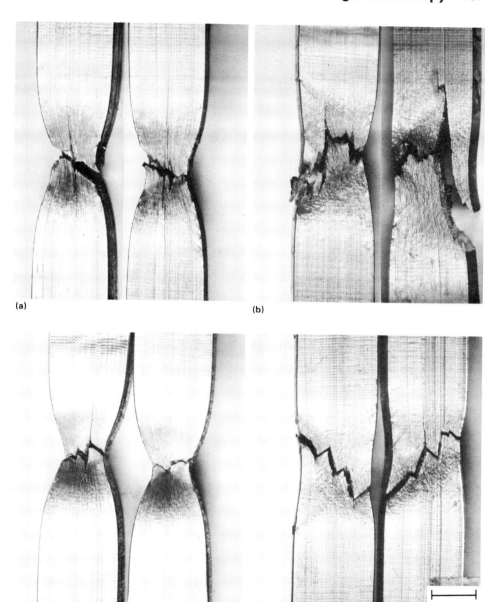

Fig. 32 Macrographs of fractured longitudinal and transverse tensile specimens from plates finish-rolled at two temperatures. (a) Longitudinal specimen finished at 315 °C (600 °F). (b) Transverse specimen finished at the same temperature. (c) Longitudinal specimen finished at 150 °C (300 °F). (d) Transverse specimen finished at the same temperature. Source: Ref 83

between chevron formation and cup-and-cone type tensile fracture. This model predicts that the angle θ (Fig. 40b) will be 72° and that this angle is independent of material thickness and properties. To construct this model, 161 measurements were made on 20 fracture surfaces, and a mean angle of 69° was observed, which varied somewhat depending on the presence and size of shear lips. Because plastic deformation is required for formation of chevron patterns, such patterns are not observed on more brittle materials.

In Fig. 40(c), locations A to E demonstrate the close resemblance between chevron fractures and the cup-and-cone tensile fracture.

Fracture begins along the centerline and spreads as a disk-shaped crack toward the surfaces. However, before reaching the surface, the fracture mode changes, and rupture occurs on the conical surfaces of maximum shear, approximately 45° from the tensile axis (dotted lines in A, B, and C, Fig. 40c).

As necking progresses, plastic deformation (shaded area, Fig. 40c) occurs ahead of the crack front, with lines of maximum shear strain approximately 45° to the plate axis. If this plastic zone is large enough, fracture will progress by shear only. However, if necking is limited, thus limiting the extent of plastic deformation, many small shear fractures of dif-

108 / Visual Examination and Light Microscopy

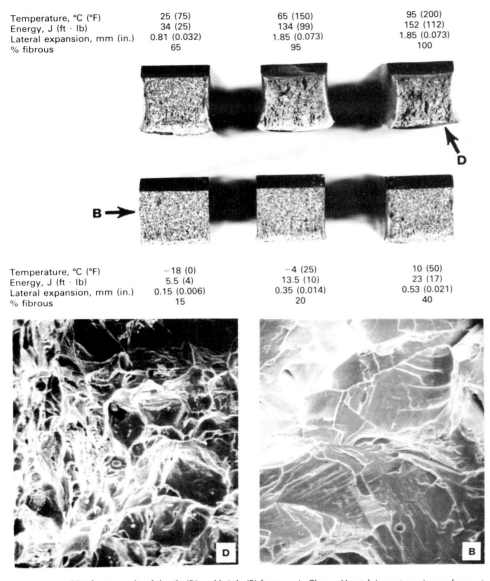

Fig. 33 SEM fractographs of ductile (D) and brittle (B) fractures in Charpy V-notch impact specimens shown at top. Both 400×

fering orientation will form, resulting in a rough or serrated appearance.

This description reveals that chevron fractures are intermediate between completely brittle and completely ductile (shear) fractures. Figure 41 shows five ship steel fractures produced by drop-weight testing at temperatures from −30 to −70 °C (−25 to −96 °F). The mating halves of each fracture are shown with the fracture starting at the notches at the top of each specimen. The specimen broken at −45 °C (−50 °F) has the best definition of the chevron pattern. Testing at higher temperatures produced shear fractures, but testing at −70 °C (−96 °F) produced a nearly flat brittle fracture with only a hint of a chevron pattern. This same trend is also demonstrated in Fig. 42, which shows sections of line pipe steel tested at four temperatures from −3 to 40 °C (27 to 104 °F).

Chevron patterns are observed in the samples tested at −3 and 16 °C (27 and 60 °F) and are most clearly developed in the latter. Testing at higher temperatures resulted in 100% shear fractures.

Figures 43 to 46, which are photographs of controlled laboratory fractures of full-size line pipes, illustrate several important macroscopic features of ductile and brittle fractures (Ref 109). When a line pipe fails catastrophically, failure occurs with rapid lengthening of an initially small crack in these large, continuously welded structures. To prevent such failures, a weldable low-strength steel that resists crack propagation is needed. The potential for a catastrophic failure in a line pipe depends on the toughness of the material (both transition temperature and upper shelf energy), the magnitude of the operating stresses, pipe dimensions, and the nature of the fluid within the pipe.

Full-scale testing of line pipe, as shown in Fig. 43 to 46, revealed a good correlation between the crack speed and the 50% shear area transition temperature as determined by the Battelle drop-weight tear test (DWTT). The fracture appearance of the pipe reflects the crack speed, for example:

- Less than 10% shear area corresponds to brittle fracture with a fast-running crack
- Greater than 40% shear area corresponds to a ductile fracture with a crack speed below 275 m/s (900 ft/s)

Figures 43 to 46 show full-scale test results of American Petroleum Institute (API) grade X-60 line pipes in which the test temperature was varied with respect to the 50% shear area DWTT transition temperature. In each test, the pipe was loaded to 40% of its yield strength, and a 30-grain explosive charge was detonated beneath a 460-mm (18-in.) long notch cut in the pipe while the entire pipe was maintained at the desired test temperature.

Figure 43 shows a completely ductile fracture in a line pipe tested at 5 °C (8 °F) above the 50% shear area DWTT transition temperature; the resultant crack speed was 85 m/s (279 ft/s). From the 460-mm (18-in.) notch, the crack propagated 840 mm (33 in.) in full shear and then 460 mm (18 in.) in tearing shear before stopping.

Figure 44 shows the result of testing at 1 °C (2 °F) below the 50% shear area DWTT transition temperature; the average crack speed was 172 m/s (566 ft/s). The crack began by brittle cleavage with a 15% shear area and progressed 200 mm (8 in.) in this manner before changing to 100% shear. The shear fracture ran 940 mm (37 in.) before stopping. The macrograph shows the initial fracture area.

Figure 45 shows the result of testing a line pipe at 6 °C (10 °F) below the 50% shear area DWTT transition temperature; the resultant crack speed was 470 m/s (1550 ft/s). From the notch, the fracture propagated by cleavage with a 15 to 18% shear area, with some small patches having shear areas as high as 70% (see macrograph of fracture, Fig. 45). The crack propagated straight along the top of the pipe for 685 mm (27 in.) and then developed a wave pattern. Only a half wave was completed before the crack changed to 100% shear and then tore circumferentially for 840 mm (33 in.) before stopping. Although the actual crack speed was 470 m/s (1550 ft/s), the straight line crack speed was 380 m/s (1250 ft/s).

Figure 46 shows the result of testing a pipe at 22 °C (40 °F) below the 50% shear area DWTT transition temperature; the crack speed was 675 m/s (2215 ft/s) (maximum straight line crack speed was 535 m/s, or 1760 ft/s). From the notch, the fracture traveled in a wave pattern for a full wave by cleavage with less than 10% shear areas present and then changed to full

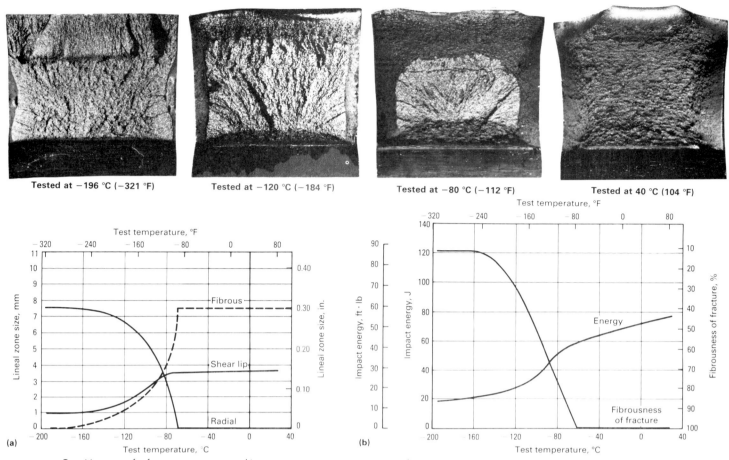

Fig. 34 Transition curves for fracture appearance and impact energy versus test temperature for specimens of 4340 steel. Light fractographs at top show impact specimens tested at various temperatures. Linear measurements were made parallel to the notch for the shear-lip zones, perpendicular to the notch for the fibrous zone, and perpendicular to the notch for the radial zone. The measurements yielded the three curves shown in (a). The curve of percentage of fibrousness of fracture (b) was constructed from visual estimates of the fibrous-plus-shear-lip zones. This curve, together with the impact energy curve shown in (b), shows that the transition temperature for fracture appearance is essentially the same as for impact energy. Source: Ref 63

shear and tore circumferentially for 480 mm (19 in.) before stopping.

The formation of sinusoidal fracture paths in line pipes has been analyzed (Ref 110). In a homogeneous (or reasonably homogeneous) material, the direction of crack propagation is perpendicular to the plane of maximum stress. Therefore, if the fracture path changes direction, it must do so as a result of the imposition of additional stresses on the circumferential stresses in the pipe due to the internal pressure of the contained gas or fluid. The sinusoidal fracture paths, as illustrated in Fig. 45 and 46, occur only in steels with substantial ductility. Also, the crack length must be considerable, as in these examples. As the crack lengthens, the gases or fluids escaping from the rupture cause the free surfaces to vibrate. This produces a tearing force normal to the crack propagation direction. As the crack grows, the vibrations produce maximum tensile stresses alternating between the outside and inside of the pipe. If the tearing stresses are sufficiently large, the crack can continue to tear circumferentially and halt further crack growth. Because the crack speed in such cases is quite high, the vibration frequency generating the wave pattern must be substantial.

Microscopically, brittle fractures have the following characteristics:

- Transgranular cleavage or quasi-cleavage
- Intergranular separation
- Features on transgranular facets, such as river marks, herringbone patterns, or tongues

Engineering alloys are polycrystalline with randomly oriented grains (or some form of preferred texture). Because cleavage occurs along well-defined crystallographic planes within each grain, a cleavage fracture will change directions when it crosses grain or subgrain boundaries. Engineering materials contain second-phase constituents; therefore, true featureless cleavage is difficult to obtain, even within a single grain. Examples of cleavage fractures, as examined with the light microscope and SEM, are shown in Fig. 1 to 5, 12, and 18, while light microscope views on cross sections are shown in Fig. 12 and 21(a). Figure 47 shows an additional example of a cleavage fracture in a low-carbon steel with a ferrite-pearlite microstructure using a light microscope cross section. The orientation of the cleavage planes can be seen to vary in the different ferrite grains.

Brittle fractures can also occur with an intergranular fracture pattern. Examples of light microscope and SEM views of intergranular brittle fractures are shown in Fig. 6 and 9. In general, it is rare for a fracture to be completely intergranular. Rather, a small amount of transgranular fracture can also be found, and it is also possible, in certain cases, for a small amount of microvoid coalescence to accompany intergranular fractures, although this is not commonly observed.

In one study, intergranular fractures were observed in Fe-0.005C material (coarse-grain) that had been reheated to 705 °C (1300 °F), quenched, and tested at −195 °C (−320 °F) (Ref 104). The intergranular fracture may have been due to strengthening of the grain boundaries by equilibrium carbon segregation, or they may have been embrittled by the segregation of some other element, such as oxygen. Inter-

110 / Visual Examination and Light Microscopy

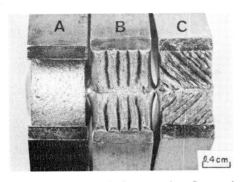

Fig. 35 Macrograph illustrating the influence of specimen orientation (with respect to the hot-working direction) on splitting observed in Charpy V-notch impact specimens. A, notch parallel to the plate surface; B, notch perpendicular to the plate surface; C, notch 45° to the plate surface. Source: Ref 83

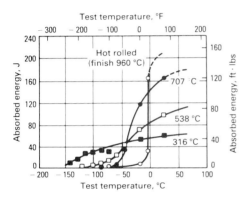

Fig. 36 Plot of absorbed-energy Charpy V-notch test data for Fe-1Mn steels finished at different temperatures (indicated on graph). Source: Ref 83

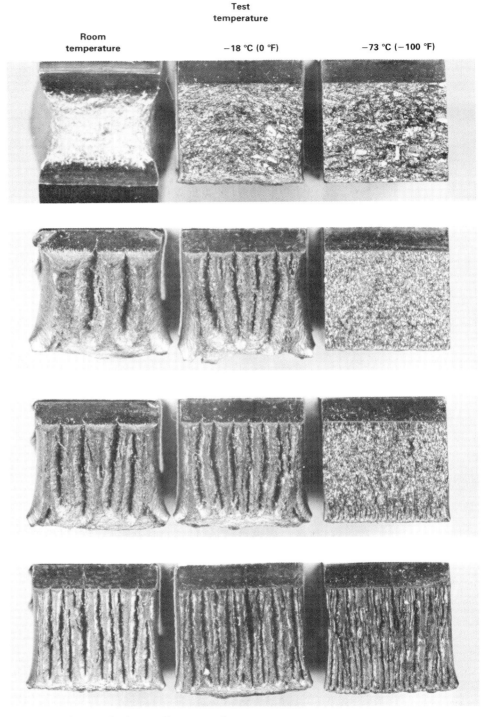

Fig. 37 Macrographs showing Charpy V-notch impact specimens from Fe-1Mn steels finish-rolled at four different temperatures and tested at three different temperatures. Top row, finished at 960 °C (1760 °F); second row, finished at 705 °C (1300 °F); third row, finished at 540 °C (1000 °F); bottom row, finished at 315 °C (600 °F). Source: Ref 83

granular fractures have also been observed at −196 °C (−321 °F) in wet hydrogen decarburized low-carbon steel. If such samples are recarburized, the embrittlement disappears (Ref 111).

Intergranular fractures can be separated into three categories: those in which a brittle second-phase film in the grain boundaries causes separation; those in which no visible film is present, with embrittlement occurring because of segregation of impurity atoms in the grain boundaries; and those caused by a particular environment, as in SCC, in which neither grain-boundary films nor segregates are present. In the case of brittle grain-boundary films, it is not necessary for the film to cover the grain boundaries completely; discontinuous films are sufficient. Some common examples of intergranular embrittlement by films or segregants include:

- Grain-boundary carbide films in steels
- Iron nitride grain-boundary films in nitrided steels
- Temper embrittlement of alloy steels by segregation of phosphorus, antimony, arsenic, or tin
- Grain-boundary carbide precipitation in austenitic stainless steels (sensitization)
- Embrittlement of molybdenum by oxygen, nitrogen, or carbon
- Embrittlement of copper by antimony

Intergranular fractures are generally quite smooth and flat unless the grain size is rather coarse. In the latter case, a rock-candy fracture appearance is observed. In a classic experiment, the microhardness was shown to increase going from the center of the grains toward the grain boundaries (Ref 112). Grain-boundary strengthening is characteristic of intergranular fractures caused by embrittlement.

Fatigue Fractures

If a component is subjected to cyclic loading involving tensile stresses below the statically determined yield strength of the material, the component may eventually fail by a process known as fatigue. Although fatigue fractures are best known in metal components, other materials, such as polymers, can also fail by fatigue.

With cyclic loading at tensile stresses below the yield strength, a crack will begin to form at the region of greatest stress concentration after some critical number of cycles. With continued cycling, the crack will grow in length in a direction perpendicular to the applied tensile stress. After the crack has progressed a certain distance, the remaining cross section can no longer support the loads, and final rupture occurs. In general, fatigue failures proceed as follows:

- Cyclic plastic deformation before crack initiation
- Initiation of microcrack(s)
- Propagation of microcrack(s) (Stage I)
- Propagation of macrocrack(s) (Stage II)
- Final rupture (overload)

Macroscopically, fatigue fractures exhibit many of the same features as brittle fractures in that they are flat and perpendicular to the stress axis with the absence of necking. However, the fracture features are quite different; part of the fracture face is cyclically grown, but the remainder occurs by overloading, that is, one-step fracture. The apparent ductility or brittleness of the overload portion varies, depending on the strength, ductility, and toughness of the material, as well as the temperature and environment.

The most distinct characteristic of fatigue failures in the field are the beach or clam shell markings on the cyclically grown portion of the fracture. It should be mentioned that similar marks on fractures can be produced under certain conditions by other fracture mechanisms that involve cyclic crack growth without cyclic loading. Also, such marks may not be visible on all materials that fail by fatigue; for example, many cast irons do not develop beach marks. Laboratory fatigue test specimens also

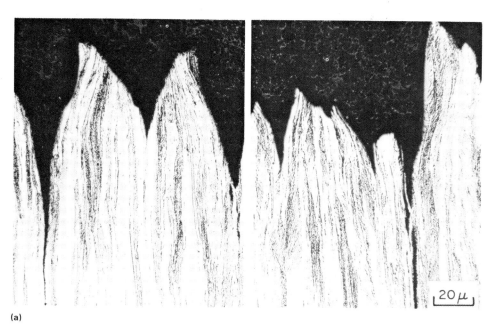

Fig. 38 The microstructure of Fe-1Mn steel finish rolled at 315 °C (600 °F) and impact tested at two temperatures. A, tested at −18 °C (0 °F); B, tested at −135 °C (−210 °F). Source: Ref 83

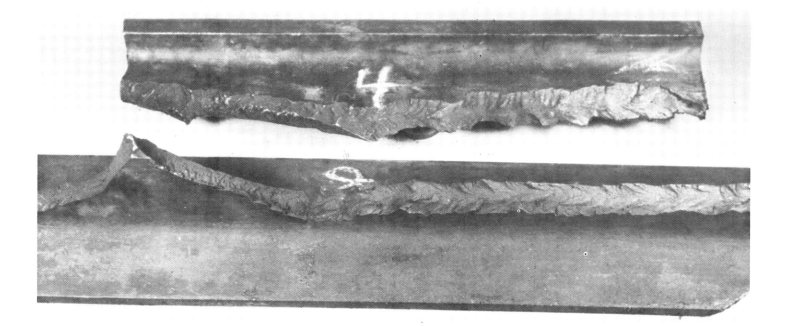

Fig. 39 Classic appearance of chevrons on a brittle fracture of a steel railroad rail. The fracture origin is not on the section shown.

do not exhibit beach marks, regardless of the material, unless the test is deliberately controlled to do so, for example, by using load blocks at widely varying loads. Beach marks document the position of the crack front at various arrest points during its growth and can reflect changes in loading that either retard or accentuate crack growth plus the influence of the environment on the fracture face. In a laboratory test conducted at constant cyclic loading in a dry environment, there is no opportunity for beach mark formation.

The presence of beach marks is fortuitous, at least for the investigator, because beach marks permit the origin to be easily determined and provide the analyst with other information concerning the manner of loading, the relative magnitude of the stresses, and the importance of stress concentration. Guides for interpreting fatigue fracture markings have been shown schematically (Ref 113, 114) and have been discussed by others (Ref 63, 115-122). The most comprehensive schematics for interpreting fatigue fractures are shown in Fig. 48 for round cross sections and in Fig. 49 for rectangular cross sections. Each consists of examples of high and low nominal stress with three degrees of stress concentration and five types of loading: tension-tension or tension-compression, unidirectional bending, reversed bending, rotating bending, and torsion.

In most cases, beach marks are concave to the failure origin. However, the notch sensitivity of the material and the residual stress patterns can influence crack propagation to produce beach marks in notch-sensitive materials that are convex to the origin (Ref 117), although such cases are rare. It is also possible for a fatigue fracture to exhibit a pattern of fanlike marks similar to the radial fracture markings present on chevron fractures. In this case, many small fatigue cracks have joined together at shear steps (Ref 123). This is an example of multiple fatigue cracks initiating from a common location.

Fatigue cracks can be initiated at a wide variety of features, such as scratches, abrupt changes in cross section, tool marks, corrosion pits, inclusions, precipitates, identification marks, and weld configuration defects. In some cases, microcracks may be present before loading begins—for example, grinding cracks, quench cracks, or hot or cold cracks from welding. All these problems increase the likelihood of early failure by fatigue, assuming the presence of alternating stresses of sufficient magnitude. The absence of surface stress raisers, smoothly polished surfaces, and a very low inclusion content will not prevent fatigue if the alternating stresses are of sufficient magnitude and are applied long enough, but these factors are all desirable features for long life. In the macroscopic examination of fatigue fractures, the analyst must carefully examine the surface at and near the origin for any contributing factors, such as those mentioned above.

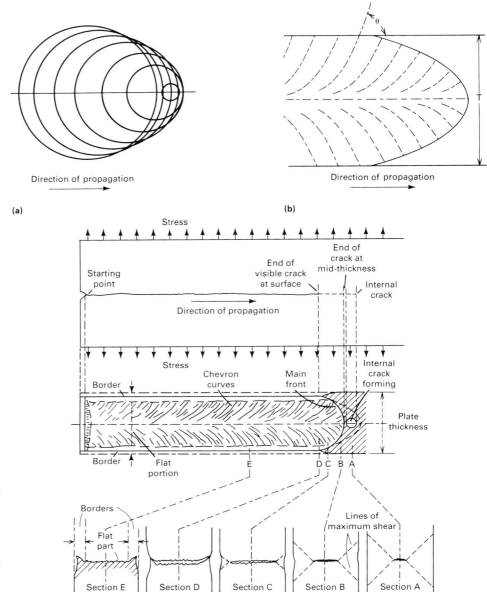

Fig. 40 Schematics showing a model for chevron crack formation. (a) Crack front propagation model. (b) Crack front at any given time. (c) Analogy between chevron formation in plate and cup-and-cone fracture. Source: Ref 108

Another macroscopic feature observed on certain fatigue fractures, particularly shafts and leaf springs, is called ratchet marks. These marks are seen where there are multiple fatigue crack origins that grow and link up. The ratchet marks are the steplike junctions between adjacent fatigue cracks.

Figure 50 shows a bolt that failed by fatigue. The arrow indicates the origin, which was the root of the first thread, the location of greatest stress concentration. The overload portion of the fracture (side opposite the arrow) is rather small compared to the overall cross section; this indicates that the loads were rather low. Comparison of Fig. 50 to the schematics in Fig. 48 suggests that loading was tension-tension or tension-compression with a mild stress concentration at low nominal applied stresses.

Figure 51 shows the fracture of a drive shaft that failed by fatigue. The fracture started at the end of the keyway (B), which was not filletted; that is, the corners were sharp. Final rupture occurred at C, a very small zone. The slight curvature of the beach marks becomes substan-

Fig. 41 Mating drop-weight tear test fractures in ship steel showing the influence of test temperature on fracture appearance. Note that chevrons are most clearly developed at −45 °C (−50 °F). The fractures were located by the notch at the top of each specimen.

tial near the final rupture zone. Comparison of Fig. 51 to the schematics shown in Fig. 48 indicates that the failure occurred by rotating-bending stresses, which would be expected in a drive shaft, with the direction of rotation clockwise looking at the fracture and with a mild stress concentration and low nominal stresses.

Figure 52 shows two fractures in AISI 9310 coupling pins that failed by fatigue. Both were subjected to reversed-bending alternating stresses. These pins actually failed by low-cycle fatigue (<10 000 cycles). High nominal stresses were applied, and stress concentration was either absent or mild. Multiple origins are frequent (arrows).

Figure 53 shows a torsion failure of a railroad spring that began at a small fatigue crack (right arrow) that was initiated at an abraded region (left arrow) at the top of the spring. The abrasion probably resulted from rubbing of the coil spring when it was fully compressed. In this case, the fatigue crack is small, and the overload portion is quite large, suggesting high nominal stresses.

Figure 54 shows the fatigue fracture of a plate of D6AC high-strength steel. The part was fractured in the laboratory under block loading, which produced a beach mark pattern. Because of the uniformity of the load cycles, the resultant beach marks are extremely uniform in appearance. The pattern increases in spacing and clarity as the crack grows because, as the crack grows, the stresses are increasing even though the applied loads are the same. The two macrographs (b and c) of the regions shown in (a) illustrate the progression of the fracture at the lower loads, that is, for the very fine marks in (a) rather than the high-load coarse marks.

Although many fatigue failures begin at a free surface, some important types of fatigue failures begin at both surface and subsurface origins. These all involve rolling-contact fatigue loading and occur in forged hardened steel rolls (Ref 125-130), bearings (Ref 131-135), railroad rails (Ref 136-139), and wheels, for example, crane wheels (Ref 140, 141) and railroad wheels. In each of these products, fatigue failures (spalling) are common modes of failure.

Spalling is the primary cause of premature failures of forged hardened steel rolls. Spalls are sections that have broken from the surface of the roll. In nearly all cases, they are observed in the outer hardened zone of the body surface, and they generally exhibit well-defined fatigue beach marks. The most common spalls are the circular spall and the line spall. Circular spalls exhibit subsurface fatigue marks in a circular, semicircular, or elliptical pattern. They are generally confined to a particular body area. A line spall has a narrow width of subsurface fatigue that extends circumferentially around the body of the roll. Most line spalls originate at or beneath the surface in the outer hardened zone.

Figure 55 shows a classic example of a circular spall that formed beneath the surface in a hardened steel roll. The very clear development of the fatigue marks out to the overload zone is evident. Radial marks can be observed all around the overload portion starting from the last fatigue mark.

Figure 56 shows a circular spall that formed at the body shoulder of a forged hardened steel roll. Two small thumbnail-shaped fatigue cracks (see enlarged view) served as the origin. These two origins were connected by subsurface cracks (not shown). The overall view of the spall shows that the region around the two origins is relatively featureless. However, a faint ring around the two origins suggests that the crack propagated outward from the two origins, stopped, and then fractured by overload.

Tested at −3 °C (27 °F) 0.96×

Tested at 16 °C (60 °F) 0.92×

Tested at 27 °C (80 °F) 1.16×

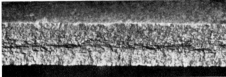
Tested at 40 °C (104 °F) 1.25×

Fig. 42 Transition of fracture appearance with change in test temperature. Light fractographs show four specimens cut from a single length of low-carbon steel pipe that were burst by hydraulic pressure at the temperatures indicated. Note the increasing size of the shear-lip zone as test temperature increases.

Figure 57 shows another circular spall—the mating fracture on the roll body rather than the spalled portion. The well-defined fatigue origin (arrow) and beach marks are visible. The overload portion of the fracture exhibits much more detail than the previous failure. The overload portion exhibits a high density of ridge marks, and it appears to have grown cyclically in that there are a number of arrest points in its growth.

Figure 58 shows a circular spall from a forged hardened steel roll with well-defined fatigue origins (areas A and B) and other regions, such as areas C and D, away from the origin that exhibit beach marks. Enlargements of areas A to D are shown in Fig. 59 and 60. It is clear that the spall originated from the multiple fatigue origins in the light-colored area (A and B) near the center. However, it appears that other fatigue origins, such as at areas C and D, were also growing when the overload fracture occurred.

Figures 61 and 62 show two views of a line spall fracture in a 545-mm (21½-in.) diam forged hardened steel roll. The term line spall comes from the linear shape of the fatigue portion of the fracture that initiated rupture. In a line spall, the linear fatigue portion often runs part way around the periphery; the overload portion and the overall shape of the spalled section need not be linear. The arrows in Fig. 61 and 62 indicate the portion of the line spall that extended under the unbroken portion of the roll body. This area was opened and is shown in Fig. 63, along with an enlarged view of the fatigue marks at the origin of the line spall. Figures 61 and 62 show that the spalled portion broke off as two pieces. The radial marks around the two spalled regions (Fig. 61 and 62) show how the overload portion of the fracture formed around the line spall.

Figure 64 shows another example of a line spall on a hardened steel roll body. In this case, the spall is rather small and did not progress far before the section spalled off. The enlarged view shows a well-defined fatigue origin (arrow). The ridge marks emanating from the fatigue area clearly reveal the crack growth direction in the overload portion.

Figure 65 shows another example of a line spall fracture from a hardened steel roll. Fine fatigue marks can be seen in the finger-shaped portion of the line spall fracture at the bottom of the macrograph. The marks indicate that the crack growth direction was from the arrow at the bottom of the picture and toward the top. After the linear portion in the center of the spall formed, final rupture occurred from the linear portion outward, as indicated by the ridge marks in the overload zone and their origin at the boundary of the linear portion of the spall. Figure 66 shows the outer roll surface of the spall, that is, the reverse side shown in Fig. 65. Macroetching of the surface revealed two intersecting craze crack patterns indicative of abusive service conditions.

Most spalls are caused by local overloading, that is, surface or near-surface damage due to either mechanical or thermal abuse in service. A common cause of roll failures is the failure to remove completely the damage by grinding from previous abusive service experience. Optical examination of replicas of roll surfaces has been implemented to study spall nucleation (Ref 142).

The literature regarding the microstructural aspects of spalling fatigue in bearings is extensive and relies heavily on light microscopy and electron microscopy (Ref 143-156). Spalling failures in bearings are caused by factors different from those in hardened steel rolls, because the service conditions are much different.

TEST TEMP. +56°F
DWTT +48°F
CRACK SPEED 279 fps

Fig. 43 Fracture of API grade X-60 line pipe tested 5 °C (8 °F) above the 50% shear area drop-weight tear test transition temperature. See also Fig. 44 to 46 and text for details.

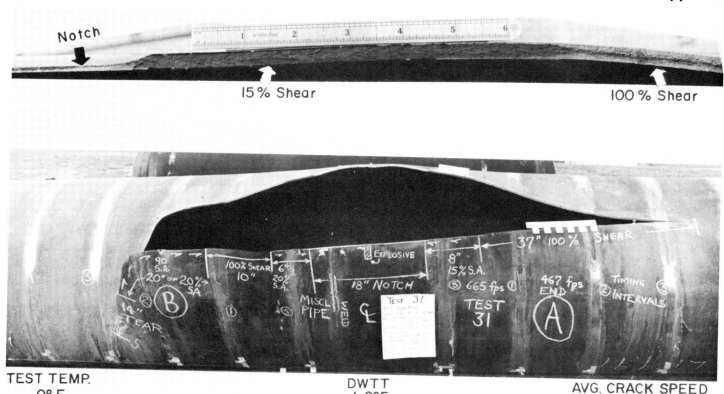

Fig. 44 Fracture of API grade X-60 line pipe tested 1 °C (2 °F) below the 50% shear area drop-weight tear test transition temperature. See also Fig. 43, 45, and 46.

Although spalls in rolls can occur after a quite limited service life, spalls in bearings generally arise after considerable service time. The most common origin for spalls in bearings is oxide inclusions. Hard carbides, nitrides, and carbonitrides are other possible sources for initiation. The initiation sources for contact fatigue in bearings, in addition to inclusions, are geometric stress concentrations, point surface origins, peeling (general superficial pitting), and subcase fatigue (carburized components) (Ref 148).

Although loaded in compression, bearings will fail by rolling-contact fatigue after long life even when the inclusion content is extremely low. The high applied loads alter the surface microstructure, generally by localized strain hardening around a hard oxide inclusion, producing a white-etching region referred to as a butterfly. Figure 67 shows an example of such a white-etching feature in a hardened steel roll. These features never form around sulfide inclusions, which are softer and more ductile than oxides. Butterflies nucleated at spherical primary carbides in M50 bearing steel have also been observed (Ref 155). In some cases, lenticular carbides form at the interface between the butterfly and the matrix, and they can also act as nucleation sites. The hardness of the white-etching butterflies is considerably greater than that of the matrix.

Detailed microstructural examinations were conducted on these white-etching regions in 52100 bearing steel (Ref 146). These areas formed in the subsurface region at a depth of about 0.3 mm (0.01 in.) and were most prominent at a depth of 0.47 to 0.55 mm (0.019 to 0.22 in.). The number and size of these areas increased with bearing service life, while thin lenticular carbides formed at the edges of these areas and thickened with time. Temper carbides were not observed in these white-etching areas, and the amount of proeutectoid carbide decreased. Tempering of specimens containing white-etching zones between 450 and 750 °C (840 and 1380 °F) caused precipitation of carbides in these zones. Electron microscopy (thin foil) examination of white-etching carbide-free areas revealed a fine cellular structure with a cell size of about 0.1 μm.

Another investigation showed that the lenticular carbides bordering white-etching areas in 52100 steel formed by carbon migration from the white-etching region (Ref 150). The white-etching areas form by solutioning of the carbides in these regions. Once the butterfly is well developed, microcracking occurs at the edges of the butterfly where the lenticular carbides are observed. These cracks grow until they reach the surface, producing a spall (Ref 151). In some cases, butterflies have been observed that did not initiate at oxide inclusions. Fatigue cracks in rolling-contact loading are not always preceded by butterfly formation (Ref 148).

The wheel spinning that occurs during the start up of movement by a locomotive can generate frictional heat and stresses in rails when traction is poor. These conditions can lead to spalling, or shelling, on the rail head surface.

An example of such a problem is shown in Fig. 68. The particular steel rail was a 67-kg/m (136-lb/yd) rail with 0.74% C and 0.90% Mn. Top and side views of the spalled regions are shown. X-ray diffraction analysis of the spalled area revealed the presence of iron oxides (Fe_2O_3 and Fe_3O_4) and sand (sand is sometimes used to improve traction). Spinning locomotive wheels can generate enough frictional heat to reaustenitize the surface of a rail head. The mass of the rail acts as a heat sink to provide cooling rates high enough to form as-quenched martensite. Microhardness testing of the white-etching layer found in the spalled region (Fig. 68) revealed a hardness of 60 HRC. The interior hardness was 28 HRC, typical of an as-rolled fully pearlitic rail.

The maximum depth of the white-etching layer was 0.4 mm (0.0175 in.). Surface regions away from the spall (Fig. 68) exhibited a heat-checked crack pattern and scorch marks. The white-etching layer consisted of two zones, the outer, featureless zone (S) and an inner zone with a ferrite network (F). These micrographs

116 / Visual Examination and Light Microscopy

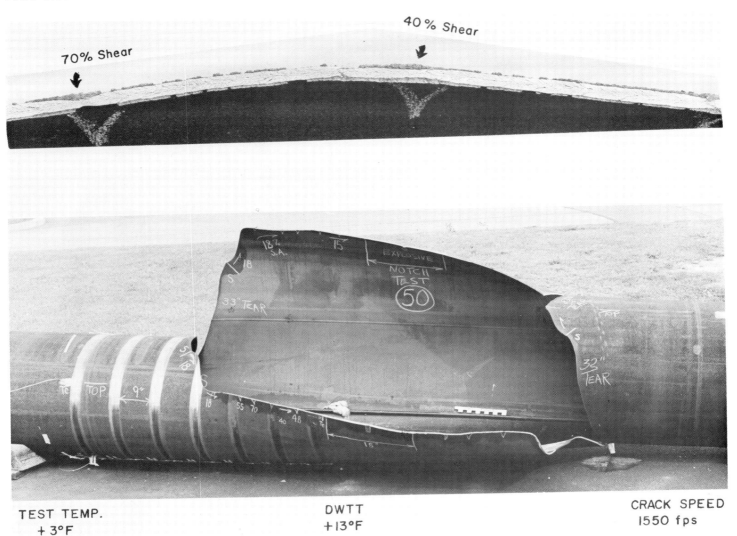

Fig. 45 Fracture of API grade X-60 line pipe tested 6 °C (10 °F) below the 50% shear area drop-weight tear test transition temperature. See also Fig. 43, 44, and 46.

Fig. 46 Fracture of API grade X-60 line pipe tested 22 °C (40 °F) below the 50% shear area drop-weight tear test transition temperature. Temperature given in °F. See also Fig. 42 to 45.

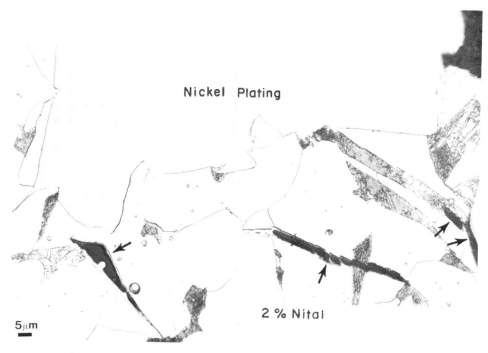

Fig. 47 Cleavage crack path in a ferritic-pearlitic low-carbon steel. Note the subsurface cracks (arrows). One crack has been partially filled by nickel plating. Etched with 2% nital. 1000×

were taken in the heat-checked region near the spall. Beneath the S and F zones is the normal pearlitic rail microstructure. Tempering of the specimen at 540 °C (1000 °F) precipitated spheroidized carbides in the white-etching layer. The distortion of the ferrite in the F zone near the rail head field side (outside) occurred before formation of the white-etching layer because of deformation of the microstructure by heavy wheel loads. This deformed structure was then austenitized by the spinning locomotive wheels. Martensite subsequently formed with cooling. Later, spalling in the white-etching layer occurred because of subsequent train traffic.

The shell-type fracture on the rail head forms on the plane of maximum residual tensile stress and may be aided by the presence of large inclusions or clusters of inclusions at this location. Another type of rail fracture, called a detail fracture, can form as a perturbation from the shell crack under cyclic loading. The detail crack is constrained and exists as an internal flaw during the early stages of its growth because it is impeded at the gage side (inside edge) of the rail and at the rail head by longitudinal compressive stresses. The detail crack forms perpendicular to the shell fracture and is connected to the shell fracture (Fig. 69). Such defects are relatively rare; only about 3½% of 30 000 of all defects were detail cracks (Ref 137). However, they can grow with cyclic loading and can lead to gross fracture of the rail (Fig. 70). Figure 70 shows the fracture face of a rail with a detail crack that was used under controlled conditions at the Facility for Accelerated Service Testing (FAST). The fracture face shows well-defined fatigue marks growing from the detail fracture.

Fatigue fractures are usually transgranular. In the absence of a stress-raiser, fatigue crack nucleation involves slip-plane fracture due to repetitive reversal of the operative slip systems. Studies have demonstrated that slip lines form as in static loading and that some of these will broaden into intense bands with further cycling. Cracks will eventually be observed at these intense slip bands. With high stress levels, the density of such bands is greater, and cracks begin earlier. It has been shown that the critical shear stress law was followed with cyclic loading and that a stress exists below which crack-free slip bands could be formed (Ref 157).

A surface roughening peculiar to fatigue loading has been observed after cyclic loading. Thin slivers of metal are seen to protrude from the surface at some of the dense, cyclically formed slip bands. Close examination reveals that there are intrusions as well as extrusions and that fatigue cracks begin from the intrusions and propagate into the slip band. This phenomenon appears to be a general characteristic of fatigue crack initiation. Indeed, studies have demonstrated that fatigue life can be dramatically improved if the metal surface is periodically removed (only a minor amount of metal need be removed); however, such a practice is not easy to apply commercially (Ref 158).

An examination of the polished surfaces of cyclically deformed copper and nickel specimens revealed slip-band formation (Ref 158). Most of the slip bands were easily removed by electropolishing, but some required more effort. Upon retesting, slip bands formed again at these locations; therefore, the term persistent slip bands was coined to describe these regions of intense plastic deformation. In another study, two-beam interferometry (white light) was used to measure the surface distortion associated with persistent slip bands (Ref 159). Typical bands were about 25 μm wide and about 0.3 μm above the surface, with a number of sharp hills and valleys with heights and depths up to about 5 μm.

Whether a fatigue crack is initiated at a slip band, a second-phase particle, or a stress riser depends on which source is the easiest for nucleation. In a study of a quenched-and-tempered medium-carbon alloy steel, cracks were observed within and around alumina (Al_2O_3) inclusions that initiated fatigue cracks (Ref 160). If the Young's modulus of the particle was greater than that of the matrix, tensile stress concentrated in the particle and caused cracking. Cracking was not observed at sulfide inclusions or cementite particles, because their Young's modulus was less than that of the matrix. Therefore, certain inclusions or precipitates act as fatigue crack initiators, but others do not.

Once nucleated, microcracks must grow to a size that can be detected. Precisely when a microcrack becomes a macrocrack is a matter of definition and of the resolving power of the observation method. The number of microcracks that form is a function of the stress or plastic-strain amplitude. At low stress, that is, near the endurance limit, the growth of a single microcrack to a macrocrack occurs; at higher loads, numerous microcracks form and link up, producing one or more macrocracks.

Macrocrack propagation has been widely studied and shown to be a function of the stress intensity range, ΔK, as defined by linear-elastic fracture mechanics. Macrocrack growth has been subdivided into three regions. In the initial growth stage, there is a critical stress intensity range, ΔK_0, required for crack growth. Once this threshold value has been exceeded, the crack growth rate da/dN increases rapidly with increasing ΔK until a steady-state condition is obtained. In this second stage of crack growth, the Paris relation is followed (Ref 161). Above this region, the maximum stress intensity K_{max} approaches the critical stress intensity for fracture K_c, and final fracture occurs.

Considerable use has been made of special straining stages for in situ direct observation of fatigue crack propagation within the SEM chamber (Ref 162-166). Electron channeling contrast is used to reveal the plastic-strain distribution at the crack tip in conductive metals. Such observations would be much more difficult with light microscopy.

Microscopic examination of fatigue fracture surfaces began with a study by Zapffe and Worden using light microscopy (Ref 167). The

118 / Visual Examination and Light Microscopy

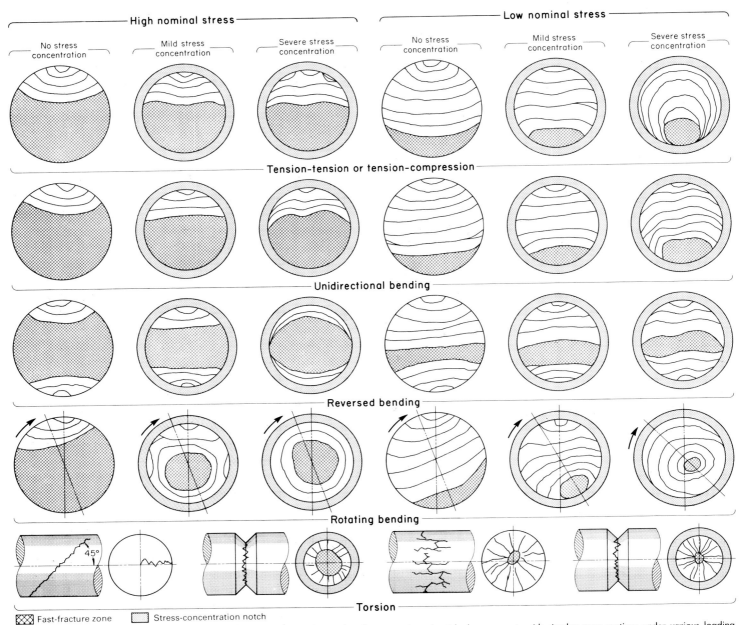

Fig. 48 Schematic representation of fatigue fracture surface marks produced on smooth and notched components with circular cross sections under various loading conditions

restricted depth of field of the light microscope limited such examinations until development of replication procedures using TEM. With the subsequent development of the scanning electron microscope, observation of fatigue fractures has become simpler. Many fractographers, however, still prefer to examine fatigue fractures with replicas because of the excellent image contrast.

The most distinguishing microscopic feature of a fatigue fracture is its striated surface appearance. If these striations are coarse enough, as in low-cycle fatigue fractures (high stress levels), they can be observed by light microscopy as demonstrated in Ref 167. At lower stress levels, however, rather high magnifications (1000 to 20 000×, for example) are required to resolve these fine marks.

Striations have been observed on the surfaces of many fatigue fractures of metals and polymers. They are more easily observed on the surfaces of more ductile metals, especially fcc metals, than on steels. Striations on low-strength steels tend to be wavy in nature and exist in patches rather than over the entire surface. Fatigue striations may be quite difficult to observe on high-strength steel fatigue fractures, depending on the manner of loading, the environment, and so on. Some researchers have been unable to observe striations on such fractures (Ref 168, 169), but others have detected them (Ref 170, 171).

Programmed fatigue loading of fcc metals, such as aluminum, copper, and austenitic stainless steel, has demonstrated that each striation represents crack extension from each load cycle. In service conditions, loads are variable in magnitude, and not all will be of sufficient magnitude to cause crack propagation. Also, after an abrupt change from a high load level to a low load level, there will be a brief period during which the crack does not grow.

The fracture surface during microcrack formation (Stage I) is generally featureless, although rub marks may be observed. The tran-

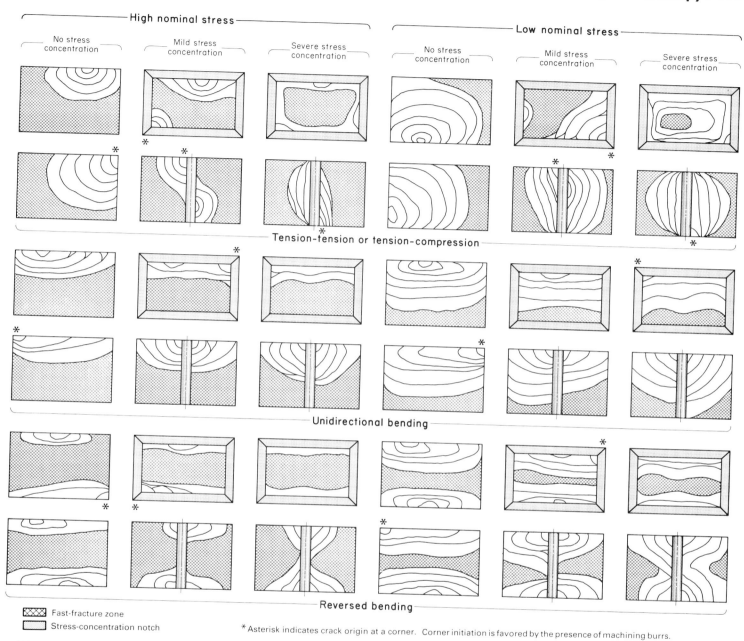

Fig. 49 Schematic representation of fatigue fracture surface marks produced in square and rectangular components and in thick plates under various loading conditions

sition from microcrack formation (Stage I) to macrocrack propagation (Stage II) often occurs at a grain boundary or a triple point. Fatigue striations are observed during Stage II crack growth, but their absence does not prove that the failure was not due to fatigue. Also, other fracture features resembling striations have been mistakenly interpreted as striations.

Two types of fatigue striations, referred to as ductile or brittle striations, have been classified (Ref 172). Ductile striations are the type most commonly observed on fatigue fractures, while brittle striations have been observed on fatigue fractures in corrosion fatigue failures and in hydrogen-embrittled steels.

Although Stage I microcracks are highly sensitive to crystallographic and microstructural features, Stage II macrocrack formation is less dependent on these features. Second-phase particles, such as inclusions, have a complex influence on striations and on the crack growth rate. The mechanism by which inclusions can either inhibit or promote crack propagation has been schematically illustrated (Ref 173). Different inclusion types and morphologies have been shown to influence the fatigue crack growth rate by using different plate steels tested with six possible orientations (Ref 174).

Fracture features that have been mistakenly identified as striations include Wallner lines (Ref 115, 168, 174, 175), slip traces (Ref 115, 168, 176), fractured pearlite lamellae (Ref 41, 115, 168), and rub marks (for example, tire-tracks) (Ref 41, 115, 168, 175). Striationlike marks have also been observed on certain metals not tested under cyclic loading conditions. These problems are discussed in the articles "Modes of Fracture" and "Scanning Electron Microscopy" in this Volume. Fatigue striations are not observed at all crack propagation rates (Ref 177). At very high crack propagation rates (high stress intensities), the fatigue fracture exhibits microvoid coalescence or cleavage, depending on the alloy. As the fatigue crack propagation rate decreases, stria-

Fig. 50 Fatigue failure of a bolt due to unidirectional cycling bending loads. The failure started at the thread root (arrow) and progressed across most of the cross section before final fast fracture. Actual size

Fig. 51 Macrograph of the fracture face of an AISI 4320 drive shaft that failed by fatigue. The failure began at the end of a keyway that was machined without fillets (B) and progressed to final rupture at (C). The final rupture zone is small, indicating that loads were low.

Fig. 52 Two examples of fatigue fractures in AISI 9310 quenched-and-tempered coupling pins caused by reversed cyclic bending loads. Arrows indicate the fracture origins. Actual size

tions are observed. At very low crack growth rates (for example, 2 to 20 × 10^{-6} mm, or 0.07 to 0.7 µin., per cycle), the fracture exhibits an appearance similar to cleavage, even in fcc metals. Facetlike fatigue fractures were observed in aluminum, copper, titanium, and austenitic stainless steels at crack growth rates below about 10^{-5} mm, or 0.3 µin., per cycle (Ref 177). Testing of specimens in a vacuum has revealed conflicting results; some observers have found striations, but others have not. Fatigue testing at high temperatures has revealed many fractures that do not exhibit striations.

Many studies have demonstrated that the striation spacing is a function of the applied stress; that is, the spacing increases with load and crack length. In most cases, light microscopy is unsuitable for such measurements, and electron metallographic techniques are required (see the articles "Scanning Electron Microscopy" and "Transmission Electron Microscopy" in this Volume). Laboratory studies have shown that the crack growth rate determined from striation spacings is within a factor of two of the macroscopic growth rate (Ref 178-183). In many cases, at low crack growth rates, the macroscopic growth rate was less than the microscopic growth rate; at high crack growth rates, the macroscopic rate exceeds the microscopic growth rate. These studies have shown that there is a relationship between the striation spacing and the stress intensity factor range ΔK (Ref 182).

Macroscopic crack growth rates are commonly measured using an optical telescope focused on the growing crack front. Because the crack front is usually not perpendicular to the specimen surfaces, the time to initiation and the crack growth rate measured macroscopically or from striation spacings will usually be somewhat different. A technique has been developed for measuring the crack propagation rate based on the observation that fracture color and roughness vary with the crack propagation rate (Ref 184, 185). The technique is usually applied to test samples using programmed fatigue loading with regularly stepped variable loads. The steps appear in different colors and roughnesses on the fracture surface. A color chart correlating known crack propagation rates with known stresses has been prepared (Ref 186). Very low crack propagation rates produce black coloration, intermediate rates produce gray colors, and high rates produce uniform coloration. Such a procedure is qualitative at best and is limited in application, especially for service failures.

The crack propagation rate and stress intensity can be estimated through measurement of

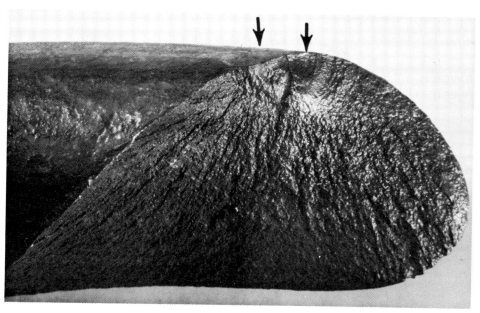

Fig. 53 Torsion failure of an AISI 51B60 railroad spring. The failure began by fatigue at the abraded area at the top (arrows).

the striation spacing. Obviously, this is limited to those fatigue fractures exhibiting striations. If the crack propagates partly by another fracture mechanism, the microscopically determined growth rate will be lower than the macroscopically determined rate. The striation measurement can provide an estimate of the number of stress cycles strong enough to grow the crack during Stage II, but can provide no information about the number of cycles required to initiate and propagate a microcrack.

The problem of fatigue striation measurement is discussed in the article "Quantitative Fractography" in this Volume and in Ref 187. The finest reported striation spacing was about 10 nm (Ref 188). The striation spacing in contiguous locations, even for specimens subjected to reasonably uniform loading, can vary by as much as a factor of five (Ref 189). Consequently, at each location from the crack origin, it is important to measure a large number of striations to ensure that the average value is representative.

Metallographic observation of either completely broken or partially broken fatigued samples is a common procedure, particularly in failure analysis. After macrofractographic and microfractographic examination of the fracture face has been completed, it is common practice to section the specimen so that the microstructure at the fatigue origin can be examined to determine if any microstructural anomaly is present.

Figure 71(a) shows the same fatigue crack in an aluminum alloy as shown in Fig. 16, but after etching. The roughness of the crack pattern and the interaction with the intermetallic particles are evident. Figure 71(b) shows a similar fatigue crack in the same aluminum alloy after etching; but in this case, the specimen was completely broken, and only one side of the fracture can be observed. Comparison of Fig. 71(a) and (b) clearly shows that the nature of the crack pattern is much easier to observe in a partially broken specimen where both sides of the fracture can be observed. Figure 72 shows a fatigue crack in a ferrite-pearlite carbon steel after etching in which both sides of the crack can be observed. The irregular path of the crack through the ferrite phase is evident in Fig. 72 as compared to the nature of cleavage cracks in ferrite shown in Fig. 47.

As a final note on the microscopic aspects of fatigue, the use of DIC illumination to study the progression of slip during cyclic loading will be briefly discussed. The sensitivity of DIC for revealing the progression of slip and microcrack formation has been demonstrated by using polished specimens of Nickel 200 examined after cyclic loading (Ref 190). When the Wollaston prism is fully inserted to produce a dark blue-green image, the surface deformation from slip is vividly revealed. Slip can also be observed by using bright-field illumination (Fig. 73), but DIC is more sensitive, particularly in the early stages of slip.

High-Temperature Fractures

As service or test temperatures are increased, the strength of metals and alloys decreases while ductility increases, although various embrittlement phenomena may be encountered. At elevated temperatures, the strength will vary with strain rate and exposure time. Also, microstructures will change with high-temperature exposure. Visual and light microscope examination are very useful for examining high-temperature fractures and for observing changes in microstructure. As with the other fracture mechanisms, electron metallographic procedures complement these techniques and are necessary tools.

The simplest measure of high-temperature properties is provided by the use of short-time tensile tests. Hot hardness tests are also quite useful. Figure 74 shows fractures of short-term high-temperature tensile specimens of type 316 stainless steel tested at 760 to 980 °C (1400 to 1800 °F). The reduction of area of the tensile fractures shown increases with temperature. Cup-and-cone fractures are not observed, however. The microstructures shown illustrate the formation of voids at the grain boundaries, which is typical of high-temperature tests. Precipitation of grain-boundary carbides (sensitization) can be clearly seen in the high-magnification view of the specimen tested at 815 °C (1500 °F). This is an example of a change in microstructure caused by high-temperature exposure.

Because metals expand with high-temperature exposure, creep tests are widely performed on alloys used at high temperature to assess the magnitude of the length change as a function of load and temperature. The creep rate will increase with load and temperature. Tests conducted at relatively high loads until fracture occurs are referred to as stress rupture tests. In this test, the time to rupture at a given load and temperature is assessed, rather than the creep rate. Creep tests are conducted at relatively low loads at which fracture may not occur for many years.

The fracture path for metals and alloys tested at high temperatures changes from transgranular to intergranular with increasing temperature (Ref 191). Transgranular fracture occurs at low temperatures, at which the slip planes are weaker than the grain boundaries. At high temperatures, the grain boundaries are weaker, and fracture is intergranular. Such observations have led to the introduction of the equicohesive temperature concept to define the temperature at which the grains and grain boundaries exhibit equal strength; that is, the temperature at which the fracture mode changes from transgranular to intergranular (Ref 192, 193). The equicohesive temperature is not a fixed temperature, but varies with the stress and strain rate for a given composition. Above the equicohesive temperature, coarse-grain specimens exhibit greater strength than fine-grain specimens because of the lower grain-boundary surface area. Transgranular fractures have been observed in high-purity metals at rather high temperatures.

Tests at temperatures above the equicohesive temperature have revealed two types of intergranular fracture. When grain-boundary sliding occurs, wedge-shaped cracks may form at grain-boundary triple points (as observed on a plane-of-polish) if the tensile stresses normal to the boundaries exceed the boundary cohesive strength. In the literature, these cracks are referred to as wedge or w-type, or as triple point or grain-corner, cracks. High stresses

122 / Visual Examination and Light Microscopy

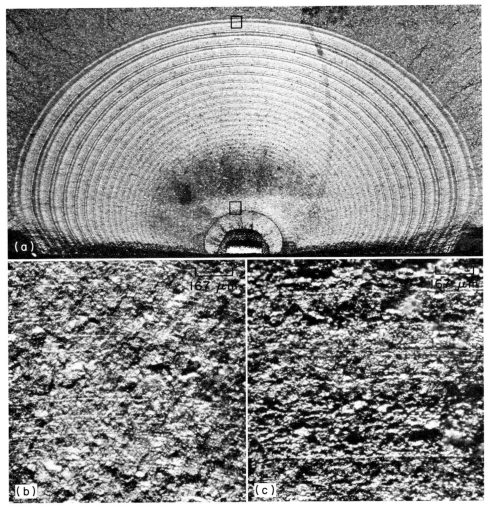

Fig. 54 Three views of a fatigue fracture in D6AC steel plate, showing beach marks. (a) Plate subjected to a series of varied loading cycles in the laboratory. The crack origin, at the bottom center, was at a starter notch formed by electrical discharge machining. (b) Area in lower square in (a), just above beach mark registering the first change in loading. (c) Area in upper square in (a), just ahead of the zone of final fast fracture. Both (b) and (c) contain faintly resolved, fine beach marks. (a) About 5×. (b) and (c) 60×. Source: Ref 124

promote this type of crack formation. One of the ways in which such cracks can form is shown in Fig. 75.

Under low-stress conditions, intergranular fractures occur by void formation at the grain boundaries (Ref 195, 196). These cavities form along grain edges rather than at grain corners. Because they appear to be round or spherical on metallographic cross sections, these voids are sometimes referred to as r-type cavities. Subsequent studies have shown that voids of more complex shapes, such as polyhedra, can be formed during cavitation. Inclusions and precipitates on grain boundaries can act as sites for void nucleation. Cavity nucleation occurs during the initial (primary) stage of creep (Ref 197). These cavities grow during the second stage (steady-state) of creep. The third (tertiary) stage of creep begins when these cavities grow to such an extent that their size and spacing are approximately the same as the grain size.

Qualitative observations of grain-boundary sliding during high-temperature tests were first made in 1913 (Ref 191). Since then, similar observations have been made for many metals under creep conditions. Quantitative measurements have been made using bicrystals and polycrystals with a variety of methods (Ref 198-211). These measurements, primarily by light microscopy, have been useful in determining the contribution that sliding makes in the overall creep extension and in understanding creep mechanisms.

Measurements of grain-boundary sliding are made on grain boundaries that intersect the specimen surface or on boundaries in the specimen interior, using either bicrystals or polycrystalline specimens. Most of the procedures fall into one of the following three types:

- Lines or grids are scribed on the specimen surface, and the displacements are measured either parallel or perpendicular to the stress axis
- Interferometry is used to measure the vertical displacement produced during sliding; alternatively, profile measurements can be made on cross sections
- The change in grain shape of the grains in the specimen interior is statistically determined by using quantitative metallographic procedures

The displacement measurements are used to calculate the contribution of grain-boundary sliding to the overall extension during creep.

Cavity formation, density, and orientations have also been quantified by using a number of procedures (Ref 212-221). These studies have demonstrated that the cavities form preferentially on grain boundaries oriented approximately perpendicular to the applied stress. Creep cavity size, shape, and density can be measured at various stages during creep testing by using standard quantitative metallographic methods. The average cavity diameter increases with time, temperature, and strain rate.

Measurements have also been made of the angular distribution of cavitated grain boundaries with respect to the stress axis, showing that the cavitated grain boundaries are oriented between 60 and 90° to the stress axis. At low strain rates, the most frequently observed cavities are on boundaries perpendicular to the stress axis. With increasing strain rates, the most frequently observed cavitated grain boundaries shift toward a 45° angle to the stress axis. These measurements are usually made on longitudinal sections cut from creep test specimens. The angles between the cavitated grain boundary and the stress axis on the sectioning plane are measured, and the distribution is statistically analyzed by using the procedure discussed in Ref 221. Metallographic studies of high-temperature test specimens and fractures have used macrophotography, optical microscopy, x-ray techniques, microradiography, TEM, and SEM, as demonstrated in the cited references and in Ref 222-232.

Figure 76 illustrates r-type cavities at grain boundaries in type 316 stainless steel tested under creep conditions. The SEM fractograph shows the r-type cavities on the fracture surface, while the etched cross sections reveal the cavities at and behind the fracture face and their relationship to the grain boundaries. Figure 77 shows w-type cracks in type 316L stainless steel tested under creep conditions. The w-type cracks and the orientation to the stress axis (vertical in the micrograph) are easily observed in the light micrograph, but are not obvious in the SEM fractograph. Figure 78 shows grain-boundary triple-point cracking and extensive deformation in type 316 stainless steel tested at a slightly lower temperature. It is interesting to compare the three SEM fractographs in Fig. 76 to 78. Figure 79 shows crack nucleation in a type 316 stainless steel specimen with heavy carbide precipitation. Cracking has occurred

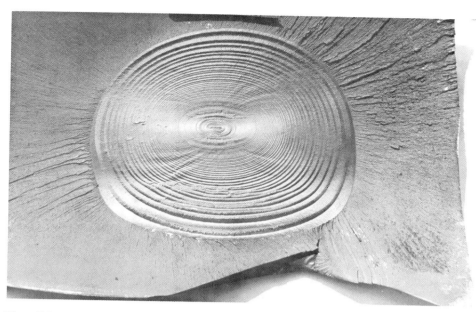

Fig. 55 Circular spall that began at a large subsurface inclusion in a hardened steel roll

along the grain-boundary carbides, and the SEM fractography reveals a heavy concentration of carbides on the intergranular fracture surface.

Because of the economic importance of creep in high-temperature service, particularly in power generation equipment, considerable emphasis has been placed on predicting the remaining life of components (Ref 233-238). This work has involved metallographic examination of the creep damage, including field metallographic procedures (Ref 239-243). Such predictions must also take into consideration the changes in microstructure that occur during the extended high-temperature exposure of metals and alloys (Ref 244-249).

Embrittlement Phenomena

The expected deformation and fracture processes can be altered by various embrittlement phenomena. These problems can arise as a result of impurity elements (gaseous, metallic, or nonmetallic), temperature, irradiation, contact with liquids, or combinations of these or other factors. Metals can become embrittled during fabrication, heat treatment, or service. If the degree of embrittlement is severe enough for the particular service conditions, premature failures will result. Some of these problems introduce rather distinctive features that may be observed by macro- or microscopic fractographic methods, and the ability to categorize these problems properly is imperative for determining cause and for selecting the proper corrective action.

It is well recognized that many metals, such as iron (Ref 250-257), are embrittled by high levels of oxygen, nitrogen, phosphorus, sulfur, and hydrogen. Of these elements, the influence of oxygen on the intergranular brittleness of iron has produced the most conflicting test results. For example, in one investigation a series of iron-oxygen alloys with up to 0.27% O was tested, and intergranular fractures were observed in all but the lowest (0.001%) oxygen sample (Ref 251). On the other hand, in a study of high-purity iron and electrolytic iron, no influence of oxygen content (up to 2000 ppm) was observed on the ductile-to-brittle transition temperature (Ref 256). Increasing the carbon content to about 40 ppm decreased the ductile-to-brittle transition temperature and decreased the intergranular brittleness, irrespective of oxygen content.

Other bcc metals, such as molybdenum, chromium, and tungsten, are embrittled by oxygen, nitrogen, and carbon (Ref 250, 258, 259). When embrittled, the fractures of these metals are intergranular. Face-centered cubic metals may also be embrittled by oxygen (Ref 260, 261) and sulfur (Ref 262-265). For example, in a study of the grain-boundary embrittlement of intermetallics with a stoichiometric excess of active metal component, the extreme brittleness of these materials was shown to be due to grain-boundary hardening through absorption of gaseous impurities (oxygen and/or nitrogen) segregated to the grain-boundary areas (Ref 266).

Metallography and fractography have played important roles in developing an understanding of embrittlement mechanisms. For example, early work on the embrittlement of copper by bismuth attributed the embrittlement to the presence of thin grain-boundary films of elemental bismuth (Ref 267, 268). However, careful metallographic preparation and examination of copper containing low amounts of bismuth (up to 0.015%) showed that the apparent films were actually steplike grooves at the grain boundaries (Ref 269). These grooves were not observed after either mechanical or electrolytic polishing, but were visible after etching. In another study, copper containing up to 4.68% Bi was tested, and the results were similar to those discussed in Ref 269; however, in alloys with high bismuth contents, either continuous grain-boundary films or discrete particles of bismuth with a lenticular shape were observed. Studies of the embrittlement of copper by antimony revealed results similar to that of the low-bismuth alloy (Ref 271, 272); that is, grain-boundary grooves, rather than discrete films, were observed after etching. The embrittled specimens fractured intergranularly.

The influence of impurity elements on the hot workability of metals is well known. Copper will be embrittled during hot working in the presence of bismuth, lead, sulfur, selenium, tellurium, or antimony (Ref 273). Lead and bismuth also degrade the hot workability of brass (Ref 273).

The hot workability of steels is degraded by sulfur (Ref 274-280) and by residual copper and tin (Ref 281-284). Sulfides have also caused intergranular cracking in alloy steel castings (Ref 285). Poor hot workability is also a problem with free-machining steels containing lead and tellurium (Ref 286). Residuals such as lead, tin, bismuth, and tellurium can cause hot cracking during hot working of stainless steels (Ref 287, 288), and residual elements such as sulfur, phosphorus, bismuth, lead, tellurium, selenium, and thallium are detrimental to nickel-base superalloys (Ref 289-291). Excessive precipitation of aluminum nitride can cause cracking in steel castings and during hot working (Ref 292-303).

Certain materials are inherently brittle because of their crystal structure, microstructure, or both. For example, gray cast iron is an inherently brittle material because of the weakness of the nearly continuous graphite phase. However, if the graphite exists in isolated, spherical particles, as in nodular cast iron, excellent ductility can be obtained. Grain-boundary cementite films in high-carbon or carburized steels produce extreme brittleness, but if the same amount of cementite exists as discrete spheroidized particles, ductility is good. As-quenched high-carbon martensite is quite brittle, but tempering improves the ductility, although at a sacrifice in strength. The normally ductile austenitic stainless steels can be embrittled by the formation of hcp ε-martensite during service (Ref 304-306).

Numerous types of embrittlement phenomena can occur in certain metals and alloys or under certain environmental conditions. These problems can be traced to compositional or manufacturing problems and/or service conditions. The more familiar embrittlement problems and their fractographic characteristics are summarized below.

Creep-Rupture Embrittlement. Under creep conditions, embrittlement can occur and result in abnormally low rupture ductility. This

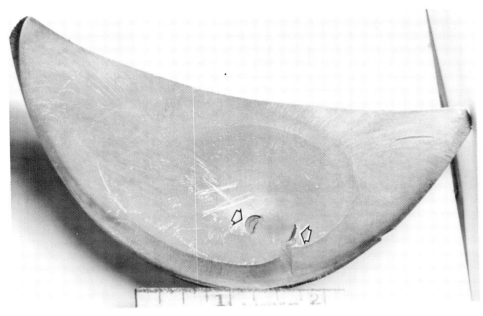

Fig. 56 Circular spall from the shoulder of a forged, hardened alloy steel mill roll with two small fatigue origins. A subsurface crack connected the two fatigue zones. Note that the spall surface is relatively featureless. (a) About actual size. (b) 5.5×

Graphitization. In the early 1940s, several failures of welded joints in high-pressure steam lines occurred because of graphite formation in the region of the weld heat-affected zone (HAZ) that had been heated during welding to the critical temperature of the steel (Ref 316-320). Extensive surveys of carbon and carbon-molybdenum steel samples removed from various types of petroleum-refining equipment revealed graphite in about one-third of the 554 samples tested (Ref 316, 319). Generally, graphite formation did not occur until about 40 000 h or longer at temperatures from 455 to 595 °C (850 to 1100 °F). Aluminum-killed carbon steels were susceptible, but silicon-killed or low-aluminum killed carbon steels were immune to graphitization. The C-0.5Mo steels were more resistant to graphitization than the carbon steels, but were similarly influenced by the manner of deoxidation. Chromium additions and stress relieving at 650 °C (1200 °F) both retarded graphitization.

Hydrogen Embrittlement. Hydrogen is known to cause various problems in many metals, most notably in steels, aluminum, nickel, and titanium alloys (Ref 321-332). Various forms of hydrogen-related problems have been observed.

- Blistering, porosity, or cracking during processing due to the lack of solubility during cooling of supersaturated material, or by cathodic charging, or other processes that form high-pressure gas bubbles
- Adsorption or absorption of hydrogen at the surface of metals in a hydrogen-rich environment producing embrittlement or cracking
- Embrittlement due to hydride formation
- Embrittlement due to the interaction of hydrogen with impurities or alloying elements

The problem of hydrogen effects in steels has been thoroughly studied. Hydrogen embrittlement is most noticeable at low strain rates and at ambient temperatures. A unique aspect of hydrogen embrittlement is the delayed nature of the failures; that is, after a specimen is charged with hydrogen, fracture does not occur instantly but only after the passage of a certain amount of time. Therefore, some researchers have used the term static fatigue to describe the phenomenon. However, this term is misleading. Tensile and bend tests have historically been used to detect and quantify the degree of embrittlement. For example, in tensile testing, it is common practice to compare the normal tensile ductility—the %RA—with the %RA in the presence of hydrogen in order to calculate an embrittlement index E showing the loss in reduction of area:

$$E = \frac{(\%RA)u - (\%RA)c}{(\%RA)u} \qquad \text{(Eq 2)}$$

where u and c indicate unchanged and changed, respectively.

problem has been encountered in aluminum (Ref 307) and steels (Ref 308-315). Iron, in amounts above the solubility limit in aluminum, has been shown to cause creep-rupture embrittlement by development of intergranular cracking (Ref 307).

The creep embrittlement of chromium-molybdenum steels has been extensively studied. Matrix precipitation strengthening has been shown to cause creep embrittlement (Ref 308). Also, coarse-grain areas in 2.25Cr-1Mo welds have been found to exhibit much lower creep ductility than fine-grain weldments (Ref 312). Impurities such as phosphorus, sulfur, copper, arsenic, antimony and tin have been shown to reduce rupture ductility, although rupture life increases. This behavior appears to be due to the grain-boundary segregants blocking grain-boundary diffusion, which reduces the cavity growth rate. High impurity contents increase the density of the cavities. Substantial intergranular cracking is observed in high-impurity material and is absent in low-impurity heats (Ref 313).

Visual Examination and Light Microscopy / 125

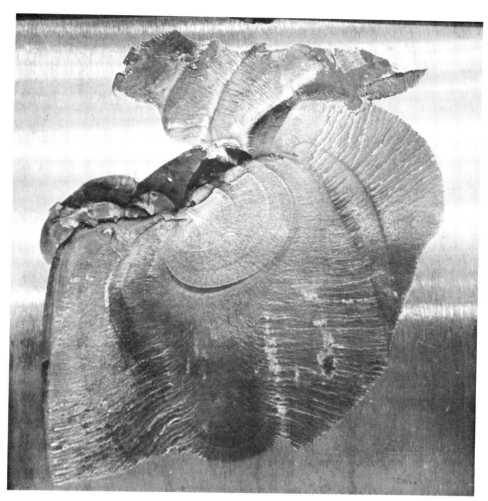

Fig. 57 Circular spall on the surface of a forged, hardened alloy steel mill roll. The arrow indicates the fracture origin. Note the fatigue marks showing the growth away from the origin, followed by brittle fracture. 0.68×

Hydrogen flaking is a well-known problem in the processing of high-carbon and alloy steels (Ref 333-339). Whether or not flaking occurs depends on several factors, such as steel composition, hydrogen content, strength, and thickness. Steels prone to flaking are made either by vacuum degassing to reduce hydrogen to a safe level or by using controlled cooling cycles. High-carbon steels are particularly susceptible to flaking. Therefore, rail steel is control cooled slowly after rolling to prevent flakes (Ref 340). When the hydrogen content is not properly controlled, flakes result (Fig. 80).

Flaking can occur in a wide variety of steels. Low-carbon steels appear to be relatively immune to flaking, but alloy steels, particularly those containing substantial nickel, chromium, or molybdenum, are quite susceptible. In general, as the strength of the material increases, less hydrogen can be tolerated. The number of flakes formed has been shown to be related to the cooling rate after hot working (Ref 338). Manufacturers have observed that if the inclusion content is quite low, flaking can occur at hydrogen levels normally considered to be safe.

Rail producers have found that the cooling cycles used in the past to prevent flaking are inadequate for this purpose when inclusion contents are very low. The flakes in such steels do not exhibit the classic appearance shown in Fig. 80.

In addition to flaking, blisters can be produced by excess hydrogen (Fig. 81). Hydrogen can also be introduced into steels during other processes, such as welding, pickling, bluing, enameling, or electroplating. Consequently, it is necessary to bake the material after such processes if the material is prone to hydrogen damage.

The influence of hydrogen on fracture appearance is complex (Ref 341-348). Studies have shown that cracks propagate discontinuously, suggesting that the crack growth rate is controlled by the diffusion of hydrogen to the triaxially stressed region ahead of the crack tip. The fracture appearance is influenced by the strength of the material. As the strength level increases, fractures are more intergranular. Impurities also influence fracture mode. For example, at low impurity levels, hydrogen-induced cracking was shown to occur by cleavage at very high stress intensities (Ref 344). At high impurity levels (grain-boundary impurities), the fracture path was intergranular, and the stress intensity required for crack growth decreased. In another study, tempering between 350 and 450 °C (660 and 840 °F) produced entirely intergranular fractures (Ref 345). Therefore, the fracture mode will be influenced by the cooperative action of temper embrittlement and hydrogen embrittlement. A common fracture feature in hydrogen-embrittled low-strength steels is flat fracture regions (100 to 200 μm in size) that are circular or elongated and centered around an inclusion or a cluster of inclusions (Ref 348). These zones are transgranular.

An investigation of crack nucleation and growth in a low-carbon ferrite-pearlite steel tested in both hydrogen and oxygen environments revealed that crack growth rates were faster in hydrogen than in oxygen and were faster when the crack growth direction was in the hot-working direction (Ref 346). Crack initiation in hydrogen was not sensitive to orientation. The specimens tested in hydrogen exhibited cleavagelike fractures with a small amount of ductile microvoids. Plastic deformation occurred near the crack tip before crack growth in hydrogen, producing voids at inclusions ahead of the crack. When the crack propagated, these voids were linked to the fracture by cracking of the matrix perpendicular to the maximum normal stress. The fracture path was transgranular with no preference, or aversion, for any microstructural feature.

Hydrogen also influences ductile fracture (Ref 349, 350). For example, a study of spheroidized carbon steels (0.16 and 0.79% C) found no significant influence of hydrogen on the initiation of voids at carbides or on the early eutectoid growth of the voids before linking (Ref 349). In the eutectoid steel, hydrogen exposure increased the dimple size and assisted void growth during linking. In the low-carbon steel, flat quasi-cleavagelike facets were observed, and the void size decreased because of hydrogen exposure. An investigation of low- and medium-carbon spheroidized steels concluded that hydrogen charging promoted void nucleation at carbides and accelerated void growth, particularly for carbides at grain or subgrain boundaries (Ref 350). Void growth acceleration was greatest in the latter stage of void growth. Quasi-cleavage facets were observed around inclusions in steels with high inclusion contents.

The influence of inclusions, particularly sulfides, on hydrogen embrittlement has been demonstrated (Ref 351-355). In one investigation of hydrogen-embrittled ultrahigh strength steels, for example, cleavage areas were observed around nonmetallic inclusions (Ref 353). Increasing the sulfur content reduced hydrogen embrittlement under certain test conditions. Sample size also influenced results. Oxides with low coefficients of thermal expansion, such as silicates, are detrimental under

126 / Visual Examination and Light Microscopy

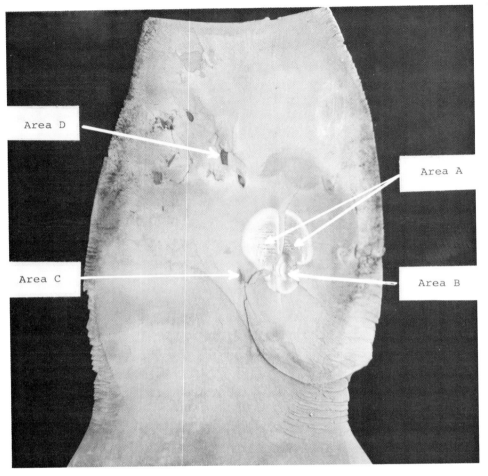

Fig. 58 Spalled section from a forged, hardened alloy steel mill roll showing three regions with fatigue beach marks. The large, shiny region appears to be the origin. See Fig. 59 and 60 for close-up views of areas A, B, C, and D. 0.73×

hydrogen charging conditions; sulfides, with high coefficients of thermal expansion, were harmless or beneficial (Ref 354). Microcracks were commonly observed at oxides but not at sulfides. Another study found that the critical concentration of hydrogen for cracking increased with increasing sulfur content and that embrittlement increased with increasing oxygen content, that is, oxide inclusions (Ref 355).

Intergranular Corrosion. Typically, metallic corrosion occurs uniformly; however, under certain conditions, the attack is localized at the grain boundaries, producing intergranular corrosion.

High-strength precipitation-hardened aluminum alloys are susceptible to intergranular corrosion. Aluminum-copper and aluminum-magnesium alloys are susceptible to intergranular corrosion in certain environments. Die-cast zinc-aluminum alloys can fail by intergranular corrosion in steam or salt water.

Austenitic stainless steels are known to be susceptible to intergranular corrosion in environments that are normally harmless if the material has been subjected to a sensitization treatment (Ref 356-361). Exposure to temperatures in the range of 480 to 815 °C (900 to 1500 °F), either isothermally or by slow cooling through this range, precipitates $M_{23}C_6$ carbides in the grain boundaries, thus sensitizing the alloy to intergranular corrosion. Ferritic stainless steels are also susceptible (Ref 362, 363). Corrosion occurs in the matrix adjacent to the sensitized grain boundary because of depletion of chromium, as demonstrated by analytical electron microscopy (Ref 364-366). The metallographic examination of specimens electrolytically etched with a 10% aqueous oxalic acid solution, as defined by Practice A of ASTM A 262, is widely used as a screening test for sensitization (Ref 367-370).

Figure 82 shows two views of an austenitic stainless steel exhibiting intergranular corrosion. Figure 82(a) shows the surface of the sample. The grain structure is visible due to the attack, and some grains have fallen out (referred to as grain dropping). The cross-sectional view (Fig. 82b) shows the depth of penetration of the attack along the grain boundaries.

Figure 83 shows the fracture of sensitized type 304 austenitic stainless steel broken in a noncorrosive environment. The micrograph of the partially broken specimen shows that the fracture is not totally intergranular. There are microvoids in the matrix beside and ahead of the crack. These often nucleate at large carbides. In the completely broken specimen, shown by SEM and with a cross section, there is little evidence for intergranular fracture in the traditional form. There are some tendencies, but the fracture surface is covered with fine dimples that form around the carbides.

Liquid-metal embrittlement (LME) is a phenomenon in which the ductility or fracture stress of a solid metal is reduced by exposure of the surface to a particular liquid metal (Ref 371-376). The phenomenon was first observed in 1914 in experiments where β-brass disintegrated intergranularly in liquid mercury. Separation is so complete that various investigators have used this process to study the three-dimension characteristics of grains. The study of LME is complicated by the existence of at least four forms of the phenomenon:

- Instantaneous fracture of a particular metal under applied or residual tensile stresses when in contact with specific liquid metals
- Delayed fracture of a particular metal in contact with a specific liquid metal after a time interval at a static load below the tensile strength of the metal
- Grain-boundary penetration of a particular solid metal by a specific liquid metal; stress does not appear to be required in all instances
- High-temperature corrosion of a solid metal by a liquid metal causing embrittlement

The first type is the classic, most common form of LME; the second type is observed in steels. Many metals and alloys are known to fail by LME when in contact with some particular liquid metal.

If the solid metal being embrittled is not notch-sensitive, as in fcc metals, the crack will propagate only when the liquid metal feeds the crack. However, in a notch-sensitive metal, as in bcc metals, the crack, once nucleated, can become unstable and propagate ahead of the liquid metal. Crack propagation rates during LME can be quite fast. Generally, the cracks are intergranular, although a few cases of transgranular fractures have been reported.

Figure 84 shows a metallographic cross section of eutectoid rail steel embrittled by liquid copper and an SEM view of the fracture surface. Tensile specimens were heated to 2110 °F (1100 °C) and loaded at 12.5 to 50% of the normal tensile strength at this temperature. Liquid copper was present at the base of a V-notch machined into the specimen. Fracture occurred in a few seconds at 50% of the tensile strength (SEM view of this sample), and the time to fracture increased with decreasing load.

Hot shortness in steels can be caused by copper segregation in steels alloyed with copper. Figure 85 shows an example of hot shortness in a structural steel section in a copper-containing grade. At the rolling temperature, the segregated, elemental copper was molten. When the section was rolled, it broke up

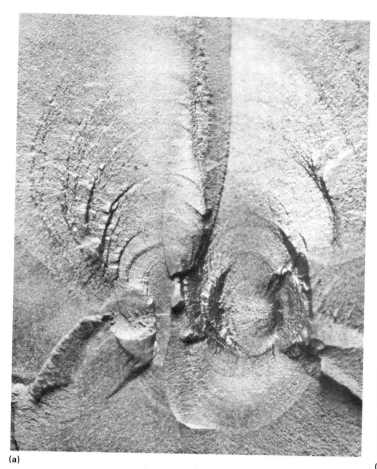 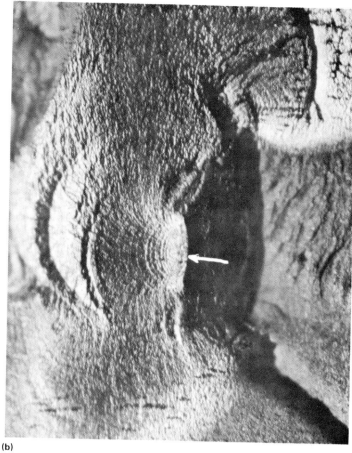

Fig. 59 Close-ups of the primary initiation site of the spall shown in Fig. 58. (a) Area A. 4×. (b) Area B. 8×. See also Fig. 60.

because the liquid copper wetted the austenite grain boundaries.

Neutron irradiation of nuclear reactor components causes a significant increase in the ductile-to-brittle transition temperature in ferritic alloys (Ref 377-385). The degree of irradiation-induced embrittlement depends on the neutron dose, neutron spectrum, irradiation temperature, steel composition, and heat treatment. Tempered martensite is less susceptible to embrittlement than tempered bainite or ferrite-pearlite microstructures (Ref 379). Impurity elements in steels can influence embrittlement; for example, phosphorus levels above 0.015% and copper levels above 0.05% are detrimental.

Radiation produces swelling and void formation following a power law dependent on fluence. Void density decreases as irradiation temperature increases, but the average void size increases. Examination of radiation-induced voids requires thin-foil TEM. Examination of fractures of irradiated ferritic materials tested at low temperatures reveals a change from cleavage fracture to a mixture of cleavage and intergranular fractures. Fractures of specimens tested at higher temperatures reveal a change in dimple size and depth. Irradiation embrittlement in austenitic stainless steels is associated with grain-boundary fracture processes, particularly for deformation at temperature above 550 °C (1020 °F). In one study, neutron irradiation of annealed aluminum alloy 1100 at high fluences at about 323 K caused a large increase in strength, a large decrease in ductility, and intergranular fracture at 478 K (Ref 385).

Overheating occurs when steels are heated at excessively high temperatures prior to hot working (Ref 386-396). Overheated steels may exhibit reduced toughness and ductility as well as intergranular fractures. The faceted grain boundaries exhibit fine ductile dimples because of reprecipitation of fine manganese sulfides at the austenite grain boundaries present during the high-temperature exposure. Heating at temperatures above 1150 °C (2100 °F) causes sulfides to dissolve, with the amount dissolved increasing as the temperature increases above 1150 °C (2100 °F).

Burning occurs at higher temperatures (generally above about 1370 °C, or 2500 °F) at which incipient melting occurs at the grain boundaries. Hot working after burning will not repair the damage. In the case of overheating, facet formation can be suppressed if adequate hot reduction, usually greater than 25% reduction, is performed. In such cases, there will be little or no change in toughness, but there may be some loss in tensile ductility. The cooling rate after overheating influences the critical overheating temperature. Low-sulfur steels are more susceptible to overheating than high-sulfur steels.

Fracture tests have been widely used to reveal facets indicative of overheating (Ref 386). The heat treatment used after overheating has a pronounced influence on the ability to reveal the facets. Quench-and-temper treatments are required. It has been shown that facets are best revealed when the sample is quenched and tempered to a hardness of 302 to 341 HB (Ref 386). The size of the plastic zone at the crack tip during fracturing of the testpiece is influenced by the yield strength of the specimen, which in turn controls the nature of the fracture surface in an overheated specimen (Ref 392). Progressively higher tempering temperatures increase the size of the plastic zone, which enhances the ability of the crack to follow the prior-austenite grain boundaries. However, if the sample is highly tempered, it is more difficult to fracture. Therefore, the optimum tempering temperature is one that permits the crack to follow the austenite grain boundaries to reveal faceting while still permitting the specimen to be broken easily. Facets are more easily observed on impact specimen fractures than on

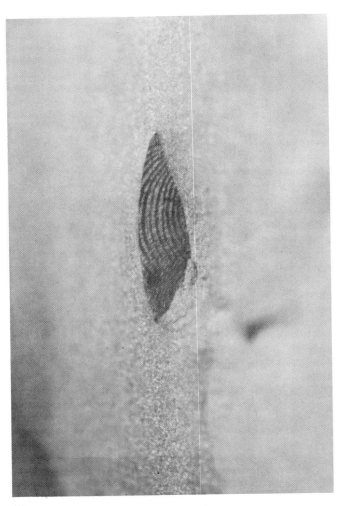 

Fig. 60 Close-ups of two regions of the spall surface shown in Fig. 58 that exhibited fatigue beach marks. (a) Area C. 10×. (b) Area D. 9×. See also Fig. 59.

tensile fractures. For impact specimens, facets are best observed when the test temperature is above the brittle-to-ductile transition temperature. Faceting is generally easier to observe when the fracture plane is transverse to the rolling direction rather than parallel to it.

Metallographers have made considerable use of special etchants to reveal overheating in suspected samples (Ref 9, 387, 393). One study documents the evaluation of over 300 different etchants in an effort to develop this technique (Ref 387). Several etchants have been found to produce different etch responses between the matrix and the grain-boundary area in overheated specimens. These procedures work reasonably well for severely overheated specimens, but are not as sensitive as the fracture test when the degree of overheating is minor, particularly in the case of low-sulfur steels.

Figure 86 shows a section through the center of an alloy steel compressor disk that cracked during forging because of overheating. A specimen was cut from the disk and normalized, quenched and tempered (to a hardness of 321 to 341 HB), and fractured, revealing facets indicative of overheating (Fig. 86). Figure 87, which shows another example of facets, illustrates the fracture of a vanadium-niobium plate steel slab due to overheating. The accompanying micrograph shows that substantial carbon segregation and grain growth were present in the overheated region.

Another investigation showed interesting features of fractures of overheated ASTM A508 class II forging steel (Ref 390). The number of facets per square inch of test fracture depended on both the overheating temperature and the tempering temperature for the fracture test (also shown in Ref 392). For a given soak temperature, the number of facets per square inch decreased as the tempered hardness decreased. For the same tempered hardness, the facet density increased as the soaking temperature increased. Figure 88 shows a series of test fractures of ASTM A508 class II material soaked at 1205 to 1370 °C (2200 to 2500 °F) and then quenched and tempered to a hardness of 37 to 39 HRC. Figure 89 shows SEM views of typical facets in samples soaked at 1205 and 1370 °C (2200 and 2500 °F). The sulfides in the dimples in the 1370-°C (2500-°F) specimen are clearly visible, but those in the 1205-°C (2200-°F) specimen are extremely fine. This difference arises from the greater dissolution of the original sulfides at the higher temperature. Because more of the sulfides are dissolved at higher temperatures, more are available for reprecipitation in the austenite grain boundaries upon cooling.

The number of facets in these specimens varied from 4/cm^2 (24/in.^{2}) at 1205 °C (2200 °F) to 121/cm^2 (783/in.^{2}) at 1370 °C (2500 °F). The facet size also increased with temperature because of grain growth at the soaking temperature. The facet fracture appearance has been referred to as intergranular microvoid coalescence due to the presence of dimples on the intergranular surfaces. Although some early researchers tried to study the facets using light fractography, subsequent investigators have used electron metallographic techniques.

Overheating will not occur if the sulfur content is below 0.001%. Increasing the sulfur content raises the temperature at which overheating begins. Low-sulfur steels are more

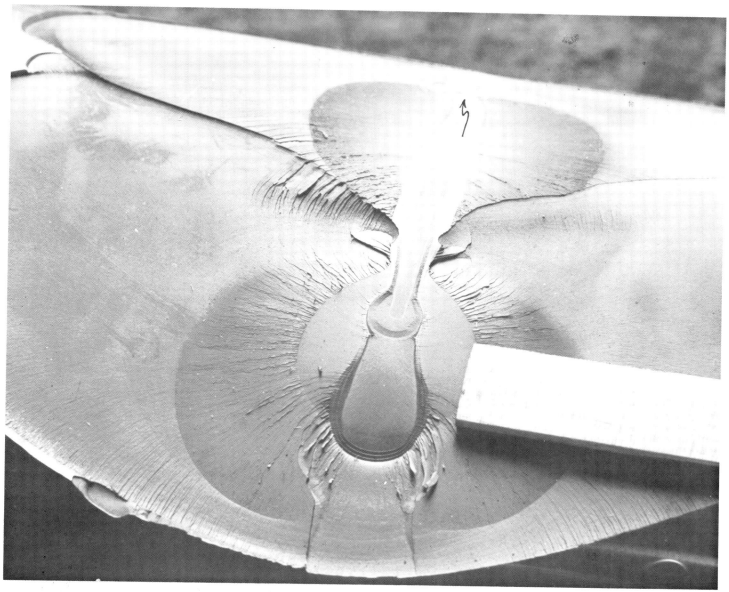

Fig. 61 Example of a line spall in a forged, hardened alloy steel mill roll. Note that the portion that spalled off broke as two sections, with the fractures propagating from the two circular fatigue regions growing from the line spall. The arrow at the top indicates that the spall continued under the unbroken roll surface. Ultrasonic examination showed that it progressed another 137 mm (5⅜ in.) under the surface at a maximum depth of 11 mm (⁷⁄₁₆ in.). The subsurface, unspalled portion of the roll is shown in Fig. 63; Fig. 62 is another view of the line spall.

prone to overheating problems than high-sulfur steels (Ref 393-396). Rare-earth additions raise the overheating temperature by reducing the solubility of the sulfides.

Quench Aging and Strain Aging. If a low-carbon steel is heated to temperatures immediately below the lower critical temperature and then quenched, it becomes harder and stronger, but is less ductile. This problem is referred to as quench aging (Ref 397-402). Brittleness increases with aging time at room temperature, reaching a maximum in about 2 to 4 weeks. The steels most prone to quench aging are those with carbon contents from about 0.04 to 0.12%. Quench aging is caused by precipitation of carbide from solid solution in α iron. Nitride precipitation can also occur, but the amount of nitrogen present is generally too low for substantial hardening. Transmission electron microscopy is the preferred technique for studying the aging phenomenon.

Strain aging occurs in low-carbon steels deformed certain amounts and then aged, producing an increase in strength and hardness and a loss of ductility (Ref 399-406). The amount of cold work is critical; about 15% reduction provides the maximum effect. The resulting brittleness varies with aging temperature and time. Room-temperature aging is very slow, and several months are generally required for maximum embrittlement. As the aging temperature increases, the time for maximum embrittlement decreases. Certain coating treatments, such as hot-dip galvanizing, can produce a high degree of embrittlement in areas that were cold worked the critical amount, leading to brittle fractures. This can be prevented if the material is annealed before coating. Additions of elements that will tie up nitrogen, such as aluminum, titanium, vanadium, or boron, also help prevent strain aging.

Strain aging can also lead to stretcher-strain formation (Lüders bands) on low-carbon sheet steels. These marks are cosmetic defects rather than cracks, but the formed parts are unacceptable (Fig. 90). During tensile loading, such a sheet steel exhibits nonuniform yielding, fol-

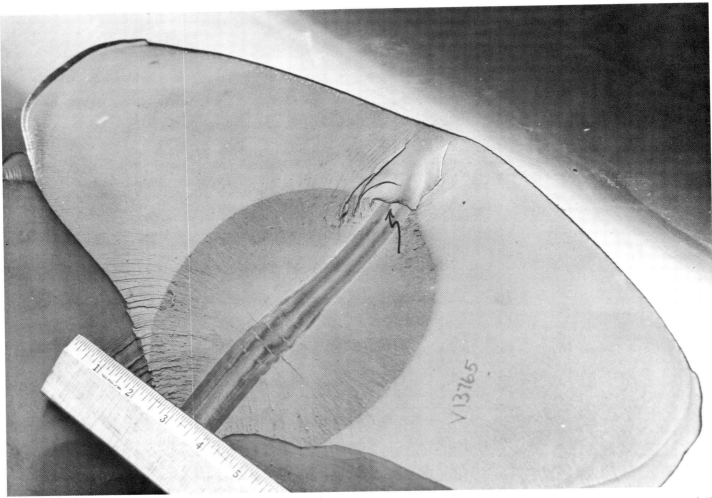

Fig. 62 Another view of the line spall shown in Fig. 61. The arrow indicates the location of the line spall under the unfractured surface. See also Fig. 61 and 63. 0.6×

lowed by uniform deformation. The elongation at maximum load and the total elongation are reduced, lessening cold formability. In non-aluminum-killed sheet steels, a small amount of deformation, about 1%, will suppress the yield-point phenomenon for several months. If the material is not formed within this safe period, the discontinuous yielding problem will eventually return and impair formability.

Strain aging occurs due to ordering of carbon and nitrogen atoms at dislocations. Strain aging results because the dislocations are pinned by the solute atmospheres or by precipitates. During tensile deformation, most dislocations remain pinned. New dislocations appear in areas of stress concentration and must intersect and cut through the pinned dislocations, resulting in higher flow stresses. Transmission electron microscopy is required to study these effects.

Quench Cracking. Production of martensitic microstructures in steels requires a heat treatment cycle that incorporates a quench after austenitization. The part size, the hardenability of the steel, and the desired depth of hardening dictate the choice of quench media. Certain steels are known to be susceptible to cracking during or slightly after quenching. This is a relatively common problem for tool steels (Ref 407), particularly those quenched in liquids.

Many factors can contribute to quench-cracking susceptibility: carbon content, hardenability, M_s temperature (the temperature at which martensite starts to form), part design, surface quality, furnace atmosphere, and heat treatment practice (Ref 407-413). As the carbon content increases, the M_s and M_f (temperatures at which martensite formation starts and finishes) temperatures decrease, and the volumetric expansion and transformation stresses accompanying martensite formation increase. Steels with less than 0.35% C are generally free from quench-cracking problems. The higher M_s and M_f temperatures permit some stress relief to occur during the quench, and transformation stresses are less severe. Alloy steels with ideal critical diameters of 4 or higher are more susceptible to quench cracking than lower-hardenability alloy steels. In general, quench crack sensitivity increases with the severity of the quench medium.

Control of the austenitizing temperature is very important in tool steels. Excessive retained austenite and coarse grain structures both promote quench cracking. Quench uniformity is important, particularly for liquid quenchants. Because as-quenched tool steels are in a highly stressed condition, tempering must be done promptly after quenching to minimize quench cracking. Surface quality is also important because seams, laps, tool marks, stamp marks, and so on, can locate and enhance cracking. These and other problems are reviewed and illustrated in Ref 407.

Quench cracking has been shown to be a statistical problem that occasionally defies prediction and is frequently difficult to diagnose. Heat treaters often experience short periods in which cracking problems are frequent. An occasional heat of steel may show an abnormally high incidence of quench cracking for no apparent reason. Instances have also been documented in which extensive cracking has been associated with material from the bottom of ingots (Ref 411).

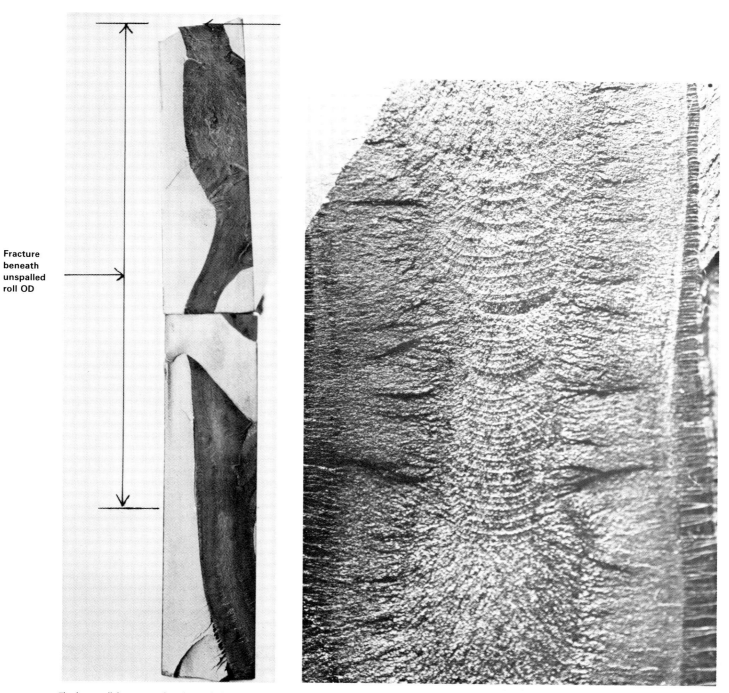

Fig. 63 The line spall fracture surface beneath the unspalled roll surface (see Fig. 61 and 62), revealed by sectioning and fracturing. The smooth, white fracture region formed when the spall was opened. The arrow at the top indicates the origin, shown at the right at higher magnification, which revealed beach marks.

The fracture surfaces of quench cracks are always intergranular. Macroscopic examples of quench cracks are shown in Ref 407. Quench crack surfaces are easiest to observe using SEM (Fig. 91). In quenched-and-tempered steels, proof of quench cracking is often obtained by opening the crack and looking (visually) for temper color typical for the tempering temperature used (Ref 414). Figure 92 shows a guide for predicting temper colors as a function of temperature and time for carbon steels. Table 1 lists temperatures at which different temper colors occur for a carbon tool steel and a stainless steel. The microstructure adjacent to the crack will not be decarburized unless a specimen with an undetected quench crack is rehardened. Quench cracks always begin at the part surface and grow inward and are most commonly oriented longitudinally or radially unless located by a change in section size.

Figure 93 shows an interesting example of quench cracking on ASTM A325 bolts heat treated in an automated, high production rate furnace system. Quench cracks in bolts usually occur longitudinally, often due to the presence of seams. These cracks, however, were circumferential, running part way around the head of the bolts, as shown in Fig. 93 (magnetic particles show the cracks). The furnace had not been used for about 2 years, and no cracking had occurred in the past. During the time the furnace was not in use, two factors had occurred that influenced the problem. First, a

132 / Visual Examination and Light Microscopy

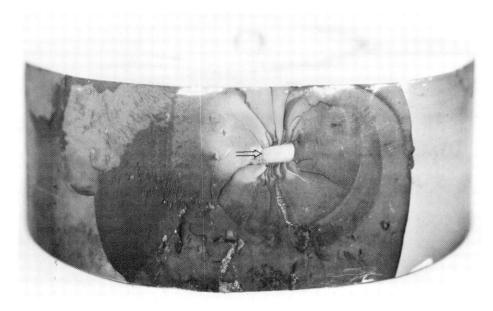

Fig. 64 Example of a line spall in a forged, hardened steel roll. (a) Section containing the spall cut from the roll. The arrow indicates the origin of the fracture, about 6 mm (0.25 in.) below the roll surface. (b) The fracture origin at 6.5×. Fatigue beach marks originate at the arrow; gross fracture marks can be seen radiating from the fatigue zone.

Table 1 Temper colors observed on steels at various temperatures

Color	Carbon tool steel, ground(b) °C	°F	Type 410, polished(c) °C	°F
First straw	195	380	230	450
Straw	215	420	...	...
Brown (bronze)	235	460	430	810
Purple	260	500	...	...
Cobalt blue	305	580	690	1275
Pale blue (gray)	350	660	...	...

(a) Samples held 60 min at heat. (b) Source: Ref 415. (c) Source: Ref 416

occurred had a reducing atmosphere, while the others had oxidizing atmospheres. This difference in atmospheres is known to reduce the cooling rate during quenching. When the furnace atmosphere was made oxidizing, cracking stopped. Higher-hardenability bolts treated in the furnace did not crack even when the reducing atmosphere was used.

Sigma-Phase Embrittlement. Sigma is a hard, brittle intermetallic phase that was discovered by Bain and Griffiths in 1927. Subsequent studies have identified σ-type compounds in over 50 different transition element alloys. Because of the influence of σ phase on the properties of stainless steels and superalloys, many studies have been performed. A few selected reviews are given in Ref 417 to 423.

In austenitic stainless steels, σ precipitates at grain and twin boundaries at temperatures between about 595 and 900 °C (1100 and 1650 °F). Sigma precipitation occurs fastest at about 845 °C (1550 °F). Cold working prior to heating in this range accelerates initiation of σ precipitation. Sigma is not coherent with the matrix. Embrittlement from σ phase is most pronounced at temperatures below 260 °C (500 °F). Therefore, σ-embrittled components present serious maintenance problems.

Sigma phase can be formed in iron-chromium alloys with chromium contents between 25 and 76%. Additions of silicon, molybdenum, nickel, and manganese permit σ to form at lower chromium levels. Carbon additions retard σ formation. Sigma forms more readily from ferrite in stainless steels than from austenite. This presents problems in the welding of austenitic stainless steels because a small amount, about 5%, of δ-ferrite is introduced to prevent hot cracking.

Sigma will slightly increase bulk hardness, but the loss in toughness and ductility is substantial. Sigma does provide increased high-temperature strength, which may prove to be beneficial if the reduced ductility is not a problem. Sigma also improves the wear resistance, and some applications have made use of this. Sigma does reduce creep resistance and has a minor influence on corrosion resistance. The magnitude of these effects depends on the amount of σ present and its size and distribution.

cooling tower was installed so that the quench water was recirculated rather than used once and discharged. Therefore, the quench water was typically about 10 to 35 °C (20 to 60 °F) warmer, depending on the time of the year. Second, basic oxygen furnace (BOF) steel (AISI 1040) was now being used rather than electric furnace (EF) steel. Basic oxygen furnace steel has a lower residual alloy content than electric furnace steel. Both of these factors reduced the hardenability.

Metallographic examination (Fig. 93) showed that the bolt heads were not uniformly hardened. The outer surface of the bolts from the wrench flats inward were hardened, but the middle of the top surface was not. Bolts from the same heats treated in other furnaces were uniformly case hardened across the heads and did not crack. The quench cracks were observed to form in the hardened region, near the interface of the unhardened surface zone. The furnace atmosphere in the line where cracking

Figure 94 shows part of a broken hook used to hold a heat treatment basket during austenitization and quenching. The hook was made from cast 25Cr-12Ni heat-resisting steel; but the composition was not properly balanced, and a higher-than-normal δ-ferrite content was present in the hook. The δ-ferrite transformed to σ during the periods that the hook was in the austenitizing furnace (temperatures from 815 to 900 °C, or 1500 to 1650 °F, generally). The micrograph shows a very heavy, nearly continuous grain-boundary σ network.

Figure 95 shows three views of impact-formed cracks in type 312 stainless steel weld metal that had been heated at 815 °C (1500 °F) for 160 h before breaking. The micrographs show both partially broken and completely broken fractures to illustrate the nature of the crack path. This sample has a rather high σ content. The SEM fractograph shows the rather brittle fracture appearance, a mixture of fine dimples around the σ and quasi-cleavage. The impact energy was only 7% of that of a similar specimen without σ present.

Stress-corrosion cracking of metals and alloys occurs from the combined effects of tensile stress and a corrosive environment (Ref 424-434). Many different metals and alloys can fail by SCC under certain specific circumstances. These failures may be catastrophic, may be due solely to SCC, or an SCC-nucleated crack may be propagated by another fracture mechanism—for example, by fatigue.

Stress-corrosion cracks may propagate transgranularly or intergranularly. In most cases, there is little evidence of the influence of corrosion, but energy-dispersive x-ray analysis can usually detect the presence of the corrosive agent on the fracture surface. Certain forms of SCC have been given other identifying names, such as season cracking of brass or caustic embrittlement of riveted carbon steel structures. In some cases, it is difficult to determine whether the failure was due to hydrogen embrittlement or SCC, or to their combined effects. There are many close parallels between these two mechanisms.

Stress-corrosion fractures exhibit many of the characteristics of brittle fractures in that little or no deformation accompanies the fracture and the fracture is macroscopically flat. The speed at which a stress-corrosion crack propagates, however, is slow compared to a brittle fracture, and crack propagation may be discontinuous. At the fracture initiation site, the metal must be stressed in tension. If the tensile stress is relieved, the crack will stop propagating by stress corrosion. Cases of SCC under compressive loading under certain circumstances have been reported, but these are not common.

There are some interesting features of SCC. For example, such failures often occur under relatively mild conditions of stress and corrosive environment. In many cases, residual stresses alone are adequate. Pure metals are immune to SCC, but the presence of minor amounts of impurity elements will make them susceptible. In some alloys, the heat treatment condition is very important.

Most commonly used alloys can fail by SCC under certain conditions. The best known case is SCC failures of austenitic stainless steels due to chloride ions. Many aluminum alloys will fail by SCC in chloride environments. Copper alloys fail by SCC in ammonia-containing environments. Carbon steels can fail by SCC in environments containing sodium hydroxide or other caustics.

The stress needed to produce SCC failures is generally low. Therefore, many studies have determined threshold stress levels below which cracking does not occur by using a fracture mechanics approach. In certain metals, a minor amount of deformation may cause a change from transgranular to intergranular fracture. Corrosion products have been shown to aid crack propagation by becoming trapped in the crack and exerting a wedging action.

The crack path can be influenced by the microstructure. Grain-boundary precipitates or grain-boundary denuded regions will promote intergranular fracture. Transgranular cracks often follow specific crystallographic planes.

Figure 96 shows a macrograph of an ASTM A325 bolt that fractured in a bridge. The fracture surface is covered by rust, but it is apparent that the fracture began at the root of the threads in the region near the arrow. Examination of the microstructure in this region revealed intergranular secondary cracks, as shown. Due to a heat treatment error, the bolt was not tempered, and the hardened surface was in the as-quenched condition (53 to 57 HRC). This made the bolt susceptible to SCC.

Figure 97 shows the fracture and microstructure of a type 304 stainless steel wire (solution annealed) that failed by SCC in boiling $MgCl_2$. The wire was bent around a 13-mm (0.5-in.) diam pin before being placed in the solution. Therefore, the region where cracking occurred was cold worked. The fracture is predominantly intergranular.

Figure 98 shows a micrograph of a predominantly transgranular stress-corrosion crack in a manganese-chromium austenitic drill collar alloy. The crack occurred at the inner diameter of

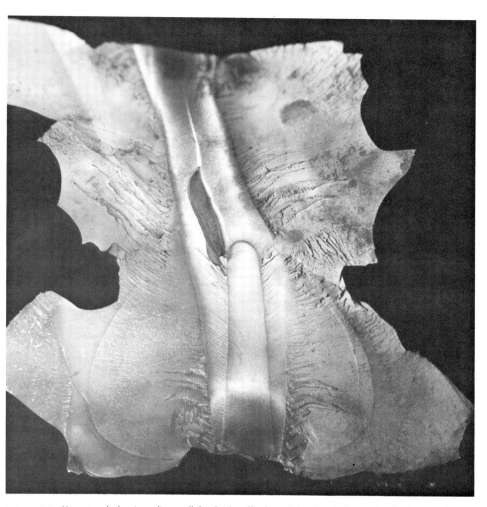

Fig. 65 Macrograph showing a line spall that broke off a forged, hardened alloy steel roll. The central region of the fracture shows evidence of crack growth by fatigue. Final rupture occurred by brittle fracture. See also Fig. 66. 0.65×

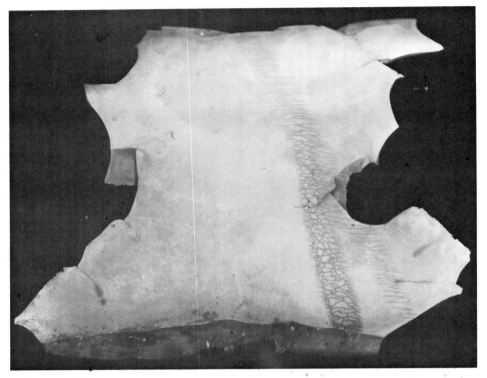

Fig. 66 The roll surface of the spall shown in Fig. 65 after macroetching with 10% aqueous HNO₃. Etching revealed a craze crack pattern similar to heat checks, caused by abusive service conditions. About 0.5×

Table 2 Properties of step-cooled embrittled AISI 4140 alloy steel

Phosphorus, %	Hardness, HRC	Tensile strength, MPa	ksi	50% FATT(a), °C	°F
0.004	33	1031	149.5	−70	−95
0.013	33.5	1071	155.4	−39	−38
0.022	35.5	1099	159.4	30	85

(a) FATT, fracture appearance transition temperature based on a temperature for a 50% ductile, 50% brittle fracture appearance

rapid cooling. Carbon steels are not susceptible to temper embrittlement. Selected reviews on temper embrittlement and its fractographic aspects are given in Ref 435 to 441.

The phenomenon of temper embrittlement has been known since 1883. Numerous service failures have been attributed to, or have been influenced by, temper embrittlement. Prior to development of electron fractographic techniques, the degree of embrittlement was identified by macroscopic fracture examination, degradation of mechanical properties, and use of grain-boundary etchants. The Charpy V-notch impact test has been widely used to assess the shift in transition temperature between embrittled and nonembrittled conditions. Electron fractography has added another tool for assessment of the degree of embrittlement by determination of the area fraction of intergranular facets on the fracture surface. The amount of grain-boundary fracture depends on the degree of impurity segregation to the grain boundaries, but is also influenced by the matrix hardness, the prior-austenite grain size, and the test temperature. Also, the amount of intergranular fracture will vary as a function of the distance from the root of the V-notch.

Phosphorus will segregate to the austenite grain boundaries during austenitization and during tempering in the critical range. Other embrittlers, such as antimony, segregate to the grain boundaries only during tempering. In embrittled martensitic steels, the fracture follows the prior-austenite grain boundaries. In nonmartensitic steels embrittled by antimony, the fracture is intergranular when tested below the transition temperature, and the crack path follows ferrite and upper bainite boundaries rather than the prior-austenite grain boundaries (Ref 440). When phosphorus is segregated, fractures follow the prior-austenite grain boundaries.

The first etchant deliberately formulated to reveal temper embrittlement was developed in 1947 (Ref 442). Other etchants have also been developed (Ref 443, 444). Use of these etchants is reviewed in Ref 9. To demonstrate the use of these etchants, three laboratory ingots of AISI 4140 alloy steel were prepared from the same melt, with the amount of phosphorus varied in the three ingots. Wrought samples were heat treated and then subjected to a stepwise embrittlement cycle; the results are given in Table 2.

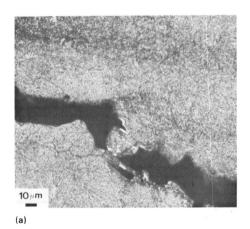
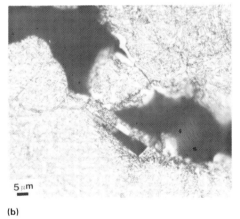

(a) (b)

Fig. 67 Two views of white-etching deformation bands, termed butterflies, that formed at a growing fatigue crack in a forged, hardened alloy steel mill roll. The crack would have led to spalling in later service. These white-etching zones form from the original tempered martensite due to cyclic service stresses and are often observed at the initiation sites of spalls.

the drill collar where residual stresses were high. The cracking was caused by the chloride ion concentration in the drilling fluid.

Temper Embrittlement. Alloy steels containing certain impurities (phosphorus, antimony, tin, and arsenic) will become embrittled during tempering in the range of 350 and 570 °C (660 and 1060 °F) or during slow cooling through this region. Embrittlement occurs because of the segregation of phosphorus, antimony, arsenic, and/or tin to the grain boundaries. The degree of embrittlement depends on the impurity content and the time within the critical tempering range. Embrittlement occurs fastest at about 455 to 480 °C (850 to 900 °F). It is most easily observed by using toughness tests, but tensile properties will be affected by severe embrittlement. The alloy content also influences embrittlement. Nickel-chromium steels are particularly prone to temper embrittlement, but molybdenum additions reduce the susceptibility. Fortunately, the embrittlement is reversible and can be removed by retempering above the critical tempering range, followed by

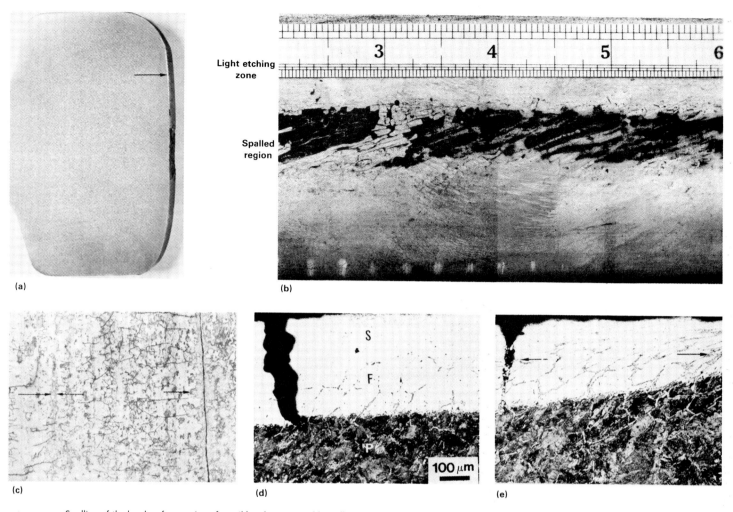

Fig. 68 Spalling of the hard surface region of a rail head was caused by rolling-contact fatigue in service. The hard area was formed by localized overheating, probably by spinning locomotive wheels. (a) Section through rail head, with field side at top and gage side at bottom. (b) Rail surface. 1.2×. (c) Rail surface. 9×. (d) Area adjacent to spall showing featureless surface zone (S), featureless zone with ferrite (F), and pearlite matrix (P). 7×. (e) Section from near field side. Etched with 4% picral. 80×

Scanning electron microscopy examination revealed no intergranular fracture in the samples containing 0.004% P, considerable intergranular fracture in the samples with 0.013% P, and predominantly intergranular fracture in the samples containing 0.022% P. Etching of all of the sample types with the reagents described in Ref 442 and 444 revealed prior-austenite grain boundaries in all samples. Etching of the samples with the saturated aqueous picric acid reagent described in Ref 443, using sodium tridecylbenzine sulfonate as the wetting agent, also revealed the grain boundaries in all samples. Micrographs of these samples are shown in Fig. 99. All samples were then deembrittled by tempering at 620 °C (1150 °F), followed by water quenching. Next, they were repolished and etched with each of the three reagents. The etheral-picral etchants described in Ref 442 and 444 did not reveal any grain boundaries. However, with the saturated aqueous picric acid etchant, grain boundaries were not visible in samples containing 0.004% P, a few were visible in samples with 0.013% P, and most were visible in the samples containing 0.022% P. This etchant is known to be very sensitive to segregated phosphorus, but will not reveal prior-austenite grain boundaries in temper-embrittled steels free of phosphorus, that is, if embrittlement is due only to segregated antimony, tin, or arsenic.

Tempered Martensite Embrittlement (TME). Ultrahigh strength steels with martensitic microstructures are susceptible to embrittlement when tempered between about 205 and 400 °C (400 and 750 °F). Tempered martensite embrittlement is also referred to as 350-°C or 500-°F embrittlement or one-step temper embrittlement. Embrittlement can be assessed with a variety of mechanical tests, but the room-temperature Charpy V-notch absorbed energy plotted against the tempering temperature is the most common procedure. These data reveal a decrease in impact energy in the embrittlement range. The ductile-to-brittle transition temperature will also increase with tempering in this range.

Depending on the test temperature, TME produces a change in the fracture mode from either predominantly transgranular cleavage or microvoid coalescence to intergranular fracture along the prior-austenite grain boundaries. In room-temperature tests, the fracture may change from microvoid coalescence to a mixture of quasi-cleavage and intergranular fracture.

Many studies have been conducted to determine the cause of TME, but the results are not as clear-cut as for temper embrittlement (Ref 445 to 453). Martensitic microstructures are of course susceptible to TME; but mixtures of martensite and lower bainite also suffer a loss in toughness while fully bainitic and pearlitic microstructures are less affected or unaffected. Tempered martensite embrittlement occurs in the tempering range in which ε-carbide changes to cementite. Early studies concluded that TME was due to precipitation of thin platelets of

Fig. 69 Macrograph of broken rail showing the interconnection between the shell fracture (across the top) and the detail fracture (on the transverse plane). 3.5×. (R. Rungta, Battelle Columbus Laboratories)

cementite at the grain boundaries. However, TME has also been found to occur in very low carbon steels (Ref 447, 448), and residual impurity elements have also been shown to be essential factors in TME (Ref 445). The decomposition of interlath retained austenite into cementite films with tempering in the range of 250 to 400 °C (480 to 750 °F) has also been found to be a factor in TME (Ref 449, 452). It appears that TME results from the combined effects of cementite precipitation on prior-austenite grain boundaries or at interlath boundaries and the segregation of impurities, such as phosphorus and sulfur, at the prior-austenite grain boundaries.

Thermal Embrittlement. Maraging steels fracture intergranularly when the toughness has been severely degraded because of improper processing after hot working. This problem, called thermal embrittlement, occurs upon heating above 1095 °C (2000 °F), followed by slow cooling or by interrupted cooling with holding in the range of 815 to 980 °C (1500 to 1800 °F) (Ref 454-458). Embrittlement has been attributed to precipitation of TiC and Ti(C,N) on the austenite grain boundaries during cooling through the critical temperature range. The severity of the embrittlement increases with decreasing cooling rate through this range.

Increases in the concentration of carbon and nitrogen render maraging steels more susceptible to thermal embrittlement. Also, as the titanium level increases, thermal embrittlement problems become more difficult to control. Auger electron spectroscopy (AES) has shown that embrittlement begins with the diffusion of titanium, carbon, and nitrogen to the grain boundaries. Precipitation of TiC or Ti(C,N) on the grain boundaries represents an advanced stage in the embrittlement.

Light microscopy can be used to reveal the nature of the fracture path in severely embrittled specimens; but the chance of observing precipitates along the fracture path is low, and pre-precipitation segregation is not detectable. Scanning electron microscopy examination of the fracture face is helpful, but the preferred approach is the use of extraction replica fractography (Ref 459). This procedure reveals the precipitates with strong contrast, and they can be easily identified with energy-dispersive spectroscopy (EDS) and electron diffraction procedures.

Figure 100 shows the fracture of a cobalt-free high-titanium maraging steel specimen that fractured because of thermal embrittlement. The light micrographs of the fracture profile and a secondary crack reveal the intergranular nature of the fracture. This is more easily observed by direct SEM examination of the fracture, but the TiC and Ti(C,N) on the intergranular fracture is more clearly revealed by the extraction fractograph. Analysis of the extracted grain-boundary precipitates is not hindered by detection of the matrix around the precipitate, as might occur with SEM-EDS analysis.

885-°F (475-°C) Embrittlement. Ferritic stainless steels containing more than about 13% Cr become embrittled with extended exposure to temperatures between about 400 and 510 °C (750 and 950 °F), with the maximum embrittlement at about 475 °C (885 °F). Therefore, this problem is referred to as 885-°F or 475-°C embrittlement (Ref 460-469). Aging at 475 °C (885 °F) increases strength and hardness, decreases ductility and toughness, and changes electrical and magnetic properties and corrosion resistance. The time at the aging temperature intensifies these changes. Embrittlement produces microstructural changes, notably a widening of the etched grain boundaries followed by a darkening of the ferrite grains.

Early x-ray studies suggested that a chromium-rich precipitate formed during embrittlement; this precipitate was incorrectly identified as σ phase. In one study, TEM extraction replicas and selected-area diffraction patterns were used, followed by x-ray diffraction and fluorescence, to show that the precipitate had a bcc crystal structure and contained about 80% Cr (Ref 460). These precipitates were coherent with the matrix and were extremely small. Aging of a 27% Cr alloy for 1000 h at 480 °C (900 °F) produced particles smaller than 5 nm in diameter, but aging for 34 000 h produced particles about 22.5 nm in diameter. This work showed that 885-°F embrittlement produces iron-rich α and chromium-rich α' ferrites. Chromium nitrides also precipitate during aging and are observed at grain boundaries, dislocations, and inclusions.

The formation of α' can occur in two ways: by nucleation and growth or by spinodal decomposition. The latter mechanism appears to be operative in higher chromium content alloys and has been the most commonly observed formation process. The reaction is reversible; heating above the embrittlement range dissolves α'. In duplex stainless steels, the embrittlement temperature range appears to be broader, with additional phases precipitating in the upper portion of the range.

As an example of 885-°F embrittlement, Figure 101 shows the fracture of a duplex alloy similar in composition to type 329 stainless steel. This material was aged for 1000 h at 370 °C (700 °F), producing a substantial hardness increase and a dramatic loss in toughness. Room-temperature half-size Charpy V-notch impact tests revealed 1.36 J (1 ft · lb) of absorbed energy compared to 64 J (47 ft · lb)

Visual Examination and Light Microscopy / 137

Fig. 70 Macrograph of a rail that contained a detail fracture (upper left, beneath the rail head surface) that was placed in service in the FAST test track. Note the fatigue fracture that grew from the detail fracture. 0.75×. (R. Rungta, Battelle Columbus Laboratories)

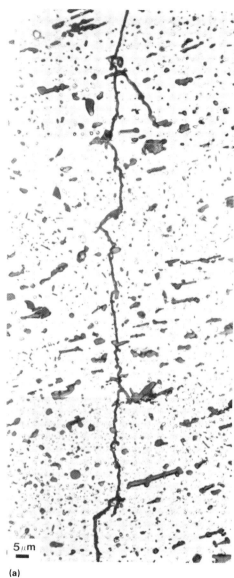

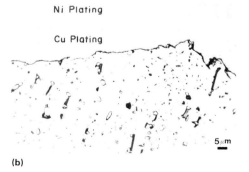

Fig. 71 Fatigue cracks in an unbroken (a) and a completely broken (b) aluminum alloy. (a) Etched with Kroll's reagent. 680× (b) Etched with Keller's reagent. 510×

for the same alloy in the hot rolled and annealed condition. The fracture is by cleavage with some dark-etching ferrite grain boundaries in the embrittled specimen, but the same etch (Vilella's reagent) does not reveal the ferrite grain boundaries in the annealed nonembrittled specimen.

Weld Cracking

Cold cracking generally occurs at temperatures below 205 °C (400 °F); however, in some cases, these cracks are initiated at higher temperatures and then grow to a detectable size at lower temperatures (Ref 470-479). Cracking may occur during cooling to room temperature or a short time after reaching room temperature, or it may occur after a considerable time delay. Cold cracks may be found in the weld metal or in the HAZ (Fig. 102). Both transgranular and intergranular fracture paths have been observed.

Three factors combine to produce cold cracks: stress, hydrogen, and a susceptible microstructure. Stresses may be due to thermal expansion and contraction, or they may be due to a phase transformation, such as the austenite-to-martensite transformation. High-carbon (plate) martensite microstructures are most susceptible to cold cracking. Low-carbon (lath) martensite microstructures are less susceptible, while ferritic and bainitic microstructures are least susceptible. Hydrogen must be present above some critical level for cracking to occur. Figure 103 shows a metallographic cross section of a cold crack in a plate steel implant specimen welded with a high-hydrogen content electrode and an SEM view of the fracture after sectioning to open the crack.

Studies of cold cracking in plate steel weldments demonstrated that at high hydrogen levels cracking depends primarily on weld metal

Fig. 72 Fatigue crack (arrows) in a ferrite-pearlite microstructure in a carbon steel. Etched with 2% nital. 800×

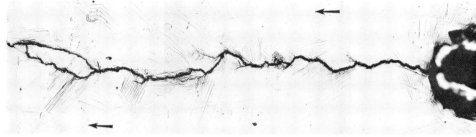

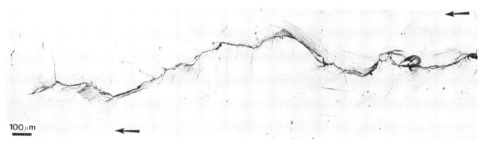

Fig. 73 Fatigue crack appearance in an austenitic stainless steel specimen polished before fatigue loading, revealing slip lines on surfaces associated with crack formation and growth. Arrows indicate the direction of crack growth. 52×

hardness (Ref 480). The risk of cold cracking increased with increasing hardness in the range of about 200 to 330 HV. At lower hydrogen levels, the resistance to cracking depends primarily upon the microstructure. Acicular ferrite provided good resistance to cracking, while ferritic structures with aligned second phases produced poor resistance to cold cracking. At hardnesses below about 270 HV, the crack path was not preferential to grain-boundary ferrite; at higher hardnesses, there was an increased tendency for the cracks to follow the ferrite grain boundaries. In the absence of grain-boundary ferrite, intergranular fracture occurred. Fracture surfaces in plate steels can exhibit a wide range of morphologies, from microvoid coalescence to cleavage to intergranular patterns.

Hot cracking, also referred to as solidification cracking, is thought to occur at or above the solidus temperature of the last portion of the metal to freeze due to the influence of stresses from shrinkage (Ref 481-491). Austenitic stainless steels are known to be susceptible to hot cracking in the fusion zone and in the base metal adjacent to the fusion zone. Cracking has been attributed to low melting point films at grain and subgrain boundaries. Analysis of these intergranular cracks has shown that the low melting point films are enriched in sulfur, phosphorus, silicon, and manganese. Therefore, one approach to reducing hot cracking is to restrict the sulfur and phosphorus levels to less than 0.002% each and the silicon content to less than 0.10%. Another approach, which is well established, is to introduce a small amount, generally about 5 to 8%, of δ-ferrite to the austenitic weld metal. There are also reports of hot cracking at temperatures about 100 to 300 °C (180 to 540 °F) below the equilibrium solidus temperature.

Several theories have been proposed to account for weld metal hot cracking. The shrinkage-brittleness theory states that cracking occurs during the final stage of solidification because the partially solidified dendritic structure is unable to tolerate the contraction strain or external restraint. The strain theory states the thin liquid films present during the final stage of solidification lower the strength and ductility of the solidifying metal. Thermal gradients present during solidification create stresses that tend to pull the solid apart, creating tears. These concepts have been refined in a generalized theory of super-solidus cracking (Ref 483). Weld solidification was considered to occur in four stages: primary dendrite formation, dendrite interlocking, grain-boundary development, and final solidification of the remaining interdendritic liquid. The third phase is critical, because cracks formed at this time will not be healed. The concept of the relative interfacial energy and dihedral angle has been used to predict hot-cracking tendencies (Ref 483). This approach demonstrates the critical influence of sulfur and copper impurities on hot-cracking susceptibility. Cracking is most likely when nearly continuous liquid films are present in the fourth stage of solidification.

Hot cracking can also occur in the base metal, and three problems have been suggested to produce liquid films in the base metal: incipient grain-boundary melting, melting of segregates, and absorption of liquid from the molten weld pool. The base metal along the fusion line is stressed in tension initially during welding. As the weld nugget cools, the HAZ becomes stressed in tension, producing separation of regions containing liquid films.

Light microscopy examination of cross sections is a very important aspect of diagnosing hot cracking. Figure 104 illustrates the use of light microscopy in the study of extensive hot cracking in an electron beam weld that was contaminated by liquid copper by accidental melting of the copper back-up plate. The cracks are clearly intergranular, and elemental copper can be found throughout the region in the prior-austenite grain boundaries. Figure 105 shows two examples of fine hot tears in the HAZ of a welded HY-80 plate steel.

Lamellar tearing occurs in the base metal because of high through-thickness strains introduced by weld metal shrinkage under conditions of high joint restraint (Ref 492-502).

Visual Examination and Light Microscopy / 139

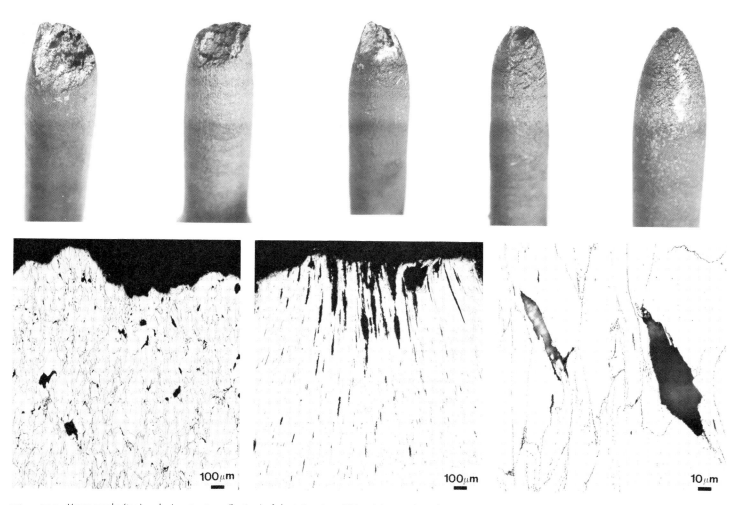

Fig. 74 Macrographs (top) and microstructures (bottom) of short-time type 316 stainless steel tensile specimens tested at various temperatures. Top, from left: specimens tested at 760 °C (1400 °F), 815 °C (1500 °F), 870 °C (1600 °F), 925 °C (1700 °F), and 980 °C (1800 °F). Bottom, from left: Specimen tested at 760 °C (1400 °F). 35×. Specimen tested at 980 °C (1800 °F). 35×. Specimen tested at 815 °C (1500 °F). 350×. Specimens at bottom etched with mixture of HCl, HNO$_3$, and H$_2$O.

Tearing occurs by decohesion and linking along the rolling direction of the plate in the base metal beneath the HAZ of multipass welds approximately parallel to the fusion line. The cracks usually have a steplike appearance indicative of discontinuous growth. The fracture surfaces appear fibrous or woody.

Steels susceptible to lamellar tearing exhibit low through-thickness ductility and toughness. The welding process must produce a weld-fusion boundary approximately parallel to the plate surface. For cracking to occur, the joint design must produce high through-thickness strains. When these conditions exist, lamellar tearing is likely, depending on the inclusion types and content in the steel.

Cracking begins at the inclusion/matrix interfaces, as shown schematically in Fig. 106. The cracked interfaces that lie in the same plane grow and link together to form terraces. The high strain level, due to thermal contraction and joint restraint, produces ductile tearing of the matrix between the inclusion/matrix interfacial cracks. Further straining connects terrace cracks on different parallel planes by ductile shearing (Fig. 106). These shear walls produce the characteristic steplike appearance of lamellar tears (Fig. 107).

Lamellar-tearing susceptibility is influenced by joint design factors that increase weld metal shrinkage strains in the through-thickness direction, that is, joint designs that orient the fusion line parallel to the hot-working direction (tee and corner joints); excessive joint restraint (larger welds than needed, full-penetration welds rather than fillet welds, and so on); excessively high weld metal strength, which concentrates the strain in the base metal; and high levels of component restraint. Material properties are also very important. Through-thickness ductility and toughness must be maximized to inhibit lamellar tearing. Anisotropy of mechanical properties is enhanced by the banding of microstructural constituents, by the type and amount of microstructural phases and constituents present, and by the number, size, shape, and types of the inclusions present (Ref 502). Elongated sulfide and silicate stringers are very detrimental. Low sulfur content and inclusion shape control have been shown to be very beneficial in providing resistance to lamellar tearing.

Stress-relief cracking, also called postweld heat treatment cracking, stress rupture cracking, and reheat cracking, usually occurs in the HAZ, and sometimes in the weld metal, during postweld heat treatments or during high-temperature service (Ref 503-512). This type of cracking occurs when crack-free weldments of a susceptible alloy composition are subjected to a thermal stress-relief heat treatment to reduce residual stresses and improve toughness. After such a treatment, cracking may be found that can be substantial in some instances. Cracking usually occurs at stress raisers, for example, at geometrical discontinuities.

Stress-relief cracking only occurs in metals that exhibit precipitation hardening during the high-temperature heat treatment or exposure. Cracking results when the creep ductility is inadequate to accommodate strains that accompany the relief of applied or residual stresses.

Stress-relief cracks are intergranular and are generally observed in the coarse-grain region of the HAZ. They usually initiate at a stress

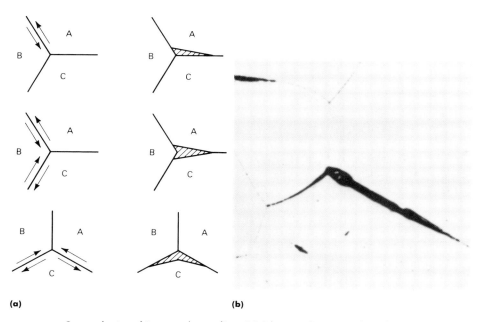

Fig. 75 One mechanism of intergranular cracking. (a) Schematic showing cracking due to grain-boundary sliding. Arrows along a grain boundary indicate that this boundary underwent sliding. (b) Cracks and voids in Al-5.1Mg that was stress rupture tested at 260 °C (500 °F). Electrolytically polished. 60×. Source: Ref 194

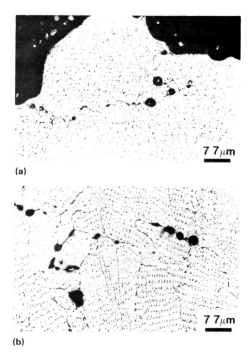

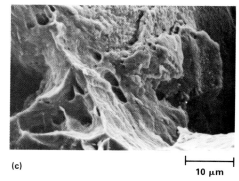

Fig. 76 Microstructure and fracture appearance of type 316 stainless steel tested in creep to fracture in air at 800 °C (1470 °F) at a load of 103 MPa (15 ksi). Time to rupture: 808 h. Light micrographs (a and b) illustrate r-type cavities caused by vacancy condensation on boundaries perpendicular to the stress axis. Both 90× and electrolytically etched with oxalic acid. The SEM fractograph (c) illustrates the formation of the r-type cavities at the grain boundaries. 1260×. (W.E. White, Petro-Canada Ltd.)

concentrator, such as a weld toe or an area of incomplete penetration or fusion. The intergranular crack path may exhibit considerable branching. The problem is most commonly encountered in thick welded sections in which restraint can produce very high residual stresses.

Steel microstructures that promote stress-relief cracking usually contain fine carbides that have precipitated within the grains in the HAZ; the grain boundaries are frequently denuded of carbides. This concentrates the creep strain at the weakened grain boundaries, producing extensive deformation, which is often accommodated by grain-boundary sliding, and intergranular cracking. Carbides may also lie along the grain boundaries, which promotes cracking. Alloy steels containing carbide-forming elements, such as chromium, molybdenum, and vanadium, are most susceptible. A metallographic example of the nature of stress-relief cracks is shown in Fig. 108.

Quality Control Applications

The examination of fractures often provides information about microstructure or quality, and fracture tests have been used for such purposes since antiquity and are invaluable in failure analysis work. Fractures can often provide the metallurgist with interesting information. For example, Figure 109 shows a fractured ingot of an iron-chromium-aluminum alloy in which the solidification structure is clearly revealed. A related example is shown in Fig. 110, which illustrates a fatigue fracture located by a shrinkage cavity. The dendritic nature of solidification is observed in the shrinkage cavity.

The Fracture Test. The examination of test sample fractures is a well recognized, simple procedure for quality evaluation. The breaking of testpieces can be a very crude operation, or it may be carefully controlled in test machines. The simplest procedure is to support the sample on its ends and strike the center with a sledge hammer. In the fracturing of hardened steel disks, a mold can be designed to support the specimen edges while a top cover is used to locate a chisel over the center of the specimen. The chisel is then struck with a sledge hammer. The closed mold prevents pieces from striking people in the area. The use of such a device to break cast coupons for inspection is discussed in Ref 513. In some cases, it may be necessary to nick the specimen in order to locate the fracture in the desired area. A fracture press is ideal for such work. A less satisfactory procedure is to place one end of the specimen in a sturdy vise in a cantilever fashion and strike it on the free end. Hardened high-hardness tool steels and high-carbon steels can usually be broken at room temperature, but more ductile alloys often require refrigeration in liquid nitrogen to facilitate breaking. Face-centered cubic materials are more difficult to fracture, particularly when section thicknesses are appreciable.

Some of the uses for test fracture examination include:

Details about composition
Detection of inclusion stringers
Assessment of degree of graphitization
Grain size measurement
Depth of hardening
Detection of overheating
Detection of segregation

Some of these applications of the fracture test are reviewed in Ref 514. The fracture appearance can be used to classify broadly the composition of certain steels and cast irons in field work (Ref 515).

Inclusion stringers of macroscopic size can be readily detected on a fractured specimen after heat tinting in the blue heat range (Ref 516-518). Several ASTM standards (A 295, A 485, A 535, and A 711) include test limitations for macroscopic inclusion stringers detected in this type of test. In most work with this test, the macroetch billet disks are hardened, fractured, and blued on a hot plate or in a laboratory furnace. Heating to 260 to 370 °C (500 to 700 °F) is usually adequate. The fracture face should be a longitudinal plane. The lengths of

Visual Examination and Light Microscopy / 141

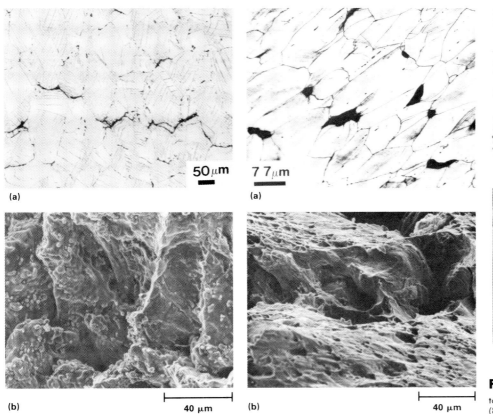

Fig. 77 Microstructure and fracture appearance of type 316L stainless steel tested in creep to fracture in air at 800 °C (1470 °F), using a 53-MPa (7.7-ksi) load. Time to rupture: 839 h. The light micrograph (a) illustrates w-crack coalescence by slow shearing along grain boundaries. The SEM fractograph (b) shows the fracture appearance characterized by poorly formed shear dimples and particle dragging. (a) Electrolytically etched with oxalic acid. 84×. (b) 420×. (W.E. White, Petro-Canada Ltd.)

Fig. 78 Microstructure and fracture appearance of type 316 stainless steel tested in creep to fracture in air at 685 °C (1265 °F) at a load of 123 MPa (17.9-ksi). Time to rupture: 710 h. The light micrograph (a) shows triple boundary cracking with extensive bulk deformation and grain elongation. The SEM fractograph (b) illustrates classic shear dimple topography. (a) Electrolytically etched with oxalic acid. 110×. (b) 380×. (W.E. White, Petro-Canada Ltd.)

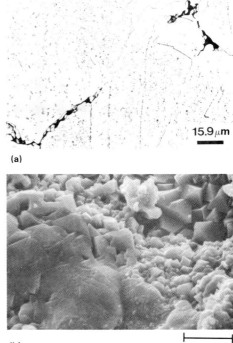

Fig. 79 Microstructure and fracture appearance of type 316 stainless steel tested in creep to fracture at 770 °C (1420 °F) using a 62-MPa (8.95-ksi) load. Time to rupture: 808 h. (a) Optical micrograph showing crack nucleation and growth by decohesion along the carbide/matrix interfaces. Etched with dilute aqua regia. 440×. (b) SEM fractograph illustrating carbide morphology at the fracture surface. 3150×. (W.E. White, Petro-Canada Ltd.)

Fig. 80 Example of the macroscopic appearance of hydrogen flakes in plate steel. 1.6×

any visible stringers, which appear white against the blue fracture, are measured. Figure 111 shows an example of inclusion stringers detected by this procedure.

Chill and wedge tests have been extensively used in the manufacture of gray and white irons to reveal carbide stability or the tendency of an iron to solidify as white iron rather than gray iron (Ref 519-521). These tests show the combined effect of melting practice and composition on carbide stability, but are not a substitute for chemical analysis. The chill test uses a small rectangular casting in which one face is a chill plate, while the wedge test employs a wedge-shaped mold in which the cooling rate varies with the wedge thickness. After a casting is made with either type of mold, it is broken and the fracture is examined. Figure 112 shows an example of a wedge test fracture. In the wedge test, the distance from the wedge tip to the end of the white fracture is measured; in the chill test, the depth of the white iron chill is measured.

In the production of tool steels, the fracture test can be used to detect graphitization in tool steels that should contain graphite, such as AISI 06, or in high-carbon tool steels that should be free of graphite. Figure 113 shows four fractured test disks of a high-carbon tool steel that contained considerable undesired graphite due to accidental addition of aluminum.

The prior-austenite grain size of high-carbon martensitic steels, such as tool steels, can be assessed by comparing a fractured specimen to a series of graded fracture specimens (Ref 9, 522, 523). This method is fast, simple, and accurate, but can be applied only to high-hardness relatively brittle martensitic (retained austenite may also be present) steels. Complete details on the use of this test and its limitations are provided in Ref 9 and 523.

The depth of hardening in case-hardened steels, such as carbon tool steels, can be easily assessed by using fractured specimens. The P-F test is commonly used for this purpose (Ref 522). Figure 114 shows P-F test specimens. A 19-mm (¾-in.) diam, 100-mm (4-in.) long specimen is austenitized at the recommended temperature and brine quenched. After fracturing, one fracture face is ground and macro-

etched, as shown. The sample is then hardness tested with the Rockwell superficial test to determine the depth to 55 HRC, that is, the location for 50% martensite. This depth correlates well with the case/core transition depth on the test fracture.

As discussed earlier, fracture tests are commonly used to detect overheating. A test specimen of appropriate size for subsequent fracturing is normalized, austenitized, quenched and tempered to about 321 HB, and fractured, usually at room temperature. The appearance of coarse-grain facets on the fracture, as shown in Figures 86 to 88, is taken as proof of overheating.

Fractures of transverse or longitudinal sections from wrought products or from castings have been used for many years to evalu-

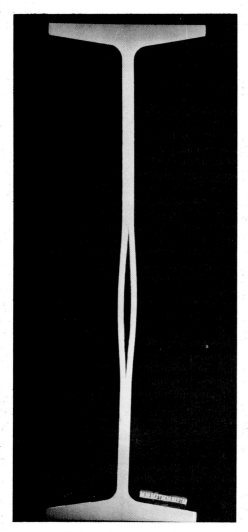

Fig. 81 Hydrogen blister in the web of a structural steel section. About 0.3×

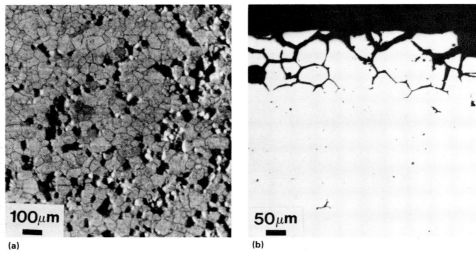

Fig. 82 Planar (a) and cross sectional (b) views of intergranular corrosion (grain dropping) in a sensitized austenitic stainless steel. As-polished. (a) 50×. (b) 100×

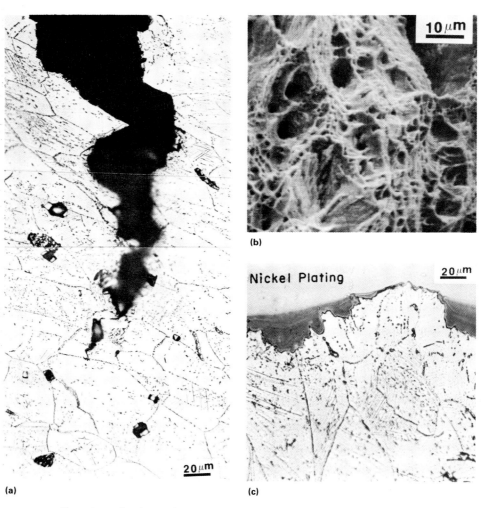

Fig. 83 Three views of a fractured sensitized specimen of type 304 stainless steel. (a) Partially broken specimen. Etched with mixed acids. (b) SEM view of fracture; specimen broken at −195 °C (−320 °F). (c) Cross section of fracture. Etched with acetic glyceregia

ate metal quality (Ref 514, 524, 525). The fracture test can reveal texture, flaking, graphitization, slag, blow holes, pipe, inclusions, and segregation. The general appearance of the test fracture is often classified as being coarse or fine, woody, fibrous, ductile, or brittle. Fibrous and woody fractures (longitudinal) result from microstructural anisotropy due to banding, segregation, or excessive inclusion stringers. Woody fractures usually result from grosser conditions compared to fibrous fractures.

Figure 115 shows an example of the transverse fracture of an AISI 1070 steel bar that broke during heat treatment due to gross chemical segregation and overheating. The transverse fracture reveals a fine-grain appearance at the hardened outer periphery of the bar. The coarse intergranular appearance of the fracture in the interior is evident in Fig. 115. Radial marks can be seen emanating from the coarse-grain central region of the bar.

As noted previously, the appearance of tensile-test fractures has long been used as an index of quality. If the tensile fracture is not affected by gross defects, its general appear-

 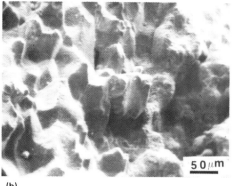 

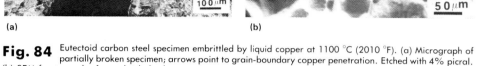

Fig. 84 Eutectoid carbon steel specimen embrittled by liquid copper at 1100 °C (2010 °F). (a) Micrograph of partially broken specimen; arrows point to grain-boundary copper penetration. Etched with 4% picral. (b) SEM fractograph of completely broken specimen

ance is classified as irregular, angular, flat, partially cupped, or a full cup-and-cone. The latter type is generally regarded as the optimum fracture appearance. An extensive study of tensile fracture appearance and its relationship to test results was reported in 1950 (Ref 58). For a given type of steel and heat treatment condition, there is an approximately linear relationship between tensile strength and tensile ductility. The data scatter for the relationship between tensile strength and percent elongation or %RA was shown to be much less for full cup-and-cone test fractures than for partial cup-and-cone fracture specimens (Ref 58).

The fracture test is widely used in the production of castings. Test fractures reveal the influence of casting temperature on the primary macrostructure. The change in fracture appearance of hypereutectic white irons with increasing casting temperature is illustrated in Ref 526. The same trend was not observed for tests of hypoeutectic white irons. One investigation found that the fracture appearance of tensile specimens of white irons varied with tensile strength (Ref 527). A specimen with a high tensile strength (490 MPa, or 71 ksi) had a fine fracture, while one with a low tensile strength (280 MPa, or 40.6 ksi) had a coarse columnar appearance.

The fracture test has been widely used in the production of copper castings (Ref 528-531). Such tests are often performed by the removal of the gating or feeding portions of the casting by fracture. In copper-base castings, the color of the fracture provides considerable information because it is influenced by alloy composition and the presence of various phases or constituents (Ref 531). For example, the α solid-solution phase in tin bronze has a reddish-brown fracture color, the tin-rich δ phase has a bluish-white fracture color, copper phosphide appears pale blue-gray, and lead appears dark blue-gray. Bronzes with low tin and phosphorus contents are reddish-brown, and the fracture color changes to gray-brown and then to gray as the tin or phosphorus contents are increased. Zinc additions to copper impart a brassy yellow color to the fracture. The texture of cast bronze test fractures varies with composition and grain size (Ref 531). Increased tin or phosphorus contents generally produce finer, silkier fractures, while lead additions produce a granular appearance. Fibrous textures are often observed in low-tin bronzes.

REFERENCES

1. G.F. Vander Voort, Conducting The Failure Examination, *Met. Eng. Q.*, Vol 15, May 1975, p 31-36
2. C.A. Zapffe and M. Clogg, Jr., Fractography—A New Tool for Metallurgical Research, *Trans. ASM*, Vol 34, 1945, p 77-107
3. C.A. Zapffe et al., Fractography: The Study of Fractures at High Magnification, *Iron Age*, Vol 161, April 1948, p 76-82
4. C.A. Zapffe and C.O. Worden, Fractographic Registrations of Fatigue, *Trans. ASM*, Vol 43, 1951, p 958-969
5. C.A. Zapffe et al., Fractography as a Mineralogical Technique, *Am. Mineralog.*, Vol 36 (No. 3 and 4), 1951, p 202-232
6. K. Kornfeld, Celluloid Replicas Aid Study of Metal Fractures, *Met. Prog.*, Vol 77, Jan 1960, p 131-132
7. P.J.E. Forsyth and D.A. Ryder, Some Results of the Examination of Aluminum Alloy Specimen Fracture Surfaces, *Metallurgia*, March 1961, p 117-124
8. K.R.L. Thompson and A.J. Sedriks, The Examination of Replicas of Fracture Surfaces by Transmitted Light, *J. Austral. Inst. Met.*, Vol 9, Nov 1964, p 269-271
9. G.F. Vander Voort, *Metallography: Principles and Practice*, McGraw-Hill, 1984
10. H.C. Rogers, The Tensile Fracture of Ductile Metals, *Trans. AIME*, Vol 218, June 1960, p 498-506
11. C. Laird and G.C. Smith, Crack Propagation in High Stress Fatigue, *Philos. Mag.*, Vol 7, 1962, p 847-857

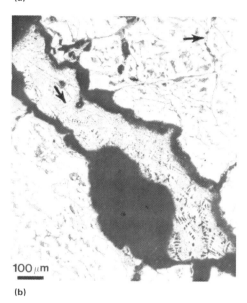

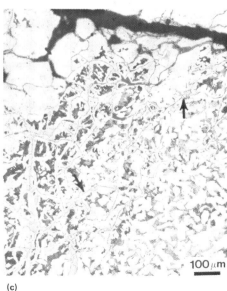

Fig. 85 Hot shortness in a structural steel caused during rolling by internal LME due to copper segregation. (a) Macrograph of section from toe of flange. 0.4×. (b) and (c) Micrographs showing grain-boundary copper films that were molten during rolling. (b) and (c) Etched with 2% nital. Both 70×

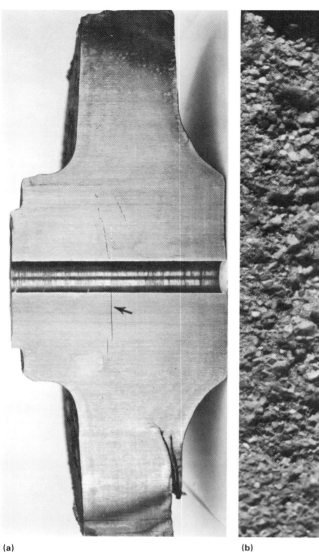 

Fig. 86 Alloy steel compressor disk that cracked from overheating during forging. (a) Macrograph of disk (cracking at arrow). 0.4×. (b) Fracture surface of a specimen from the disk that was normalized, quenched, and tempered to 321 to 341 HB. The treatment revealed facets indicative of overheating. 5×

12. C. Laird, The Influence of Metallurgical Structure on the Mechanisms of Fatigue Crack Propagation, in *Fatigue Crack Propagation*, STP 415, American Society for Testing and Materials, 1967, p 131-180
13. D.P. Clausing, The Development of Fibrous Fracture in a Mild Steel, *Trans. ASM*, Vol 60, 1967, p 504-515
14. I.E. French and P.F. Weinrich, The Shear Mode of Ductile Fracture in a Spheroidized Steel, *Metall. Trans.*, Vol 10A, March 1979, p 297-304
15. R.H. Van Stone and T.B. Cox, Use of Fractography and Sectioning Techniques to Study Fracture Mechanisms, in *Fractographic-Microscopic Cracking Processes*, STP 600, American Society for Testing and Materials, 1976, p 5-29
16. R.W. Staehle et al., Mechanism of Stress Corrosion of Austenitic Stainless Steels in Chloride Waters, *Corrosion*, Vol 15, July 1959, p 51-59 (373t-381t)
17. K.W. Burns and F.B. Pickering, Deformation and Fracture of Ferrite-Pearlite Structures, *J. Iron Steel Inst.*, Vol 202, Nov 1964, p 899-906
18. J.H. Bucher et al., Tensile Fracture of Three Ultra-High-Strength Steels, *Trans. AIME*, Vol 233, May 1965, p 884-889
19. U. Lindborg, Morphology of Fracture in Pearlite, *Trans. ASM*, Vol 61, 1968, p 500-504
20. C.T. Liu and J. Gurland, The Fracture Behavior of Spheroidized Carbon Steels, *Trans. ASM*, Vol 61, 1968, p 156-167
21. D. Eylon and W.R. Kerr, Fractographic and Metallographic Morphology of Fatigue Initiation Sites, in *Fractography in Failure Analysis*, STP 645, American Society for Testing and Materials, 1978, p 235-248
22. W.R. Kerr et al., On the Correlation of Specific Fracture Surface and Metallographic Features by Precision Sectioning in Titanium Alloys, *Metall. Trans.*, Vol 7A, Sept 1976, p 1477-1480
23. D.E. Passoja and D.J. Amborski, Fracture Profile Analysis by Fourier Transform Methods in *Microstructural Science*, Vol 6, Elsevier, 1978, p 143-158
24. W.T. Shieh, The Relation of Microstructure and Fracture Properties of Electron Beam Melted, Modified SAE 4620 Steels, *Metall. Trans.*, Vol 5, May 1974, p 1069-1085
25. J.R. Pickens and J. Gurland, Metallographic Characterization of Fracture Surface Profiles on Sectioning Planes, in *Proceedings of the 4th International Congress for Stereology*, NBS 431, National Bureau of Standards, 1976, p 269-272
26. S.M. El-Soudani, Profilometric Analysis of Fractures, *Metallography*, Vol 11, July 1978, p 247-336
27. E. Rabinowicz, Taper Sectioning. A Method for the Examination of Metal Surfaces, *Met. Ind.*, Vol 76, 3 Feb 1950, p 83-86
28. L.E. Samuels, An Improved Method for the Taper Sectioning of Metallographic Specimens, *Metallurgia*, Vol 51, March 1955, p 161-162
29. W.A. Wood, Formation of Fatigue Cracks, *Philos. Mag.*, Vol 3 (No. 31), Series 8, July 1958, p 692-699
30. W.A. Wood and H.M. Bendler, Effect of Superimposed Static Tension on the Fatigue Process in Copper Subjected to Alternating Torsion, *Trans. AIME*, Vol 224, Feb 1962, p 18-26
31. W.A. Wood and H.M. Bendler, The Fatigue Process in Copper as Studied by Electron Metallography, *Trans. AIME*, Vol 224, Feb 1962, p 180-186
32. P.J.E. Forsyth, Some Metallographic Observations on the Fatigue of Metals, *J. Inst. Met.*, Vol 80, 1951-1952, p 181-186
33. K. Kitajima and K. Futagami, Fractographic Studies on the Cleavage Fracture of Single Crystals of Iron, in *Electron Microfractography*, STP 453, American Society for Testing and Materials, 1969, p 33-59
34. R.M.N. Pelloux, Mechanisms of Formation of Ductile Fatigue Striations, *Trans. ASM*, Vol 62, 1969, p 281-285
35. W.D. Hepfer et al., A Method for Determining Fracture Planes in Beryllium Sheet by the Use of Etch Pits, *Trans. ASM*, Vol 62, 1969, p 926-929
36. A. Inckle, Etching of Fracture Surfaces, *J. Mater. Sci.*, Vol 5 (No. 1), 1970, p 86-88

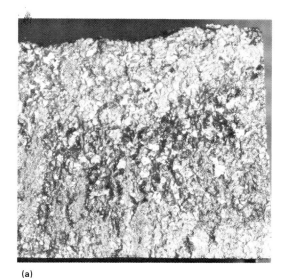 

Fig. 87 Fracture (a) of a slab that was rolled from a vanadium-niobium plate steel. The coarse fracture facets indicate overheating before hot rolling. 1.25×. The etched section (b, with fracture along the top edge) shows carbon segregation in the center, as evidenced by the greater amount of pearlite. Note also the coarseness of the pearlite in the center. Etched with 2% nital. 3.5×

37. Y. Mukai et al., Fractographic Observation of Stress-Corrosion Cracking of AISI 304 Stainless Steel in Boiling 42 Percent Magnesium-Chloride Solution, in *Fractography in Failure Analysis*, STP 645, American Society for Testing and Materials, 1978, p 164-175
38. D. McLachlan, Jr., Extreme Focal Depth in Microscopy, *Appl. Opt.*, Vol 3 (No. 9), 1964, p 1009-1013
39. M.D. Kelly and J.E. Selle, The Operation and Modification of the Deep Field Microscope, in *Proceedings of the 2nd Annual Technical Meeting*, International Metallographic Society, 1969, p 171-177
40. J.H. Waddell, Deep-Field Low-Power Microscope, *Res. Dev.*, Vol 19, May 1968, p 34
41. A. Phillips et al., *Electron Fractography Handbook*, AFML-TDR-64-416, Air Force Materials Laboratory, 31 Jan 1965
42. A. Phillips et al., *Electron Fractography Handbook*, MCIC-HB-08, Air Force Materials Laboratory and the Metals and Ceramics Information Center, June 1976
43. B.V. Whiteson et al., Electron Fractographic Techniques, in *Techniques of Metals Research*, Vol II, Pt. I, Interscience, 1968, p 445-497
44. C.D. Beachem, The Effects of Crack Tip Plastic Flow Directions Upon Microscopic Dimple Shapes, *Metall. Trans.*, Vol 6A, Feb 1975, p 377-383
45. B.J. Brindley, The Mechanism of Ductile Fracture in an Fe-21% Cr-0.5% C Alloy, *Acta Metall.*, Vol 16, April 1968, p 587-595
46. A.W. Thompson and P.F. Weihrauch, Ductile Fracture: Nucleation at Inclusions, *Scr. Metall.*, Vol 10, Feb 1976, p 205-210
47. E.R. Parker et al., A Study of the Tension Test, *Proc. ASTM*, Vol 46, 1946, p 1159-1174
48. I.E. French and P.F. Weinrich, The Tensile Fracture Mechanisms of F.C.C. Metals and Alloys—A Review of the Influence of Pressure, *J. Austral. Inst. Met.*, Vol 22 (No. 1), March 1977, p 40-50
49. H.C. Rogers, The Effect of Material Variables on Ductility, in *Ductility*, American Society for Metals, 1968, p 31-56
50. L.D. Kenney et al., Effect of Particles on the Tensile Fracture of Aluminum Alloys, in *Microstructural Science*, Vol 8, Elsevier, 1980, p 153-156
51. B.I. Edelson and W.W. Baldwin, Jr., The Effect of Second Phases on the Mechanical Properties of Alloys, *Trans. ASM*, Vol 55, 1962, p 230-250
52. J. Gurland and J. Plateau, The Mechanism of Ductile Rupture of Metals Containing Inclusions, *Trans. ASM*, Vol 56, 1963, p 442-454
53. T. Inoue and S. Kinoshita, Mechanism of Void Initiation in the Ductile Fracture Process of a Spheroidized Carbon Steel, in *Microstructure and Design of Alloys*, Vol 1, Institute of Metals and the Iron and Steel Institute, 1973, p 159-163
54. C. Wells and R.F. Mehl, Transverse Mechanical Properties in Heat Treated Wrought Steel Products, *Trans. ASM*, Vol 41, 1949, p 715-818
55. E.A. Loria, Transverse Ductility Variations in Large Steel Forgings, *Trans. ASM*, Vol 42, 1950, p 486-498
56. E.G. Olds and C. Wells, Statistical Methods for Evaluating the Quality of Certain Wrought Steel Products, *Trans. ASM*, Vol 42, 1950, p 845-899
57. C. Wells et al., Effect of Composition on Transverse Mechanical Properties of Steel, *Trans. ASM*, Vol 46, 1954, p 129-156
58. H.H. Johnson and G.A. Fisher, Steel Quality as Related to Test Bar Fractures, *Trans. AFS*, Vol 58, 1950, p 537-549
59. J. Welchner and W.G. Hildorf, Relationship of Inclusion Content and Transverse Ductility of a Chromium-Nickel-Molybdenum Gun Steel, *Trans. ASM*, Vol 42, 1950, p 455-485
60. H.D. Shephard and E.A. Loria, The Nature of Inclusions in Tensile Fractures of Forging Steels, *Trans. ASM*, Vol 41, 1949, p 375-395
61. F.L. Carr et al., Correlation of Microfractography and Macrofractography of AISI 4340 Steel, in *Application of Electron Microfractography to Materials Research*, STP 493, American Society for Testing and Materials, 1971, p 36-54
62. J.A. Kies et al., Interpretation of Fracture Markings, *J. Appl. Phys.*, Vol 21, July 1950, p 716-720
63. J. Nunes et al., Macrofractographic Techniques, in *Techniques of Metals Research*, Vol 2, Pt. 1, John Wiley & Sons, 1968, p 379-444
64. J.H. Hollomon, Temper Brittleness, *Trans. ASM*, Vol 36, 1946, p 473-541
65. W.R. Clough et al., The Rosette, Star, Tensile Fracture, *J. Basic Eng. (Trans. ASME)*, Vol 90, March 1968, p 21-27
66. A.S. Shneiderman, Star-Type Fracture in the Tensile Testing of Testpieces, *Ind. Lab.*, Vol 41, July 1974, p 1061-1062
67. F.R. Larson and F.L. Carr, Tensile Fracture Surface Configurations of a Heat-Treated Steel as Affected by Tempera-

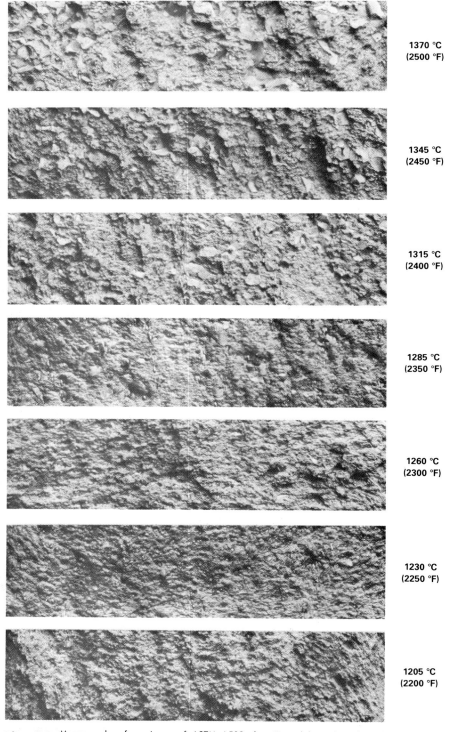

Fig. 88 Macrographs of specimens of ASTM A508 class II steel heated to the indicated temperatures, normalized, quenched, tempered to about 37 to 39 HRC, and fractured. Overheating facets are observed in all samples, but are not excessively large until the steel is heated to 1260 °C (2300 °F) or above. Source: Ref 390

ture, *Trans. ASM*, Vol 55, 1962, p 599-612

68. F.R. Larson and J. Nunes, Low Temperature Plastic Flow and Fracture Tension Properties of Heat-Treated SAE 4340 Steel, *Trans. ASM*, Vol 53, 1961, p 663-682

69. F.L. Carr and F.R. Larson, Fracture Surface Topography and Toughness of AISI 4340 Steel, *J. Mater.*, Vol 4, Dec 1969, p 865-875

70. J.H. Bucher et al., Tensile Fracture of Three Ultra-High-Strength Steels, *Trans. AIME*, Vol 233, May 1965, p 884-889

71. F.L. Carr et al., Mechanical Properties and Fracture Surface Topography of a Thermally Embrittled Steel, in *Temper Embrittlement in Steel*, STP 407, American Society for Testing and Materials, 1968, p 203-236

72. R.J. Hrubec et al., The Split, Layered, Cup-and-Cone Tensile Fracture, *J. Basic Eng. (Trans. ASME)*, Vol 90, March 1968, p 8-12

73. B.M. Kopadia et al., Influence of Mechanical Fibering on Brittle Fracture in Hot Rolled Steel Plate, *Trans. ASM*, Vol 55, 1962, p 289-398

74. E.A. Almond et al., Fracture in Laminated Materials, in *Interfaces in Composites*, STP 452, American Society for Testing and Materials, 1969, p 107-129

75. E.A. Almond, Delamination in Banded Steels, *Metall. Trans.*, Vol 1, July 1970, p 2038-2041

76. D.F. Lentz, Factors Contributing to Split Fractures, in *Mechanical Working and Steel Processing*, Meeting XII, American Institute of Mining, Metallurgical, and Petroleum Engineers, 1974, p 397-411

77. D.M. Fegredo, The Effect of Rolling at Different Temperatures on the Fracture Toughness Anisotropy of a C-Mn Structural Steel, *Can. Metall. Q.*, Vol 14 (No. 3), 1975, p 243-255

78. H. Hero et al., The Occurrence of Delamination in a Control Rolled HSLA Steel, *Can. Metall. Q.*, Vol 14 (No. 2), 1975, p 117-122

79. P. Brozzo and G. Buzzichelli, Effect of Plastic Anisotropy on the Occurrence of "Separations" on Fracture Surfaces of Hot-Rolled Steel Specimens, *Scr. Metall.*, Vol 10, 1976, p 233-240

80. D.N. Hawkins, Cleavage Separations in Warm-Rolled Low-Carbon Steels, *Met. Technol.*, Sept 1976, p 417-421

81. A.J. DeArdo, On Investigation of the Mechanism of Splitting Which Occurs in Tensile Specimens of High Strength Low Alloy Steels, *Metall. Trans. A*, Vol 8, March 1977, p 473-486

82. B.L. Bramfitt and A.R. Marder, Splitting Behavior in Plate Steels, in *Toughness Characterization and Specifications for HSLA and Structural Steels*, American Institute of Mining, Metallurgical, and Petroleum Engineers, 1977, p 236-256

83. B.L. Bramfitt and A.R. Marder, A Study of the Delamination Behavior of a Very Low-Carbon Steel, *Metall. Trans.*, Vol 8A, Aug 1977, p 1263-1273

84. J.E. Ryall and J.G. Williams, Fracture Surface Separations in the Charpy V-Notch Test, *BHP Tech. Bull.*, Vol 22, Nov 1978, p 38-45

85. B. Engl and A. Fuchs, Macroscopic and Microscopic Features of Separations in

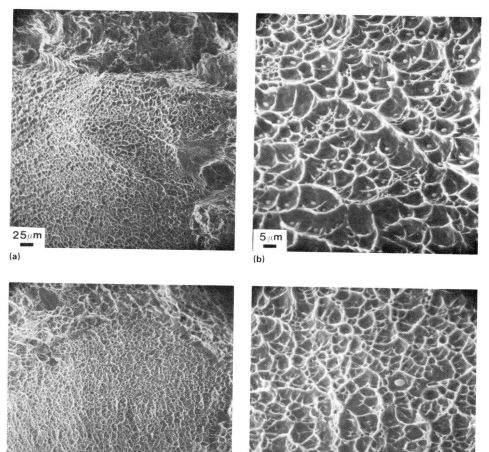

Fig. 89 SEM fractographs of the appearance of facets in specimens heated to 1205 and 1370 °C (2200 and 2500 °F). (a) and (b) Specimens heated to 1370 °C (2500 °F). (c) and (d) Specimens heated to 1205 °C (2200 °F). (a) 135×. (b) 680×. (c) 135×. (d) 680×. Source: Ref 390

Structural Steels, *Prakt. Metallogr.*, Vol 17, Jan 1980, p 3-13
86. G.D. Fearnehough Fracture Propagation Control in Gas Pipelines: A Survey of Relevant Studies, *Int. J. Pressure Vessels Piping*, Vol 2, 1974, p 257-282
87. R. Schofield et al., "Arrowhead" Fractures in Controlled-Rolled Pipeline Steels, *Met. Technol.*, July 1974, p 325-331
88. T. Yamaguchi et al., Study of Mechanism of Separation Occurring on Fractured Surface of High Grade Line Pipe Steels, *Nippon Kokan Tech. Rep. (Overseas)*, Dec 1974, p 41-53
89. D.S. Dabkowski et al., "Splitting-Type" Fractures in High-Strength Line-Pipe Steels, *Met. Eng. Q.*, Feb 1976, p 22-32
90. M. Iino et al., On Delamination in Linepipe Steels, *Trans. ISIJ*, Vol 17, 1977, p 450-458
91. A. Gangulee and J. Gurland, On the Fracture of Silicon Particles in Aluminum-Silicon Alloys, *Trans. AIME*, Vol 239, Feb 1967, p 269-272
92. C.J. McMahon, Jr., The Microstructural Aspects of Tensile Fracture, in *Fundamental Phenomena in the Material Sciences*, Vol 4, Plenum Press, 1967, p 247-284
93. J.C.W. Van De Kasteele and D. Broek, The Failure of Large Second Phase Particles in a Cracking Aluminum Alloy, *Eng. Fract. Mech.*, Vol 9 (No. 3), 1977, p 625-635
94. A.S. Argon and J. Im, Separation of Second Phase Particles in Spheroidized 1045 Steel, Cu-0.6 Pct Cr Alloy, and Maraging Steel in Plastic Straining, *Metall. Trans. A*, Vol 6, April 1975, p 839-851
95. T. Kunio et al., An Effect of the Second Phase Morphology on the Tensile Fracture Characteristics of Carbon Steels, *Eng. Fract. Mech.*, Vol 7, Sept 1975, p 411-417
96. W.A. Spitzig, Effect of Sulfides and Sulfide Morphology on Anistropy of Tensile Ductility and Toughness of Hot-Rolled C-Mn Steels, *Metall. Trans.*, Vol 14A, March 1983, p 471-484
97. J.E. Croll, Factors Influencing the Through-Thickness Ductility of Structural Steels, *BHP Tech. Bull.*, Vol 20, April 1976, p 24-29
98. I.D. Simpson et al., Effect of the Shape and Size of Inclusions on Through-Thickness Properties, *BHP Tech. Bull.*, Vol 20, April 1976, p 30-36
99. I.D. Simpson et al., The Effect of Non-Metallic Inclusions on Mechanical Properties, *Met. Forum*, Vol 2 (No. 2), 1979, p 108-117
100. H. Takada et al., Effect of the Amount and Shape of Inclusions on the Directionality of Ductility in Carbon-Manganese Steels, in *Fractography in Failure Analysis*, STP 645, American Society for Testing and Materials, 1978, p 335-350
101. W.A. Spitzig and R.J. Sober, Influence of Sulfide Inclusions and Pearlite Content on the Mechanical Properties of Hot-Rolled Carbon Steels, *Metall. Trans.*, Vol 12A, Feb 1981, p 281-291
102. W.C. Leslie, Inclusions and Mechanical Properties, in *Mechanical Working & Steel Processing*, Meeting XX, American Institute of Mining, Metallurgical, and Petroleum Engineers, 1983, p 3-50
103. C.J. McMahon, Jr. and M. Cohen, Initiation of Cleavage in Polycrystalline Iron, *Acta Metall.*, Vol 13, June 1965, p 591-604
104. C.J. McMahon, Jr. and M. Cohen, The Fracture of Polycrystalline Iron, in *Proceedings of the First International Conference on Fracture*, Sendai, Japan, 12-17 Sept 1965, p 779-812
105. E.O. Hall, The Deformation and Aging of Mild Steel: III Discussion of Results, *Proc. Phys. Soc. (London) B*, Vol 64, 1 Sept 1951, p 747-753
106. N.J. Petch, The Cleavage Strength of Polycrystals, *J. Iron Steel Inst.*, Vol 173, May 1953, p 25-28
107. C.F. Tipper, The Study of Fracture Surface Markings, *J. Iron Steel Inst.*, Vol 185, Jan 1957, p 4-9
108. G.M. Boyd, The Propagation of Fractures in Mild Steel Plates, *Engineering*, Vol 175, 16 Jan 1953, p 65-69; 23 Jan 1953, p 100-102
109. J.B. Cornish and J.E. Scott, Fracture Study of Gas Transmission Line Pipe, in *Mechanical Working & Steel Processing*, Vol II, American Institute of Mining, Metallurgical, and Petroleum Engineers, 1969, p 222-239
110. D.E. Babcock, Brittle Fracture: An Inter-

Fig. 90 Stretcher-strain marks (Lüders bands) on the surface of a range component after forming 0.25×.

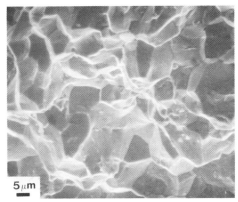

Fig. 91 SEM fractograph of a quench crack surface in AISI 5160 alloy steel showing a nearly complete intergranular fracture path. 680×

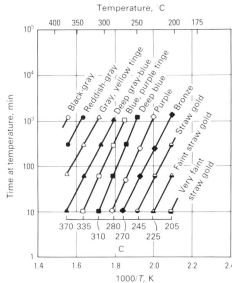

Fig. 92 Temper colors as a function of time at heat for AISI 1035 steel. Source: Ref 414

pretation of Its Mechanism, in *AISI Yearbook*, American Iron and Steel Institute, 1968, p 255-278
111. J.R. Low, Jr. and R.G. Feustel, "Inter-Crystalline Fracture and Twinning of Iron at Low Temperature, *Acta Metall.*, Vol 1, March 1953, p 185-192
112. J.H. Westbrook and D.L. Wood, Embrittlement of Grain Boundaries by Equilibrium Segregation, *Nature*, Vol 192, 30 Dec 1961, p 1280-1281
113. C. Lipson, *Basic Course in Failure Analysis*, Penton (reprinted from *Mach. Des.*)
114. G. Jacoby, Fractographic Methods in Fatigue Research, *Exp. Mech.*, March 1965, p 65-82
115. D.A. Ryder, *The Elements of Fractography*, NATO Report AGARD-AG-155-71, NTIS AD-734-619, National Technical Information Service, Nov 1971
116. G.F. Vander Voort, Macroscopic Examination Procedures for Failure Analysis, in *Metallography in Failure Analysis*, Plenum Press, 1978, p 33-63
117. D.J. Wulpi, *How Components Fail*, American Society for Metals, 1966
118. D.J. Wulpi, *Understanding How Components Fail*, American Society for Metals, 1985
119. J. Mogul, Metallographic Characterization of Fatigue Failure Origin Areas, in *Metallography in Failure Analysis*, Plenum Press, 1978, p 97-120
120. R.D. Barer and B.F. Peters, *Why Metals Fail*, Gordon & Breach, 1970
121. J.A. Bennett and J.A. Quick, "Mechanical Failures of Metals in Service," NBS Circular 550, National Bureau of Standards, 27 Sept 1954
122. V.J. Colangelo and F.A. Heiser, *Analysis of Metallurgical Failures*, John Wiley & Sons, 1974
123. D. McIntyre, Fractographic Analysis of Fatigue Failures, *J. Eng. Mater. Technol. (Trans. ASME)*, July 1975, p 194-205
124. C.E. Feddersen, "Fatigue Crack Propagation in D6AC Steel Plate for Several Flight Load Profiles in Dry Air and JP-4 Fuel Environments," AFML-TR-72-20, Battelle Memorial Institute, Jan 1972
125. M.K. Chakko and K.N. Tong, Evaluation of Resistance to Spalling of Roll Materials, *Iron Steel Eng.*, Vol 42, Oct 1965, p 141-154
126. J.D. Keller, Effect of Roll Wear on Spalling, *Iron Steel Eng.*, Vol 37, Dec 1960, p 171-178
127. F.K. Naumann and F. Spies, Working Roll With Shell-Shaped Fractures, *Pract. Metallogr.*, Vol 13, 1976, p 440-443
128. J.M. Chilton and M.J. Roberts, Factors Influencing the Performance of Forged Hardened Steel Rolls, in *AISE Yearbook*, Association of Iron and Steel Engineers, 1981, p 85-90
129. M. Nakagawa et al., Causes and Countermeasures of Spalling of Cold Mill Work Rolls, in *AISE Yearbook*, Association of Iron and Steel Engineers, 1981, p 134-139
130. S.J. Manganello and D.R. Churba, Roll Failures and What to Do When They

Visual Examination and Light Microscopy / 149

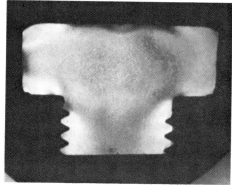

Fig. 93 Example of quench cracks on the head of AISI 1040 steel bolts. Cracks were caused by incomplete development of the case. (a) Bolt heads at 0.72×; cracks accentuated using magnetic particles. (b) Quench crack near a corner. Etched with 2% nital. 54×. (c) Opened quench crack, with arrows indicating temper color. 1.5×. (d) Macrograph showing lack of complete case hardening around head. Actual size

Occur, in *Mechanical Working & Steel Processing*, Meeting XVIII, American Institute of Mining, Metallurgical, and Petroleum Engineers, 1980, p 204-230

131. A.F. Kaminskas, Antidotes for Sleeve Bearing Failures, in *Iron and Steel Engineering Yearbook*, Association of Iron and Steel Engineers, 1955, p 717-724

132. W.E. Duckworth and G.H. Walter, Fatigue in Plain Bearings, in *Proceedings of the International Conference on Fatigue of Metals*, The Institute of Mechanical Engineering and The American Society of Mechanical Engineers, 1956, p 585-592

133. W.R. Good and A.J. Gunst, Bearing Failures and Their Causes, *Iron Steel Eng.*, Vol 43, Aug 1966, p 83-93

134. R.L. Widner and J.O. Wolfe, Valuable Results From Bearing Damage Analysis, *Met. Prog.*, Vol 93, April 1968, p 79-86

135. S. Borgese, An Electron Fractographic Study of Spalls Formed in Rolling Contact, *J. Basic Eng. (Trans. ASME) D*, Vol 89, Dec 1967, p 943-948

136. R.J. Henry, The Cause of White Etching Material Outlining Shell-Type Cracks in Rail-Heads, *J. Basic Eng. (Trans. ASME) D*, Vol 91, Sept 1969, p 549-551

137. R. Rungta *et al.*, An Investigation of Shell and Detail Cracking in Railroad Rails, in *Corrosion, Microstructure & Metallography*, Vol 12, *Microstructural Science*, American Society for Metals and the International Metallographic Society, 1985, p 383-406

138. C.G. Chipperfield and A.S. Blicblau, Modelling Rolling Contact Fatigue in Rails, *Railw. Gaz. Int.*, Jan 1984, p 25-31

139. H. Masumoto *et al.*, Some Features and Metallurgical Considerations of Surface Defects in Rail Due to Contact Fatigue, in *Rail Steels—Developments, Processing, and Use*, STP 644, American Society for Testing and Materials, 1976, p 233-255

140. T. Mitsuda and F.G. Bauling, Research on Shelling of Crane Wheels, in *Iron and Steel Engineering Yearbook*, Association of Iron and Steel Engineers, 1966, p 272-281

141. S. Neumann and L.E. Arnold, Prediction and Analysis of Crane Wheel Service Life, in *Iron and Steel Engineering Yearbook*, Association of Iron and Steel Engineers, 1971, p 102-110

142. L.E. Arnold, Replicas Enable New Look at Roll Surfaces, *Iron Steel Eng.*, Vol 43, Aug 1966, p 129-133

143. J.J. Bush *et al.*, Microstructural and Residual Stress Changes in Hardened Steel Due to Rolling Contact, *Trans. ASM*, Vol 54, 1961, p 390-412

144. A.J. Gentile *et al.*, Phase Transformations in High-Carbon, High-Hardness Steels Under Contact Loads, *Trans. AIME*, Vol 233, June 1965, p 1085-1093

145. A.H. King and J.L. O'Brien, Microstructural Alterations in Rolling Contact Fatigue, in *Advances in Electron Metallography*, STP 396, American Society for Testing and Materials, 1966, p 74-88

146. J.A. Martin *et al.*, Microstructural Alterations of Rolling Bearing Steel Undergoing Cyclic Stressing, *J. Basic Eng. (Trans. ASME) D*, Vol 88, Sept 1966, p 555-567

147. J.L. O'Brien and A.H. King, Electron Microscopy of Stress-Induced Structural Alterations Near Inclusions in Bearing Steels, *J. Basic Eng. (Trans. ASME) D*, Vol 88, Sept 1966, p 568-572

148. W.E. Littmann and R.L. Widner, Propagation of Contact Fatigue From Surface and Subsurface Origins, *J. Basic Eng. (Trans. ASME) D*, Vol 88, Sept 1966, p 624-636

149. J.A. Martin and A.D. Eberhardt, Identification of Potential Failure Nuclei in Rolling Contact Fatigue, *J. Basic Eng. (Trans. ASME) D*, Vol 89, Dec 1967, p 932-942

150. J. Buchwald and R.W. Heckel, An Analysis of Microstructural Changes in 52100 Steel Bearings During Cyclic Stressing, *Trans. ASM*, Vol 61, 1968, p 750-756

151. R. Tricot *et al.*, How Microstructural Alterations Affect Fatigue Properties of 52100 Steel, *Met. Eng. Q.*, Vol 12, May 1972, p 39-47

152. H. Swahn *et al.*, Martensite Decay During Rolling Contact Fatigue in Ball Bearings, *Metall. Trans.*, Vol 7A, Aug 1976, p 1099-1110

153. R. Österlund and O. Vingsbo, Phase Changes in Fatigued Ball Bearings, *Metall. Trans.*, Vol 11A, May 1980, p 701-707

154. P.C. Becker, Microstructural Changes Around Non-Metallic Inclusions Caused by Rolling-Contact Fatigue of Ball-Bearing Steels, *Met. Technol.*, Vol 8, June 1981, p 234-243

155. R. Österlund *et al.*, Butterflies in Fatigued Ball Bearings—Formation Mechanisms and Structure, *Scand. J. Metall.*, Vol 11 (No. 1), 1982, p 23-32

156. K. Tsubota and A. Koyanagi, Formation of Platelike Carbides during Rolling Contact Fatigue in High-Carbon Chromium Bearing Steel, *Trans. ISIJ*, Vol 25, p 496-504

157. H.J. Gough, Crystalline Structure in Relation to Failure of Metals—Especially by Fatigue, *Proc. ASTM*, Vol 33, Pt. II, 1933, p 3-114

158. N. Thompson *et al.*, The Origin of Fatigue Fracture in Copper, *Philos. Mag.*, Series 8, Vol 1, 1959, p 113-126

159. J.M. Finney and C. Laird, Strain Localization in Cyclic Deformation of Copper Single Crystals, *Philos. Mag.*, Series 8, Vol 31 (No. 2), Feb 1975, p 339-366

160. N.M.A. Eid and P.F. Thomason, The Nucleation of Fatigue Cracks in a Low-Alloy Steel Under High-Cycle Fatigue Conditions and Uniaxial Loading, *Acta Metall.*, Vol 27, July 1979, p 1239-1249

161. P.C. Paris, The Fracture Mechanics Ap-

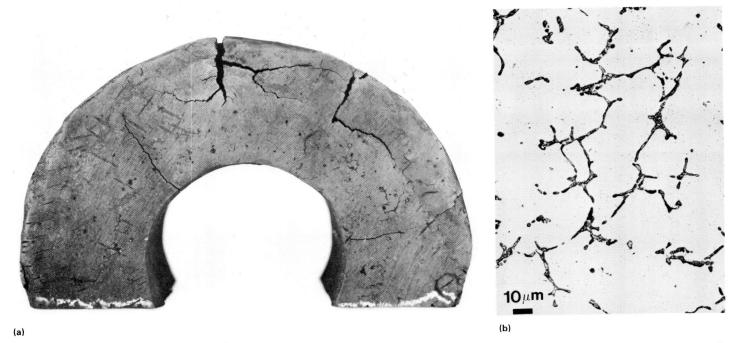

Fig. 94 Cracked 25Cr-12Ni cast stainless steel quenching fixture. (a) Macrograph of part of the fixture. (b) Microstructure showing substantial σ phase. Electrolytically etched with 10 N KOH. 500×

proach to Fatigue, in *Fatigue—An Interdisciplinary Approach*, 10th Sagamore Army Materials Research Conference, Syracuse University Press, 1964, p 107-132

162. W.L. Morris, Microcrack Closure Phenomena for Al 2219-T851, *Metall. Trans.*, Vol 10A, Jan 1979, p 5-11
163. M. Kikukawa et al., Direct Observation and Mechanism of Fatigue Crack Propagation, in *Fatigue Mechanisms*, STP 675, American Society for Testing and Materials, 1979, p 234-253
164. D.L. Davidson and J. Lankford, Dynamic, Real-Time Fatigue Crack Propagation at High Resolution as Observed in the Scanning Electron Microscope, in *Fatigue Mechanisms*, STP 675, American Society for Testing and Materials, 1979, p 277-284
165. D.L. Davidson and J. Lankford, Fatigue Crack Propagation: New Tools for the Study of an Old Problem, *J. Met.*, Vol 31, Nov 1979, p 11-16
166. D.L. Davidson, The Study of Fatigue Mechanisms With Electron Channeling, in *Fatigue Mechanisms*, STP 675, American Society for Testing and Materials, 1979, p 254-275
167. C.A. Zapffe and C.O. Worden, Fractographic Registrations of Fatigue, *Trans. ASM*, Vol 43, 1951, p 958-969
168. W.J. Plumbridge and D.A. Ryder, The Metallography of Fatigue, *Met. Rev.*, No. 136, Aug 1969, p 119-142
169. G.A. Miller, Fatigue Fracture Appearance and the Kinetics of Striation Formation in Some High-Strength Steels, *Trans. ASM*, Vol 62, 1969, p 651-658
170. I. LeMay and M.W. Lui, Fractographic Observations of Fatigue Fracture in High-Strength Steels, *Metallography*, Vol 8, 1975, p 249-252
171. M.W. Lui and I. LeMay, Fatigue Fracture Surface Features: Fractography and Mechanisms of Formation, in *Microstructural Science*, Vol 8, Elsevier, 1980, p 341-352
172. P.J.E. Forsyth et al., Cleavage Facets Observed on Fatigue-Fracture Surfaces in an Aluminum Alloy, *J. Inst. Met.*, Vol 90, 1961-1962, p 238-239
173. M. Sumita et al., Fatigue Fracture Surfaces of Steels Containing Inclusions, *Trans. Natl. Res. Inst. for Metals*, Vol 14 (No. 4), 1972, p 146-154
174. R.M.N. Pelloux, "The Analysis of Fracture Surfaces by Electron Microscopy," Report Dl-82-0169-R1, Boeing Scientific Research Laboratory, Dec 1963; see also Technical Report P19-3-64, American Society for Metals, Oct 1964
175. C.D. Beachem, "Electron Microscope Fracture Examination to Characterize and Identify Modes of Fracture," Report 6293 (AFML-TR-64-408), Naval Research Laboratory, 28 Sept 1965
176. C.D. Beachem and D.A. Meyn, "Illustrated Glossary of Fractographic Terms," NRL Memorandum, Report 1547, Naval Research Laboratory, June 1964
177. R.W. Hertzberg and W.J. Mills, Character of Fatigue Fracture Surface Micromorphology in the Ultra-Low Growth Rate Regime, in *Fractographic-Microscopic Cracking Processes*, STP 600, American Society for Testing and Materials, 1976, p 220-234
178. R. Koterazawa et al., Fractographic Study of Fatigue Crack Propagation, *J. Eng. Mater. Technol. (Trans. ASME)*, Oct 1973, p 202-212
179. B.V. Whiteson et al., Special Fractographic Techniques for Failure Analysis, in *Electron Fractography*, STP 436, American Society for Testing and Materials, 1968, p 151-178
180. A. Yuen et al., Correlations Between Fracture Surface Appearance and Fracture Mechanics Parameters for Stage II Fatigue Crack Propagation in Ti-6Al-4V, *Metall. Trans.*, Vol 5, Aug 1974, p 1833-1842
181. A.J. Brothers and S. Yukawa, Engineering Applications of Fractography, in *Electron Fractography*, STP 436, American Society for Testing and Materials, 1968, p 176-195
182. R.C. Bates and W.G. Clark, Jr., Fractography and Fracture Mechanics, *Trans. ASM*, Vol 62, 1969, p 380-389
183. R.C. Bates et al., Correlation of Fractographic Features With Fracture Mechanics Data, in *Electron Microfractography*, STP 453, American Society for Testing and Materials, 1969, p 192-214
184. E. Gassner, Fatigue Strength. A Basis for Measuring Construction Parts With Random Loads Under Actual Usage, *Konstruction*, Vol 6, 1954, p 97-104
185. E. Gassner, Effect of Variable Load and Cumulative Damage in Vehicle and Airplane Structures, in *International Conference on Fatigue of Metals*, 1956, p

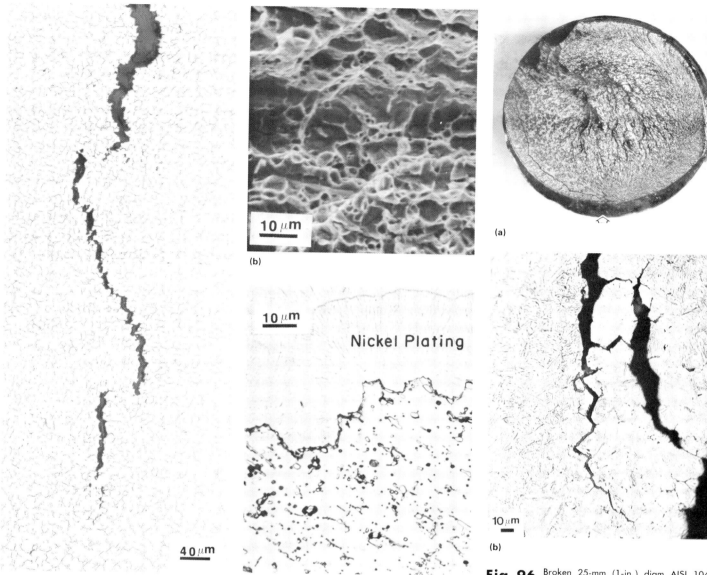

Fig. 95 Three views of a fractured specimen of type 312 weld metal that was exposed to high temperatures to transform the δ ferrite to σ phase. The specimen was subsequently broken by impact at room temperature. (a) Partially broken specimen. Etched with mixed acids. (b) SEM view of fracture. (c) Cross section of fracture. Etched with acetic glyceregia

Fig. 96 Broken 25-mm (1-in.) diam AISI 1040 steel bolt. (a) Macrograph of fracture surface; corrosion products obscure most of the surface. 2×. Intergranular secondary cracks (b) were observed in the region near the surface of the bolt shown by the arrow in (a). The bolt was not tempered (surface hardness was 53 to 57 HRC) and probably failed by SCC. (b) Etched with 2% nital. 340×

304-309
186. G. Quest, Quantitative Determination of the Load and the Number of Cycles from the Surface of Fatigue Fractures, *Der Maschineenschaden*, Vol 33, 1960, p 4-12, 33-44
187. E.E. Underwood and E.A. Starke, Jr., Quantitative Stereological Methods for Analyzing Important Microstructural Features in Fatigue of Metals and Alloys, in *Fatigue Mechanisms*, STP 675, American Society for Testing and Materials, 1979, p 633-682
188. J.C. McMillan and R.W. Hertzberg, Application of Electron Fractography to Fatigue Studies, in *Electron Fractography*, STP 436, American Society for Testing and Materials, 1968, p 89-123
189. P.J.E. Forsyth and D.A. Fyder, Fatigue Fracture. Some Results Derived From the Microscopic Examination of Crack Surfaces, *Aircr. Eng.*, Vol 32, April 1960, p 96-99
190. C.E. Price and D. Cox, Observing Fatigue With The Nomarski Technique, *Met. Prog.*, Vol 123, Feb 1983, p 37-39
191. W. Rosenhain and D. Ewen, The Intercrystalline Cohesion of Metals, *J. Inst. Met.*, Vol 10, 1913, p 119-149
192. Z. Jeffries, The Amorphous Metal Hypothesis and Equicohesive Temperatures, *J. Am. Inst. Met.*, Vol 11 (No. 3), Dec 1917, p 300-324
193. Z. Jeffries, Effect of Temperature, Deformation, and Grain Size on the Mechanical Properties of Metals, *Trans. AIME*, Vol 60, 1919, p 474-576
194. H.C. Chang and N.J. Grant, Mechanisms of Intercrystalline Fracture, *Trans. AIME*, Vol 206, 1956, p 544-551
195. J.N. Greenwood, Intercrystalline Cracking of Metals, *J. Iron Steel Inst.*, Vol 171, Aug 1952, p 380
196. J.N. Greenwood, Intercrystalline Cracking of Metals, *Bull. Inst. Met.*, Vol 1, Pt. 12, Aug 1952, p 104-105; Intercrystalline Cracking of Brass, *Bull. Inst. Met.*, Vol 1, Pt. 14, Oct 1952, p 120-121

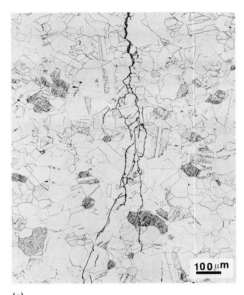

(a)

(b)

Fig. 97 Type 304 stainless steel specimen after testing in boiling MgCl$_2$. (a) Cross section of partially broken specimen. Etched with mixed acids. (b) SEM fractograph of completely broken specimen

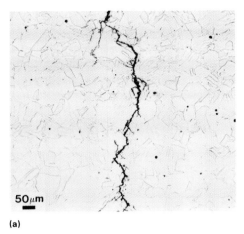

(a)

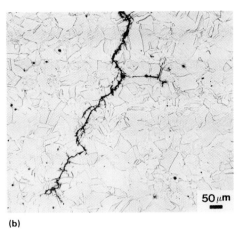

(b)

Fig. 98 Two views of the crack path, which was predominantly transgranular, in an austenitic manganese-chromium stainless steel drill collar alloy. The SCC was caused by chlorides in the drilling fluid. Cracking began at the inside bore surface. Etched with acetic glyceregia. Both 65×

197. W. Pavinich and R. Raj, Fracture at Elevated Temperatures, *Metall. Trans.*, Vol 8A, Dec 1977, p 1917-1933
198. W.A. Rachinger, Relative Grain Translations in the Plastic Flow of Aluminum, *J. Inst. Met.*, Vol 81, 1952-1953, p 33-41
199. J.A. Martin *et al.*, Grain-Boundary Displacement Vs. Grain Deformation as the Rate-Determining Factor in Creep, *Trans. AIME*, Vol 209, Jan 1957, p 78-81
200. D. McLean and M.H. Farmer, The Relation During Creep Between Grain-Boundary Sliding, Sub-Crystal Size, and Extension, *J. Inst. Met.*, Vol 85, 1956-1957, p 41-50
201. H.C. Chang and N.J. Grant, Observations of Creep of the Grain Boundary in High Purity Aluminum, *Trans. AIME*, Vol 194, June 1952, p 619-625
202. D. McLean and M.H. Farmer, Grain-Boundary Movement, Slip, and Fragmentation During Creep of Aluminum-Copper, Aluminum-Magnesium and Aluminum-Zinc Alloys, *J. Inst. Met.*, Vol 83, 1954-1955, p 1-10
203. D. McLean, Deformation at High Temperatures, *Met. Rev.*, Vol 7 (No. 28), 1962, p 481-527
204. R.C. Gifkins and T.G. Langdon, On The Question of Low-Temperature Sliding at Grain Boundaries, *J. Inst. Met.*, Vol 93, 1964-1965, p 347-352
205. C.M. Sellars, Estimation of Slip Strain of Interior Grains During Creep, *J. Inst. Met.*, Vol 93, 1964-1965, p 365-366
206. Y. Ishida *et al.*, Internal Grain Boundary Sliding During Creep, *Trans. AIME*, Vol 233, Jan 1965, p 204-212
207. F.N. Rhines *et al.*, Grain Boundary Creep in Aluminum Bicrystals, *Trans. ASM*, Vol 48, 1956, p 919-951
208. R.N. Stevens, Grain Boundary Sliding in Metals, *Met. Rev.*, Vol 11, Oct 1966, p 129-142
209. R.L. Bell and T.G. Langdon, An Investigation of Grain-Boundary Sliding During Creep, *J. Mater. Sci.*, Vol 2, 1967, p 313-323
210. R.L. Bell *et al.*, The Contribution of Grain Boundary Sliding to the Overall Strain of a Polycrystal, *Trans. AIME*, Vol 239, Nov 1967, p 1821-1824
211. T.G. Langdon and R.L. Bell, The Use of Grain Strain Measurements in Studies of High-Temperature Creep, *Trans. AIME*, Vol 242, Dec 1968, p 2479-2484
212. P.W. Davies and B. Wilshire, An Experiment on Void Nucleation During Creep, *J. Inst. Met.*, Vol 90, 1961-1962, p 470-472
213. R.V. Day, Intercrystalline Creep Failure in 1%Cr-Mo Steel, *J. Iron Steel Inst.*, Vol 203, March 1965, p 279-284
214. A. Gittins and H.D. Williams, The Effect of Creep Rate on the Mechanism of Cavity Growth, *Philos. Mag.*, Vol 16 (No. 142), Oct 1967, p 849-851
215. P.W. Davies *et al.*, On the Distribution of Cavities During Creep, *Philos. Mag.*, Vol 18 (No. 151), July 1968, p 197-200
216. T. Johannesson and A. Tholen, Cavity Formation in Copper and in a Steel During Creep, *J. Inst. Met.*, Vol 97, 1969, p 243-247
217. D.M.R. Taplin, A Note on the Distribution of Cavities During Creep, *Philos. Mag.*, Vol 20 (No. 167), Nov 1967, p 1079-1982
218. V.V.P. Kutumbarao and P. Rama Rao, On the Determination of the Distribution of Creep Cavities, *Metallography*, Vol 5, 1972, p 94-96
219. B.J. Cane, Creep-Fracture Initiation in 2-1/4% Cr-1%Mo Steel, *Met. Sci.*, Vol 10, Jan 1976, p 29-34
220. D.A. Miller and R. Pilkington, The Effect of Temperature and Carbon Content on the Cavitation Behavior of a 1.5 Pct Cr-0.5 Pct V Steel, *Metall. Trans.*, Vol 9A, April 1978, p 489-494
221. R.A. Scriven and H.D. Williams, The Derivation of Angular Distributions of Planes by Sectioning Methods, *Trans. AIME*, Vol 233, Aug 1965, p 1593-1602
222. D.M.R. Taplin and L.J. Barker, A Study of the Mechanism of Intergranular Creep Cavitation by Shadowgraphic Electron Microscopy, *Acta Metall.*, Vol 14, Nov 1966, p 1527-1531
223. G.J. Cocks and D.M.R. Taplin, An Appraisal of Certain Metallographic Techniques for Studying Cavities, *Metallurgia*, Vol 75 (No. 451), May 1967, p 229-235
224. D.M.R. Taplin and A.L. Wingrove, Study of Intergranular Cavitation in Iron by Electron Microscopy of Fracture Surfaces, *Acta Metall.*, Vol 15, July 1967, p 1231-1236
225. K. Farrell and J.O. Stiegler, Electron Fractography for Studying Cavities, *Metallurgia*, Vol 79 (No. 471), Jan 1969, p

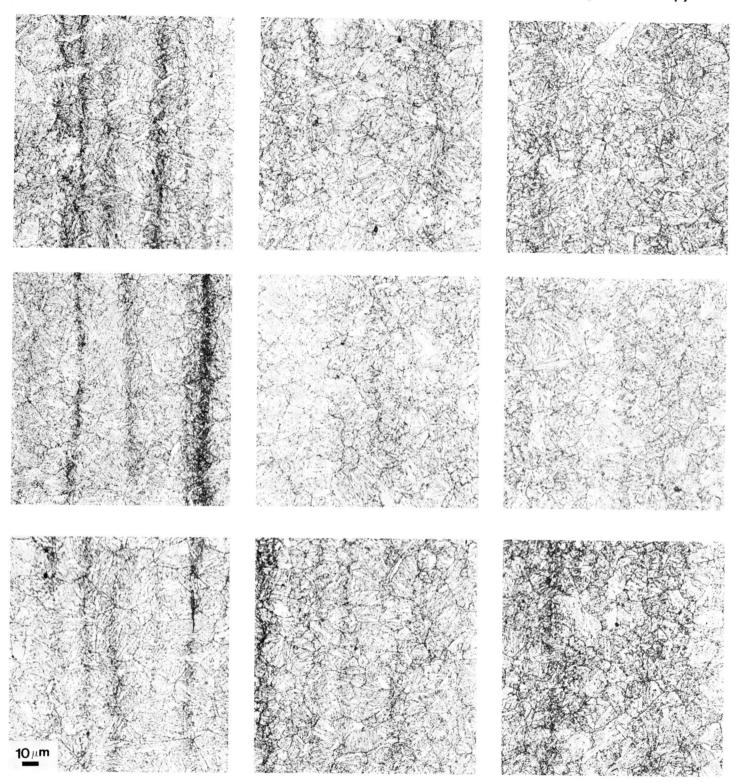

Fig. 99 Microstructures of AISI 4140 steel with 0.004% P (left column), 0.013% P (center column), and 0.022% P (right column). Specimens were etched with the etheral-picral etchant described in Ref 442 (top row) and Ref 444 (middle row) and with saturated aqueous picric acid plus a wetting agent (bottom row). All 425×

35-37
226. A.L. Wingrove and D.M.R. Taplin, The Morphology and Growth of Creep Cavities in α-Iron, *J. Mater. Sci.*, Vol 4, Sept 1969, p 789-796
227. H.R. Tipler *et al.*, Some Direct Observations on the Metallography of Creep-Cavitated Grain Boundaries, *Met. Sci. J.*, Vol 4, Sept 1970, p 167-170
228. W.E. White and I. LeMay, Metallographic and Fractographic Analyses of Creep Failure in Stainless Steel Weld

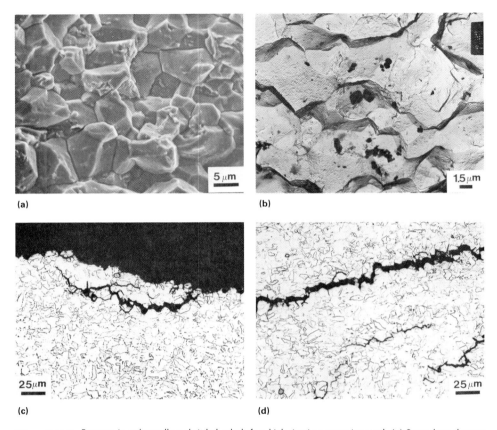

Fig. 100 Fracture in a thermally embrittled cobalt-free high-titanium maraging steel. (a) Secondary electron image of fracture surface. 1300×. (b) TEM extraction fractograph. 2150×. (c) Light micrograph of fracture edge. 260×. (d) Light micrograph of internal cracks. 260×. Light micrograph specimens etched with modified Fry's reagent

ments, in *Microstructural Science*, Vol 5, Elsevier, 1977, p 145-160
229. V.K. Sikka et al., Twin-Boundary Cavitation During Creep in Aged Type 304 Stainless Steel, *Metall. Trans.*, Vol 8A, July 1977, p 1117-1129
230. D.G. Morris and D.R. Harries, Wedge Crack Nucleation in Type 316 Stainless Steel, *J. Mater. Sci.*, Vol 12, Aug 1977, p 1587-1597
231. W.M. Stobbs, Electron Microscopical Techniques for the Observation of Cavities, *J. Microsc.*, Vol 116, Pt. 1, May 1979, p 3-13
232. R.J. Fields and M.F. Ashby, Observation on Wedge Cavities in the SEM, *Scr. Metall.*, Vol 14 (No. 7), July 1980, p 791-796
233. A.J. Perry, Cavitation in Creep, *J. Mater. Sci.*, Vol 9, June 1974, p 1016-1039
234. B.F. Dyson and D. McLean, A New Method of Predicting Creep Life, *Met. Sci. J.*, Vol 6, 1972, p 220-223
235. B. Walser and A. Rosselet, Determining the Remaining Life of Superheater-Steam Tubes Which Have Been in Service by Creep Tests and Structural Examinations, *Sulzer Res.*, 1978, p 67-72
236. N.G. Needham and T. Gladman, Nucleation and Growth of Creep Cavities in a Type 347 Steel, *Met. Sci.*, Vol 14, Feb 1980, p 64-72
237. Y. Lindblom, Refurbishing Superalloy Components for Gas Turbines, *Mater. Sci. Technol.*, Vol 1, Aug 1985, p 636-641
238. J. Wortmann, Improving Reliability and Lifetime of Rejuvenated Turbine Blades, *Mater. Sci. Technol.*, Vol 1, Aug 1985, p 644-650
239. C.J. Bolton et al., Metallographic Methods of Determining Residual Creep Life, *Mater. Sci. Eng.*, Vol 46, Dec 1980, p 231-239
240. R. Sandstrom and S. Modin, "The Residual Lifetime of Creep Deformed Components. Microstructural Observations for Mo- and CrMo-Steels," Report IM-1348, Swedish Institute for Metals Research, 1979
241. C. Bengtsson, "Metallographic Methods for Observation of Creep Cavities in Service Exposed Low-Alloyed Steel," Report IM-1636, Swedish Institute for Metals Research, March 1982
242. J.F. Henry and F.V. Ellis, "Plastic Replication Techniques for Damage Assessment," Report RP2253-01, Electric Power Research Institute, Sept 1983
243. J.F. Henry, Field Metallography. The Applied Techniques of In-Place Analysis, in *Corrosion, Microstructure, & Metallography*, Vol 12, *Microstructural Science*, American Society for Metals and the International Metallographic Society, 1985, p 537-549
244. M.C. Murphy and G.D. Branch, Metallurgical Changes in 2.25 CrMo Steels During Creep-Rupture Test, *J. Iron Steel Inst.*, Vol 209, July 1971, p 546-561
245. J.M. Leitnaker and J. Bentley, Precipitate Phases in Type 321 Stainless Steel After Aging 17 Years at ~600 °C, *Metall. Trans.*, Vol 8A, Oct 1977, p 1605-1613
246. M. McLean, Microstructural Instabilities in Metallurgical Systems—A Review, *Met. Sci.*, Vol 12, March 1978, p 113-122
247. S. Kihara et al., Morphological Changes of Carbides During Creep and Their Effects on the Creep Properties of Inconel 617 at 1000 °C, *Metall. Trans.*, Vol 11A, June 1980, p 1019-1031
248. S.F. Claeys and J.W. Jones, Role of Microstructural Instability in Long Time Creep Life Prediction, *Met. Sci.*, Vol 18, Sept 1984, p 432-438
249. Y. Minami et al., Microstructural Changes in Austenitic Stainless Steels During Long-Term Aging, *Mater. Sci. Technol.*, Vol 2, Aug 1986, p 795-806
250. J.R. Low, Jr., Impurities, Interfaces and Brittle Fracture, *Trans. AIME*, Vol 245, Dec 1969, p 2481-2494
251. W.P. Rees and B.E. Hopkins, Intergranular Brittleness in Iron-Oxygen Alloys, *J. Iron Steel Inst.*, Vol 172, Dec 1952, p 403-409
252. J.R. Low, Jr. and R.G. Feustel, Inter-Crystalline Fracture and Twinning of Iron at Low Temperatures, *Acta Metall.*, Vol 1, March 1953, p 185-192
253. B.E. Hopkins and H.R. Tipler, Effect of Heat-Treatment on the Brittleness of High-Purity Iron-Nitrogen Alloys, *J. Iron Steel Inst.*, Vol 177, May 1954, p 110-117
254. B.E. Hopkins and H.R. Tipler, The Effect of Phosphorus on the Tensile and Notch-Impact Properties of High-Purity Iron and Iron-Carbon Alloys, *J. Iron Steel Inst.*, Vol 188, March 1958, p 218-237
255. A.R. Troiano, The Role of Hydrogen and Other Interstitials in the Mechanical Behavior of Metals, *Trans. ASM*, Vol 52, 1960, p 54-80
256. C. Pichard et al., The Influence of Oxygen and Sulfur on the Intergranular Brittleness of Iron, *Metall. Trans.*, Vol 7A, Dec 1976, p 1811-1815
257. M.C. Inman and H.R. Tipler, Grain-Boundary Segregation of Phosphorus in an Iron-Phosphorus Alloy and the Effect Upon Mechanical Properties, *Acta Metall.*, Vol 6, Feb 1958, p 73-84
258. G.T. Hahn et al., "The Effects of Solutes

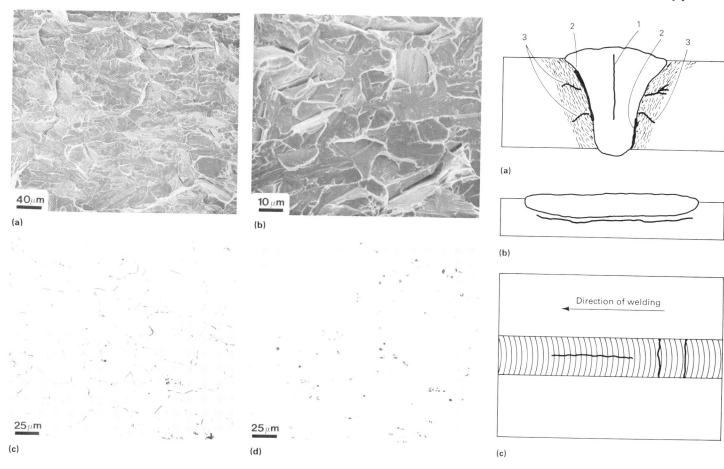

Fig. 101 Example of the fracture appearance and microstructure of a duplex stainless steel (similar to type 329) in the embrittled condition after heating at 370 °C (700 °F) for 1000 h. (a) and (b) SEM fractographs at 165 and 650×, respectively, showing cleavage fracture with some splitting. (c) Light micrograph showing dark-etching ferrite-ferrite grain boundaries in a specimen subjected to the embrittling treatment. (d) Light micrograph of a hot-rolled and annealed nonembrittled specimen, which shows only carbides. (c) and (d) Etched with Vilella's reagent. Both at 260×

Fig. 102 Schematics illustrating the nomenclature used to describe weldment cracks. (a) Weld metal (1), fusion line (2), and HAZ (3) cracks. (b) Underbead crack. (c) Longitudinal and transverse weld metal cracks. Source: Ref 480

on the Ductile-to-Brittle Transition in Refractory Metals," DMIC Memorandum 155, Battelle Memorial Institute, 28 June 1962
259. R.E. Maringer and A.D. Schwope, On the Effects of Oxygen on Molybdenum, *Trans. AIME*, Vol 200, March 1954, p 365-366
260. T.G. Nieh and W.D. Nix, Embrittlement of Copper Due to Segregation of Oxygen to Grain Boundaries, *Metall. Trans.*, Vol 12A, May 1981, p 893-901
261. R.H. Bricknell and D.A. Woodford, The Embrittlement of Nickel Following High Temperature Air Exposure, *Metall. Trans.*, Vol 12A, March 1981, p 425-433
262. K.M. Olsen et al., Embrittlement of High Purity Nickel, *Trans. ASM*, Vol 53, 1961, p 349-358
263. S. Floreen and J.H. Westbrook, Grain Boundary Segregation and the Grain Size Dependence of Strength of Nickel-Sulfur Alloys, *Acta Metall.*, Vol 17, Sept 1969, p 1175-1181
264. W.C. Johnson et al., Confirmation of Sulfur Embrittlement in Nickel Alloys, *Scr. Metall.*, Vol 8, Aug 1974, p 971-974
265. C. Loier and J.Y. Boos, The Influence of Grain Boundary Sulfur Concentration on the Intergranular Brittleness of Nickel of Different Purities, *Metall. Trans.*, Vol 12A, July 1981, p 1223-1233
266. J.H. Westbrook and D.L. Wood, A Source of Grain Boundary Embrittlement in Intermetallics, *J. Inst. Met.*, Vol 91, 1962-1963, p 174-182
267. E. Voce and A.P.C. Hallowes, The Mechanism of the Embrittlement of Deoxidized Copper by Bismuth, *J. Inst. Met.*, Vol 73, 1947, p 323-376
268. T.H. Schofield and F.W. Cuckow, The Microstructure of Wrought Non-Arsenical Phosphorus-Deoxidized Copper Containing Small Quantities of Bismuth, *J. Inst. Met.*, Vol 73, 1947, p 377-384
269. L.E. Samuels, The Metallography of Copper Containing Small Amounts of Bismuth, *J. Inst. Met.*, Vol 76, 1949-1950, p 91-102
270. C.W. Spencer et al., Bismuth in Copper Grain Boundaries, *Trans. AIME*, Vol 209, June 1957, p 793-794
271. D. McLean and L. Northcott, Antimonial 70:30 Brass, *J. Inst. Met.*, Vol 72, 1946, p 583-616
272. D. McLean, The Embrittlement of Copper: Antimony Alloys at Low Temperatures, *J. Inst. Met.*, Vol 81, 1952-1953, p 121-123
273. R. Carlsson, Hot Embrittlement of Copper and Brass Alloys, *Scand. J. Metall.*, Vol 9 (No. 1), 1980, p 25-29
274. H.K. Ihrig, The Effect of Various Elements on the Hot-Workability of Steel, *Trans. AIME*, Vol 167, 1946, p 749-790
275. J.M. Middletown and H.J. Protheroe, The Hot-Tearing of Steel, *J. Iron Steel Inst.*, Vol 168, Aug 1951, p 384-400
276. C.T. Anderson et al., Effect of Various Elements on Hot-Working Characteristics and Physical Properties of Fe-C Alloys, *J. Met.*, Vol 5, April 1953, p 525-529
277. C.T. Anderson et al., Forgeability of Steels with Varying Amounts of Manga-

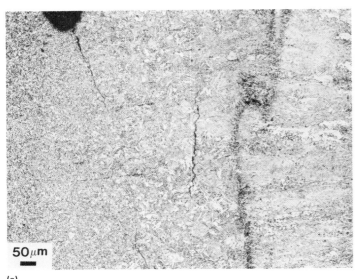
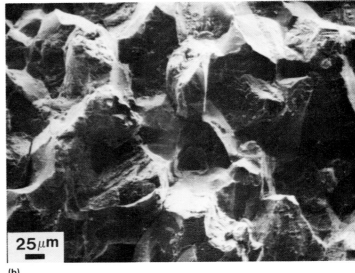

Fig. 103 Cold cracks in an RQC-90 steel plate welded with a high-hydrogen electrode. The sample was an implant specimen loaded to 193 MPa (28 ksi) during solidification. (a) Light micrograph showing cracking. Etched with nital. 80×. (b) SEM fractograph showing the intergranular nature of the cracks. 200×. (J.P. Snyder, Bethlehem Steel Corporation)

nese and Sulfur, *Trans. AIME*, Vol 200, July 1954, p 835-837

278. D. Smith *et al.*, Effects of Composition on the Hot Workability of Resulphurized Free-Cutting Steels, *J. Iron Steel Inst.*, Vol 210, June 1972, p 412-421

279. W.J. McG. Tegart and A. Gittins, The Role of Sulfides in the Hot Workability of Steels, in *Sulfide Inclusions in Steel*, American Society for Metals, 1975, p 198-211

280. A. Josefsson *et al.*, The Influence of Sulphur and Oxygen in Causing Red-Shortness in Steel, *J. Iron Steel Inst.*, Vol 191, March 1959, p 240-250

281. P. Bjornson and H. Nathorst, A Special Type of Ingot Cracks Caused by Certain Impurities, *Jernkontorets Ann.*, Vol 139, 1955, p 412-438

282. W.J.M. Salter, Effect of Mutual Additions of Tin and Nickel on the Solubility and Surface Energy of Copper in Mild Steel, *J. Iron Steel Inst.*, Vol 207, Dec 1969, p 1619-1623

283. W.J. Jackson and D.M. Southall, Effect of Copper and Tin Residual Amounts on the Mechanical Properties of 1.5Mn-Mo Cast Steel, *Met. Technol.*, Vol 5, Pt. 11, Nov 1978, p 381-390

284. K. Born, Surface Defects in the Hot Working of Steel, Resulting from Residual Copper and Tin, *Stahl Eisen*, Vol 73 (No. 20), BISI 3255, 1953, p 1268-1277

285. I.S. Brammar *et al.*, The Relation Between Intergranular Fracture and Sulphide Precipitation in Cast Alloy Steels, in *ISI 64*, Iron and Steel Institute, 1959, p 187-208

286. D. Bhattacharya and D.T. Quinto, Mechanism of Hot-Shortness in Leaded and Tellurized Free-Machining Steels, *Metall. Trans.*, Vol 11A, June 1980, p 919-934

287. R.A. Perkins and W.O. Binder, Improving Hot-Ductility of 310 Stainless, *J. Met.*, Vol 9, Feb 1957, p 239-245

288. L.G. Ljungström, The Influence of Trace Elements on the Hot Ductility of Austenitic 17Cr13NiMo-Steel, *Scand. J. Metall.*, Vol 6, 1977, p 176-184

289. W.B. Kent, Trace-Element Effects in Vacuum-Melted Superalloys, *J. Vac. Sci. Technol.*, Vol 11, Nov/Dec 1974, p 1038-1046

290. R.T. Holt and W. Wallace, Impurities and Trace Elements in Nickel-Base Superalloys, *Int. Met. Rev.*, Vol 21, March 1976, p 1-24

291. A.R. Knott and C.H. Symonds, Compositional and Structural Aspects of Processing Nickel-Base Alloys, *Met. Technol.*, Vol 3, Aug 1976, p 370-379

292. C.H. Lorig and A.R. Elsea, Occurrence of Intergranular Fracture in Cast Steels, *Trans. AFS*, Vol 55, 1947, p 160-174

293. B.C. Woodfine, "First Report on Intergranular Fracture in Steel Castings," BSCRA Report 38/54/FRP.5, British Steel Casting Research Association, March 1954

294. B.C. Woodfine, Effect of Al and N on the Occurrence of Intergranular Fracture in Steel Castings, *J. Iron Steel Inst.*, Vol 195, Aug 1960, p 409-414

295. R.F. Harris and G.D. Chandley, High Strength Steel Castings. Aluminum Nitride Embrittlement, *Mod. Cast.*, March 1962, p 97-103

296. J.A. Wright and A.G. Quarrell, Effect of Chemical Composition on the Occurrence of Intergranular Fracture in Plain Carbon Steel Castings Containing Aluminum and Nitrogen, *J. Iron Steel Inst.*, Vol 200, April 1962, p 299-307

297. N.H. Croft *et al.*, Intergranular Fracture of Steel Castings, in *Advances in the Physical Metallurgy and Applications of Steels*, Publication 284, The Metals Society, 1982, p 286-295

298. E. Colombo and B. Cesari, The Study of the Influence of Al and N on the Susceptibility to Crack Formation of Medium Carbon Steel Ingots, *Metall. Ital.*, Vol 59 (No. 2), 1967, p 71-75

299. S.C. Desai, Longitudinal Panel Cracking in Ingots, *J. Iron Steel Inst.*, Vol 191, March 1959, p 250-256

300. R. Sussman *et al.*, Occurrence and Control of Panel Cracking in Aluminum Containing Steel Heats, in *Mechanical Working & Steel Processing*, Meeting XVII, American Institute of Mining, Metallurgical, and Petroleum Engineers, 1979, p 49-78

301. L. Ericson, Cracking in Low Alloy Aluminum Grain Refined Steels, *Scand. J. Metall.*, Vol 6, 1977, p 116-124

302. F. Vodopivec, Influence of Precipitation and Precipitates of Aluminum Nitride on Torsional Deformability of Low-Carbon Steel, *Met. Technol.*, Vol 5, April 1978, p 118-121

303. G.D. Funnell and R.J. Davies, Effect of Aluminum Nitride Particles on Hot Ductility of Steel, *Met. Technol.*, Vol 5, May 1978, p 150-153

304. T. Lepistö and P. Kettunen, Embrittlement Caused by ε-Martensite in Stainless Steels, *Scand. J. Metall.*, Vol 7, 1978, p 71-76

305. F.W. Schaller and V.F. Zackay, Low Temperature Embrittlement of Austenitic Cr-Mn-N-Fe Alloys, *Trans. ASM*, Vol

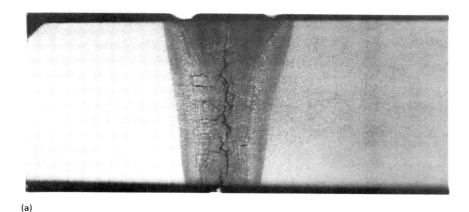

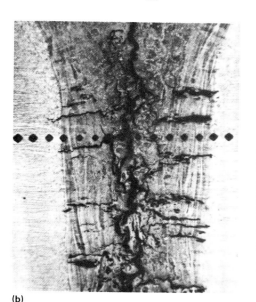

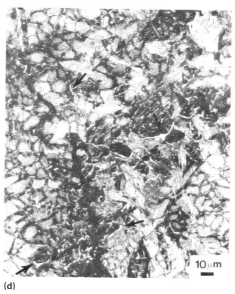

Fig. 104 Hot cracking of an electron beam weld due to accidental melting of the copper backing plate. Note the extensive intergranular cracking and the grain-boundary copper film. (a) 4×. (b) 13×. (c) 68×. (d) 340×

51, 1959, p 609-628

306. D. Hennessy et al., Phase Transformation of Stainless Steel During Fatigue, *Metall. Trans.*, Vol 7A, March 1976, p 415-424

307. H.H. Bleakney, The Creep-Rupture Embrittlement of Metals as Exemplified by Aluminum, *Can. Metall. Q.*, Vol 2 (No. 3), 1963, p 391-315

308. D.E. Ferrell and A.W. Pense, Creep Embrittlement of 2-1/4% Cr-1% Mo Steel, in *Report to Materials Division*, Pressure Vessel Research Council, May 1973

309. H.R. Tipler, "The Role of Trace Elements in Creep Embrittlement and Cavitation of Cr-Mo-V Steels," National Physical Laboratory, 1972

310. R. Bruscato, Temper Embrittlement and Creep Embrittlement of 2-1/4Cr-1Mo Shielded Metal Arc Weld Deposits, *Weld. J.*, Vol 49, April 1970, p 148s-156s

311. R.A. Swift and H.C. Rogers, Study of Creep Embrittlement of 2-1/4Cr-1Mo Steel Weld Metal, *Weld. J.*, Vol 55, July 1976, p 188s-198s

312. R.A. Swift, The Mechanism of Creep Embrittlement in 2-1/4Cr-1Mo Steel, in *2-1/4 Chrome-1 Molybdenum Steel in Pressure Vessels and Piping*, American Society of Mechanical Engineers, 1971

313. L.K.L. Tu and B.B. Seth, Effect of Composition, Strength, and Residual Elements on Toughness and Creep Properties of Cr-Mo-V Turbine Rotors, *Met. Technol.*, Vol 5, March 1978, p 79-91

314. S.H. Chen et al., The Effect of Trace Impurities on the Ductility of a Cr-Mo-V Steel at Elevated Temperatures, *Metall. Trans.*, Vol 14A, April 1983, p 571-580

315. M.P. Seah, Impurities, Segregation and Creep Embrittlement, *Philos. Trans. R. Soc. (London) A*, Vol 295, 1980, p 265-278

316. H.J. Kerr and F. Eberle, Graphitization of Low-Carbon and Low-Carbon-Molybdenum Steels, *Trans. ASME*, Vol 67, 1945, p 1-46

317. S.L. Hoyt et al., Summary Report on the Joint E.E.I.-A.E.I.C. Investigation of Graphitization of Piping, *Trans. ASME*, Aug 1946, p 571-580

318. R.W. Emerson and M. Morrow, Further Observations of Graphitization in Aluminum-Killed Carbon-Molybdenum Steel Steel Steam Piping, *Trans. AIME*, Aug 1946, p 597-607

319. J.G. Wilson, Graphitization of Steel in Petroleum Refining Equipment, *Weld. Res. Counc. Bull.*, No. 32, Jan 1957, p 1-10

320. A.B. Wilder et al., Stability of AISI Alloy Steels, *Trans. AIME*, Vol 209, Oct 1957, p 1176-1181

321. I.M. Bernstein and A.W. Thompson,

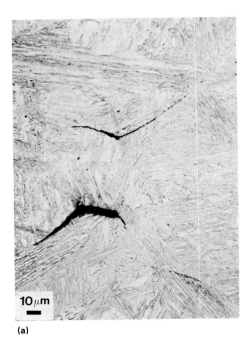
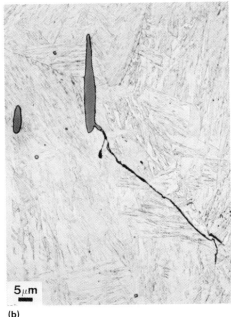
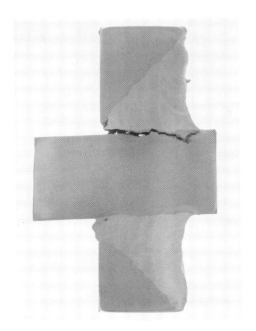

Fig. 105 Two examples of hot tears in the HAZ of gas-metal arc welded HY-80 steel. Note the crack associated with the manganese sulfide inclusion (b). Both etched with 1% nital. (a) 370×. (b) 740×. Courtesy of C.F. Meitzner, Bethlehem Steel Corporation

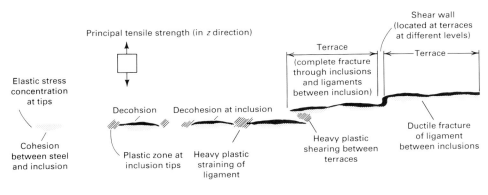

Fig. 106 Schematic illustrating the formation of lamellar tears. Source: Ref 499

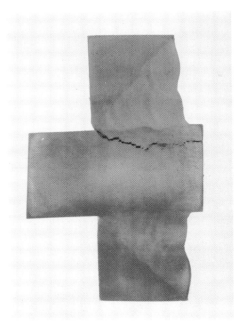

Fig. 107 Macrographs showing typical lamellar tears in structural steel

Ed., *Hydrogen in Metals*, American Society for Metals, 1974
322. C.D. Beachem, Ed., *Hydrogen Damage*, American Society for Metals, 1977
323. R.W. Staehle et al., Ed., *Stress Corrosion Cracking and Hydrogen Embrittlement in Iron Base Alloys*, NACE Reference Book 5, National Association of Corrosion Engineers, 1977
324. A.E. Schuetz and W.D. Robertson, Hydrogen Absorption, Embrittlement and Fracture of Steel, *Corrosion*, Vol 13, July 1957, p 437t-458t
325. I.M. Bernstein, The Role of Hydrogen in the Embrittlement of Iron and Steel, *Mater. Sci. Eng.*, Vol 6, July 1970, p 1-19
326. M.R. Louthan, Jr. et al., Hydrogen Embrittlement of Metals, *Mater. Sci. Eng.*, Vol 10, 1972, p 357-368
327. I.M. Bernstein and A.W. Thompson, Effect of Metallurgical Variables on Environmental Fracture of Steels, *Int. Met. Rev.*, Vol 21, Dec 1976, p 269-287
328. J.P. Hirth, Effects of Hydrogen on the Properties of Iron and Steel, *Metall. Trans.*, Vol 11A, June 1980, p 861-890
329. I.M. Bernstein et al., Effect of Dissolved Hydrogen on Mechanical Behavior of Metals, in *Effect of Hydrogen on Behavior of Materials*, American Institute of Mining, Metallurgical, and Petroleum Engineers, 1976, p 37-58
330. L.H. Keys, The Effects of Hydrogen on the Mechanical Behavior of Metals, *Met. Forum*, Vol 2 (No. 3), 1979, p 164-173
331. S.P. Lynch, Mechanisms of Hydrogen-Assisted Cracking, *Met. Forum*, Vol 2 (No. 3), 1979, p 189-200
332. H.G. Nelson, Hydrogen Embrittlement, in *Embrittlement of Engineering Alloys*, Academic Press, 1983, p 275-359
333. C.A. Zapffe, Hydrogen Flakes and Shatter Cracks, *Metals and Alloys*, Vol 11, May 1940, p 145-151; June 1940, p 177-184; Vol 12, July 1940, p 44-51; Aug 1940, p 145-148
334. C.A. Zapffe, Defects in Cast and Wrought Steel Caused by Hydrogen, *Met. Prog.*, Vol 42, Dec 1942, p 1051-1056

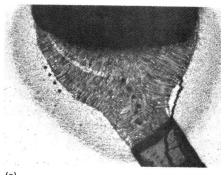

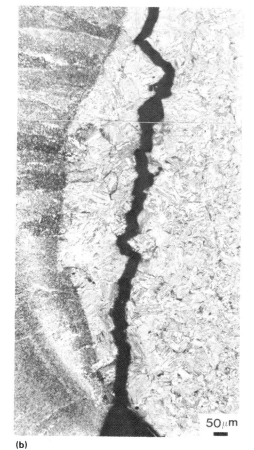

Fig. 108 Stress-relief crack in plate steel Lehigh restraint sample. The crack is intergranular, but the high restraint level suppresses crack branching. The square marks in the weld HAZ in (a) are hardness impressions. (a) 9×. (b) Etched with 2% nital. 74×

335. E.R. Johnson et al., Flaking in Alloy Steels, in *Open Hearth Conference*, 1944, p 358-377
336. A.W. Dana et al., Relation of Flake Formation in Steel to Hydrogen, Microstructure, and Stress, *Trans. AIME*, Vol 203, Aug 1955, p 895-905
337. J.D. Hobson, The Removal of Hydrogen by Diffusion from Large Masses of Steel, *J. Iron Steel Inst.*, Vol 191, April 1959, p 342-352
338. J.M. Hodge et al., Effect of Hydrogen Content on Susceptibility to Flaking, *Trans. AIME*, Vol 230, 1964, p 1182-1193
339. J.E. Ryall et al., The Effects of Hydrogen in Rolled Steel Products, *Met. Forum*, Vol 2 (No. 3), 1979, p 174-182
340. R.E. Cramer and E.C. Bast, The Prevention of Flakes by Holding Railroad Rails at Various Constant Temperatures, *Trans. ASM*, Vol 27, 1939, p 923-934
341. F. Terasaki and S. Okamoto, Fractography of Hydrogen-Embrittled Steels, *Tetsu-to-Hagané (J. Iron Steel Inst. Jpn.)*, Vol 60, 1974, p S576 (HB 9225)
342. A.P. Coldren and G. Tither, Metallographic Study of Hydrogen-Induced Cracking in Line Pipe Steel, *J. Met.*, Vol 28, May 1976, p 5-10
343. M.R. Louthan, Jr. and R.P. McNitt, The Role of Test Technique in Evaluating Hydrogen Embrittlement Mechanisms, in *Effects of Hydrogen on Behavior of Materials*, American Institute of Mining, Metallurgical, and Petroleum Engineers, 1976, p 496-506
344. C.L. Briant et al., Embrittlement of a 5 Pct. Nickel High Strength Steel by Impurities and Their Effects on Hydrogen-Induced Cracking, *Metall. Trans.*, Vol 9A, May 1978, p 625-633
345. D. Hardie and T.I. Murray, Effect of Hydrogen on Ductility of a High Strength Steel in Hardened and Tempered Conditions, *Met. Technol.*, Vol 5, May 1978, p 145-149
346. T.A. Adler et al., Metallographic Studies of Hydrogen-Induced Crack Growth in A-106 Steel, in *Microstructural Science*, Vol 8, Elsevier, 1980, p 217-230
347. J.E. Costa and A.W. Thompson, Effect of Hydrogen on Fracture Behavior of a Quenched and Tempered Medium-Carbon Steel, *Metall. Trans.*, Vol 12A, May 1981, p 761-771
348. B.D. Craig, A Fracture Topographical Feature Characteristic of Hydrogen Embrittlement, *Corrosion*, Vol 37, Sept 1981, p 530-532
349. R. Garber et al., Hydrogen Assisted Ductile Fracture of Spheroidized Carbon Steels, *Metall. Trans.*, Vol 12A, Feb 1981, p 225-234
350. H. Cialone and R.J. Asaro, Hydrogen Assisted Ductile Fracture of Spheroidized Plain Carbon Steels, *Metall. Trans.*, Vol 12A, Aug 1981, p 1373-1387
351. T. Boniszewski and J. Moreton, Effect of Micro-Voids and Manganese Sulphide Inclusion in Steel on Hydrogen Evolution and Embrittlement, *Br. Weld. J.*, Vol 14, June 1967, p 321-336
352. P.H. Pumphrey, Effect of Sulphide Inclusions on Hydrogen Diffusion in Steels, *Met. Sci.*, Vol 16, Jan 1982, p 41-47
353. K. Farrell and A.G. Quarrell, Hydrogen Embrittlement of an Ultra-High-Tensile Steel, *J. Iron Steel Inst.*, Vol 202, Dec 1964, p 1002-1011
354. A. Ciszewski et al., Effect of Nonmetallic Inclusions on the Formation of Microcracks in Hydrogen Charged 1% Cr Steel, in *Stress Corrosion Cracking and Hydrogen Embrittlement of Iron Base Alloys*, NACE Reference Book 5, National Association of Corrosion Engineers, 1977, p 671-679
355. G.M. Presouyre and C. Zmudzinski, Influence of Inclusions on Hydrogen Embrittlement, in *Mechanical Working & Steel Processing*, Meeting XVIII, American Institute of Mining, Metallurgical, and Petroleum Engineers, 1980, p 534-553
356. R.L. Cowan II and C.S. Tedmon, Jr., Intergranular Corrosion of Iron-Nickel-Chromium Alloys, in *Advances in Corrosion Science and Technology*, Vol 3, Plenum Press, 1973, p 293-400
357. E.M. Mahla and N.A. Nielson, Carbide Precipitation in Type 304 Stainless Steel—An Electron Microscope Study, *Trans. ASM*, Vol 43, 1951, p 290-322
358. R. Stickler and A. Vinckier, Morphology of Grain-Boundary Carbides and its Influence on Intergranular Corrosion of 304 Stainless Steel, *Trans. ASM*, Vol 54, 1961, p 362-380
359. R. Stickler and A. Vinckier, Precipitation of Chromium Carbide on Grain Boundaries in a 302 Austenitic Stainless Steel, *Trans. AIME*, Vol 224, Oct 1962, p 1021-1024
360. K.T. Aust et al., Intergranular Corrosion and Electron Microscopic Studies of Austenitic Stainless Steels, *Trans. ASM*, Vol 60, 1967, p 360-372
361. S. Danyluk et al., Intergranular Fracture, Corrosion Susceptibility, and Impurity Segregation in Sensitized Type 304 Stainless Steel, *J. Mater. Energy Syst.*, Vol 7, June 1985, p 6-15
362. A.P. Bond and E.A. Lizlovs, Intergranular Corrosion of Ferritic Stainless Steels, *J. Electrochem. Soc.*, Vol 116, Sept 1969, p 1305-1311
363. A.P. Bond, Mechanisms of Intergranular Corrosion in Ferritic Stainless Steels, *Trans. AIME*, Vol 245, Oct 1969, p 2127-2134
364. E.L. Hall and C.L. Briant, Chromium Depletion in the Vicinity of Carbides in Sensitized Austenitic Stainless Steels, *Metall. Trans.*, Vol 15A, May 1984, p 793-811
365. J.B. Lee et al., An Analytical Electron Microscope Examination of Sensitized AISI 430 Stainless Steel, *Corrosion*, Vol 41, Feb 1985, p 76-80
366. E.P. Butler and M.G. Burke, Chromium Depletion and Martensite Formation at Grain Boundaries in Sensitized Austenitic

Fig. 109 Fracture surface of a small ingot of iron-chromium-aluminum alloy. The as-cast solidification pattern is clearly revealed. 0.57×

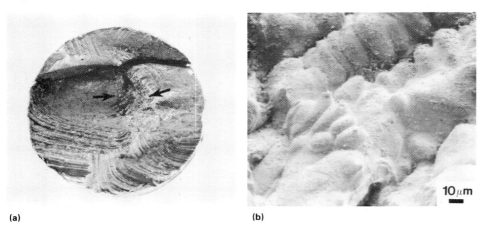

Fig. 110 Macrograph (a) of defect in test bar that initiated fatigue failure. 1.5×. (b) SEM fractograph showing that the defect was a shrinkage cavity. Note the dendritic appearance. 375×

Stainless Steel, *Acta Metall.*, Vol 34, March 1986, p 557-570

367. M.A. Streicher, Screening Stainless Steels from the 240-Hr Nitric Acid Test by Electrolytic Etching in Oxalic Acid, *ASTM Bull.*, No. 188, Feb 1953, p 35-38; July 1953, p 58-59

368. M.A. Streicher, Theory and Application of Evaluation Tests for Detecting Susceptibility to Intergranular Attack in Stainless Steels and Related Alloys—Problems and Opportunities, in *Intergranular Corrosion of Stainless Alloys*, STP 656, American Society for Testing and Materials, p 3-84

369. A.P. Majidi and M.A. Streicher, "The Effect of Methods of Cutting and Grinding on Sensitization in Surface Layers in Type 304 Stainless Steel," Paper 25, presented at Corrosion/83, National Association of Corrosion Engineers, 1983

370. J.M. Schluter and J.A. Chivinsky, Surface Preparation Requirements for ASTM A262, in *Laboratory Corrosion Tests and Standards*, STP 866, American Society for Testing and Materials, 1985, p 455-464

371. W. Rostoker *et al.*, *Embrittlement by Liquid Metals*, Reinhold, 1960

372. N.S. Stoloff, Liquid Metal Embrittlement, in *Surfaces and Interfaces*, Pt. II, Syracuse University Press, 1968, p 157-182

373. A.R.C. Westwood *et al.*, Adsorption—Induced Brittle Fracture in *Liquid-Metal Environments*, in Fracture, Vol III, Academic Press, 1971, p 589-644

374. M.H. Kamdar, Embrittlement by Liquid Metals, *Prog. Mater. Sci.*, Vol 15, Pt. 4, 1973, p 289-374

375. M.H. Kamdar, Liquid Metal Embrittlement, in *Embrittlement of Engineering Alloys*, Academic Press, 1983, p 361-459

376. M.H. Kamdar, Ed., *Embrittlement by Liquid and Solid Metals*, American Institute of Mining, Metallurgical, and Petroleum Engineers, 1984

377. M.B. Reynolds, Radiation Effects in Some Engineering Alloys, *J. Mater.*, Vol 1, March 1966, p 127-152

378. K. Ohmae and T.O. Ziebold, The Influence of Impurity Content on the Radiation Sensitivity of Pressure Vessel Steels: Use of Electron Microprobe for Irregular Surfaces, *J. Nucl. Mater.*, Vol 43, 1972, p 245-257

379. J.R. Hawthorne and L.E. Steele, Metallurgical Variables as Possible Factors Controlling Irradiation Response of Structural Steels, in *The Effects of Radiation on Structural Metals*, STP 426, American Society for Testing and Materials, 1967, p 534-572

380. *Irradiation Effects on the Microstructure and Properties of Metals*, STP 611, American Society for Testing and Materials, 1976

381. J.O. Stiegler and E.E. Bloom, The Effects of Large Fast-Neutron Fluences on

Fig. 111 Example of inclusion stringers (white streaks against the dark background) on a longitudinal test fracture. The specimen was blued to reveal the stringers.

Fig. 112 Example of a fractured wedge test specimen used to assess the carbide stability in a cast iron melt. Note the change in fracture appearance in the tip (white fracture due to the presence of carbide) compared to the base of the wedge (dark due to the presence of graphite). Actual size

the Structure of Stainless Steel, *J. Nucl. Mater.*, Vol 33, 1969, p 173-185

382. S.D. Harkness and C.Y. Li, A Study of Void Formation in Fast Neutron-Irradiated Metals, *Metall. Trans.*, Vol 2, May 1971, p 1457-1470

383. H.R. Brager et al., Irradiation-Produced Defects in Austenitic Stainless Steel, *Metall. Trans.*, Vol 2, July 1971, p 1893-1904

384. H.R. Brager and J.L. Straalsund, Defect Development in Neutron Irradiated Type 316 Stainless Steel, *J. Nucl. Mater.*, Vol 46, 1973, p 134-158

385. K. Farrell and R.T. King, Radiation-Induced Strengthening and Embrittlement in Aluminum, *Metall. Trans.*, Vol 4, May 1973, p 1223-1231

386. J.R. Strohm and W.E. Jominy, High Forging Temperatures Revealed by Facets in Fracture Tests, *Trans. ASM*, Vol 36, 1946, p 543-571

387. A. Preece et al., The Overheating and Burning of Steel, *J. Iron Steel Inst.*, Vol 153, Pt. 1, 1946, p 237p-254p

388. A. Preece and J. Nutting, The Detection of Overheating and Burning in Steel by Microscopical Methods, *J. Iron Steel Inst.*, Vol 164, Jan 1960, p 46-50

389. B.J. Schulz and C.J. McMahon, Fracture of Alloy Steels by Intergranular Microvoid Coalescence as Influenced by Composition and Heat Treatment, *Metall. Trans.*, Vol 4, Oct 1973, p 2485-2489

390. J.A. Disario, "Overheating and Its Effect on the Toughness of ASTM A508 Class II Forgings," MS thesis, Lehigh University, 1973

391. T.J. Baker and R. Johnson, Overheating and Fracture Toughness, *J. Iron Steel Inst.*, Vol 211, Nov 1973, p 783-791

392. T.J. Baker, Use of Scanning Electron Microscopy in Studying Sulphide Morphology on Fracture Surfaces, in *Sulfide Inclusions in Steel*, American Society for Metals, 1975, p 135-158

393. G.E. Hale and J. Nutting, Overheating of Low-Alloy Steels, *Int. Met. Rev.*, Vol 29 (No. 4), 1984, p 273-298

394. R.C. Andrew et al., Overheating in Low-Sulphur Steels, *J. Austral. Inst. Met.*, Vol 21, June-Sept 1976, p 126-131

395. R.C. Andrew and G.M. Weston, the Effect of Overheating on the Toughness of Low Sulphur ESR Steels, *J. Austral. Inst. Met.*, Vol 22, Sept-Dec 1977, p 171-176

396. S. Preston et al., Overheating Behavior of a Grain-Refined Low-Sulphur Steel, *Mater. Sci. Technol.*, Vol 1, March 1985, p 192-197

397. A.L. Tsou et al., The Quench-Aging of Iron, *J. Iron Steel Inst.*, Vol 172, Oct 1952, p 163-171

398. T.C. Lindley and C.E. Richards, The Effect of Quench-Aging on the Cleavage Fracture of a Low-Carbon Steel, *Met. Sci. J.*, Vol 4, May 1970, p 81-84

399. A.S. Keh and W.C. Leslie, Recent Observations on Quench-Aging and Strain Aging of Iron and Steel, in *Materials Science Research*, Vol 1, Plenum Press, 1963, p 208-250

400. E.R. Morgan and J.F. Enrietto, Aging in Steels, in *AISI 1963 Regional Technical Meeting*, American Iron and Steel Institute, 1964, p 227-252

401. W.C. Leslie and A.S. Keh, Aging of Flat-Rolled Steel Products as Investigated by Electron Microscopy, in *Mechanical Working of Steel*, Pt. II, Gordon & Breach, 1965, p 337-377

402. E. Stolte and W. Heller, The State of Knowledge of the Aging of Steels, I. Fundamental Principles, *Stahl Eisen*, Vol 90 (No. 16), BISI 8920, 1970, p 861-868

403. L.R. Shoenberger and E.J. Paliwoda, Accelerated Strain Aging of Commercial Sheet Steels, *Trans. ASM*, Vol 45, 1953, p 344-361

404. F. Garofalo and G.V. Smith, The Effect of Time and Temperature on Various Mechanical Properties During Strain Aging of Normalized Low Carbon Steels, *Trans. ASM*, Vol 47, 1955, p 957-983

405. J.D. Baird, Strain Aging of Steel—A Critical Review, *Iron Steel*, Vol 36, May 1963, p 186-192; June 1963, p 326-334; July 1963, p 368-374; Aug 1963, p 400-405; Sept 1963, p 450-457

406. J.D. Baird, The Effects of Strain-Aging Due to Interstitial Solutes on the Mechanical Properties of Metals, *Metall. Rev.*, Vol 16, Feb 1971, p 1-18

407. G.F. Vander Voort, Failures of Tools and Dies, in *Failure Analysis and Prevention*, Vol 11, 9th ed., *Metals Handbook*, American Society for Metals, 1986, p 563-585

408. L.D. Jaffee and J.R. Hollomon, Hardenability and Quench Cracking, *Trans. AIME*, Vol 167, 1946, p 617-626

409. M.C. Udy and M.K. Barnett, A Laboratory Study of Quench Cracking in Cast Alloy Steels, *Trans. ASM*, Vol 38, 1947, p 471-487

410. J.W. Spretnak and C. Wells, An Engineering Analysis of the Problem of Quench Cracking in Steel, *Trans. ASM*, Vol 42, 1950, p 233-269

411. C. Wells, Quench Cracks in Wrought Steel Tubes, *Met. Prog.*, Vol 65, May 1954, p 113-121

412. T. Kunitake and S. Sugisawa, The Quench-Cracking Susceptibility of Steel, *Sumitomo Search*, No. 5, May 1971, p 16-25

413. K.E. Thelning, Why Does Steel Crack on Hardening, *Härt.-Tech.-Mitt.*, Vol 4, BISI 12602, 1970, p 271-281

Fig. 113 Four fractured, hardened etch disks of a high-carbon (1.3% C) tool steel that contained excessive graphite (dark regions) due to an accidental aluminum addition. About 0.5×

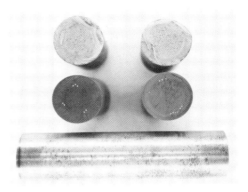

Fig. 114 Example of specimens used in the Shepherd P-F test. The 19-mm (¾-in.) diam, 100-mm (4-in.) long specimen is shown at the bottom. The test consists of fracturing an austenitized and brine-quenched specimen. Case depth can be observed on the fracture. After fracture grain size is rated, the fracture surface is carefully ground off. Macroetching then reveals the depth of hardening. Superficial hardness tests are used to determine the case depth to a particular hardness.

414. P. Gordon, The Temper Colors on Steel, *J. Heat Treat.*, Vol 1, June 1979, p 93
415. *Modern Steels*, 6th ed., Bethlehem Steel Corporation, 1967, p 149
416. D.J. McAdam and G.W. Geil, Rate of Oxidation of Steels as Determined From Interference Colors of Oxide Films, *J. Res. NBS*, Vol 23, July 1939, p 63-124
417. A.J. Lena, Effect of Sigma Phase on Properties of Alloys, *Met. Prog.*, Vol 66, Aug 1954, p 94-99
418. J.I. Morley and H.W. Krikby, Sigma-Phase Embrittlement in 25Cr-20Ni Heat-Resisting Steels, *J. Iron Steel Inst.*, Vol 172, Oct 1952, p 129-142
419. A.M. Talbot and D.E. Furman, Sigma Formation and Its Effect on the Impact Properties of Iron-Nickel-Chromium Alloys, *Trans. ASM*, Vol 45, 1953, p 429-442
420. E.O. Hall and S.H. Algie, The Sigma Phase, *Met. Rev.*, Vol 11, 1966, p 61-88
421. G. Matern et al., The Formation of Sigma Phase in Austenitic Ferrite Stainless Steels and Its Influence on Mechanical Properties, *Mem. Sci. Rev. Met.*, BISI 13972, 1974, p 841-851
422. W.J. Boesch and J.S. Slaney, Preventing Sigma Phase Embrittlement in Nickel Base Superalloys, *Met. Prog.*, Vol 86, July 1964, p 109-111
423. J.R. Mihalisin et al., Sigma—Its Occurrence, Effect, and Control in Nickel-Base Superalloys, *Trans. AIME*, Vol 242, Dec 1968, p 2399-2414
424. H.L. Logan, *The Stress Corrosion of Metals*, John Wiley & Sons, 1966
425. V.V. Romanov, *Stress-Corrosion Cracking of Metals*, Israel Program for Scientific Translations and the National Science Foundation, 1961
426. H.L. Craig, Jr., Ed., *Stress Corrosion—New Approaches*, STP 610, American Society for Testing and Materials, 1976
427. R.W. Staehle et al., Ed., *Fundamental Aspects of Stress Corrosion Cracking*, National Association of Corrosion Engineers, 1969
428. C. Edeleanu, Transgranular Stress Corrosion in Chromium-Nickel Stainless Steels, *J. Iron Steel Inst.*, Vol 173, Feb 1953, p 140-146
429. N.A. Neilsen, Environmental Effects on Fracture Morphology, in *Electron Fractography*, STP 436, American Society for Testing and Materials, 1968, p 124-150
430. J.C. Scully, Scanning Electron Microscope Studies of Stress-Corrosion Cracking, in *Scanning Electron Microscopy*, IIT Research Institute, 1970, p 313-320
431. J.C. Scully, Failure Analysis of Stress Corrosion Cracking with the Scanning Electron Microscope, in *Scanning Electron Microscopy*, Pt. IV, IIT Research Institute, 1974, p 867-874
432. B. Poulson, The Fractography of Stress Corrosion Cracking in Carbon Steels, *Corros. Sci.*, Vol 15, Sept 1975, p 469-477
433. H. Okada et al., Scanning Electron Microscope Observation of Fracture Faces of Austenitic Stainless Steels by Stress Corrosion Cracking, *Nippon Kinzoku Gakkai-shi*, Vol 37, 1973, p 197-203
434. Y. Mukai et al., Fractographic Observation of Stress-Corrosion Cracking of AISI 304 Stainless Steel in Boiling 42 Percent Magnesium-Chloride Solution, in *Fractography in Failure Analysis*, STP 645, American Society for Testing and Materials, 1978, p 164-175
435. B.C. Woodfine, Temper-Brittleness: A Critical Review of the Literature, *J. Iron Steel Inst.*, Vol 173, March 1953, p 229-240
436. J.R. Low, Jr., Temper Brittleness—A Review of Recent Work, in *Fracture of Engineering Materials*, American Society for Metals, 1964, p 127-142
437. C.J. McMahon, Jr., Temper Brittleness—An Interpretive Review, in *Temper Embrittlement in Steel*, STP 407, American Society for Testing and Materials, 1968, p 127-167
438. I. Olefjord, Temper Embrittlement, *Int. Met. Rev.*, Vol 23 (No. 4), 1978, p 149-163
439. R.G.C. Hill and J.W. Martin, A Fractographic Study of Some Temper Brittle Steels, *Met. Treat. Drop Forg.*, Vol 29, Aug 1962, p 301-310
440. H. Ohtani and C.J. McMahon, Jr., Modes of Fracture in Temper Embrittled Steels, *Acta Metall.*, Vol 23, March 1975, p 377-386
441. J. Yu and C.J. McMahon, Jr., Variation of the Fracture Mode in Temper Embrittled 2.25Cr-1Mo Steel, *Metall. Trans.*, Vol 16A, July 1985, p 1325-1331
442. J.B. Cohen et al., A Metallographic Etchant to Reveal Temper Brittleness in Steel, *Trans. ASM*, Vol 39, 1947, p 109-138
443. D. McLean and L. Northcott, Micro-Examination and Electrode-Potential Measurements of Temper-Brittle Steels, *J. Iron Steel Inst.*, Vol 158, 1948, p 169-177
444. J.P. Rucker, "Improved Metallographic Technique for Revealing Temper Brittleness Network in Ordnance Steels," NPG

Fig. 115 AISI 1070 steel bar that broke during heat treatment due to gross chemical segregation. (a) Macrograph at actual size. (b) Fracture near the outside surface of the bar. 8×. (c) Fracture near the center of the bar. 8×. Note the coarse grain structure of the center region.

1555, United States Naval Proving Ground, 28 Aug 1957
445. J.M. Capus and G. Mayer, The Influence of Trace Elements on Embrittlement Phenomena in Low-Alloy Steels, *Metallurgia*, Vol 62, 1960, p 133-138
446. E.B. Kula and A.A. Anctil, Tempered Martensite Embrittlement and Fracture Toughness in SAE 4340 Steel, *J. Mater.*, Vol 4, Dec 1969, p 817-841
447. G. Delisle and A. Galibois, Tempered Martensite Brittleness in Extra-Low-Carbon Steels, *J. Iron Steel Inst.*, Vol 207, Dec 1969, p 1628-1634
448. G. Delisle and A. Galibois, Microstructural Studies of Tempered Extra-Low Carbon Steels and Their Effectiveness in Interpreting Tempered Martensite Brittleness, in *Microstuctural Science*, Vol 1, Elsevier, 1974, p 91-112
449. G. Thomas, Retained Austenite and Tempered Martensite Embrittlement, *Metall. Trans.*, Vol 9A, March 1978, p 439-450
450. R.M. Horn and R.O. Ritchie, Mechanisms of Tempered Martensite Embrittlement in Low Alloy Steels, *Metall. Trans.*, Vol 9A, Aug 1978, p 1039-1053
451. C.L. Briant and S.K. Banerji, Tempered Martensite Embrittlement in a High Purity Steel, *Metall. Trans.*, Vol 10A, Aug 1979, p 1151-1155
452. M. Sarikaya *et al.*, Retained Austenite and Tempered Martensite Embrittlement in Medium Carbon Steels, *Metall. Trans.*, Vol 14A, June 1983, p 1121-1133
453. N. Bandyopadhyay and C.J. McMahon, Jr., The Micro-Mechanisms of Tempered Martensite Embrittlement in 4340-Type Steels, *Metall Trans.*, Vol 14A, July 1983, p 1313-1325
454. G.J. Spaeder, Impact Transition Behavior of High-Purity 18Ni Maraging Steel, *Metall. Trans.*, Vol 1, July 1970, p 2011-2014
455. D. Kalish and H.J. Rack, Thermal Embrittlement of 18Ni (350) Maraging Steel, *Metall. Trans.*, Vol 2, Sept 1971, p 2665-2672
456. W.C. Johnson and D.F. Stein, A Study of Grain Boundary Segregants in Thermally Embrittled Maraging Steel, *Metall. Trans.*, Vol 5, March 1974, p 549-554
457. E. Nes and G. Thomas, Precipitation of TiC in Thermally Embrittled Maraging Steels, *Metall. Trans.*, Vol 7A, July 1976, p 967-975
458. H.J. Rack and P.H. Holloway, Grain Boundary Precipitation in 18Ni-Maraging Steels, *Metall. Trans.*, Vol 8A, Aug 1977, p 1313-1315
459. A.J. Birkle *et al.*, A Metallographic Investigation of the Factors Affecting the Notch Toughness of Maraging Steels, *Trans. ASM*, Vol 58, 1965, p 285-301
460. R.M. Fisher *et al.*, Identification of the Precipitate Accompanying 885 °F Embrittlement in Chromium Steels, *Trans. AIME*, Vol 197, May 1953, p 690-695
461. B. Cina and J.D. Lavender, The 475 °C Hardening Characteristics of Some High-Alloy Steels and Chromium Iron, *J. Iron Steel Inst.*, Vol 174, June 1953, p 97-107
462. A.J. Lena and M.F. Hawkes, 475 °C (885 °F) Embrittlement in Stainless Steels, *Trans. AIME*, Vol 200, May 1954, p 607-615
463. R.O. Williams and H.W. Paxton, The Nature of Aging of Binary Iron-Chromium Alloys Around 500 °C, *J. Iron Steel Inst.*, Vol 185, March 1957, p 358-374
464. R. Lagneborg, Metallography of the 475C Embrittlement in an Iron-30% Chromium Alloy, *Trans. ASM*, Vol 60, 1967, p 67-78
465. P.J. Grobner, The 885 °F (475 °C) Embrittlement of Ferritic Stainless Steels, *Metall. Trans.*, Vol 4, Jan 1973, p 251-260
466. H.D. Solomon and E.F. Koch, High

Temperature Precipitation of α' in a Multicomponent Duplex Stainless Steel, *Scr. Metall.*, Vol 13, 1979, p 971-974
467. H.D. Solomon and L.M. Levinson, Mössbauer Effect Study of "475 °C Embrittlement" of Duplex and Ferritic Stainless Steels, *Acta Metall.*, Vol 26, 1978, p 429-442
468. J. Chance et al., Structure-Property Relationships in a 25Cr-7Ni-2Mo Duplex Stainless Steel Casting Alloy, in *Duplex Stainless Steels*, American Society for Metals, 1983, p 371-398
469. P. Jacobsson et al., Kinetics and Hardening Mechanism of the 475 °C Embrittlement in 18Cr-2Mo Ferritic Steels, *Metall. Trans.*, Vol 6A, Aug 1975, p 1577-1580
470. A.L. Schaeffler et al., Hydrogen in Mild-Steel Weld Metal, *Weld. J.*, Vol 31, June 1952, p 283s-309s
471. H.G. Vaughan and M.E. deMorton, Hydrogen Embrittlement of Steel and Its Relation to Weld Metal Cracking, *Br. Weld. J.*, Vol 4, Jan 1957, p 40-61
472. T.E.M. Jones, Cracking of Low Alloy Steel Weld Metal, *Br. Weld. J.*, Vol 6, July 1959, p 315-323
473. N. Christensen, The Role of Hydrogen in Arc Welding With Coated Electrodes, *Weld. J.*, Vol 40, April 1961, p 145s-154s
474. E.P. Beachum et al., Hydrogen and Delayed Cracking in Steel Weldments, *Weld. J.*, Vol 40, April 1961, p 155s-159s
475. F. Watkinson et al., Hydrogen Embrittlement in Relation to the Heat-Affected Zone Microstructure of Steels, *Br. Weld. J.*, Vol 10, Feb 1963, p 54-62
476. F. Watkinson, Hydrogen Cracking in High Strength Weld Metals, *Weld. J.*, Vol 48, Sept 1969, p 417s-424s
477. T. Boniszewski and F. Watkinson, Effect of Weld Microstructures on Hydrogen-Induced Cracking in Transformable Steels, *Met. Mater.*, Vol 7, Feb 1973, p 90-96; March 1973, p 145-151
478. J.M.F. Mota and R.L. Apps, "Chevron Cracking"—A New Form of Hydrogen Cracking in Steel Weld Metals, *Weld. J.*, Vol 61, July 1982, p 222s-228s
479. P.H.M. Hart, Resistance to Hydrogen Cracking in Steel Weld Metals, *Weld. J.*, Vol 65, Jan 1986, p 14s-22s
480. P.A. Kammer et al., "Cracking in High-Strength Steel Weldments—A Critical Review," DMIC 197, Battelle Memorial Institute, 7 Feb 1964
481. P.W. Jones, Hot Cracking of Mild Steel Welds, *Br. Weld. J.*, Vol 6, June 1959, p 269-281
482. J.C. Borland, Some Aspects of Cracking in Welded Cr-Ni Austenitic Steels, *Br. Weld. J.*, Vol 7, Jan 1960, p 22-59
483. J.C. Borland, Generalized Theory of Super-Solidus Cracking in Welds (and Castings), *Br. Weld. J.*, Vol 7, Aug 1960, p 508-512
484. J.C. Borland, Hot Cracking in Welds, *Br. Weld. J.*, Vol 7, Sept 1960, p 558-559
485. J.H. Rogerson and J.C. Borland, Effect of the Shapes of Intergranular Liquid on the Hot Cracking of Welds and Castings, *Trans. AIME*, Vol 227, Feb 1963, p 2-7
486. W.F. Savage et al., Copper-Contamination Cracking in the Weld Heat-Affected Zone, *Weld. J.*, Vol 57, May 1978, p 145s-152s
487. H. Homma et al., A Mechanism of High Temperature Cracking in Steel Weld Metals, *Weld. J.*, Vol 58, Sept 1979, p 277s-282s
488. M.J. Cieslak et al., Solidification Cracking and Analytical Electron Microscopy of Austenitic Stainless Steel Weld Metals, *Weld. J.*, Vol 61, Jan 1982, p 1s-8s
489. T. Ogawa and E. Tsunetomi, Hot Cracking Susceptibility of Austenitic Stainless Steels, *Weld. J.*, Vol 61, March 1982, p 82s-93s
490. G. Rabensteiner et al., Hot Cracking Problems in Different Fully Austenitic Weld Metals, *Weld. J.*, Vol 62, Jan 1983, p 21s-27s
491. J.A. Brooks et al., A Fundamental Study of the Beneficial Effects of Delta Ferrite in Reducing Weld Cracking, *Weld. J.*, Vol 63, March 1984, p 71s-83s
492. J.C.M. Farrar and R.E. Dolby, An Investigation Into Lamellar Tearing, *Met. Constr. Br. Weld. J.*, Vol 1, Feb 1969, p 32-39
493. D.N. Elliott, Fractographic Examination of Lamellar Tearing in Multi-Run Fillet Welds, *Met. Constr. Br. Weld. J.*, Vol 1, Feb 1969, p 50-57
494. J.C.M. Farrar et al., Lamellar Tearing in Welded Structural Steels, *Weld. J.*, Vol 48, July 1969, p 274s-282s
495. D.N. Elliott, Lamellar Tearing in Multi-Pass Fillet Joints, *Weld. J.*, Vol 48, Sept 1969, p 409s-416s
496. S. Hasebe et al., Factors for Lamellar Tearing of Steel Plate, *Sumitomo Search*, No. 13, May 1975, p 19-27
497. S. Ganesh and R.D. Stout, Material Variables Affecting Lamellar Tearing Susceptibility in Steels, *Weld. J.*, Vol 55, Nov 1976, p 341s-355s
498. J.C.M. Farrar, The Effect of Hydrogen on the Formation of Lamellar Tearing, *Weld. Res. Int.*, Vol 7 (No. 2), 1977, p 120-142
499. J. Sommella, "Significance and Control of Lamellar Tearing of Steel Plate in the Shipbuilding Industry," SSC-290, United States Coast Guard, May 1979
500. E. Holby and J.F. Smith, Lamellar Tearing. The Problem Nobody Seems to Want to Talk About, *Weld. J.*, Vol 59, Feb 1980, p 37-44
501. A.D. Hattangadi and B.B. Seth, Lamellar Tearing in Fillet Weldments of Pressure Vessel Fabrications, *Weld. J.*, Vol 62, April 1983, p 89s-96s
502. J. Heuschkel, Anisotropy and Weldability, *Weld. J.*, Vol 50, March 1971, p 110s-126s
503. T. Boniszewski and N.F. Eaton, Electron Fractography of Weld-Reheat Cracking in CrMoV Steel, *Met. Sci. J.*, Vol 3, 1969, p 103-110
504. R.A. Swift, The Mechanism of Stress Relief Cracking in 2-1/4Cr-1Mo Steel, *Weld. J.*, Vol 50, May 1971, p 195s-200s
505. D. McKeown, Re-Heat Cracking in High Nickel Alloy Heat-Affected Zones, *Weld. J.*, Vol 50, May 1971, p 201s-206s
506. A.W. Pense et al., Stress Relief Cracking in Pressure Vessel Steels, *Weld. J.*, Vol 50, Aug 1971, p 374s-378s
507. R.A. Swift and H.C. Rogers, Embrittlement of 2-¼Cr-1Mo Steel Weld Metal by Postweld Heat Treatment, *Weld. J.*, Vol 52, April 1973, p 145s-153s, 172s
508. C.F. Meitzner, Stress-Relief Cracking in Steel Weldments, *Weld. Res. Counc. Bull.*, No. 211, Nov 1975
509. J. Myers, Influence of Alloy and Impurity Content on Stress-Relief Cracking in Cr-Mo-V Steels, *Met. Technol.*, Vol 5, Nov 1978, p 391-396
510. A. Vinckier and A. Dhooge, Reheat Cracking in Welded Structures During Stress Relief Heat Treatments, *J. Heat Treat.*, Vol 1, June 1979, p 72-80
511. C.P. You et al., Stress Relief Cracking Phenomena in High Strength Structural Steel, *Met. Sci.*, Vol 18, Aug 1984, p 387-394
512. J. Shin and C.J. McMahon, Jr., Comparison of Stress Relief Cracking in A508 2 and A533B Pressure Vessel Steels, *Met. Sci.*, Vol 18, Aug 1984, p 403-410
513. A.G. Fuller, Apparatus for Breaking Test Castings, *BCIRA J.*, Vol 8, July 1960, p 586-587
514. G.M. Enos, Fractures, in *Visual Examination of Steel*, American Society for Metals, 1940 p 37-54
515. B. Ostrofsky, Materials Identification in the Field, *Mater. Eval.*, Vol 36, Aug 1978, p 33-39, 45
516. Blue-Brittleness Test for Assessing Macroscopic Inclusion Contents in Steels, *Stahl-Eisen-Prufblatt 1584*, BISI 12397, Dec 1970
517. "Wrought Steels—Macroscopic Methods for Assessing the Content of Non-Metallic Inclusions," ISO 3763, International Standards Organization
518. "Determination of Slag Inclusion Content of Steel. Macroscopic Methods," SIS 11111105, Swedish Institute of Standards
519. D.E. Krause, Chill Test and the Metallurgy of Gray Iron, *Trans. AFS*, Vol 59, 1951, p 79-91
520. A.T. Batty, The Wedge Test and Its Use in the Ironfoundry, *SEAISI Q.*, Vol 4, April 1975, p 55-59

521. W.H. Moore, Melting Engineering Cast Iron in the Electric Furnace, *Cast. Eng./Foundry World*, Vol 16, Summer 1984, p 40, 43-46
522. B.F. Shepherd, The P-F Characteristic of Steel, *Trans. ASM*, Vol 22, Dec 1934, p 979-1016
523. G.F. Vander Voort, Grain Size Measurement, in *Practical Applications of Quantitative Metallography*, STP 839, American Society for Testing and Materials, 1984, p 85-131
524. W.J. Priestly, Fracture Test on Steel To Determine its Quality, *Trans. ASST*, Vol 2, April 1922, p 620-622
525. J.H. Hruska, Fracture Tests and Ingot Defects, *Blast Furn. Steel Plant*, Vol 22, Dec 1934, p 705, 707
526. A.L. Norbury, The Effect of Casting Temperature on the Primary Microstructure of Cast Iron, *J. Iron Steel Inst.*, Vol 140 (No. II), 1939, p 161P-180P
527. W.J. Williams, A Relationship of Microstructure to the Mechanical Properties of White Iron, *BCIRA J.*, Vol 5, Dec 1953, p 132-134
528. F.M. Baker *et al.*, Melt Quality and Fracture Characteristics of 85-5-5-5 Red Brass, *Trans. AFS*, Vol 58, 1950, p 122-132
529. Fracture Test for Determining Melt Quality of 85-5-5-5 Red Brass, *Am. Foundryman*, Vol 24, Oct 1953, p 68-69, 71
530. A.R. French, Melt-Quality Tests for Copper-Base Alloys, *Foundry Trade J.*, Vol 98, 10 March 1955, p 253-257; 17 March 1955, p 281-293
531. D. Hanson and W.T. Pell-Walpole, Methods of Assessing the Quality of Cast Bronzes, in *Chill-Cast Tin Bronzes*, Edward Arnold, 1957, p 22-55

Scanning Electron Microscopy

Barbra L. Gabriel, Packer Engineering Associates, Inc.

THE SCANNING ELECTRON MICROSCOPE has unique capabilities for analyzing surfaces. A beam of electrons moves in an x-y pattern across a conductive specimen, which releases various data signals containing structural and compositional information. Because electrons are used as the radiation source instead of light photons, resolution is improved. Simultaneously, because the specimen is irradiated in a time-sequenced mode, high depth of field is attained, and the images appear three dimensional. In addition, a broad range of magnifications (10 to 30 000×) facilitates the correlation of macro- and microscopic images.

The scanning electron microscope also has analytical capabilities. Among the data signals released during examination are x-rays that characterize the elemental composition of the specimen. When x-ray and structural information are combined, a unique description of the specimen emerges. More recent developments in scanning electron microscopy (SEM) include thermal-wave imaging, which is used to detect subsurface defects. Devices are also available for *in situ* fracture studies and have application in the kinematic analysis of deformation.

These features make SEM an ideal tool for the study of fracture surfaces. Different fracture modes exhibit unique features that are easily documented by SEM.

This article will discuss the basic principles and practice of SEM, with emphasis on applications in fractography. The topics include an introduction to SEM instrumentation, imaging and analytical capabilities, specimen preparation, and the interpretation of fracture features. A discussion of the historical development of the scanning electron microscope and its application to fracture studies can be found in the article "History of Fractography" in this Volume. Detailed information on the interpretation of SEM fractographs and the correlation between fracture appearance and properties of various metals and alloys can be found in the article "Modes of Fracture" in this Volume.

SEM Instrumentation

The scanning electron microscope (Fig. 1) can be subdivided into four systems. The illuminating/imaging system consists of an electron source and a series of lenses that generate the electron beam and focus it onto the specimen. The information system comprises the specimen and data signals released during irradiation as well as a series of detectors that discriminates among and analyzes the data. The display system is simply a cathode ray tube (CRT) synchronized with the electron detectors such that the image can be observed and recorded on film. Lastly, the vacuum system removes gases that would otherwise interfere

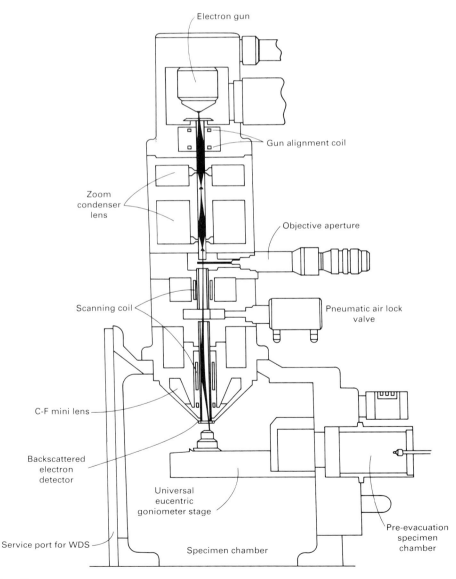

Fig. 1 Schematic cross section of a commercially available scanning electron microscope. Courtesy of JEOL

with operation of the scanning electron microscope column. These four systems are described below in more detail. Supplementary information on the principles and instrumentation associated with SEM can be found in the article "Scanning Electron Microscopy" in Volume 10 of the 9th Edition of *Metals Handbook*.

Illuminating/Imaging System

This system contains an electron gun that generates electrons as well as a series of convergent magnetic lenses that reduces electron beam diameter and focuses the beam at the level of the specimen. The conventional electron gun consists of a tungsten filament that generates electrons when heated to incandescence, an apertured shield centered over the tip of the filament, and the anode, which is held at high positive potential relative to the filament. All three components act as an electrostatic lens; heating the filament generates electrons that are accelerated by the potential difference between the filament and anode into the imaging system. Alternate electron sources, such as the lanthanum hexaboride and field-emission guns, are discussed in Ref 1.

In the imaging system, a series of magnetic lenses reduces the beam diameter from roughly 4000 to 10 nm at the specimen level (Ref 2). Simultaneously, stray electrons are intercepted by apertures such that a collimated electron beam strikes the specimen. Associated with the final lens is a scanning coil that deflects the electron beam in an *x-y* pattern; this activity is reproduced on the observation screen as a raster pattern.

The illuminating/imaging system is responsible for several factors that ultimately define instrument performance, including accelerating voltage, beam diameter, and levels of spherical aberration and astigmatism. Within instrumental specifications for resolution, these factors are subject to operator control. As will be discussed below, the microscopist rarely receives a perfect specimen for examination; consequently, to obtain the highest quality information from any specimen, the scanning electron microscope must be maintained and operated at peak performance (Ref 3).

The accelerating voltage, variable from about 5 to 30 keV on most scanning electron microscopes, is the difference in potential between the filament and anode. Accelerating voltage, V_0, is related to atomic number, Z, and depth of penetration of the incident beam into the specimen, d_p, by:

$$d_p \propto \frac{W_a V_0^2}{Z\rho}$$

where W_a is atomic weight and ρ is density. A secondary effect is the formation of an excitation volume considerably larger than the beam diameter. Consequently, metal specimens are examined at high voltages (25 to 30 keV),

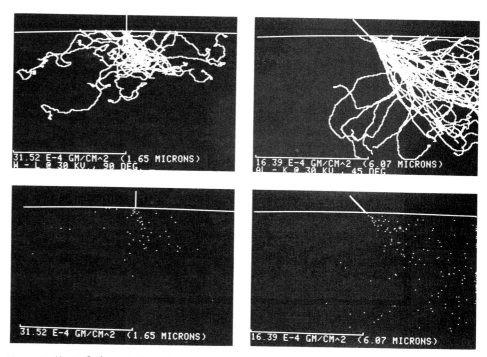

Fig. 2 Monte Carlo projections of the trajectory of incident electrons (top) and emitted x-rays (bottom). Projections are for tungsten (left) and aluminum (right). Note the effect of specimen tilt on the location of the excitation volume.

nonconductive but coated specimens at moderate voltages (~15 keV), and nonconductive, uncoated specimens at low voltages (~5 keV). Figure 2 shows a Monte Carlo projection of electron trajectories in tungsten and aluminum (note the differing sizes of the volumes). The excitation volume is an important quantity because it is the source of data signals used for imaging and analysis. The location of the excitation volume depends on the angle of incidence of the electron beam relative to the specimen surface. This geometry must be known for correct interpretation of x-ray data.

Beam diameter, or spot size, is the width of the beam incident upon the specimen surface. As shown in Fig. 2, it is considerably smaller than the excitation volume. A general rule is that smaller spot sizes always produce higher resolution images. However, at very small spot sizes and beam currents, the signal-to-noise ratio may increase, causing a loss in resolution. Smaller spot sizes are used for image recording; larger spot sizes may be required for x-ray analysis, backscattered electron imaging, and TV mode operation.

Focus and magnification are also controlled by the magnetic lenses. Focus is achieved by varying the current passing through the objective lens. Magnification is the ratio of the size of the display area on the CRT to the area of the specimen scanned. Because a change in magnification involves simply scanning a larger or smaller area, the image should always be focused at least two magnification steps higher than the desired level. This ensures that photographic enlargements will exhibit the same clarity as the original micrograph.

Digital readouts of magnification are not very sensitive; a better indicator is the micron bar imprinted directly onto the micrograph. Because the micron bar is sensitive to both focus and magnification settings, dimensions in enlargements can be measured. However, serious errors arise if very accurate measurements are required, as in the analysis of fatigue crack growth rates. Excessive parallax and other factors complicate the issue. Where high levels of sensitivity are required, internal calibration with commercially available grating replicas is a good starting point, followed by quantitative analysis of stereo pairs, as discussed in the section "Display System" in this article.

Astigmatism is an optical aberration caused by minute flaws in the magnetic-lens coilings. It is manifested as a distortion in shape as focus is varied; for example, a circle forms an ellipse on either side of focus. This asymmetry is compensated for by incorporating weak lenses called stigmators into the lens. The stigmators are of variable amplitude and direction, which oppose and thus cancel the lens asymmetry. Astigmatism must be regularly corrected at a magnification level (~20 000×) roughly double the typical operating magnification.

Spherical aberration arises because an electromagnetic field is strongest along the center of the optical axis and becomes progressively weaker at its periphery. Electrons passing through these different zones are influenced at different magnitudes. This aberration is re-

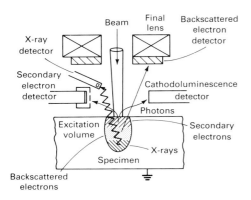

Fig. 3 Origin and detection of data signals

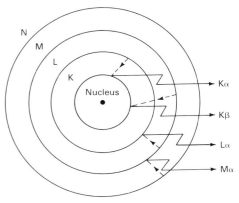

K shell
Kα x-rays are the most intense x-rays and originate from L-shell → K-shell transitions.
Kβ x-rays are the most energetic x-rays and originate from M-shell → K-shell transitions.
L shell
Lα x-rays originate from M-shell → L-shell transitions.
M shell
Mα x-rays originate from N-shell → M-shell transitions.
Energy levels: M < L < Kα < Kβ; intensity levels: Kα > Kβ > L > M

Fig. 4 Origin of x-rays as shown in the Bohr model of the atom

lieved by intercepting peripheral electrons with apertures. In general, smaller apertures (50-μm bore size) are used closest to the specimen level, and larger apertures (~200 μm) are used closest to the electron gun. Image clarity and depth of field are both enhanced with small final apertures. As expected, the apertures must be centered in the optical axis and must be regularly replaced because of the accumulation of contaminants.

Maintenance of the illumination/imaging system requires replacement of filaments (average service life, 40 h), apertures, and the column liner tube as well as alignment of the column. Manufacturer operating manuals should be consulted for maintenance procedures. Additional information can be found in Ref 4.

Information System

Electron Signals. Various data signals are simultaneously released by an irradiated specimen, and in the presence of appropriate detectors, the signals can be analyzed (Fig. 3). Data signals arise from either elastic (electron-nucleus) or inelastic (electron-electron) collisions. Elastic collisions produce backscattered electrons carrying topographic and compositional data (Ref 5, 6). Inelastic collisions deposit energy within the specimen, which then returns to the ground state by releasing secondary electrons, x-rays, and heat phonons.

The conventional SEM image consists of more secondary electrons than backscattered electrons. An important difference between these types of electrons is their relative energy; backscattered electrons retain 80% of the incident beam energy, whereas secondary electrons are of low energy (~4 eV). Therefore, backscattered electrons follow a line-of-sight trajectory and are detected only if they intersect the electron detector. In comparison, secondary electrons are attracted toward the detector by a positively charged Faraday cage and can follow a curved trajectory. The conventional Everhart-Thornley electron detector is ideal for analyzing secondary electrons, but its geometry within the scanning electron microscope is such that it detects only a fraction of the backscattered electrons emitted by the specimen (Ref 7). This secondary electron detector is positioned 90° relative to the optical axis, and the specimen is tilted 10 to 30° to enhance electron collection. In contrast, backscattered electron detectors, such as the Robinson detector (Ref 8), are located immediately beneath the final pole piece, and the specimen is perpendicular to the optical axis (Fig. 3).

The advantage of distinguishing between secondary and backscattered electrons is that the latter can be used for atomic number imaging; the number of backscattered electrons reflected by a specimen increases with atomic weight. This is a powerful technique when used in conjunction with x-ray analysis. As shown in Fig. 3, backscattered electrons originate from a zone closest to the x-ray excitation volume. In failure analysis, atomic number contrast is used in the analysis of segregation, plating defects, and composite failures.

Effect of Specimen/Instrument Geometry. The geometry of the specimen, optical axis, and detector influences data collection. The specimen is manipulated with x,y,z, tilt, and rotational controls. The z-axis controls specimen height, also known as working distance. A large working distance increases depth of field and decreases the lower limit of magnification; the converse is true for small working distances. A compromise for imaging is to position the specimen surface immediately at or slightly below the level of the secondary electron detector. For x-ray analysis, the specimen should be at the level of the detector because x-rays follow a line-of-sight trajectory. Usually, only minor adjustments of the z-axis are required to optimize detection of both signals.

Image clarity is also affected by specimen tilt. Secondary electron and x-ray collection can be maximized by tilting the specimen toward the detector. The optimum angle depends on specimen topography; in general, larger angles are required for smoother specimens. As shown in Fig. 2, the degree of tilt will affect the position of the excitation volume. This is crucial for valid interpretation of point x-ray analysis because the data may originate from a position that does not correspond exactly to the SEM image.

X-Ray Signals. Characteristic x-rays are distinct quanta of energy released from excited atoms. Specimen composition is analyzed by measuring x-ray energy or wavelength. X-rays arise from electron transitions within the orbitals of an atom (Fig. 4). Although there is some overlap among x-ray energies, all atoms generally possess at least one x-ray, or spectrum of x-rays, that is unique to that element.

Energy-dispersive spectroscopy (EDS) is the more common method of x-ray analysis used in SEM. The conventional system can quantitatively analyze elements with Z exceeding or equal to 11 (sodium) (Ref 9). Windowless detectors permit light element detection (Ref 10, 11). All x-rays ranging from about 0.7 to 13 keV are simultaneously detected. Standard tables of x-ray energy are available for manual data reduction, but modern spectrometers automatically identify each peak and its relative intensities (Ref 12).

Wavelength-dispersive spectroscopy (WDS) uses a crystal spectrometer for the detection of specific x-rays. Unlike EDS, a specific wavelength is tuned in and analyzed. This is a higher-resolution technique, but is more frequently associated with electron probe x-ray microanalysis than with SEM (Ref 13-15).

The advantage of conducting an x-ray analysis with the scanning electron microscope is that the area to be analyzed is visualized directly on the CRT; that is, at low magnification, one may analyze the bulk specimen, then increase magnification and selectively analyze smaller areas. However, the effects of geometry on location of the excitation volume, as shown in Fig. 2 and 3, should be considered. To identify the sources of x-ray emission, x-ray or dot maps are produced by feeding the x-ray data for a given element back into the scanning electron microscope (Ref 16). Direct correlations between structure and composition can be made by recording the x-ray spectrum, dot map, and electron image. Dot maps are very useful for sorting inclusions, demonstrating corrosion sites, and illustrating any type of atomic number difference, especially in conjunction with backscattered electron imaging.

A great deal of information is available on x-ray analysis. Manufacturer's publications are good sources, as are Ref 4, 12, 14, and 17 and the articles "Scanning Electron Microscopy" and "Electron Probe X-Ray Microanalysis" in Volume 10 of the 9th Edition of *Metals Handbook*.

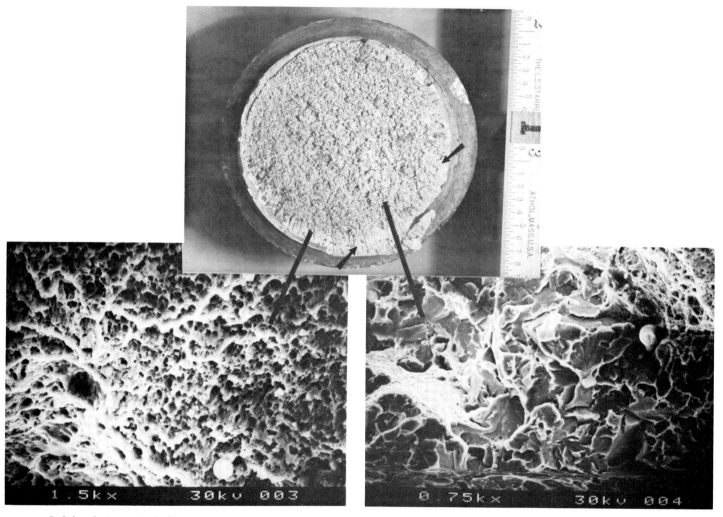

Fig. 5 Radial marks (arrows) in the fibrous zone of a bolt fractured under conditions of tensile overload. The morphologies of the different texture zones are shown in the SEM fractographs: ductile fracture (left) and transgranular fracture (right).

Thermal-wave imaging is a near-surface high-resolution technique that produces images resulting from localized changes in thermal parameters (Ref 18). Although this technique is more widely used to analyze microelectronic devices, it has application in metallurgy for the detection of subsurface (5 to 10 μm) defects and the imaging of metallographic features of unpolished samples.

In Situ **Studies.** The advent of large scanning electron microscope specimen chambers has permitted design of devices for the *in situ* analysis of mechanical behavior, such as fatigue crack initiation and propagation studies. Fatigue crack initiation has been studied (Ref 19), and fatigue cracks near the threshold value have been analyzed (Ref 20). Other devices include those for the analysis of wear (Ref 21, 22), high-temperature *in situ* oxidation (Ref 23, 24), fiber-reinforced metal matrix composites (Ref 25), and *in situ* evaluation of ductile material behavior (Ref 26, 27). Videotaping of such experiments provides a microscopic view of fracture mechanisms.

Display System

Scanning electron microscopy images are displayed on a CRT synchronized with the imaging system. Micrographs are recorded from a high-resolution CRT, usually onto Polaroid film. Very slow scan rates (30 to 120 s) are used to improve the signal-to-noise ratio. Contrast and brightness are modulated by the operator of the scanning electron microscope. A good micrograph exhibits a range of gray levels; as the number of gray levels increases, so does the information content of the micrograph. The operating parameters that influence the quality of the micrograph include correct accelerating voltage, small beam spot size, optimum specimen geometry, and column alignment. The specimen itself must be clean and conductive. Detailed information on SEM photography can be found in Ref 4 and 28 to 30.

Most scanning electron microscopes have various signal-processing devices that modulate the image. Gamma modulation suppresses very dark or light levels, thus intensifying intermediate gray levels; it is used for specimens having very rough surfaces. Other devices include split screens for display of dual magnification or different imaging modes, for example, the side-by-side display of secondary electron and backscattered electron images of the same area.

The most crucial aspect of image recording for fractography is to maintain orientation and perspective. In general, only selected areas of a fracture surface are examined in depth, for example, the fracture origin. If the specimen exhibits multiple fracture modes, usually visible with a binocular microscope, the different areas are documented (Fig. 5). Consequently, to maintain orientation, the microscopist should use a macrophotograph or detailed sketch to identify sites where SEM photos are recorded. The fractographs should progress from low to

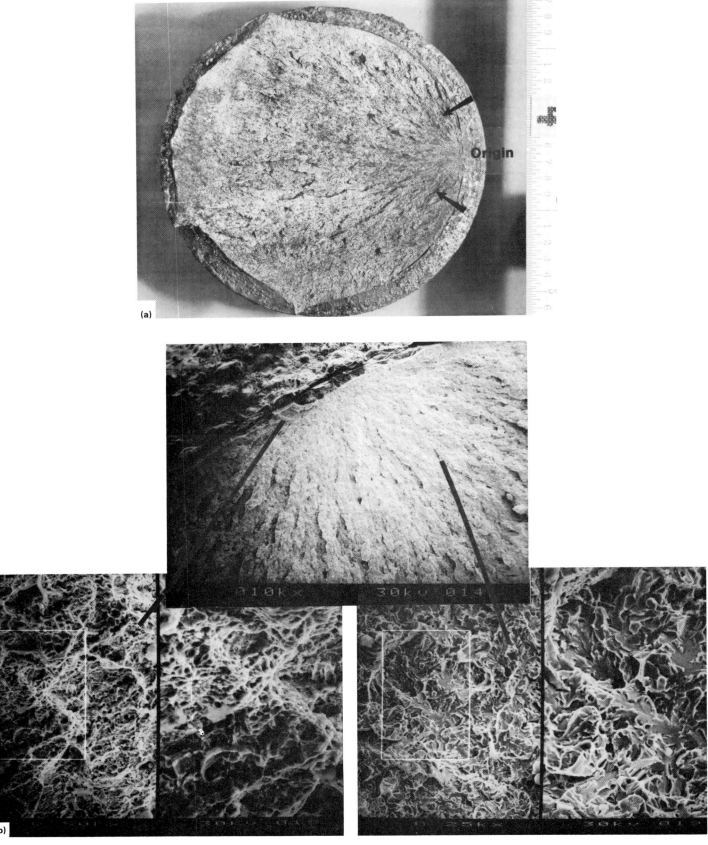

Fig. 6 (a) Chevrons (arrows) emanating from the fracture origin in a bolt that failed under conditions of bending overload. (b) SEM fractographs of the origin and fracture surface shown in (a)

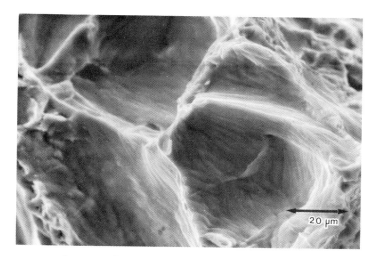 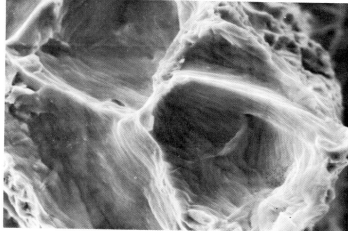

Fig. 7 Stereo pair showing deep dimples in the fracture surface of commercially pure titanium. Average grain size is 46 μm. Large dimples originated at grain-boundary triple points. Note small dimples at rim of large dimples that nucleated at dislocation cell walls. (M. Erickson-Natishan, University of Virginia)

high magnification, with identifiable features present in the series (Fig. 6). Such a correlation of macroscopic and microscopic features provides an excellent record of fracture morphology and is invaluable for interpretation. A similar approach is used if the images are videotaped.

Stereo Imaging. A serious problem often encountered in SEM fractography is perspective distortion due to incorrect perception of the direction of illumination. This artifact is eliminated by stereo imaging, which involves recording the same field of view twice, each at slightly different orientations, then simultaneously viewing the stereo pair. The correct relationships are restored, and valid spatial judgments replace subjective impressions.

The tilt method of stereo recording can be used with any scanning electron microscope as follows:

- Select and record the desired field of view, noting the tilt value of the specimen stage
- Mark the location of a prominent surface feature on the observation screen with a wax pencil
- Tilt the specimen about 7° (the stereo angle), and realign the prominent features beneath the wax pencil mark
- Refocus the image using the z-axis control; do not refocus with the lens controls
- Adjust brightness and contrast, and record the image

Figure 7 illustrates the tilt method for stereo SEM. Stereo pairs are viewed using simple pocket viewers, double-prism viewers, or a mirror stereoscope (Ref 31). Methods of stereo projection are discussed in Ref 32 and 33.

Quantitative stereoscopy, which involves stereoscopic imaging and photogrammetric methods, is used for conducting spatial measurements on stereo pairs (Ref 31, 34-37). Detailed information on stereoscopic imaging and photogrammetric methods can be found in the article "Quantitative Fractography" in this Volume.

In stereo photogrammetry, calibrated topographic maps of fracture surfaces can be generated by using a newly developed adaptation of a Hilger-Watts stereoscope interfaced to a microcomputer. Transducers are mounted so as to follow the motion of the viewing table and the motion of the micrometer used to superimpose the image of the light spot onto the three-dimensional image of the surface below. The light spot (generated by two light sources mounted on either side of the stereoscope and seen through half-silvered mirrors) is raised and lowered in order to appear to lie along the surface below. The micrographs are translated, and the apparent height of the light spot is recorded with each trigger event, generating a matrix of x-, y-, and z-coordinates when processed.

The voltage signals of the transducers are processed through an analog-digital conversion board and recorded on the microcomputer. The arrays are then calibrated and normalized using user-supplied information on magnification and parallax angle. The array of calibrated x-, y-, and z-files can then be used to generate graphical output in various forms: carpet plots, hidden line plots, or contour plots, depending on the need. Figure 8 shows a stereo pair of the fracture surface of a Ti-10V-2Fe-3Al alloy and the corresponding carpet plot and contour plot.

Vacuum System

The scanning electron microscope optical column and specimen chamber are operated under high-vacuum conditions ($\leq 10^{-4}$ torr), to improve the quality of imaging, minimize contamination, and, in general, extend the service lives of all components. A typical scanning electron microscope is equipped with a high-vacuum diffusion pump backed by a rotary pump. Some manufacturers market turbomolecular pumps, which relieve the contamination problems sometimes associated with conventional systems. Because vacuum technology is standard regardless of the equipment it is associated with, the scanning electron microscope vacuum system will not be discussed further in this article. Additional information on vacuum pumping systems can be found in the article "Scanning Electron Microscopy" in Volume 10 of the 9th Edition of *Metals Handbook*.

Specimen Preparation

The microscopist must know the objectives of an SEM examination before preparing a specimen. Different preparation protocols are used, depending on whether SEM is required alone or in combination with x-ray analysis, particularly when the specimen is too large for the specimen chamber or is nonconductive. In some litigation cases, use of an inappropriate preparation method can be disastrous. The least aggressive method of preparation should be selected for any fracture specimen.

The major criteria for SEM specimen preparation are that the specimen be conductive, clean, and small enough to enter the specimen chamber. If the specimen is too large, replicas composed of cellulose acetate or dental-impression media are prepared and coated with a conductive thin film (Ref 4, 38). Cellulose acetate replicas are also used to remove and simultaneously preserve oxidation products that obscure the specimen surface. The fracture surface morphology can be analyzed by direct examination of the fracture, and the products held within the replica can be identified by coating the replica with thin carbon film. The handling and cleaning of fracture surfaces are the most important aspects of fracture specimen preparation. Methods for handling, sectioning, and cleaning fractographic specimens are de-

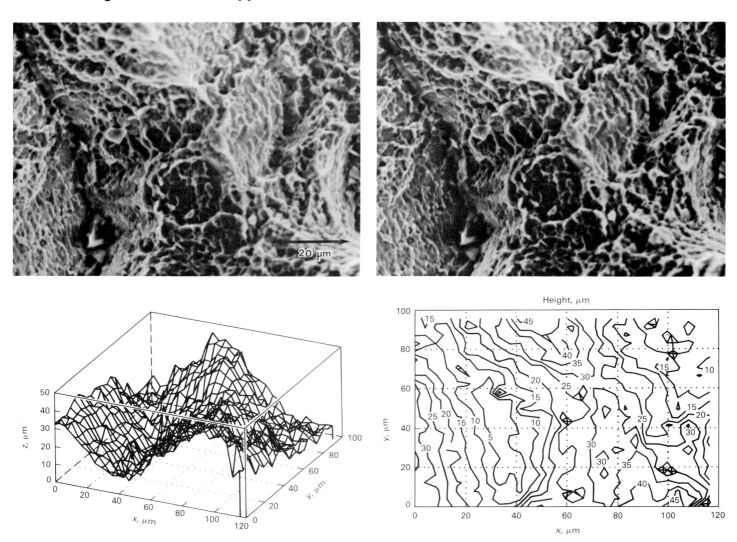

Fig. 8 Stereo pair (top left and right) of a fractured Ti-10V-2Fe-3Al alloy that was heat treated at 780 °C (1435 °F) for 3 h, water quenched, and aged for 1 h at 500 °C (930 °F). The corresponding carpet plot (bottom left) and contour plot (bottom right) of the fracture surface are also shown. (J.D. Bryant, University of Virginia)

scribed in the article "Preparation and Preservation of Fracture Specimens" in this Volume.

In most cases, a metal fracture can be directly examined in the scanning electron microscope. After cleaning, the specimen is mounted in a specimen holder or on a substrate using conductive paint or tape (Ref 39). Substrates include aluminum stubs and carbon planchets; the latter is preferred for x-ray analysis. The paint or tape must be positioned such that the area of interest is not obscured. With large specimens, it is helpful to identify the area of interest with small arrows cut from metallic tape; their position and orientation can be indicated on both the macrophotograph and low-magnification micrographs to facilitate correlations. At higher magnifications, these overviews can be used as maps to pinpoint location and orientation.

Replicas and other nonconductive specimens are coated with a conductive thin film for SEM examination. Nonconductive specimens accumulate a net negative charge that interferes with imaging unless examined at very low accelerating voltages (~5 keV). Coating the specimen permits use of higher voltages (15 to 20 keV), which significantly enhances image quality. Metallic coatings (gold or chromium) are preferred for imaging purposes because they increase the image-forming electron yield; such coatings are prepared by thermal evaporation or sputter coating. Carbon coatings prepared by evaporation are used for x-ray analysis because the surface film is nearly transparent to x-rays.

Thermal Evaporation. Evaporated thin films are prepared in a bell jar under high-vacuum conditions by resistance heating of a metal wire or basket, which holds the evaporant above the specimen to be coated. At the vaporization temperature of the metal, atoms are released and follow a line-of-sight trajectory until they strike the specimen surface. As more metal vaporizes, a thin film will gradually adhere and eventually coat the specimen. If the specimen is held stationary and at an angle relative to the source, the deposition is oblique. This is the technique of shadowing, which is used to highlight surface features. In fractography, shadowing is used to enhance the fidelity of very fine fatigue striations and the contrast of faint river patterns in cleavage fracture; the shadow is deposited in the direction of crack propagation. If the specimen is mounted on a planetary stage in motion during evaporation, a continuous thin film is deposited. The latter is preferred for coating nonconductive specimens, because film continuity is required for conductivity.

Assuming that all other factors are constant, the metal used for evaporation determines the structure of the coating. In general, the higher the vaporization temperature of the metal, the finer the thin film. Gold with a vaporization temperature of 1465 °C (2670 °F) produces a coarse-grain film, while platinum (2090 °C, or 3795 °F) produces a finer film. Alloys such as platinum-carbon form very fine-grain films. The latter are required for transmission electron microscopy (TEM), but coarser films are adequate for routine SEM fractography because the grain rarely becomes objectionable and interferes with image quality. Finer films are required only when resolution exceeds approximately 8 nm and magnification is greater than 40 000×.

Carbon thin films are used for x-ray analysis or as a preliminary coating to enhance adhesion of metal films. The vacuum bell jar is used, but is modified such that two carbon electrodes are used in place of a tungsten substrate for a metal wire. The carbon is evaporated by passing an alternating current of 20 A at 30 V through the electrodes. More detailed descriptions of this technique and of thermal evaporation are available in Ref 4 and 40. Shadowing is also discussed in the article "Transmission Electron Microscopy" in this Volume.

Sputter coating involves the erosion of metal atoms from a target by an energetic plasma under low-vacuum conditions. This technique is preferred over evaporation for coating rough-surface specimens, because metal atoms released from the target are deflected by gas molecules within the chamber and thus approach the specimen from all directions. For example, replicas of ductile fracture surfaces often have an exaggerated topography, and it is difficult to coat the cavities of dimples without increasing film thickness when thermal evaporation is used. With sputter coating, the cavities will be coated without increasing thickness.

The diode sputter coater consists of the specimen stage (anode) and a small bell jar containing a metal target (usually gold) that functions as a cathode. Under low-vacuum conditions ($\sim 10^{-2}$ torr, or 1.3 Pa), argon or nitrogen is bled into the chamber and forms a plasma during glow discharge. These energetic ions strike the metal target, and a transfer of momentum causes metal atoms to be ejected from the target. The metal atoms are attracted toward the specimen stage by the potential difference between the target and stage. Because heat is generated during sputtering, some diode coaters are equipped with cooled specimen stages (Ref 41) or are modified into triode units (Ref 42). Sputter coating is also discussed in Ref 4 and 43 to 45.

SEM Fractography

The general features of ductile and brittle fracture modes are summarized in this section. More detailed information on fracture modes and the effect on fracture morphologies of environmental factors, such as corrosion, temperature, stress state, and strain rate, can be found in the article "Modes of Fracture" in this Volume. Overviews on fractography (Ref 46-51) and various fractographic atlases (Ref 52-55) should also be consulted.

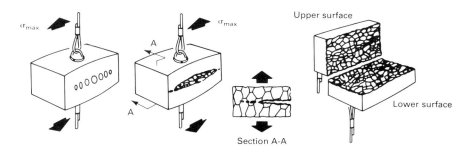

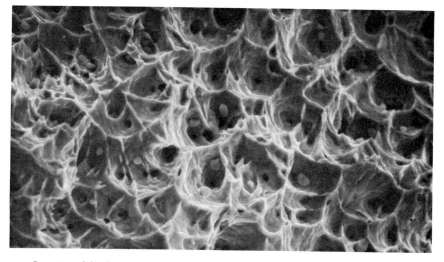

Fig. 9 Formation of dimples under conditions of tension using a copper test specimen. Note that the dimples are equiaxed. 750×

Ductile and brittle are terms that describe the amount of macroscopic plastic deformation that precedes fracture. Ductile fractures are characterized by tearing of metal accompanied by appreciable gross plastic deformation and expenditure of considerable energy. Ductile tensile fractures in most materials have a gray, fibrous appearance and are classified on a macroscopic scale as either flat (perpendicular to the maximum tensile stress) or shear (at a 45° slant to the maximum tensile stress) fractures.

Brittle fractures are characterized by rapid crack propagation with less expenditure of energy than with ductile fractures and without appreciable gross plastic deformation. Brittle tensile fractures have a bright, granular appearance and exhibit little or no necking. They are generally of the flat type, that is, normal (perpendicular) to the direction of the maximum tensile stress. A chevron pattern may be present on the fracture surface, pointing toward the origin of the crack, especially in brittle fractures in flat platelike components. Fractographic features that can be observed without magnification or at low magnifications are discussed in the article "Visual Examination and Light Microscopy" in this Volume.

These terms can also be applied, and are applied, to fracture on a microscopic level. Ductile fractures are those that occur by microvoid formation and coalescence, whereas brittle fractures can occur by either transgranular (cleavage, quasicleavage, or fatigue) or intergranular cracking. Intergranular fractures are specific to certain conditions that induce embrittlement. These include embrittlement by thermal treatment or elevated-temperature service and embrittlement by the synergistic effect of stress and environmental conditions. Both types are discussed below. Additional information can also be found in the article "Ductile and Brittle Fractures" in Volume 11 of the 9th Edition of *Metals Handbook*.

Ductile Fracture. Examination of ductile fracture surfaces by SEM reveals information about the type of loading experienced during fracture, the direction of crack propagation, and the relative ductility of the material (Ref 56-58). The shape of the dimples produced is determined by the type of loading the component experienced during fracture, and the orientation of the dimples reveals the direction of crack extension.

Equiaxed or hemispheroidal dimples are cupshaped, and they form under conditions of uniform plastic strain in the direction of applied tensile stress; equiaxed dimples are typically produced under conditions of tensile overload (Fig. 9). In comparison, elongated dimples shaped like parabolas result from nonuniform plastic-strain conditions, such as bending or shear overloads (Fig. 10). These dimples are

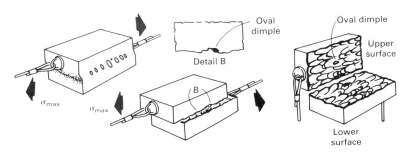

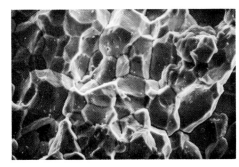

Fig. 11 Grain-boundary separation induced by atmospheric stress-corrosion cracking of a high-strength aluminum alloy. 130×

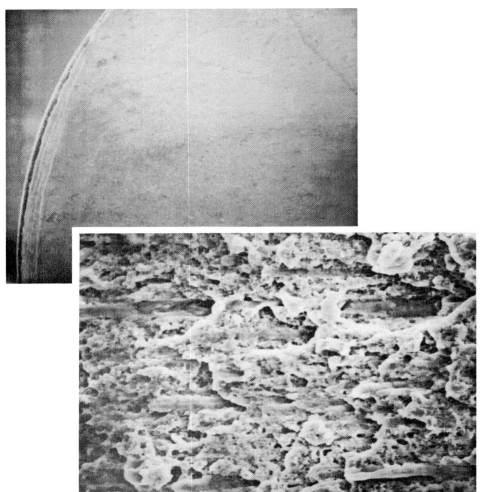

Fig. 10 Fracture of a high-strength steel under conditions of transverse shear overload. Fractographs at 25× (top) and 1000× (bottom)

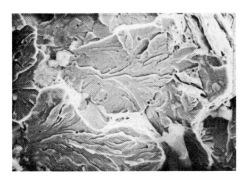

Fig. 12 Cleavage in a low-carbon steel specimen that was impact fractured at liquid-nitrogen temperatures. 385×

elongated in the direction of crack extension and reveal the origin of the fracture. Orientation is critical in this fracture mode; the microscopist should observe the conditions described earlier regarding mapping macro- and microfractographs. Similar conditions should be observed when examining fractures due to torsional shear and bending overloads.

Because the size of the dimples is largely a function of the relative ductility of the material, one magnification level cannot be specified for all ductile fractures. Fractographs should be recorded up to magnification at which the shape and orientation of the dimples are clearly revealed. If there is any confusion about the orientation of the dimples, stereo fractographs should be recorded to ensure that no perspective distortion was introduced by the complex geometry.

If x-ray analysis of the precipitates or inclusions held within the dimple is desired, the parameters on the location of the excitation volume and accelerating voltage discussed in the section "Illuminating/Imaging System" in this article should be considered. The excitation volume may include the area beneath the inclusion. It may be necessary to reduce the accelerating voltage for analysis of low molecular weight inclusions and to compare this spectrum with one at a higher voltage. This is adequate for most purposes, although a more sophisticated approach is to strip the spectrum of the bulk specimen from that of the inclusion. Modern electron probe microanalysis systems readily manipulate the spectra and greatly simplify this type of analysis. The mechanism of dimple rupture fracture and the effect of environment on dimple size and shape are discussed in the article "Modes of Fracture" in this Volume.

Intergranular brittle fracture, also referred to as grain-boundary separation or decohesive rupture, is characterized by a rock-candy or faceted appearance (Fig. 11). It is promoted by the synergistic effect of environmental conditions and sustained stress. Although it is easy to recognize intergranular fracture, identification of the cause of fracture is much more complex. Consequently, SEM provides a means to identify the mode of fracture, but yields little other information.

These fractures are generally characterized at magnifications under 1000×. The relationships among grains must be demonstrated in the fractographs because small zones of microvoid

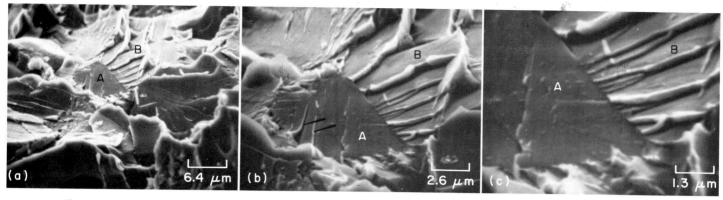

Fig. 13 Cleavage fracture in a notched impact specimen of hot-rolled 1040 steel broken at −196 °C (−320 °F), shown at three magnifications. The specimen was tilted at an angle of 40° to the electron beam. The cleavage planes followed by the crack show various alignments, as influenced by the orientations of the individual grains. Grain A, at center in fractograph (a), shows two sets of tongues (see arrowheads in fractograph b) as a result of local cleavage along the {112} planes of microtwins created by plastic deformation at the tip of the main crack on {100} planes. Grain B and many other facets show the cleavage steps of river patterns. The junctions of the steps point in the direction of crack propagation from grain A through grain B, at about 22° to the horizontal plane. The details of these forks are clear in fractograph (c).

Fig. 14 Forged aluminum alloy 2014-T6 aircraft component that failed by fatigue. Characteristic beach marks are evident. See also Fig. 15.

coalescence may be observed on grain facets or interfaces. Recording the images of these very small areas is misleading; again, a range of magnifications is required to portray the surface accurately.

Transgranular fracture modes include cleavage and fatigue. The features of each are discussed below. See the sections "Cleavage" and "Fatigue" in the article "Modes of Fracture" for a description of the fracture mechanisms involved in transgranular fractures. The effects on fatigue of gaseous environments, liquid environments, vacuum, temperature, and loading are described in the section "Effect of Environment" in the aforementioned article.

Cleavage. A visual examination of a cleavage fracture reveals brightly reflecting facets, which appear in the scanning electron microscope as very flat surfaces (Fig. 12). At higher magnifications, the facets reveal features related to the direction of local crack propagation, which can in turn be related to the origin of the primary crack (Fig. 13). River patterns represent steps between different local cleavage facets at slightly different heights but along the same general cleavage plane. Because local crystallographic structure can modify the local direction of crack propagation, the overall direction is assigned only after confirming the orientation of the river patterns in several areas on the fracture surface. The same criteria used for characterizing ductile fractures apply in this case.

Fatigue is a time-dependent mechanism that can be separated into three stages that exhibit different features. Stage I is crack initiation, stage II is crack propagation, and stage III is unstable fast fracture by overload. The fatigue crack initiation zone is a point (or points, producing multiple origins) that is usually at or near the surface, where the cyclic strain is greatest or where material defects or residual stresses lessen the fatigue resistance of the component. The crack typically initiates at a small zone and propagates by slip-line fracture, extending inward from the surface at roughly 45° to the stress axis (Ref 59).

The location of the origin is defined by interpreting features of the stage II crack propagation zone. Macroscopic beach marks, or clamshell markings, radiate away from the origin in concentric semicircles (Fig. 14). These are a special form of progression mark commonly associated with fatigue. When the propagation zone is examined by SEM at progressively higher magnifications, the beach marks can be resolved into hundreds or thousands of fatigue striations (Fig. 15).

Striations are characteristically mutually parallel and at right angles to the local direction of crack propagation. They vary in striation-to-striation spacing with cyclic stress intensity, they are equal in number to the number of load cycles (under cyclic stress-loading conditions), and they are generally grouped into patches within which all markings are continuous (Ref 60). Also, fatigue striations do not cross one another, but may join and form a new zone of local crack propagation.

Because beach marks and fatigue striations radiate away from the origin as a series of concentric arcs, the crack initiation site(s) can be identified by drawing an imaginary radius perpendicular to their direction and centered at the origin.

If the component has been subjected to uniformly applied loads of sufficient magnitude, a

Fig. 15 A series of low- to high-magnification micrographs of the specimen shown in Fig. 14. Note that as magnification is increased, progressively finer striations are resolved. (a) 80×. (b) 800×. (c) 4000×. (d) 8000×

single advance of the crack front, that is, the distance between two adjacent striations, is a measure of the rate of propagation per stress cycle (Ref 61, 62). Therefore, the appearance of a fatigue fracture surface can be directly related to a given stress cycle (Ref 63-66).

However, if the loading is nonuniform, there are wide variations between a given stress cycle series and the spacing of the striations; each stress cycle does not necessarily produce a striation. For example, overload cycles can produce zones of microvoid coalescence interspersed among bands of striations (Fig. 16). Because the area of the fracture surface occupied by dimples does not exhibit striations, there are wide variations between the pattern of striations and the applied cyclic stress (Ref 67, 68). Further, under nonuniform loading conditions, the lower-amplitude stress cycles may not be of sufficient magnitude to produce resolvable striations. *In situ* fracture devices are one possible solution to this problem (Ref 69, 70).

A different type of complication is that not all fatigue fractures exhibit striations. Although the presence of striations establishes fatigue as the mode of failure, their absence does not eliminate fatigue as a possibility. For example, striations are usually well defined in aluminum alloys fatigued in air, but do not form if the component is tested under vacuum (Ref 71); the same holds true for titanium alloys (Ref 72). Further, the fidelity of the striations changes with composition. For example, striations are often prominent in aluminum alloys (Fig. 15), but are often poorly defined in ferrous alloys (Fig. 17). Oxidation, corrosion, or mechanical damage can obliterate striations.

Fatigue fractures are examined in the scanning electron microscope at progressively higher magnifications at the origin and at least two other areas in the crack propagation zone to confirm the location of the origin. It is obviously important to maintain orientation and to record fractographs at several levels of magnification. If any obvious defect is observed at the origin, its composition should be compared to that of the bulk material.

Mixed Fracture Modes. In practice, very clear-cut cases are often encountered; however, more complicated cases occasionally arise. For example, in mixed modes of fracture, the predominant fracture mode is normally assigned. In other cases, such as combined stress-corrosion cracking and fatigue, both failure modes contribute to the fracture. This illustrates the importance of collecting data during a failure analysis and combining all sources with fractographic evidence before defining modes of fracture.

REFERENCES

1. A.N. Broers, *IITRI/SEM Proceedings*, 1975, p 661
2. B. Siegel, *IITRI/SEM Proceedings*, 1975, p 647
3. G.F. Pfefferkorn et al., *SEM, Inc.*, Vol 1, 1978, p 1
4. B.L. Gabriel, *SEM: A User's Manual for Materials Science*, American Society for Metals, 1985
5. D.E. Newbury, *IITRI/SEM Proceedings*, Vol 1, 1977, p 553
6. V.N. Robinson and E.P. George, *SEM, Inc.*, Vol 1, 1978, p 859

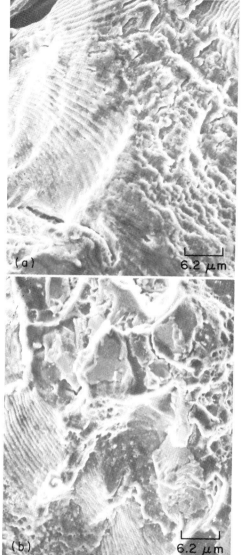

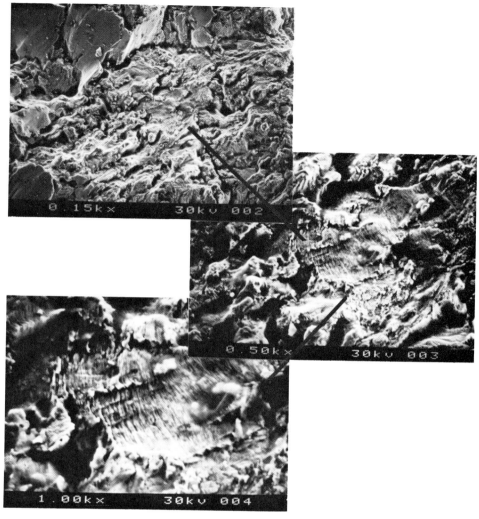

Fig. 17 Poorly defined fatigue striations in a low-carbon steel specimen

Fig. 16 Intermingled dimples and fatigue striations in low-cycle fatigue test fractures in aluminum alloy 2024-T851 at a high range of stress intensity (ΔK) at the crack tip. Orientation of fatigue striations differs from patch to patch, particularly in fractograph (a). Dimples in fractograph (b) are associated with inclusions. Both 1600×

7. T.E. Everhart and R.E.M. Thornley, *J. Sci. Inst.*, Vol 37, 1960, p 246
8. V.N.E. Robinson, *J. Phys. E, Sci. Instrum.*, Vol 7, 1974, p 650
9. R. Woldseth, *X-ray Energy Spectrometry*, Kevex Corporation, 1973
10. R.G. Musket, in *Energy Dispersive X-ray Spectrometry*, NBS 604, National Bureau of Standards, 1981, p 97
11. J.C. Russ and A.O. Sandborg in *Energy Dispersive X-ray Spectrometry*, NBS 604, National Bureau of Standards, 1981, p 71
12. N.C. Barbi, *Electron Probe Microanalysis Using Energy Dispersive X-ray Spectroscopy*, PGT, Inc., 1981
13. S.J.B. Reed, *Electron Microprobe Analysis*, Cambridge University Press, 1975
14. J.I. Goldstein et al., *Scanning Electron Microscopy and X-ray Microanalysis*, Plenum Press, 1981
15. D.T. Quinto et al., *Low-Z Element Analysis in Hard Materials*, Plenum Press, 1983
16. J.J. McCarthy et al., in *Proceedings of the Microbeam Analysis Society*, 1981, p 30
17. D.E. Newbury, *SEM, Inc.*, Vol 2, 1979, p 1
18. A. Rosenscwaig, *Science*, Vol 218, 1982, p 223
19. K. Wetzig et al., *Pract. Metallogr.*, Vol 21, 1984, p 161
20. M. Schaper and D. Boesel, *Prakt. Metallogr.*, Vol 22 (No. 4), 1985, p 197
21. G. Gille and K. Wetzig, *Thin Solid Films*, Vol 110 (No. 1), 1983, p 37
22. S.V. Prasad and T.H. Kosel, in *Wear of Materials*, American Society of Mechanical Engineers, 1983, p 121
23. E. Kny et al., *J. Vac. Sci. Technol.*, Vol 17 (No. 5), 1980, p 1208
24. S.K. Verma et al., *Oxid. Met.*, Vol 15 (No. 5-6), 1981, p 471
25. D.L. Davidson et al., in *Mechanical Behavior of Metal/Matrix Composites*, American Institute of Mining, Metallurgy, and Petroleum Engineers, 1982, p 117
26. Z.Q. Hu et al., in *In Situ Composites IV*, Elsevier, 1981
27. F. Mousy, in *Advances in Fracture Research*, Vol 5, Pergamon Press, 1982, p 2537
28. H. Horenstein, *Black and White Photography: A Basic Manual*, Little, Brown and Co., 1974
29. H. Horenstein, *Beyond Basic Photography*, Little, Brown and Co., 1977
30. C.B. Neblette et al., *Photography: Its Materials and Processes*, Van Nostrand Reinhold, 1976
31. A. Boyde, *SEM, Inc.*, Vol 2, 1979, p 67

32. V.C. Barber and C.J. Emerson, *Scanning*, Vol 3, 1980, p 202
33. W.P. Wergin and J.B. Pawley, *SEM, Inc.*, Vol 1, 1980, p 239
34. P.G.T. Howell and A. Boyde, *IITRI/SEM Proceedings*, 1972, p 233
35. A. Boyde, *IITRI/SEM Proceedings*, 1974, p 101
36. A. Boyde, *SEM, Inc.*, Vol 1, 1981, p 91
37. P.G.T. Howell, *Scanning*, Vol 4, 1981, p 40
38. C.H. Pameijer, *SEM, Inc.*, Vol 2, 1978, p 831
39. J.A. Murphy, *SEM, Inc.*, Vol 2, 1982, p 657
40. C.C. Shiflett, in *Thin Film Technology*, R.W. Berry *et al.*, Ed., Van Nostrand Reinhold, 1968, p 113
41. P.N. Panayi *et al.*, *IITRI/SEM Proceedings*, Vol 1, 1977, p 463
42. P. Ingram *et al.*, *IITRI/SEM Proceedings*, Vol 1, 1976, p 75
43. P. Echlin, *IITRI/SEM Proceedings*, 1975, p 217
44. P. Echlin, *SEM, Inc.*, Vol 1, 1978, p 109
45. P. Echlin, *SEM, Inc.*, Vol 1, 1981, p 79
46. C.D. Beachem, in *Fracture*, Vol 1, H. Liebowitz, Ed., Academic Press, 1968, p 243
47. C.D. Beachem, Ed., *Electron Fractography*, STP 436, American Society for Testing and Materials, 1968
48. V.J. Colangelo and F.A. Heiser, *Analysis of Metallurgical Failures*, John Wiley & Sons, 1974
49. B.M. Strauss and W.H. Cullen, Ed., *Fractography in Failure Analysis*, STP 645, American Society for Testing and Materials, 1978
50. P.P. Tung *et al.*, Ed., *Fracture and Failure: Analyses, Mechanisms, and Applications*, American Society for Metals, 1981
51. D.J. Wulpi, *Understanding How Components Fail*, American Society for Metals, 1985
52. A. Phillips *et al.*, *Electron Fractography Handbook*, sponsored by Air Force Materials Laboratory, AFML-TR-64-416, Wright-Patterson Air Force Base, published by Battelle Columbus Laboratories, June 1976
53. G.F. Pittinato *et al.*, *SEM/TEM Fractography Handbook*, sponsored by Air Force Materials Laboratory under Contract No. F33615-74-C-5004, published by Battelle Columbus Laboratories, 1975
54. S. Bhattacharyya *et al.*, *IITRI Fracture Handbook*, IIT Research Institute, 1979
55. L. Engel and H. Klingele, *An Atlas of Metal Damage*, Prentice-Hall, 1981
56. C.D. Beachem, *Trans. ASM*, Vol 56 (No. 3), 1963, p 318
57. C.D. Beachem, *Met. Trans. A*, Vol 6, 1975, p 377
58. C.D. Beachem and G.R. Yoder, *Met. Trans.*, Vol 4, 1973, p 1145
59. D. Eylon and W.R. Kerr, in *Fractography in Failure Analysis*, B.M. Strauss and W.H. Cullen, Ed., STP 645, American Society for Testing and Materials, 1978, p 235
60. D.A. Meyn, Memorandum Report 707, Naval Research Laboratory, 1966
61. P.C. Paris, *Proceedings of the 10th Sagamore Conference*, Syracuse University Press, 1964, p 107
62. J.C. McMillan and R.W. Hertzberg, in *Electron Fractography*, C.D. Beachem, Ed., STP 436, American Society for Testing and Materials, 1968, p 89
63. R.W. Hertzberg and W.J. Mills, in *Fractography—Microscopic Cracking Processes*, STP 600, American Society for Testing and Materials, 1976, p 220
64. P.R. Abelkis, in *Fractography in Failure Analysis*, B.M. Strauss and W.H. Cullen, Ed., STP 645, American Society for Testing and Materials, 1978, p 213
65. A. Madeysik and L. Albertin, in *Fractography in Failure Analysis*, B.M. Strauss and W.H. Cullen, Ed., STP 645, American Society for Testing and Materials, 1978, p 73
66. A.J. Krasowsky and V.A. Stepanenko, *Int. J. Fract.*, Vol 15, 1979, p 203
67. W. Wiebe and R.V. Dainty, *Can. Aero. Space J.*, Vol 27 (No. 2), 1981, p 107
68. C.R. Morin and B.L. Gabriel, *Microscope*, Vol 30, 1982, p 139
69. K. Schulte *et al.*, *AGARD (NATO)*, Vol 16, 1984, p 1-10
70. J. Lankford and D.E. Davidson, in *Advances in Fracture Research*, Vol 2, Pergammon Press, 1981, p 899
71. D.A. Meyn, *Trans. ASTM*, Vol 61 (No. 1), 1968, p 52
72. D.A. Meyn, *Met. Trans.*, Vol 2, 1971, p 853

Transmission Electron Microscopy

THE APPLICATION of the transmission electron microscope to the study of fracture surfaces and related phenomena made it possible to obtain magnifications and depths of field much greater than those possible with the light (optical) microscope. As a result, new information regarding the micromechanisms of fracture processes was obtained during the 1960s, when the transmission electron microscope was the principal tool for the study of fractures. However, because of the problems introduced by the necessity of preparing a replica of the fracture surface (see discussion below) and because of improvements in the scanning electron microscope, fracture studies using the transmission electron microscope and associated replicas are currently confined to the following cases:

- Where higher resolution (better than 1 nm) is required, such as the examination of extremely fine fatigue striations
- Where the surface of a large component must be examined without sectioning the part
- Where extraction of particles for identification is necessary

This article will review methods for preparing replicas, will discuss artifacts in replicas, and will compare transmission electron and scanning electron fractographs. Information on the historical development of the transmission electron microscope and its application to fractography can be found in the article "History of Fractography" in this Volume. A detailed review of the instrumentation, principles, and applications associated with the transmission electron microscope is provided in the article "Analytical Transmission Electron Microscopy" in Volume 10 of the 9th Edition of *Metals Handbook*.

Specimen Preparation

Specimens for transmission electron microscopy (TEM) must be reasonably transparent to electrons, must have sufficient local variations of thickness or density, or both, to provide adequate contrast in the image, and must be small enough to fit within the specimen-holder chamber of the microscope. Transparency to electrons is provided by plastic or carbon replicas of the fracture surface. Fractures are usually too rough to permit electrolytic thinning although alternative and more direct techniques for the examination of ion-milled, thin-foil transverse sections of fracture surfaces have been utilized (Ref 1, 2) as will be discussed below. Thin-foil preparation from bulk samples is discussed in the article "Analytical Transmission Electron Microscopy" in Volume 10 of the 9th Edition of *Metals Handbook*.

A fracture surface can be directly examined using scanning electron microscopy (SEM) (see the article "Scanning Electron Microscopy" in this Volume), but replicas offer the unique capability of transposing topographic information from the actual fracture surface to a high-resolution facsimile that can be conveniently handled and transported and can be readily examined in the transmission electron, scanning electron, or light microscope (see the article "Visual Examination and Light Microscopy" in this Volume). Use of a replica is of great importance when the fracture surface to be studied is located on a large structure, vessel, or machine that cannot be moved to the laboratory or that cannot undergo sectioning.

Replicas used for fractography are broadly classified as single-stage and two-stage replicas, with the two-stage plastic-carbon replica technique being the most commonly used for examination of rough fracture surfaces. A third technique, extraction replication, is used to examine and characterize small particles embedded in a matrix, such as small second-phase particles in a steel.

Direct Observation of Thin-Foil Specimens. As mentioned above, there have been reported cases where TEM thin-foil fracture specimens were examined directly. In one study, information on the process of crack initiation for cleavage fracture in a pearlitic eutectoid steel was obtained by examining thin-foil transverse sections (Ref 1). The technique involved electroplating the fracture surface with nickel to provide a deposit ~1.5 mm (0.06 in.) thick. Following plating, 0.2 mm (0.008 in.) thick slices were cut from the specimen so as to include the fracture surface. These slices were mechanically ground to 50 μm and then disks of 3 mm (0.12 in.) diam were cut out such that the fracture surface was located across the center of the disk. The disks of half nickel and half steel were thinned by argon ion bombardment in an ion miller. The thinning was performed until a perforation crossed the fracture surface. It was then possible to examine the fracture surface and the region directly adjacent to it by TEM, yielding information on both crack initiation and propagation.

Transverse sectioning, using diamond knife ultramicrotomy, has also been applied with considerable success to the study of stress-corrosion cracking of brass (Ref 2).

Cleaning of Fracture Surfaces for Replication

The important first step in all replication methods is cleaning of the fracture surface to remove contaminants. Initial cleaning of fracture surfaces is discussed in detail in the article "Preparation and Preservation of Fracture Specimens" in this Volume.

The final stage of cleaning usually consists of successively applying and mechanically stripping several plastic films prior to formation of the replica to be used for examination. The films applied for cleaning should be relatively thick (at least 0.125 mm, or 0.005 in.), and they should be allowed to dry thoroughly before removal. Otherwise, small regions of the film that adhere tightly to the fracture surface may tear out when the film is stripped off, leaving bits of film on the surface that are difficult to detect and remove and that effectively represent artifacts that can be picked up in the final replica.

Single-Stage Replicas. Three methods are available for producing single-stage replicas:

- A film of plastic can be applied or formed on the fracture surface
- A film of carbon can be formed directly on the fracture surface by vacuum vapor deposition
- A conversion oxide film can be formed on the fracture surface by chemical or chemical-plus-thermal treatment of the surface

Single-stage replicas can be examined directly or after they have been shadowed (shadowing will be discussed in the section "Shadowing of Replicas" in this article). Because the replicas are examined in transmission, it is difficult to visualize the actual topographical shapes; shadowing the replica usually eliminates this difficulty.

A replica can be directly examined in a transmission electron microscope if it is sufficiently thin (of the order of 150 nm in thickness). For examination by reflection in a scanning electron microscope (with either the secondary electron emission or the backscattered electron mode of operation) or in a light microscope, it is made thicker.

Thin Single-Stage Plastic Replicas. The formation of a single-stage replica is shown sche-

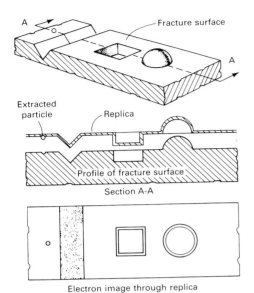

Fig. 1 Schematic views illustrating the single-stage replication technique

matically in Fig. 1. The contrast obtained with single-stage replicas in a transmission electron microscope derives solely from the greater effective thickness (the greater distance the electrons travel perpendicularly through the film) of the replica in regions corresponding to the sides of hills and valleys in the fracture surface, resulting in relatively more absorption or scattering of the electrons and darkening in the electron image. An inadequate degree of contrast is usually provided by the use of plastic only; therefore, a shadowing technique involving the vacuum vapor deposition of a heavy metal is implemented to enhance contrast.

Because thin single-stage plastic replicas are fragile and difficult to remove from a rough fracture surface, their use is generally limited to polished-and-etched specimens. On the few occasions when they are used on fracture surfaces, they usually cannot be stripped without chemical assistance, and they are likely to collapse or break open at local regions corresponding to sharp hills and valleys on the fracture surface. In fractography, thin single-stage plastic replicas are useful only for investigation of relatively smooth fracture surfaces, when the greatest resolution is needed to reveal fractographic details.

In the preparation of thin (100 to 150 nm) single-stage plastic replicas, the plastic, in a suitable solvent, is applied in the form of a dilute solution, ranging in concentration from one to a few percent, depending on the film thickness desired. The plastic-solvent combinations most widely used are:

- Cellulose acetate (acetyl cellulose) with acetone
- Cellulose nitrate with amyl, ethyl, or methyl acetate
- Polyvinyl formal with ethylene dichloride
- Polystyrene with benzene

Because moisture collects in droplets, causing the formation of holes and artifacts in the replica, it should be avoided in the plastic solution and in condensate form on the fracture surface.

The plastic-solvent solution is flowed onto the clean fracture surface. Thickness of the film is controlled by tilting the specimen to drain off some of the solution. The film should be allowed to dry thoroughly before any attempt is made to strip it from the fracture surface. Heating of the film is not recommended, because this may cause the formation of a solvent vapor bubble, followed by rupture of the film when the bubble bursts.

The replica can be removed from the fracture surface by either of two methods. The simpler method consists of mechanical stripping by lifting an edge of the replica away from the surface in the presence of water; the capillary action of the water may be sufficient to free a portion of it. The more effective method consists of dissolving a very small amount of the underlying metal in the fracture surface by chemical etching or electropolishing. However, care must be taken to avoid formation of gas bubbles, which can deform and fragment the replica. This technique alters or destroys the fine fractographic features in the area replicated, preventing a meaningful second replication.

After it is stripped, the replica is washed in distilled water. It is then floated facedown on the surface of distilled water, and replica grids are brought in contact with its top surface. The replica is cut around the grids, and then, with grids attached, it is placed faceup on a glass slide and allowed to dry. After this, the replica is shadowed.

Thick single-stage plastic replicas are used in fractography for the first stage in the two-stage replication technique (see the section "Two-Stage Replicas" in this article) and for scanning electron microscope and light microscope examinations, particularly for field investigations of fractures that occur in service. Thick replicas can be more readily stripped, even from relatively rough fracture surfaces; they are more durable and do not easily tear or collapse; and they provide sufficient resolution for viewing fine fractographic features.

Thick single-stage plastic replicas usually range in thickness from about 0.025 to 0.305 mm (0.001 to 0.01 in.), or about 250 times thicker than the thin single-stage plastic replicas. This thickness corresponds to the thickness range of the commercially available plastic tape and sheet used for replication. The thinner replicas (0.025 mm, or 0.001 in.) in this category are preferred because they dry more quickly, but use of thicker replicas may be dictated by the roughness of the fracture surface. In a seldom-used technique, an initially thin replicating film is built up and reinforced by additional plastic layers until the composite is strong enough to maintain its integrity during mechanical stripping. Because thick replicas

are intended for reflection viewing or for the first stage in two-stage replication, there is no upper limitation on thickness such as that imposed on replicas for direct examination in a transmission electron microscope.

To prepare a thick replica, a strip is cut from the plastic tape or sheet, and one side is softened by applying a few drops of solvent. The plastic-solvent combinations most commonly used are cellulose acetate with acetone, and cellulose nitrate with amyl acetate. Excess solvent is drained from the plastic strip a few seconds after the solvent has been applied. The softened face of the strip is placed on the fracture surface by contacting one edge and then slowly and progressively draping the remainder down on the surface, allowing for the escape of air and vapor. (As an additional means of removing air and vapor bubbles, the fracture surface can be wetted with the solvent immediately before the plastic strip is placed onto the surface.) The plastic strip is then pressed down firmly for 10 to 15 s with a finger or a resilient pad.

For a properly softened and drained plastic strip, the hardening time varies from 10 to 60 min, depending on the thickness. The degree of softening, the amount of solvent draining, and the time of pressure application and subsequent drying are critical. Optimum combinations are best determined by experiment and observation. If the plastic strip is insufficiently softened, the plastic will not conform accurately and completely to the contour of the fracture surface; this will result in the generation of artifacts. If the plastic strip is too soft, entrapped solvent vapor will cause bubble artifacts, and the strip will tear when pressure is applied. The surface of the plastic strip should be semifluid on the softened side and firm on the back.

Application of additional layers of plastic strip will facilitate removal of a replica from a rough surface. The application procedure is essentially the same as that for the initial replica layer, but the amount of softening of the contact surface of the second layer, which is necessary for good adhesion, is less. The stripped replica is shadowed and then coated with an electrically conductive film for viewing in the scanning electron microscope, or it is shadowed before proceeding to second-stage replication.

An alternative method of preparing thick plastic replicas consists of brushing or flowing a concentrated (40 to 60%) solution of cellulose acetate in acetone or of cellulose nitrate in amyl, ethyl, or methyl acetate onto the fracture surface. Because this method does not produce a highly-detailed replica, it is primarily used for replicating rough fracture surfaces that are to be examined for gross fracture features only. The replicating solution has the consistency of molasses. After it has been applied to the fracture surface, it is allowed to dry and harden to a tacky consistency, and a strip of the same type of plastic is pressed into it. Some fractographers prefer to apply a dry plastic strip immediately on the solution, using firm thumb or finger pressure for about 3 to 5 min while the solvent of the solution partly dissolves the strip and the replica partly dries. Long drying times (about 30 min or more) are required before stripping. Application of a stream of warm air may facilitate drying, but entrapment of vapor bubbles is a problem. The replica is mechanically stripped and then shadowed.

Another method consists of pressing a wafer of polystyrene onto the fracture surface while simultaneously heating the wafer and the specimen. This thermal molding method, in addition to having the obvious advantage of being a dry method, is capable of replicating rough surfaces and of providing good resolution. Its unfavorable feature is the multiple-step time-consuming procedure required. Commercially available pellets of polystyrene are melted down between glass platens, such as microscope slides, in an oven at 165 °C (330 °F) to form flat wafers upon cooling. A wafer is pressed onto the fracture surface by means of a platen and weight or a clamp, and the assembly is placed in an oven at 165 °C (330 °F). The assembly is heated to temperature and held for about 5 min. It is then cooled to room temperature, and the replica is mechanically stripped. The replica can be shadowed or can be shadowed and coated with a conductive film, depending on how it is to be viewed.

Single-Stage Direct Carbon Replicas. The direct carbon replication method is usually superior to the plastic replication method for obtaining thin replicas for examination in the transmission electron microscope. Furthermore, carbon replicas are stronger and more stable under electron beam bombardment in the transmission electron microscope. They are durable and have high resolution. They are free of many of the artifacts that can be present in plastic replicas. Carbon replicas are used to investigate the most delicate fracture features by TEM.

Single-stage carbon replicas are prepared directly from the fracture surface. Deposition of the carbon film on the surface is carried out in a laboratory vacuum evaporator operating at a pressure of 10^{-4} torr (1.3×10^{-2} Pa) or less. The carbon source consists of two spectrographic carbon rods that are heated by electrical contact resistance. The rods are held horizontal in spring-loaded holders to maintain contact. Their contact tips are specially shaped—one rod is sharpened to a point and the other is machined to a small (about 1-mm, or 0.04-in.) uniform diameter. The specimen is placed faceup on a small, rotating stage (driven by an externally controlled variable-speed motor), which is below and vertically in line with the carbon source (contacting tips of the rod electrodes). The spacing between source and specimen should be no less than about 100 mm (4 in.) to avoid heating of the specimen by radiation.

The area outside the region to be replicated is usually masked with adhesive tape or stop-off lacquer before carbon deposition; this will facilitate stripping of the replica. The fracture surface is often preshadowed with carbon or a heavy metal before deposition of the uniform carbon film. The preshadowing is also conducted in the vacuum evaporator.

The thickness of the carbon film can be controlled by completely evaporating a premeasured length of the 1-mm (0.04-in.) diam electrode; a length of 10 mm (0.4 in.) is usually optimum. Alternatively, a strip of dead-white paper or card stock can be placed near the specimen, and the darkening caused by carbon deposition can be visually calibrated and plotted against deposition time or total amount of carbon evaporated (or both) or can be plotted against the degree of transparency of a sequence of carbon deposits of increasing thickness. The specimen should be rotated slowly during carbon deposition to provide a uniformly thin and continuous film that will resist breakup during stripping.

Stripping of carbon replicas usually requires dissolution of some underlying metal from the fracture surface, either by chemical or electrochemical etching. The solutions and conditions used are the same as those used for preparation of metallographic specimens of the same metals, particularly for making extraction replicas for investigation in the transmission electron microscope (see the section "Extraction Replicas" in this article).

To facilitate the release of the carbon film, the replica is scribed into a pattern of small squares, about the same size as the piece of grid to be used, before etching or electropolishing. The time to effect release of a replica may vary from minutes to hours, depending on the etching rate of the metal and the topography of the fracture surface. After the replica is released, it floats on the surface of the etchant; it is lifted off and rinsed in a series of alcohol or distilled water baths. Dilute acids—for example, a 5 to 10% solution of hydrochloric acid in alcohol—can be used to dissolve reaction products and metallic debris, provided the acid does not attack the metal used for shadowing. Carbon films have a strong tendency to curl when stripped. Curling is minimized by shadowing with platinum; curled replicas can sometimes be salvaged by alternate immersion in distilled water and alcohol. After application of the grid, the grid and replica are dried by carefully touching an edge of a piece of filter paper to the junction of the tweezer points that hold the grid and replica and the grid itself. Drying is conducted at room temperature.

Conversion Oxide Film Replicas. The fracture surfaces of some metals and alloys, notably aluminum and aluminum alloys, nickel and nickel alloys, titanium, high-alloy refractory metals, and austenitic stainless steels, can be oxidized to provide a thin, microscopically structureless oxide film of controlled thickness. This film is strong, dimensionally stable, and rigid enough to resist deformation and flattening during stripping and subsequent handling. The main requirement is that the oxide film be essentially structureless. Microcrystalline, porous, or granular films do not provide adequate detail in replicating the fine features of the fracture surface. Local variations in fracture surface composition, including the presence of segregate phases, solid-state precipitates, and nonmetallic inclusions, cause local differences in replica thickness and topography. These effects may not be relevant to the processes involved in crack propagation and fracture and therefore may interfere with fractographic interpretation. On the other hand, they may yield additional useful information. In general, those metals and alloys that have useful corrosion passivity and that form uniform, thin, continuous oxide films can be replicated by this method. Except for the austenitic stainless steels, the method is not applicable to ferrous alloys.

Oxidation of the clean fracture surface is achieved either electrolytically in an aqueous solution at room temperature or by immersion in a molten salt bath at elevated temperature. Aluminum and its alloys can be suitably oxidized within 5 to 10 min in a 3% tartaric acid electrolyte adjusted to a pH of 5.5 with ammonium hydroxide, using the specimen as the anode and high-purity aluminum as the cathode, at a potential of 20 V. Austenitic stainless steels, nickel, and nickel alloys can be thermally oxidized by heat tinting in a molten salt bath composed of 50 wt% each of sodium nitrate and potassium nitrate at 425 °C (795 °F). The time of immersion is determined by intermittently withdrawing the specimen and visually inspecting it; when the oxide film has grown to a thickness that produces a light-yellow interference color, it constitutes a suitable replica, and the specimen is removed from the bath.

Conversion oxide film replicas are more difficult to strip from the fracture surface than single-stage carbon replicas. The oxide films on aluminum and its alloys can be electrolytically removed in a solution of 20% perchloric acid in alcohol, using the specimen as the anode and high-purity aluminum as the cathode; a potential of about 12 V is applied until the oxide film is released.

Some solutions of perchloric acid and organic materials can be ignited or exploded, but solutions of perchloric acid in alcohol are believed to be safe to mix and use if the following precautions are strictly observed. The alcohol-perchloric acid solution should be prepared in small quantities, with the acid being added to the alcohol, and should be stored in full, glass-stoppered bottles. Evaporated alcohol should be promptly replaced, keeping the bottles filled. While being used, the solution must not be allowed to become heavily contaminated with dissolved metal, and spent or exhausted solution should be promptly discarded. No departure should be allowed from the prescribed formula, the method of mixing, or the strength of the acid used. The solution should always be protected from heat or fire.

Undermining of the oxide film by dissolution of some of the underlying metal in the fracture surface is also required for austenitic stainless steels, nickel, and nickel alloys. The more aggressive action of a solution of bromine in methyl alcohol is needed. Use of the bromine solution is hazardous in several respects. In contact with the skin, bromine will cause severe third-degree burns, even with immediate first aid. Bromine vapor is dangerous to inhale. If the bromine solution becomes overheated during mixing or use, it may erupt violently. For these reasons, the entire operation should be carried out in a well-ventilated hood, the solution should be cooled in an ice bath during mixing, and bromine concentration should not exceed 10%.

Before dissolving the underlying metal, the oxide film should be cut into small squares, the size of the piece of grid to be used, to facilitate lifting of the film and thus to shorten the time of treatment. Solution concentrations of 2 to 3 mL of bromine per 100 mL of methyl alcohol are adequate for nickel and nickel alloys; higher concentrations (7 to 10 mL of bromine per 100 mL of methyl alcohol) are required for austenitic stainless steels.

After the film squares are released, they are collected on 200-mesh corrosion-resistant wire screen and are washed by several immersions in fresh methyl alcohol. The cleaned replicas are picked up on replica grids, with the side of the replicas that was next to the fracture surface facing up. The replicas are then dried. Because they provide adequate image contrast in the transmission electron microscope, oxide replicas usually are not shadowed.

Two-Stage Replicas. Two-stage replication can yield either thin second-stage replicas for transmission viewing in a transmission electron microscope or thick second-stage replicas for viewing by reflection in a scanning electron microscope or a light microscope.

Two methods are commonly used to prepare a final replica from a first-stage plastic replica. In the first method, a carbon replica is formed by vacuum vapor deposition of carbon on a first-stage plastic replica. In the second method, a plastic replica is formed by applying to a first-stage plastic replica a softened strip (or a solution) of a second, chemically different plastic; this plastic should not be soluble in the solvent or solvents used to prepare the first plastic replica, and its solvent should not affect the first plastic replica. Because of the superior characteristics of carbon replicas in transmission electron microscope examinations and because the equipment for producing them is usually available, carbon replicas have replaced plastic replicas for the second stage of replication; a plastic second stage is used only in very special applications. Figure 2 shows schematically the formation of a two-stage replica.

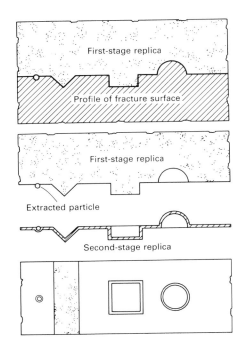

Fig. 2 Schematic views illustrating the two-stage replication technique

The procedure for preparing two-stage plastic-carbon replicas is a combination of thick single-stage plastic replication and single-stage carbon replication of that plastic replica, with an intervening shadowing step. The two-stage method is applicable to analysis of service fractures, where portability and integrity of the replica are of paramount importance. It is more susceptible to the generation of artifacts than the single-stage carbon replication method is.

After a thick first-stage plastic replica has been prepared in accordance with the procedure outlined earlier in this article, the replica is stretched out flat and faceup on a clean glass microscope slide. The replica is taped down at the edges to keep it flat. The slide bearing the replica is placed in a vacuum evaporator operating at a pressure of less than 10^{-4} torr (1.3×10^{-2} Pa) for shadowing and for formation of a carbon second-stage replica. The carbon for the second step is deposited at about 45° to the replica surface. During deposition, replicas of rough fracture surfaces can be continuously rotated on a small motor-driven platform to ensure uniform carbon deposition, especially on steeply tilted facets of hills and valleys and in sharp declivities at an angle to the surface, and to preclude confusing double shadows.

The thickness of the carbon film deposited is important. For high resolution, a thinner carbon film (yellowish interference color) is preferable, but when integrity of the film is a problem, such as in the replication of very rough surfaces, a thicker film (grayish interference color) is needed to provide resistance to film breakup during subsequent dissolution of the first-stage plastic replica. The thickness is controlled by the amount of carbon evaporated, which can be judged by the length of carbon rod consumed or by the degree of darkening of a piece of white paper located alongside the replica.

The separation of the carbon replica from the first-stage plastic replica is best done by dissolution of the plastic in successive baths of fresh acetone. The replica "sandwich" is first examined under a bench microscope to select regions of interest, which are removed as small squares by gentle slicing with a razor blade. The size should be about 0.5 mm (0.02 in.) larger than the grid, which is usually 2.3 or 3.0 mm (0.09 or 0.12 in.) square, to provide ultimate support. The overlap will fold over the edges of the grid by surface tension and hold the replica in place.

Up to ten squares, carbon-side up, are placed in a 50-mL petri dish and carefully covered with about 5 mm (0.2 in.) of acetone. Gentle heat from a lamp to stimulate circulation and evaporation will assist in freeing the carbon replica, which should float to the top in about 5 min. A gentle jet of acetone slowly expelled from a medicine dropper will aid in freeing replicas that are not too fragile. When free, each replica is sucked up by the partly filled dropper and transferred to the next dish of acetone. The carbon replica actually curls up during entry and uncurls during ejection, but if either entry or ejection is performed forcefully, the replica will be torn. The dropper is then cleaned with fresh acetone, and after 2 to 3 min, the squares are transferred to the last dish of acetone, where they remain for 5 to 15 min of final cleaning.

The following technique is used to position the replica on the grid. Grip an edge of a grid with tweezers and bend the edge slightly upward so that when the grid is immersed in the last dish of acetone, the grid plane is still horizontal. If the acetone is not too shallow, the replicas will be moving slowly about. Position the grid beneath a replica and raise the grid to the surface; the replica will be held to the grid by surface tension and mechanical friction. If positioned properly, the edges of the replica will overhang the edges of the grid (except at the side next to the tweezers; the carbon should not touch the tweezers, because it will adhere). If positioned incorrectly, lower the grid a bit to free the replica and relocate it; do this quickly, because motion of the acetone at the surface is rapid.

Raise the grid from the acetone and begin drying by carefully contacting the junction between the tweezers and the grid with absorbent paper. Maintain pressure on the tweezers; otherwise, the grid may be drawn up between the tweezer points by surface tension and the replica destroyed. Hold the grid until the replica is dry; then place it, replica-side up, in a clean, dry receptacle.

Extraction Replicas. This technique finds frequent application for identifying precipitates

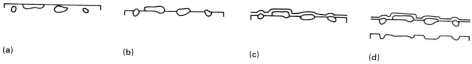

Fig. 3 Procedure used to produce single-stage extraction replicas for the analysis of small particles originally embedded in a matrix. (a) The original material with particles embedded in a matrix. (b) The particles have been etched to stand in relief. (c) The specimen is coated with a carbon film, usually by vacuum evaporation. (d) The specimen is immersed in an aggressive solution to release the particles from the matrix. Source: Ref 3

in metallic, ceramic, or glass systems. If the particles (precipitates) to be studied are large enough (>100 nm in diameter), it should be possible to examine them directly in a thin-foil specimen. If the particles are smaller, it is best to extract them from the matrix by using extraction replicas (Fig. 3).

The first step in producing an extraction replica is to etch the alloy heavily to leave the particles of interest in relief. A carbon film is then deposited on the surface. A second etch is then implemented to dissolve the matrix further, leaving only the particles attached to the carbon film. The film, which contains the particles, can then be examined in the analytical electron microscope. The replicating procedure may alter the extracted particles and may produce spurious oxide particles. To assist in distinguishing detail from artifacts induced by specimen preparation, it is advisable to prepare the extraction replicas with several etchants and to compare the results.

An alternative procedure that has proved successful for investigating the notch toughness of steels (Ref 4) involves cleaning the fracture surface in an ultrasonic cleaner by using an Alconox-water solution and an alcohol solution (Ref 5). Next, the specimen is heavily carbon coated at various angles. The surface, except the fracture area to be examined, is then painted with an acid-resistant lacquer. After the lacquer has been dried, the specimen is extracted in a 10% hydrochloric acid-90% methanol electrolyte, freeing the carbon film from the surface of the fracture for subsequent examination. Figure 4 shows the range of particle sizes and a typical distribution of particles from a fracture surface of an Fe-0.03C-18.5Ni-3Mo-1.4Ti-0.1Al maraging steel specimen that was prepared by this method.

Shadowing of Replicas

Most untreated replicas do not provide enough contrast to permit full realization of maximum detail and resolution of the features of fracture surfaces. This deficiency may result in failure to detect or recognize small but important features, such as fine fatigue striations, even though their dimensions are well within the resolution limits of the replication method and the viewing instrument. Also, in two-dimensional viewing of replicas, questions frequently arise concerning the degree of vertical relief, that is, whether certain fracture features are higher or lower than others and by how much. For these reasons, most replicas are shadowed (or shadow cast) by directional vapor deposition in vacuum of a heavy metal and an oblique angle to the replica surface. Contrast is enhanced by the buildup of more electron-opaque material (compared to the replicating material) at the leading edges or frontal slopes of hills and at the trailing edges or upward slopes of valleys in the replica (snowdrift effect). The length of the shadows provides an estimate of the relative height or depth of such features, and more accurate measurement is possible by suitable calibration of shadow length. Figure 5 shows the shadowing processes for single-stage plastic replicas and for two-stage plastic-carbon replicas, together with one method for shadow length calibration. Although shadowing can give useful indications of vertical dimensions, stereography is the most reliable method of obtaining depth and height perception and measurement.

When viewing a fractograph of a shadowed replica, it should be remembered that the dark portions in the print represent regions where the shadowing layer is thick, and shadows on the print represent regions where less shadowing metal was deposited. Because shadowing metal will not deposit in thick layers at the bottoms of depressions, it can be deduced that features that appear as light regions were depressions in the surface.

If the shadowing metal is deposited directly on the fracture surface, as in single-stage direct carbon replicas, a light region observed in the replica was a depression on the fracture surface. On the other hand, if the shadowing metal is deposited on a thick first-stage plastic replica, a light region inside a feature on a second-stage carbon replica shows that it was a hill on the actual fracture surface. For this reason, sound

Fig. 4 Transmission electron stereo pair at 3300× showing grain-boundary precipitates in an extraction replica taken from the fracture surface of a cobalt-free maraging steel that was solution annealed at 815 °C (1500 °F) for 1 h and air cooled, followed by aging at 480 °C (900 °F) for 3 h and air cooling. The specimen fractured several hours after aging. The particles shown ranged from 0.1 to 4 μm. The smallest were titanium-rich and also contained a small amount of molybdenum. These are probably TiC carbides. A few small (0.2 to 0.3 μm) iron-titanium-molybdenum-rich particles were also observed. The larger particles were either titanium- or zirconium-rich, and these may also be an MC-type precipitate. Such grain-boundary precipitates lower the notch toughness of steels. (G.F. Vander Voort, Carpenter Technology Corporation)

184 / Transmission Electron Microscopy

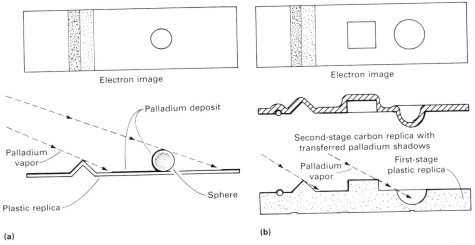

Fig. 5 Shadowing processes for single-stage plastic replicas (a) and two-stage plastic-carbon replicas (b). A sphere of known size can be placed on a replica, as in (a), to calibrate shadow length.

Fig. 6 Four types of artifacts found on two-stage plastic-carbon replicas. The appearance of the bottom half of the fractograph (below arrow A) is the result of tears in the first-stage plastic replica and is not representative of the fracture surface. Arrow B indicates a crack in the carbon second-stage replica; arrow C, a particle from the improperly cleaned fracture surface; and arrow D, residual plastic from incomplete dissolution of the first stage. 3000×

interpretation of a fractograph depends on knowing the replication procedure used.

The shadowing of replicas and of fracture surfaces is carried out in a laboratory vacuum evaporator fitted with holders for tilting and rotating the replica or specimen and with fixtures for supporting the evaporation sources, which are connected to electrical vacuum leadthroughs for resistance heating. Special spring-loaded devices are used for continuously feeding opposed carbon rods (electrodes) into contact. Spiral tungsten filaments, in the form of baskets, are used for resistance heating and evaporation of the metals used for shadowing (except for platinum, which alloys readily with tungsten). Evaporation is carried out at a pressure of less than 10^{-4} torr (1.3×10^{-2} Pa) to prevent oxidation of the vaporized metal and to avoid deposition of a granular shadow.

The vaporized metal atoms travel in essentially straight lines from the evaporation source, impinge at an oblique angle on the tilted replica, and condense. Generally, a 45° angle is used, but a more oblique angle (15 to 30°) produces longer shadows and more clearly delineates small elevations and depressions—for example, fine fatigue striations. To use shallow deposition angles satisfactorily, the replica should be generally flat; otherwise, some areas will be masked. If the nature and orientation of the surface features of interest are known or can be surmised before shadowing, the direction of shadowing that will produce the most definitive shadows can be selected. For example, to accentuate fatigue striations most effectively, shadowing metal should be deposited in a direction oriented 90° to the striations, which will make the shadowing direction parallel to the direction of crack propagation. If the crack propagation direction on the microscopic scale differs from that on the macroscopic scale, it may be necessary to make and shadow more than one replica.

Depending on their tilt and orientation, some facets on fracture surfaces will receive a thicker deposit than others. For example, on the side of a protuberance facing the evaporation source, the deposit will be thicker than on the side away from the source (the opposite is true for depressions). This difference in deposit thickness will result in a difference in the degree of electron absorption or scattering in a transmission electron microscope, which enhances the image contrast of the feature. The degree of scattering increases with the atomic number of the shadowing element; for this reason, heavy metals (those with high atomic numbers, namely chromium, platinum, palladium, germanium, and gold-palladium alloys) are commonly used. They minimize the thickness of shadow needed to provide adequate image contrast. Thick deposits are generally undesirable because they may obscure fine topographic details, and the shadows may be difficult to interpret.

The thickness of the shadow deposits is a function of the amount of shadowing material vaporized, which can be controlled by completely evaporating a premeasured amount. Alternatively, if an excess of shadowing material is placed in the evaporator, a piece of white paper (or card) having an upright tab can be placed near the replica, with the tab facing the evaporation source, so that the relative darkening of the paper in front of the tab, as compared to the shadow behind it, can be visually monitored and evaporation stopped at a suitable stage.

Artifacts in Replicas

The disadvantage most often encountered with any replication process is the presence of features on the replica that do not accurately reflect the details of the fracture surface, such as the artifacts shown in Fig. 6.

Poor technique in the formation of the first-stage plastic replica is a principal cause of artifacts in two-stage replicas. If the fracture surface is not cleaned properly, foreign material is likely to appear in the replica (arrow C, Fig. 6). If the plastic solution for the first-stage replica is not flowed into all recesses of the fracture surface, air or solvent vapor may become trapped and will appear in the replica as smooth, rounded areas. Many fractographers avoid this problem by applying solvent to the fracture surface immediately before forcing a piece of presoftened plastic strip onto it. During the 10- to 15-s interval before the plastic begins to set, the dry top of the piece of plastic is pressed, using tweezers, to remove visible bubbles from the area to be replicated.

The most frequently encountered artifact is the small-bubble artifact, that is, a bubble embedded in the plastic (see Fig. 24, 30, and 32 in the section "Comparison of TEM and SEM Fractographs" in this article). Small-bubble artifacts result from release of solvent vapors at the fracture surface. They occur at irregularities on the fracture surface and resemble bubbles that arise from scratches on the bottom of a beaker when water is being boiled. They are caused by the high vapor pressure of excess solvent as it attempts to diffuse out of the plastic. They can be avoided by using plastic strip that is thick enough to allow the dry back (or top) side to act as an absorbent sink for the solvent. Ideally, the plastic strip is softened on one side only and no more than necessary to make it conform to the fracture surface.

Another artifact is encountered in the deposition of carbon for single-stage direct

Fig. 7 Scraping artifacts (as at arrows) which resemble linear striations, on the sides of dimples. TEM plastic-carbon replica. 3000×

Fig. 8 Palladium-shadowed plastic-carbon replica of a fracture in nickel showing reticulated shadowing metal on dimples. Reticulation was caused by the melting of the shadowing metal in the microscope and the formation of globules. 22 500×

with the two-stage plastic-carbon technique, artifacts, in the form of cracks, can develop at the bottom of troughs or the tops of ridges when the replica is cut into grid-size pieces. The problem can be avoided by cutting the plastic replica into grid-size pieces before the carbon is deposited. In general, replicas should be cut by gently slicing them, using several light cuts, rather than pushing the cutting blade straight through.

A further problem encountered in making replicas by the two-stage plastic-carbon technique arises form the fact that the plastic expands as it soaks up solvent during the dissolution process. Expansion exerts tensile forces on the thin carbon film that are usually sufficient to stretch it and possibly to produce a crack (arrow B, Fig. 6). This can be avoided by using a relatively thin layer of plastic, thereby limiting the tensile forces, or by using the so-called wax technique. The wax technique involves forming a frame of paraffin wax around the replica, which prevents the plastic from swelling laterally in the solvent.

The most severe damage often occurs during drying of the final carbon replica, when surface tensions are very high. The folding and tearing that occur during drying are major limitations of the replication technique. They can be avoided by freeze-drying the replica, although this is a complicated procedure. The smoother the surface, the less damage that occurs in drying. Replicas of flat, fatigue fracture surfaces and of the intergranular surfaces of fine-grain metals are barely affected, but replicas of irregular cleavage fracture surfaces and surfaces including large ridges and valleys undergo severe distortion (see Fig. 10, 20, 22, and 24 in the following section for examples of folding and tearing artifacts).

Artifacts that arise after a replica is inserted in the microscope include contamination, reticulation, and melting of materials contained in the replica. Contamination resulting from the action of the electron beam is unavoidable and increases with exposure time. Reticulation occurs when the shadowing metal gets hot enough to melt and form globules, such as those shown in Fig. 8. Reticulation can be avoided by using a less finely focused electron beam or by using platinum as the shadowing metal. Use of a less finely focused beam also avoids melting of other materials contained in the replica.

Comparison of TEM and SEM Fractographs (Ref 6)

The following comparisons of TEM and SEM fractographs (Fig. 9 to 32) represent point-by-point comparisons of identical fields of view, giving a one-to-one correspondence. Additional TEM replicas are shown in the Atlas of Fractographs in this Volume. Additional information is available in the article "Interpretation of Transmission-Electron-Microscope Fractographs" in Volume 9 of the 8th Edition of *Metals Handbook*.

The two stereofractographs of each fracture surface presented in the following pages are mounted with the SEM view at the top and the TEM view at the bottom of each set. When viewing in three dimensions, the eyes may be shifted from the SEM fractographs to the TEMs and back at will without losing the three-dimensional appearance. It will be observed that the conformity of contour between the SEM and the TEM pairs is surprisingly close, considering the disparity in appearance between the two types of fractographs. Artifacts in the TEM fractographs will be indicated when present.

Method of Preparation. In each case, the preparation of these fractographs began with a careful scan of the fracture surface under a low-power binocular microscope to select an area of maximum interest. This was then marked by an identifying scratch in the center of the selected area.

A plastic replica of each marked area was prepared with replica tape, which was softened in acetone and then pressed in place to lessen the occurrence of air bubbles and to replicate all crevices. The tape was allowed to harden in place before being stripped from the fracture surface. The first replica of each area, which was made to clean the fracture surface, was discarded, and a second replica was prepared for use in fractographic examination.

(continued after Fig. 32)

carbon replicas. Frequently, features that are 2 to 5 nm in major dimension are to be resolved. If too much carbon is deposited, the fine features are distorted by enlargement or are totally hidden. If fine features are to be retained, the replica must be made as thin as possible. Although most fractographers prefer replicas that are strong enough to remain intact over an entire 75-mesh grid, others, in the interest of fine detail, prefer a replica that is so thin that it will not remain intact over more than half the grid openings of a 200-mesh grid.

Artifacts that arise during removal of the replica from the fracture surface are of three types. In single-stage direct carbon replicas, holes occur if the carbon adheres to the fracture surface. In plastic single-stage or first-stage replicas, some of the plastic is usually trapped in deep crevices and remains there when the remainder of the replica is removed. This results in a characteristic torn-plastic feature, such as that shown in the lower half of Fig. 6. Another common artifact develops when the plastic scrapes against the fracture surface as it is pulled off. This gives rise to linear marks such as those indicated by arrows in Fig. 7. These linear marks usually occur at several neighboring features. When viewed in stereo, they are seen to be on the sides of hills on the replica (all on one side or the other, depending on the direction of stripping and scraping). This type of replica artifact is particularly troublesome because it closely resembles valid linear striations.

With the single-stage direct carbon technique, artifacts can result from use of a washing or freeing chemical that attacks the extracted particles on the shadowing metal. In contrast,

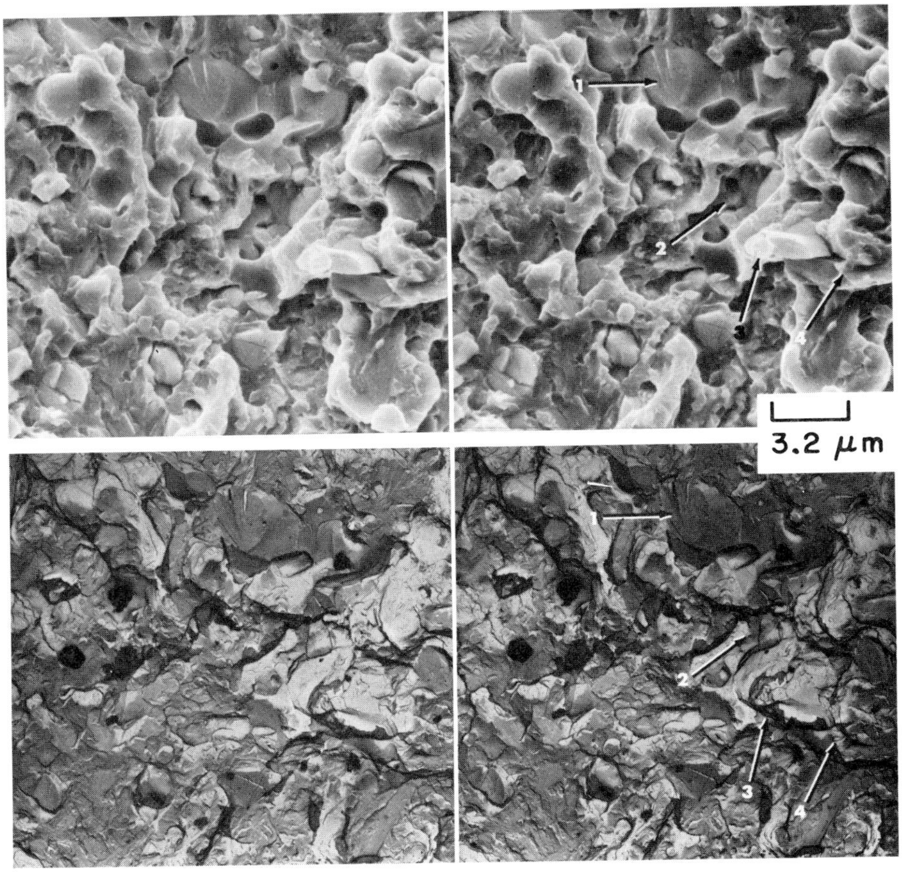

Fig. 9, 10 Surface of a room-temperature impact fracture in a specimen of AISI T2 high-speed tool steel having a nominal composition of 18W-4Cr-2V-1C. The specimen had been heat treated for 5 min at 1200 °C (2190 °F) and oil quenched, and tempered for 30 min at 500 °C (930 °F) and air cooled, yielding a hardness of 64 HRC. Surface shows quasi-cleavage facets and scattered dimples. In some of the features, the two fractographic techniques provide exact duplicates; in others, some details are missing from the TEM view. The region around the quasi-cleavage facet at arrow 1 is an example of quite good, detailed conformance between SEM and TEM, including the deep hole below (arrow 2). At arrow 3, stereo viewing of the SEM pair reveals a very high prong that is not found in the TEM; this bears evidence that the replica tore very locally. Compare, however, the fine details that appear in both pairs at arrow 4, as well as at other points of duplication. See also Fig. 11 and 12. Fig. 9 (top): SEM stereo pair. Fig. 10 (bottom): TEM stereo pair. Both 3100×

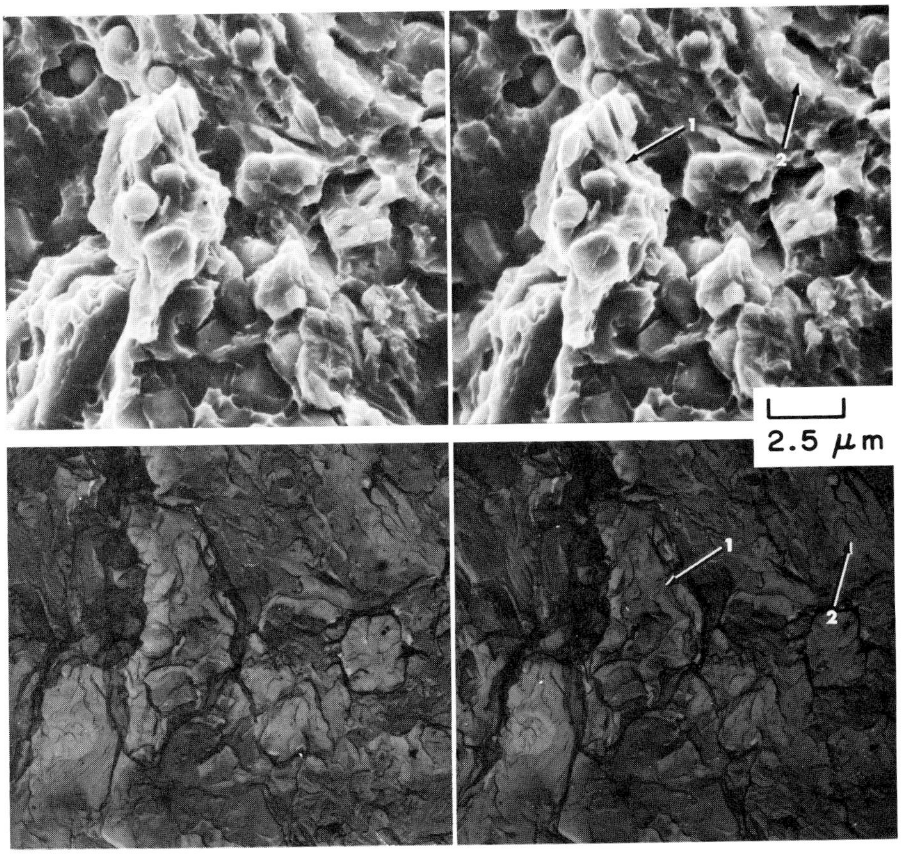

Fig. 11, 12 Views of a different area, and at higher magnification, of the fracture surface shown in Fig. 9 and 10. This is a remarkably complex area, containing many small and varied quasi-cleavage facets and exhibiting a number of spheroidal inclusions. The most prominent feature is the high peak at arrow 1, which is reproduced in very faithful and identical detail in both stereo pairs. Note that the two inclusions between the twin tips of the peak are shown in both pairs but that similar inclusions slightly nearer the top of the SEM pair are missing in the TEM because of damage to the replica. The top center area of the SEM view appears to record a greater number of surface complexities than the same area in the TEM view; at arrow 2, however, the TEM view displays very fine fracture marks that cannot be found in the same location in the SEM view. Fig. 11 (top): SEM stereo pair. Fig. 12 (bottom): TEM stereo pair. Both 4000×

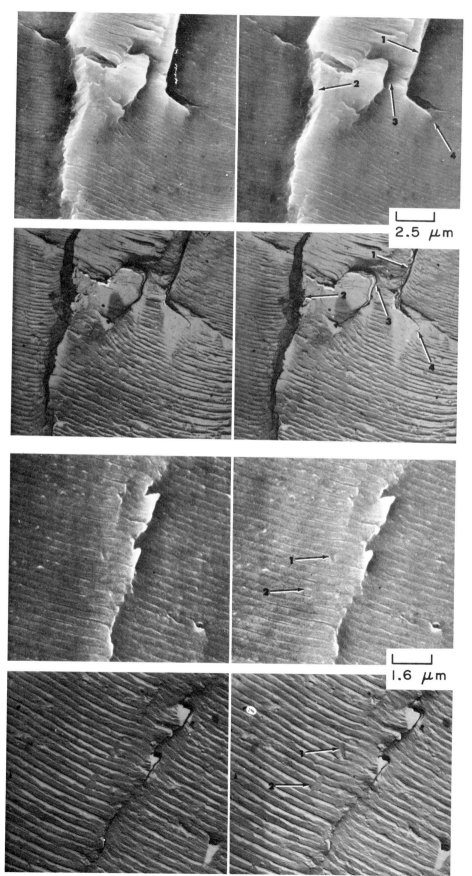

Fig. 13, 14 The first of two sets of views at increasing magnification showing four areas of the surface of a low-cycle fatigue test fractured in aluminum alloy 7075-T6 obtained with a maximum loading of 117 MPa (17 ksi) at 1900 cpm. Fracture occurred at 600 000 cycles. The tensile strength of the material was 586 MPa (85 ksi). In this area, the fracture surface shows well-defined fatigue striations that are clearly visible in both the SEM and the TEM pairs but with more detail and contrast in the TEM. Note the excellent match of the reproductions made by the two techniques. The vertical surface at arrow 1 is quite accurately registered in the TEM fractograph, but at the other side (arrow 2), the replica has been torn. The detail in the crevices marked by arrows 1 and 3 is more clearly displayed in the TEM pair, but could be matched in the SEM if the exposure were adjusted for this purpose. The sharp line at arrow 4 in the TEM fractograph bears some resemblance to a replica tear, but it is also visible in the SEM and is actually a secondary crack in the aluminum. See also Fig. 15 and 16. Fig. 13 (top): SEM stereo pair. Fig. 14 (bottom): TEM stereo pair. Both 4000×

Fig. 15, 16 A second set of views, again at higher magnification, showing a fourth area of the surface of the low-cycle fatigue test fracture in aluminum alloy 7075-T6 in Fig. 13 and 14. This region of the fracture surface shows portions of two fatigue patches separated by a sizable step. At this magnification, much of the detail of the striations is visible in both the SEM and the TEM fractographs. Note the slight indentation marks at arrows 1 and 2; these are of course much more clearly defined in the TEM pair. The gray mark below the cavity at lower right in the TEM was produced during shadowing of the replica, a negative surface in which the cavity in the specimen was reproduced as a projection. The height of the step at the juncture of the two fatigue patches is shown to be much greater in the SEM than in the TEM pair; this is the result of replica collapse. Fig. 15 (top): SEM stereo pair. Fig. 16 (bottom): TEM stereo pair. Both 6400×

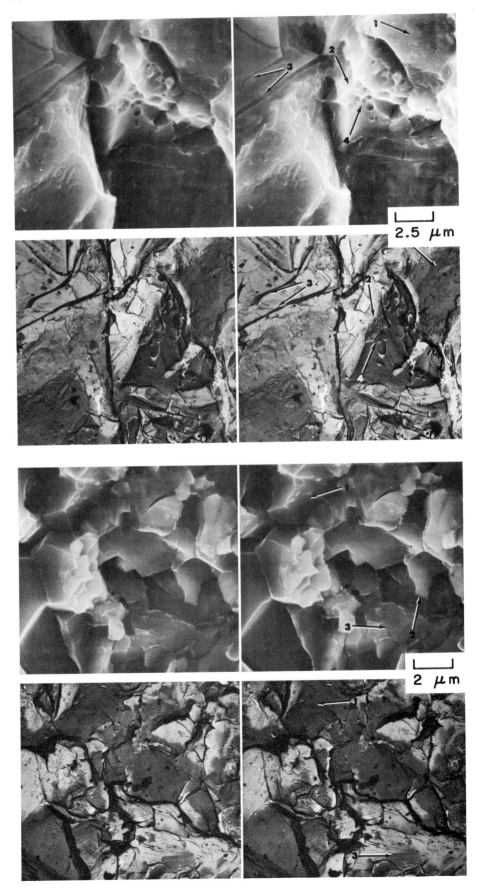

Fig. 17, 18 The first of two sets of views at increasing magnification of four areas of the surface of a fracture resulting from stress-corrosion cracking in aluminum alloy 7075-T6 exposed to water at room temperature and to stress (maximum not reported) by hydrostatic pressure. The tensile strength of the material was 586 MPa (85 ksi). In this area, the fracture surface is markedly intergranular, contains several secondary cracks (some of which appear to be transgranular), exhibits some corrosion pits but little corrosion debris, and contains local regions of moderate-size dimples. In general, the reproduction of surface contours in the TEM pair matches very closely that in the SEM. However, mismatch is evident at regions affected by replica collapse; for example, the region at arrow 1 (as viewed in three dimensions) appears to be higher than the region at arrow 2 in the SEM pair but lower than the region at arrow 2 in the TEM. Note the excellent agreement between the two techniques in registering the secondary cracks at arrow 3 and the details of the dimples at arrow 4. However, very few corrosion pits are resolved in the SEM pair. See also Fig. 19 and 20. Fig. 17 (top): SEM stereo pair. Fig. 18 (bottom): TEM stereo pair. Both 4000×

Fig. 19, 20 A second set of views at higher magnification showing another area of the stress-corrosion cracking fracture in aluminum alloy 7075-T6 in Fig. 17 and 18. Similar to the preceding set, this set of views shows a variety of grain surfaces resulting from intergranular rupture. The TEM fractograph shows many minute corrosion pits on these separated-grain surfaces, none of which are resolved in the SEM pair. The dark zones between grains in the TEM pair are caused by folding of the replica at secondary cracks. Observe the excellent agreement between the SEM and TEM pairs as to contour and detail, particularly in the region at arrow 1. Arrow 2 marks a feature in the SEM that does not appear in the TEM, which is unusual. Again, differences in elevation are minimized in the TEM; the region at arrow 3 appears to be at the same level as the ridge above arrow 1, but the SEM shows that the region at arrow 3 is much lower. Fig. 19 (top): SEM stereo pair. Fig. 20 (bottom): TEM stereo pair. Both 5000×

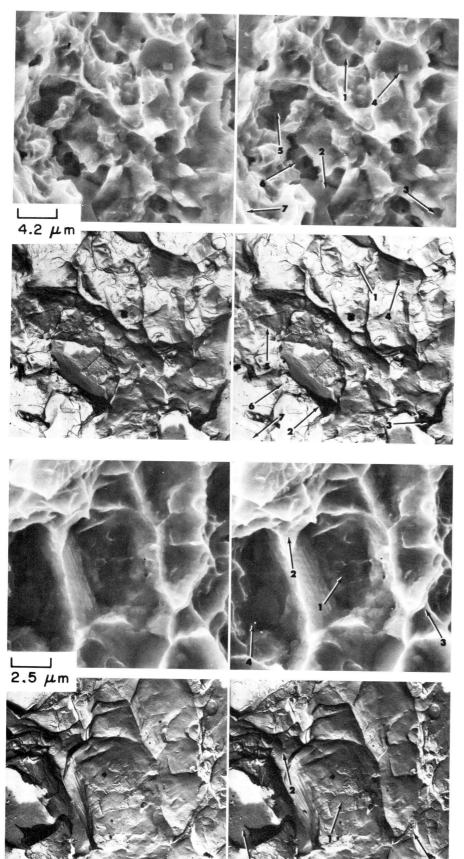

Fig. 21, 22 Surface of a tension overload fracture in a notched specimen of annealed Ti-8Al-1Mo-1V alloy that was broken at room temperature. The tensile strength of the material was 1000 MPa (145 ksi). The fracture surface consists of dimples of various sizes, a few of which show some sidewall stretching. The SEM and TEM pairs are in close agreement as to details of contour; compare, for example, the contours in and around the depression at arrow 1. In the TEM fractograph, tears and folds in the replica obscure the surface in the crevices at arrows 2 and 3. Note that the cube-shaped second-phase particle at arrow 4 in the SEM is missing entirely from the TEM. Many superficial surface features clearly visible in the TEM pair are not resolved in the SEM. Some replica collapse is evident when the elevations at arrows 5 to 7 in the TEM fractograph are compared with those in the SEM. See also Fig. 23 and 24. Fig. 21 (top): SEM stereo pair. Fig. 22 (bottom): TEM stereo pair. Both 2400×

Fig. 23, 24 A more highly magnified set of views of a different area of the surface of the tension overload fracture in annealed Ti-8Al-1Mo-1V alloy shown in Fig. 21 and 22. This area shows rather large dimples. For the most part, the topographic agreement between the SEM and TEM pairs is very close, as can be seen by comparing the contours at arrows 1 to 3. Note that the stretch marks on the left side of the cavity at arrow 2 are clearly visible in both pairs. Elsewhere, however, similar marks are largely unresolved in the SEM fractograph. The feature at arrow 4 in the TEM is not a fracture contour but an artifact caused by a bubble entrapped in the plastic replica. The dark regions above and below this artifact are the result of folds in the carbon replica. The SEM fractograph shows the true surface at arrow 4. Fig. 23 (top): SEM stereo pair. Fig. 24 (bottom): TEM stereo pair. Both 4000×

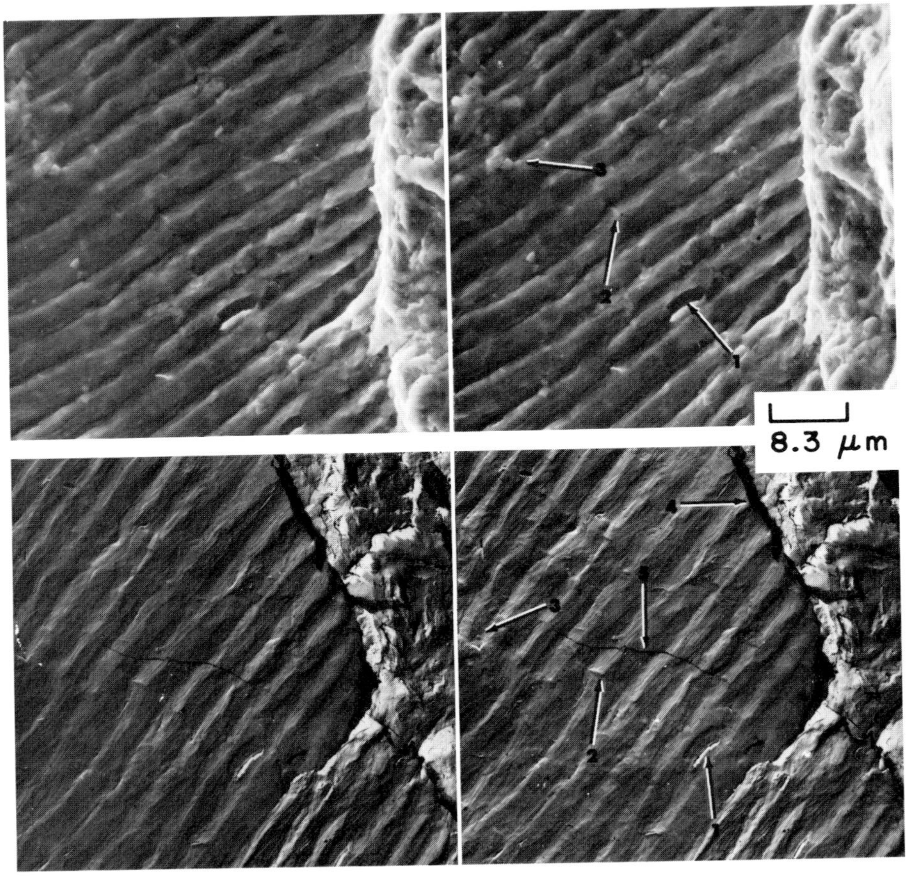

190 / Transmission Electron Microscopy

Fig. 25, 26 Surface of high-cycle fatigue fracture produced at room temperature in a specimen of the same annealed Ti-8Al-1Mo-1V alloy shown in Fig. 21 to 24. Shown here is a portion of a fatigue patch with widely spaced well-defined fatigue striations bounded on the right by a high step to the next patch. There is good agreement between the SEM and TEM pairs, as evidenced by the details in the two views at arrows 1 to 3, which are readily recognized in both fractographs. The TEM replica has been torn and folded at the step at right, resulting in the dark band at arrow 4 and the simulation of a crack at arrow 5. The apparent elevation of the step is much reduced in the TEM fractograph compared to the true elevation shown in the SEM pair. Note also that the contours of the striations appear much flatter in the TEM than in the SEM but that the fine marks on the ridges, which are in sharp focus in the TEM, are not well resolved in the SEM. See also Fig. 27 and 28. Fig. 25 (top): SEM stereo pair. Fig. 26 (bottom): TEM stereo pair. Both 1200×

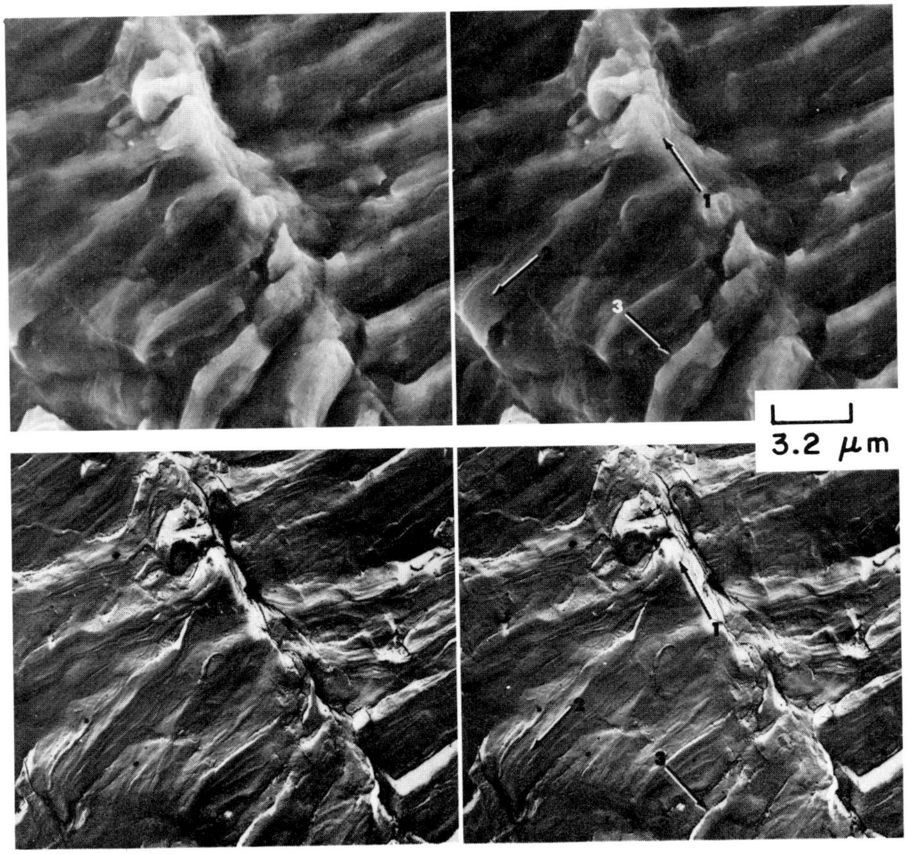

Fig. 27, 28 A more highly magnified set of views of another area of the surface of the high-cycle fatigue fracture in annealed Ti-8Al-1Mo-1V alloy in Fig. 25 and 26. These stereofractographs show the junction of two adjacent fatigue striation patches that exhibit considerable irregularity in striation orientation in the neighborhood of the step separating the patches. Comparison of the SEM and TEM pairs shows good agreement in registry of contours, as can be seen at arrows 1 to 3. The fine marks along the ridges of the striations are sufficiently pronounced for many to be shown in the SEM fractograph, but others are not resolved there and are shown much more distinctly in the TEM pair. Fig. 27 (top): SEM stereo pair. Fig. 28 (bottom): TEM stereo pair. Both 3100×

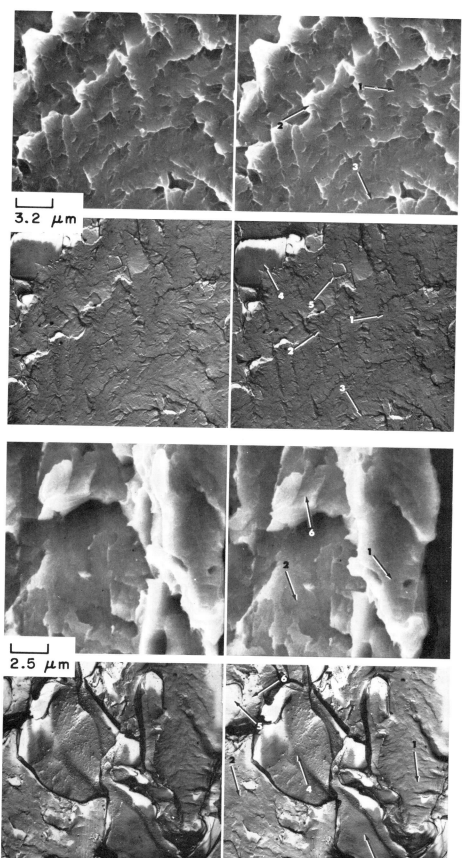

Fig. 29, 30 Stress-corrosion fracture in a specimen of the same annealed titanium alloy Ti-8Al-1Mo-1V shown in Fig. 21 to 28. This specimen was subjected to a maximum stress of 170 MPa (25 ksi) while exposed to an environment of 3.5% NaCl in water at room temperature. The fracture surface possesses many small and slightly offset facets, such as those at arrows 1 and 2, that appear to be quasi-cleavage facets. Many of the features along the edges of these facets, however, show dimple-like characteristics. Serrated ranks of the facets are separated by large steps, although the fracture surface area here is not large enough to show how repetitive these tiers can be. Agreement in contour details between the SEM view and the TEM view is very good, as can be seen by comparing the surfaces at arrows 1 to 3. The feature in the upper-left corner of the TEM stereo pair (arrow 4) is an artifact created by entrapment of a gas bubble in the plastic replica medium. Another feature not common to both the SEM and the TEM views is the apparent artifact at arrow 5 in the TEM. See also Fig. 31 and 32. Fig. 29 (top): SEM stereo pair. Fig. 30 (bottom): TEM stereo pair. Both 3100×

Fig. 31, 32 A higher-magnification set of views in another area of the stress-corrosion fracture of the annealed Ti-8Al-1Mo-1V specimen shown in Fig. 29 and 30. This area is somewhat similar to that shown at the lower magnification in the above fractographs, exhibiting rows of modest-size quasi-cleavage facets at different levels. Some of these facets, as at right, are separated by steps of appreciable height. There is excellent agreement in surface features between the SEM and the TEM pairs in the high-level region at arrow 1 (at right) and in the lower-level region at arrow 2. The surface details at arrows 3 to 5 in the TEM view, however, are artifacts resulting either from bubbles or foreign material trapped by the replica. The small facet at arrow 6 is the portion of this region that is common to both the SEM and the TEM fractographs. The surfaces that are real show essentially no accumulation of corrosion debris. Fig. 31 (top): SEM stereo pair. Fig. 32 (bottom): TEM stereo pair. Both 4000×

Each replica was shadowed with platinum at a 45° angle with a layer about 2 to 3 nm thick; this was followed by evaporation of carbon onto the plastic at a 90° angle to a thickness of about 10 nm. Finally, the replica was trimmed to a triangle centered on the identifying scratch mark, with the points of the triangle serving to orient the piece for tilting to produce the stereofractographs. The triangular replica was mounted on a 150-mesh copper grid and immersed in acetone to dissolve the plastic.

Each stereo pair of TEM fractographs was taken at a 6° tilt from the normal to the replica plane and at a selected location acceptably distant from the identifying scratch. The outcome, however, because of the procedure used in preparing the replica, was a negative surface in which each hill was a valley and each valley a hill. This meant that during the shadowing process each valley cast a shadow outside itself. This negative surface made an exact comparison between the TEM and the SEM fractographs difficult, because the SEMs show a positive surface. For the purpose of this article, therefore, it was decided to ignore the nature of the shadowing and to abandon the original arrangement of the stereo pairs by transposing the prints in each TEM pair so that the view originally on the left was moved to become the view on the right. This resulted in a positive image in three-dimensions, with the only unusual aspect being the casting of an external shadow by a cavity. This disadvantage is outweighed by providing conformity of contours in the TEM view with those in the SEM, thus making possible a close visual comparison of the two types of fractographs.

After preparation of the replica, the sample was sectioned to produce a specimen suitable for mounting in the scanning electron microscope, with the specimen containing the identifying scratch mark in the center of the fracture area preserved. The specimen was scanned in the microscope to locate the scratch. Then, using the TEM fractographs previously prepared, the exact location of the replica was found and the magnification and orientation were adjusted to give an exact counterpart of the TEMs. The stereo pair of the SEM fractographs was taken, tilting the fracture surface to an angle of 6° on both sides of the position normal to the electron beam.

For stereo viewing, two photographs are taken. The first is taken with the replica tilted 5 or 6° from the horizontal in one direction, such as the left, and the second is taken with the replica tilted the same angle to the right, producing an included angle of 10 or 12° between the two positions. When the fractographs are printed and placed under a stereo viewer, which are commercially available, the features are readily apparent in three dimensions. By interchanging the two prints under the viewer, relief is reversed—hills become depressions, and conversely. Additional information on stereofractography is available in the article "Scanning Electron Microscopy" in this Volume.

REFERENCES

1. Y.J. Park and I.M. Bernstein, The Process of Crack Initiation and Effective Grain Size for Cleavage Fracture in Pearlitic Eutectoid Steel, *Met. Trans.*, Vol 10A, Nov 1979, p 1653-1664
2. A.J. Morris, Olin Corporation, private communication, 18 Nov 1986
3. K.C. Thompson-Russell and J.W. Edington, *Electron Microscope Specimen Preparation Techniques in Materials Science*, Macmillan, 1977
4. A.J. Birkle, D.S. Dabkowski, J.P. Paulina, and L.F. Porter, A Metallographic Investigation of the Factors Affecting the Notch Toughness of Maraging Steels, *Trans. ASM*, Vol 58, 1965, p 285-301
5. G.F. Vander Voort, Carpenter Technology Corporation, private communication, 19 Sept 1986
6. J.L. Hubbard, " A Comparison Atlas of Electron and Scanning Electron Fractography," Masters thesis, Georgia Institute of Technology, June 1971

Quantitative Fractography

Ervin E. Underwood and Kingshuk Banerji, Fracture and
Fatigue Research Laboratory, Georgia Institute of Technology

THE PRINCIPAL OBJECTIVE of quantitative fractography is to express the characteristics of the features in the fracture surface in quantitative terms, such as the true area, length, size, spacing, orientation, and location, as well as distributions of these, as required. As will be discussed in this article, the more prominent techniques for studying the fracture surface are based on the projected images (the picture obtained with the scanning electron microscope), stereoscopic viewing (using stereophotogrammetry), and sectioning (to generate profiles). The use of an automatic image analysis system with a digitizing tablet is extremely helpful, although manual methods offer efficient alternatives in some cases. Instrumentation, principles, and applications associated with image analysis systems are discussed in the article "Image Analysis" in Volume 10 of the 9th Edition of *Metals Handbook*.

Treatment of the basic data includes the choice of triangulation methods, stereophotogrammetry, or the angular distribution of elements along the fracture profile. Although these procedures allow the fracture surface area to be estimated, they are essentially only approximations of the complex and irregular fracture surfaces found in metals.

Another approach to this problem invokes the statistically exact assumption-free relationships of stereology, which yield a linear equation between the surface, R_S, and profile, R_L, roughness parameters. The parameter R_S is defined as the true surface area divided by its projected area, and R_L is the true profile length divided by its projected length. Because R_L is experimentally available, calculation of R_S gives the quantity sought, the fracture surface area, with an accuracy determined by the accuracy with which R_L was determined. The mathematical relationships developed in quantitative stereology are described in the article "Quantitative Metallography" in Volume 9 of the 9th Edition of *Metals Handbook*.

Knowing the roughness parameters enables simple relationships to be set up for features in the fracture surface (see the section "Analytical Procedures" in this article for a discussion of roughness parameters). Errors of over 100% are found when values of dimple size or facet size are calculated directly from the scanning electron microscopy (SEM) fractographs without corrections (see Example 5). Other examples of the application of these new quantitative methods are given in this article for striation spacings (Example 1), precision matching (Example 2), crack path tortuosity (Example 4), and the use of statistical methods as opposed to individual measurements on features in the fracture surface (Example 7).

Development of Quantitative Fractography

There has been an increased effort recently toward developing more quantitative geometrical methods for characterizing the nonplanar surfaces encountered in fractures (Ref 1-8). A concerted effort has produced gratifying advances in the capabilities to quantify the fracture surface and its features (Ref 9-24). Although experimental methods are still being explored and efficient analytical procedures are being developed, the results achieved to this point have significantly advanced the progress toward a general treatment of the problem.

The earliest attempts at a quantitative approach to fractography were concerned with individual features. Simple measurements of height, depth, or separation could be performed by shadowing, stereoscopy, or interferometry (Ref 25-27). Clearly, these methods were applicable only to the simplest geometric configurations, such as a hillock raised above a flat background. Carl Zapffe's early optical examinations of fracture surfaces contributed significantly to the qualitative knowledge of fracture geometry (Ref 28). However, low magnifications and an unsatisfactory depth of field limited the usefulness of this technique. With the advent of fracture replicas and the transmission electron microscope, great strides were achieved in the resolution of fine detail and in higher magnifications (Ref 29-31). Quantification techniques were still limited to the more primitive measurements and were applied mostly to the projected image (or photograph) of the replica (Ref 32). The general availability of the scanning electron microscope opened new avenues toward the understanding of fracture structures in three dimensions. Researchers increasingly used quantitative measurements on SEM photomicrographs. Although direct measurements on these projected images do not yield correct spatial information, a large step forward was being taken toward quantification. A detailed review of Zapffe's contributions to modern fractography and the development of the transmission and scanning electron microscopes and their application to fracture studies is presented in the article "History of Fractography" in this Volume.

The basic problem in quantifying a fracture surface is a very general one. Briefly put, the area of a fracture surface must be known in order to apply numbers to the components that constitute the nonplanar surface. Once the area has been determined, preferably by a fairly simple and direct procedure, the other quantities of interest in the fracture surface can be determined.

As stated above, the current, more prominent techniques for studying the fracture surface are based on projected images (the SEM picture), stereoscopic methods (including photogrammetry), and profile generation (obtained from the fracture surface). Each method has its advantages and disadvantages. The metallurgist needs a procedure that will accommodate the complex and irregular fracture surfaces found in metals and alloys. This overriding requirement immediately narrows down the selection of suitable experimental techniques.

It would also be beneficial to use the basic equations of stereology to the utmost (Ref 33). These statistically exact assumption-free relationships apply equally well to the flat features in the plane of polished specimens as they do to the spatial objects in three-dimensional sample space. If a planar cut is made through the fracture surface, a profile is obtained. Its length, angular distribution of its segments, orientation characteristics, and so on, can be measured as if it were a line in a plane.

On the other hand, it may also be necessary to determine the area of the fracture surface. Unfortunately, the facets of a fracture surface are usually preferentially, rather than randomly, oriented. This presents a difficult sampling problem if the stereological relationships based on random sampling and measurements are to be used. For example, to determine the spatial angular distribution of facets, a prohibitive number of sampling planes at many angles and locations are theoretically required (Ref 34). However, alternatives to random testing are available, such as those that express the surface area in terms of roughness parameters, the degree of orientation, and so on. Fortunately, an investigation of sampling requirements (Ref 17) on a computer-simulated fracture surface (CSFS) (Ref 35) indicates that relatively simple serial sectioning may be adequate for determining the fracture surface area.

Because of the interest in sectioning procedures, methods for characterizing the resulting profiles have proliferated. Several profile parameters have emerged that show considerable promise (Ref 18, 19, 36, 37), including the roughness parameters R_L and R_P and the fractal dimension $\mathcal{D}$ (Ref 38) (these are discussed in the section "Roughness Parameters" in this article and in the article "Fractal Analysis of Fracture Surfaces" in this Volume). A number of equations have been derived that relate the profile roughness parameter R_L to the surface roughness parameter R_S and therefore to the fracture surface area (Ref 9, 10, 13, 17, 23).

Another approach is being pursued to exploit the profile characteristics. Mathematical relationships are available that transform the angular distribution of linear elements in a profile to the angular distribution of facets in three-dimensional sample space (Ref 34, 39, 40). This type of method will be discussed in terms of the analysis described in Ref 39 and in the section "Profile Angular Distributions" in this article.

At the same time, some proposed relationships involving roughness parameters and fractographic measurements or models are obviously incorrect (Ref 41, 42). Moreover, other equations that invoke randomly oriented surface elements should be used with care unless the assumption of randomness can be justified (Ref 10, 13). Such relationships represent only a trivial solution to an unrealistic condition. It should also be noted that previous measurements from SEM fractographs that were not corrected for roughness in the z-direction may have errors of more than ±50%, depending on the quantity measured (Ref 43, 44).

Considerable work remains to be done before facet and dimple areas, striation spacings, interparticle distances, crack path length, and other attributes of features in the fracture surface can be efficiently estimated (Ref 23). However, the use of roughness parameters appears to offer a viable solution to this quantification problem.

Experimental Techniques

As indicated above, the experimental procedures currently being used the most can be grouped under three categories: projected images, stereoscopic methods, and profile generation. These techniques will be briefly discussed in terms of their advantages and disadvantages. Also presented in this section are tabulations of quantitative relationships applicable to an assumed flat fracture surface (as represented by the SEM fractograph) and to an assumed randomly oriented fracture surface.

Projected Images

Although stereo pairs are readily obtained with SEM and transmission electron microscopy (TEM), single SEM images are also extensively used in fractographic studies. The vertical photographs permit some relative assessments to be made, and numerical (counting) results are quite common (Ref 21, 45). Single SEM photographs are also widely used for detailed qualitative interpretations concerning the fracture mechanisms and features involved in the fracture process (Ref 29, 30). Other studies have attempted to extract more quantitative information from these planar projected images (Ref 46, 47). The analyses were largely two dimensional because general fractographic relationships bearing on the third dimension were not available.

These early attempts are commendable in the absence of better procedures. However, with the development of the relationships of projection stereology, additional tools are now available (Ref 33, 48). Even so, this is not enough. Measurements restricted to just the SEM photomicrograph are limited in the amount of information they can provide about the nonplanar surface. An additional degree of freedom, such as that provided by stereometry or profile analysis, is necessary to generalize the three-dimensional spatial analysis.

Single SEM Fractograph. Limited information can be obtained from the flat TEM or SEM picture. Basically, in a two-dimensional surface, there can only be areal, lineal, or numerical quantities. These include projected areas, lengths, sizes, spacings, numbers, and more sophisticated parameters dealing with distributions, locational analyses, shape, and orientational tendencies. No information is possible about the features in three-dimensional sample space, except by making *ad hoc* assumptions.

Figure 1 shows schematically the measured quantities and the basic equations applicable to individual convex figures in a projection plane (Ref 33, 48). Table 1 lists the working equations for features in a projection plane, both for individual features (Part I) and for systems of features (Part II). All the quantities in Table 1 can be obtained by simple counting measurements, which are defined at the bottom of the table. If image analysis equipment with operator-interactive capability is available, most of the quantities listed in Table 1 can be measured directly.

When many individual measurements are required on a system of features, such as for a size distribution, more labor is involved. If appropriate or available, semiautomatic image analysis equipment can be used in such cases. For example, the areal size distribution of the

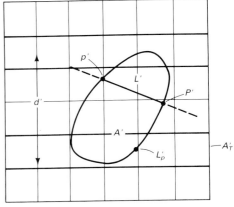

Measured quantities(a)	Basic equations(a)
A' Area of figure	$A' = L_p' \dfrac{\overline{L'}}{\pi}$
L' Intercept length	$L' = \dfrac{A'}{\overline{d'}}$
L_p' Perimeter length	$L_p' = \pi \overline{d'}$
d' Tangent diameter	$\overline{d'} = \dfrac{2}{\overline{k_L'}}$ (curvature)
A_T' Test area	
P' Intersection points	

(a) Primes signify projected quantities.

Fig. 1   Basic quantities for a convex figure in a projection plane

facets shown in Fig. 2 was determined by tracing the perimeter of each facet in the SEM photomicrograph with an electronic pencil (Ref 18). With a typical semiautomatic image analysis system, a printout can be produced with the data in the form of a histogram, along with other statistics.

It should be emphasized that the equations listed in Table 1 describe only the images of features in the projection plane and do not give spatial information. Nevertheless, due to the lack of anything better, the two-dimensional results are valid for that particular projected surface and may even be compared with other fractographs from fracture surfaces of comparable roughness.

Assumption of Randomness. It would be better if the true magnitudes of areas, lengths, sizes, distances, and so on, were obtainable instead of the projected quantities. If this is not possible, however, order of magnitude calculations can be made by assuming that the fracture surface is composed of randomly oriented elements. With this assumption, the stereological equations can be used with only one (or a limited number) of projection planes (Ref 33, 48).

This is so because the stereological equations are valid for structures with any degree of randomness, provided randomness of sampling is achieved. If the structure (in this case, the

Table 1 Stereological relationships for features in a projection plane
Primes indicate projected quantities or measurements made on projection planes.

Variable	Explanation	Equation
Part I: Individual projected features(a)		
L'	Length of a linear feature	$L' = \left(\dfrac{\pi}{2}\right)\overline{P'_L}A'_T$
L'_P	Perimeter length of a closed figure	$L'_P = \left(\dfrac{\pi}{2}\right)\overline{P'_L}A'_T$
$\overline{L'_2}$	Mean intercept length of a closed figure	$\overline{L'_2} = \dfrac{2\overline{P'_P}}{\overline{P'_L}}$
A'	Area of a closed figure	$A' = \overline{P'_P}A'_T$
Part II: Systems of projected features(b)		
$\overline{L'}$	Mean length of discrete linear features	$\overline{L'} = \dfrac{L'_A}{N'_A}$
$\overline{L'_P}$	Mean perimeter length of closed figures	$\overline{L'_P} = \dfrac{L'_A}{N'_A}$
$\overline{L'_2}$	Mean intercept length of closed figures	$\overline{L'_2} = \dfrac{L'_L}{N'_L}$
$\overline{A'}$	Mean area of closed figures	$\overline{A'} = \dfrac{A'_A}{N'_A}$
A'_A	Area fraction of closed figures	$A'_A = L'_L = P'_P$
L'_A	Line length per unit area	$L'_A = \left(\dfrac{\pi}{2}\right)\overline{P'_L}$
λ'	Mean free distance	$\lambda' = \dfrac{(1 - A'_A)}{N'_L}$
$\overline{d'}$	Mean tangent diameter of convex figures	$\overline{d'} = \dfrac{N'_L}{N'_A}$
$\overline{k'_L}$	Mean curvature of convex figures	$\overline{k'_L} = \dfrac{2N'_A}{N'_L}$

(a) A'_T: the selected test area in the projection plane; $\overline{P'_L}$: the number of intersections of a linear feature per unit length of test line, averaged over several directions of the test grid; $\overline{P'_P}$: the number of point hits in areal features per grid test point, averaged over several angular placements of test grid. (b) N'_A: the number of features per unit area of the test plane; N'_L: the number of interceptions of a feature per unit length of test line; L'_A: the length of linear features per unit area of the test plane

Fig. 2 SEM projection of facets in fractured Al-4Cu alloy

lationship of particles and dimples among these three planes has been thoroughly investigated (Ref 49).

If it is permissible to assume an absence of correlation, then measurements of P'_P, L'_L, or A'_A in the flat SEM fractograph should yield the same values as from the plane of polish (Ref 50). That is, for a two-phase structure:

$$A'_A = A_A,$$
$$L'_L = L_L, \text{ and}$$
$$P'_P = P_P \quad \text{(Eq 1)}$$

where A_A, L_L, and P_P refer to quantities measured in the plane of polish. The three projected quantities in Eq 1 are dimensionless ratios and are therefore independent of magnification and distortion in the SEM image.

If it can also be assumed that the surface elements of the fracture surface are randomly oriented, then the relationships given in Table 2 can also be used. For example:

$$N_L = \left(\dfrac{2}{\pi}\right)N'_L \text{ and } N_S = \dfrac{N'_A}{2} \quad \text{(Eq 2)}$$

Under these circumstances, ratios from the fracture surface—for example, N_L or N_S—have the same values, respectively, as their counterparts (N_L or N_A) in the plane of polish.

It is evident from the above discussion that considerable quantification is possible in several ways from SEM photomicrographs. First, calculations can be made in the plane of projection alone, without any attempt to convert to three dimensions (Table 1). Second, spatial quantities in the fracture surface can be calculated if random orientation of surface elements can be reasonably assumed (Table 2). Correlation effects will determine whether additional relationships are possible with the plane of polish (Ref 51).

The assumption of angular randomness in the fracture surface is admittedly somewhat tenuous, especially considering the strongly oriented nature of a fracture surface. This problem will be addressed in the section "Analytical Procedures" in this article, in which the subject

fracture surface) happens to possess complete angular randomness, the randomness of sampling is not required. Instead, directed measurements are adequate because for a random structure the measured value should be the same in any direction. Accordingly, with the assumption of randomness, the standard equations of stereology, based only on vertical (directed) measurements from the SEM picture, can be used.

Basic equations that are valid under these conditions are listed in Table 2. The quantities to the left-hand side of Table 2 pertain to the fracture surface. Thus, instead of L_A, for example, L_S is used, which represents the length of the line per unit area of the (curved) fracture surface. On the right-hand side of Table 2 the working equations are expressed in terms of projected quantities.

Figure 3 shows the differences between L_ϕ and L_c in the fracture surface. The value L_ϕ corresponds to a straight line L'_ϕ with fixed direction in the projection plane, while L_c relates to a curved line L'_c with variable direction in the projection plane. Statistically, the angular dependence of L_ϕ is averaged in a vertical plane defined by L'_ϕ, while L_c must be averaged in three dimensions, amounting to a difference of up to 23%. This elementary distinction has not been recognized in work purported as three-dimensional fractography (Ref 42). Other quantities in Table 2 can be used as they are or can be combined into more complicated functions as required.

Correlation of Fracture Path and Microstructure. It is possible to relate bulk microstructural properties to the fracture path only if there is no correlation between the configuration of the surface and the underlying structure. However, in general, a correlation does exist. Frequently, the fracture path passes preferentially through particles, or some feature of the microstructure, rather than along an independent path through the material. This correlation results in a statistically higher concentration of particles in the fracture surface, N_S, than in the metallographic plane of polish, N_A. Figure 4 shows the relationship of the fracture surface to a horizontal plane of polish and to the SEM projection plane. The interre-

Table 2 Stereological relationships between spatial features and their projected images

Primes indicate projected quantities or measurements made on projection planes.

Variable	Explanation	Equation
Part I: Individual features in the fracture surface		
L_ϕ	Length of linear feature (fixed direction ϕ in the projection plane)	$L_\phi = \left(\frac{\pi}{2}\right)\overline{L'_\phi}$
L_c	Length of curved linear feature (variable directions in the projection plane)	$L_c = \left(\frac{4}{\pi}\right)\overline{L'_c}$
$\overline{L_3}$	Mean intercept length of a closed figure (averaged over all directions in the projection plane)	$\overline{L_3} = \left(\frac{\pi}{2}\right)\overline{L'}$
L_p	Perimeter length of a closed figure	$L_p = \left(\frac{4}{\pi}\right)\overline{L'_p}$
S	Area of a curved two-dimensional feature (no overlap)	$S = 2\overline{A'}$
Part II: Systems of features in fracture surface(a)		
$\overline{L_\phi}$	Mean length of linear features (fixed direction ϕ in projection plane)	$\overline{L_\phi} = \left(\frac{\pi}{2}\right)\frac{L'_A}{N'_A}$
$\overline{L_c}$	Mean length of curved lines (variable directions in the projection plane)	$\overline{L_c} = \left(\frac{4}{\pi}\right)\frac{L'_A}{N'_A}$
$\overline{L_3}$	Mean intercept length of closed features (averaged over all directions in the projection plane)	$\overline{L_3} = \left(\frac{\pi}{2}\right)\frac{L'_L}{N'_L}$
$\overline{L_p}$	Mean perimeter length of closed figures	$\overline{L_p} = \left(\frac{4}{\pi}\right)\frac{L'_A}{N'_A}$
$\overline{S}$	Mean area of curved two-dimensional features (no overlap)	$\overline{S} = \frac{2A'_A}{N'_A}$
S_S	Area fraction of curved two-dimensional features (no overlap)	$S_S = A'_A = L'_L = P'_P$
L_S	Length of linear features per unit area of fracture surface (variable directions in projection plane)	$L_S = \left(\frac{2}{\pi}\right)L'_A$

(a) $L'_A = (\pi/2)\overline{P'_L(\phi)}$; $\overline{P'_L(\phi)}$ is the number of intersections of linear features per unit length of test line (variable directions in the projection plane).

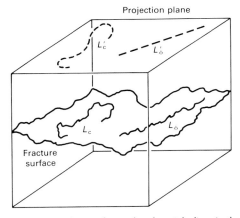

Fig. 3 Relation of curved and straight lines in the projection plane to lines in the fracture surface. $L_c = (4/\pi)\,\overline{L'_c}$ and $L_\phi = (\pi/2)\,\overline{L'_\phi}$.

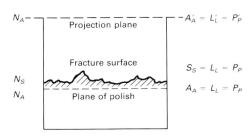

Fig. 4 Correlation among plane of polish, fracture surface, and projection plane

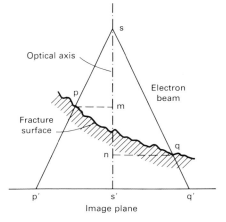

Fig. 5 Geometry of image formation in the scanning electron microscope

of partially oriented surfaces is treated in a more quantitative manner.

Stereoscopic Methods

In this section, conventional stereoscopic imaging and photogrammetric methods will be considered, as well as a geometric method that requires no instrumentation. Basically, these methods measure the locations of points.

Stereoscopic Imaging. Stereoscopic pictures can be readily taken by SEM and TEM (Ref 52). In any SEM picture, there are two main types of distortion: perspective error due to tilt of the surface and magnification error arising from surface irregularities. The first type of error can be minimized by keeping the beam close to perpendicular to the fracture surface. The second error can be understood by reference to Fig. 5, which is the rectilinear optical equivalent of the SEM image (Ref 50).

The magnification is defined as the ratio of the image distance to the object distance. In the case of an irregular surface, the object distance is not constant. Consequently, high points have higher magnification than low points on the surface. For example, at point p in Fig. 5, the magnification is proportional to ss'/sm, but at point q, it is proportional to ss'/sn.

The coordinates of the points in the fracture surface are usually measured by stereo SEM pairs, that is, two photographs of the same field taken at small tilt angles with respect to the normal. The geometry of this case is shown in Fig. 6, in which the points A,B appear at A',B' and A",B" in stereo pictures taken at tilt angles $\pm\alpha$. The lengths A'B' and A"B" can be measured either from the two photos separately or from the stereo image directly (Ref 53). The difference A'B' − A"B" is called the parallax, Δx.

According to the geometry shown in Fig. 6:

$$\Delta z = \left[\frac{1}{2M \sin \alpha}\right]\Delta x \qquad (Eq\ 3)$$

where the height difference Δz between the two points is proportional to the measured parallax Δx, and M is the average magnification (Ref 30, 53). Because the magnification and tilt angle are fixed for one pair of photographs, the terms in the square brackets are constant. If $\alpha = 10°$, for example, the value of Δz should be about 2.88 times greater than the corresponding measurement along the x- or y-direction. The x- and y-coordinate points can be measured directly with a superimposed grid or can be obtained automatically with suitable equipment (Ref 54). Equation 3 is strictly correct for an orthogonal projection; that is, the point S in Fig. 5 is situated at infinity. This is a reasonable assumption at higher magnifications ($>1000\times$); however, at lower magnifications, there are induced errors (Ref 50).

Once the (x,y,z) coordinates have been obtained at selected points in the fracture surface, elementary calculations can be made, such as the equation of a straight line or a planar surface, the length of a linear segment between two points in space, and the angle between two lines or two surfaces (the dihedral angle) (Ref 55). Some of these basic equations are given in Ref 16. Also available is a computerized graphical method for analyzing stereophotomicrographs (Ref 56).

Photogrammetric Methods. Another procedure for mapping fracture surfaces uses stereoscopic imaging with modified photogrammetry equipment (Ref 14, 42, 57). Several reports from the Max-Planck-Institut in Stuttgart have described the operation in detail (Ref 14, 54, 58, 59). Their instrument is a commercially available mirror stereometer with parallax-measuring capability (adjustable light point type). It is linked to an image analysis system and provides semiautomatic measurement of up to 500 (x,y,z) coordinate points over the fracture surface. The output data generate height profiles or contours at selected locations, as well as the angular distribution of profile

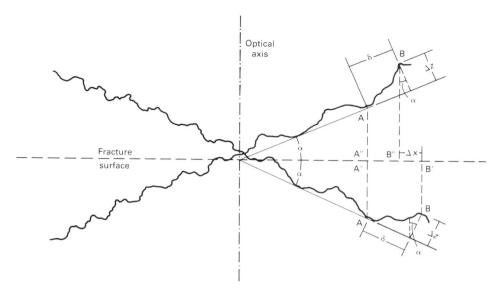

Fig. 6 Determination of parallax Δx from stereo imaging

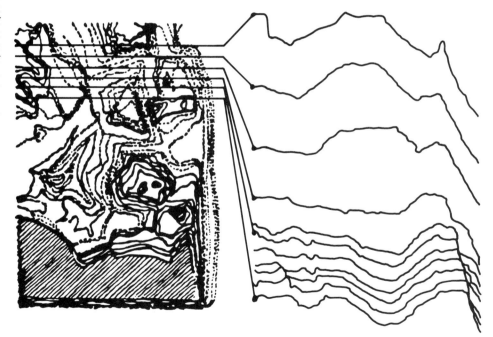

Fig. 7 Contour map and profiles obtained by stereophotogrammetry. Source: Ref 58

elements (Fig. 7). The accuracy of the z-coordinates is better than 5% of the maximum height difference of a profile, and the measurement time is about 3 s per point.

Stereo pair photographs with symmetrical tilts about the normal incident position are used to generate the stereoscopic image. The built-in floating marker (a point light source) of the stereometer is adjusted to lie at the level of the fracture surface at the chosen (x',y') position. Small changes in height are recognized when the marker appears to float out of contact with the fracture surface. The accuracy with which the operator can place the marker depends on his stereo acuity and amount of practice. When the floating marker is positioned in the surface, a foot switch sends all three spatial coordinates to the computer. Thus, the operator can take a sequence of observations without interrupting the stereo effect. A FORTRAN program then calculates the three-dimensional coordinates and produces the corresponding profile or contour map on the plotter or screen.

In another research program having the objective of mapping fracture surfaces, a standard Hilger-Watts stereophotogrammetry viewer was modified so that the operator does not have to take his eyes from the viewer to record the

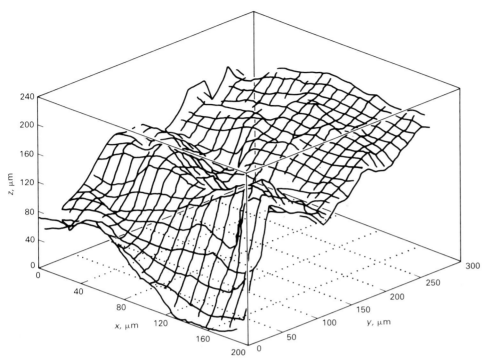

Fig. 8 Fracture surface map (carpet plot) of a Ti-10V-2Fe-3Al specimen by stereophotogrammetry. Source: Ref 57

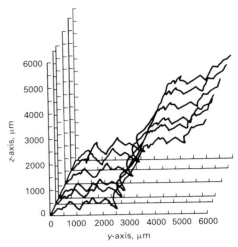

Fig. 9 Profiles obtained by serial sectioning of a fractured Al-4Cu alloy

micrometer readings (Ref 57). The graphic output is based on a matrix of 19 by 27 data points and is in the form of contour plots, profiles, or carpet plots. Figure 8 shows an example of the latter for a fractured Ti-10V-2Fe-3Al tensile specimen.

These photogrammetric procedures are nondestructive with regard to the fracture surface. They also allow detailed scrutiny of some areas and wider spacings in regions of less interest. It is also possible to obtain a roughness profile from a fracture surface along a curved or meandering path, a task that is very difficult to accomplish by sectioning methods. However, this is still a point-by-point method that relies heavily on operator skill and training. No information is provided on any subsurface cracking or possible interactions between microstructure and fracture path. Overlaps and complex fracture surfaces are difficult, if not impossible, to handle. Moreover, smooth or structureless areas cannot be measured unless a marker is added (Ref 59). This method is best suited to relatively large facets and to samples that must be preserved.

Geometrical Methods. The stereoscopic and photogrammetric methods discussed above usually require special equipment and a visually generated stereo effect. Three-dimensional information can also be obtained from stereo pair photomicrographs by a computer graphical method (Ref 56) or a microcomputer-based system (Ref 60). A geometric method has also been described in which the three dimensional data are obtained by analytical geometry (Ref 61). Special stereoscopic or photogrammetric equipment or visual stereo effects are not required. The three-dimensional information is derived from three micrographs taken at different tilting angles, without regard to conditions for a good stereo effect. The x- and y-coordinates are measured from the micrographs, while the z-coordinate is calculated from a simple angle formula. A computer program in BASIC calculates the metric characteristics.

In all these methods that utilize stereoscopic viewing or photography, the basic limitations are the inherent inefficiency of mapping areas with points, the inability to see into reentrances, and the difficulty of obtaining the fine detail required to characterize complex, irregular fracture surfaces. However, in some cases, the requirement of retaining the specimen in the original condition precludes all other considerations.

Profile Generation

The analysis of fracture surfaces by means of profiles appears more amenable to quantitative treatment than by other methods. Profiles are essentially linear in nature, as opposed to the two-dimensional sampling of SEM pictures and the point sampling using photogrammetry procedures. Several types of profiles can be generated, either directly or indirectly, but this article will discuss only three categories of profiles. Those selected are profiles obtained by metallographic sectioning, by nondestructive methods, and by sectioning of fracture surface replicas.

Metallographic Sectioning Methods. Although many kinds of sections have been investigated—for example, vertical (Ref 16, 62, 63), slanted (Ref 64-66), horizontal (Ref 22, 47, 49, 67), and conical (Ref 20, 64, 68)—this discussion will be confined to planar sections. The major experimental advantages of planar sections are that they are obtained in ordinary metallographic mounts and that any degree of complexity or overlap of the fracture profile is accurately reproduced. Moreover, serial sectioning (Ref 18, 45, 47) is quite simple and direct (Fig. 9). In addition, planar sections reveal the underlying microstructure and its relation to the fracture surface (Ref 16, 62, 63), the standard equations of stereology are rigorously applicable on the flat section (Ref 33), and the angular characteristics of the fracture profile (Ref 36) can be mathematically related to those of the surface facets (Ref 39).

Some objections to planar sectioning have been raised on the grounds that it is destructive of the fracture surface and sample. Also, the fracture surface must be coated with a protective layer before sectioning to preserve the trace. Moreover, in retrospect, additional detailed scrutiny of special areas on the fracture surface is not possible once it is coated and cut. However, bearing these objections in mind, if the fracture surface is carefully inspected by SEM in advance, areas of interest can be photographed before coating and cutting take place (Ref 69, 70).

The preliminary experimental procedures are straightforward. When the specimen is fractured, two (ostensibly matching) nonplanar surfaces are produced. After inspection and photography by SEM, one or both surfaces can be electrolytically coated to preserve the edge upon subsequent sectioning (Ref 71). The coated specimen is then mounted metallographically and prepared according to conventional metallographic procedures (Ref 71). One fracture surface can be sectioned in one direction, and the other at 90° to the first direction, if desired.

Once the profile is clearly revealed and the microstructure underlying the crack path properly polished and etched, the measurements can begin. Photographs of the trace or microstructure can be taken in the conventional manner for subsequent stereological measurements. However, with the currently available commercial image analysis equipment, one can bypass photography and work directly from the specimen, greatly improving sampling statistics, costs, and efficiency (Ref 19).

Data that can be obtained with a typical semiautomatic image analysis system include coordinates at preselected intervals along the trace (Ref 36), angular distributions (Ref 18, 59), true profile length (Ref 15, 19), and fractal data (Ref 19, 24, 36). If the unit is interfaced with a large central computer, printouts and graphs are also readily available (Ref 19). After the desired basic data have been acquired, analysis of the profile and calculation of three-dimensional properties can proceed (Ref 18, 19, 36).

Nondestructive Profiles. The sectioning methods described above are basically destructive with regard to fracture surface. What is desired, then, is a nondestructive method of generating profiles representative of the fracture surface. In this way, the wealth of detail available in the flat SEM photomicrograph could be supplemented by the three-dimensional information inherent in the profile at any time.

Comparatively simple procedures for generating profiles nondestructively across the fracture surface are available. One attractive method provides profiles of light using a light-profile microscope (Ref 72) or a modification (Ref 73). A narrow illuminated line is projected on the fracture surface and reveals the profile of the surface in a specified direction. The line can be observed through a light microscope and photographed for subsequent analysis. The resolution obtained depends on the objective lens optics. Profile roughness using a light-section microscope was reported as the mean peak-to-trough distance, with values of approximately 50 ± 5 μm determined for a 1080 steel fractured in fatigue (Ref 20). Unfortunately, other quantitative profile roughness parameters were not investigated.

In a second method, a narrow contamination line is deposited on the specimen across the area of interest, using the SEM linescan mode (Ref 74). The specimen is then tilted, preferably around an axis parallel to the linescan direction, by an angle α. The contamination line then appears as an oblique projection of the profile that can be evaluated (point by point) by the relationship:

$$\Delta z = \left[\frac{1}{M \cos (90° - \alpha)}\right] \Delta k \quad \text{(Eq 4)}$$

where Δk is the displacement of the contamination line at the point of interest, and Δz is the corresponding height of the profile at that point. The tilt angle should be as large as possible (Ref 75). The major source of error comes from broadened contamination lines with diffuse contours (Ref 58). Fortunately, the lines appear considerably sharper in the tilted position. Another disadvantage of this method is that it cannot be used on surfaces that are too irregular. Other than these restrictions, however, it appears that both the contamination line and light beam profiles possess useful attributes that should be investigated more thoroughly for metallic fracture surfaces. Other methods for generating a profile nondestructively include a focusing technique with the SEM for height differences (Ref 76), an interferometric fringe method that yields either areal or lineal elevations (Ref 26, 77), and use of the Tallysurf (Ref 78) or stylus profilometer (Ref 79) for relatively coarse measurements of surface roughness.

Profiles From Replicas. Another method for generating profiles nondestructively is to section replicas of the fracture surface (Ref 30, 58, 80). The primary concern is to minimize distortion of the slices during cutting of the replica. Recent experimental studies have established a suitable foil material, as well as procedures for stripping and coating the replicas. After embedding the replica, parallel cuts are made with an ultramicrotome. In one investigation, a slice of about 2 μm thick was found to be optimum with regard to resolution (Ref 58). Thicker slices deform less; however, the resolution is also less.

The slices obtained by serial sectioning can be analyzed to give the spatial coordinates of a fracture surface (Ref 58). In one study, coordinates along several profiles were obtained using stereogrammetry as described above. The right-hand side of Fig. 7 shows six vertical sections spaced 25 μm apart and seven vertical sections spaced 5 μm apart. On the left-hand side of Fig. 7, the locations of the wider spaced sections are superimposed over a contour map of the same surface. In another investigation, the dimples in a fractured 2024 aluminum alloy were studied using the profiles of microtomed slices from the replica (Ref 80). Sketches of the dimpled profiles show that the depth-to-width ratios are low and that the dimples in this case are relatively shallow holes.

This method of obtaining profiles nondestructively from a fracture surface seems to be useful. It is of course limited to relatively smooth surfaces in order to avoid tearing the replica. The replica-sectioning method appears to be a viable way to quantify the usual measurements from a flat SEM photomicrograph without destroying the specimen.

Analytical Procedures

This section will discuss various roughness and materials-related parameters. The outstanding ones to emerge are the profile, R_L, and surface, R_S, roughness parameters. Based on these parameters, it is possible to develop general relationships for the features in partially oriented fracture surfaces. This section will also examine and compare direct methods for estimating the fracture surface area. An alternative to these methods is the parametric equation based on the relationship between R_L and R_S.

Roughness Parameters

Several types of roughness parameters have been proposed for profiles and surfaces. The major criterion for their use is their suitability for characterizing irregular curves and surfaces. Because profiles are easily obtained experimentally, it is natural that considerable attention has centered on their properties. Those profile parameters that express roughness well, that relate readily to the physical situation, and that equate simply to spatial quantities are particularly favored. Surface roughness studies encounter considerably more difficulty; therefore, descriptive parameters are generally too simplified for meaningful quantitative purposes. Based on the experimental measurements required, the types of derived quantities that are possible, and the ease of interrelating profile to surface roughness, a few parameters have emerged that possess outstanding attributes for quantitative fractography (Ref 23, 37).

Profile Parameters. Most profile parameters consist basically of ratios of length or points of intersection. Thus, they are dimensionless and do not vary with size for curves of the same shape. For definition and further discussion, four profile parameters are described below.

Profile Parameter 1. The lineal roughness parameter, R_L, is defined as the true profile

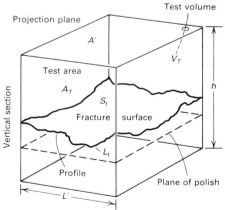

Profile roughness parameter, $R_L = \frac{L_t}{L'}$

Surface roughness parameter, $R_S = \frac{S_t}{A'}$

S_t is true surface area, L_t is true profile length
$A_T = L'h$
$V_T = A'h$

Fig. 10 Arbitrary test volume enclosing a fracture surface

Table 3 Tabulation of published values of profile roughness parameters, R_L

Material	Range of R_L values	Ref
Al-4Cu	1.480-2.390	18, 69
X7091 aluminum alloy	1.12-1.31	81
Ti-24V	1.064	82
Ti-28V	1.060-1.147	16
4340 steels	1.447-1.908	24, 69
Al_2O_3 + 3% glass	1.23-1.62	14
Prototype faceted surface	1.016	44
Computer-simulated fracture surface	1.0165-2.8277	82

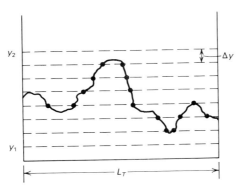

Fig. 11 Profile configuration parameter, R_P, which is the ratio of average peak height to average peak spacing

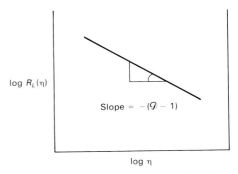

Fig. 12 Fractal dimension $\mathcal{D}$ and its relationship to slope of log-log plot of $R_L(\eta)$ versus η

length divided by the projected length (Ref 12), or:

$$R_L = \frac{L_t}{L'} \quad \text{(Eq 5)}$$

These quantities can be seen in the vertical section plane shown in Fig. 10. A simple, direct way to measure the profile length is to use a digitizing tablet (Ref 11, 36). If this is not available, the true length can be measured manually (Ref 15). Theoretically, for a profile consisting of randomly oriented linear segments, $R_L = \pi/2$ (Ref 33). Experimental R_L values between 1.06 and 2.39 have been reported for a variety of materials. Actual values from the literature are summarized in Table 3.

Profile Parameter 2. A profile configuration parameter, R_P, described in Ref 83 is equal to the average height divided by the average spacing of peaks, or:

$$R_P = \frac{1}{2L_T} \int_{y_1}^{y_2} P(y) dy \quad \text{(Eq 6)}$$

where L_T is the (constant) length of the test line, $P(y)$ is the intersection function, and y_1, y_2 are the bounds of the profile envelope. The working equation can be expressed as:

$$R_P = \frac{\Delta y}{2L_T} \sum_i P_i \quad \text{(Eq 7)}$$

where Δy is the constant displacement of the test line of length L_T, and ΣP_i is the total number of intersections of the test line with the profile (Fig. 11). Applications of this parameter are reviewed in Ref 37, and R_P is compared experimentally with R_L in Ref 36. The parameter R_P is also related to R_V, the vertical roughness parameter, by a factor of two because R_V measures the heights of both sides of each peak (Ref 13). It can be expressed as:

$$R_V = \sum \frac{h_i}{L'} = \left(\sum \frac{h_i}{n}\right)\left(\frac{L'}{n}\right) \quad \text{(Eq 8)}$$

which is equivalent to twice the average peak height over the average peak spacing; thus, $R_V = 2R_P$.

Profile Parameter 3. A simple form of roughness parameter proposed for profiles is merely the average peak-to-trough height, $\bar{h}$ (Ref 20). As such, $\bar{h}$ is closely related to the well-known arithmetic average $\mu(AA)$ used by machinists (Ref 37). This fracture surface roughness can be expressed as:

$$\bar{h} = \left(\frac{1}{n}\right)\sum_i^n h_i \quad \text{(Eq 9)}$$

where h_i is the vertical distance between the ith peak and associated trough. Because peaks and troughs are measured separately from a central reference line in the arithmetic average method, $\bar{h}$ is twice as large as $\mu(AA)$. Because two profiles can have identical values of $\bar{h}$ and yet one curve could have half the number of peaks, it would probably be better to normalize $\bar{h}$ by dividing by the mean peak spacing, L'/n, where n is the number of peaks. This then becomes R_P.

Profile Parameter 4. The symbol $\mathcal{D}$ represents the fractal dimension of an irregular planar curve as described in Ref 38. Its value is obtained from the slope of the log-log plot of the expression:

$$L(\eta) = L_0 \eta^{-(\mathcal{D}-1)} \quad \text{(Eq 10)}$$

where $L(\eta)$ is the apparent length of the profile, and L_0 is a constant with dimensions of length. The apparent profile length varies inversely with the size of the measurement unit η. The symbol $\mathcal{D}$ can have fractional values equal to or greater than 1. Dividing both sides of Eq 10 by L', the projected length of the $L(\eta)$ curve, produces the more useful expression:

$$R_L(\eta) = \frac{L_0}{L'} \eta^{-(\mathcal{D}-1)} \quad \text{(Eq 11)}$$

or the linear form:

$$\log R_L(\eta) = \log\left(\frac{L_0}{L'}\right) - (\mathcal{D}-1)\log \eta \quad \text{(Eq 12)}$$

where the (negative) slope is equal to $-(\mathcal{D} - 1)$. Figure 12 shows the essential relationship of these quantities. This approach has been reanalyzed for the case of fracture surfaces and profiles (Ref 24).

These parameters are interrelated to some extent. For example, R_L and $\mathcal{D}$ give qualitatively similar experimental curves (Ref 24), while $R_L = \pi R_P$ for a profile composed of randomly oriented linear elements. Under some circumstances, the stepped fracture surface model described in Ref 13 yields $R_V = R_L - 1$ and, as demonstrated above, $R_V = 2R_P$.

Surface Parameters. The major interest in surface parameters lies in their possible relationships to the fracture surface area. Surface parameters are not as plentiful as profile roughness parameters, primarily because they are more difficult to evaluate. In fact, most so-called surface parameters are expressed in terms of linear quantities (Ref 84). A natural surface roughness parameter of great importance that parallels the profile roughness parameter dimensionally is R_S*, which is defined in terms of true surface area S_t divided by its projected area A' according to:

$$R_S = \frac{S_t}{A'} \quad \text{(Eq 13)}$$

or, in an alternate form, as:

$$R_S = \frac{S_V}{A'_V} \quad \text{(Eq 14)}$$

Figure 10 also defines and illustrates the surface roughness parameter. For a fracture surface with randomly oriented surface elements and no overlap, $R_S = 2$ (Ref 33).

Other surface parameters for characterizing some aspect of a rough surface have been proposed, but the motivation has been mostly for applications other than fractography. However, the concepts may prove useful for a particular fractographic problem. For example, the slit-island technique uses horizontal serial sections (Ref 22). As the intersection plane moves into the rough surface, islands (and islands with

---
*The earlier literature used other symbols for R_S, for example, K_A, R_A, and S_A. The symbol R_S is preferred because of its position relative to R_L and R_P.

lakes) appear. The perimeter-to-area ratios are measured at successive levels of the test plane, and the analysis is based on the fractal characteristics of the slit-island perimeters. Correlation with the fracture toughness of steels (Ref 22) and brittle ceramics (Ref 67) is claimed.

Another method for determining the topography of an irregular surface also uses horizontal sectioning planes. The analysis, although general in nature, was proposed for the study of paper surfaces (Ref 85). The method is based on experimental measurements of the area exposed as the peaks are sectioned at increasingly lower levels. A graphical solution yields a frequency distribution curve, the mode of which defines the Index of Surface Roughness.

Two other parameters sensitive to surface roughness are the Surface Volume and the Topographic Index (Ref 84). Surface volume is defined as the volume per unit surface area enclosed by the surface and a horizontal reference plane located at the peak of the highest summit. Experimentally, this parameter is evaluated as the distance between the reference plane and the mean plane through the surface. The Topographic Index, ξ, is defined in terms of the peaks nipped off by a horizontal sectioning plane. Experimentally, $\xi = \Delta/\rho$, where Δ equals the average separation of asperity contacts and ρ equals the average asperity contact spot radius. An average ξ, obtained from different levels of the sectioning plane, would probably be more useful.

Other Roughness Parameters. In addition to the rather general profile and surface roughness parameters discussed above, other parameters have been proposed for specific application to a particular requirement. The motivation has often been to connect fracture characteristics to microstructural features.

A linear Fracture Path Preference Index, Q_i, has been proposed for each microstructural constituent i, according to (Ref 63):

$$Q_i = \frac{\Sigma(\ell_i)_{\text{profile}}}{\Sigma(L_i)_{\parallel}} \quad \text{(Eq 15)}$$

where $\Sigma(\ell_i)_{\text{profile}}$ is the total length of the fracture profile that runs through the ith constituent, and $\Sigma(L_i)_{\parallel}$ is the total intercept length through the microstructure across the ith constituent, along a straight test line parallel to the effective fracture direction. As such, this parameter can be recognized as a form of R_L. Furthermore, $Q_i/\overline{Q}$, where $\overline{Q}$ is the average for all constituents, is a measure of the preference of the fracture path for microstructural constituent i (Ref 86). Because deviations from randomness are part of the effect being characterized, the presence or absence of correlation does not affect the accuracy of the result (see the section "Correlation of Fracture Path and Microstructure" in this article).

Another profile parameter, $\mathcal{P}_i(k)$, represents the probability of a microstructural constituent i being associated with a fracture mode k. The probability parameter is defined as (Ref 63):

$$\mathcal{P}_i(k) = \frac{\Sigma(\ell_i)_k}{\Sigma(\ell_i)_{\text{profile}}} \quad \text{(Eq 16)}$$

where the denominator denotes the total profile length through constituent i, and the numerator represents only that portion of the profile length through constituent i that is associated with fracture mode k. Three-dimensional analogs to the above linear parameters are also proposed. The index k for fracture modes is assigned according to $k = 0$, cleavage fracture; $k = 1$, quasi-cleavage fracture; $k = 2$, quasi-dimple fracture; and $k = 3$, dimple fracture.

In an early study, the effects of microstructure on fatigue crack growth were investigated in a low-carbon steel (Ref 86). Inspection of crack traces in the plane of polish showed that the crack generally preferred a path through the ferrite phase, especially at higher ΔK, the stress intensity factor range. At lower ΔK values, however, the crack would penetrate a martensitic region if directly in its path, giving rise to a relatively flat fracture surface. To characterize this crack path preference numerically, a connectivity parameter ψ was proposed for the martensite, whereby:

$$\psi = \frac{(P_L)_{\text{M}\alpha}}{(P_L)_{\text{M}\alpha} + (P_L)_{\alpha\alpha}} \quad \text{(Eq 17)}$$

where the P_L terms signify the number of intersections per unit length of test lines with the Mα-phase boundaries and the $\alpha\alpha$-ferrite grain boundaries. Thus, ψ is equivalent to the fraction of martensite phase boundary area in the microstructure. The ψ parameter is closely related to the contiguity parameter (Ref 87), which also represents the fractional part of total boundary area shared by two grains of the same phase. A more direct crack path preference index is the ratio of the crack length through a selected phase compared to total crack length (Ref 15). Such parameters, together with the profile roughness parameter R_L, should be very useful in quantifying crack path characteristics for most purposes.

A unique approach toward characterizing a fracture profile was described in Ref 88 and 89. A discrete Fourier transform was used to analyze the profiles in ferrous and nonferrous alloys. The energy spectra of fracture profiles show an inverse relationship between the Fourier amplitudes and wave numbers. The spectral lines in the energy spectra can be indexed by assuming that the fundamental fracture unit has a triangular shape. Thus, the results were rationalized on the basis that the profile was built up from a random shifting of a basic triangular element. The resulting profile configuration leads automatically to a bimodal angular distribution, which was reported experimentally by direct angular measurements (Ref 23) (see Fig. 16 and the section "Profile Angular Distributions" in this article).

An essential component in all these specialized parametric studies is the image analysis equipment used to measure the lengths of irregular lines and the areas of irregular planar regions. Such quantities can be measured manually (Ref 15, 16, 33), but automated procedures are generally preferred.

Parametric Relationships for Partially Oriented Surfaces. Among the various profile and surface roughness parameters discussed above, two have emerged as having outstanding characteristics. They are the dimensionless ratios R_L and R_S, the first being a ratio of lengths and the second a ratio of areas. With these parameters, simple expressions can be written for the average value of features in partially oriented fracture surfaces.

Because directionality is important in this case, it is necessary to distinguish between linear features having a fixed direction in the projection plane and those having variable directions in the projection plane. For example, a profile with fixed direction ϕ in the projection plane would have its roughness parameter written as $[R_L]_\phi$, and would probably have different values in different directions. Other quantities are averages of measurements made at several angles ϕ in the projection plane; for example, $\overline{P'_P(\phi)}$ is the average of several angular placements of a point count grid.

At this point, it may be useful to summarize expressions for the more common quantities in partially oriented surfaces. Table 4 lists these relationships as functions of R_L and R_S. In general, R_L appears with lineal terms, and R_S with areal terms. If R_L is known, R_S can be calculated (see the section "Parametric Relationships Between R_S and R_L" in this article).

Estimation of Fracture Surface Area

There are several ways to estimate the actual fracture surface area. A real surface can be approximated with arbitrary precision by triangular elements of any desired size. Another method is based on the relationship of the angular distribution of linear elements along a profile to the angular distribution in space of the surface elements of the fracture surface. A third method involves the stereological relationship between profile and surface roughness parameters. These procedures will be discussed below.

Triangular Elements. This approach uses either vertical or horizontal serial sections, which generate either profiles or contours, respectively. Similar results can be achieved through stereophotogrammetry. A triangular network is constructed between adjacent profiles (or contours). The summation of the areas of the triangular elements constitutes the estimate of the fracture surface area. The accuracy of these triangular approximation methods depends on the closeness of the serial sections and on the number of coordinate points in relation to the complexity of the profile.

In the case of a series of adjacent profiles, one procedure forms the triangular facets by two consecutive points in one profile and the closest point in the adjacent profile, giving about 50 triangular facets per profile (Ref 14). The estimated total facet surface area is given by:

$$S_\triangle = \sum_{i,j} a_{i,j} \quad \text{(Eq 18)}$$

where a is the area of the individual triangular facets, and subscripts i and j refer to the triangles along a profile, and the profiles, respectively. A similar procedure is being evaluated in which the triangular facets between adjacent profiles are formed automatically at 500 regularly spaced coordinate points along each profile (Ref 90).

Another approximate surface composed of triangular elements has been proposed, but this approach uses a contour map rather than serial sections to form the triangles (Ref 13). The triangles are connected between adjacent contour lines, then divided into right angle subtriangles (A'B'C') as shown in Fig. 13. The area of the triangular element ABC can be calculated because the sides $\overline{AB}$ and $\overline{BC}$ can be determined. The area of the entire model surface is then obtained by summation.

The roughness parameter $(R_S)_i$ for the triangular element ABC is given by:

$$(R_S)_i = \sqrt{1 + c^2} \quad \text{(Eq 19)}$$

where $c = d_3/d_1$. Thus, the overall R_S is easily determined by a computer program by knowing only the coordinates of the triangular nodes. The method is simple and direct. However, experimentation has shown that contour sections ("horizontal" cuts) tend to yield wide lines that cannot be accurately evaluated, especially with flat fracture surfaces (Ref 58).

Profile Angular Distributions. A promising analytical procedure, although not specifically developed for fracture surfaces, provides a methodology for transforming the angular distribution of linear elements along a trace to the angular distribution of surface elements in space. From this spatial distribution, the fracture surface area can be obtained.

Three versions of this procedure have been proposed for grain boundaries (Ref 34, 39, 40). There are problems in applying these analyses directly to the case of fracture surfaces, however. The transform procedures require an axis of symmetry in the angular distribution of facets, with angular randomness around the axis, as well as facets that are planar, equiaxed, and finite. Unfortunately, most fracture surfaces do not possess the angular randomness exhibited by grain boundaries. It does appear, however, that there may be an axis of symmetry.

One of the first analyses of the angular distribution of grain-boundary facets is discussed in Ref 39. This method also requires an axis of symmetry so that all sections cut through the axis have statistically identical properties. A recursive formula and tables of coefficients are given, from which the three-dimensional facet angular distribution Q_r can be obtained from the experimental profile angular distribution, G_r. The essential working equations are:

$$Q_r = Q(rh) - Q[(r-1)h] \quad \text{(Eq 20)}$$

where $r = 1, 2, 3 \ldots 18$ (for 0 to 90°), and h is a constant interval value of 5°, and $Q(0) = 0$. Moreover:

$$Q(rh) = \frac{\sum_{s=1}^{r} b_{rs} G_s}{\sum_{s=1}^{18} b_{18,s} G_s} \quad \text{(Eq 21)}$$

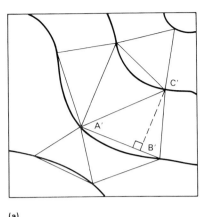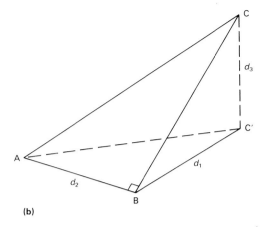

Fig. 13 Representation of a fracture surface with triangular elements between contour lines. (a) Projection of contours and triangles. (b) Three-dimensional view of triangle ABC

Table 4 Parametric relationships between features in the fracture surface and their projected images

Primes indicate projected quantities or measurements made on projection planes

Variable	Explanation	Equation
Part I: Individual features in the fracture surface(a)		
L_ϕ	Length of a linear feature (fixed direction ϕ in the projection plane)	$L_\phi = [R_L]_\phi L'_\phi$
L_c	Length of curved linear feature (variable directions in the projection plane)	$L_c = \overline{R_L(\phi)} L'_c$
L_3	Mean intercept length of a closed figure (averaged over all directions in the projection plane)	$L_3 = \overline{R_L(\phi)} L'_3$
L_p	Perimeter length of a closed figure	$L_p = \overline{R_L(\phi)} L'_p$
S	Area of a curved two-dimensional feature (no overlap)	$S = R_S A'$
Part II: Systems of features in fracture surface		
$\overline{L_\phi}$	Mean length of linear features (fixed direction ϕ in projection plane)	$\overline{L_\phi} = [R_L]_\phi \overline{L'_\phi}$
$\overline{L_c}$	Mean length of curved lines (variable directions in projection plane)	$\overline{L_c} = \overline{R_L(\phi)} \overline{L'_c}$
$\overline{L_3}$	Mean intercept length of closed features (averaged over all directions in the projection plane)	$\overline{L_3} = \overline{R_L(\phi)} \overline{L'_3}$
$\overline{L_p}$	Mean perimeter length of closed figures	$\overline{L_p} = \overline{R_L(\phi)} \overline{L'_p}$
$\overline{S}$	Mean area of curved two-dimensional features (no overlap)	$\overline{S} = R_S \overline{A'}$
$(S_S)_f$	Area fraction of curved two-dimensional features (no overlap)	$(S_S)_f = \dfrac{(R_S)_f (A'_A)_f}{R_S}$
$\overline{L_S}$	Length of linear features per unit area of fracture surface (variable directions in projection plane)	$L_S = \dfrac{\overline{R_L(\phi)} L'_A}{R_S}$

(a) $[R_L]_\phi = L_{\text{true}}/L_{\text{proj}}$ (in ϕ direction in the projection plane); $\overline{R_L(\phi)} = L_{\text{true}}/L_{\text{proj}}$ (over all directions in the projection plane); $R_S = S_{\text{true}}/A_{\text{proj}}$ (total area in selected region of fracture plane); $(R_S)_f = S_f/(A')_f$ (for selected features in fracture plane)

where the coefficients b_{rs} are tabulated for each successive value of r, and the values of G_s are experimentally obtained from $s = r$. The denominator is a constant, with the coefficients $b_{18,s}$ tabulated in Table III in Ref 39 under $r = 18$. The values of Q_r give the distribution of elevation angle θ, per unit angular interval in θ, in the form of a histogram. The total surface area is obtained simply by multiplying the Q_r values by the cosine in each class interval and summing over all class intervals (Ref 18, 69). Thus, for the surface roughness parameter:

$$R_S = \frac{1}{\Sigma Q_r (\cos \theta)_r} \quad \text{(Eq 22)}$$

To illustrate this procedure, two profiles representing two important types of fracture surfaces are selected (Ref 36). The first profile (Fig. 14a) comes from an actual dimpled frac-

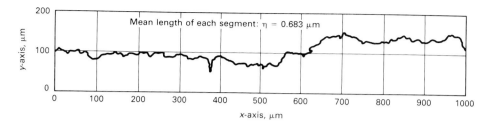

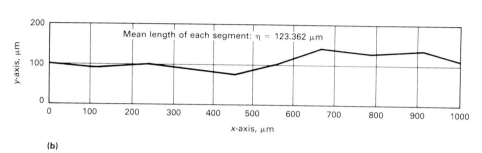

Fig. 14 Profiles for 4340 steel. (a) Dimpled fracture surface. (b) Prototype faceted fracture surface. Compare with Fig. 15.

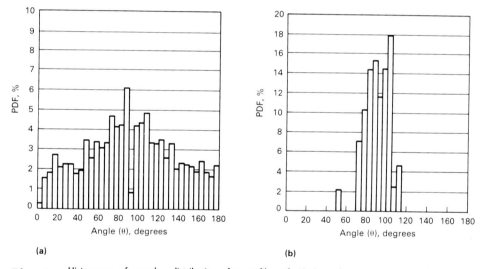

Fig. 15 Histograms of angular distributions for profiles of 4340 steel. (a) Dimpled fracture surface. (b) Prototype faceted fracture surface. PDF, probability density function. Compare with Fig. 14.

Table 5 Values of surface roughness parameter R_S for dimpled and prototype faceted 4340 steel

Fracture mode		Calculated values of R_S	
	Eq 22	Eq 32	Eq 26
Dimpled rupture			
$R_L(\eta = 0.68 \, \mu m) = 1.4466\ldots$	1.6639	1.5686	1.8419
Prototype facets			
$R_L(\eta = 123.4 \, \mu m) = 1.0163\ldots$	1.0482	1.0207	1.2940

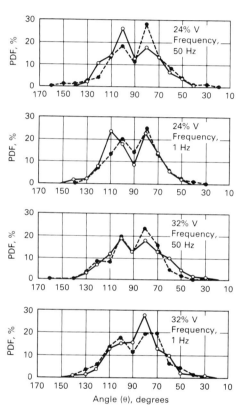

Fig. 16 Experimental angular distribution curves for profiles of Ti-24V and Ti-32V alloys fractured in fatigue. Longitudinal sections made along edge (solid lines) or through center (dashed lines) of compact tensile specimens. Angles measured counter-clockwise from horizontal to segment normals

ture surface for 4340 steel. Figure 14(b) shows a prototype profile of a fracture surface representing 100% cleavage or intergranular facets. The experimental data for these two extreme cases were obtained simply by tracing the profile image (once) with a cursor on a digitizing tablet. The measuring distance η between coordinates on the dimple fracture profile was 0.68 μm. A very large value of η = 123.4 μm was selected to generate the prototype facet profile.

The two histograms for the angular distribution of segment normals along the two profiles are shown in Fig. 15(a) and (b). As expected, there is a much broader angular distribution in Fig. 15(a) than in (b). The distribution for the dimple fracture profile is extremely broad, which might seem to justify the assumption of angular randomness. To verify this assumption, the surface areas were calculated according to the procedure described in Ref 39 as well as Eq 26, and results are given in Table 5. The R_S values fall short of the value of 2 required for a completely random orientation of surface elements. Thus, instead of complete randomness, the dimpled fracture surface belongs in the category of partially oriented surfaces (Ref 33).

Another major requirement of this procedure is that there should be an axis of symmetry in the facet orientations. Surprisingly, an axis of symmetry has been observed in both two-dimensional (profile) (Ref 35, 36) and three-dimensional (facet) (Ref 35) angular distributions, for real fracture surfaces (Ref 18, 35, 36) and a computer-simulated fracture surface (Ref 35), and in compact tensile specimens (Ref 35) and round tensile specimens (Ref 18, 36).

The type of angular distribution observed is essentially bimodal, showing symmetry about the normal to the macroscopic fracture plane. The experimental angular distribution curves for two titanium-vanadium alloys (compact tensile specimens) shown in Fig. 16 illustrate this bimodal behavior clearly. Similar distributions were obtained for both longitudinal and transverse sections, which establishes the three-dimensional angular symmetry about the macroscopic normal. Three-dimensional bimodality was also confirmed clearly with the computer-simulated fracture surface (Ref 35). Bimodal behavior also appears to a greater or lesser extent in other surface studies, in which histograms of two-dimensional angular distributions reveal a relative decrease in the density function at or near 90° (Ref 14, 59). The conclusion to be drawn from this combined evidence is that there is a symmetry axis about the normal to the macroscopic fracture plane. Thus, the mathematical transform procedures that require an axis of symmetry may be more applicable to metallic fracture surfaces than originally thought.

It is interesting to speculate briefly on the physical reasons for such an unexpected type of angular distribution. One of the simplest explanations appears to be that the fracture surface (and profile) consists statistically of alternating up- and down-facets (or linear segments) rather than a basic horizontal component with slight tilts to the left and right (which would give a unimodal distribution curve centered about the normal direction). The explanation of a zig-zag profile is also invoked in the studies of fracture in HY180 and low-carbon steels described in Ref 88.

Parametric Relationships Between R_S and R_L. Most methods for determining the true area of an irregular surface are basically approximations. The smaller the scale of measurement, the closer the true surface area can be approached. This truism is apparent in both the triangulation and angular distribution methods discussed above. A third method for estimating the surface area takes a different approach. The statistically correct equations of stereology between surfaces and their traces are invoked, suitable boundary conditions are determined, and an analytical expression that covers the range of possible fracture surface configurations is then constructed (Ref 17).

The basic geometrical relationship between surface and profile roughness parameters derives from the general stereological equation:

$$S_V = \left(\frac{4}{\pi}\right) L_A \quad \text{(Eq 23)}$$

where, according to Fig. 10, the surface area per unit volume is:

$$S_V = \frac{S_t}{V_T} = \frac{S_t}{A'h} \quad \text{(Eq 24)}$$

and the trace length per unit area is:

$$L_A = \frac{L_t}{A_T} = \frac{L_t}{L'h} \quad \text{(Eq 25)}$$

Substituting for S_V and L_A in Eq 23 yields:

$$R_S = \left(\frac{4}{\pi}\right) R_L \quad \text{(Eq 26)}$$

Equation 26 has been misinterpreted in several papers (Ref 10, 13). It does not refer exclusively to surfaces with random angular orientation of surface elements, nor does it represent an end-point in a progression of surfaces with various configurations between perfectly flat and perfectly random. It does represent surfaces of any configuration (including randomly oriented surfaces) provided the surface is sampled randomly (Ref 91). Thus, for any surface with a value of R_S between 1 and ∞, there should be a corresponding value of R_L between $\pi/4$* and ∞.

From a practical point of view, it should be noted that SEM fractographs represent only one direction of viewing. Thus, when analyzing SEM pictures, the equations that apply to directed measurements may be required. Three structures that frequently use directed measurements are the perfectly flat surface, the ruled surface, and the perfectly random surface.

For the perfectly oriented surface (the flat fracture surface), there is an associated straight-line trace, and:

$$(R_S)_{or} = (R_L)_\perp = 1 \quad \text{(Eq 27)}$$

for a sectioning plane perpendicular to the plane of the surface. Both parameters in this case equal unity and can have no other value.

The ruled surface also represents a low-area surface, depending on the complexity of the trace in the vertical section. For a partially oriented ruled surface:

$$(R_S)_{ruled} = (R_L)_\perp \geq 1 \quad \text{(Eq 28)}$$

where the sectioning plane is perpendicular to the elements of the ruled surface. Values for ruled surfaces can increase indefinitely from 1.

For a randomly oriented surface, R_S is always related to R_L by the coefficient $4/\pi$ regardless of the measurement direction, because a random surface should give the same value (statistically speaking) from any direction. Accordingly, for a directed measurement perpendicular to the effective fracture surface plane:

$$(R_S)_{ran} = \left(\frac{4}{\pi}\right) (R_L)_\perp \quad \text{(Eq 29)}$$

Thus, only a randomly oriented surface has the $4/\pi$ coefficient with a simple directed measurement.

The above three cases are fairly straightforward. The problems arise with partially oriented surfaces that fall between the perfectly oriented surface and the surface with infinitely large surface area. Several attempts have been made to relate R_S and R_L for nonrandom surface configurations (Ref 9, 10, 13, 17). Most derivations relate R_S and R_L linearly, but have different slope constants. One relationship is based on a stepped-surface model (Ref 9), that

*For the random sampling dictated by Eq 23, $\overline{L_2}$ for a square area a^2 equals $(\pi/4)a$ (Ref 33, p 42)

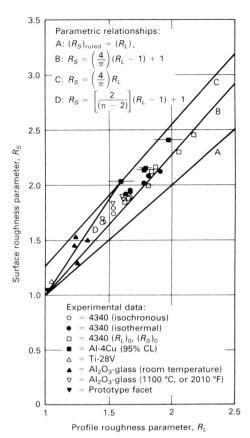

Fig. 17 Plot of all known experimental pairs of R_S, R_L with respect to Eq 32 and upper and lower limits. CL, confidence limits

is, a profile with square-cornered hills and valleys:

$$R_L = \left(\frac{4}{\pi}\right) \frac{[R_S(2-R_S)]^{1/2}}{\tan^{-1}[(2-R_S)/R_S]^{1/2}} \quad \text{(Eq 30)}$$

Other equations (Ref 10, 13) are based on a linear change between a perfectly flat and a perfectly random surface:

$$R_S = \frac{2}{\pi-2}(R_L-1) + 1 \quad \text{(Eq 31)}$$

with a postulated cutoff at $R_S = 2$. A third model proposes a linear change in the degree of orientation between a completely oriented surface and a surface with an unlimited degree of roughness (Ref 17), or:

$$R_S = \left(\frac{4}{\pi}\right)(R_L-1) + 1 \quad \text{(Eq 32)}$$

A plot of these curves on R_S, R_L coordinates has been made, on which all known experimental data points have been superimposed (Ref 23). An updated version of this plot is shown in Fig. 17. The curves labeled A to D represent the following: curve A, Eq 28; curve B, Eq 32; curve C, Eq 26; and curve D, Eq 31. Experi-

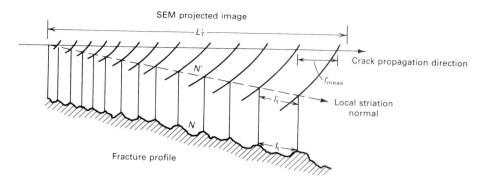

Fig. 18 Relationship of fatigue striation spacings to their projected images. L_T': test line length; N: number of striations; ℓ_t: true striation spacing; ℓ'_{meas}: measured striation spacing in crack propagation direction. Primes indicate quantities in projection plane.

mental roughness data are entered for 4340 steels (Ref 43, 69), Al-4Cu alloys (Ref 18, 69), Ti-28V alloys (Ref 16), and an $Al_2O_3 + 3\%$ glass ceramic (Ref 14). The horizontal bars intersecting the data points in Fig. 17 represent the 95% confidence limits of R_L values from six serial sections for each of the Al-4Cu alloys studied (see Fig. 9).

Almost all the experimental data points fall between curves B and C. However, the locations of these points depend on the accuracy of R_L, which in turn depends on the accuracy with which L_t, the true length of the profile, is determined. The same considerations apply to R_S, which can be considered a dependent variable of R_L. The true length of the profile, L_t, is measured by coordinate points spaced a distance η apart. Most values of R_L are based on the minimum value of $\eta = 0.68$ µm. As an example, see the data points for the isochronous 4340 alloy data shown in Fig. 17.

The data for these six alloy conditions were used to assess the influence of η spacing on the calculated values of L_t (and R_L). Extrapolated values were obtained for R_L as $\eta \to 0$, designated $(R_L)_0$, along with the corresponding calculated values of $(R_S)_0$ (Ref 24). The six points for $(R_L)_0, (R_S)_0$ are plotted in Fig. 17 as open squares. They all fall closely around curve B, which greatly enhances the credibility of Eq 32. If these findings are generally true, then Eq 32 can be used with confidence to obtain values of R_S from measured values of R_L.

Applications

The measurement procedures and equations developed to this point remain incomplete. In some cases, there are special features that require individual measurements or additional techniques (Ref 75, 92). Even so, most of the basic quantities needed to characterize common fracture surfaces are now available. To illustrate what can be done quantitatively, some specific examples will be examined below.

Example 1: Fatigue Striation Spacings. The problem of determining the actual striation spacing in the nonplanar fracture surface was examined previously (Ref 15). Corrections for orientation and roughness effects were formulated, but with separate equations. Figure 18 shows the geometrical relationships involved. Striation spacings are measured in the SEM picture in the crack propagation direction according to:

$$\overline{\ell'_{meas}} = \frac{L_T'}{N'} = \frac{1}{N_L'} \quad \text{(Eq 33)}$$

where L_T' is the distance over which measurements are made, and N' is the number of striations within this distance. Because the directions of the striations vary widely from the crack propagation direction, an angular correction is required. From basic stereology (Ref 33):

$$\overline{\ell'_{meas}} = \left(\frac{\pi}{2}\right)\overline{\ell'_t} \quad \text{(Eq 34)}$$

where $\overline{\ell'_t}$ is the mean normal striation spacing.

In addition, a roughness factor must be considered, and this can be evaluated if the lineal roughness parameter, R_L, is known. It is assumed that R_L over the entire profile length can be equated to the local ratio of mean true striation spacing, $\overline{\ell_t}$, to its mean projected length, $\overline{\ell'_t}$. That is:

$$R_L = \frac{L_{profile}}{L'_{profile}} = \frac{\overline{\ell_t}}{\overline{\ell'_t}} \quad \text{(Eq 35)}$$

Equations 33 to 35 can now be incorporated to yield Eq 36, an overall relationship that combines corrections for both orientation and roughness:

$$\ell_t = R_L\overline{\ell'_t} = R_L\left(\frac{2}{\pi}\right)\overline{\ell'_{meas}} = \left(\frac{2}{\pi}\right)\frac{R_L}{N_L'} \quad \text{(Eq 36)}$$

Equation 36 gives the mean true striation spacing in terms of measurable quantities.

A further modification is possible if warranted by the additional work. The $(\pi/2)$ coefficient in Eq 34 represents a random angular distribution of local striation normals with respect to the crack propagation direction. Although the actual angular distribution can be determined only by measuring it, there is another way to eliminate this problem.

If a second set of measurements is made (Ref 15), in which the striation spacings are first measured along the local striation normals and then projected along the crack propagation direction, the result is:

$$\overline{\ell'_{proj}} = \left(\frac{2}{\pi}\right)\overline{\ell'_t} \quad \text{(Eq 37)}$$

Multiplication of Eq 34 by Eq 37 permits $\overline{\ell'_t}$ to be expressed as the geometric mean of the two measurements and eliminates the two numerical coefficients. Accordingly:

$$\overline{\ell'_t} = \left[\overline{\ell'_{meas}} \cdot \overline{\ell'_{proj}}\right]^{1/2} \quad \text{(Eq 38)}$$

where the actual coefficients may cancel out for real fatigue striation structures.

The close correlation among fatigue striation spacings (FSS), the macroscopic growth rate (da/dN), and an effective stress intensity range (ΔK_{eff}) is well established (Ref 93). It is interesting to note that in this case the conventionally measured FSS is the correct quantity to use (rather than the true striation spacing), because both the crack extension and FSS are projected quantities.

Example 2: A Check on Precision Matching. This technique is occasionally implemented and requires extremely precise experimental controls (Ref 80, 94, 95). An unusually careful and well-documented fractographic investigation is described in Ref 96, which discusses a precision matching study of fatigue striations in nickel. Two pairs of matching profiles were obtained using a stereoscopic method. Illustrations of these four curves are shown in Fig. 19, with the pairs of matching profiles identified as I′I′ and II′II′ in (a) and I″I″ and II″II″ in (b).

To assess the degree of matching, R_L was measured on all profiles. The results are given in Table 6, including duplicate values for the same curves that appeared elsewhere in Ref 97. Of major interest is the correspondence between (ostensibly) matching profile curves. Other than one pair, matching values of R_L agreed within about 2%, which can be considered quite good, especially because measurements were made on illustrations of the curves. Another profile parameter, R_P, was also used in the analysis of these curves, with percent deviations about three times greater than those for R_L (Ref 23). This is probably due to the fact that only the line length is involved with R_L, while R_P depends more on the tortuosity of the profile (Ref 37).

Example 3: Measurements of Features in the Fracture Surface of an Al-4Cu Alloy. Three procedures suggest themselves for obtaining information from the SEM fractograph. The simplest assumption (Case 1) is that the planar SEM projection is a picture of a flat fracture surface. Another approach (Case 2)

Table 6 Analysis of matching profiles of fatigue striations in nickel

	Profile roughness parameters, R_L			
	Profiles in Fig. 19(a)		Profiles in Fig. 19(b)	
	I'I'	II'II'	I"I"	II"II"
Fig. 19(a) and (b)....	1.164	1.251	1.251	1.231
(percent deviation)....	(7.5)		(1.6)	
Duplicate curves.....	1.215	1.242	1.265	1.240
(percent deviation)....	(2.2)		(2.0)	

Source: Ref 37, 96

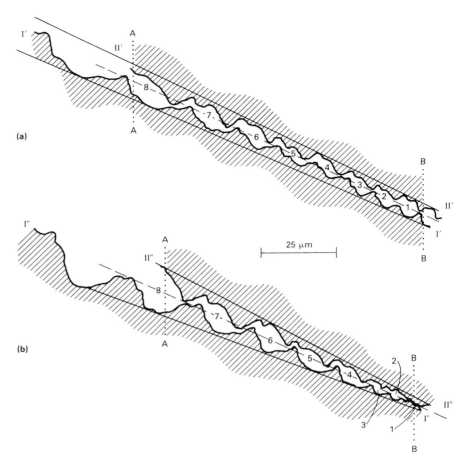

Fig. 19 Precision matching of fatigue striation profiles in nickel. (a) Matching profile of I'I' and II'II'. (b) Matching profile of I"I" and II"II". Source: Ref 96

is to assume that the fracture surface has a uniform angular distribution of surface element normals in sample space, that is, a randomly oriented surface. A third procedure (Case 3) is to analyze the actual fracture surface without any special assumptions about the surface configuration. In this case, roughness parameters R_L and R_S are needed to augment measurements made on the SEM fractograph.

The equations suitable for each of these three cases can be found in Tables 1, 2, and 4. To put numbers on the calculations, SEM fractographs from a fractured Al-4Cu alloy, such as that shown in Fig. 2, are used (Ref 18, 44). Four features are selected for calculation: they are $\overline{S}_f$, the mean facet area; $\overline{L}_p$, the mean facet perimeter length; N_L, the number of facets intercepted per unit length of test line in the fracture surface; and N_S, the number of facets per unit area of the fracture surface.

The results of the calculations are listed in Table 7. From the magnitude of the entries in the last column of Table 7, it can be seen that considerable error is involved if calculations are made directly from the SEM fractograph without corrections. However, to take advantage of the Case 3 estimates, the roughness parameters must be known. If this is not possible, then calculations based on an assumed random fracture surface may be a satisfactory alternative.

Example 4: Crack Path Tortuosity. References 97 and 98 deal with models for characterizing crack path tortuosity. The idealized model of a deflected crack profile consists of a horizontal segment S and a deflected portion D at an angle θ to the horizontal. The crack deflection ratio, D/(D + S), is used to signify the extent of crack deflection.

It can be appreciated that a tremendous amount of experimental labor is involved in evaluating the crack deflection ratio step by step and that many subjective decisions would have to be made by the operator. A parameter comparable to the crack deflection ratio is Ω_{12}, the degree of orientation of a linear element 1 in a plane 2 (Ref 15, 16, 33). The value Ω_{12} is defined as the length of the completely oriented elements of the profile divided by the total length of the profile, regardless of profile complexity. This useful parameter varies between 0 and 1 as the shape of the profile progresses from completely random ($\Omega_{12} = 0$) to completely oriented ($\Omega_{12} = 1$).

In terms of the deflection model discussed in Ref 97, $\Omega_{12} \approx S/(S + D)$ and approximates the coefficient [D cos θ + S)/(S + D)], because both fractions represent oriented line length over total line length. Thus, in order to link the average measured crack growth rate ($\overline{da/dN}$) to the undeflected crack growth rate $(da/dN)_0$, Ω_{12}, or some function of Ω_{12}, could be used as the coefficient. This gives:

$$\left(\overline{\frac{da}{dN}}\right) = \Omega_{12} \left(\frac{da}{dN}\right)_0 \quad \text{(Eq 39)}$$

For a completely undeflected crack, $\Omega_{12} = 1$; as the crack becomes increasingly jagged, $\Omega_{12} \rightarrow 0$ and the calculated value of ($\overline{da/dN}$) will decrease accordingly. It is likely that for the variety of crack configurations observed in fatigue crack propagation studies, the degree of orientation would serve a useful purpose.

Example 5: Quantitative Measurements of Two-Dimensional Dimple Characteristics. There is an abundance of literature on various aspects of dimple size and shape, both with and without particles (Ref 99). Measurements have been made on the SEM fractograph (Ref 62, 100), on stereopictures using stereophotogrammetry (Ref 42, 95), and on profiles, either from the fractured specimen itself (Ref 63) or from replicas of the fracture surface (Ref 80). Detailed information on dimple size and shape can also be found in the article "Modes of Fracture" in this Volume.

What is frequently reported as dimple size is some selected dimension from an arbitrary shape representing some feature in the projection plane. Because of the multiplicity of choices available to express a size, the word is meaningless unless defined. It could be the dimple area, A (Ref 42, 43), the edge of a square ($\sqrt{A}$ or $\sqrt{1/N_A}$) (Ref 42, 80), the diameter of a circle ($D = (4/\pi)\overline{L}$) (Ref 102), the mean intercept length ($\overline{L} = 1/N_L$) (Ref 49, 62, 102), or the maximum intercept length (L_{max}) (Ref 100, 101). Generally, the reported quantity is the projected value. Sometimes, in an effort to estimate the magnitude of the figure in three dimensions, an incorrect coefficient has been applied to the projected value (Ref 42, 46, 103).

One approach toward a quantitative study of dimple characteristics was made recently with

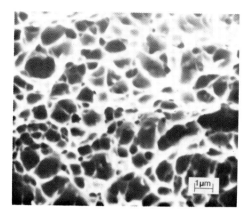

Fig. 20 SEM fractograph of dimples in fractured 4340 steel

Table 7 Calculation of features in SEM fractographs of a peak-aged Al-4Cu alloy

Feature in fracture surface	Equivalent projected quantity	Case 1: SEM (projected)	Case 2: SEM (random)	Case 3: SEM (R_L, R_S)	Deviation: Case 3 from Case 1, %
$\overline{S}_f$ (mm²)	$\overline{A}'_f$	290	580	512(c)	77
$\overline{L}_p$ (mm)	$\overline{L}'_p$	66.6	84.8	106.7(b)	60
N'_L (mm⁻¹)	$[N'_L]_\phi$(a)	0.0731	0.0465	0.037	−49
N'_S (mm⁻²)	N'_A	3.45×10^{-3}	1.73×10^{-3}	1.95×10^{-3}	−43

(a) $[R_L]_\phi = 1.976$, in ϕ direction in projection plane. (b) $\overline{R_L(\phi)} = 1.602$, averaged over many directions ϕ in projection plane. (c) $R_S = 1.767$, based on $\overline{R_L(\phi)}$ using Eq 32. Source: Ref 44, 69

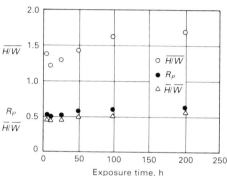

Fig. 21 Height and width measurements of dimples in fractured 4340 steel

4340 steel (Ref 43, 69). The outlines of 514 dimples were traced individually on SEM fractographs, using the digitizing tablet of an image analysis system that records areas and perimeter lengths automatically. Also measured were N'_A and N'_L, the areal and lineal dimple densities, respectively. An SEM fractograph of dimples in the fractured 4340 steel is shown in Fig. 20.

After coating, sectioning, and mounting the fractured specimen to show the profile, the x-y coordinates were obtained by tracing the profile image with the cursor on the digitizing tablet. These data allow the profile roughness parameter, R_L, to be calculated, and then R_S, the surface roughness parameter, follows according to Eq 32.

The basic quantities, calculated from the SEM measurements, gave $\overline{A}' = 0.583$ μm² and $\overline{L}' = 0.638$ μm. Using the roughness parameters, the true mean dimple area is calculated:

$$\overline{S}_{\text{dimples}} = R_S \overline{A}' = 2.0384(0.583)$$
$$= 1.188 \text{ μm}^2 \quad \text{(Eq 40)}$$

and the true mean intercept length is:

$$\overline{L}_{\text{dimples}} = R_L \overline{L}' = 1.8155(0.638)$$
$$= 1.158 \text{ μm} \quad \text{(Eq 41)}$$

according to the equations in Table 4. These values are greater than the corresponding uncorrected quantities from the SEM photomicrographs by 104 and 82%, respectively. Thus, significant improvements in the estimated values are realizable when roughness parameters are available to correct SEM measurements.

Example 6: Quantitative Measurements of Three-Dimensional Dimple Characteristics. Other dimple characteristics of interest are the depth (or height, H) of the dimple and the height-to-width ratio, $\overline{H/W}$. In one case, the average of the ratios, $\overline{H/W}$, was obtained in addition to the customary ratio of the averages, $\overline{H}/\overline{W}$ (Ref 69). As reported in the literature, the heights of the half-dimples were measured from profiles (Ref 63), sectioned replicas (Ref 80), or by stereophotogrammetry (Ref 42, 95). Moreover, in several recent investigations, profiles were digitized in order to measure $\overline{H}$ and $\overline{W}$ (Ref 42, 43, 69).

A new method for obtaining $\overline{H}/\overline{W}$ and $\overline{H/W}$ was developed based solely on digitized profiles (Ref 69). Six 4340 steel specimens, held for different times at 650 °C (1200 °F), were fractured at room temperature. Profiles were prepared and mounted metallographically as described above, then digitized by tracing the image with the cursor of an image analysis system (Ref 43). Dimensions extracted directly from the coordinate file by a suitable filtering algorithm gave values of $\overline{H/W}$ and $\overline{H}/\overline{W}$, as well as R_P, the profile configuration parameter. The results are plotted in Fig. 21 versus exposure time at 650 °C (1200 °F). It is seen that $\overline{H/W}$ is about three times greater than $\overline{H}/\overline{W}$ and R_P, which also represents the mean peak height divided by the mean spacing. The latter two parameters differ by only about 13% and bear a linear relationship to each other. Because of the large discrepancy between the two types of ratios, models that incorporate dimple dimensions should explicitly identify which ratio is being used.

Example 7: Statistical Rather Than Individual Measurements. Investigations are frequently reported in which laborious and time-consuming measurements of individual features have been performed. The distributions of properties are sometimes needed, but often only the mean values are plotted. Three typical examples will be discussed in this example, with proposed alternative procedures that yield quantitative microstructural and fractographic information in a relatively short time.

In Ref 104, an excellent study of fatigue failure, a particle fracture model is set up in which the major microstructural quantities are D, the slip distance from a matrix particle to the grain boundary, and w, which is related to the particle size. Instead of individual measurements of D, it seems that $\overline{L}$, the mean grain intercept length, is the ideal choice to represent three-dimensional slip paths in randomly oriented equiaxed polycrystals. One-half this mean path through the grains could be equated to the D in the model in Ref 104. Also, the w parameter, or the projected length of a crack through a particle, described in Ref 104 can be measured rapidly and accurately with an electronic pencil, such as would be found in any automatic image analysis system having operator-interactive capability. The straight line distance between the two crack ends is obtained simply by touching the two ends. If the true crack length is desired, the operator merely traces the crack from one end to the other. These quantitative measurements can be performed quickly and efficiently, the sampling is unbiased, and the printouts of the individual measurements are available if required.

Another investigation concerns fatigue damage in the form of surface cracks as well as fatigue life prediction (Ref 105). The researchers painstakingly measured the lengths of individual surface cracks and obtained distribution curves of the crack lengths. In some cases, the mean values of the surface crack lengths were plotted versus N/N_f, the ratio of load cycles to load cycles to fracture.

The mean crack length can be obtained manually by simple intersection counts of a test line (or parallel grid of lines) placed at various angles θ over the microstructure (Ref 33), that is:

$$L_A = \left(\frac{\pi}{2}\right) \overline{P_L(\theta)} \quad \text{(Eq 42)}$$

where L_A is the crack length per unit area, and $\overline{P_L(\theta)}$ is the average number of grid intersections with cracks per unit length of test lines. Then, a simple count of discrete cracks in selected test areas gives N_A, the number of cracks per unit test area. Dividing L_A by N_A gives the mean crack length $\overline{L}$. The two counting measurements, P_L and N_A, are performed quickly and efficiently and are free of any assumptions concerning crack size, location, or complexity. Although this procedure applies equally to random or oriented crack traces, there are special procedures for oriented linear traces in a plane that give additional information. For example, the degree of orientation of Ω_{12} is a powerful, yet simple parameter for quantitatively expressing the fraction of a linear system that lies in a particular direction.

A third study examines the facet areas reported for Charpy specimens of an Fe-8Ni-2Mn-0.1Ti ferritic steel (Ref 106). The projected areas of about 1000 selected facets were obtained for each heat treatment by digitizing the images in SEM photomicrographs. Tilt angles were obtained from stereofractograms, allowing a tilt correction to be applied to the measured SEM facet areas. The results were presented in the form of histograms of percent facets of a certain area versus facet area. Examination of the QL histogram (Ref 106, Fig. 12) reveals that the data are grouped into 92 class intervals, or an average of about 10 measurements per class interval. Thus, the considerable background noise in the histogram plot results from too many class intervals for the available data.

The medians of the histograms were selected because of concern about the possible influence of anomalous data on the means. However, a conventional plot on probability paper would immediately indicate whether a size distribution fits the chosen graph paper and would reveal any anomalous data. To check this point, the QL histogram data were plotted on log-normal probability paper, revealing good conformance to a log-normal distribution and a satisfactory linear region in the central part of the plot. The geometric mean of facet areas was 530 μm^2, and the arithmetic mean was 703 μm^2, which can be compared with the median quoted of 560 μm^2. The three histograms presented appear to be typical log-normal size distributions and should be amenable to standard analyses. Even so, after making all these measurements for the histogram plots, only the median values were used.

A further point concerns the method used to correct for tilt angles. Instead of being a three-dimensional technique as claimed, the correction is in reality a two-dimensional correction in the θ plane only, and not in three-dimensional (θ,ϕ) space. As discussed above, the averaging statistics for these two cases differ by more than 23%. The best estimate of true mean facet area $\overline{S}_f$ requires only the surface roughness parameter, R_S, and mean facet projected area $\overline{A'_f}$, as indicated in Table 4.

ACKNOWLEDGMENT

The support of the National Science Foundation, Division of Materials Research, under Grant No. DMR 8504167, is gratefully acknowledged.

REFERENCES

1. Fractography and Materials Science symposium, Williamsburg, VA, American Society for Testing and Materials, Nov 1979
2. Quantitative Fractography, at the Fall Meeting, Pittsburgh, PA, American Institute of Mining, Metallurgical and Petroleum Engineers, Oct 1980
3. Quantification of Nonplanar Surfaces, at the 3rd European Symposium on Stereology, Ljubljana, Yugoslavia, June 1981
4. Fracture and the Role of Microstructure, at the 4th European Conference on Fracture, Leoben, Austria, Sept 1982
5. Fractures and Other Nonplanar Surfaces, at the 6th International Congress on Stereology, Gainesville, FL, Dec 1983
6. Fractography, Failure Analysis, and Microstructural Studies, at the 17th Annual Meeting, Philadelphia, PA, International Metallographic Society, July 1984
7. Nonplanar Surfaces, at the Fourth European Symposium on Stereology, Göteborg, Sweden, Aug 1985
8. Stochastic Aspects of Fracture, at the Annual Meeting, New Orleans, LA, American Institute of Mining, Metallurgical, and Petroleum Engineers, Feb 1986
9. S.M. El-Soudani, Profilometric Analysis of Fractures, *Metallography*, Vol 11, 1978, p 247-336
10. M. Coster and J.L. Chermant, Recent Developments in Quantitative Fractography, *Int. Met. Rev.*, Vol 28 (No. 4), 1983, p 228-250
11. M. Coster and J.-L. Chermant, Fractographie Quantitative, in *Précis D'Analyse D'Images*, Éditions du Centre Nat. de la Recherche Sci., 1985, p 411-462
12. J.R. Pickens and J. Gurland, Metallographic Characterization of Fracture Surface Profiles on Sectioning Planes, in *Proceedings of the Fourth International Conference on Stereology*, E.E. Underwood, R. de Wit, and G.A. Moore, Ed., NBS 431, National Bureau of Standards, 1976, p 269-272
13. K. Wright and B. Karlsson, Topographic Quantification of Nonplanar Localized Surfaces, *J. Microsc.*, Vol 130, Part 1, 1983, p 37-51
14. H.E. Exner and M. Fripan, Quantitative Assessment of Three-Dimensional Roughness, Anisotropy, and Angular Distributions of Fracture Surfaces by Stereometry, *J. Microsc.*, Vol 139, Part 2, 1985, p 161-178
15. E.E. Underwood and E.A. Starke, Jr., Quantitative Stereological Methods for Analyzing Important Microstructural Features in Fatigue of Metals and Alloys, in *Fatigue Mechanisms*, STP 675, J.T. Fong, Ed., American Society for Testing and Materials, 1979, p 633-682
16. E.E. Underwood and S.B. Chakrabortty, Quantitative Fractography of a Fatigued Ti-28V Alloy, in *Fractography and Materials Science*, STP 733, L.N. Gilbertson and R.D. Zipp, Ed., American Society for Testing and Materials, 1981, p 337-354
17. E.E. Underwood and K. Banerji, Statistical Analysis of Facet Characteristics in Computer Simulated Fracture Surface, in *Acta Stereologica*, M. Kališnik, Ed., Proceedings of the 6th International Congress on Stereology, Gainesville, FL, 1983, p 75-80
18. K. Banerji and E.E. Underwood, On Estimating the Fracture Surface Area of Al-4% Cu Alloys, in *Microstructural Science*, Vol 13, S.A. Shiels, C. Bagnall, R.E. Witkowski, and G.F. Vander Voort, Ed., Elsevier, 1985, p 537-551
19. K. Banerji and E.E. Underwood, Fracture Profile Analysis of Heat Treated 4340 Steel, in *Advances in Fracture Research*, Vol 2, S.R. Valluri, D.M. Taplin, P.R. Rao, J.F. Knott, and R. Dubey, Ed., Proceedings of the 6th International Conference on Fracture, New Delhi, India, 1984, p 1371-1378
20. G.T. Gray III, J.G. Williams, and A.W. Thompson, Roughness-Induced Crack Closure: An Explanation for Microstructurally Sensitive Fatigue Crack Growth, *Metall. Trans.*, Vol 14, 1983, p 421-433
21. R.H. Van Stone, T.B. Cox, J.R. Low, Jr., and J.A. Psioda, Microstructural Aspects of Fracture by Dimpled Rupture, *Int. Met. Rev.*, Vol 30 (No. 4), 1985, p 157-179
22. B.B. Mandelbrot, D.E. Passoja, and A.J. Paullay, The Fractal Character of Fracture Surfaces of Metals, *Nature*, Vol 308, 19 April 1984, p 721-722
23. E.E. Underwood, Quantitative Fractography, chapter 8, in *Applied Metallography*, G.F. Vander Voort, Ed., Van Nostrand Reinhold, 1986, p 101-122
24. E.E. Underwood and K. Banerji, Fractals in Fractography, *Mater. Sci. Eng.*, Vol 80, 1986, p 1-14
25. E.K. Brandis, Comparison of Height and Depth Measurements With the SEM and TEM Using a Shadow Casting Technique, *Scan. Elec. Microsc.*, Vol 1, 1972, p 241
26. R.C. Gifkins, *Optical Microscopy of Metals*, Pitman, 1970
27. H. Dong and A.W. Thompson, The Effect of Grain Size and Plastic Strain on Slip Length in 70-30 Brass, *Metall. Trans. A*, Vol 16, 1985, p 1025-1030
28. C.A. Zapffe and G.A. Moore, A Micrographic Study of the Cleavage of Hydrogenized Ferrite, *Trans. AIME*, Vol 154, 1943, p 335-359

29. C.D. Beachem, Microscopic Fracture Processes, in *Fracture*, Vol 1, H. Liebowitz, Ed., Academic Press, 1969, p 243-349
30. D. Broek, Some Contributions of Electron Fractography to the Theory of Fracture, *Int. Met. Rev.*, Vol 19, 1974, p 135-182
31. R.M. Pelloux, The Analysis of Fracture Surfaces by Electron Microscopy, *Met. Eng. Quart.*, Vol 5 (No. 4), Nov 1965, p 26-37
32. D. Broek, A Study on Ductile Fracture, *Nat. Lucht. Ruimteraartlab. Rep.*, NLR (TR71021), 1971, p 98-108
33. E.E. Underwood, *Quantitative Stereology*, Addison-Wesley, 1970
34. J.E. Hilliard, Specification and Measurement of Microstructural Anisotropy, *Trans. AIME*, Vol 224, 1962, p 1201-1211
35. E.E. Underwood and E.S. Underwood, Quantitative Fractography by Computer Simulation, in *Acta Stereologica*, M. Kališnik, Ed., Proceedings of the 3rd European Symposium on Stereology, Ljubljana, Yugoslavia, 1982, p 89-101
36. K. Banerji and E.E. Underwood, Quantitative Analysis of Fractographic Features in a 4340 Steel, in *Acta Stereologica*, M. Kališnik, Ed., Proceedings of the 6th International Congress on Stereology, Gainesville, FL, 1983, p 65-70
37. E.E. Underwood, Practical Solutions to Stereological Problems, in *Practical Applications of Quantitative Metallography*, STP 839, J.L. McCall and J.H. Steele, Ed., American Society for Testing and Materials, 1984, p 160-179
38. B.B. Mandelbrot, *The Fractal Geometry of Nature*, W.H. Freeman, 1982
39. R.A. Scriven and H.D. Williams, The Derivation of Angular Distributions of Planes by Sectioning Methods, *Trans. AIME*, Vol 233, 1965, p 1593-1602
40. V.M. Morton, The Determination of Angular Distributions of Planes in Space, *Proc. R. Soc. (London) A*, Vol 302, 1967, p 51-68
41. A.W. Thompson and M.F. Ashby, Fracture Surface Micro-Roughness, *Scr. Metall.*, Vol 18, 1984, p 127-130
42. G.O. Fior and J.W. Morris, Jr., Characterization of Cryogenic Fe-6Ni Steel Fracture Modes: A Three-Dimensional Quantitative Analysis, *Metall. Trans. A*, Vol 17, 1986, p 815-822
43. E.E. Underwood, Estimating Feature Characteristics by Quantitative Fractography, *J. Met.*, Vol 38 (No. 4), 1986, p 30-32
44. E.E. Underwood, Recent Developments in Quantitative Fractography, in *Proceedings of the 1985 Metals Congress*, (Ballarat, Australia), 1985, p D7-D11
45. T.B. Cox and J.R. Low, Jr., An Investigation of the Plastic Fracture of AISI 4340 and 18 Nickel-200 Grade Maraging Steels, *Metall. Trans.*, Vol 5, 1974, p 1457-1470
46. D.E. Passoja and D.C. Hill, On the Distribution of Energy in the Ductile Fracture of High Strength Steels, *Metall. Trans.*, Vol 5, 1974, p 1851-1854
47. R.H. Van Stone and T.B. Cox, Use of Fractography and Sectioning Techniques to Study Fracture Mechanisms, in *Fractography—Microscopic Cracking Processes*, STP 600, C.D. Beachem and W.R. Warke, Ed., American Society for Testing and Materials, 1976, p 5-29
48. E.E. Underwood, The Stereology of Projected Images, *J. Microsc.*, Vol 95, Part 1, 1972, p 25-44
49. D.E. Passoja and D.C. Hill, Comparison of Inclusion Distributions on Fracture Surfaces and in the Bulk of Carbon-Manganese Weldments, in *Fractography—Microscopic Cracking Processes*, STP 600, C.D. Beachem and W.R. Warke, Ed., American Society for Testing and Materials, 1976, p 30-46
50. J.E. Hilliard, Quantitative Analysis of Scanning Electron Micrographs, *J. Microsc.*, Vol 95, Part 1, 1972, p 45-58
51. D.J. Widgery and J.F. Knott, Method for Quantitative Study of Inclusions Taking Part in Ductile Fracture Process, *Met. Sci.*, Vol 12, Jan 1978, p 8-11
52. J.I. Goldstein, D.E. Newbury, P. Echlin, D.C. Joy, C. Fiori, and E. Lifshin, *Scanning Electron Microscopy and X-Ray Microanalysis*, Plenum Press, 1981, p 143-146
53. A. Boyd, Quantitative Photogrammetric Analysis and Quantitative Stereoscopic Analysis of SEM Images, *J. Microsc.*, Vol 98, Part 3, 1973, p 452-471
54. P.G.T. Howell, Stereometry as an Aid to Stereological Analysis, *J. Microsc.*, Vol 118, Part 2, 1980, p 217-220
55. L.P. Eisenhart, *Coordinate Geometry*, Dover, 1960
56. F.J. Minter and R.C. Pillar, A Computerized Graphical Method for Analyzing Stereo Photo-micrographs, *J. Microsc.*, Vol 117, 1979, p 305-311
57. D. Bryant, Semi-Automated Topographic Mapping of Fracture Surfaces Through Stereo-Photogrammetry, *Micron Microsc. Acta*, to be published
58. B. Bauer and A. Haller, Determining the Three-Dimensional Geometry of Fracture Surfaces, *Pract. Metallogr.*, Vol 18, 1981, p 327-341
59. B. Bauer, M. Fripan, and V. Smolej, Three Dimensional Fractography, in *Fracture and the Role of Microstructure*, K.L. Maurer and F.E. Matzer, Ed., Proceedings of the 4th European Conference on Fracture, Leoben, Austria, 1982, p 591-598
60. S.G. Roberts and T.F. Page, A Microcomputer-Based System for Stereogrammetric Analysis, *J. Microsc.*, Vol 124, 1981, p 77-88
61. S. Simov, E. Simova, and B. Davidkov, Electron Microscope Study of Surface Topography by Geometrical Determination of Metric Characteristics of Surface Elements, *J. Microsc.*, Vol 137, Part 1, 1985, p 47-55
62. F.L. Bastian and J.A. Charles, Mechanism of Fibrous Fracture of Powder Forged Steels, in *Advances in Fracture Research*, Vol 1, Proceedings of the 5th International Conference on Fracture, Cannes, France, 1981, Pergamon Press, 1982, p 209-216
63. W.T. Shieh, The Relation of Microstructure and Fracture Properties of Electron Beam Melted, Modified SAE 4620 Steels, *Metall. Trans.*, Vol 5, 1974, p 1069-1085
64. D. Shechtman, Fracture-Microstructure Observations in the SEM, *Metall. Trans. A*, Vol 7, 1976, p 151-152
65. W.R. Kerr, D. Eylon, and J.A. Hall, On the Correlation of Specific Fracture Surface and Metallographic Features by Precision Sectioning in Titanium Alloys, *Metall. Trans. A*, Vol 7, 1976, p 1477-1480
66. E.A. Almond, J.T. King, and J.D. Embury, Interpretation of SEM Fracture Surface Detail Using a Sectioning Technique, *Metallography*, Vol 3, 1970, p 379-382
67. K.S. Feinberg, "Establishment of Fractal Dimensions for Brittle Fracture Surfaces," B.S. thesis, Pennsylvania State University, 1984
68. J.C. Chestnut and R.A. Spurling, Fracture Topography-Microstructure Correlations in the SEM, *Metall. Trans. A*, Vol 8, 1977, p 216
69. K. Banerji, "Quantitative Analysis of Fracture Surfaces Using Computer Aided Fractography," Ph.D. thesis, Georgia Institute of Technology, 1986
70. P. Nenonen, K. Törrönen, M. Kemppainen, and H. Kotilainen, Application of Scanning Electron Microscopy for Correlating Fracture Topography and Microstructure, in *Fractography and Materials Science*, STP 733, L.N. Gilbertson and R.D. Zipp, Ed., American Society for Testing and Materials, 1981, p 387-393
71. G.F. Vander Voort, *Metallography: Principles and Practice*, McGraw-Hill, 1984, p 86-90, 538-540
72. S. Tolansky, A Light-Profile Microscope for Surface Studies, *Z. Elektrochemie*, Vol 56 (No. 4), 1952, p 263-267
73. V.R. Howes, An Angle Profile Technique for Surface Studies, *Metallography*, Vol 7, 1974, p 431-440
74. J.A. Swift, Measuring Surface Variations With the SEM Using Lines of Evaporated Metal, *J. Phys. E., Sci. Instrum.*, Vol 9, 1976, p 803

75. R. Wang, B. Bauer, and H. Mughrabi, The Study of Surface Roughness Profiles of Fatigued Metals by Scanning Electron Microscopy, *Z. Metallkd.*, Vol 73, 1982, p 30-34
76. D.M. Holburn and D.C.A. Smith, Topographical Analysis in the SEM Using an Automatic Focusing Technique, *J. Microsc.*, Vol 127, Part 1, 1982, p 93-103
77. L.H. Butler and A.A. Juneco, Use of Micro-Interferometry to Study Surface Topography during Cavitation, *J. Inst. Met.*, Vol 99, 1971, p 163-166
78. W.H.L. Hooper and J. Holden, Some Methods of Measuring Surface Topography as Applied to Stretcher-Strain Markings on Metal Sheet, *J. Inst. Met.*, 1954, p 161-165
79. D.H. Park and M.E. Fine, Origin of Crack Closure in the Near-Threshold Fatigue Crack Propagation of Fe and Al-3%Mg, in *Fatigue Crack Growth Threshold Concepts*, D.L. Davidson and S. Suresh, Ed., American Institute of Mining, Metallurgical, and Petroleum Engineers, 1984, p 145-161
80. D. Broek, The Role of Inclusions in Ductile Fracture and Fracture Toughness, *Eng. Fract. Mech.*, Vol 5, 1973, p 55-66
81. V.W.C. Kuo and E.A. Starke, Jr., The Development of Two Texture Variants and Their Effect on the Mechanical Behavior of a High Strength P/M Aluminum Alloy, X7091, *Metall. Trans. A*, Vol 16, 1985, p 1089-1103
82. E.E. Underwood, Georgia Institute of Technology, unpublished research, 1986
83. E.W. Behrens, private communication, 1977
84. H.C. Ward, "Profile Description," Notes to "Surface Topography in Engineering," Short Course offered at Teeside Polytechnic
85. B.-S. Hsu, Distribution of Depression in Paper Surface: A Method of Determination, *Br. J. Appl. Phys.*, Vol 13, 1962, p 155-158
86. H. Suzuki and A.J. McEvily, Microstructural Effects on Fatigue Crack Growth in a Low Carbon Steel, *Metall. Trans. A*, Vol 10, 1979, p 475-481
87. J. Gurland, The Measurement of Grain Contiguity in Two-Phase Alloys, *Trans. AIME*, Vol 212, 1958, p 452
88. D.E. Passoja and D.J. Amborski, Fracture Profile Analysis by Fourier Transform Methods, in *Microstructural Science*, J.E. Bennett, L.R. Cornwell, and J.L. McCall, Ed., Elsevier, 1978, p 143-158
89. D.E. Passoja and J.A. Psioda, Fourier Transform Techniques—Fracture and Fatigue, in *Fractography and Materials Science*, STP 733, L.N. Gilbertson and R.D. Zipp, Ed., American Society for Testing and Materials, 1981, p 355-386
90. K. Banerji, Georgia Institute of Technology, unpublished research, 1984
91. C.S. Smith and L. Guttman, Measurement of Internal Boundaries in Three-Dimensional Structures by Random Sectioning, *Trans. AIME*, Vol 197, 1953, p 81-87
92. W.W. Gerberich and K. Jatavallabhula, Quantitative Fractography and Dislocation Interpretations of the Cyclic Cleavage Crack Growth Process, *Acta Metall.*, Vol 31, 1983, p 241-255
93. R.W. Hertzberg and J. von Euw, Crack Closure and Fatigue Striations in 2024-T3 Aluminum Alloy, *Metall. Trans.*, Vol 4, 1973, p 887-889
94. C.D. Beachem, An Electron Fractographic Study of the Influence of Plastic Strain Conditions Upon Ductile Rupture Processes in Metals, *Trans. ASM*, Vol 56, 1963, p 318-326
95. O. Kolednik and H.P. Stüwe, Abschätzung der Risszähigkeit eines duktilen Werkstoffes aus der Gestalt der Bruchfläche, *Z. Metallkd.*, Vol 73, 1982, p 219-223
96. A.J. Krasowsky and V.A. Stepanenko, Quantitative Stereoscopic Fractographic Study of the Mechanism of Fatigue Crack Propagation in Nickel, *Int. J. Fract.*, Vol 15 (No. 3), 1979, p 203-215
97. S. Suresh, Crack Deflection: Implications for the Growth of Long and Short Fatigue Cracks, *Metall. Trans. A*, Vol 14, 1983, p 2375-2385
98. A.K. Vasudévan and S. Suresh, Lithium-Containing Aluminum Alloys: Cyclic Fracture, *Metall. Trans. A*, Vol 16, 1985, p 475-477
99. A.S. Argon and J. Im, Separation of Second Phase Particles in Spheroidized 1045 Steel, Cu-0.6Pct Cr Alloy, and Maraging Steel in Plastic Straining, *Metall. Trans. A*, Vol 6, 1975, p 839-851
100. W.W. Gerberich and C.E. Hartbower, Some Observations on Dimple Size in Cyclic-Load-Induced Plane-Stress Fracture, *Trans. ASM*, Vol 61, 1968, p 184-187
101. A.W. Thompson, Ductile Fracture Topography: Geometrical Contributions and Effects of Hydrogen, *Metall. Trans. A*, Vol 10, 1979, p 727-731
102. A.W. Thompson and J.A. Brooks, Hydrogen Performance of Precipitation-Strengthened Stainless Steels Based on A-286, *Metall. Trans. A*, Vol 6, 1975, p 1431-1442
103. T.G. Nieh and W.D. Nix, A Comparison of the Dimple Spacing on Intergranular Creep Fracture Surfaces With the Slip Band Spacing for Copper, *Scr. Metall.*, Vol 14, 1980, p 365-368
104. W.L. Morris and M.R. James, Statistical Aspects of Fatigue Failure Due to Alloy Microstructure, in *Fatigue Mechanisms: Advances in Quantitative Measurement of Physical Damage*, STP 811, J. Lankford, D.L. Davidson, W.L. Morris, and R.P. Wei, Ed., American Society for Testing and Materials, 1983, p 170-206
105. H. Kitagawa, Y. Nakasone, and S. Miyashita, Measurement of Fatigue Damage by Randomly Distributed Small Cracks Data, in *Fatigue Mechanisms: Advances in Quantitative Measurement of Physical Damage*, STP 811, J. Lankford, D.L. Davidson, W.L. Morris, and R.P. Wei, Ed., American Society for Testing and Materials, 1983, p 233-263
106. D. Frear and J.W. Morris, Jr., A Study of the Effect of Precipitated Austenite on the Fracture of a Ferritic Cryogenic Steel, *Metall. Trans. A*, Vol 17, 1986, p 243-252

Fractal Analysis of Fracture Surfaces

Ervin E. Underwood and Kingshuk Banerji, Fracture and Fatigue Research Laboratory, Georgia Institute of Technology

RESEARCH into the field of quantitative fractography has progressed considerably during the past few years (Ref 1-4). The main thrust has been directed toward ways to augment the two-dimensional information in the scanning electron microscope fractograph. Elevation information is needed in order to quantify the true magnitudes of features in the nonplanar fracture surface. The essential key that unlocks these quantitative data is the true fracture surface area. Although several experimental procedures have been proposed for obtaining the surface area, one of the more promising techniques involves vertical sections through the fracture surface (Ref 5) (see the section "Profile Generation" in the article "Quantitative Fractography" in this Volume). The resulting profiles are statistically related to the fracture surface Thus, characterization of the profile has been investigated in detail (Ref 6, 7).

As part of the intensive studies of profile characteristics, the fractal properties of profiles have also been investigated (Ref 8, 9). (The use of fractals is discussed in the *Metals Handbook* in Volume 7, *Powder Metallurgy*. See the section "Fractals as Descriptors of P/M Systems" in the article "Particle Shape Analysis.") The reported experimental results on the fractal behavior of numerous materials—both organic and inorganic—have been unexpected. Basically, the theoretical linear fractal plot, whose slope yields the fractal dimension, is not obtained in many experimental studies. Thus, the current fractal theory does not appear to apply to natural, irregular curves, such as profiles from fracture surfaces.

An alternative procedure has been developed for linearizing the fractal plot (Ref 10). New fractal equations have been proposed, not only for profiles but also for fracture surfaces. Modified fractal dimensions that result from the analysis appear to possess some generality for natural irregular nonplanar surfaces and their profiles. These developments are described below, leading to a procedure that overcomes some of the problems with the current theory.

Experimental Background

The mathematical concept of fractals is based on the change in apparent length of an irregular profile as a function of the size of the measuring unit (Ref 11). This idea is illustrated in Fig. 1, which shows that the apparent measured length of the same curve obviously increases as the size of η decreases (Ref 12). The estimated lengths $L(\eta)$ are a function of the measuring unit η according to the theoretical fractal equation (Ref 13):

$$L(\eta) = L_0 \, \eta^{-(\mathcal{D}-1)} \quad \text{(Eq 1)}$$

where L_0 is a constant with dimensions of length, and $\mathcal{D}$, the fractal dimension, is a constant related to the slope of the linear form of Eq 1, which is:

$$\log L(\eta) = \log L_0 - (\mathcal{D} - 1) \log \eta \quad \text{(Eq 2)}$$

The resulting curve of $L(\eta)$ versus η on log-log coordinates is termed the fractal plot. Equation 1 predicts an infinitely large value of $L(\eta)$ as $\eta \to 0$. This has been explained by postulating self-similitude in rough planar curves, by which the same apparent roughness and configuration are always observed regardless of the magnification (Ref 11). This behavior has been found in only a few cases, such as the pathological Koch figures, whose boundaries can be increased without limit (Ref 11, 14).

However, nonfractal behavior is quite common in natural, irregular curves. The linearity predicted by Eq 2 is not observed; instead, there is a trend toward asymptotic curvature in the fractal curve. This tendency has been seen in many microstructural features that have rough, irregular boundaries (both closed loops and open curves). Papers on the fractal behavior of metal particle profiles (Ref 15), open pores (Ref 16), lung cells (Ref 17), liver membranes (Ref 12), minerals (Ref 18), clusters of overlapping circles (Ref 19) and fracture profiles (Ref 3, 20) have inadvertently revealed noncompliance with the postulate of self-similitude. Asymptotic behavior is understandable in real curves, of course, because of the limitations of atomic size or, at other levels, the size of the picture element (pixel) in automatic image analysis or the thickness of the line in a curve. The deviations from linearity were not detected in early work, primarily because the range of η values was too limited. In later work, researchers noticed curvature in their fractal plots, giving rise to such terms as semifractal (Ref 17) or critical resolution point (Ref 12). Later, more extensive data obtained from fracture profiles largely confirmed the inapplicability of Eq 2 (Ref 8).

A recent development has recognized the actual fractal behavior of real fracture surfaces

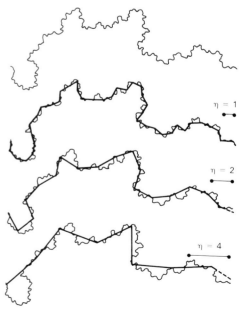

Fig. 1 Interaction of an irregular curve and the measuring unit η. (Ref 12)

and their profiles (Ref 10). Both profile and surface fractal equations have been proposed, based on the linearization of reversed sigmoidal curves (RSCs). The modified profile fractal curve is represented by:

$$L(\eta) = L_0 \eta^{-(\mathcal{D}_\beta - 1)} \qquad \text{(Eq 3)}$$

where the subscript β emphasizes that the fractal constant $\mathcal{D}_\beta$ applies to RSC behavior. A parallel expression for irregular surfaces is:

$$S(\eta^2) = S_0[\eta^2]^{-(\mathcal{D}_\gamma - 2)/2} \qquad \text{(Eq 4)}$$

where $S(\eta^2)$ is expressed as a function of an area-measuring unit η^2 rather than the linear η used for profiles. The value $\mathcal{D}_\gamma$ is the fractal constant for surfaces that conform to an RSC fractal plot. An equation comparable to Eq 4 has also been proposed for the fractal surface irregularities of a silicic acid absorbent (Ref 21).

Applications based on these developments will be discussed below. First, however, the important relationship between the profile and surface roughness parameters that yields the surface area of irregular fracture surfaces will be described, followed by the experimental procedures required to obtain profiles and the measurements that are made. Further experimental details are also available in the article "Quantitative Fractography" in this Volume.

Profile and Surface Roughness Parameters

It has become evident in the study of fracture profiles that R_L, the profile roughness parameter, occupies a central position in expressing the characteristics of the profile (Ref 22). The parameter R_L is defined by $L(\eta)$ divided by L', the (constant) profile projected length. It also bears a direct relationship to R_S, the surface roughness parameter, which is defined by the surface area S divided by its projected area, A' (Ref 5). Although R_S is the quantity sought because it gives the fracture surface area, it is relatively difficult to obtain. The parameter, R_L, on the other hand, is an experimental quantity that is measured directly from a vertical section.

Several analytical methods have been proposed for determining R_S (see the article "Quantitative Fractography" in this Volume) (Ref 23, 24). One of the simpler ways is through the linear parametric equation that links R_S directly to R_L according to:

$$R_S = \left(\frac{4}{\pi}\right)(R_L - 1) + 1 \qquad \text{(Eq 5)}$$

Equation 5 provides the best fit to all known experimental data (Ref 7). It is based on a realistic model, that is, a fracture surface that can have any configuration between the limits of completely oriented ($R_S = 1$) and extremely complex ($R_S \to \infty$).

In analyzing the fractal behavior of irregular fracture surfaces, data point pairs (R_L, R_S) are needed for each configuration. Equation 5 supplies the inaccessible quantity, R_S, with a high degree of reliability and a minimum of calculation.

Equations 3 and 4 are easily modified in terms of the more useful roughness parameters. Dividing both sides of Eq 3 by L' gives:

$$R_L(\eta) = C_1 \eta^{-(\mathcal{D}_\beta - 1)} \qquad \text{(Eq 6)}$$

and dividing both sides of Eq 4 by A' results in:

$$R_S(\eta^2) = C_2[\eta^2]^{-(\mathcal{D}_\gamma - 2)/2} \qquad \text{(Eq 7)}$$

where $\mathcal{D}_\beta$ is the RSC fractal dimension for profiles, $\mathcal{D}_\gamma$ is the RSC fractal dimension for surfaces, and C_1 and C_2 are constants. Equation 6 is the experimental counterpart to Eq 1.

Experimental Procedure

A very useful experimental procedure for determining the fracture surface area is based on vertical sections through the fracture surface (Ref 5, 7). The sections generate profiles that sample continuously across the entire surface as opposed, for example, to the point-by-point registration obtained by stereophotogrammetry measurements (Ref 4). With the sample in a metallographic mount, the fracture path and underlying microstructure are revealed in relation to each other, and the standard equations of stereology are applicable to both. Moreover, serial sectioning is readily performed merely by grinding down the face of the sample parallel to the previous location. This procedure yields a new profile each time and systematically samples through sample space.

In addition to the sectioning procedure, a sampling plan must be chosen. Theoretically, in calculating the surface area, profiles are required from sections at all possible locations, angles of rotation, and angles of tilt (Ref 25). This was investigated by using a computer-simulated fracture surface, having a known surface area for any configuration, that can be sectioned at will (Ref 23, 26). It was found that simple serial sectioning was equally as effective as sections taken over all possible orientations. This finding, if generally applicable, should greatly reduce the number of sections required for adequate sampling.

The profile characteristics are readily measured with a commercial image analysis system connected to a suitable optical metallographic system (Ref 9, 27). The image of the curve is traced with the cursor, which is on the digitizing tablet. Measurements are taken directly from the metallographic samples (eliminating photography and darkroom time) by using a drawing tube to link the specimen image and

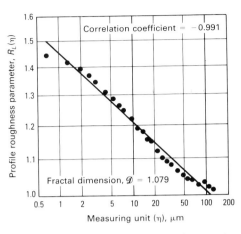

Fig. 2 Fractal plot for AISI 4340 steel tempered at 700 °C (1290 °F) for 1.5 h and fractured at room temperature showing RSC behavior

digitizing tablet. The basic data obtained are the (x,y) coordinates along the profile at preselected intervals. These values are then transmitted to a main-frame computer for analysis, graphs, and printouts. Calculated quantities may include the "true" length of the profile, its projected length, and the angular distribution of elementary segments along the curve.

Figure 2 shows typical fractal data obtained from the profile of an AISI 4340 steel fractured at room temperature (Ref 8). The profile image is first digitized as described above. More than 12 000 coordinate points are registered along this profile at a spacing of 0.683 μm. This value represents the minimum η size. The successively larger values of η are obtained merely by skipping more coordinate points each time, until the largest size of 123.5 μm is reached.* The retained coordinate points are used to calculate the 26 values of apparent profile lengths $L(\eta)$ shown in Fig. 2.

It is clear that a curve through the data points shown in Fig. 2 is not linear as postulated, even though the fitted regression line shows a correlation exceeding 99%. In fact, a tendency toward asymptotic behavior is evident at both termini of the curve. In an effort to clarify this situation, fracture profiles were obtained from a series of six AISI 4340 steel specimens tempered for 1.5 h at various temperatures between 200 and 700 °C (390 and 1290 °F). The upper limit of η was extended to 426.5 μm, making the spread in values greater than 645 times. The fractal plots for the six profiles are shown in Fig. 3, which clearly reveals the RSC shape. The curves are definitely not linear.

Conventional $\mathcal{D}$ values were calculated for each of the six curves shown in Fig. 3. Average

---

*The way in which η is determined with automatic image analysis equipment differs slightly from the way in which η is determined with dividers, as used in Fig. 1. With automatic image analysis, the size of each successively larger η is based on the original coordinate points along the curve. With the dividers technique, a new set of point intersections of the curve is established for each size.

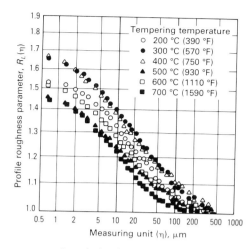

Fig. 3 Fractal plots for AISI 4340 steel specimens tempered at temperatures between 200 and 700 °C (390 and 1290 °F) for 1.5 h and fractured at room temperature showing RSC behavior

Table 1 Summary of fractal dimensions and roughness parameters for AISI 4340 steel specimens

Tempering temperature		$\mathcal{D}$				
°C	°F	(conventional)	$\mathcal{D}_\beta$	$\mathcal{D}_\gamma$	R_L^0	R_S^0
200	390	1.085	1.512	2.5064	1.650	1.850
300	570	1.091	1.446	2.4020	1.850	2.100
400	750	1.090	1.392	2.3290	2.175	2.450
500	930	1.072	1.351	2.3152	1.800	2.000
600	1110	1.084	1.398	2.3372	2.050	2.300
700	1290	1.079	1.440	2.3670	1.875	2.150

slopes were determined by linear regression using the central portion of each curve. The $\mathcal{D}$ values obtained are given in Table 1 for comparison with other results. A minimum is apparent at 500 °C (930 °F). The significance of this minimum has been discussed in relation to microstructural and fractographic factors (Ref 8, 9).

Unfortunately, $\mathcal{D}$ values calculated in this manner depend on subjective operator decisions. Fractal equations that linearize all the experimental data and provide constant fractal dimensions are developed below. Note that the fractal analysis for profiles has been extended to include a comparable fractal equation for irregular surfaces.

Linearization of RSC Fractal Curves

The extensive fractal data shown in Fig. 3 reveal the essential characteristics of the entire experimental fractal curve. An RSC is found instead of the straight line predicted in Ref 12. In an effort to rationalize this nonlinear behavior, it is useful to recall the treatment of sigmoidal curves frequently encountered in phase transformation studies (Ref 28). That is, if the fraction transformed in the forward direction is y_F, then as a function of x:

$$y_F = 1 - e^{-\alpha x^\beta} \quad \text{(Eq 8)}$$

is the forward sigmoidal curve with constants α and β. It follows that the fraction untransformed, y_R, is simply $1 - y_F$. The general linear form for y_R is given by:

$$\log\log\left(\frac{1}{y_R}\right) = \log\left(\frac{\alpha}{2.3}\right) + \beta \log x \quad \text{(Eq 9)}$$

Equation 9 requires a plot of $\log(1/y_R)$ versus x on log-log coordinates to yield a straight line of slope β.

In order to evaluate y_R, the limits to the log $R_L(\eta)$ versus log η curve shown in Fig. 4 are used (Ref 27). The asymptotic limit as $\eta \to 0$ is denoted by R_L^0, while the limit for $\eta \to \infty$, R_L^∞, is 1. Thus, the fractional representation of y_R is given by (Ref 10):

$$y_R = \frac{R_L(\eta) - R_L^\infty}{R_L^0 - R_L^\infty} = \frac{R_L(\eta) - 1}{R_L^0 - 1} \quad \text{(Eq 10)}$$

Substituting the latter value for y_R into Eq 9 and replacing x by η gives the linear equation:

$$\log\log\left\{\frac{R_L^0 - 1}{R_L(\eta) - 1}\right\} = C_1 + \beta \log \eta \quad \text{(Eq 11)}$$

for the RSC profile fractal plot in terms of the experimental variables $R_L(\eta)$ and η, where β is the slope and C_1 and R_L^0 are constants (R_L^0 is a disposable constant whose value may be determined by an iterative optimization procedure, or graphically, until the data are best linearized). It should be noted that the profile length term in R_L^0 represents the "true" length of the profile, that is at $\eta \to 0$.

In addition to the linear fractal equation for RSC profiles, a similar expression for irregular surfaces is also available:

$$\log\log\left\{\frac{R_S^0 - 1}{R_S(\eta^2) - 1}\right\} = C_2 + \gamma \log \eta^2 \quad \text{(Eq 12)}$$

where γ is the (constant) slope, and R_S^0 and C_2 are constants. Values of R_S are obtained from Eq 5. Here, of course, the left-hand side is plotted versus $\log \eta^2$. If desired, the left-hand side can be plotted against $\log \eta$, but γ will then equal the experimental slope divided by 2. The value R_S^0 is determined by an iterative optimization procedure similar to that used to determine R_L^0. Also, the surface area term in R_S^0 represents the "true" area of the surface, independent of fractal variations.

Modified Fractal Dimensions

Plots of experimental fractal data show that Eq 11 and 12 are linear over a wide range of η (Ref 10). Thus, the β and γ slopes are satisfactory measures of the fractal characteristics of profiles and surfaces. However, if desired, β can be expressed in terms of the modified

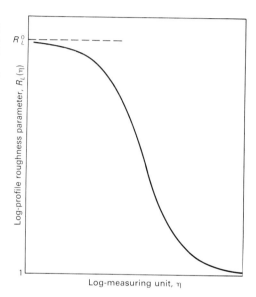

Fig. 4 Upper and lower asymptotic limits to fractal plot showing RSC behavior

fractal dimension $\mathcal{D}_\beta$, thereby providing values comparable to the conventional Mandelbrot fractal dimension $\mathcal{D}$. Comparing the slope of Eq 6 to β gives:

$$\mathcal{D}_\beta = \beta + 1 \quad \text{(Eq 13)}$$

Similarly, the slope of Eq 7 is equated to γ, resulting in:

$$\mathcal{D}_\gamma = 2\gamma + 2 \quad \text{(Eq 14)}$$

A typical β value of 0.512 gives $\mathcal{D}_\beta = 1.512$, and a typical γ value of 0.253 gives $\mathcal{D}_\gamma = 2.506$. The difference $\mathcal{D}_\gamma - \mathcal{D}_\beta$ is close to unity, which is very satisfying from a fractal point of view. Also, $\mathcal{D}_\beta > 1$ and $\mathcal{D}_\gamma > 2$, as required by fractal theory. Calculated values of $\mathcal{D}_\beta$ and $\mathcal{D}_\gamma$ for the series of AISI 4340 specimens are given in Table 1.

Fractal Analysis of AISI 4340 Steels

Data reported previously (Ref 7, 8, 9, 27), as well as the results of fractal analyses along lines discussed above, have been brought together in

214 / Fractal Analysis of Fracture Surfaces

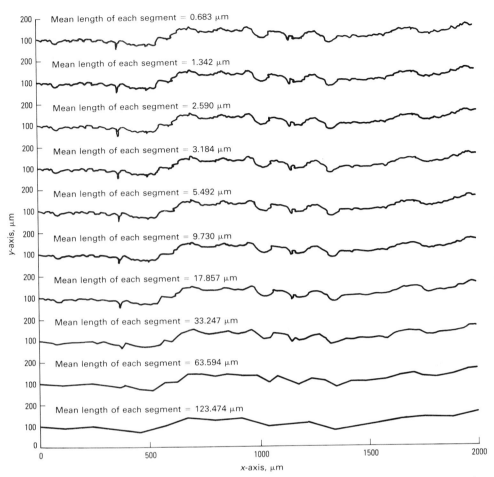

Fig. 5 Fracture profile of AISI 4340 steel specimen plotted with increasing lengths of measuring unit η. The numbers on the curves indicate the value of η.

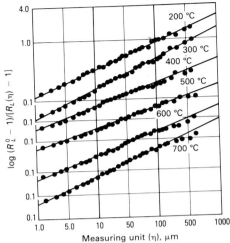

Fig. 6 Linearized RSCs as a function of $R_L(\eta)$ for AISI 4340 steel specimens fractured at room temperature

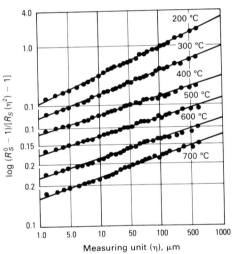

Fig. 7 Linearized RSCs as a function of $R_S(\eta^2)$ for AISI 4340 steel specimens fractured at room temperature

this article. Six AISI 4340 steel specimens were fractured at room temperature after tempering for 1.5 h between 200 and 700 °C (390 and 1290 °F). At 500 °C (930 °F), there is a well-known embrittlement phenomenon associated with these steels (see the section "Thermally Induced Embrittlement" in the article "Ductile and Brittle Fractures" in Volume 11 of the 9th Edition of *Metals Handbook*). However, both tensile strength and hardness curves decrease smoothly as tempering temperature increases. Moreover, the microstructures, scanning electron microscope fractographs, and profiles for each condition do not indicate any special behavior at 500 °C (930 °F).

Figure 5 shows ten profiles generated from the same profile from an AISI 4340 steel specimen (Ref 9). Varying the size of η, as indicated on the curves, reveals the gradual loss of detail as the value of η increases from 0.683 to 123.474 µm. Each curve represents a different length $L(\eta)$ as a function of a particular η size. The data pairs are then plotted as shown in Fig. 3.

Next, the experimental data for the series of six AISI 4340 fractured specimens are plotted according to Eq 11 and 12 (Ref 10). The RSCs for profiles and surfaces are linearized quite well, as shown in Fig. 6 and 7. These curves yield the β and γ slopes used to calculate $\mathcal{D}_\beta$ and $\mathcal{D}_\gamma$ according to Eq 13 and 14. Values of $\mathcal{D}_\beta$ and $\mathcal{D}_\gamma$ are given in Table 1, along with corresponding values of R_L^0 and R_S^0. In addition, both $\mathcal{D}_\beta$ and $\mathcal{D}_\gamma$ are plotted versus tempering temperature in Fig. 8, showing minima at 500 °C (930 °F).

Minima at 500 °C (930 °F) are apparent from all the data listed in Table 1. It can be seen that $\mathcal{D}_\beta$ is an order of magnitude larger than $\mathcal{D}$. Moreover, $\mathcal{D}_\beta$ is based on all data points, but $\mathcal{D}$ involves only a selected central portion of the fractal plot. Thus, $\mathcal{D}_\beta$ is considered more reliable than the conventional $\mathcal{D}$. The modified fractal dimensions and the roughness parameters appear capable of detecting subtle differences in fracture behavior not readily apparent by more conventional means.

Summary

In order to investigate the fractal characteristics of fracture surfaces, it is necessary to determine the fracture surface areas. These can be calculated from the parametric equation:

$$R_S = \left(\frac{4}{\pi}\right)(R_L - 1) + 1$$

where R_L, the profile roughness parameter, is experimentally obtained from vertical sections through the fracture surface. The surface roughness parameter, R_S, is relatively inaccessible and yields the true fracture surface area.

Conventional fractal plots are not linear, and the conventional fractal dimension $\mathcal{D}$ is not constant. Experimental fractal plots display the shape of an RSC and are linearized by:

$$\log \log \left\{\frac{R_L^0 - 1}{R_L(\eta) - 1}\right\} = C_1 + \beta \log \eta$$

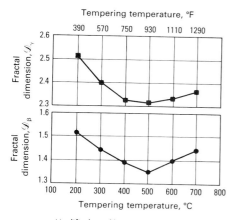

Fig. 8 Modified profile and surface fractal dimensions $\mathscr{D}_\beta$ and $\mathscr{D}_\gamma$ versus tempering temperature from RSCs for AISI 4340 steel specimens fractured at room temperature

and

$$\log \log \left\{ \frac{R_S^0 - 1}{R_S(\eta^2) - 1} \right\} = C_2 + \gamma \log \eta^2$$

for profiles and surfaces, respectively. The slopes β and γ are related to modified fractal dimensions $\mathscr{D}_\beta$ for profiles and $\mathscr{D}_\gamma$ for surfaces.

Results obtained from a series of tempered AISI 4340 steel specimens confirm that the differences $\mathscr{D}_\gamma - \mathscr{D}_\beta$ are close to unity, $\mathscr{D}_\beta > 1$ and $\mathscr{D}_\gamma > 2$, as would be expected from fractal theory. This analysis appears to possess general validity for the irregular, complex curves and surfaces of fractured materials.

ACKNOWLEDGMENT

This work was performed under the auspices of the Division of Materials Research, Metallurgy Section, National Science Foundation on Grant DMR 8504167.

REFERENCES

1. S.M. El-Soudani, Profilometric Analysis of Fractures, *Metallography*, Vol 11, 1978, p 247-336
2. M. Coster and J.L. Chermant, Recent Developments in Quantitative Fractography, *Int. Met. Rev.*, Vol 28 (No. 4), 1983, p 228-250
3. K. Wright and B. Karlsson, Topographic Quantification of Nonplanar Localized Surfaces, *J. Microsc.*, Vol 130, 1983, p 37-51
4. H.E. Exner and M. Fripan, Quantitative Assessment of Three-Dimensional Roughness, Anisotropy, and Angular Distributions of Fracture Surfaces by Stereometry, *J. Microsc.*, Vol 139, Part 2, 1985, p 161-178
5. E.E. Underwood and S.B. Chakrabortty, Quantitative Fractography of a Fatigued Ti-28V Alloy, in *Fractography and Materials Science*, STP 733, L.N. Gilbertson and R.D. Zipp, Ed., American Society for Testing and Materials, 1981, p 337-354
6. E.E. Underwood, Practical Solutions to Stereological Problems, in *Practical Applications of Quantitative Metallography*, STP 839, J.L. McCall and J.H. Steele, Ed., American Society for Testing and Materials, 1984, p 160-179
7. E.E. Underwood, Quantitative Fractography, chapter 8, in *Applied Metallography*, G.F. Vander Voort, Ed., Van Nostrand Reinhold, 1986, p 101-122
8. K. Banerji and E.E. Underwood, Fracture Profile Analysis of Heat Treated 4340 Steel, in *Advances in Fracture Research*, Vol 2, S.R. Valluri, D.M. Taplin, P.R. Rao, J.F. Knott, and R. Dubey, Ed., Proceedings of the 6th International Conference on Fracture, New Delhi, India, 1984, p 1371-1378
9. K. Banerji and E.E. Underwood, Quantitative Analysis of Fractographic Features in a 4340 Steel, in *Acta Stereologica*, M. Kališnik, Ed., Proceedings of the 6th International Congress on Stereology, Gainesville, FL, 1983, p 65-70
10. E.E. Underwood and K. Banerji, Fractals in Fractography, *Mater. Sci. Eng.*, Vol 80, 1986, p 1-14
11. B.B. Mandelbrot, *The Fractal Geometry of Nature*, W.H. Freeman, 1982
12. D. Paumgartner, G. Losa, and E.R. Weibel, *J. Microsc.*, Vol 121 (No. 1), 1981, p 51-63
13. B.B. Mandelbrot, *Fractals: Form, Chance and Dimension*, W.H. Freeman, 1977, p 32
14. E. Kasner and J. Newman, *Mathematics and the Imagination*, Simon and Schuster, 1967, p 344
15. B.H. Kaye, Chapter 10, in *Direct Characterization of Fine Particles*, John Wiley & Sons, 1981, p 367-375
16. H. Pape, L. Riepe, and J.R. Schopper, Abstract 32, in *Abstracts of the 4th European Symposium on Stereology*, Göteborg, 1985
17. J.P. Rigaut, P. Berggren, and B. Robertson, Abstract 14, in *Abstracts of the 4th European Symposium on Stereology*, Göteborg, 1985
18. W.B. Whalley and J.D. Orford, *Scan. Elec. Microsc.*, Vol 11, 1982, p 639-647
19. H. Schwarz and H.E. Exner, *Powder Technol.*, Vol 27, 1980, p 207-213
20. M. Coster and A. Deschanvres, *Pract. Metallogr.*, Special Issue, Vol 8, 1977, p 61-73
21. D. Avnir and P. Pfeifer, Fractal Dimension in Chemistry. An Intensive Characteristic of Surface Irregularity, *Nouveau J. de Chimie*, Vol 7 (No. 2), 1983, p 71-72
22. J.R. Pickens and J. Gurland, Metallographic Characterization of Fracture Surface Profiles on Sectioning Planes, in *Proceedings of the 4th International Congress for Stereology*, NBS 431, E.E. Underwood, R. de Wit, and G.A. Moore, Ed., National Bureau of Standards, 1976, p 269-272
23. E.E. Underwood and K. Banerji, Statistical Analysis of Facet Characteristics in Computer Simulated Fracture Surface, in *Acta Stereologica*, M. Kališnik, Ed., Proceedings of the 6th International Congress for Stereology, Gainesville, FL, 1983, p 75-80
24. E.E. Underwood, *Stereological Analysis of Fracture Roughness Parameters*, 25th Anniversary Volume, International Society for Stereology, 1987
25. R.A. Scriven and H.D. Williams, The Derivation of Angular Distributions of Planes by Sectioning Methods, *Trans. AIME*, Vol 233, 1965, p 1593-1602
26. E.E. Underwood and E.S. Underwood, Quantitative Fractography by Computer Simulation, in *Acta Stereologica*, M. Kališnik, Ed., Proceedings of the 3rd European Symposium for Stereology, 2nd Part, 1982, p 89-101
27. K. Banerji, "Quantitative Analysis of Fracture Surfaces using Computer Aided Fractography," Ph.D. thesis, Georgia Institute of Technology, June 1986
28. J. Burke, *The Kinetics of Phase Transformations in Metals*, Pergamon Press, 1965, p 36, 46

Atlas of Fractographs

THIS ATLAS provides a collection of fractographs that is intended to assist the reader in recognizing the features of fracture surfaces and in understanding the causes and mechanisms of fracture of industrial alloys and engineered materials. With its 1343 illustrations, the Atlas both complements and supplements the preceding nine articles to provide a Handbook designed to describe and illustrate fractographic techniques, to identify and interpret fracture morphologies by visual examination, light microscopy, or electron microscopy (particularly SEM), and to elucidate the benefits of fractography in determining the relationship of the mode of fracture to the microstructure, evaluating the responses of materials to mechanical, chemical, and thermal environments, and performing failure analyses.

Organization and Presentation

As shown in Table 1, the Atlas contains photographic illustrations for 30 categories of materials. The fractographs of metallic materials (Fig. 1 to 1249) are arranged by alloy grade, following an order similar to that in Volume 9 of the 9th Edition of *Metals Handbook* (*Metallography and Microstructures*), and the fractographs of engineered materials (Fig. 1250 to 1343) are categorized according to type. To assist the reader in locating a material of interest, Table 1 lists the figure numbers of the fractographs within each category. Further, headlines at the top of each Atlas page identify the material categories presented. Specific alloy designations can be located by consulting the Index in this Volume.

Table 1 Distribution content of illustrations for various materials in the Atlas of Fractographs

Material	Figure numbers	Photographs	Macrographs/micrographs	Light fractographs	SEM fractographs	TEM fractographs	Total illustrations
Irons	1-28	...	...	2	26	...	28
Gray irons	29-35	1	2	1	4	...	7
Ductile iron	36-97	...	14	4	44	...	62
Malleable irons/white irons	98-108	...	10	...	1	...	11
Low-carbon steels	109-164	3	20	8	23	2	56
Medium-carbon steels	165-244	9	10	55	6	...	80
High-carbon steels	245-318	5	6	28	27	8	74
AISI/SAE alloy steels	319-572	23	22	100	73	36	254
ASTM/ASME alloy steels	573-610	2	4	3	28(a)	...	38(b)
Austenitic stainless steels	611-697	9	20	5	40	11	87(c)
Martensitic stainless steels	698-719	2	4	7	...	9	22
Precipitation-hardening stainless steels	720-747	1	2	7	18	...	28
Tool steels	748-797	1	2	26	21	...	50
Maraging steels	798-822	...	...	3	18	4	25
Superalloys(d)	823-866	1	9(e)	8	20	3	44(f)
Nickel alloys	867-880	...	...	...	14	...	14
Cobalt alloys	881-885	...	...	2	3	...	5
Copper alloys	886-920	4	4	1	26(g)	...	35
Cast aluminum alloys	921-963	11	1	8	23	...	43
Wrought aluminum alloys	964-1095	12	11	25	76	8	132
P/M aluminum alloys	1096-1102	...	...	...	7	...	7
Titanium alloys	1103-1195	1	7	8	68	9	93
Miscellaneous metals and alloys(h)	1196-1249	1	...	9	27	17	54
Metal-matrix composites	1250-1272	...	4	...	19	...	23
Cemented carbides	1273-1277	...	...	...	1	4	5
Ceramics	1278-1280	1	...	...	2	...	3
Concrete and asphalt	1281-1293	...	1	...	12(j)	...	13
Resin-matrix composites	1294-1316	...	6	...	17	...	23
Polymers	1317-1325	...	3	...	6	...	9
Electronic materials	1326-1343	1	...	...	17	...	18
Total	...	88	162	310	667	111	1343

(a) Total includes phosphorus and sulfur maps obtained in a scanning Auger microprobe (Fig. 604 and 605). (b) Total includes a schematic of a fatigue fracture surface (Fig. 577). (c) Total includes a graph showing the effect of hydrogen on ductility (Fig. 639) and a drawing of a compression hip screw (Fig. 670). (d) Total includes both iron-base (Fig. 823 to 827) and nickel-base (Fig. 828 to 866) superalloys. (e) Total includes two sulfur dot maps obtained by energy-dispersive x-ray spectroscopy (Fig. 825 and 827). (f) Total includes a graph of fracture toughness versus neutron irradiation that contains three superimposed electron fractographs (Fig. 823), a graph charting fatigue crack growth rate versus the stress intensity factor range that contains four superimposed electron fractographs (Fig. 846), and a schematic illustrating the selective area polishing technique (Fig. 847). (g) Total includes a tin distribution map obtained by energy-dispersive x-ray spectroscopy (Fig. 916). (h) Miscellaneous metals and alloys include tungsten, iridium, magnesium-base, iron-base, molybdenum-base, and tantalum-base materials. (j) Total includes two sulfur dot maps obtained by energy-dispersive x-ray spectroscopy (Fig. 1290 and 1292).

Table 2 Causes of fractures illustrated in the Atlas of Fractographs for various ferrous and nonferrous alloys

Material	Fracture source			Fracture mode				
	Parts	Test specimens	Total	Dimple rupture	Cleavage	Fatigue(a)	Decohesive rupture(b)	Miscellaneous(c)
Irons	...	19	19	5	6	1	3	4
Gray irons	1	1	2	...	1	1	...	...
Ductile irons	4	45	49	8	17	11	...	13
Malleable irons/white irons	1	4	5	1	...	...	...	4
Low-carbon steels	18	11	29	1	2	11	7	8
Medium-carbon steels	40	1	41	2	...	34	...	5
High-carbon steels	34	...	34	...	1	21	3	9
AISI/SAE alloy steels	66	66	132	15	4	46	15	52
ASTM/ASME alloy steels	4	10	14	...	...	6	5	3
Austenitic stainless steels	18	11	29	3	1	8	12	5
Martensitic stainless steels	3	...	3	...	...	2	1	...
Precipitation-hardening stainless steels	4	6	10	2	...	1	3	4
Tool steels	7	14	21	3	...	7	4	7
Maraging steels	...	7	7	2	...	3	...	2
Superalloys(d)	3	18	21	2	...	11	6	2
Nickel alloys	...	3	3	...	...	...	...	3
Cobalt alloys	1	2	3	...	...	3	...	...
Copper alloys	7	8	15	3	...	5	2	5
Cast aluminum alloys	5	1	6	1	...	1	...	4
Wrought aluminum alloys	18	34	52	6	1	32	5	8
P/M aluminum alloys	...	4	4	...	...	3	1	...
Titanium alloys	3	36	39	15	4	8	8	4
Miscellaneous metals and alloys(e)	1	25	26	3	9	2	5	7
Total	238	326	564	72	46	217	80	149

(a) This category includes 21 parts and test specimens that failed because of corrosion fatigue. (b) This category includes creep rupture, stress-corrosion cracking, hydrogen embrittlement, thermally induced embrittlement, and any other combination of applied stress and environment that results in an intergranular fracture path. Fractures resulting from the synergistic interaction of a corrosive environment and applied stress that are entirely transgranular would be listed under "Cleavage." (c) This category includes quasi-cleavage; mixed-mode fractures; overload failures involving tensile, shear, or torsion stresses that do not result in an entirely dimpled rupture fracture surface; fractures due to such material defects as laps, seams, cold shuts, pre-existing cracks, inclusions, porosity, and grain-boundary discontinuities; photographs illustrating microcrack initiation and propagation; fractures due to design deficiencies; and radiation-induced failures. (d) Superalloys include both iron-base (Fig. 823-827) and nickel-base (Fig. 828-866) alloys. (e) Miscellaneous metals and alloys include tungsten, iridium, magnesium-base, iron-base, molybdenum-base, and tantalum-base materials.

Each fractograph or group of fractographs is accompanied by a caption that provides (1) the alloy grade, the chemical composition of the part or specimen, or the commercial or proprietary name of the material, (2) details of the fabrication or heat treatment history of the part or specimen, (3) material properties, and (4) a brief summary of the fracture surface features displayed, the fracture mechanisms, and when possible, the cause of fracture. The etchant used with each micrograph is also noted in the caption and identified by its common name or composition. Detailed information on the composition selection, preparation, and handling of etchants is available in the article "Etching" in Volume 9 of the 9th Edition of *Metals Handbook* and in the Section "Metallographic Techniques and Microstructures: Specific Metals and Alloys" in Volume 9.

Magnifications are given for each illustration in the Atlas and are usually placed at the end of the caption. Magnification markers are also provided wherever possible. Electron fractographs are differentiated by the acronyms SEM and TEM. The type of replication technique used for TEM fractographs is also included in the caption. For example, "TEM p-c replica, 10 000×" indicates a plastic-carbon two-stage replica examined in a transmission electron microscope at a magnification of 10 000×. Replication techniques are discussed in the articles "History of Fractography" and "Transmission Electron Microscopy" in this Volume.

A number of the fractographs show the fracture surface first at low magnification and then show the details visible at higher magnification by light microscopy, scanning electron microscopy, or transmission electron microscopy. Photographs of the part, macrographs of important sections, photomicrographs of the structure, and elemental distribution maps obtained by scanning Auger electron spectroscopy (AES) or energy-dispersive x-ray spectroscopy (EDS) have been included, when appropriate, to enhance the reader's understanding of the fracture. The Atlas also includes five line drawings, two of which are graphs containing superimposed electron fractographs. Information on the principles, instrumentation, and applications associated with AES and EDS is provided in the articles "Auger Electron Spectroscopy" and "Electron Probe X-Ray Microanalysis," respectively, in Volume 10 of the 9th Edition of *Metals Handbook* (*Materials Characterization*).

Table 3 Causes of fractures or failures illustrated in the Atlas of Fractographs for various engineered materials

Material	Fracture source			Fracture mode				
	Parts	Test specimens	Total	Ductile	Brittle	Fatigue	Mixed mode	Miscellaneous(a)
Metal-matrix composites	...	12	12	4	3	1	3	1
Cemented carbides	...	3	3	...	3	...	...	...
Ceramics	1	1	2	...	1	...	...	1
Concrete and asphalt	...	4	4	...	4	...	...	...
Resin-matrix composites	...	16	16	...	8	...	...	8
Polymers	3	3	6	1	...	2	...	3
Electronic materials	11	...	11	...	...	3	...	8
Total	15	39	54	5	19	6	3	21

(a) This category includes fractures or failures due to design deficiencies, contamination by foreign particles, splitting of reinforcing fibers, cracks originating at corrosion pits during four-point bending, cracks initiating at internal defects or voids, overload by three-point bending resulting in dramatic differences between the tensile and compressive sides of the fracture surface, fiber/matrix debonding, fiber buckling, interfacial melting due to arcing, and cracking due to the combination of improperly applied metallized layer and high voltage.

Distribution of Content

The totals for each type of illustration featured in this Atlas are listed in Table 1. As expected, SEM fractographs constitute the highest percentage:

Type of illustration	Approximate percentage
SEM fractographs	50
Light fractographs	23
Macrographs/micrographs	12
TEM fractographs	8
Photographs	7

Table 2 summarizes the number of fractures from the various causes represented in the Atlas for metallic materials, both as fractured parts or fractured test specimens. The categorization of the fracture mode follows the same convention used in the article "Modes of Fracture" in this Volume, which should be referred to for detailed explanations. Fatigue fractures represent the most common cause, amounting to more than one-third of the total.

The causes of fractures or failures of engineered materials are listed in Table 3. Additional information is available in the articles "Failure Analysis of Continuous Fiber Reinforced Composites," "Failure Analysis of Ceramics," "Failure Analysis of Polymers," and "Failure Analysis of Integrated Circuits" in Volume 11 of the 9th Edition of *Metals Handbook* (*Failure Analysis and Prevention*).

ACKNOWLEDGMENTS

Names and affiliations of contributors of illustrations in the Atlas of Fractographs are listed parenthetically at the end of each caption, where applicable. To these individuals, whose time, efforts, and considerable talents made this compilation possible, ASM and the Handbook staff owe special thanks. Unlisted, but equally deserving of mention, are the numerous individuals who contributed anonymously through their participation in the preparation and interpretation of fractographs as well as the contributors of material for the 8th Edition Handbook whose illustrations warranted inclusion in this Volume. Special acknowledgment and sincere thanks are also due to Editorial Assistant Robert T. Kiepura for his persistence during the many months spent collecting fractographs and to Consulting Editor Donald F. Baxter for his exacting approach to writing the hundreds of captions new to this Volume.

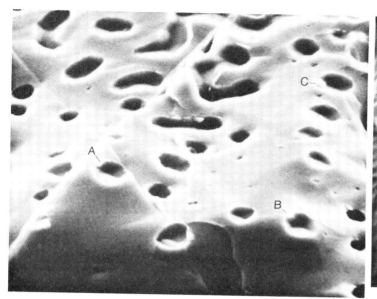

Fig. 1 Grain-boundary cavitation in iron. This is the mechanism by which metals typically fail when subjected to elevated temperatures and low strain rates. Composition, in parts per million: 70 C, 60 S, 54 O, 11 N, 40 P. Rod, 13 mm (0.5 in.) in diameter, was made by vacuum induction melting, chill casting, and swaging. Heat treatment: recrystallize for 30 min at 850 °C (1560 °F), austenitize for 1 h at 1100 °C (2010 °F), air cool. Sample was tensile tested at 700 °C (1290 °F) and an initial strain rate of 1.1×10^{-6}/s. Test was interrupted after 12 h, and the sample was then broken by impact at -100 °C (-150 °F). Fracture surface reveals cavitated grain boundaries. The intergranular cavities nucleated on second-phase (iron sulfide) particles, examples of which are shown at A, B, and C. SEM, 4500× (E.P. George and D.P. Pope, University of Pennsylvania)

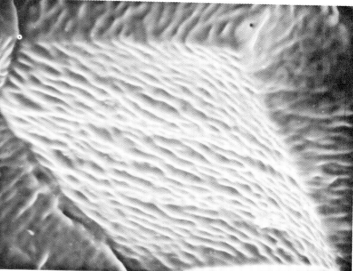

Fig. 2 Slip lines in iron. Composition, in parts per million: 160 C, 40 S, 13 O, 6 N, 30 P. Rod, 13 mm (0.5 in.) in diameter, was made by vacuum induction melting, chill casting, and swaging. Heat treatment: recrystallize for 30 min at 850 °C (1560 °F), austenitize for 1 h at 1100 °C (2010 °F), air cool. Sample was tensile tested to failure at 700 °C (1290 °F) and an initial strain rate of 4.4×10^{-5}/s. Slip lines, smoothened by diffusional flow, are visible on grain boundaries of the elevated-temperature fracture surface. Failure was intergranular and resulted from the nucleation, growth, and eventual coalescence of grain-boundary cavities (see Fig. 1). In this case, however, cavity outlines were masked by the slip steps created on grain boundaries due to severe plastic deformation within the grains. The wavy nature of the slip lines is a characteristic of body-centered cubic iron. SEM, 2180× (E.P. George and D.P. Pope, University of Pennsylvania)

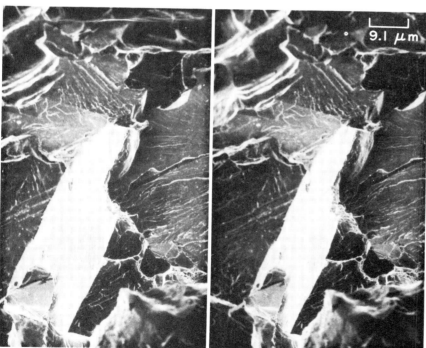

Fig. 3 Stereo pair of scanning electron microscope views of the fracture surface of a Charpy impact test bar of high-purity iron. The specimen was broken after being cooled to equilibrium in liquid nitrogen (-196 °C, or -321 °F). Flat cleavage has taken place on a variety of sharply divergent crystal planes. There are fine river patterns evident on nearly all of the facets. The cleavage steps that delineate the "river" systems are, however, very minute in height, providing only very small departures from a single crystallographic plane of crack growth. Characteristic of iron when fractured at low temperature is the formation of "tongues," one of which is located at the lower end of the very bright facet. There even appears to be a river pattern on the left-hand portion of the "tongue." A second, smaller tongue projects from the facet shown obliquely at top left. 1100×

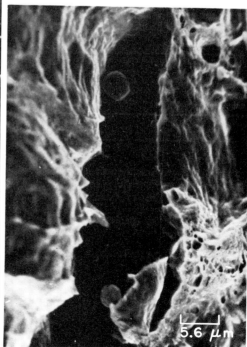

Fig. 4 Surface of a room-temperature tensile-test fracture in a specimen taken from an ingot prepared by adding Fe_2O_3 to pure iron in a vacuum melt equilibrated at 1550 °C (2820 °F) in a silica crucible. The ingot contained 0.07% O in the form of FeO. The fracture surface contains dimples that initiated at globular FeO inclusions averaging 5.2 μm in diameter. SEM, 1800×

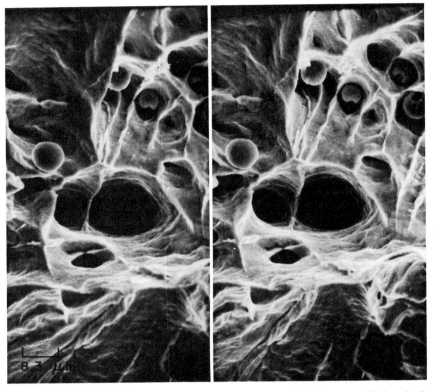

Fig. 5 Stereo pair of scanning electron microscope fractographs of the surface of a tensile-test fracture obtained at room temperature. The alloy was a low-carbon iron to which an appreciable amount of Fe_2O_3 had been added to form an aggregate of FeO inclusions. The dimples that are characteristic of ductile rupture are evident here, and many of these dimples contain one or more globular oxide inclusions that are readily apparent. There appear to be two sizes of these oxide inclusions— some being about 6 μm in diameter and others about 3 μm in diameter. Unlike many inclusions displayed in other fractographs, which have relatively smooth, unbroken contours several of the particles shown here possess sizable surface defects. Some of these defects may be exposed internal shrinkage cavities. The surfaces of the dimples show contours that vaguely resemble fatigue-striation marks. The differences in topographic contours of the dimples displayed in this stereo pair of fractographs can be appreciated by viewing the fractographs stereographically, which provides a three-dimensional effect. It then becomes apparent that the dimples are chimneylike cavities with nearly vertical walls in many instances and with bottoms at great depth that appear black and without detail. The FeO inclusions appear to cling to the cavity walls, many at a point part way to the bottom of the "chimney." Most of the separating walls between adjacent chimneys are extremely thin, which makes it surprising that these walls did not rupture at a point closer to the bottom of the chimney. 1200×

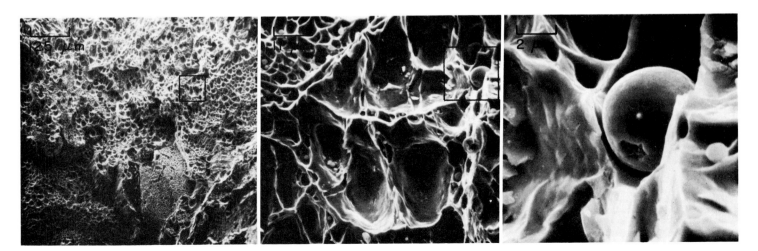

Fig. 6, 7, 8 Sequence of SEM fractographs, at increasing magnifications (80×, 950×, and 5000×, respectively), that show a fracture in an iron alloy containing 0.14% S and 0.04% O. The fracture was obtained by bending at room temperature. Several spheroidal oxide inclusions are visible, most of them having diameters in the range of 1 to 3 μm. The rectangle in Fig. 6 (left) indicates the area that is shown at higher magnification in Fig. 7 (center), and the rectangle in Fig. 7 indicates the area that is shown at still higher magnification in Fig. 8 (right). The 6-μm-diam oxysulfide particle in Fig. 8 shows a shrinkage cavity plus a white spot from an electron beam impingement in fluorescent x-ray analysis. It is quite evident that, during the process of microvoid coalescence, the iron matrix has become detached from the globular inclusions at the metal-to-oxide and metal-to-sulfide interfaces, leaving these inclusions unaffected by the applied stresses and severe deformation taking place around them.

Pure Irons / 221

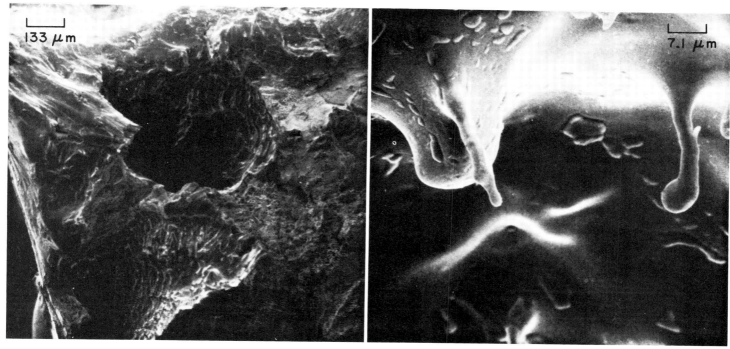

Fig. 9 Fracture produced at room temperature by bending iron containing 0.02% C, 0.14% S, and 0.04% O cast in a 7 × 7 × 20 cm (2.75 × 2.75 × 8 in.) ingot mold. A carbon-FeO reaction caused blowholes such as shown at top. Note the fine lamellar structure at bottom. See also Fig. 10. SEM, 75×

Fig. 10 Higher-magnification view of the blowhole shown at the top in Fig. 9, showing the interior of the blowhole that resulted from the carbon-FeO reaction. The pendants are droplets of a liquid oxysulfide that spread over the surface of the blowhole during freezing of the ingot. SEM, 1400×

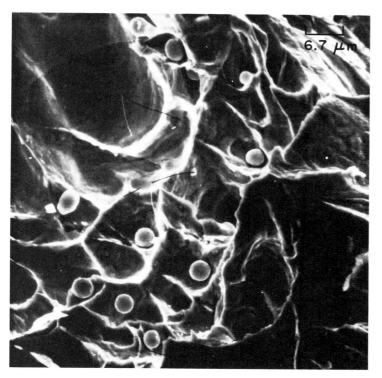
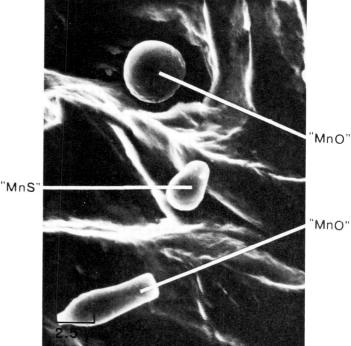

Fig. 11 Fracture by room-temperature bending in casting of similar composition to that of casting in Fig. 9, but containing 1.1% Mn. The numerous inclusions contained within the dimples are particles of manganese oxide and manganese sulfide trapped between growing dendrite branches. SEM, 1500×

Fig. 12 Higher-magnification view of the same fractured specimen shown in Fig. 11. X-ray fluorescent analysis of the inclusions indicates that some of them are Mn(Fe)O and that others are Mn(Fe)S, but the exact amount of contained iron was not determined. SEM, 4000×

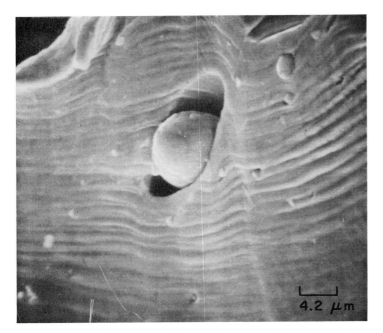

Fig. 13 Low-carbon iron containing a high percentage of oxygen, fractured in fatigue at room temperature. A large oxide inclusion has been nearly completely disengaged from its original pocket. Fatigue striations detour around, or extend into, the pocket. Crack propagation was from bottom to top. SEM, 2400×

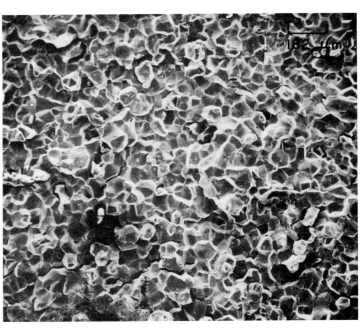

Fig. 14 Intergranular fracture that was generated in a specimen of oxygen-embrittled Armco iron by a Charpy impact test at room temperature. The grain facets appear sharp and clean. Note the secondary cracks, which follow grain boundaries. See also Fig. 15 and 16 for views of other regions of this fracture. SEM, 55×

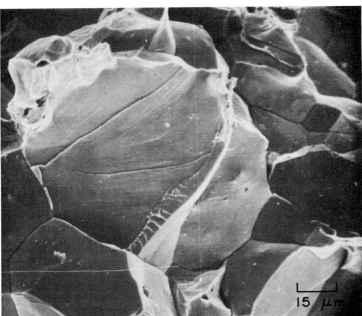

Fig. 15 View of another region of the surface of the impact fracture shown in Fig. 14, showing facets that resulted from a combination of intergranular rupture and transcrystalline cleavage. Note the array of small river patterns at the bottom edge of the large facet at center. See also Fig. 16. SEM, 655×

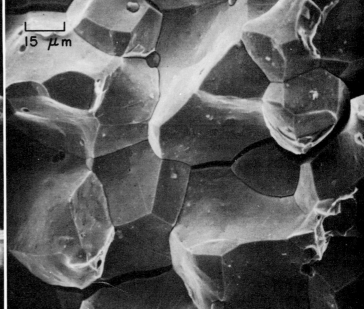

Fig. 16 View of a third region of the surface of the impact fracture shown in Fig. 14 and 15. Note the almost perfect grain-boundary surfaces and the sharp edges and points at which the separated-grain facets meet. The secondary cracks are equally clean separations. SEM, 670×

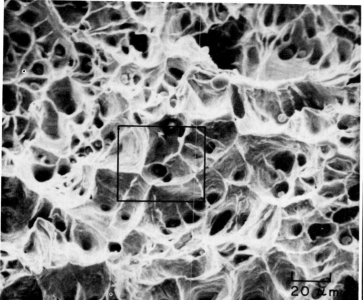

Fig. 17 Surface of tensile-test fracture in specimen of low-carbon, high-oxygen iron that was broken at room temperature. Many of the equiaxed dimples contain spheroidal particles of FeO. The rectangle marks the area shown at higher magnification in Fig. 18 and 19. SEM, 500×

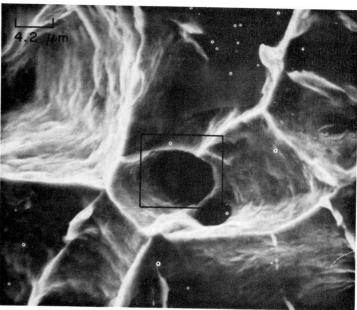

Fig. 18 Enlargement of the area within the rectangle in Fig. 17, showing the surface contours of the dimple cavities of the very ductile fracture. Dimly visible in the central dimple is a globular particle of FeO; the particle is shown more clearly in Fig. 19. SEM, 2400×

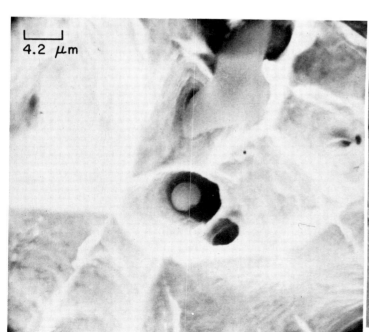

Fig. 19 Same fracture-surface area as that shown in Fig. 18, but processed with a different exposure to bring out the shape and size of the globular particles of FeO in the central dimple. The small dark spot at the lower right on the FeO particle is a shrinkage area. SEM, 2400×

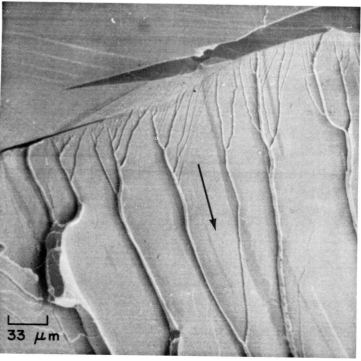

Fig. 20 Fe, 0.01% C, 0.24% Mn, and 0.02% Si, heat treated at 950 °C (1740 °F) ½ h, air cooled. The structure is ferrite. Hardness, 62 HV. The fracture was generated by impact at −196 °C (−321 °F). Cleavage steps beginning at the twin at top form a sharply defined river pattern. Crack propagation was in direction of arrow. SEM, 300×

Fig. 21 Surface of an impact fracture in a notched specimen of wrought iron. The longitudinal stringers of slag in the material are parallel to the direction of fracture, which gives the surface this typically "woody" appearance. Compare with Fig. 22. 6×

Fig. 22 A companion notch-impact fracture to the one shown in Fig. 21, but here the longitudinal stringers of slag are normal to the fracture surface. Individual particles of slag are not readily visible at this magnification, but they cause the speckled appearance. 6×

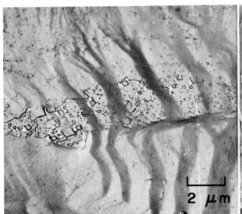

Fig. 23 External surface of a specimen of Armco iron that had been etch-pitted and then subjected to tension during one cycle of a 2.5° bend test. Note the slip steps. See also Fig. 24. TEM p-c replica, 5000×

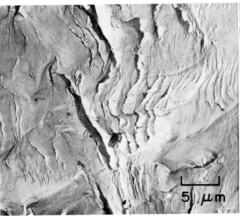

Fig. 24 Same specimen as in Fig. 23 after 5.5 cycles of 2.5° bending. At center is the same area as in Fig. 23 (reversed left to right in printing), showing slip-band cracks that have grown at slip steps. TEM p-c replica, 2000×

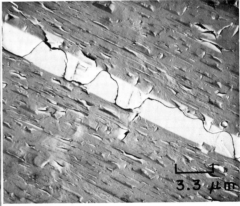

Fig. 25 Cleavage fracture in Armco iron broken at dry-ice temperature (−78.5 °C, or −109.3 °F). The light band shows where cleavage followed a twin-matrix interface. The black meandering line is a shear step through the thickness of the twin. TEM p-c replica, 3000×

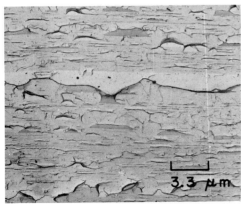

Fig. 26 Cleavage fracture in Armco iron broken at dry-ice temperature (−78.5 °C, or −109.3 °F), showing facets of which most have the same orientation. Facets that depart from the general orientation appear lighter or darker than the majority. TEM direct carbon replica, 3000×

Fig. 27 Cleavage fracture in Armco iron broken at −45 °C (−49 °F). Instead of cleavage steps, tear ridges (occasionally forming river patterns) were produced here by microscopic plastic flow. TEM p-c replica, 2000×

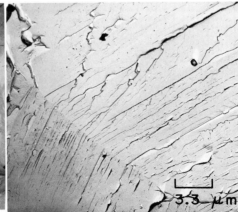

Fig. 28 Cleavage fracture in Armco iron broken at −196 °C (−321 °F), showing river patterns, tongues, and (from bottom right to top left) a grain boundary. TEM p-c replica, 3000×

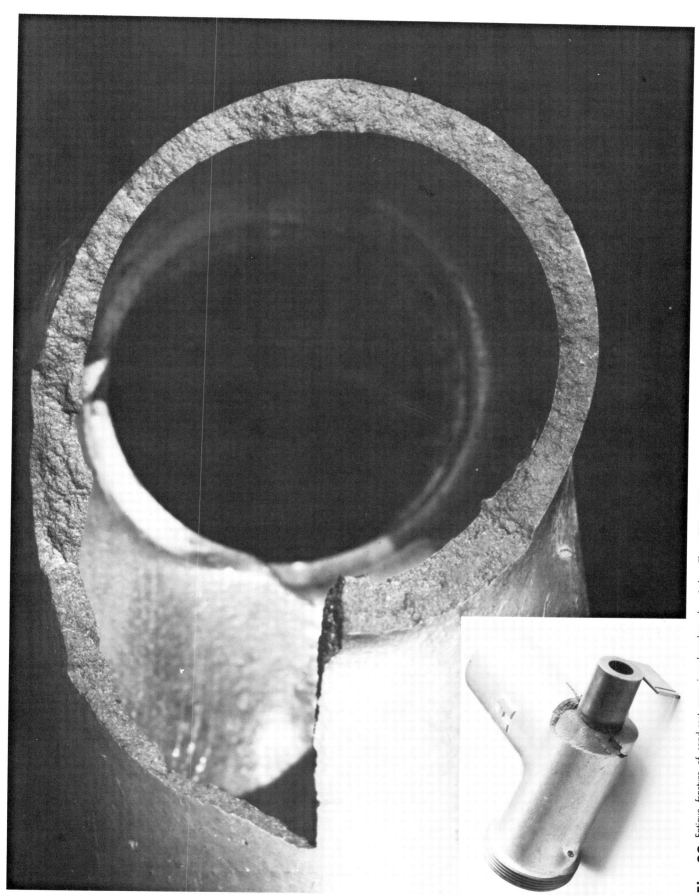

Fig. 29 Fatigue fracture of sand-cast gray iron bread-crumb grinder. The ASTM A159 part was machined and hot dip galvanized after casting. Also, a lockpin hole (not visible) was drilled through the casting in a region containing type C, size 3 graphite flakes. Cause of failure was traced to a crack that initiated at one of these flakes. Inset: broken grinder in as-received condition. 1.5× (C.-A. Baer, California Polytechnic State University)

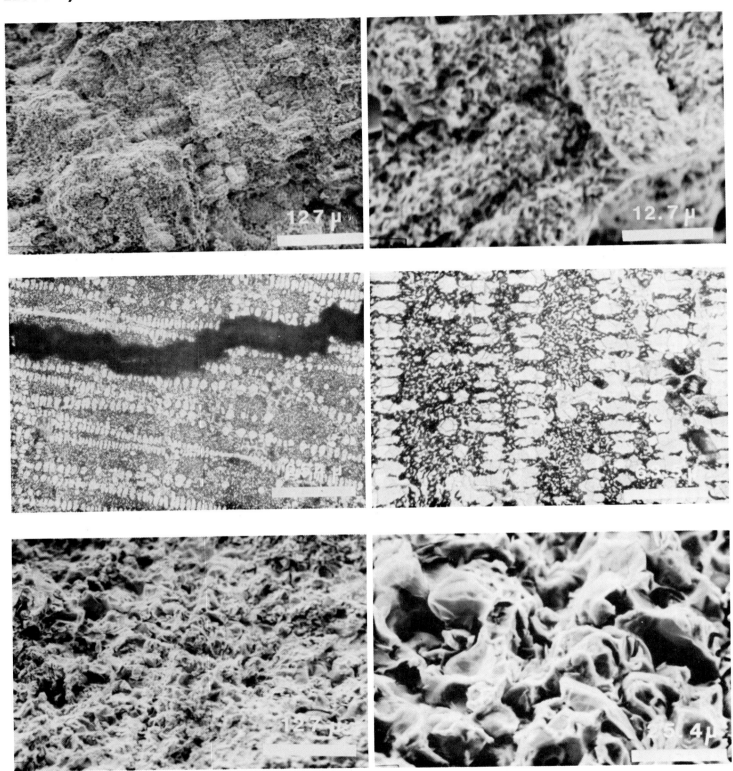

Fig. 30 to 35 Effect of graphite morphology on fracture of permanent mold cast, hypereutectic gray iron (4.5 to 4.8% carbon equivalent). Permanent mold castings are typically small—15 kg (30 lb) max—and thus their as-cast mechanical properties are particularly sensitive to solidification structure and size, shape, and distribution of graphite. In this study, cylindrical samples were cast in a cold, 230 °C (450 °F) mold and allowed to air cool. Typical microstructure: undercooled (type D) graphite in a ferrite matrix. Initially cracked (nil ductility) and crack-free (normal ductility) samples were compared. Fig. 30 and 31 (top row): Fracture surface of nil-ductility sample that cracked during casting or machining. Fracture initiated at and propagated from graphite-ferrite interfaces. SEM, 190× and 1900×. Fig. 32 and 33 (center row): Solidification structure and graphite morphology of permanent mold cast, gray iron sample with nil ductility. The crack propagated along the preferred orientation of solidification. Dendrite spacing was a narrow 50 μm. Graphite was fine tipped, roughly cylindrical, and isolated from ferrite cells—a morphology that apparently has an adverse effect on ductility and is also a major contributor to graphite-ferrite interface cracking. 2% nital, 82× and 330×. Fig. 34 and 35 (bottom row): Fracture surface of sample having normal ductility. Fracture was artificially generated by impact. Note how material resisted graphite-ferrite interface cracking. Microstructure (not shown) was more isotropic than that of the nil-ductility casting, with a wider dendrite spacing (85 μm). Graphite was medium sized, interconnected, and penetrated ferrite cells. SEM, 200× and 1000× (D.C. Wei, Kelsey-Hayes Company)

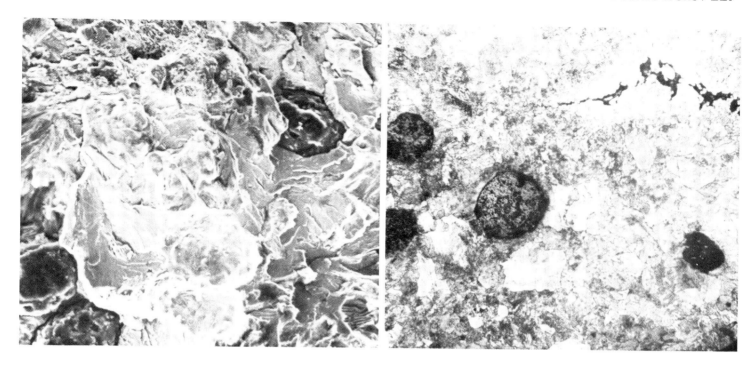

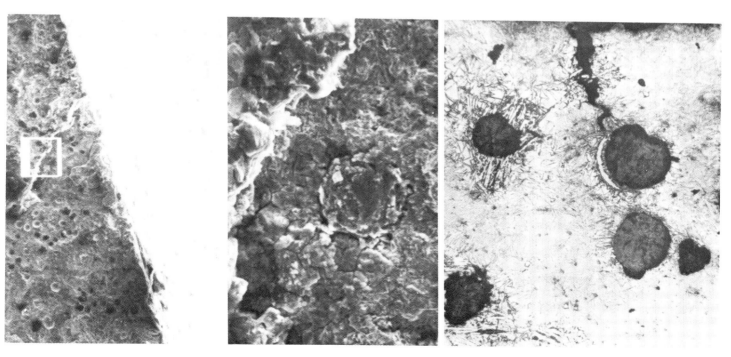

Fig. 36 to 40 Brittle cleavage fracture of ductile iron spur gear (ASTM A536, grade 100-70-03) due to improper heat treatment. Tensile strength was 544 MPa (78.9 ksi), much less than the 690 MPa (100 ksi) required by the specification. Elongation was nil; the specification called for 3% min. The induction-hardened case on the teeth was shallower and harder than specified (50+ HRC versus 46 HRC), and the martensitic microstructure had not been tempered as specified. Direct cause of fracture was the presence of inverse chill (carbides in thick sections of the casting) associated with microporosity. This carbidic material, which formed at thermal centers due to segregation of carbide-forming elements, increased hardness, decreased tensile strength, and promoted brittle fracture. Proper heat treatments would have corrected the deficiencies in case structure and properties and would also have prevented the occurrence of inverse chill. Fig. 36 (top left): Fracture surface at core of gear directly below origin. Note transcrystalline (cleavage) mode of fracture. SEM, 200×. Fig. 37 (top right): Photomicrograph of core. Matrix is 100% pearlite. Note presence of inverse chill (carbides) and associated porosity. 2% nital, 200×. Fig 38 (bottom left): Fracture face at casting surface and in the induction-hardened case. SEM, 20×. Fig. 39 (bottom center): Boxed area in Fig. 38. Note cleavage fracture appearance and nodule surrounded by cracked material believed to be carbidic. SEM, 200×. Fig. 40 (bottom right): Microstructure directly below fracture surface and in case. There is a carbide-appearing envelope around each nodule of temper carbon that formed as the inverse chill decomposed. The two nodules at upper left also have retained austenite around them. Note the secondary crack extending between the casting surface and the carbide envelope of a temper nodule. 2% nital, 200×. (G.M. Goodrich, Taussig Associates Inc.)

228 / Ductile Irons

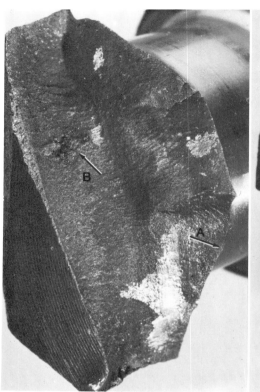

Fig. 41 Surface of a fatigue-test fracture in an experimental crankshaft of induction-hardened 80-60-03 ductile iron with a hardness of 197 to 225 HB. Fatigue-crack origin is at arrow A. Porosity at arrow B was unrelated to fracture initiation. 2.5×

Fig. 42 Surface of a fatigue-test fracture in an experimental crankshaft of ductile-iron with a hardness of 241 to 255 HB. Note the multiple fatigue-crack origins at the journal edge (at right). Fatigue beach marks are evident, which is unusual in cast iron. Actual size

Fig. 43 Surface of a fatigue fracture in an experimental crankshaft broken in a fatigue test. The material is ductile iron with a hardness of 241 HB. The origin of the fatigue crack is at the edge of the journal, at arrow. Actual size

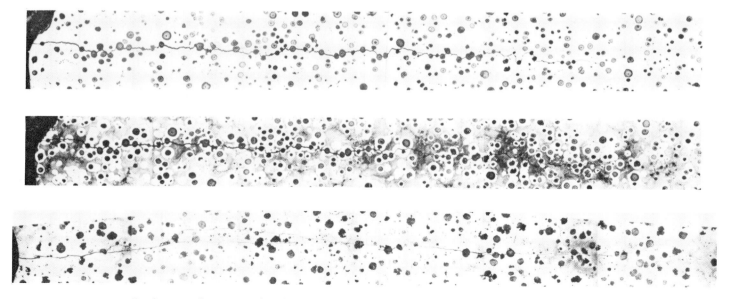

Fig. 44, 45, 46 How fatigue cracks propagate through ductile irons. Fig. 44 and 45 (top and center): In an as-cast, commercial pearlitic ductile iron, the crack changes direction from grain to grain, usually following nodule-matrix interfaces. Fig. 45 etched in 2% nital, both at 80×. Fig. 46 (bottom): Crack propagation through the tempered martensite of a heat-treated ductile iron is more matrix controlled than in either a pearlitic (Fig. 44 and 45) or ferritic microstructure. 100× (F.J. Worzala, University of Wisconsin)

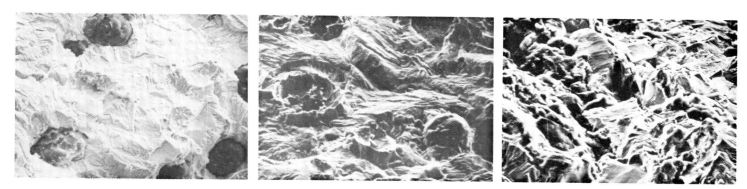

Fig. 47, 48, 49 Fatigue fracture surfaces of pearlitic and ferritic ductile irons. Compositions of pearlitic irons: 3.63 to 3.80% C, 0.34% Mn, 2.02 to 2.66% Si. Compositions of ferritic iron: 3.75 to 2.82% C, 0.34% Mn, 2.30 to 2.66% Si. Striations noted in all cases. Fig. 47 and 48 (left and center): Mixture of striations and fractured pearlite lamellae on fracture surfaces of commercial pearlitic ductile irons. Striations are the fine steplike features, not the macroscopic waviness or undulations. SEM, 207× and 198×. Fig. 49 (right): A high load fatigue fracture surface of a ferritic ductile iron. SEM, 375× (F.J. Worzala, University of Wisconsin)

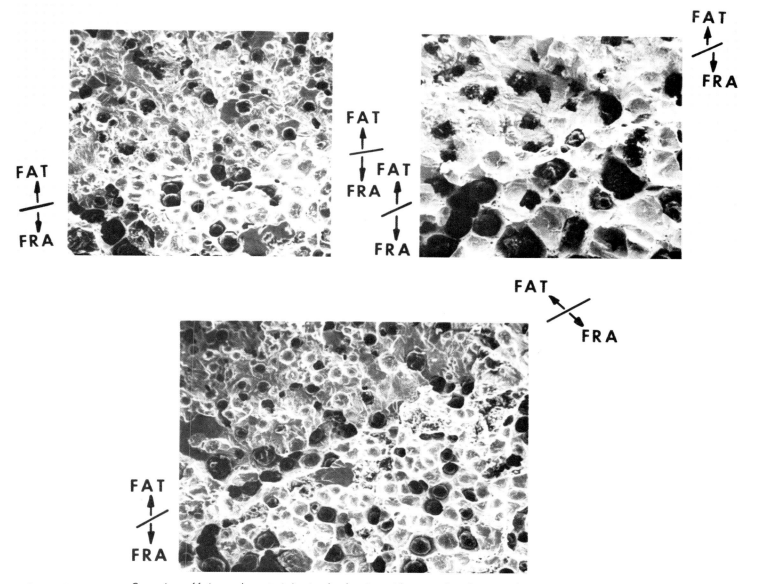

Fig. 50, 51, 52 Comparisons of fatigue and monotonic (tension, bending, impact) fracture surfaces for various ferritic ductile iron microstructures. Fatigue fractures are characterized by striations (see Fig. 47 to 49), by relatively little opening up or stretching of matrix material around nodules, and by nodules that appear to have had large pieces of graphite broken off of them. Mode of crack propagation in monotonic fractures is very ductile with considerable stretching of nodule-bearing cavities. The demarcation line between fatigue (FAT) and monotonic (FRA) fracture is noted in each of the three fractographs. Fig. 50 (top left): Commercial ferritic iron (2.30 to 2.66% Si) tested at room temperature. SEM, 90×. Fig. 51 (top right): Same material as in Fig. 50, but during tensile portion of test at −40 °C (−40 °F). SEM, 90×. Fig. 52 (bottom): High-silicon ferritic ductile iron (3.5% Si). SEM, 90× (F.J. Worzala, University of Wisconsin)

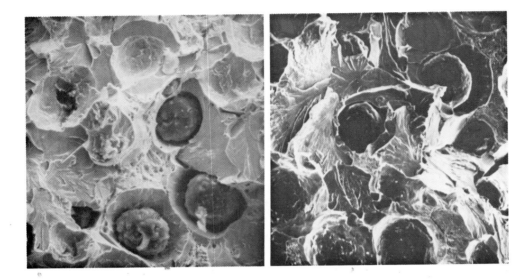

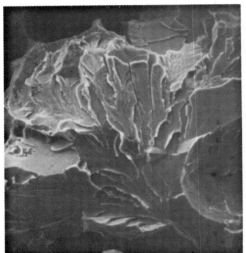
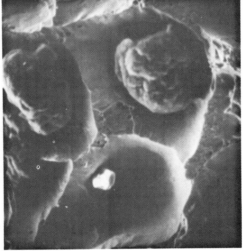

Fig. 53 to 56 Fracture modes in slow monotonic loading (tension or bending) of pearlitic ductile irons. These cast irons consist of a pearlitic matrix with ferritic rings of varying thickness surrounding graphite nodules. They exhibit predominantly brittle fracture with river patterns in pearlitic areas. However, ductile fracture can occur in areas where several graphite nodules are closely spaced. Fig. 53 (top left): Typical slow monotonic fracture surface of pearlitic ductile iron. Note river patterns and both constrained and opened ferritic rings (areas of brittle and ductile fracture, respectively). SEM, 230×. Fig. 54 (top right): Fracture surface with isolated graphite nodules. Brittle fracture results because the ferritic rings are under severe mechanical constraint due to the surrounding stronger and nondeformable matrix. SEM, 260×. Fig. 55 (bottom left): Fracture surface at cluster of graphite nodules. Note the considerable deformation of the unrestrained ferrite. Fracture occurs here by ductile tearing and microvoid coalescence. SEM, 385×. Fig. 56 (bottom right): River patterns indicative of brittle failure on fracture surface of a pearlitic ductile iron. SEM, 385× (F.J. Worzala, University of Wisconsin)

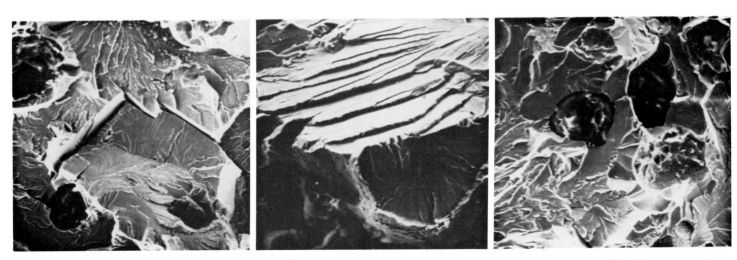

Fig. 57, 58, 59 Slow monotonic fracture of pearlitic ductile iron at temperatures below ambient is 100% brittle. There is no evidence of ductile fracture at graphite nodule clusters (compare with Fig. 55). Fig. 57 (left): Fracture surface of pearlitic ductile iron tested at −40 °C (−40 °F). SEM, 249×. Fig. 58 (center): Higher-magnification view of matrix of material in Fig. 57. SEM, 603×. Fig. 59 (right): Fracture surface of pearlitic ductile iron tested at −75 °C (−100 °F). SEM, 285× (F.J. Worzala, University of Wisconsin)

Fig. 60, 61 Fracture modes in slow monotonic loading (tension or bending) of ferritic ductile irons. The only modes operative at ambient temperature are ductile tearing and microvoid coalescence. The propagating crack will cut through any "lumps" of nonspheroidal graphite that may be present. Fig. 60 (left): Fracture surface of ferritic ductile iron subjected to slow monotonic loading is characterized by ductile tearing and microvoid coalescence. Internodular ferrite undergoes a substantial amount of deformation. SEM, 300×. Fig. 61 (right): Fracture surface of ferritic ductile iron with nonspheroidal graphite shows how crack cuts through the irregularly shaped graphite lumps while leaving graphite nodules intact. SEM, 534× (F.J. Worzala, University of Wisconsin)

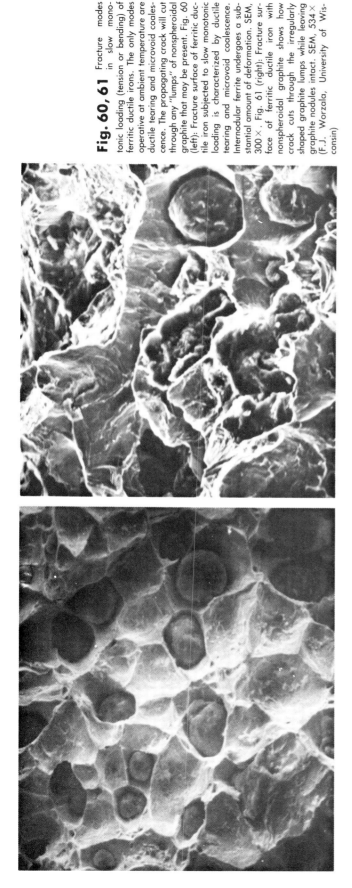

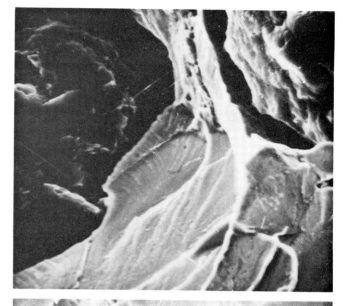

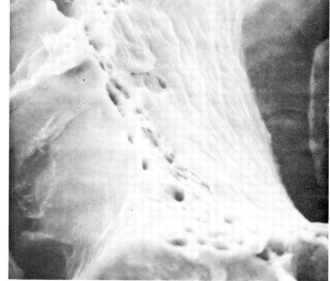

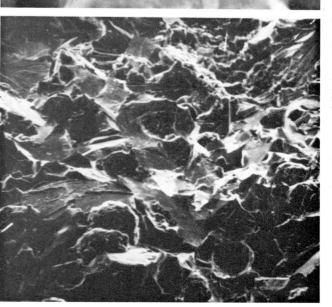

Fig. 62, 63, 64 As test temperature falls below ambient, the slow monotonic fracture mode for ferritic ductile irons undergoes a gradual ductile-to-brittle transition. Fig. 62 (left): Fracture surface of ferritic ductile iron tested at −40 °C (−40 °F). Note river patterns and plateaus characteristic of brittle fractures. SEM, 150×. Fig. 63 and 64 (center and right): High-magnification views compare internodular areas of ferritic ductile irons tested at room temperature and at −40 °C (−40 °F), respectively. SEM, ∼1900× (F.J. Worzala, University of Wisconsin)

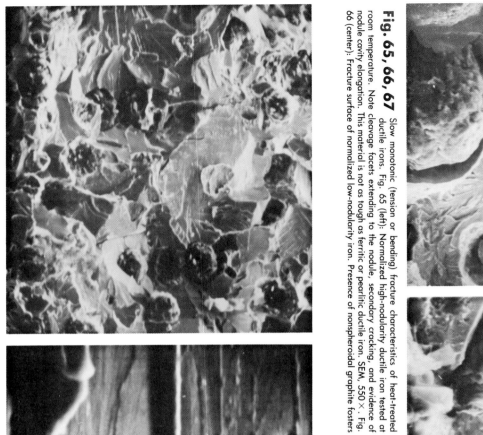
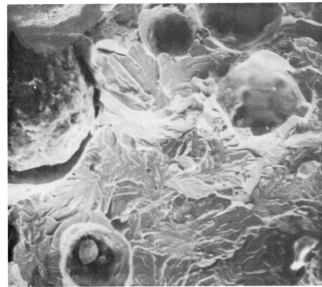
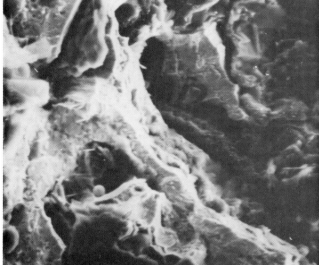
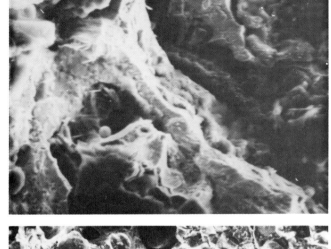
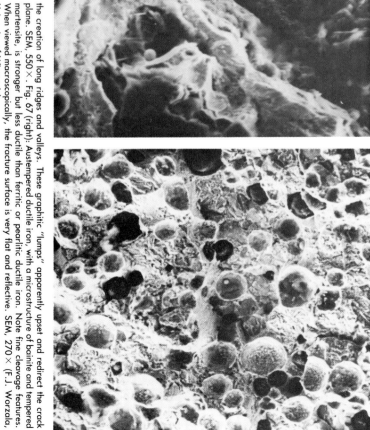

Fig. 65, 66, 67 Slow monotonic (tension or bending) fracture characteristics of heat-treated ductile irons. Fig. 65 (left): Normalized high-nodularity ductile iron tested at room temperature. Note cleavage facets extending to the nodule, secondary cracking, and evidence of nodule cavity elongation. This material is not as tough as ferritic or pearlitic ductile iron. SEM, 550×. Fig. 66 (center): Fracture surface of normalized low-nodularity iron. Presence of nonspheroidal graphite fosters the creation of long ridges and valleys. These graphitic "lumps" apparently upset and redirect the crack plane. SEM, 550×. Fig. 67 (right): Austempered ductile iron, with a microstructure of bainite and tempered martensite, is stronger but less ductile than ferritic or pearlitic ductile iron. When viewed macroscopically, the fracture surface is very flat and reflective. SEM, 270×. (F.J. Worzala, University of Wisconsin)

Fig. 68, 69 Fracture modes in impact (fast monotonic) loading of a ferritic ductile iron tested at −45 °C (−50 °F). Fig. 68 (left): Fracture surface reveals little deformation with no stretching around graphite nodules. Several nodules appear to have been broken. Also, the matrix failed in an entirely brittle mode, as evidenced by the numerous river patterns and cleavage and quasi-cleavage steps. Compare with the slow monotonic (tension or bending) fracture in Fig. 62. Although they generally resemble each other, the fast fracture exhibits no graphite cavity stretching or microvoids, and nodules remain intact in the slow fracture. SEM, 220×. Fig. 69 (right): Internodular area of impact-loaded specimen in Fig. 68. Fracture in this region was entirely brittle. Microtongues lie on the iron {112} plane and intersect the cleavage plane along [110] directions. SEM, 5600×. (F.J. Worzala, University of Wisconsin)

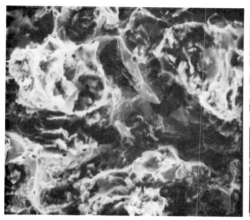 

Fig. 70, 71 Presence of nonspheroidal graphite affects the impact fracture of ductile irons by producing a rough fracture surface with continuous change in crack path direction even at low test temperatures. Fig. 70 (left): Fracture surface of ferritic ductile iron tested at −18 °C (0 °F). SEM, 240×. Fig 71 (right): Fracture surface of pearlitic ductile iron impact tested at −75 °C (−100 °F). SEM, 100× (F.J. Worzala, University of Wisconsin)

 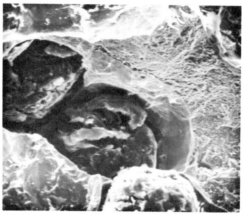

Fig. 72, 73 Impact fracture of ductile irons at test temperatures above the nil-ductility transition (NDT) temperature is accompanied by gross plastic deformation (ductile behavior). Fig. 72 (left): Fracture surface of low-nodularity ferritic ductile iron tested at 65 °C (150 °F). SEM, 80×. Fig. 73 (right): Pearlitic ductile iron tested at 205 °C (400 °F). Note microvoid coalescence and ductile tearing. SEM, ∼400× (F.J. Worzala, University of Wisconsin)

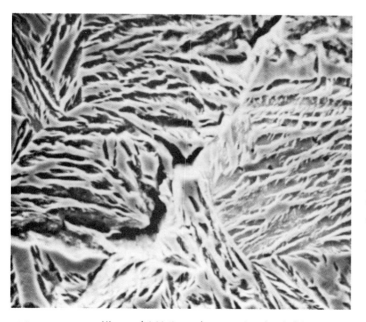 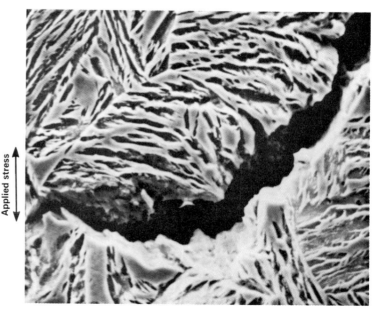

Fig. 74, 75 Microcrack initiation and propagation ahead of the primary crack front in an austempered ductile iron (3.6% C, 2.3% Si, 0.9% Mn). Heat treatment: austenitize for 2 h at 870 °C (1600 °F), austemper for 40 min at 400 °C (750 °F). Samples were polished, etched in 2% nital, and then strained (arrow indicates direction of applied stress). Austempered matrix (shown here) consists of ferrite (dark) and stabilized austenite (light). Microcracks initiate and propagate in the low-strength ferrite, avoiding regions of strong and tough austenite. SEM, 2500× (R.C. Voigt and L.M. Eldoky, University of Kansas)

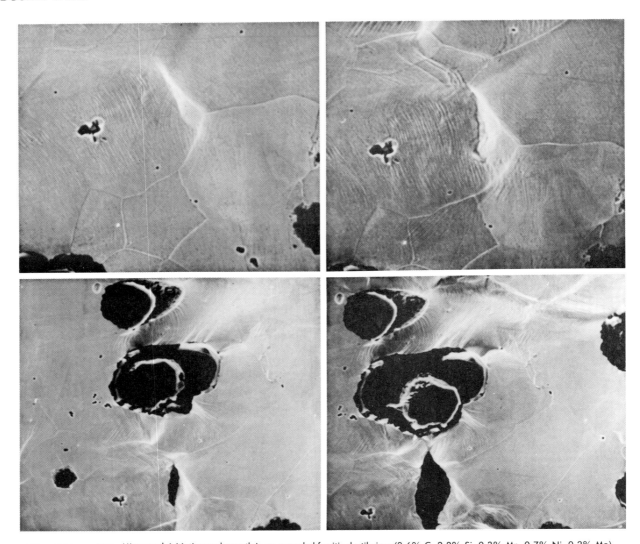

Fig. 76 to 79 Microcrack initiation and growth in an annealed ferritic ductile iron (3.6% C, 2.2% Si, 0.3% Mn, 0.7% Ni, 0.2% Mo). Sample polished, etched in 2% nital, and then plastically deformed by amount indicated. Tensile load applied horizontally. Fig. 76 and 77 (top row): After 610 and 670 strain units, respectively. SEM, both at 1200×. Fig. 78 and 79 (bottom row): After 700 and 760 strain units, respectively. SEM, both at 600× (R.C. Voigt and L.M. Eldoky, University of Kansas)

Fig. 80, 81 Effects of plastic deformation on internodule bridges in ferritic ductile iron. Material and sample preparation same as in Fig. 76 to 79. Fig. 80 (left): Surface plastic deformation and microcracking ahead of the primary crack front. Note strain concentration at narrow internodule bridge. Tensile load applied horizontally. SEM, 950×. Fig. 81 (right): Severe plastic deformation associated with tearing of internodule bridge. Note separation of secondary graphite ring from both the primary graphite and ferrite matrix. SEM, 900× (R.C. Voigt and L.M. Eldoky, University of Kansas)

Fig. 82 Microvoids and microtearing common to fractures of wide internodule bridges in ferritic ductile iron. Material same as in Fig. 76 to 79. Surface perpendicular to fracture surface and polished and etched in 2% nital. SEM, 300× (R.C. Voigt and L.M. Eldoky, University of Kansas)

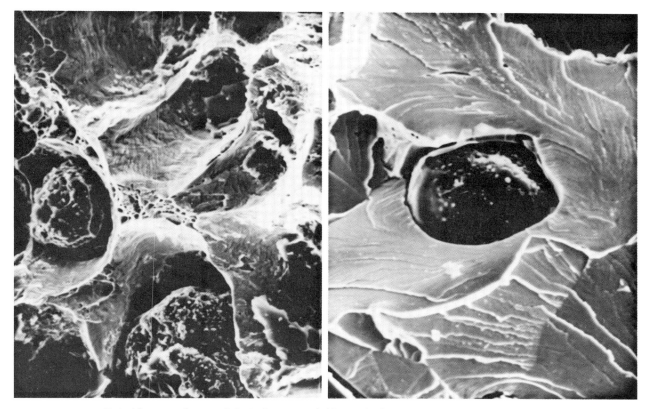

Fig. 83, 84 Typical fracture surface morphologies for an annealed ferritic ductile iron. Composition: 3.6% C, 2.2% Si, 0.3% Mn, 0.7% Ni, 0.2% Mo (same as in Fig. 76 to 82). Fig. 83 (left): Dimpled rupture (ductile fracture) at room temperature. SEM, 800×. Fig. 84 (right): Quasi-cleavage (brittle fracture) at low temperature. SEM, 1600× (R.C. Voigt and L.M. Eldoky, University of Kansas)

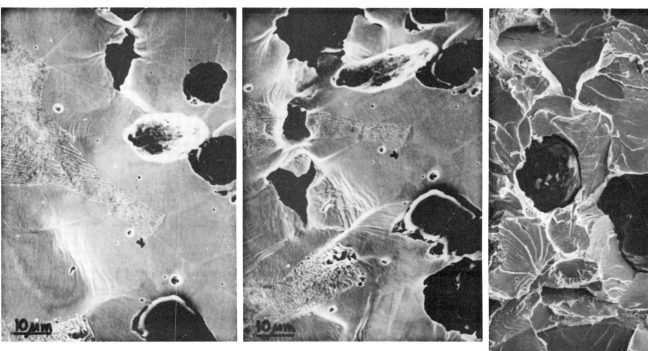

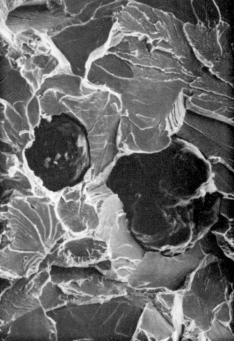

Fig. 85, 86 Microcracking in an as-cast ferritic-pearlitic ductile iron (3.6% C, 2.2% Si, 0.3% Mn, 0.7% Ni, 0.2% Mo). SEM fractographs show localized plastic deformation and microcrack initiation and propagation in the plastic zone ahead of the primary crack. Samples were polished, etched in 2% nital, and then strained. (Tensile stress applied horizontally; direction of crack growth is toward the bottom.) The high-strength pearlite initially resists microcrack initiation. With sufficient plastic strain, however, the pearlite fractures, and the resultant microcracks are then blunted by the surrounding ferrite. 1030× (R.C. Voigt and L.M. Eldoky, University of Kansas)

Fig. 87 Fracture surface of ferritic-pearlitic ductile iron in Fig. 85 and 86. The low-temperature fracture occurred via a brittle, quasi-cleavage mode. SEM, 715× (R.C. Voigt and L.M. Eldoky, University of Kansas)

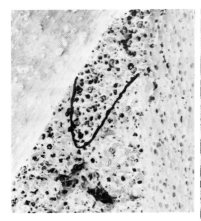

Fig. 88 Ductile-to-brittle transition in an annealed ferritic ductile iron (same alloy as in Fig. 83 and 84). Above demarcation line is region of dimpled rupture (the ductile fracture surface of the test sample after partial fracture at room temperature). Below line is region of quasi-cleavage (the brittle fracture surface of the test sample after final fracture at low temperature). SEM, 30×. (R.C. Voigt and L.M. Eldoky, University of Kansas)

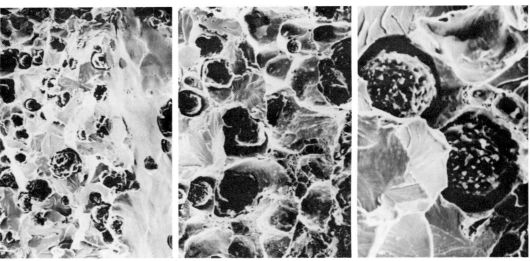

Fig. 89, 90, 91 Fracture transition zone at center of the ferritic ductile iron sample in Fig. 88. Fig. 89 (left): Subsurface fracture transition zone. SEM, 140×. Fig. 90 (center): Higher magnification view of Fig. 89. Note how far microcracks formed at room temperature had propagated by time of final low-temperature fracture. SEM, 275×. Fig. 91 (right): High-magnification view of transition zone. Visible features include nodule decohesion and microplastic deformation and tearing ahead of the primary crack front. SEM, 1100×. (R.C. Voigt and L.M. Eldoky, University of Kansas)

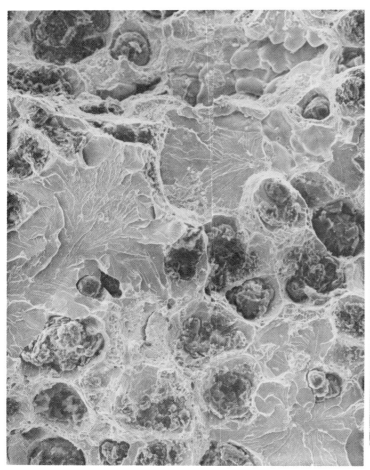

Fig. 92 Fracture surface of a ferritic-pearlitic ductile iron. Note ductile fracture of ferrite in matrix around nodules and cleavage (brittle) fracture of pearlite in matrix. SEM, 50×. (W.L. Bradley, Texas A&M University)

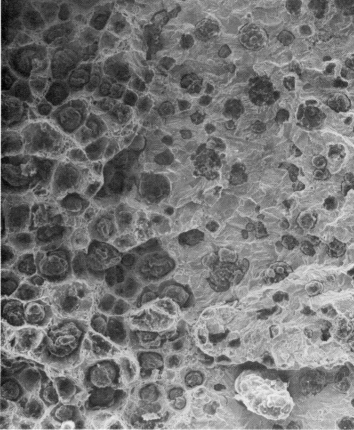

Fig. 93 Regions of fatigue precracking (at right) and crack extension or fracture (at left) in the fracture surface of a ferritic ductile iron compact tension specimen. Note how crack ignores nodules in fatigue and grows almost exclusively through nodule-nucleated voids during ductile fracture. ~80×. (W.L. Bradley, Texas A&M University)

Ductile Irons / 237

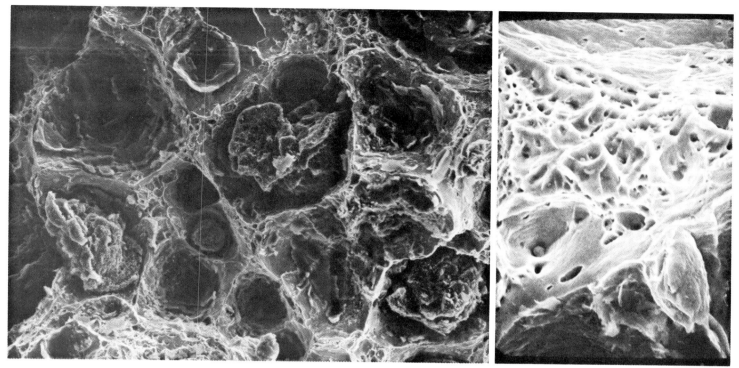

Fig. 94, 95 Ductile fracture in ferritic ductile iron. Fig. 94 (left): SEM, 160×. Fig. 95 (right): Example of the finely dimpled ductile fracture that occurs between the much larger voids nucleated at graphite nodules. SEM, 2000× (W.L. Bradley, Texas A&M University)

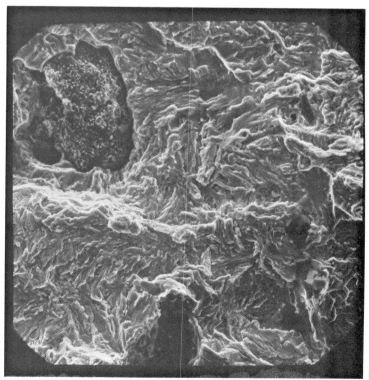

Fig. 96 Fatigue precracked region on the fracture surface of a ferritic ductile iron compact tension specimen. Morphology is typical of fatigue fractures in this material. SEM, 500× (W.L. Bradley, Texas A&M University)

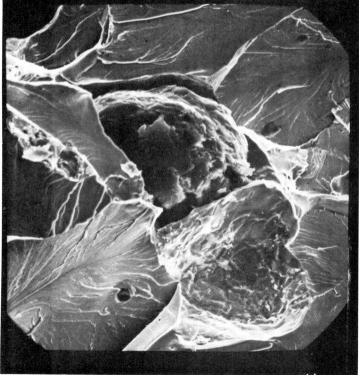

Fig. 97 Brittle cleavage fracture in ferritic ductile iron. SEM, 1000× (W.L. Bradley, Texas A&M University)

238 / Malleable Irons/White Irons

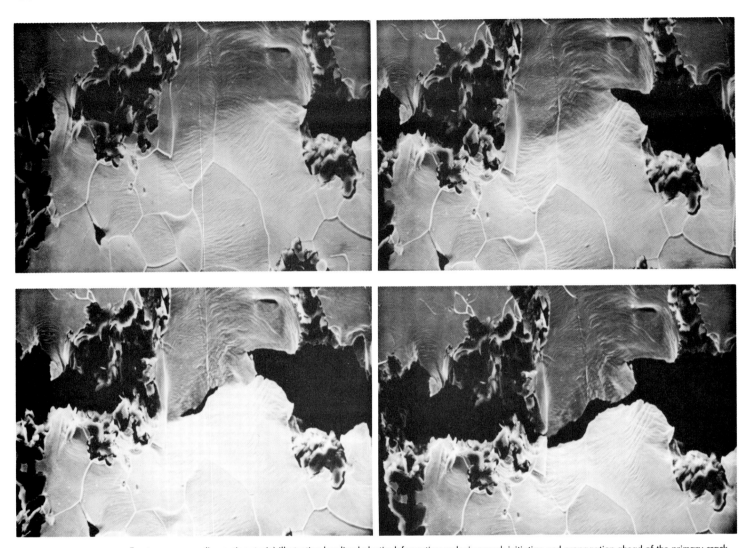

Fig. 98 to 101 Fracture sequence (increasing strain) illustrating localized plastic deformation and microcrack initiation and propagation ahead of the primary crack in a ferritic malleable iron (ASTM A47, grade 32510). Heat treatment was standard (graphitizing anneal, furnace cool). Hardness: 140 HB. Samples were polished, etched in 4% nital, and then strained. Figures 98 to 101 are in order of increasing amounts of strain. Note how cracks propagate through the graphite rather than around it as is the case for ductile iron (see Fig. 76 to 79 and Fig. 85 and 86). SEM, all at 580× (R.C. Voigt and B. Pourlaidian, University of Kansas)

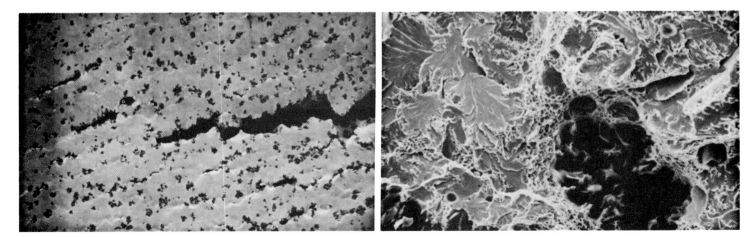

Fig. 102 Crack initiation and propagation in a ferritic malleable iron. Same material and test conditions as in Fig. 98 to 101, but at low magnification. Note that preferential sites for crack initiation and propagation are in interdendritic regions defined by the observed alignment of temper carbon (graphite). SEM, 4% nital, 40× (R.C. Voigt and B. Pourlaidian, University of Kansas)

Fig. 103 Fracture surface of ferritic malleable iron after impact loading at −196 °C (−320 °F). Same material as in Fig. 98 to 102. Regions of dimpled rupture can still be observed even at this low temperature. Primary fracture mode is brittle quasi-cleavage. SEM, 830× (R.C. Voigt and B. Pourlaidian, University of Kansas)

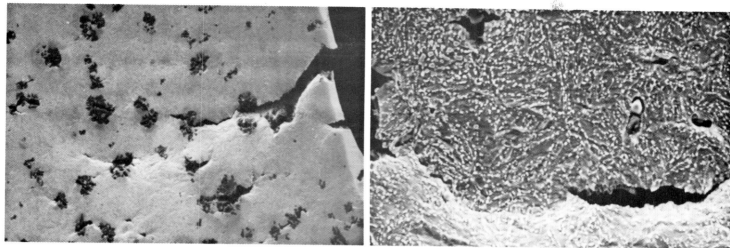

Fig. 104, 105 Microcracking in a pearlitic malleable iron (ASTM A220, grade 50005). Heat treatment: graphitize anneal, then quench and temper to 223 HB. Samples polished, etched in 4% nital, and then strained. Fig. 104 (left): Low-magnification view shows crack propagating through temper carbon, and localized plastic deformation and microcracking ahead of the primary crack front. SEM, 150×. Fig. 105 (right): Microcracking occurring between adjacent temper carbon sites. Also note small microcracks through sulfide inclusions. SEM, 1150×. (R.C. Voigt and B. Pourlaidian, University of Kansas)

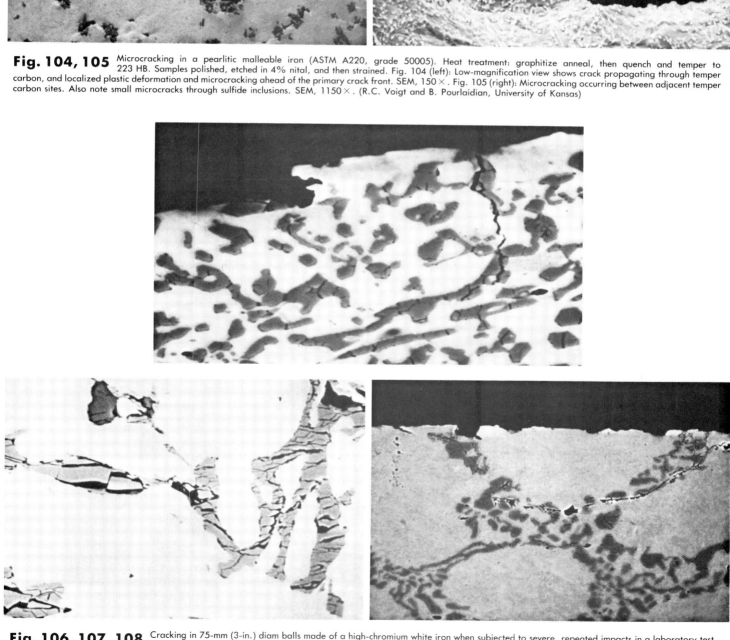

Fig. 106, 107, 108 Cracking in 75-mm (3-in.) diam balls made of a high-chromium white iron when subjected to severe, repeated impacts in a laboratory test. Fig. 106 (top): Cracks through eutectic M_7C_3 carbides near the surface (dark region) of the ball. SEM, 2000×. Fig. 107 (bottom left): Cracks parallel to the surface of the ball and extending through carbides located several millimeters beneath the surface. These cracks were caused by a combination of shear stress and the tensile stress due to rebounding. SEM, 2000×. Fig. 108 (bottom right): Crack path of an imminent near-surface spall. SEM, 1000× (H.W. Leavenworth, Jr., U.S. Bureau of Mines)

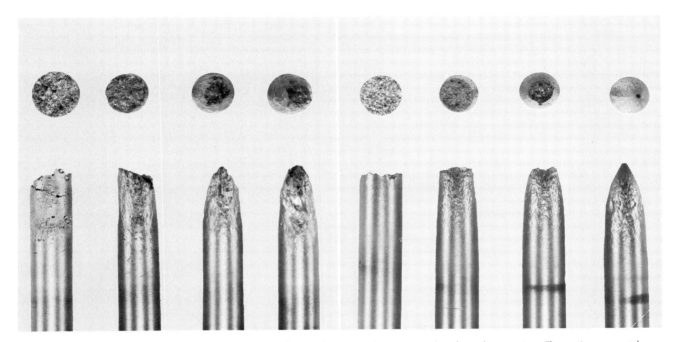

Fig. 109, 110 Fractures obtained in tension in low-carbon steel specimens that were tested at elevated temperature. The specimens were taken from a series of 225-kg (500-lb) laboratory heats melted to yield a range of carbon contents (0.04 to 0.08%) and a range of Mn:S ratios (11:1 to 68:1). The ingots were rolled to 19- and 16-mm (¾- and ⅝-in.) plate at 1175 °C (2150 °F). The specimens shown in Fig. 109 (left) were machined, placed within a silica sleeve in a Gleeble test unit, and remelted in the central region of the gage length. They were then solidified *in situ*, cooled to 1425 °C (2600 °F) at 14 °C (25 °F) per second, cooled to either 1095 °C (2000 °F) or 925 °C (1700 °F) at 5.5 °C (10 °F) per second, and then pulled to rupture. From left to right, the Mn:S ratios were 11, 22, 38, and 63 to 1, and the reductions of area were 6, 37, 63, and 83%. The specimens shown in Fig. 110 (right) were "sensitized" in the Gleeble test unit by heating to 1425 °C (2600 °F), holding for 300 s, and then cooling at 5.5 °C (10 °F) per second to 1095 °C (2000 °F), and were broken in tension at the latter temperature. From left to right, the Mn:S ratios were 14, 22, 36, and 62 to 1, and the reductions of area were 4, 47, 74, and 99%. Both at 0.75×

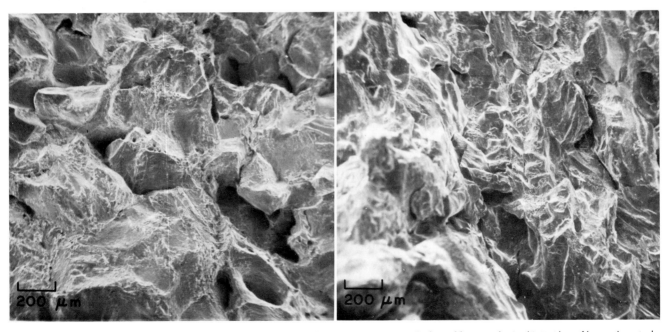

Fig. 111 Fracture in low-carbon steel containing 0.065% C, 0.52% Mn, 0.01% P, 0.022% S, and 0.076% Si (Mn:S ratio, 23.6:1). The specimen was "sensitized," and broken in tension, as described for Fig. 110. Under these conditions, a low Mn:S ratio would result in a completely intergranular fracture. However, because of the intermediate ratio here (23.6:1), the grain-boundary facets show dimples in many areas, indicating ductility. Reduction of area was 33%, which agrees with the appearance of the fracture surface. SEM, 50×

Fig. 112 Surface of fracture obtained in test bar of low-carbon steel containing 0.041% C, 0.62% Mn, 0.008% P, 0.028% S, and 0.032% Si (Mn:S ratio, 22.1:1). The test bar was remelted in its central zone in a Gleeble test unit, solidifying *in situ* in its quartz sleeve. Following controlled cooling from the freezing point, the test bar was pulled in tension at 925 °C (1700 °F). Although the path of this fracture was largely intergranular, there is also transgranular cleavage, and some facets show very fine dimples. SEM, 50×

Fig. 113, 114 In-service fatigue fracture of SAE 1010 tie rod adjusting sleeve for automobile steering system. Fig 113 (top): Tubular part was made by cold forming cold-rolled sheet. 1.8×. Fig. 114 (bottom): Macrograph reveals fracture progressed at 90° to surface. (General Motors Research Laboratories)

242 / Low-Carbon Steels

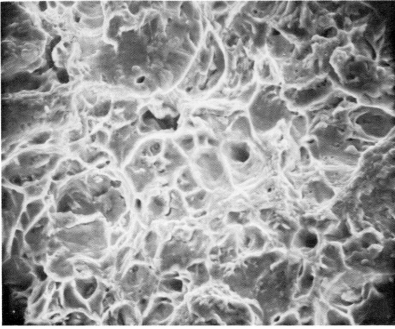

Fig. 115, 116 Bending impact fracture of SAE 1010 tie rod adjusting sleeve for automobile steering system. This part, unlike that in Fig. 113 and 114, was made of cold-drawn seamless tubing. Compressive forces produced when wheel was struck during an accident caused sleeve failure. Fig. 115 (top): Macrograph of mating fracture surfaces shows impact fracture progressed at 45° to surface. 2×. Fig. 116 (left): Photomicrograph near point of origin reveals dimpled structure characteristic of a ductile, single load fracture. SEM, 1700× (General Motors Research Laboratories)

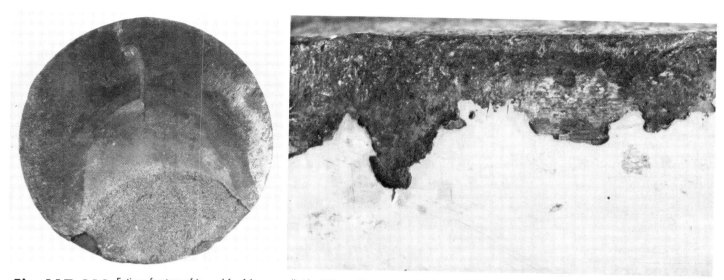

Fig. 117, 118 Fatigue fracture of journal for felt press roll. The 180-mm (7-in.) diam AISI 1019 shaft (88 HRB) had been in service 2 years at 20 °C (70 °F) and 100% relative humidity. Rust-preventive coating flaked off in fillet, resulting in pitting corrosion. Fatigue cracks initiated at pit bottoms. Fig. 117 (left): Fracture surface. Fig. 118 (right): Macrograph of shaft outer surface. Regions visible, top to bottom: fracture surface, corroded portion of shaft where coating had delaminated, and coated surface. 2.5×. (Z. Flanders, Packer Engineering Associates Inc.)

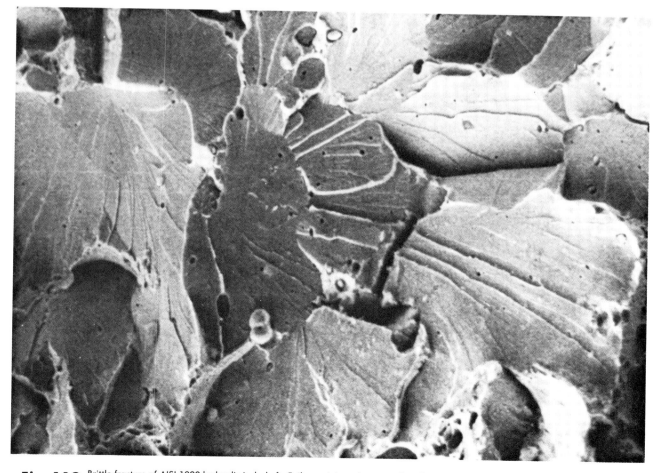

Fig. 119 Brittle fracture of AISI 1020 hydraulic jack shaft. Failure originated at root of machined thread. Corrosion (evident on part) and fatigue (due to repeated loading of shaft) may also have played roles in the failure. Photomicrograph of fracture surface shows transgranular cleavage with individual grains identified by changes in fracture plane orientation. The apparent grain size of 100 to 125 μm corresponds to a ductile-to-brittle transition temperature of approximately 100 °C (212 °F). SEM, 450×. (J.M. Rigsbee, University of Illinois)

244 / Low-Carbon Steels

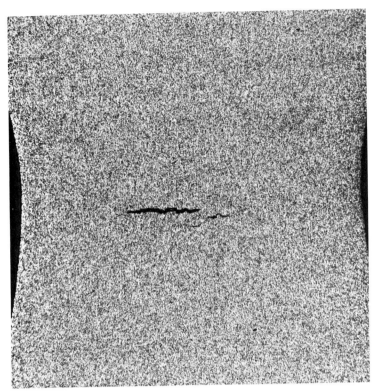

Fig. 120 Polished and etched section taken longitudinally through a 13-mm (0.505-in.) diam tensile-test specimen of AISI 1020 steel. The tensile test was halted after the centerline cracks shown here were formed, but before complete fracture occurred. See also Fig. 121 and 122. Nital etch, 10×

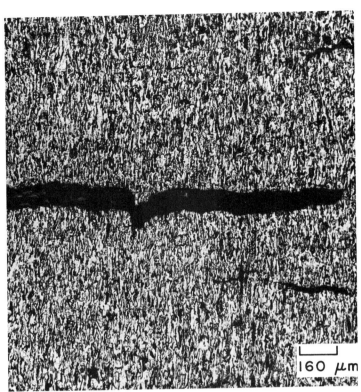

Fig. 121 Higher-magnification view of one of the centerline cracks in the specimen shown in Fig. 120, demonstrating that the cracks are transgranular. Secondary cracks also are visible at right, above and below the main crack. See Fig. 122. Nital etch, 62.5×

Fig. 122 A further enlargement of a centerline crack in the AISI 1020 steel tensile-test specimen shown in Fig. 120 and 121. Note that the crack crosses alternate plates of ferrite and pearlite; note also the longitudinal deformation of the original equiaxed structure. Nital etch, 312.5×

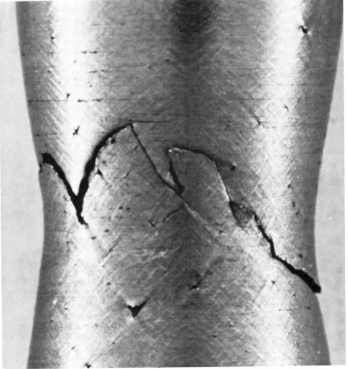

Fig. 123 Tensile-test fracture in a 13-mm (0.505-in.) diam specimen of cast 0.20% C steel with hardness of 255 HB. Note that pronounced 45° shear deformation has produced shear lips and also numerous secondary cracks, which formed at pores. 7.5×

Low-Carbon Steels / 245

 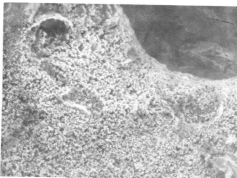 

Fig. 124, 125, 126 Brittle intergranular fracture of cast AISI 1025 collar for the emergency crank handle on a dam crane. Part failed after 63 years of infrequent use in a brackish water environment (industrial canal). Severe metal degradation promoted by bacterial action caused the collar to fracture when the crank was called on to help manually elevate dam logs. The culprits were aerobic and anaerobic thiobacillus bacteria, which can both oxidize sulfur and reduce sulfur compounds to eventually form corrosive sulfuric acid. Fig. 124 (left): Fracture surface of handle collar. SEM, 6.5×. Fig. 125 (center): Close-up crater near circular cavity in Fig. 124. Morphology indicates presence of calcareous thiobacilli carcasses. SEM, 65×. Fig. 126 (right): Same as Fig. 125, but at higher magnification. SEM, 650× (H.J. Snyder, Snyder Technical Laboratory)

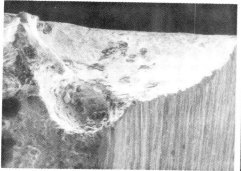 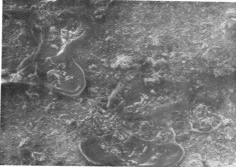 

Fig. 127, 128, 129 Other views of crank handle collar failure shown in Fig. 124 to 126. Fig. 127 (left): Craters and cavities on fractured collar. Note dark, roughly circular markings in large cavity. SEM, 8×. Fig. 128 (center): Close-up of dark patches. SEM, 80×. Fig. 129 (right): High-magnification view of dark patch revealing thiobacilli carcasses (nodules) and sulfur-bearing compounds (smooth area). SEM, 800× (H.J. Snyder, Snyder Technical Laboratory)

  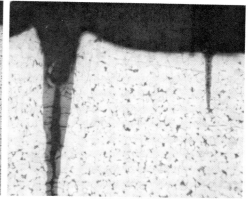

Fig. 130, 131, 132 Internal cracking of ASME SA178 grade A boiler tube due to corrosion fatigue. Failure occurred at the inner radius of a tube bend near the steam drum. Tube OD, 65 mm (2.5 in.); wall thickness, 3 mm (0.12 in.). Boiler operating pressure, 1.4 kPa (200 psi). Cyclic stresses at the tube bend were produced by thermal expansion and contraction of the entire tube. Mechanical vibration also may have played a role in crack initiation. Corrosion products formed within cracks acted as wedges, increasing stress levels at crack tips and aiding propagation. Fig. 130 (left): Close-up of inside surface of tube section. Note cracks and presence of corrosion products. Actual size. Fig. 131 (center): Cross section shows three cracks. Paths of propagation are very straight and perpendicular to the tube surface. 2% nital. 22×. Fig. 132 (right): Same as Fig. 131, but at higher magnification. Transgranular nature of cracks is evident. Note corrosion products within cracks. They prevented cracks from completely closing during a compression cycle (or when tensile stress was reduced), and caused the metal surface to bulge at the edges of larger cracks. 2% nital. 90× (F.W. Tatar, Factory Mutual Research Corporation)

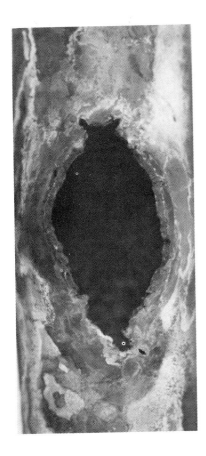 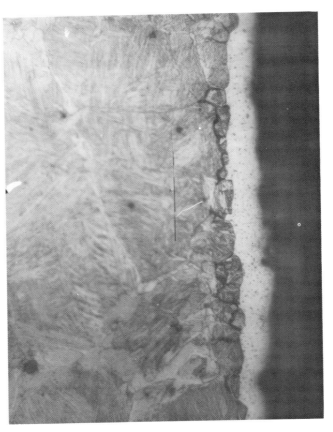

Fig. 133, 134 Failure of ASTM A178 grade B boiler tubes due to grain-boundary embrittlement by copper. Thick waterside deposits on tube inner-diameter surfaces impaired heat transfer, causing overheating of the steel during boiler start-up. Underneath the scale was a layer of metallic copper, an element often present as a feedwater impurity and which can plate out on the inner-diameter surfaces of boiler tubes. Overheating of the tubes to above 815 °C (1500 °F) caused accelerated diffusion of the copper along grain boundaries in the steel, in some cases reaching the tube outer-diameter surface. The copper-embrittled grain boundaries could not withstand both the high temperature and internal tube pressure. Either a longitudinal rupture (Fig. 133, left) or circumferential cracking (Fig. 134, right) resulted, depending on the orientation of maximum copper infiltration. (F.W. Tatar, Factory Mutual Research Corporation)

Fig. 135, 136 SEM photomicrographs at inner-diameter surface of circumferentially cracked ASTM A178 grade B boiler tube in Fig. 134. Fig. 135 (left): Note intergranular penetration by copper; some oxidation is present near surface. 100×. Fig. 136 (right): Higher-magnification photomicrograph reveals a 25-μm (1-mil) thick layer of metallic copper on the tube inner diameter. 2% nital. 300× (F.W. Tatar, Factory Mutual Research Corporation)

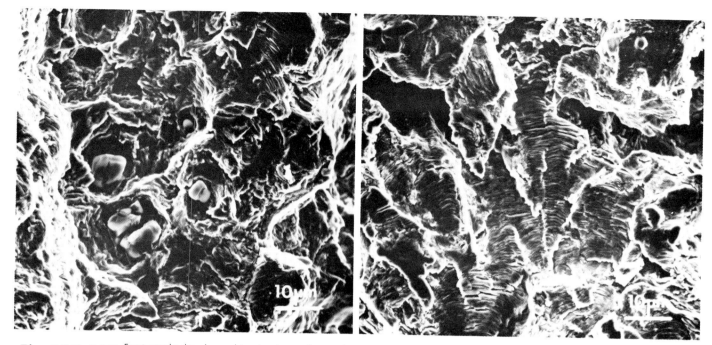

Fig. 137, 138 Fractographs show how calcium treatment affects inclusion morphology and fatigue crack propagation (FCP) in ASTM A516-70. Fracture surfaces are of FCP test specimens of calcium-treated (43 ppm Ca) electric furnace steel normalized at 900 °C (1650 °F) and air cooled. Both are in the S-L (through-thickness) orientation, $\Delta K = 43$ MPa$\sqrt{m}$ (39 ksi$\sqrt{in.}$). Calcium treatment reduces the number and size of inclusions and minimizes the formation of groups of Type II manganese sulfide inclusions and alumina inclusions via desulfurization (to 0.010% S max) and inclusion shape control. Positive results include improved levels and isotropy of FCP resistance. Fig. 137 (left): Note round, calcium-modified, duplex inclusions of sulfide and aluminate phases. Fig. 138 (right): Another fracture surface shows formation of ductile fatigue striations, the principal mode of fracture in this steel and in other calcium-treated grades. This mode is also exhibited by conventionally melted steels when fracture occurs remote from inclusion formations. Direction of macroscopic FCP is from bottom to top in both figures. SEM, 1200× (A.D. Wilson, Lukens Steel Company)

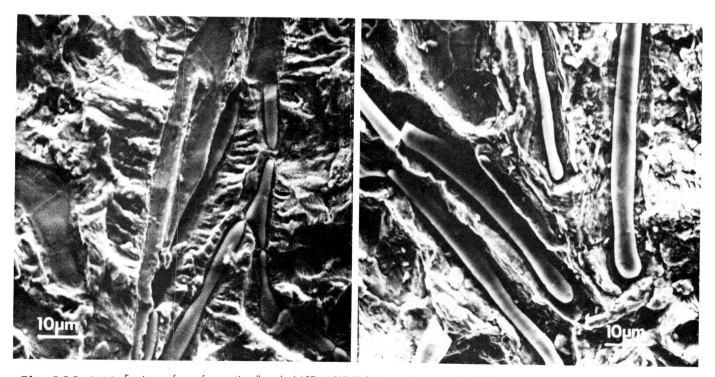

Fig. 139, 140 Fracture surfaces of conventionally melted ASTM A517-70 fatigue crack propagation (FCP) test specimens, S-L orientation, at two levels of ΔK. Fig. 139 (left): 22 MPa$\sqrt{m}$ (20 ksi$\sqrt{in.}$). Fig. 140 (right): 55 MPa$\sqrt{m}$ (50 ksi$\sqrt{in.}$). Steel normalized at 900 °C (1650 °F) and air cooled. Fracture occurred in regions containing inclusion formations (compare with Fig. 137 and 138). Failure in both cases was by fatigue striation formation in the interinclusion material. In higher strength steels at the higher ΔK level, interinclusion material undergoes ductile fracture (microvoid coalescence). Direction of macroscopic FCP is from bottom to top in both figures. SEM, 1200× (A.D. Wilson, Lukens Steel Company)

248 / Low-Carbon Steels

Fig. 141, 142 Delayed fracture of martensitic low-carbon boron steel (similar to AISI 15B22) automotive bolts due to corrosion-induced, hydrogen-assisted cracking. Application: secure forward ends of rear suspension lower control arms to vehicle frame. Class 12.8 bolts (SAW J1199, Appendix) were selected because the cold formability of the steel permitted a bolt head design with integral washer. Bolts were quenched and tempered to 39 to 44 HRC. Failed bolts had a hardness near or above the 44-HRC maximum and a soft, decarburized surface layer (permitted by the specification) typically 0.2 mm (8 mils) thick. Fig. 141 (left): Actual size. Fig. 142 (right): 4× (T.J. Hughel, General Motors Research Laboratories)

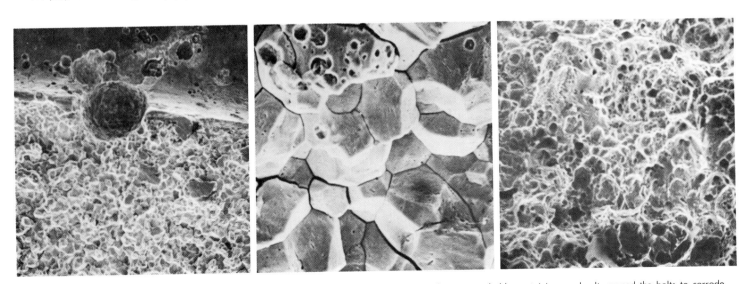

Fig. 143, 144, 145 Fractographic study of bolt failure in Fig. 141 and 142. Entry of water, probably containing road salt, caused the bolts to corrode, liberating hydrogen. Corrosion was most severe in thread roots, and eventually pits formed. However, cracks could not readily propagate in the soft decarburized layer. This layer represented a "slow fuse," delaying the onset of failure by more than 2 years. Once the pit penetrated to the hard subsurface material, the hydrogen promoted rapid crack propagation along prior-austenite grain boundaries. The intergranular hydrogen-assisted crack continued to grow until the bolt could no longer carry the clamp load. Failure then occurred by a ductile, microvoid coalescence mechanism. Fig. 143 (left): Origin of fracture in corrosion pit at bottom of thread root. SEM, 20×. Fig. 144 (center): Region of intergranular hydrogen-assisted fracture surrounding crack origin. SEM, 1000×. Fig. 145 (right): Balance of fracture surface characterized by dimpled appearance associated with microvoid coalescence. Results of laboratory tests indicated that hardness was the key factor in susceptibility to hydrogen-assisted cracking. Subsequently, the class 12.8 bolts were replaced with less formable, but lower hardness (39 HRC max) class 10.9 fasteners that used a separate washer. SEM, 500× (T.J. Hughel, General Motors Research Laboratories)

Low-Carbon Steels / 249

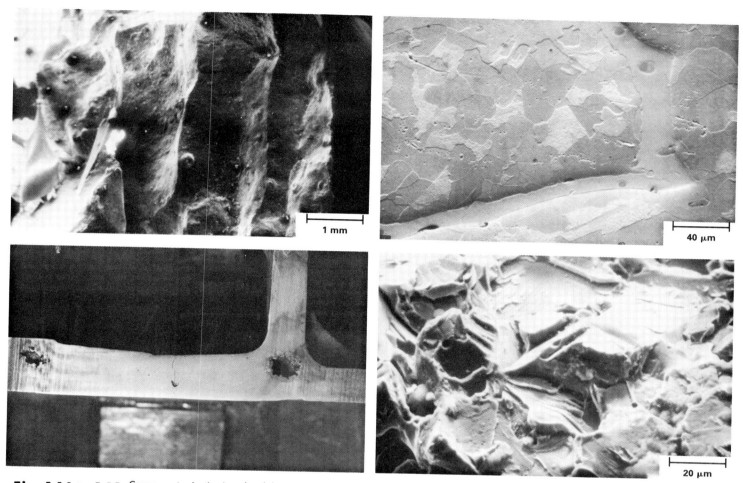

Fig. 146 to 149 Copper contamination introduced during casting led to brittle fracture of a normalized 0.25% C carbon steel part. The boxcar side frame failed during a freight train derailment. The foundry mold had accidentally been contaminated with copper. During solidification of the casting, some of the copper dissolved in the steel while the remainder was entrapped as liquid metal and either penetrated austenite grain boundaries or solidified as copper inclusions. Dissolved oxygen in the grain-boundary copper reacted with iron to form a black scale consisting largely of Fe_3O_4 and Fe_2O_3. Fracture initiated during the derailment, and upon reaching the copper-contaminated region it followed the scale network, delineating prior-austenite grain boundaries. Under repeated impacts, the fracture advanced further by low-cycle fatigue until part failure occurred by cleavage. Fig. 146 (top left): "Rock candy" portion of fracture surface where fracture followed scale network, defining copper-contaminated prior-austenite grain boundaries. SEM, 1100×. Fig. 147 (top right): Vein of copper along prior-austenite grain boundary near rock candy portion of fracture surface. Section polished and etched with 2% nital. SEM, 275×. Fig. 148 (bottom left): Transverse crack emanating from copper inclusion is visible at center of photo. Large shrinkage cavities at left and right played no role in fracture. Fig. 149 (bottom right): Region on fracture surface exhibiting cleavage. SEM, 550× (A. Johnson, University of Louisville)

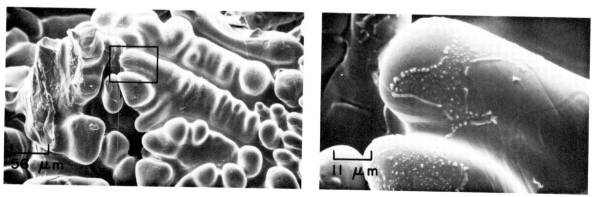

Fig. 150, 151 Surface of a brittle fracture in a specimen of low-carbon steel whose central zone was remelted in a Gleeble test unit, resolidified, cooled *in situ* to 1425 °C (2600 °F) and held for 20 s at that temperature, and then fractured in tension. This specimen contained 0.024% P. Figure 150 (left, shown at 180×) indicates that the fracture followed an interdendritic path through regions that had not yet solidified at 1425 °C (2600 °F). The rounded dendrite surfaces thus exposed are obvious, and the only evidence of a torn surface is in a small area at extreme left. Figure 151 (right), which is a higher-magnification view (900×) of the area within the rectangle in Fig. 150, shows patterns of what were confirmed by x-ray fluorescent analysis to have been, at the moment of fracture, liquid droplets of manganese oxysulfide on the dendrite surfaces. These patterns were the only evidence of segregation that was discovered within the fracture.

250 / Low-Carbon Steels

Fig. 152 Fracture surface of a carbon steel drive shaft of a 180-Mg (200-gross-ton) marine tug. The shaft failed by corrosion fatigue after seawater displaced the lubricant between the shaft and the stern tube. Beach marks indicate that the fracture origin was at left (arrow). See also Fig. 153 to 156. 0.8×

Fig. 153 Side view of the fractured drive shaft in Fig. 152. Note the serrated profile (retouched for clarity) of this edge of the fracture surface and the diagonal cracks on the shaft surface. The cracks were caused by corrosion fatigue from the combination of reversed shaft torsion, perhaps some bending, and seawater attack. 0.8×

Fig. 154, 155 Figure 154 (left) shows a macroetched section of the fractured shaft in Fig. 152. Fig. 155 (right) is a view at 90° to that in Fig. 154 and shows the shaft surface, containing diagonal cracks (Fig. 154 shows depths of these cracks). Note in Fig. 155 that the seawater produced not only corrosion-fatigue cracking but also general surface corrosion. Both at 0.6×

Fig. 156 Section of the fractured shaft shown in Fig. 151, revealing the bottom of one of the corrosion-fatigue cracks. Note that corrosion blunted the tip of the main crack but not the tip of the sharp secondary crack at left, which formed later. Picral etch, 67×

Fig. 157 Mating fracture surfaces of carburized and hardened low-carbon (0.08% C) steel bicycle pedal axle. The in-service fatigue failure initiated at quench cracks in the thin carburized case. 5× (R. Goco, California Polytechnic State University)

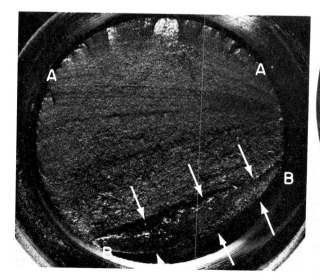

Fig. 158 Surface of a fracture in a lift-truck hydraulic-piston rod of low-carbon steel that broke at a fillet by fatigue. Several fatigue-crack nuclei are evident at top edge (between A's), and there is another at the bottom edge (between B's). Fracture was by reversed bending. Final fast fracture was in the small region indicated by the arrows. See also Fig. 159. 3×

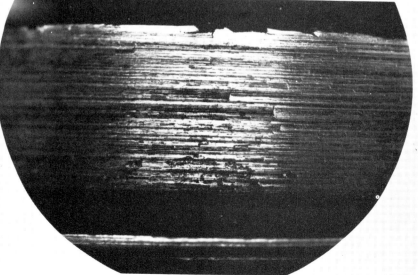

Fig. 159 A view of coarse machining marks that were found in the fillet in Fig. 158 between the larger-diameter portion of the piston rod and the threaded end. A secondary crack that follows a machining mark is visible at top center. These surface marks are considered to be areas of local stress concentrations, which led to fatigue cracking. 10×

252 / Low-Carbon Steels

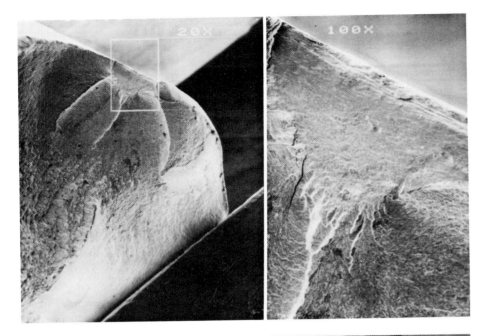

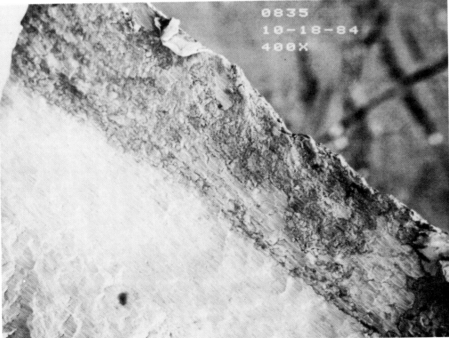

Fig. 160, 161, 162 Fracture of low-carbon steel spring due to a lap. The defect was most likely introduced during rolling of the wire rod. Unlike seams, penetration of a lap below the wire surface is more tangential than radial. Like seams, laps tend to induce failures in springs by first initiating a longitudinal shear crack. Fig 160 (top left): Fracture surface. SEM, 70×. Fig. 161 (top right): Boxed area in Fig. 160. SEM, 100×. Fig. 162 (bottom): Higher-magnification view shows the dark, nonuniformly textured lap and the lighter, more uniformly textured shear crack. SEM, 400× (J.H. Maker, Associated Spring, Barnes Group Inc.)

Fig. 163 Fractograph of cleavage fracture in low-carbon steel shows tongue (arrow) formed by local fracture along twin-matrix interfaces. Tongue formation occurs as a result of the high velocity at which a cleavage crack propagates (it has limiting velocity between 0.4 and 0.5 of the speed of sound). The high-speed crack produces a local strain rate too high for slip processes to provide all the accommodation required; thus, many twins are normally formed just ahead of the moving crack tip. TEM replica, 6500× (I. Le May, Metallurgical Consulting Services Ltd.)

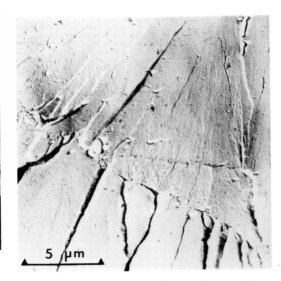

Fig. 164 River pattern on cleavage fracture surface of low-carbon steel bolt. When a crack crosses a twist boundary, many small parallel cracks may form with cleavage steps between them. These steps run together, forming larger ones and leading to the river patterns characteristic of cleavage fracture in polycrystalline metals. TEM replica, 4000× (I. Le May, Metallurgical Consulting Services Ltd.)

Medium-Carbon Steels / 253

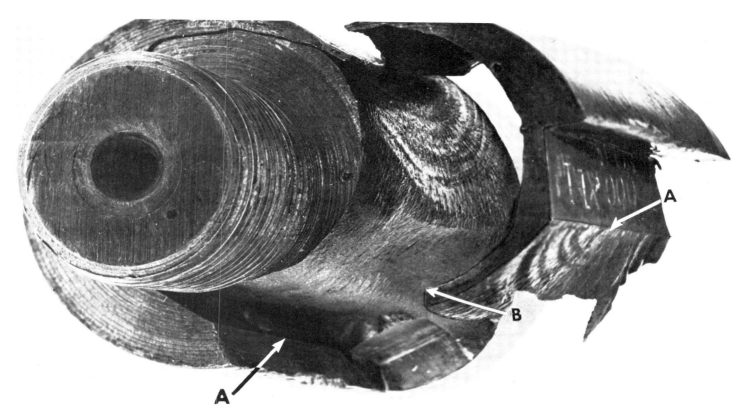

Fig. 165 Torsional fatigue fracture in an 86-mm (3⅜-in.) diam keyed tapered shaft of 1030 steel, commonly termed a "peeling" type of fracture. A loose nut had reduced the frictional force on the tapered portion of the shaft, transferring the torsional load to the key. The fatigue crack originated at a corner in the keyway (arrows A) and progressed completely around the shaft in a clockwise circular path, returning beneath the keyway to the location marked by arrow B, where final fracture occurred. 1.2×

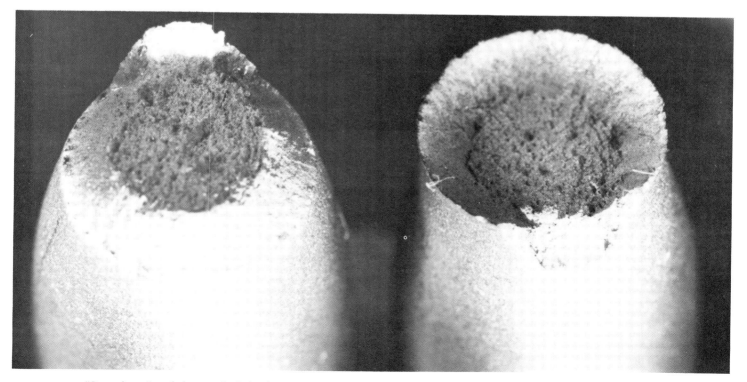

Fig. 166 "Cup and cone" tensile fracture of cylindrical test specimen is typical for ductile metals; in this case, annealed AISI 1035. Fracture originates near the center of the section with multiple cracks that join and spread outward. When cracks reach a region near the surface, the stress state changes from tension to macroshear, forming a fracture at approximately 45° to the plane of the major fracture—the familiar "shear lip." (D.J. Wulpi, Consultant)

254 / Medium-Carbon Steels

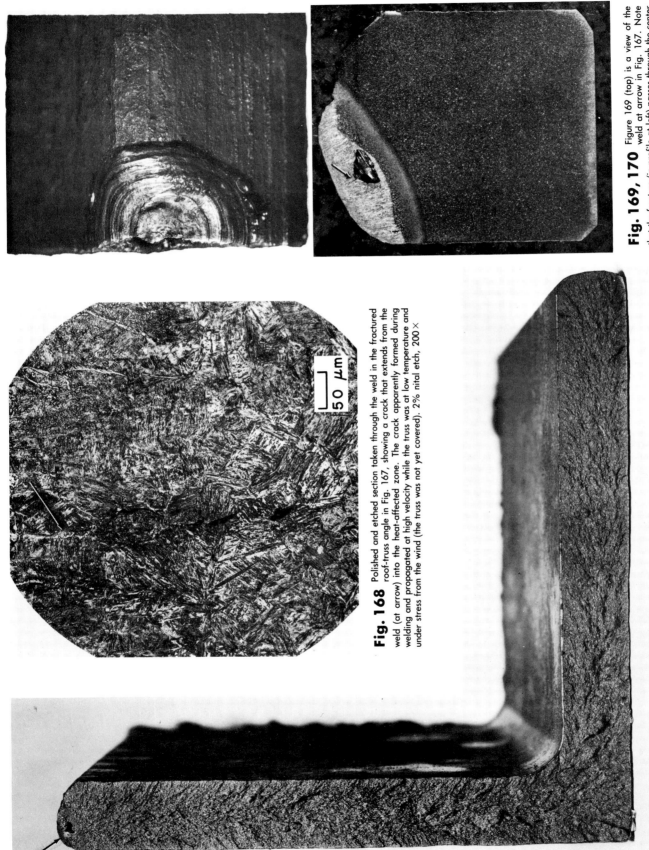

Fig. 168 Polished and etched section taken through the weld in the fractured roof-truss angle in Fig. 167, showing a crack that extends from the weld (at arrow) into the heat-affected zone. The crack apparently formed during welding and propagated at high velocity while the truss was at low temperature and under stress from the wind (the truss was not yet covered). 2% nital etch, 200×

Fig. 167 Surface of a fracture in a 1033 steel angle 100 by 100 by 13 mm (4 by 4 by ½ in.) that formed part of a truss of a roof under construction. Fracture occurred when the temperature was below freezing and the wind velocity was 50 to 65 km/h (30 to 40 mph). Visible are chevron marks, which clearly point toward the fracture origin (at arrow) at a weld deposit in the upper edge of the vertical leg. Note the large cavity in the weld zone. See also Fig. 168 to 170. 1.5×

Fig. 169, 170 Figure 169 (top) is a view of the weld at arrow in Fig. 167. Note that the fracture (in profile at left) passes through the center of the weld. Figure 170 (bottom) is a portion of the fracture surface at the weld, after polishing and etching (2% nital). A crack is visible (at arrow) at the tip of the cavity. Hardness was 187 HB in the base metal, 516 HB in the heat-affected zone, and 363 to 534 HB in the weld. Both at 5×

Medium-Carbon Steels / 255

Fig. 171 Mating surfaces of a crack in a shell plate of an ASTM A515, grade 70, steel pressure vessel used in dehydration of natural gas. Normally, operating pressure was 1.5 MPa (220 psi) and operating temperature was −12 to +4 °C (10 to 40 °F). The crack was exposed by sectioning after a leak was discovered. See Fig. 172 to 176 for views of the portion near arrow at left. 0.4×

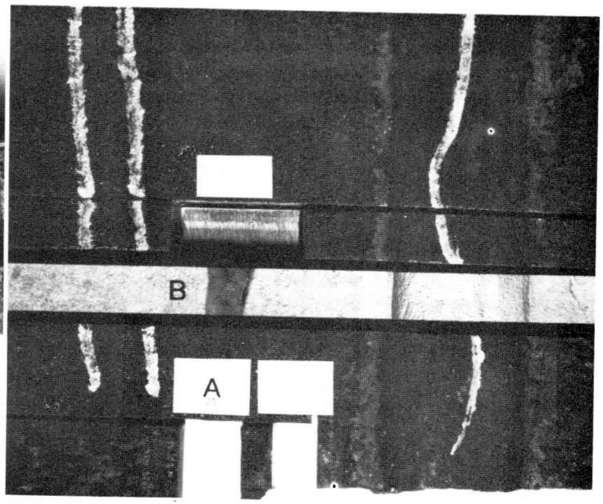

Fig. 172 Portion of the fracture shell plate in Fig. 171 near arrow. A and B indicate where sections were cut for further study. Fracture surface is shown in profile at bottom. 0.57×

Fig. 173 Same portion of the shell plate as in Fig. 172, but before sectioning and at a different orientation. The two dark crescents partly visible at top edge of the fracture surface (at arrows) were caused by arc strikes and served as crack nuclei. 0.7×

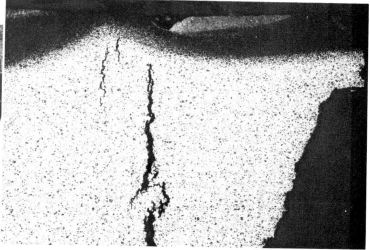

Fig. 174 Polished and etched section taken from location A in Fig. 172, showing the arc strike visible at the left arrow in Fig. 173. The large cracks present here are extensions of the fracture, which is visible in profile at right. Nital etch, 7×

Fig. 175 Another view of the fracture surface in Fig. 173, displaying clearly the two dark arc-strike crescents that are indicated in both views by arrows at the top edge of the fracture surface. Chevron marks pointing to these crescents (which are the crack nuclei) are visible both here and in Fig. 171. ~0.9×

Fig. 176 Polished and etched shell-plate section marked B in Fig. 172. Note that the smaller arc strike, visible at the arrow at right in Fig. 173 and 175, does not extend this far from the fracture. Note also the surface damage caused by the arc strike, in contrast to the excellent condition of the weld deposit. 0.95×

256 / Medium-Carbon Steels

Fig. 177 Fractured shell of an 865-mm (34-in.) diam pressure vessel fabricated from a 32 by 2440 by 9145 mm (1¼ by 96 by 360 in.) plate of ASTM A515, grade 70, steel for pressure vessels. The shell broke during testing at an internal gage pressure of 8.3 MPa (1.2 ksi). The fracture, which originated adjacent to a flange (arrow at S), followed a very complex path. See also Fig. 178 to 186. ~0.17×

Fig. 178 Polished and etched section through the shell in Fig. 177, showing an as-rolled microstructure consisting of a mixture of ferrite and fine pearlite, which resulted in a maximum hardness of 86 HRB. Nital etch, 500×

Fig. 179 View of the flange in Fig. 177, showing an adjacent area of the fracture surface of the pressure-vessel shell. Although not as discernible here as in Fig. 184, chevron marks throughout the fracture surface clearly point toward the area at S, and subsequent examination identified the toe of the flange weld at S as the crack nucleus. ~0.3×

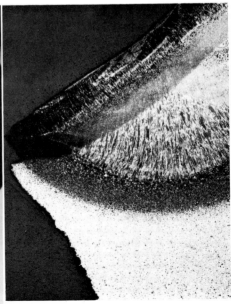

Fig. 180 Polished and etched section through the toe of the flange weld at S (crack nucleus) in Fig. 179, taken normal to the fracture surface (at left). The toe of the weld had a maximum hardness of 45 HRC. Nital etch, 6×

Medium-Carbon Steels / 257

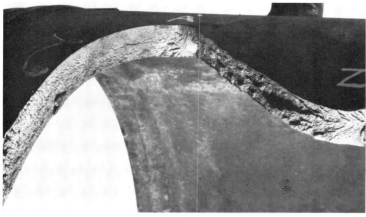

Fig. 181 View of the far left end of the top segment of the fractured pressure-vessel shell in Fig. 177, showing chevron marks. The chevron marks were smeared subsequent to fracture, but can be clearly identified as pointing to the right. 0.3×

Fig. 182 View of the top segment of the fractured pressure-vessel shell in Fig. 177, showing an area just to the right of the area in Fig. 181. Again, chevron marks are visible and consistently point to the right. 0.25×

Fig. 183 View of the bottom segment of the fractured shell in Fig. 177, showing an area opposite the one in Fig. 182. This view also reveals sharply defined chevron marks that point to the right. 0.45×

Fig. 184 View of the top segment of the fractured shell in Fig. 177, showing an area just to the left of the fracture origin. The continuity of the chevron marks that point toward location S (at right end) is much more perceptible here than in Fig. 179 because of a difference in illumination. 0.28×

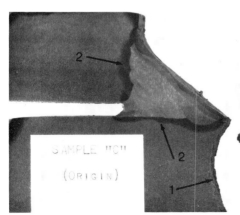

Fig. 185 Section through the flange weld in Fig. 179 and 184, taken at the fracture origin and normal to the fracture surface, shown in profile at bottom right (arrow 1). Heat-affected zones (arrows 2) appear as shadows here. ~0.9×

Fig. 186 View of the top segment of the fractured pressure-vessel shell in Fig. 177, showing an area of the fracture surface just to the right of the crack nucleus. Note that here the chevron marks clearly point to the left, toward the fracture origin. The crack ends at far right, where a section was removed for testing. 0.28×

258 / Medium-Carbon Steels

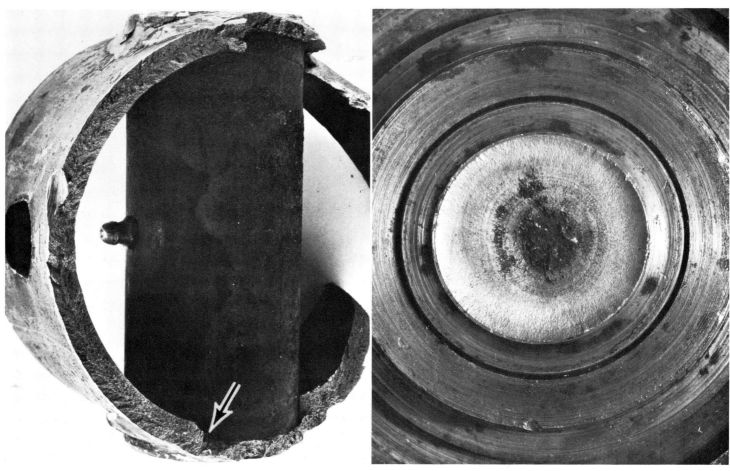

Fig. 187 Surface of a brittle fracture in an axle made of cold-drawn and stress-relieved AISI 1035 steel tubing. Fracture originated at a weld defect (at arrow) during very cold weather. Visible are chevron marks, which show that the fracture progressed clockwise along the left wall before the remainder of the axle wall (cross section shown incompletely here) ruptured in a ductile manner. Actual size

Fig. 188 Surface of a fatigue fracture in a shaft of medium-carbon steel (composition approximately that of AISI 1035) with a hardness of 143 HB. The fracture occurred under rotating bending. Fatigue cracks began at several locations within a sharp tool mark in a fillet that had an adequate radius and joined around the circumference as they penetrated to the center. 1.13×

Fig. 189, 190 Figure 189 (left) is a view of a drive shaft (actual size shown) of AISI 1035 steel, with a hardness of 34 HRC, that fractured under torsional overload, twisting off in a fashion that produced a surface almost as flat as if it had been machined. The threaded portion, at right, contains a fracture surface. A thin slice containing the mating fracture surface was cut from the portion at left; this slice is shown at center, with the fracture surface up. Shaft portion at left, including cut fracture surface, was hot etched in equal parts of hydrochloric acid and water to reveal deformation flow and forging flash lines. The plastic deformation is characteristic of torsional overload in ductile metals and is not typical of rotating fatigue. Figure 190 (right) is a higher-magnification view (2×) of the threaded portion of the fractured drive shaft in Fig. 189 and the mating surface of the fracture.

Fig. 191, 192 Fatigue fracture of automotive axle shaft due to improper wheel bearing removal. Material: modified AISI 1038. Heat treatment: induction hardened and tempered to a surface hardness of 50 to 58 HRC. Replacement of a press-fit wheel bearing was accomplished with the aid of a cutting torch. Fatigue failure initiated in subsequent service at one of the pitlike defects created by the intense flame of the torch. Fig. 191 (top): Fracture surface of axle shaft. Note extent of fatigue and relatively small final fracture area, indicating that operating loads were normal. Fig. 192 (bottom): Axle surface with bearing removed (edge of fracture surface at right). Visible are several of the burn/melt defects caused by use of a cutting torch to assist in bearing removal. The procedure is not recommended. (Z. Flanders, Packer Engineering Associates, Inc.)

260 / Medium-Carbon Steels

Fig. 193 Surface of a fatigue fracture in a shaft of AISI 1040 steel with a hardness of approximately 30 HRC. The shaft was fractured when subjected to rotating bending. The oval shape of the central region of final fast fracture indicates that two mutually perpendicular, unequal bending stresses were present. The ratchet marks around the perimeter of the shaft indicate that fracture began at several points. 1.63×

Fig. 194 Fatigue-fracture surface of a keyed shaft of AISI 1040 steel (~30 HRC). The fatigue crack originated at the left bottom corner of the keyway and extended almost through the entire cross section before final fast rupture occurred. Beach marks are visible; these swing counterclockwise because the direction of rotation was clockwise. 1.88×

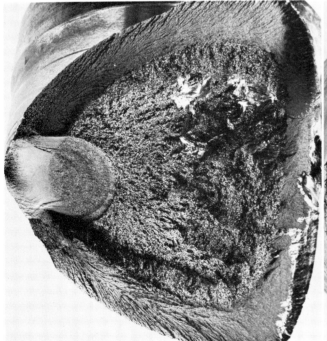
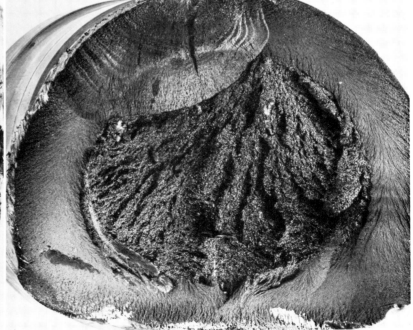

Fig. 195 Surface of a fatigue fracture in an induction-hardened axle of AISI 1041 steel with a hardness of 46 HRC in the hardened zone. The fatigue crack originated at a fillet (with a radius smaller than specified) at a change in shaft diameter near a keyway runout. Note the two sets of chevron marks along the hardened perimeter, each set pointing to the crack origin at left. Actual size

Fig. 196 Surface of a bending-plus-torsional-fatigue fracture in an experimental 89-mm (3½-in.) diam tractor axle of AISI 1041 steel that had been induction hardened. Fracture occurred after 1212 h on an endurance-test track. Note beach marks fanning out from the fatigue-crack origin (slightly to left of center, at top). Beyond the beach marks, cracking progressed by fast fracture along the hardened perimeter, producing two sets of chevron marks pointing toward the crack origin (like the chevron marks in Fig. 195). See also Fig. 197. 1.25×

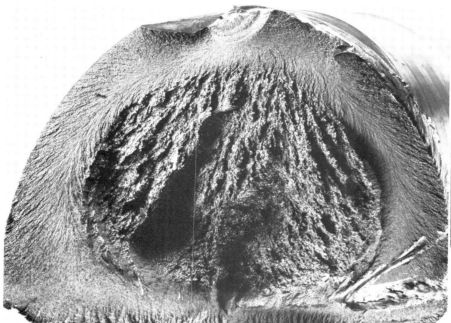

Fig. 197 Surface of a torsional-fatigue fracture in an induction-hardened AISI 1041 steel experimental tractor axle same as in Fig. 196. Hardness in the hardened zone was 50 HRC at 11 to 12 mm (7/16 to 15/32 in.) beneath the axle surface at the crack origin. This axle fractured after 450 h of endurance testing. In this axle, as in the axle in Fig. 196, crack origin is near top, at a sharp corner of a keyway runout. 1.25×

Fig. 198 Surface of a fatigue fracture in a truck axle of AISI 1041 steel induction hardened to 55 HRC. The axle broke after 490 h of service. Crack origin was subsurface, near a keyway. Note that fatigue crack progressed in steps. Actual size

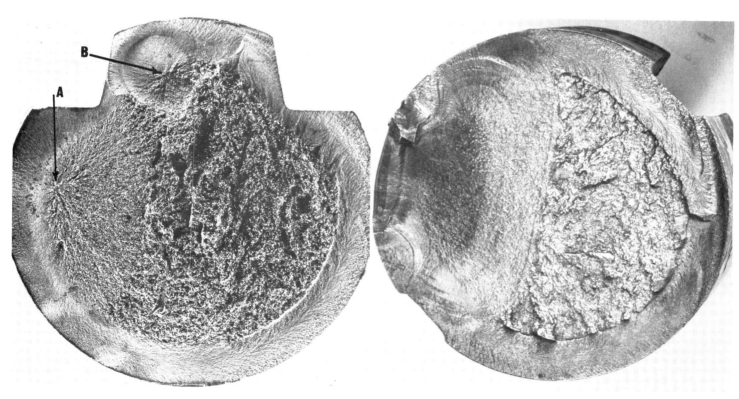

Fig. 199 Fatigue fracture in an AISI 1041 steel axle induction hardened to 50 HRC and tested in rotating bending. Two cracks are evident; one began at A and progressed around most of the perimeter by fatigue, then by fast fracture, before meeting a smaller crack, which began by fatigue at B. Chevron marks in hardened zone point to both origins. 1.2×

Fig. 200 Surface of a fatigue fracture in a keyed axle of AISI 1041 steel induction hardened to 50 HRC. An accidental arc strike on the axle surface, at left, initiated a fatigue crack, which grew about halfway through the cross section before fast fracture occurred in the hardened zone; fast fracture progressed from both sides of the fatigue crack, as shown by chevron marks. 1.3×

262 / Medium-Carbon Steels

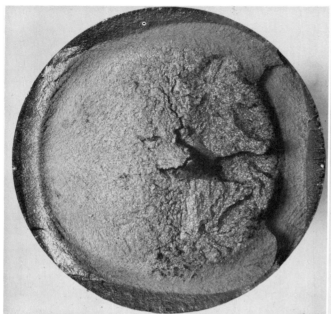

Fig. 201 Fatigue-fracture surface of a case-hardened AISI 1039 steel shaft. Case hardness, 50 HRC; core, 19 HRC. Note that fracture of the case was nearly complete before the fatigue cracks penetrated into the core. 2×

Fig. 202 Surface of a fatigue fracture in a shaft of AISI 1041 steel with a hardness of 302 HB. The shaft was broken by reversed stressing. Note the two fatigue-crack origins at right edge and the third origin at left. 2×

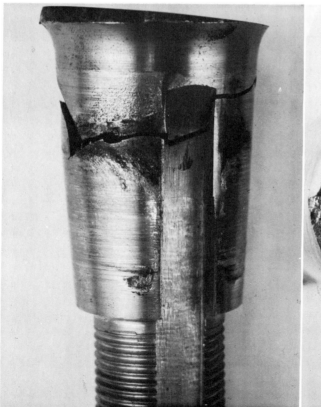

Fig. 203 View of a broken keyed spindle of AISI 1041 steel with a hardness of 333 HB. A large amount of fretting is visible on both sides of the fracture. Note that the radius of the keyway edge seems small near the fracture. See also Fig. 204. 1.5×

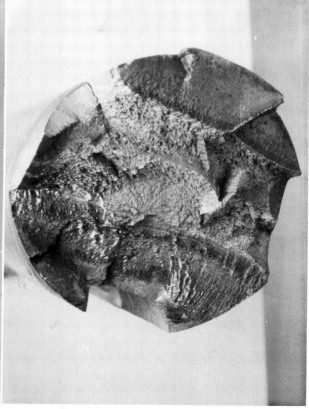

Fig. 204 Surface of the fracture in Fig. 203. Note that the radii of the keyway edges have acted as origins for fatigue cracks. A second fatigue zone is visible on the side opposite the keyway. The region between the fatigue zones underwent fast fracture. 2×

Medium-Carbon Steels / 263

Fig. 205 Fatigue-fracture surface of the shaft of a forging hammer of AISI 1144 steel containing 0.45% C, 1.60% Mn, and 0.28% S. The fatigue crack began at a heavily abraded surface area (out of view to the left), penetrated to the center of the shaft, then turned 90° and propagated longitudinally to the point of termination at right (about 635 mm, or 25 in.). See also Fig. 206. ~0.5×

Fig. 206 Mating fracture surface to the one shown in Fig. 205. This piece of the shaft broke away during service. It is believed that the crack propagated a long distance in the longitudinal direction because of weaknesses associated with the presence of longitudinal bands of elongated manganese sulfide inclusions. ~0.5×

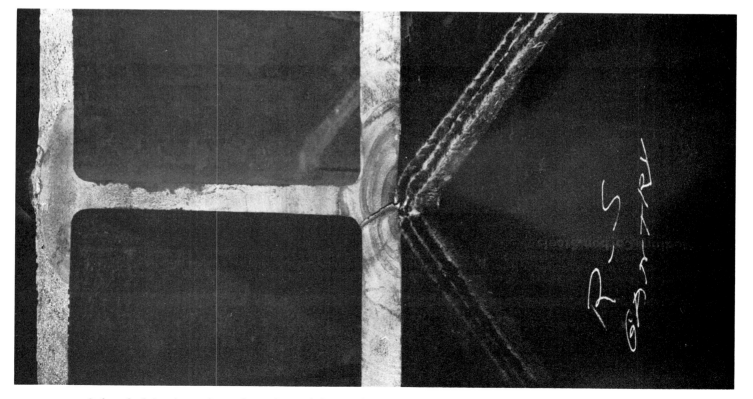

Fig. 207 Surface of a fatigue fracture in a medium-carbon steel I-beam, with a flange 400 by 400 by 40 mm (15¾ by 15¾ by 1⁹⁄₁₆ in.) and a 25-mm (1-in.) thick web, that was part of a dragline-excavator A-frame. Each flange of the I-beam was reinforced by a plate (such as that at right) joined by a double-pass weld. There are two fatigue-crack nuclei, separated by a cleavage step, at the toe of the weld at the outer edge of the flange at right. See also Fig. 208 to 210. 0.3×

Fig. 209 View of the weld joining the flange and reinforcement plate at top in Fig. 208, showing a crack at the toe of the weld. This crack was present before the fatigue crack, which led to fracture, formed at the toe of the weld in the other flange. See also Fig. 210. 2×

Fig. 210 Polished and etched section through the weld in Fig. 209, at a position just to the left of the area shown there. Like that in Fig. 209, the crack shown here was present prior to formation of the fatigue crack in the opposing flange. The weld metal (light) is at upper left. 2% nital etch, 100×

Fig. 208 Another view of the fatigue-fracture surface of the medium-carbon steel I-beam in Fig. 207. The fracture origins are clearly evident in the flange at bottom, where two nuclei are situated at the toe of the weld joining a reinforcement plate to the flange. It is remarkable that the fatigue crack penetrated the entire flange at bottom, the entire web, and part of the flange at top before final fast fracture occurred. See also Fig. 209 and 210. 0.5×

Fig. 211, 212 Figure 211 (top) is a polished and etched cross section of a partially flame-hardened crane gear of AISI 1045 steel in which fatigue cracks formed at the roots of many teeth after one year of service. The cracks have been sharply delineated by etching. Note that hardening did not extend even to the root fillet of any tooth. Figure 212 (bottom) is a polished and etched cross section of a portion of a gear that was subjected to full-contour flame hardening; note the better tooth form, with fully radiused tooth roots. Both etched in 5% nital. Fig. 211: actual size. Fig. 212: 2×

266 / Medium-Carbon Steels

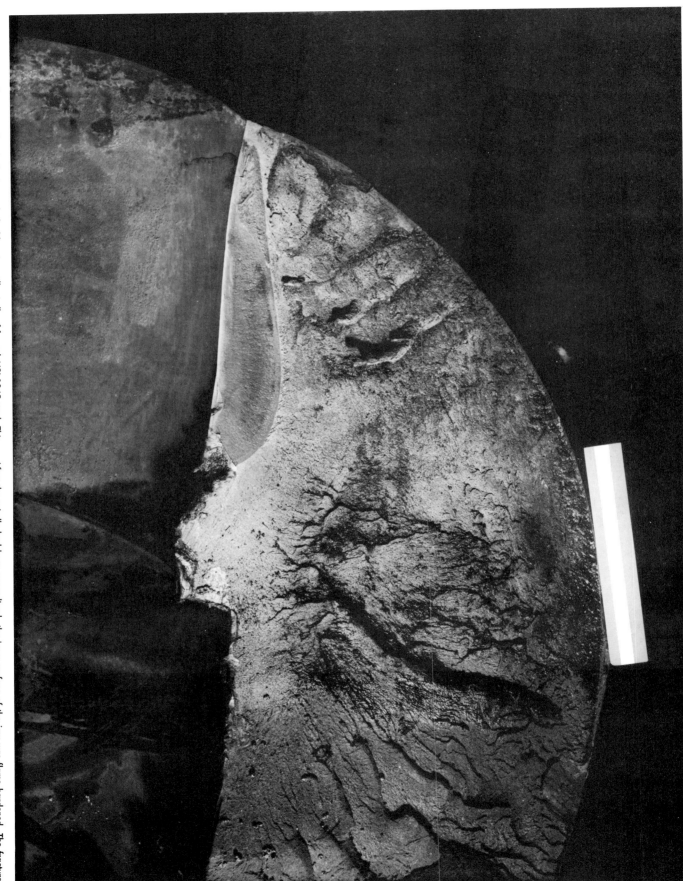

Fig. 213 Fracture surface in the jaw end of a blooming-mill spindle of forged AISI 1045 steel. This spindle was 8.2 m (27 ft) long, 890 mm (35 in.) in diameter at the jaw end, and 1065 mm (42 in.) in diameter at the drive end. The white 6-in. rule at top indicates the massive size of the spindle. After the spindle had been normalized, the inner surface of the jaw was flame hardened. The fracture originated at the jaw surface, at the lower edge of the fatigue zone (left of center). See also Fig. 214. 0.4×

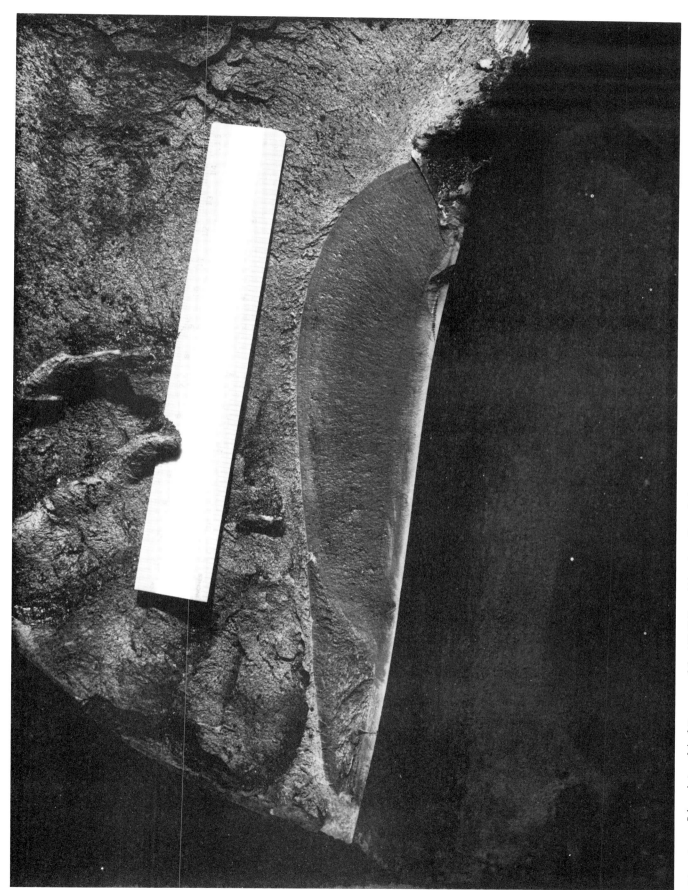

Fig. 214 Enlarged view of the fatigue zone of the AISI 1045 steel blooming-mill spindle fracture shown in Fig. 213. Note the radial marks originating at the surface of the jaw. A few beach marks are faintly visible at the top edge of the fatigue zone. Note the sharpness of the terminating edge of the fatigue zone at the transition to final fast fracture. Note also the tongues of metal near the rule in the zone of final fast fracture. 0.83×

268 / Medium-Carbon Steels

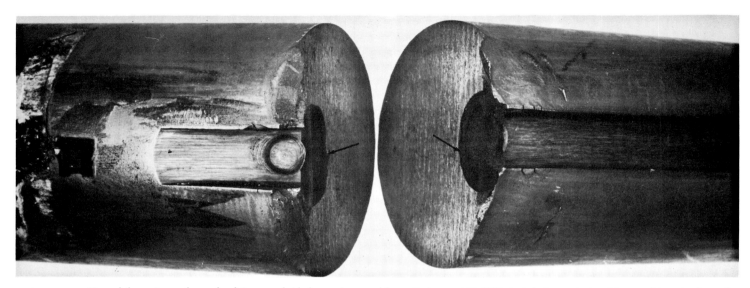

Fig. 215 View of the mating surfaces of a fatigue crack (dark areas) exposed by sectioning an AISI 1045 steel shaft quenched and tempered to a hardness of approximately 40 HRC. The crack, which was discovered before it could penetrate the entire cross section of the shaft, originated at a weld deposit that served as a stop in a keyway. 1.7×

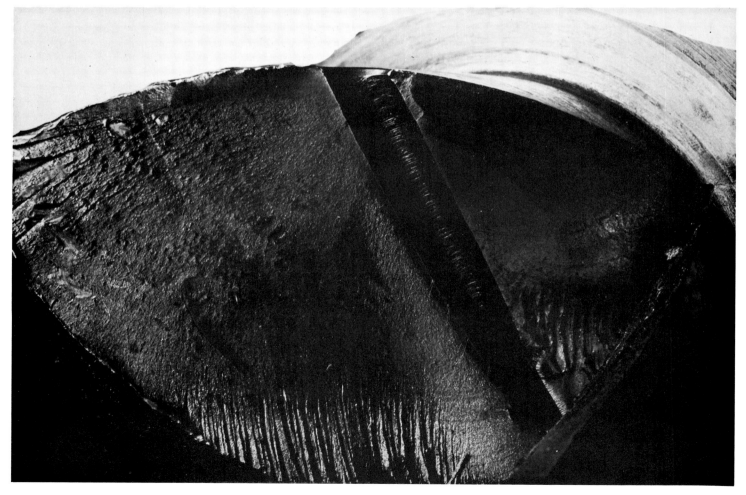

Fig. 216 Surface of a torsional-fatigue fracture in an AISI 1045 steel crankshaft induction hardened to 55 HRC. The crack originated at the edge of an oil hole. Although it is not clearly evident in this view, the crack grew at a 45° angle to the axis of the crankshaft because of tensile-stress components caused by the torsional loading. 2.5×

Medium-Carbon Steels / 269

Fig. 217 Reversed bending fatigue of a 40-mm (1.6-in.) diam AISI 1046 steel shaft (30 HRC). Note the symmetrical fatigue pattern of beachmarks on each side, with the final rupture on the diameter. This indicates that each side of the shaft was subjected to the same maximum stress and to the same number of load applications. (D.J. Wulpi, Consultant)

Fig. 218 Fatigue-fracture surface of an AISI 1050 shaft (35 HRC) subjected to rotating bending. Numerous ratchet marks (small shiny areas at surface) indicate that fatigue cracks were initiated at many locations along a sharp snap-ring groove. The eccentric pattern of oval beachmarks indicates that the load on the shaft was not balanced. The final rupture area is near the left (low stress) side, where there may have been no fatigue action. (D.J. Wulpi, Consultant)

Fig. 219 Fractograph of fatigue failure in SAE 1050 pin, induction hardened to a depth of 5 mm (3/16 in.) and surface hardness of 55 HRC. Core hardness: 21 HRC. Fatigue initiated inside the grease hole at the metallurgical notch created by the very sharp case-core hardness gradient. 1.8× (K. Marden, California Polytechnic State University)

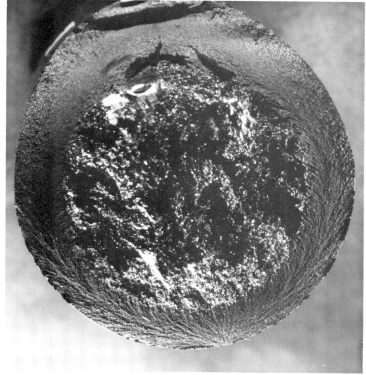

Fig. 220 Fracture surface of the modified SAE 1050 axle shaft shown in Fig. 221. Note chevrons pointing to fracture origin in the fine structured case. See also Fig. 222 to 227. 2.75×

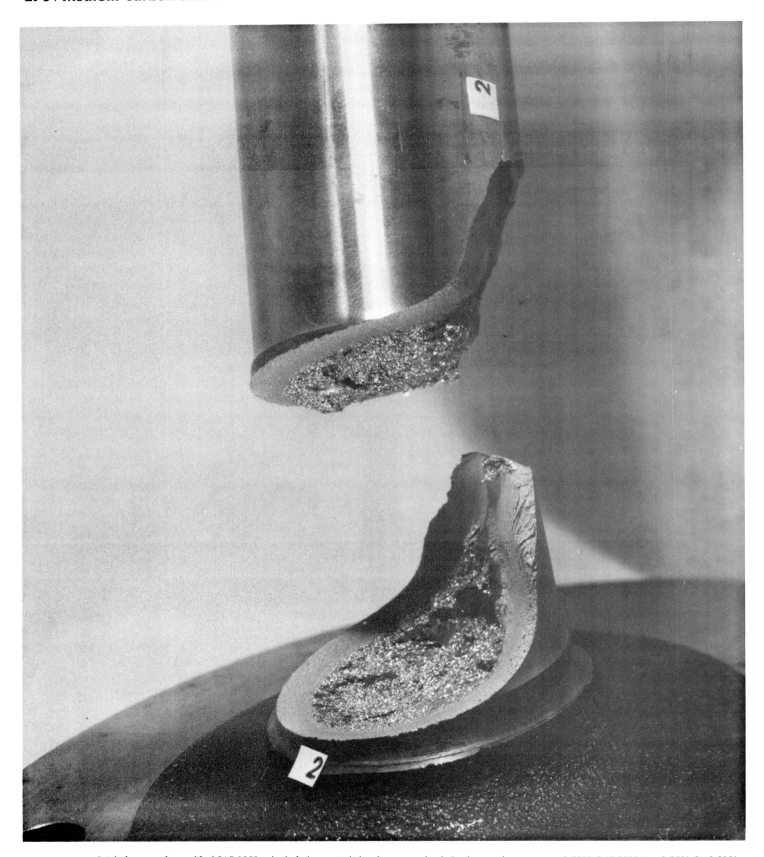

Fig. 221 Brittle fracture of a modified SAE 1050 axle shaft due to single bending impact load. Steel nominal composition: 0.50% C, 0.95% Mn, 0.25% Si, 0.01% S, and 0.01% P. The hot-rolled and upset shaft had an induction-hardened martensitic case (60 HRC) and a pearlite-ferrite core (20 HRC). Failure occurred at the flange end. Part was broken in the laboratory. Shown in this fractograph are the fracture surfaces, including a compression chip. See also Fig. 220 and 222 to 227. Actual size (General Motors Research Laboratories)

Medium-Carbon Steels / 271

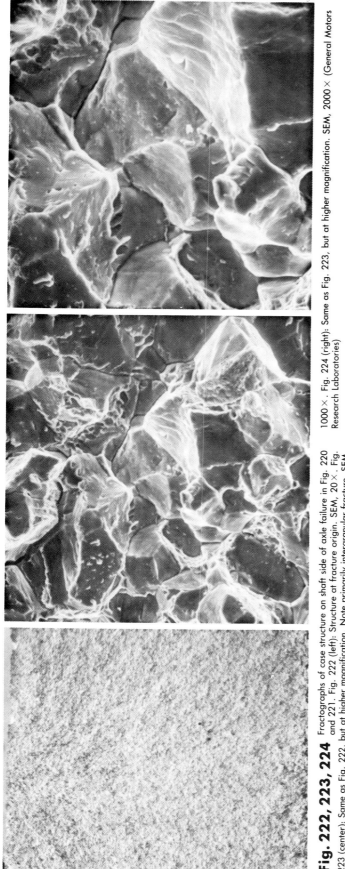

Fig. 222, 223, 224 Fractographs of case structure on shaft side of axle failure in Fig. 220 and 221. Fig. 222 (left): Structure at fracture origin. SEM, 20×. Fig. 223 (center): Same as Fig. 222, but at higher magnification. SEM, 1000×. Fig. 224 (right): Same as Fig. 223, but at higher magnification. SEM, 2000×. (General Motors Research Laboratories)

Fig. 225, 226, 227 Fractographs of core structure on shaft side of axle failure in Fig. 220 and 221. Fig. 225 (left): Case-core transition zone. SEM, 50×. Fig. 226 (center): Core structure at transition zone (see Fig. 225) showing primarily cleavage fracture. SEM, 1000×. Fig. 227 (right): Same as Fig. 226, but at higher magnification. SEM, 2000×. (General Motors Research Laboratories)

272 / Medium-Carbon Steels

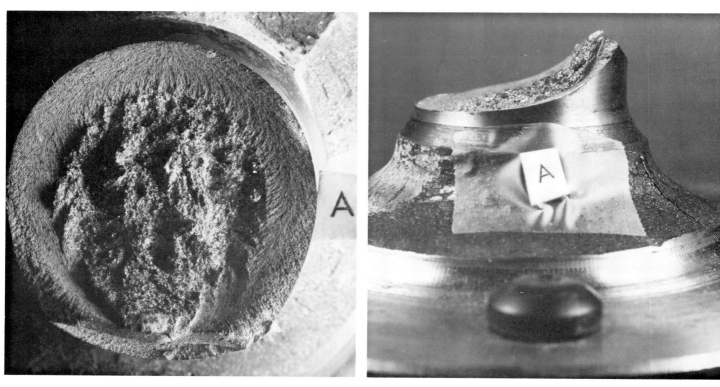

Fig. 228, 229 Classic bending overload fracture of automotive axle shaft (direct-on bearing design). Material and heat treatment: modified AISI 1050, induction hardened and tempered to 60 HRC at the bearing diameter. The part was broken in the laboratory. Fig. 228 (left): Fracture surface with chevrons in case-hardened zone pointing to single origin at top. Fine- and coarse-grained structures in case and core regions, respectively, indicate that heat treatment was properly performed. Fig. 229 (right): Side view of fractured axle shaft in Fig. 228. (Z. Flanders, Packer Engineering Associates, Inc.)

Fig. 230, 231 In-service rotary bending fatigue fracture of automotive axle shaft (direct-on bearing design) due to improper heat treatment. The modified AISI 1050 part was induction hardened and tempered to 60 HRC min at the bearing diameter. However, the hardened zone did not extend into the flange fillet radius. Fig. 230 (left): Fracture surface. Note ratchet marks at several locations around circumference indicating multiple fatigue origins. Fig. 231 (right): Side view of axle shaft in Fig. 230. (Z. Flanders, Packer Engineering Associates, Inc.)

Fig. 232 Surface of a fatigue fracture in a tooth of an AISI 1050 steel gear with a hardness of 55 HRC. The fatigue crack originated at a forging defect (at arrow). Fatigue beach marks are visible near the fatigue-crack origin. Beyond these marks are the steps of a river pattern, caused by cleavage. ~2.5×

Fig. 233 Surface of a fatigue fracture in an AISI 1050 steel shaft, with a hardness of about 35 HRC, that was subjected to rotating bending, showing ratchet marks indicating that fatigue cracks were initiated at many locations along a sharp snap-ring groove. An eccentric pattern of oval beach marks is visible, the presence of which suggests an imbalance in loading. ~2.4×

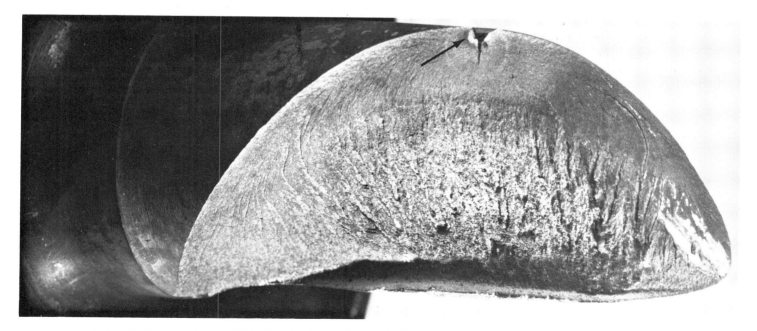

Fig. 234 Surface of a fatigue fracture in a full-floating axle of an AISI 1050 steel induction hardened to about 50 HRC. The axle was tested in reversed torsion. The arrow indicates the origin of the fatigue crack, which grew by shear fatigue to the small, circular beach mark. Subsequent crack growth was by tension fatigue until final fracture (brittle tensile) occurred under a single torsional load. ~2×

274 / Medium-Carbon Steels

Fig. 235, 236, 237 Fatigue failure of a heat-treated automotive bolt (30 HRC). The medium-carbon steel part had been slightly bent prior to onset of fatigue. Fig. 235 (top): Fractured bolt. 0.9×. Fig. 236 (bottom left) and 237 (bottom right): Matching fracture faces. 3× (General Motors Research Laboratories)

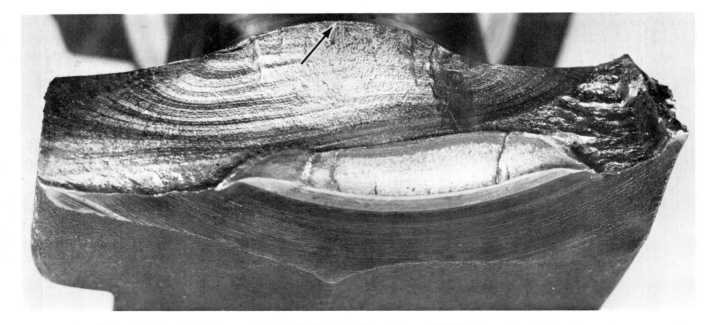

Fig. 238 Surface of a fatigue fracture in a crankshaft of AISI 1046 steel with a hardness of 25 HRC in the region of fracture. Two cracks were generated under a bending stress; one crack started in the journal fillet at the arrow and progressed through the cheek until it met the second crack, which began in the fillet on the opposite side of the cheek. Note the sharply defined beach marks in the upper portion of the fracture surface. ~0.75×

Medium-Carbon Steels / 275

Fig. 239, 240, 241 Fatigue failure of crane wheel by spalling of a section of its selectively hardened tread. Failure of the quenched and tempered medium-carbon steel forging occurred after 2 years of operation on a crane serving an open-hearth furnace. Fig. 239 (left): Spalled section of tread. Fatigue crack originated at a forging defect (arrow). 3×. Fig. 240 (center): Close-up of failure origin. 0.75×. Fig. 241 (right): Oblique lighting reveals texture of fracture surface (compare with Fig. 240). Note beach marks associated with fatigue crack propagation. 0.75× (R.K. Bhargava, Xtek Inc.)

Fig. 242 Single-overload torsional fracture on transverse shear plane of a medium-carbon steel shaft of moderate hardness. Degree of torsional deformation preceding fracture is indicated by counterclockwise twisting of originally straight splines. Final rupture was slightly off-center because of the presence of a relatively low bending force in addition to the torsional force. Fracture face has been severely rubbed and distorted in a circular direction due to contact with the mating fracture surface at the moment of separation. (D.J. Wulpi, Consultant)

276 / Medium-Carbon Steels

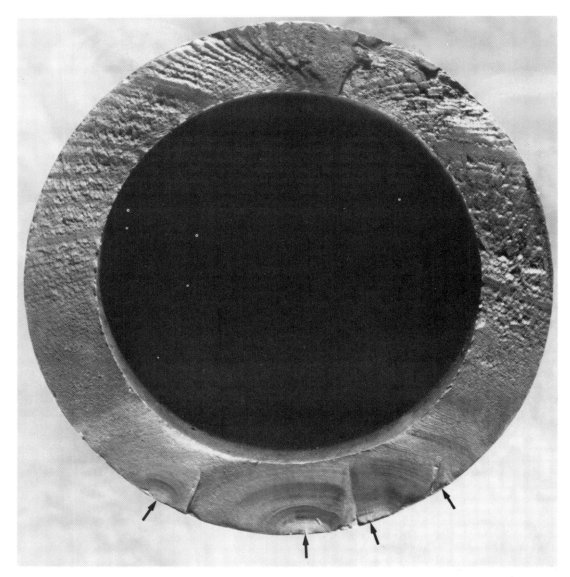

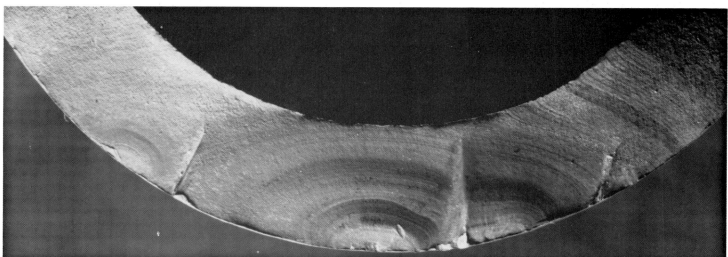

Fig. 243, 244 Fatigue fracture of 90-mm (3.6-in.) diam medium-carbon steel axle housing (217 to 229 HB). The part was subjected to unidirectional bending stresses normal for the application. Fig. 243 (top): Fracture surface showing four major fatigue crack origins (arrows). Cracks progressed up both sides of the tube and joined at the small final rupture area at top. Note increasing coarseness of fracture surface as cracks grew. Fig. 244 (bottom): Higher-magnification view of origin area in Fig. 243. Note beach marks. A small amount of postfracture damage is also present. (D.J. Wulpi, Consultant)

Fig. 245 Surface of a fatigue fracture that occurred, after 732 h of service, in a tooth of the induction-hardened AISI 1053 steel left-side bull gear of a tractor. Hardness of the tooth was 57 HRC at the tip and 50 HRC at a depth of 13.3 mm (0.52 in.). Hardness in the root fillet was approximately 56 HRC near the surface and 50 HRC at a depth of 11.5 mm (0.45 in.). The arrow indicates the origin of the fatigue crack. See also Fig. 246 to 248. ~2.5×

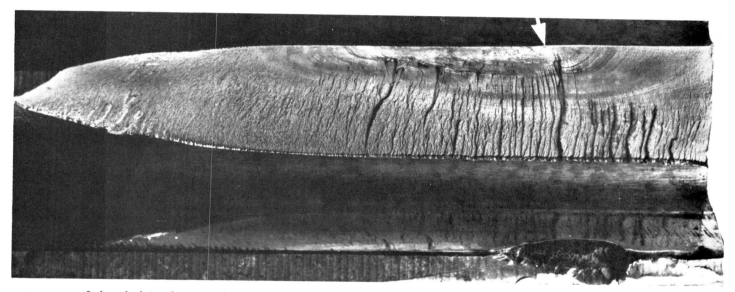

Fig. 246 Surface of a fatigue fracture (similar to that in Fig. 245) that occurred, after 841 h of service, in a tooth of the right-side bull gear of the same tractor. This gear also was made of AISI 1053 steel and induction hardened. Hardness of this gear was similar to that of the gear in Fig. 245 except that the depths at which hardness was 50 HRC were 12.2 mm (0.48 in.) below the tip of the tooth and 9.5 mm (0.37 in.) below the surface of the root fillet. The arrow indicates the fatigue-crack origin. The damage to the tooth just below the one that fractured probably resulted from jamming by a fragment from the fracture. ~2.5×

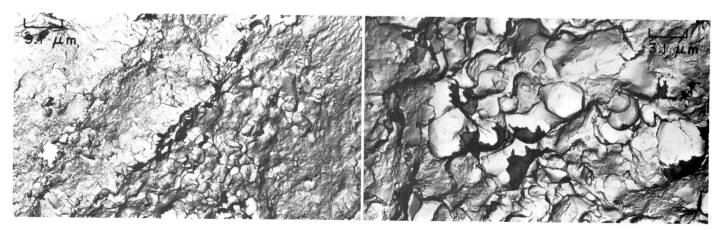

Fig. 247 Area of the fracture surface of the left-side bull gear in Fig. 245, taken near the fatigue-crack origin. Visible are several inclusions believed to have contributed to the formation of the fatigue crack. See also Fig. 248. TEM p-c replica, 1100×

Fig. 248 Portion of the area in Fig. 247, but at higher magnification, showing the inclusions more clearly. Similar inclusions were found near the crack origin in the fracture surface of the right-side bull gear in Fig. 246. TEM p-c replica, 3200×

Fig. 249 to 252 Torsional overload fracture of AISI 1060 drive shaft for power boiler stoker grate. Service temperature: 370 °C (700 °F). Boiler had a history of fractured shafts. This particular grate had jammed twice before, and each time a new key had to be machined due to twisting of the shaft. Mechanical properties at surface of shaft: yield strength, 348 MPa (50.5 ksi); tensile strength, 556 MPa (80.6 ksi); elongation, 31%. Core properties: yield, 300 MPa (43.5 ksi); tensile, 620 MPa (89.9 ksi); elongation, 25%. Fig. 249 and 250 (top row): One half of broken shaft and its fracture surface. Fig. 251 and 252 (bottom row): Mating half of shaft (two of three keyways are visible) and its fracture surface (note twisting and smearing). All at ~0.5× (Z. Flanders, Packer Engineering Associates, Inc.)

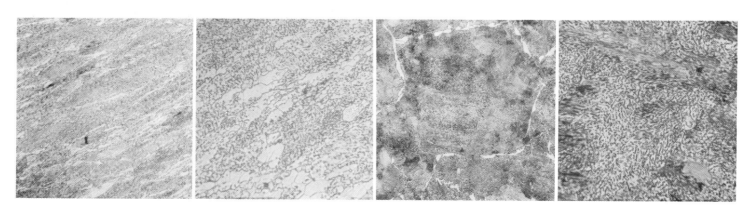

Fig. 253 to 256 Ferrite at prior-austenite grain boundaries and spheroidized pearlite characterize microstructure of drive shaft in Fig. 249 to 252. All etched in 2% nital. Fig. 253 and 254 (left side): Structure at surface of shaft (radial orientation). 100× and 400×. Fig. 255 and 256 (right side): Transverse microstructure. 100× and 400× (Z. Flanders, Packer Engineering Associates, Inc.)

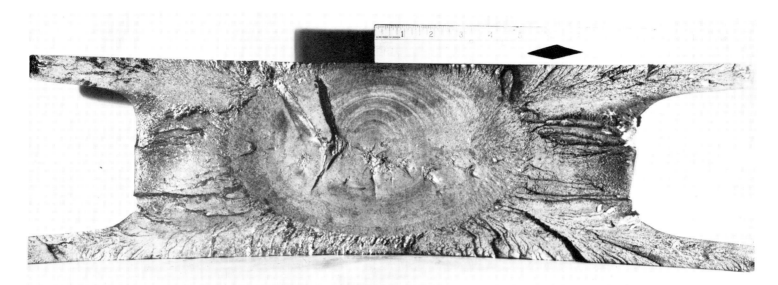

Fig. 257 Surface of a fatigue fracture that occurred, after approximately 1 year of service, in an AISI 1055 steel wheel of a 545-Mg (600-ton) stripper crane. Visible in the center of the fracture is a fatigue zone showing beach marks that are concentric around the fatigue-crack origin, which evidently was an internal flaw approximately 38 mm (1½ in.) beneath the wheel-tread surface. The chemical analysis of the steel was 0.60% C, 0.82% Mn, 0.009% P, 0.024% S, and 0.24% Si, which was acceptably close to specification. Note the chevron patterns in the area of final fast fracture that surrounds the fatigue zone. 0.3×

Fig. 258 Surface of a fatigue fracture that developed in the flange of an AISI 1055 steel wheel of an open-hearth crane after nearly 20 years of service. The fracture surface was exposed by arc cutting the adjacent metal. This fatigue fracture, like the one shown in Fig. 257, was initiated by an internal flaw and propagated outward radially; no external defect was found to be involved in either this fracture or the fracture in Fig. 257. 0.44×

Fig. 259 Fatigue fracture in AISI 1060 steel spring wire, 4.6-mm (0.18-in.) diam, originating at two or more crack nuclei at and above lower shoulder at left edge. 7×

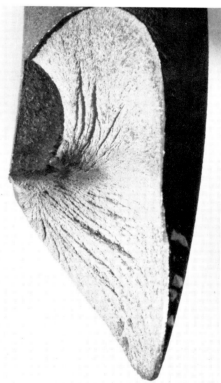

Fig. 260 Surface of a fatigue fracture in 16-mm (⅝-in.) diam spring wire of AISI 1060 steel. The crack origin, which undoubtedly was a surface flaw, generated a fatigue zone (dark area at upper left). The fatigue crack grew slowly, and beach marks became obscured by oxide. ~3×

280 / High-Carbon Steels

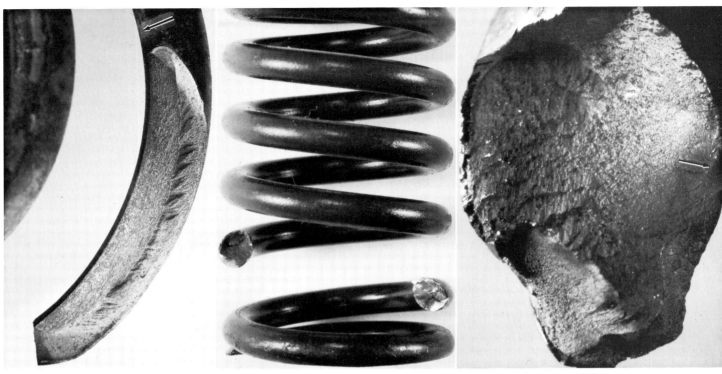

Fig. 261 Fatigue-fracture surface of a suspension spring of AISI 10B62 steel wire with a hardness of 460 HB. Note the fine seam (at arrow), which is the fatigue-crack origin. 2×

Fig. 262 Fatigue-fractured valve spring of 4.8-mm (3/16-in.) diam AISI 1060 steel wire. The wire appears to be free of surface defects, a conclusion that is supported by the evidence in Fig. 263. 2×

Fig. 263 Higher-magnification view of fracture surface in Fig. 262. The fatigue crack began at subsurface origin (at arrow), not at a surface defect. Final fast fracture formed shear lips at top, at left, and at bottom. 15×

Fig. 264 Fatigue fracture in a 16-mm (5/8-in.) diam suspension spring of AISI 10B62 steel wire. The fatigue crack probably originated at a surface defect and grew so slowly that oxidation occurred (note darkness of fatigue zone at left). 3×

Fig. 265 Fractured 13-mm (1/2-in.) diam spring of AISI 10B62 steel wire with a hardness of 477 HB. Note the spiral gouges, which are screw marks that were generated during coiling. See also Fig. 266. 2×

Fig. 266 Fracture surface of the spring in Fig. 265. The crack originated at a screw mark. The small size of the fatigue zone (at left) indicates that final fast fracture occurred soon after crack initiation. 3×

High-Carbon Steels / 281

Fig. 267 Fatigue-fractured spring formed from 16-mm (⅝-in.) diam AISI 1060 steel wire. The seam on the back side of the top turn enters the fracture surface at the dark spot at left edge. See Fig. 268 for a view of the fracture surface. 0.5×

Fig. 268 Fracture surface of the spring in Fig. 267. The fatigue-crack origin is at the seam at left edge. The surface of the wire was gouged after fracture. 3×

Fig. 269 Fatigue-fractured spring of 5-mm (0.200-in.) diam AISI 1060 steel wire (hardness, 43 to 48 HRC). This fracture, unlike that of the similar spring in Fig. 261, originated at the surface (see Fig. 270). 1.7×

Fig. 270 Fracture surface of spring at right side in Fig. 269, shown at higher magnification. The nucleus of the fatigue crack is clearly visible on the wire surface at top, with a succession of beach marks fanning out below it. The surface of the zone of final fast fracture appears to be quite woody. 15×

Fig. 271 Surface of a complex fatigue fracture in a valve spring formed from AISI 1070 steel wire, showing two facets at right angles. 8×

Fig. 272, 273 Fatigue fracture in a spring of AISI 1070 steel wire with a hardness of 54 HRC. Fig. 272 (left): Side view of one fracture surface, deep etched to show a long seam (at arrow). 10×. Fig. 273 (right): Frontal view of the mating fracture surface, showing the fatigue-crack origin (at right), which is at a second seam. 10×

282 / High-Carbon Steels

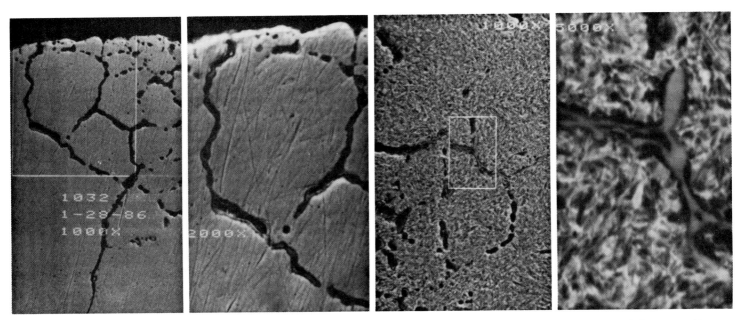

Fig. 274 to 277 Cause of brittle behavior in a small percentage of flat spring parts made of annealed AISI 1070 was traced to a nitrided/nitrogenized layer, 0.25 mm (0.010 in.) thick, produced during heat treating. The heat-treating operation included a 5-min stay in a shaker hearth furnace having a dissociated ammonia atmosphere. However, some of the parts stuck to the furnace hearth, which increased their residence time to as long as 30 min or more. The extended heating time, plus the presence of a small amount of undissociated ammonia, led to formation of the embrittling layer. Fig. 274 and 275 (left side): Grain-boundary damage at surface of AISI 1070 spring part. SEM, 1000× and 2000×. Fig. 276 and 277 (right side): Nitrides (or carbonitrides) in grain boundaries. SEM, 1000× and 5000× (J.H. Maker, Associated Spring, Barnes Group Inc.)

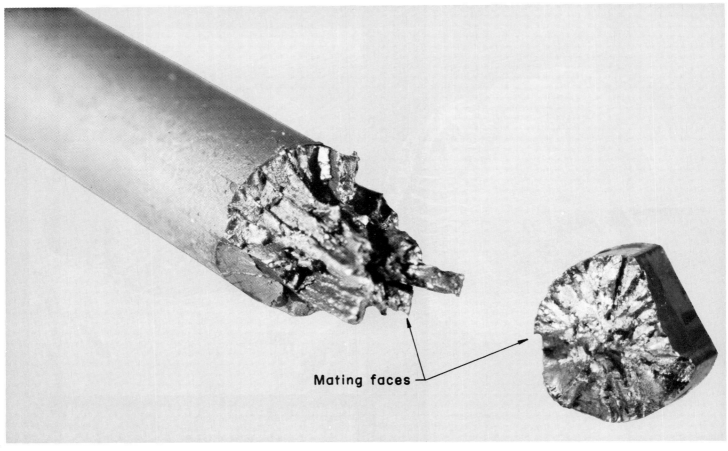

Fig. 278 Torsional fatigue fracture in an 8-mm (5/16-in.) diam torsion bar of AISI 1070 steel that was quenched and tempered to a hardness of 45 HRC. Cracking occurred along both longitudinal and transverse shear planes. 5.6×

High-Carbon Steels / 283

Fig. 279, 280 Power spring made of tempered AISI 1074 strip that fractured during coiling. Note significant surface damage. These defects were probably caused by mechanical action or lubrication breakdown during strip production while the material was in the annealed condition. Fig. 279 (left): SEM, 25×. Fig. 280 (right): Same as Fig. 279, but at higher magnification. SEM, 100×. (J.H. Maker, Associated Spring, Barnes Group Inc.)

Fig. 281, 282 Low-cycle fatigue fracture of AISI 1074 strip. Failure initiated at defects that were probably formed during hot rolling of the band. Binocular examination of defects revealed dark (oxidized) areas. Metallographic sections contained decarburized regions intimately associated with the defects. Fig. 281 (left): SEM, 50×. Fig. 282 (right): Same as Fig. 281, but at higher magnification. SEM, 250×. (J.H. Maker, Associated Spring, Barnes Group Inc.)

284 / High-Carbon Steels

Fig. 283, 284, 285 Fracture of power spring made of tempered AISI 1074. Initially, a "scab" or some other type of mechanical damage was suspected as the cause of failure because of the surface appearance of the steel. However, SEM examination revealed that fracture was due to a large nonmetallic inclusion (probably alumina, type B) located within 6 to 13 μm (0.25 to 0.50 mils) of the surface. Fig. 283 (left): Surface of fractured power spring and portion of fracture area showing origin. Note surface damage near "C" and prominent feature, "D," at edge near fracture origin. SEM, 50×. Fig. 284 (center): Close-up of feature at fracture edge in Fig. 283 reveals the subsurface inclusion at "D" and its effects on surface appearance ("C"). SEM, 400×. Fig. 285 (right): Longitudinal microsection near origin. SEM, 200× (J.H. Maker, Associated Spring, Barnes Group Inc.)

Fig. 286, 287, 288 Hydrogen embrittlement fracture of AISI 1074 part. The part was formed in the annealed condition, quenched and tempered to 48 HRC, and then electroplated. A surface imperfection in the steel was initially thought to have played a key role in the fracture. However, SEM examination revealed an intergranular fracture mode indicative of hydrogen embrittlement due, in this case, to insufficient postplate baking. Fig. 286 (left): Fractured part after stripping of electroplated coating. Note surface imperfection (boxed) near edge. SEM, 30×. Fig. 287 (center): Close-up of boxed area in Fig. 286. SEM, 150×. Fig. 288 (right): High-magnification view of fracture surface reveals that part actually failed intergranularly due to hydrogen embrittlement. Note secondary cracking. SEM, 1500× (J.H. Maker, Associated Spring, Barnes Group Inc.)

High-Carbon Steels / 285

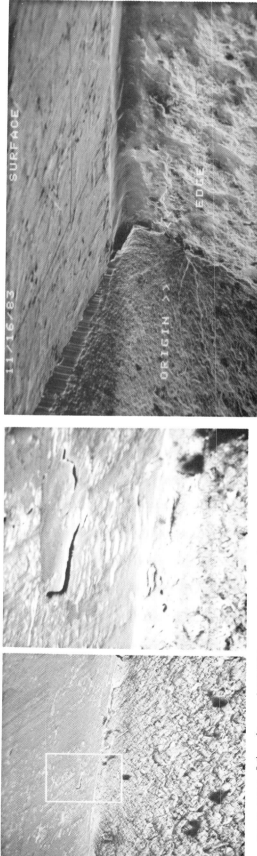

Fig. 289, 290 Failure of tempered AISI 1074 power spring due to galling. Power springs should be lubricated between coils and should not be wound too tight. When these requirements are not met, the resultant abrasion between overlapping coils can produce galling and, occasionally, frictionally induced, untempered martensite. Maximum stress also rises sharply under these conditions. In this case, galling is the only visible cause of failure. Fig. 289 (left): SEM, 80×. Fig. 290 (right): SEM, 400× (J.H. Maker, Associated Spring, Barnes Group Inc.)

Fig. 291 Fatigue failure of flat, cantilever-type AISI 1070 spring due to inadequate removal of blanking fracture. Failure initiated at a point on the edge of the spring. SEM, 100× (J.H. Maker, Associated Spring, Barnes Group Inc.)

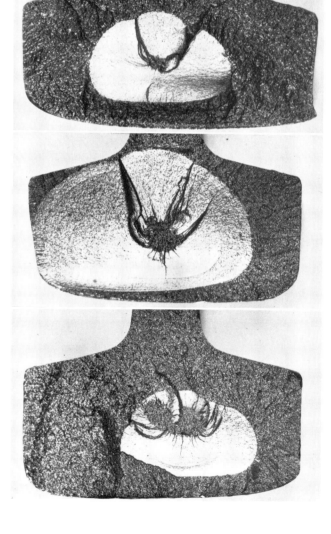

Fig. 292 to 295 Figures 292 (left), 293 (left center), and 294 (right center) show surfaces of "transverse fissures" that occurred in the heads of AREA 59-kg (131-lb) railroad rails made of AISI 1075 steel. All three fissures were caused by the presence of hydrogen-induced flakes (or "fisheyes"), which appear as dark gray areas within the bright zones. These bright zones are fatigue cracks that were initiated by the flakes. Beach marks are visible in the fatigue zones, particularly near the initiations of final fast fracture. Figure 295 (at far right) shows the surface of a longitudinal fracture in an AREA 59-kg (131-lb) rail of AISI 1075 steel that had been hydrogen embrittled. The white arrowhead points to an unusually large nonmetallic stringer that acted as a site for hydrogen segregation. All shown at actual size

286 / High-Carbon Steels

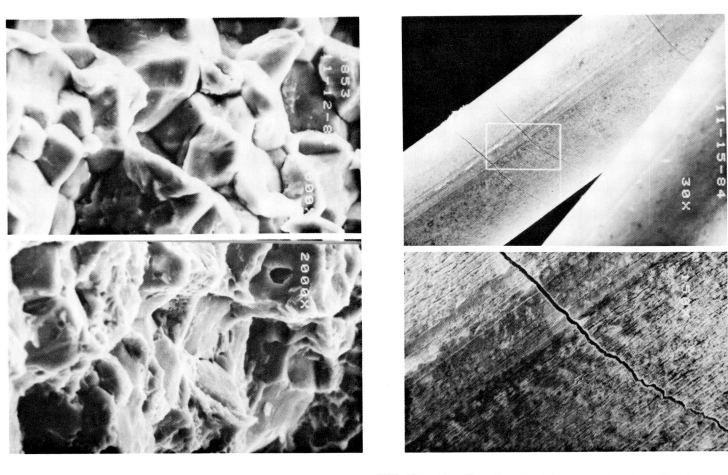

Fig. 296 to 300 Premature fatigue failures of springs due to hydrogen embrittlement. Cause was traced to a combination of inadequate stress relieving and use of an acid cleaning process. Material: ASTM A228 (0.70 to 1.0% C). Fig. 296 (top left): Three hydrogen/residual tensile stress induced cracks on wire surface. SEM, 30×. Fig. 296 (top center): Boxed area in Fig. 296. SEM, 150×. Fig. 298 (top right): Surface of one wire fracture. Hydrogen crack is at upper left and is 0.35 mm (0.014 in.) deep. At lower right is region of fatigue. SEM, 100×. Fig. 299 (bottom left): High-magnification view of intergranular hydrogen crack on fracture surface of ASTM A228 wire in Fig. 298 (compare with Fig. 300). SEM, 2000×. Fig. 300 (bottom right): Morphology of fatigue crack on fracture surface of ASTM A228 wire in Fig. 298 (compare with Fig. 299). SEM, 2000× (J.H. Maker, Associated Spring, Barnes Group Inc.)

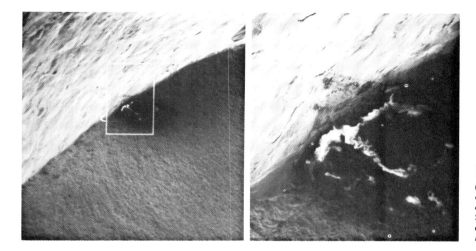

Fig. 301, 302 Fatigue failure of spring made of oil-tempered carbon valve spring wire (ASTM A230). Failure occurred after 10 000 cycles of engine operation. Fracture surface morphology is atypical. One possible cause of failure: a ribbonlike subsurface inclusion. Fig. 301 (left): SEM, 40×. Fig. 302 (right): Higher-magnification view of boxed area in Fig. 301. SEM, 200×. See also Fig. 303 and 304. (J.H. Maker, Associated Spring, Barnes Group Inc.)

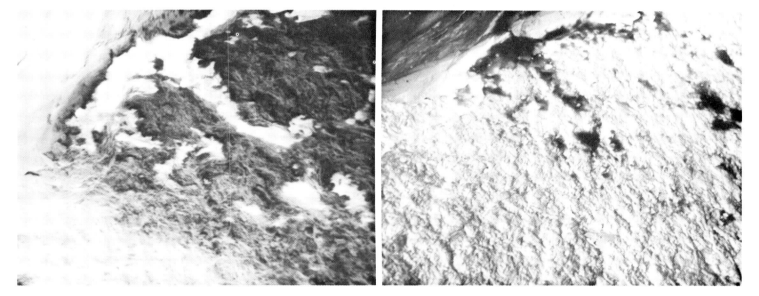

Fig. 303, 304 Higher-magnification views of the fractured ASTM A230 spring wire shown in Fig. 301 and 302. Fig. 303 (left): Secondary electron SEM image, 500×. Fig. 304 (right): Backscattered electron SEM image, 300×. See the article "Scanning Electron Microscopy" in this Volume for additional information on secondary and backscattered electron images. (J.H. Maker, Associated Spring, Barnes Group Inc.)

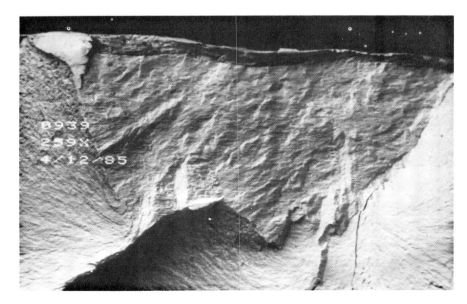

Fig. 305 Fatigue failure of an automotive engine valve spring made of a steel similar to ASTM A230. The spring was shot peened; service stresses were very high. Cause of fracture was a seam, 15 μm (0.5 mils) deep. The surface defect initiated a longitudinal shear crack that propagated to a depth of 75 to 100 μm (3 to 4 mils), where the crack root then induced stage I fatigue. SEM, 250× (J.H. Maker, Associated Spring, Barnes Group Inc.)

288 / High-Carbon Steels

Fig. 306 Surface of a fatigue fracture, which occurred after about 1 year of service, in an AISI 1095 steel spring of a 225-Mg (250-ton) shear. A fatigue zone is evident at the inner (right) edge of the fracture surface (at arrow). This is a normal location for a fatigue-crack origin in a coiled spring, and the fracture may be rated as typical. The fatigue crack penetrated a distance of approximately 13 mm (½ in.) into the spring before final fast fracture occurred. See also Fig. 307. Actual size

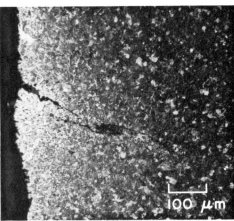

Fig. 307 Polished and etched section through the spring in Fig. 306, taken adjacent to the fracture. Visible is a cracklike defect that probably nucleated the fatigue crack. Decarburization (white zone) around the defect indicates that the defect is a seam, which existed prior to service. Nital etch, 100×

Fig. 308 Deep "shell crack" and "detail fracture" in head of railroad rail removed from service. Composition of standard carbon rail steel: 0.60 to 0.82% C, 0.70 to 1.00% Mn, and 0.10 to 0.23% Si. Gage side of rail is at left. A 5- to 10-mm (¼- to ⅜-in.) thick layer has been removed from the running surface to expose the shell crack (top). Deep shells such as this one typically nucleate at stringers or microinclusions and propagate approximately parallel to the running surface. Some shells form a branch crack, or detail fracture, in the transverse plane (lower portion of figure) that can cause a service break if not detected in time. Detail fractures typically have featureless crack propagation surfaces, but may exhibit beach marks (as in this example) that reflect the traffic pattern or temperature (thermal stress) changes during service. See also Fig. 309 and 310. 4.25× (J.M. Morris, U.S. Department of Transportation)

Fig. 309, 310 Detail fracture (see Fig. 308) in head of railroad rail that had been removed from revenue track and installed in test track to determine crack growth rate under heavy train traffic. Material: high-carbon rail steel (0.69 to 0.82% C). Fig. 309 (top): Featureless region inside contour 1 outlines crack at time of removal from service. Contours 1 through 9 and the unlabeled contours that follow identify crack propagation steps during testing. Contours separate inclined ridges that result from the unique traffic pattern on the test track. The detail fracture grew in fatigue to 80% of rail head area before breaking under the test train. 2.4×. Fig. 310 (bottom): Longitudinal section along diagonal line in Fig. 309. Contours 1 through 9 are identified. Note ridge structure of crack surface that results from combined stage I and stage II fatigue crack propagation. The less prominent, horizontal ridges in Fig. 309 result from combined stage I/stage III propagation. 13× (J.M. Morris, U.S. Department of Transportation)

290 / High-Carbon Steels

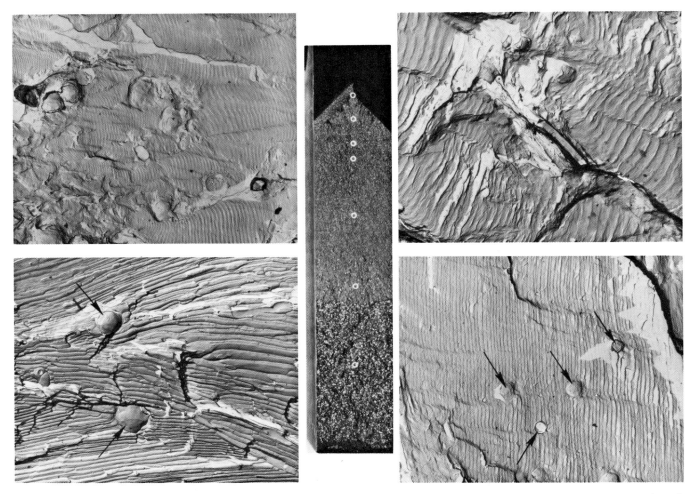

Fig. 311 to 315 Fatigue crack propagation in compact tension test specimen of rail steel (0.69 to 0.82% C). Fig. 311 (center): Fracture surface of typical specimen from rail head. Crack plane lies in the cross section. First two circles at top identify precracked region. Fatigue crack propagation extends from second to sixth circle (from top to bottom). Bottom one third of surface is region of fast fracture. 1.66×. Fig. 312 (top left) and 313 (top right): Fractographs of fatigue propagation region in Fig. 311 illustrate typical fatigue striations. Note wavelike crests and variations of wavelength near inclusion sites. TEM replicas, 4900× and 4000×. Fig. 314 (bottom left) and 315 (bottom right): Two additional fractographs of fatigue crack propagation region in Fig. 311. Sharp crest outlines and lack of wavelength change near inclusion sites (arrows) identify these features as lamellar pearlite structure (see Fig. 316). TEM replicas, both at 4000× (J.M. Morris, U.S. Department of Transportation)

Fig. 316 Photomicrograph shows typical lamellar pearlite structure in railroad rail steel (0.69 to 0.82% C). The interlamellar spacing is approximately 300 nm. This structure is sometimes mistaken for fatigue crack growth striations in SEM and TEM fractographs (see Fig. 311 to 315). 5000× (J.M. Morris, U.S. Department of Transportation)

Fig. 317, 318 Fractographs of fast fracture region in rail steel (0.69 to 0.82% C) compact tension specimen similar to that in Fig. 311. Fig. 317 (left): Typical transgranular cleavage. At right-center is a cleavage region that follows steps along pearlite lamellae. TEM replica, 5000×. Fig. 318 (right): Comparison of transgranular cleavage (left) with intergranular cleavage (right). TEM replica, 5000× (J.M. Morris, U.S. Department of Transportation)

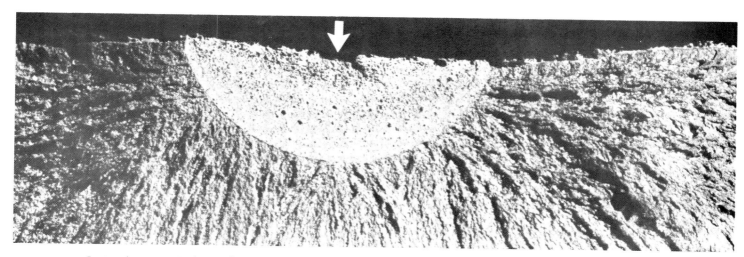

Fig. 319 Fracture due to corrosion fatigue of a 115-mm (4.5-in.) diam API 5A, grade E, seamless drill pipe. The fine-grained 0.4C-Mn alloy steel (~AISI 1340) was used in the normalized condition in an oil-field application where chlorides and sulfides were present in the strata being drilled. Fracture initiated at a corrosion pit (arrow) on the inside diameter of the pipe. 5.5× (E.V. Bravenec, Anderson & Associates, Inc.)

Fig. 320, 321, 322 Brittle fractures of AISI 4130 hitch post shafts due to thermal stresses from welding imposed on an improperly heat-treated region. The tapered shaft had a maximum diameter of about 305 mm (12 in.) and a length of approximately 1 m (40 in.). It was forged, rough machined, heat treated—845 °C (1550 °F), water quench, temper at 595 °C (1100 °F)—and a small bracket was welded to the larger diameter end. Fracture occurred after a short time in service or, in some cases, immediately after welding. Cause: Incomplete austenitization left a coarse structure at the shaft center characterized by a high residual stress level and low ductility. Thermal stresses accompanying welding initiated brittle cracks in this region, which then propagated to the surface. Fig. 320 (top): Transverse section of field failure. Fig. 321 (bottom left): Fracture surface near center of shaft. Failure mode is brittle fracture. Note river patterns. SEM, 100×. Fig. 322 (bottom right): Fractograph of near-surface region of shaft. Failure mode: dimple rupture. SEM, 100× (R.W. Bohl, University of Illinois)

292 / AISI/SAE Alloy Steels

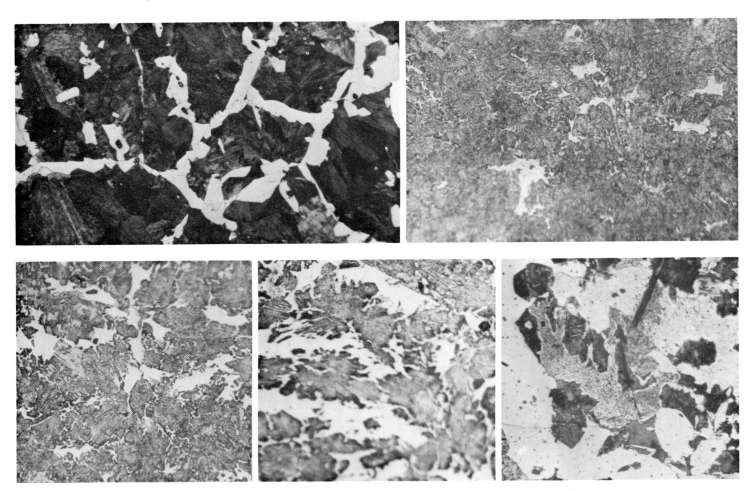

Fig. 323 to 327 Metallographic study of hitch post shaft failure in Fig. 320 to 322. All etched in 2% nital. Fig. 323 (top left): Coarse, as-forged structure of AISI 4130 shaft. 100×. Fig. 324 (top right) and 325 to 327 (bottom row): Microstructures at increasing depths beneath surface of fractured shaft. Gradual coarsening is evidence of superficial austenitization. Fig. 324: 4 mm (0.15 in.). 150×. Fig. 325: 11 mm (0.45 in.). 150×. Fig. 326: 30 mm (1.2 in.). 150×. Fig. 327: 75 mm (3 in.). 225× (R.W. Bohl, University of Illinois)

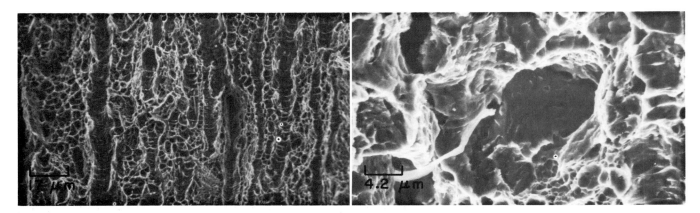

Fig. 328, 329 Two SEM views, at different magnifications, of the surface of a fracture in a specimen of AISI 4130 steel that was austenitized at 845 °C (1550 °F), oil quenched, and tempered for 2 h at 250 °C (480 °F). The fracture was produced in a toughness test in air; the rate of crack growth was in the range of rapid, unstable fracture. Dimples in an extremely wide variety of sizes and channels formed by linear coalescence of voids are evident. See Fig. 330 to 335 for the fracture behavior of this material when tested in a hydrogen atmosphere. Fig. 328: 600×. Fig. 329: 2400×

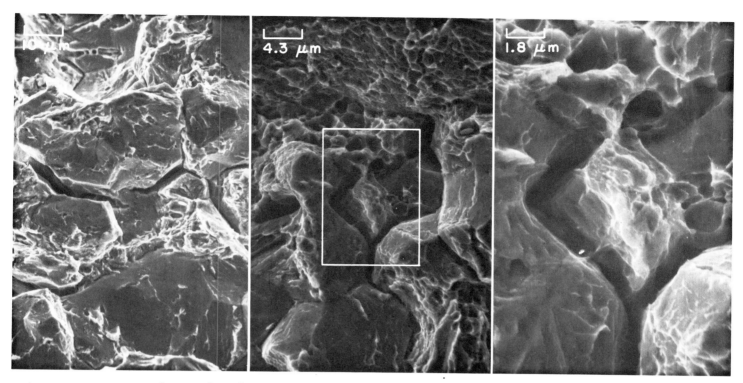

Fig. 330, 331, 332 Fracture surfaces of two specimens of AISI 4130 steel that were given the same heat treatment as described in Fig. 328 and 329. The specimen shown in Fig. 330 (left) was tested in 1 atm of hydrogen at 25 °C (75 °F); the specimen shown in Fig. 331 (center) and Fig. 332 (right) was tested in 1 atm of hydrogen at 80 °C (175 °F). Note that the area within the rectangle in Fig. 331 is shown at higher magnification in Fig. 332. In contrast to those in Fig. 328 and 329, these fracture surfaces show completely intergranular fractures with deep secondary cracks between the grains. The fracture in Fig. 331 and 332 also shows a considerable number of dimples that were formed by microvoid coalescence. See Fig. 333 to 335 for the fracture behavior of this material in an atmosphere of partially dissociated hydrogen. Fig. 330: SEM, 1000×. Fig. 331: SEM, 2350×. Fig. 332: SEM, 5600×

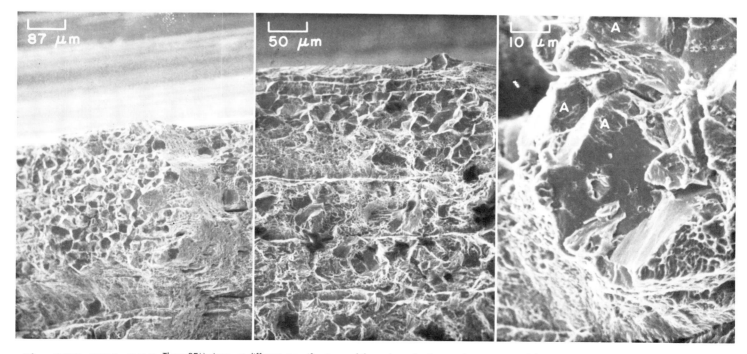

Fig. 333, 334, 335 Three SEM views, at different magnifications, of the surface of a fracture in a specimen of the same steel and heat treament as in Fig. 328 to 332, tested for stable crack-growth rate in 0.1 atm of partially dissociated hydrogen at 80 °C (175 °F). Similar to the fracture in Fig. 331 and 332, this fracture shows facets of intergranular rupture intermingled with areas of dimples. The bottom of the notch in the specimen is visible at top in Fig. 333 (left) and Fig. 334 (center). The horizontal marks in the fracture surface parallel with the notch could be indications of temporary arrests of crack penetration. The hairline indications (marked A) visible in Fig. 335 (right) are fine tear ridges typical of hydrogen embrittlement. Fig. 333: 115×. Fig. 334: 200×. Fig. 335: 1000×

Fig. 336 Fracture surface of a broken AISI 4140 steel load cell. The radial marks indicate that two crack origins formed at the bottom surface of the cell, apparently near the toe of a weld. Note the continuous shear lip around the top edge. See also Fig. 337 and 338. Actual size

Fig. 337 Mating fracture surface to that shown in Fig. 336. There appears to be no evidence of fatigue marks; however, fatigue cracking may have taken place within a very small area at the crack origins. Fracture was evidently caused by a sudden overload. See also Fig. 338. Actual size

Fig. 338 Magnified view of the fracture surface in Fig. 336, taken at an area away from the edge, and showing the dimples that are typical of transgranular fast fracture. Vertical marks are sulfide stringers. SEM, 100×

Fig. 339 Fracture surface of an AISI 4140 steel specimen that was heat treated to high strength and broken in fatigue. Deep secondary cracks are present. Features that appear to be fatigue striations (such as at arrows) may actually be fissures. SEM, 720×

AISI/SAE Alloy Steels / 295

Fig. 340 Surface of a fatigue fracture in a stud of AISI 4140 steel with hardness of 34 HRC in the core and 45 HRC at the surface. The crack originated at the bottom in this view, in a well-rubbed fatigue zone, and penetrated nearly 90% of the cross section of the stud before final fast fracture occurred. 2×

Fig. 341 Surface of a fatigue fracture in a suspension component of AISI 4140 steel heat treated to a hardness of 273 HB. Two fatigue zones are evident, at lower left and at upper right; the remainder of the fracture surface is the zone of final fast fracture. 5×

Fig. 342 Fatigue fracture in an AISI 4140 steel tail-rotor drive-pinion shaft of a helicopter, heat treated to a hardness of 35 HRC. See also Fig. 343 to 346. Actual size

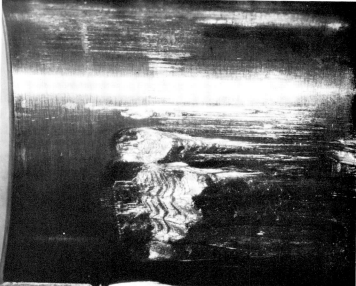

Fig. 343, 344 Figure 343 (left) is a view of the threaded end of the fractured shaft in Fig. 342, after removal of the pinion gear. Figure 344 (right) is a view of the bore of the pinion gear. Both photographs display severe galling damage of the contacting surfaces, which resulted in the gear being frozen on the shaft; the gear was removed by cutting through it from both sides. The appearance of the gouge marks indicates that most of the galling damage was done in forcing the gear onto the shaft. Figure 343 also shows a keyway and an oblique view of a fracture surface; the fracture originated at a location on the keyway edge that is not visible here. Both at 5×

296 / AISI/SAE Alloy Steels

Fig. 345, 346 Two views of a surface of the fatigue fracture in the tail-rotor drive-pinion shaft in Fig. 342, taken with the key and the pinion gear in place. The fatigue-crack origin (arrows) is in the keyway, slightly to the right of the corner formed by the edge of the keyway and the outside surface of the shaft. Not visible here are deep tool marks and machining tears on the inside of the keyway, which were the sources of the stress concentrations responsible for initiation of the fatigue crack. The smooth fracture surface shown in these views is typical of those produced by high-cycle fatigue in combined bending and torsion. Fig. 345 (left): 2×. Fig. 346 (right): 10×

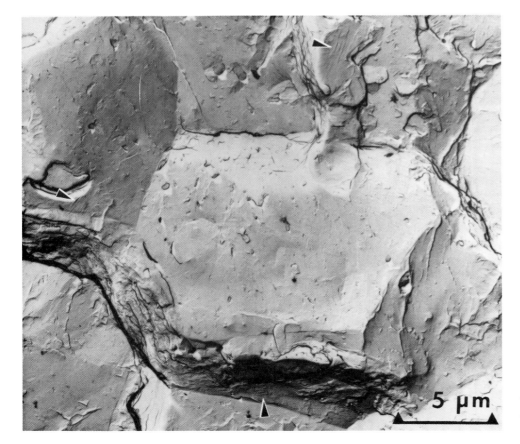

Fig. 347 Fatigue fracture along prior-austenite grain boundaries in as-quenched AISI 4140. Note striations (arrows) at places where the crack met grain-boundary carbides. TEM replica, 5000× (I. Le May, Metallurgical Consulting Services Ltd.)

AISI/SAE Alloy Steels / 297

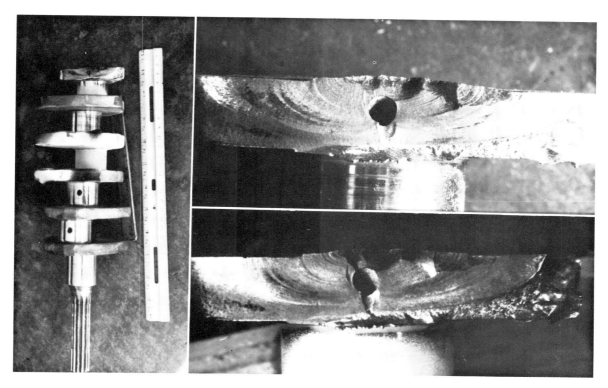

Fig. 348 Outboard-motor crankshaft, forged of AISI 4140 steel and heat treated to a hardness of 35 to 40 HRC, that fractured in service through a throw (at top). Chromium plating had been used to build up worn journals. See also Fig. 349 to 353. ~0.25×

Fig. 349, 350 Figure 349 (top) is a view of the surface of the fracture in Fig. 348, showing several small crack nuclei above the oil hole. Fatigue cracks from these nuclei merged to form a single crack front, which nearly penetrated the entire cross section of the throw before final fast fracture occurred. Figure 350 (bottom) is another view of the surface of the fracture in Fig. 348, photographed using illumination that reveals more of the details of the area of final fast fracture than does Fig. 349. The fatigue crack was initiated at a bottom fillet of the No. 2 main-bearing journal, which was chromium plated. Fig. 349: ~1.7×. Fig. 350: ~1.3×

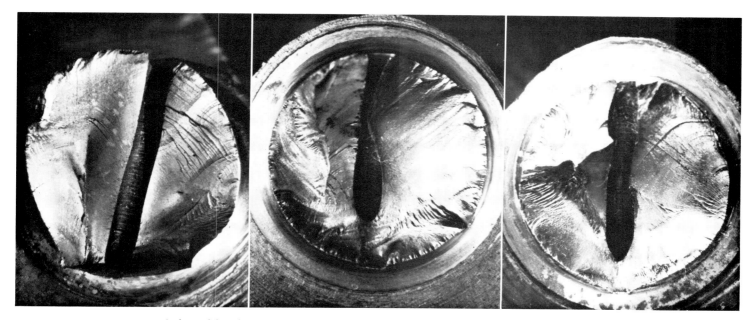

Fig. 351, 352, 353 Surfaces of three fatigue fractures in crankshafts the same as that in Fig. 348; in each of these crankshafts, however, fracture progressed through a plated connecting-rod journal, rather than through a throw as in Fig. 348. Each of these fractures originated at multiple nuclei along a fillet. Fatigue cracks began at these nuclei and then joined to form a single front along a portion of the journal periphery. Final separation was by fast fracture. In the fracture at left (Fig. 351), there are three or four crack nuclei at the right edge; the fatigue-crack front penetrated past the oil hole before final fast fracture occurred. In the fractures at center (Fig. 352) and right (Fig. 353), fatigue cracks formed at many nuclei at the bottom edge, and the fatigue-crack front followed the oil hole toward the opposite edge. In Fig. 353, a second crack front, independent of the main front, advanced from upper left. All at ~1.6×

Fig. 354 Ductile fracture of quenched and tempered AISI 4140 showing elongated dimples on fracture surface. Dimple shape is related to mode of fracture—in this case, either shearing or tearing. Matching replicas from adjacent surfaces are needed to distinguish between the two. TEM replica, 8000×. (I. Le May, Metallurgical Consulting Services Ltd.)

Fig. 355 Slip band cracks on quenched and tempered AISI 4140. The cracks form most often in more ductile metals during fatigue crack growth on multiple slip planes. Their formation is due to the restraint imposed on reverse slip by large carbides or other second-phase particles in conjunction with the large amount of forward slip taking place at high stress levels. TEM replica, 2000×. (I. Le May, Metallurgical Consulting Services Ltd.)

Fig. 356 Tire tracks on fatigue fracture surface of AISI 4140 quenched and tempered at 700 °C (1290 °F). This feature is unique to fatigue fractures and is frequently observed in cases where crack propagation rate was high and few or no striations formed (as in low-cycle fatigue). In these instances, tire tracks represent valuable evidence that fatigue was indeed the operative failure mode. TEM replica, 2080×. (I. Le May, Metallurgical Consulting Services Ltd.)

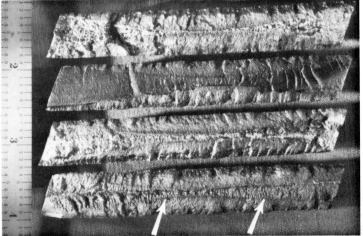

Fig. 357, 358 Fatigue fracture of AISI 4140 bull gear due to improper heat treatment. The one-of-a-kind replacement gear had a service life of just 2 weeks. Heat treatment did not produce full hardness in gear teeth. Hardness at tooth face, 15 HRC; at tooth core, 82 HRB. Fig. 357 (left): Outside diameter of gear showing where teeth broke off. Fatigue pattern on fracture surfaces is typical of reversed bending even though the gear was driven in only one direction. 0.6×. Fig. 358 (right): Fracture surfaces of four teeth. Fatigue cracks propagated from both roots of each tooth. Arrows point to area of final overload on one tooth. 0.84×. (Z. Flanders, Packer Engineering Associates, Inc.)

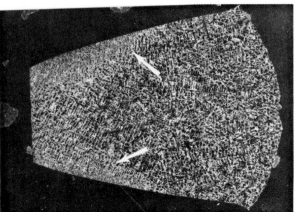
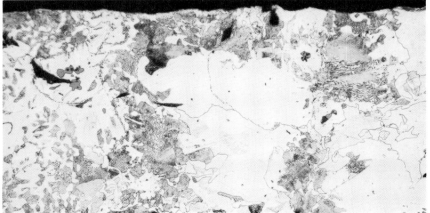

Fig. 359, 360 Structure of fractured tooth from AISI 4140 bull gear in Fig. 357 and 358. Fig. 359 (left): Tooth cross section. Arrows show depth to which temperature exceeded Ac_1 during heat treatment. Fig. 360 (right): Near-surface hardened region of tooth. Large ferrite indicates that complete austenitization was not obtained during heat treating. 2% nital, 200×. (Z. Flanders, Packer Engineering Associates, Inc.)

AISI/SAE Alloy Steels / 299

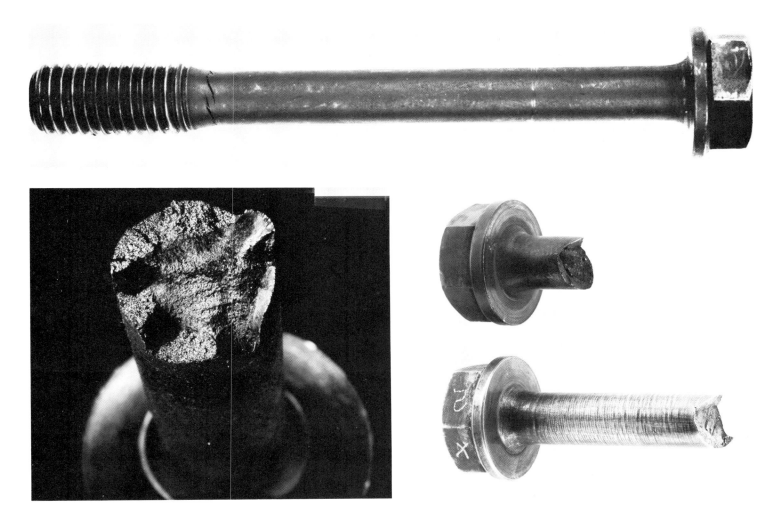

Fig. 361, 362, 363 Spontaneous rupture of high-strength diesel engine bearing cap bolts due to sulfide stress-corrosion cracking. Bolts were made of 42 CrMo 4 (~AISI 4140), quenched and tempered to 38 to 40 HRC. Parts were supplied by two vendors. Only bolts from one vendor failed, and all failures were bolts from main bearing caps 2 and 4 of the four-cylinder engine (those removed from bearing inspection after engine run-in and before final assembly). The vendor whose bolts were failure-free finish machined the parts after heat treating. This produced high compressive residual stresses on their surface and rendered them resistant to stress corrosion by effectively reducing the applied tensile stress on the shank. The vendor whose bolts failed heat treated the parts after machining—bolt surfaces were essentially free of beneficial residual stresses. The aggressive environment needed for stress-corrosion cracking was provided by water containing hydrogen sulfide. The H_2S derived from chemical breakdown of an engine oil additive. The moisture was admitted to the normally sealed bearing cap chamber surrounding the bolt shank when the easy-to-access caps 2 and 4 were removed from hot engines after run-in. Fig. 361 (top): A reduced shank bearing cap bolt. Location of cracks and fracture on bolt shank was random. In this case, circumferential cracks were found adjacent to the threaded section. 1.5×. Fig. 362 (bottom left): Fracture surface of a failed bolt. Several crack origins can be seen at left on outer edge. These multiple origins occur in a stepwise fashion and have bright, faceted surfaces. Central region is fibrous and resulted from ductile overloading. At the right edge are shear lips formed during final separation. 4×. Fig. 363 (bottom right): Comparison of sulfide stress-corrosion failure produced in laboratory (left) with broken bolt removed from an engine (right). Note similarity of fracture surfaces (origins are at right in both cases). 1.5× (R.D. Zipp, J.I. Case, a Tenneco Company)

Fig. 364, 365, 366 Fractographic analysis of sulfide stress-corrosion failures of bearing cap bolts in Fig. 361 to 363. Although this analysis is of a failure induced in the laboratory, the mode is very similar to that of bolts that failed in engines. Fig. 364 (left): Failure origin area with intergranular fracture continuous to the surface. SEM, 620×. Fig. 365 (center): Approximately 0.4 mm (0.015 in.) from origin. Mode of crack propagation is intergranular. Note corrosion products and secondary cracking. TEM replica, 1560×. Fig. 366 (right): Nonembrittled area in overload region exhibits primarily a dimpled-rupture mode of fracture. TEM replica, 3240× (R.D. Zipp, J.I. Case, a Tenneco Company)

300 / AISI/SAE Alloy Steels

Fig. 367 to 371 Fracture of AISI 4146 axle due to improper induction hardening that caused burning (incipient melting). A surface hardness of 50 to 58 HRC was specified. Actual hardness: approximately 15 HRC. Fig. 367 (top left): Fracture surface of shaft. Fig. 368 (top right): Fractography of intergranular portion of fracture surface. SEM, 15×. Fig. 369 (bottom left): Higher-magnification fractograph of intergranular portion of fracture surface (see Fig. 368) reveals partial melting of grain boundaries. Also note very large grain boundaries. SEM, 90×. Fig. 370 (bottom center): Microstructure of intergranular region of fracture surface. 2% nital, 165×. Fig. 371 (bottom right): High-magnification view of microstructure at burned grain boundary. Note decarburization, intergranular oxidation, and presence of spheroidal oxides. 2% nital, 920×. (Z. Flanders, Packer Engineering Associates, Inc.)

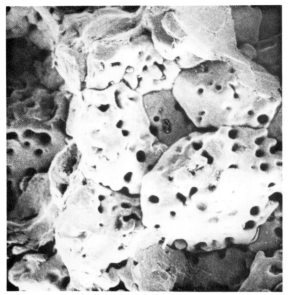

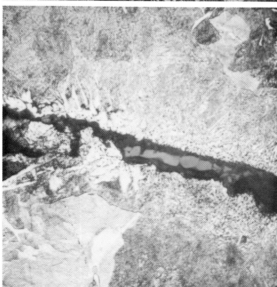

Fig. 372, 373, 374 Figure 372 (left) shows a fractured AISI 4150 steel splined shaft heat treated to a hardness of 43 HRC. Note that there is a secondary transverse crack below the main fracture. Figure 373 (center) is a view of one surface of the fracture in Fig. 372, showing a very complicated starlike pattern. Close examination of the many arms of the "star" reveals the presence of fatigue beach marks produced by crack fronts that propagated separately but simultaneously. Particularly evident are the beach marks on the arms at bottom, which are shown at higher magnification in Fig. 374 (right). Sharp radii at the bases of the spline appear to have been the origins of the fatigue cracks. Fig. 372: 0.5×. Fig. 373: 1.5×. Fig. 374: 3.5×

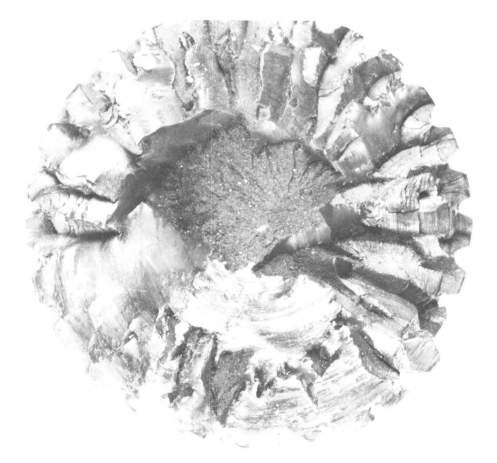

Fig. 375 Reversed torsional fatigue fracture of splined shaft due to overtempering. The SAE 4150 part was oil quenched and tempered to 34 HRC throughout—a hardness too soft for the application. Note the "starry" pattern characteristic of multiple fatigue cracks. 3.5× (D. Roche and H.H. Honegger, California Polytechnic State University)

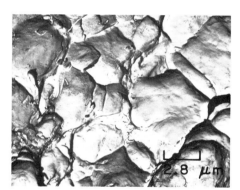

Fig. 376 Surface of a tension-overload fracture in a specimen of AISI 4315 steel that was quenched from 980 °C (1800 °F) and tempered at 315 °C (600 °F). Note the large, equiaxed dimples. TEM p-c replica, 3600×

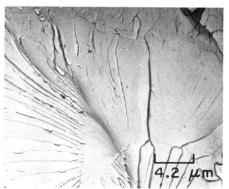

Fig. 377 Surface of a transgranular cleavage fracture, caused by hydrogen embrittlement, in AISI 4315 steel heat treated same as Fig. 376. Note cleavage steps that originated at tilt boundaries. TEM p-c replica, 2400×

302 / AISI/SAE Alloy Steels

Fig. 378, 379 How hydrogen damage affects fracture in AISI 4340. Shown are side views (Fig. 378, left) and fracture surfaces (Fig. 379, right) of 13-mm (0.505-in.) diam tensile specimens. Specimen A was tempered at 400 °C (750 °F) for 114 min in argon. Note specimen B's more ductile behavior in the tensile test. See also Fig. 380 to 384. Both at actual size (S. Harding and S.W. Stafford, University of Texas)

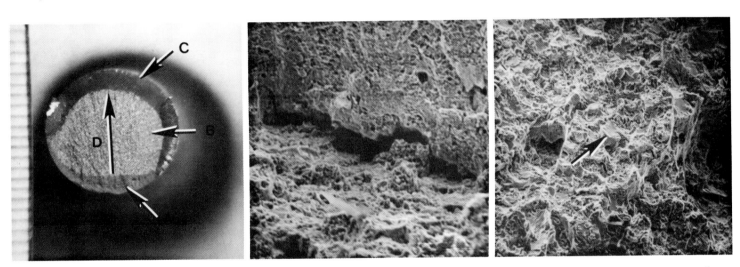

Fig. 380, 381, 382 Fractographic analysis of AISI 4340 specimen tempered in hydrogen (specimen A) in Fig. 378 and 379. Fig. 380 (left): Fracture surface showing region of ductile crack growth and crack initiation site (A), region of rapid crack growth indicated by presence of radial lines (B), a shear lip (C), and the direction of crack propagation in the region of rapid crack growth (arrow D). 2.5×. Fig. 381 (center): Fractograph of region of subsurface crack initiation (B in Fig. 380). Note large transgranular crack and very small dimples. SEM, 550×. Fig. 382 (right): Fractograph reveals predominantly cleavage (transgranular) rupture (facet indicated by arrow) in hydrogen-embrittled specimen of AISI 4340. SEM, 220× (S. Harding and S.W. Stafford, University of Texas)

Fig. 383, 384 Fractographic analysis of AISI 4340 specimen tempered in argon (specimen B in Fig. 378 and 379). Fig. 383 (left): Fracture surface showing an area of microvoid coalescence and crack initiation at the central fibrous zone (A), radial lines emanating from the center and indicating rapid crack growth (B), and a shear lip (C). This was a typical cup-and-cone fracture. 3×. Fig. 384 (right): Fractograph reveals equiaxed dimples characteristic of ductile fracture in tension. 224× (S. Harding and S.W. Stafford, University of Texas)

Fig. 385 Surface of a fatigue fracture that occurred in tension-tension ($R = 0.1$) in a test bar of electroslag remelt AISI 4340 steel heat treated to a hardness of 55 HRC. The origin of the fracture is an inclusion at the center of the "star" at right. See also Fig. 386 to 390. SEM, 20×

Fig. 386 A higher-magnification view of the center of the "star" in Fig. 385, showing the inclusion (dark, at center) that initiated the crack. Fine radial marks are evident. See also Fig. 387. SEM, 200×

Fig. 387 Higher-magnification view of the inclusion (fracture origin) shown at center in Fig. 386. Little evidence of fatigue striations is found this close to the origin. See also Fig. 388. SEM, 2000×

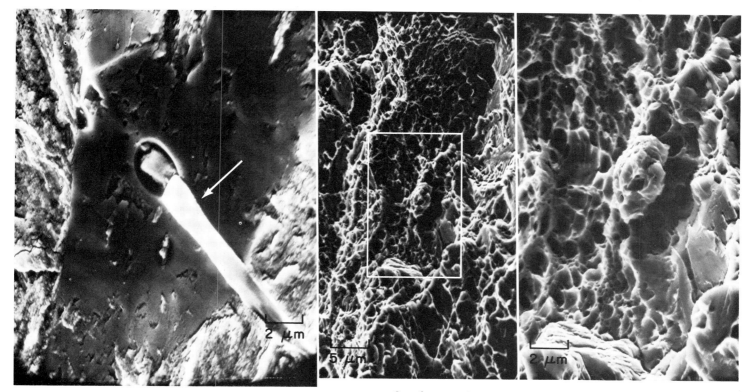

Fig. 388 View, at still higher magnification, of the inclusion in Fig. 386 and 387, showing that the inclusion was fractured by cleavage. The cleavage surface bears a bright metal sliver (at arrow). SEM, 5000×

Fig. 389 Same fracture surface as in Fig. 385, but a view from the region of fast fracture. Typical dimples and local facets of quasi-cleavage. See also Fig. 390. SEM, 2000×

Fig. 390 Higher-magnification view of the area within the rectangle in Fig. 389, showing more distinctly that the dimples are present in a wide variety of sizes. SEM, 5000×

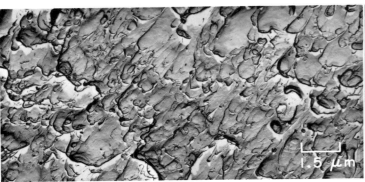

Fig. 391 Surface of a fracture in an unnotched specimen of AISI 4340 steel that was broken in tension overload, displaying elongated and quite flat shear dimples. Note that some crude steps have formed, which suggests that there may have been some degree of pulsation in the applied stress during testing. TEM p-c replica, 6500×

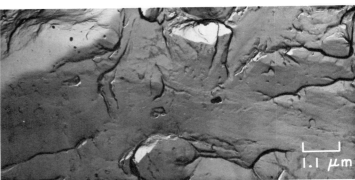

Fig. 392 Surface of a fracture in a notched specimen of AISI 4340 steel that was broken in tension overload, showing scattered dimples. It is not certain whether the remainder of the surface in this area consists of quasi-cleavage facets or of large dimples that have undergone appreciable stretching. TEM p-c replica, 9300×

Fig. 393 Surface of a fracture in a specimen of AISI 4340 steel that was broken by impact at room temperature, showing small dimples of quite uniform size, some of which (at right) are equiaxed. At left are elongated tear dimples pointing toward a tear ridge (at bottom). At upper left are regions in which considerable local stretching has occurred. Some secondary cracking may be present. TEM p-c replica, 3000×

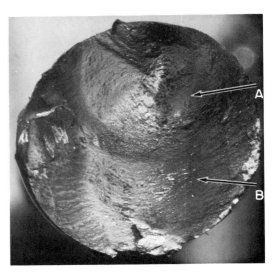

Fig. 394 Torsion-overload fracture in AISI 4340 steel oil quenched from 845 °C (1550 °F) and tempered at 635 °C (1175 °F) to a tensile strength of 931 MPa (135 ksi). In general, the fracture surface is flat and shiny. See Fig. 395 and 396 for TEM views of areas A and B.

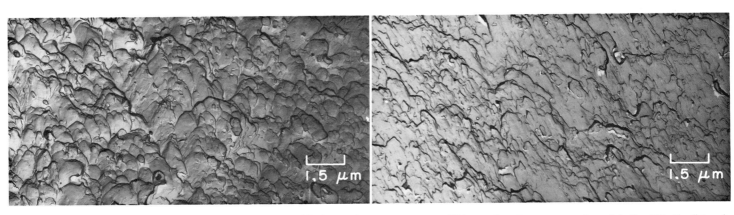

Fig. 395 TEM p-c replica of an area near letter A (in region of final rupture) in the torsion-overload fracture shown in Fig. 394, showing dimples, some of which are slightly elongated but many of which are essentially equiaxed. 6500×

Fig. 396 TEM p-c relica of an area near letter B in Fig. 394. The flat and noticeably elongated shear dimples present here are to be expected in this region, which in Fig. 394 (at lower magnification) shows an appreciable amount of torsional deformation. 6500×

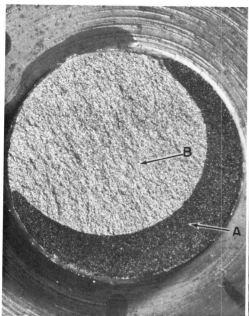

Fig. 397 Impact fracture in AISI 4340 steel machined to a 50-mm (2-in.) diam, 38-mm (1½-in.) long cylinder with a 13-mm (½-in.) diam, 25-mm (1-in.) long central projection from one end. Specimen was heat treated 45 min at 815 °C (1500 °F) and water quenched, but not tempered; then projection was broken by hammer blow. The dark crescent (A) is a quench crack; light area (B) is the impact fracture. See also Fig. 398 to 402. 5×

Fig. 398 TEM p-c replica taken from area A (the quench-crack crescent) in the impact fracture surface shown in Fig. 397. Here, the crack is completely intergranular in nature, having faithfully followed the grain interfaces. Traces of oxidation are evident. See also Fig. 399 and 400. 6500×

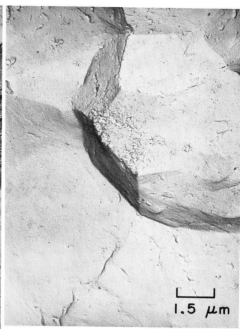

Fig. 399 TEM p-c replica from a different region in area A (the quench-crack crescent) in the fracture surface shown in Fig. 397. Here, as well as in Fig. 398, the crack path has been along the grain interfaces. Scattered marks here resemble the hairline indications associated with either hydrogen embrittlement or stress-corrosion cracking. 6500×

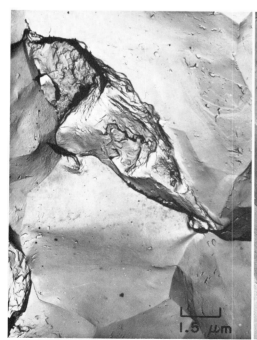

Fig. 400 TEM p-c replica taken near the root of the quench crack, marked A, in Fig. 397. In addition to separated-grain facets, localized regions of transgranular tearing are visible. The grain facets are clean of oxidation, although they display a few hairline indications. 6500×

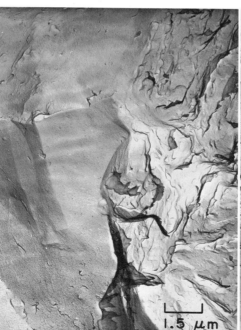

Fig. 401 TEM p-c replica taken at the root of the quench crack, marked A, in Fig. 397, and the beginning of the impact fracture, marked B there. At left are the last vestiges of intergranular cracking, and at right is the beginning of quasi-cleavage cracking in this very brittle specimen. 6500×

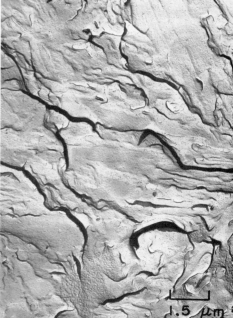

Fig. 402 TEM p-c replica from deep within the impact-fracture area, B, in Fig. 397, showing the quasi-cleavage characteristics of this brittle, untempered specimen. No large facets are visible; this suggests that the specimen had a fine grain size. Many steps exist in the crack path. 6500×

Fig. 403 View of an aircraft horizontal tail-actuator shaft that was found during a preflight inspection to contain several cracks (one is shown at A). The shaft was forged from AISI 4340 steel, was heat treated to a minimum tensile strength of 1793 MPa (260 ksi) and a hardness of 50 to 53 HRC, and then was vacuum cadmium plated. Cracks were attributed to hydrogen embrittlement. See also Fig. 404 to 412. 0.5×

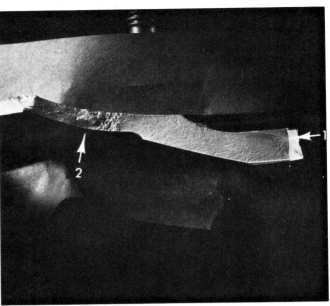

Fig. 404 Surface of the crack visible at A in Fig. 403, opened for examination, showing two internal points of crack nucleation, which are marked 1 and 2 and are shown at higher magnification in Fig. 405 and 406. Neither this crack nor any other crack found in this shaft or in companion shafts showed evidence of the corrosion products (red rust) that would be expected if stress corrosion had been involved. Actual size

Fig. 405, 406 Enlarged views of the internal points of crack nucleation shown in Fig. 404. In Fig. 405 (top), which includes point 1, chevron marks indicate that direction of fracture was right to left; shear lips are present at the borders. Figure 406 (bottom), which includes point 2, shows secondary cracking. See also Fig. 407 to 409. Both at 7×

Fig. 407 TEM p-c replica of a region near the right edge of the fracture surface in Fig. 405, showing a transition from intergranular facets (at right) to dimples (at left). This transition is typical of all cracks in the actuator shaft that originated near holes; in each instance, transition occurred at 0.5 to 1.8 mm (0.02 to 0.07 in.) from the edge of the hole. 3000×

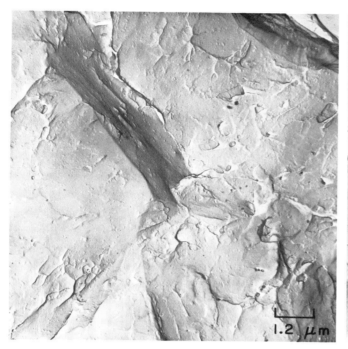

Fig. 408 TEM p-c replica of a region centered on the crack-nucleation point visible near the right edge of the fracture surface shown in Fig. 405. The surface is intergranular and free of corrosion products, which is consistent with fracture caused by hydrogen embrittlement. 8000×

Fig. 409 TEM p-c replica of a region containing the crack-nucleation site near point 2 in Fig. 404. The features that are visible in this region are the same as those in the region shown in Fig. 408, which further indicates that hydrogen embrittlement caused the cracks. 6000×

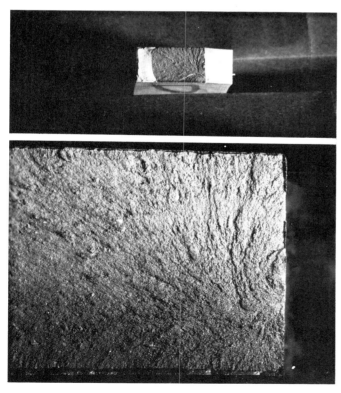

Fig. 410, 411 Views, at two magnifications, of a crack 135° around the shaft from the crack at A in Fig. 403, opened for examination. The point of crack nucleation was internal. The crack then propagated both inward and outward, as shown by the chevron marks and shear lips. Fig. 410 (top): Actual size. Fig. 411 (bottom): 7×

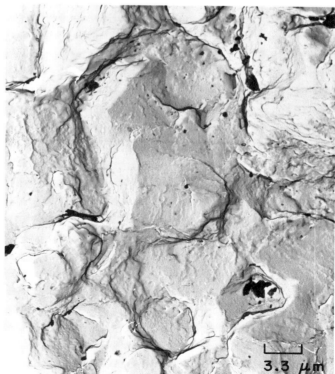

Fig. 412 TEM p-c replica of a region containing a point of nucleation in a crack very similar to those shown in Fig. 404 and 411, in a companion actuator shaft. This surface shows all of the intergranular characteristics of fracture by hydrogen embrittlement seen in Fig. 408 and 409. 3000×

Fig. 413 Surface of a low-cycle fatigue fracture in an aircraft landing-gear cylinder made of AISI 4340 steel that was hardened and tempered to a tensile strength of 1793 to 1931 MPa (260 to 280 ksi). The cylinder was stressed in the laboratory to its tensile strength four times and broke at a stress of 1103 MPa (160 ksi) during the final loading. Note the secondary crack (arrow) to the right of the crack origin (at center), representative of many that extended completely through the wall of the cylinder. See also Fig. 414 to 417. 2×

Fig. 414 Higher-magnification view of the crack-origin area of the fracture surface in Fig. 413. Fracture originated at a small surface crack (visible at arrow) that was present prior to testing. The small flat-topped projections (at A's) and flat-bottomed depressions (at B's) were not found elsewhere in the fracture surface. See also Fig. 415 to 417. 14×

Fig. 415 TEM p-c replica of a portion of the surface crack in Fig. 414, where fracture originated. Note rubbed areas between arrows. The facets are intergranular; the opaque particles lie 0.05 to 0.075 mm (0.002 to 0.003 in.) below the outer surface of the cylinder. 2000×

Fig. 416 TEM p-c replica of the fracture surface in Fig. 414, showing a region 0.25 mm (0.01 in.) below the outer surface. The mixture of intergranular facets and dimples shown here is typical of some plane-strain fracture surfaces in AISI 4340 steel. 2100×

Fig. 417 TEM p-c replica of the fracture surface in Fig. 414, showing a region at the bottom of a depression directly below the surface crack. In this entire area of projections and depressions (see Fig. 414), fracture was completely intergranular. 1000×

Fig. 418, 419, 420 Fatigue failure of shaft on bucket elevator used to handle asphalt. Fracture of the AISI 4340 part (330 HB) initiated at an area of fretting wear in the keyway. The bending stress was calculated at just 131 MPa (19 ksi), but the stress riser due to fretting increased this by a factor of 2 to 4. Fig. 418 (left): Fracture surface of shaft with key in place. 0.35×. Fig. 419 (center): Origin of the fatigue crack was an area of fretting located halfway up the keyway. 0.98×. Fig. 420 (right): Sidewall of keyway. Arrow points to fretting. 1.05× (Z. Flanders, Packer Engineering Associates, Inc.)

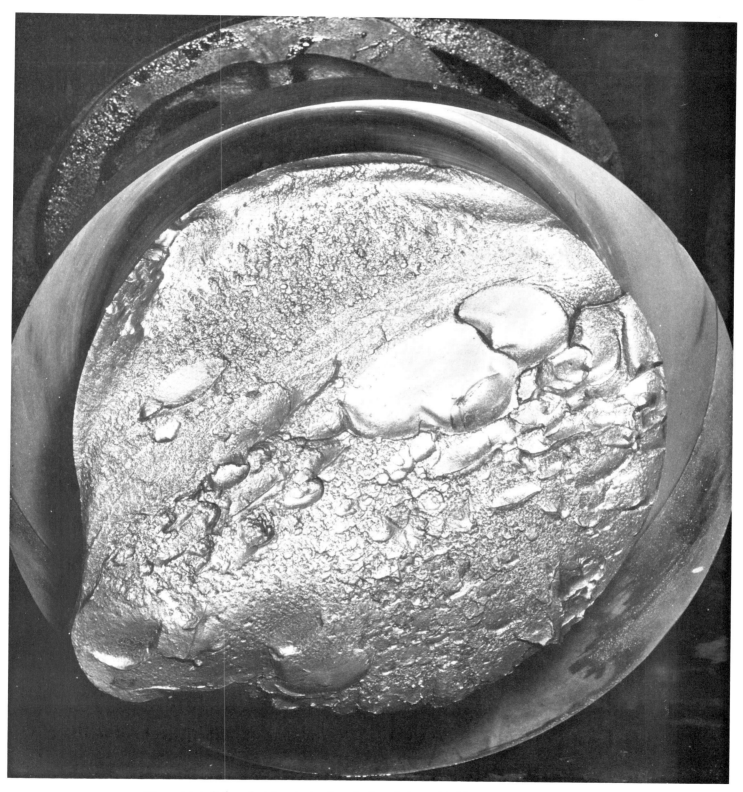

Fig. 421 Surface of a fatigue fracture through a journal of a crankshaft for a 150-mm (6-in.) upsetting machine. The crankshaft was forged of AISI 4340 steel and was normalized to a yield strength of 593 MPa (86 ksi), a tensile strength of 807 MPa (117 ksi), 20.5% elongation, and a hardness of 232 HB. Specimens cut from the crankshaft and reheat treated in exact accordance with prescribed practice had a tensile strength of 945 MPa (137 ksi), 18% elongation, and a hardness of 277 HB. The increased tensile strength and hardness indicate that the original heat treatment of the crankshaft was improper and led to a short fatigue life. See Fig. 422 for a view of the mating fracture surface. Actual size

310 / AISI/SAE Alloy Steels

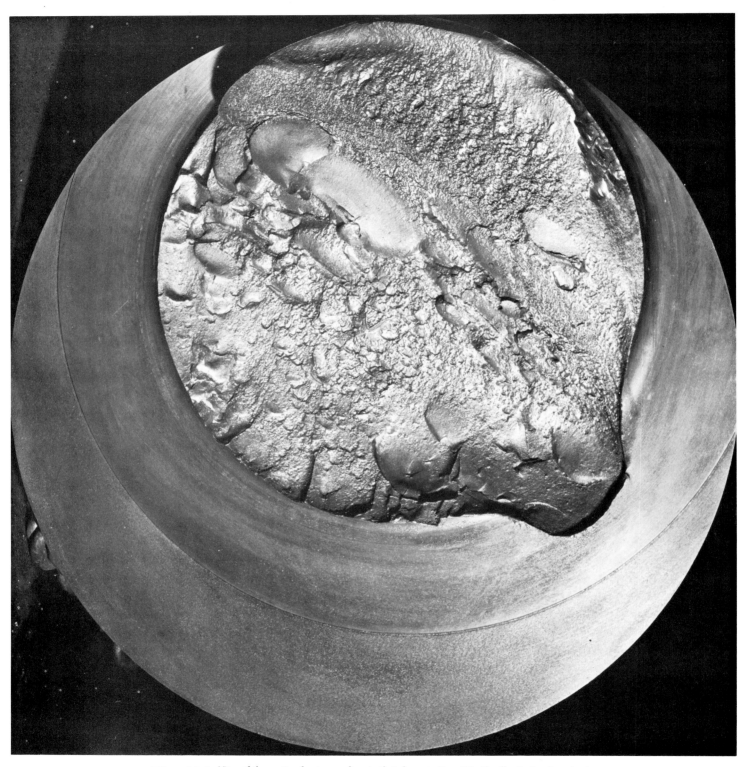

Fig. 422 View of the mating fracture surface to that shown in Fig. 421. The illumination here is at a different angle from that in Fig. 421, with the result that in some locations it is difficult to determine which of the features are projections and which are depressions. It appears likely that there were several fatigue-crack nuclei around the journal perimeter; the principal nucleus probably was in the rubbed area at upper right here (upper left in Fig. 421). There is a possibility that the crankshaft contained forging flakes and that some of these flakes were exposed during machining of the journal and initiated the fatigue cracks. 0.88×

Fig. 423, 424, 425 Surfaces of tension-overload fractures produced at different temperatures in three unnotched specimens of AISI 4340 steel, heat treated to a hardness of 35 HRC. The specimen in Fig. 423 (left) was tested at 160 °C (320 °F). A shear lip is visible, surrounding a fibrous region. The specimen in Fig. 424 (center) was tested at 90 °C (195 °F). At this lower test temperature, a radial fracture zone formed around the fibrous region, which formed first; also, the shear lip here is smaller than that in Fig. 423. The specimen in Fig. 425 (right) was tested at −80 °C (−110 °F). At this still lower test temperature, no fibrous region formed. Instead, fracture formed a radial zone that extends nearly to the specimen surface and terminates in a narrow shear lip. All at 15×

Fig. 426 Tension-overload fracture in a notched specimen of AISI 4340 steel, heat treated to a hardness of 27 HRC and tested at −40 °C (−40 °F), showing a completely fibrous surface. See Fig. 427 and 428 for fractures in similar specimens tested at different temperatures. 17×

Fig. 427 Surface of a tension-overload fracture in another notched specimen of AISI 4340 steel, the same as in Fig. 426, but tested at −120 °C (−185 °F). Two distinctly different fracture-surface zones are visible; a fibrous zone, which originated at the notch, surrounds a radial region of final fast fracture. 17×

312 / AISI/SAE Alloy Steels

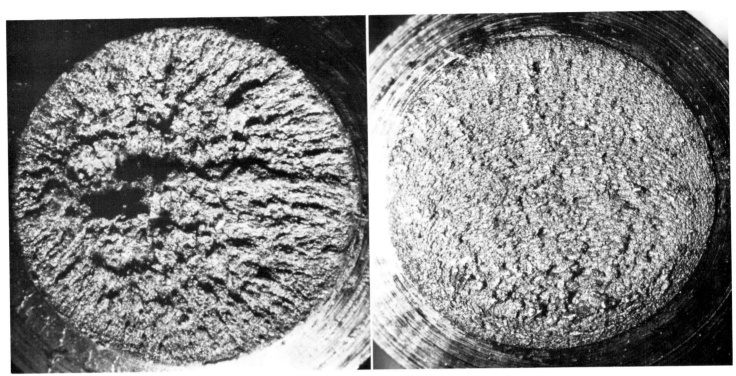

Fig. 428 Surface of tension-overload fracture in a third notched specimen of AISI 4340 steel, the same as those in Fig. 426 and 427, but tested at −155 °C (−245 °F). At this low test temperature, a completely radial fracture was produced. The surface shows no fibrous zone, only radial marks. 17×

Fig. 429 Tension-overload fracture in a notched specimen similar to those in Fig. 426 to 428, but heat treated to a hardness of 35 HRC. Tested at −40 °C (−40 °F). The surface shows only fibrous marks. See Fig. 430 and 431 for surfaces of fractures in similar specimens at different test temperatures. 17×

Fig. 430 Surface of a tension-overload fracture in a notched specimen of AISI 4340 steel the same as the specimen shown in Fig. 429, but tested at −80 °C (−110 °F). A fibrous zone, which originated at the notch, surrounds a radial zone, which is off-center because of nonsymmetrical crack propagation. 17×

Fig. 431 Tension-overload fracture in a notched specimen same as in Fig. 429 and 430, but tested at −155 °C (−245 °F). Even at this very low temperature, a small annular zone of fibrous fracture was formed next to the notch. Final fast fracture produced the radial marks in the central region. 17×

AISI/SAE Alloy Steels / 313

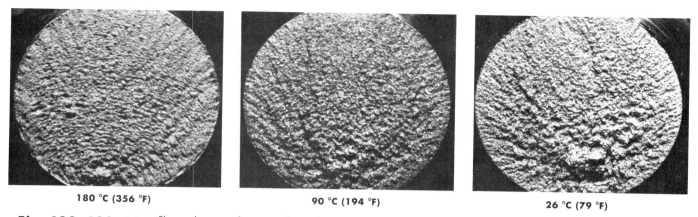

180 °C (356 °F)   90 °C (194 °F)   26 °C (79 °F)

Fig. 432, 433, 434 Shown above are fracture surfaces of three notched tensile-test specimens prepared from AISI 4340 steel that had been austenitized, quenched, and tempered at 315 °C (600 °F) to provide the following room-temperature properties: yield strength, 1558 MPa (226 ksi); unnotched tensile strength, 1758 MPa (255 ksi); notched tensile strength, 1765 MPa (256 ksi). The specimens were broken by tension overload at the temperatures indicated beneath the fractographs. The fracture surface shown in Fig. 432 (left) is completely fibrous. Lowering the test temperature from 180 °C (356 °F) to 90 °C (194 °F) resulted in the formation of a very narrow, obscure fibrous rim enclosing a zone of radial marks (see Fig. 433, center). Lowering the test temperature further, to 26 °C (79 °F), produced a fracture made up entirely of radial features (see Fig. 434, right). All at 8×

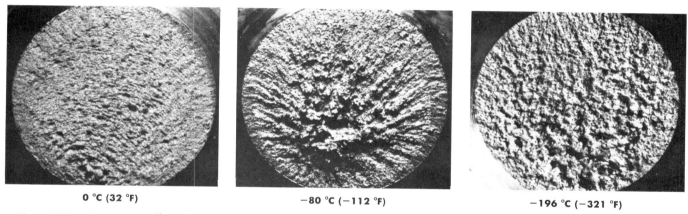

0 °C (32 °F)   −80 °C (−112 °F)   −196 °C (−321 °F)

Fig. 435, 436, 437 Shown above are fracture surfaces of three notched tensile-test specimens prepared from AISI 4340 steel that had been austenitized, quenched, and tempered at 480 °C (895 °F) to provide the following room-temperature properties: yield strength, 1234 MPa (179 ksi); unnotched tensile strength, 1310 MPa (190 ksi); notched tensile strength, 1903 MPa (276 ksi). The specimens were broken by tension overload at the temperatures indicated beneath the fractographs. The specimen shown in Fig. 435 (left) was broken at 0 °C (32 °F) but, in contrast to the specimen in Fig. 434, which was tempered at 315 °C (600 °F), yielded a completely fibrous fracture. The fracture shown in Fig. 436 (center), which was produced at −80 °C (−112 °F), shows a fibrous rim surrounding a zone of radial marks. The fracture shown in Fig. 437 (right), which was produced at −196 °C (−321 °F), shows only radial marks. All at 8×

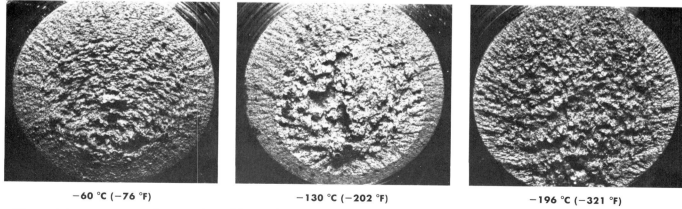

−60 °C (−76 °F)   −130 °C (−202 °F)   −196 °C (−321 °F)

Fig. 438, 439, 440 Fracture surfaces of three notched tensile-test specimens prepared from AISI 4340 steel that had been austenitized, quenched, and tempered at 650 °C (1200 °F) to provide the following room-temperature properties: yield strength, 917 MPa (133 ksi); unnotched tensile strength, 1014 MPa (147 ksi); notched tensile strength, 1489 MPa (216 ksi). The specimens were broken by tension overload at the temperatures indicated beneath the fractographs. The specimen shown in Fig. 438 (left) was broken at −60 °C (−76 °F), but displays a completely fibrous surface. The fracture in Fig. 439 (center) was produced at −130 °C (−202 °F); note the appreciable increase in the size of the fibrous zone, compared with that in Fig. 436, caused by the lower strength that resulted from tempering at 650 °C (1202 °F). The specimen in Fig. 440 (right), which was broken at −196 °C (−321 °F), exhibits a fracture surface containing only radial marks. All at 8×

314 / AISI/SAE Alloy Steels

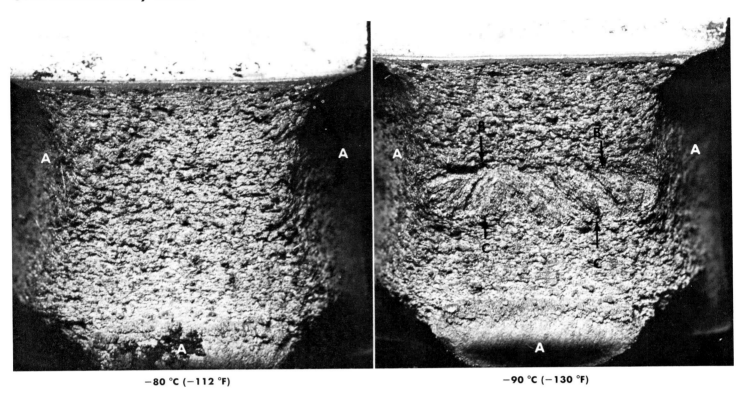

−80 °C (−112 °F)

−90 °C (−130 °F)

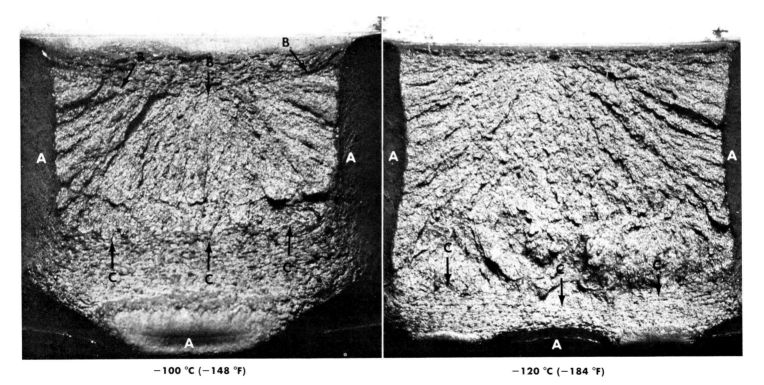

−100 °C (−148 °F)

−120 °C (−184 °F)

Fig. 441 to 444 These four light fractographs show the surfaces of impact fractures in Charpy specimens of AISI 4340 steel heat treated to a hardness of 27 HRC and depict the changes in surface characteristics as test temperature was decreased. The specimen in Fig. 441 (top left), which was broken at −80 °C (−112 °F), shows a fibrous fracture surface containing shear lips (at A's). The specimen in Fig. 442 (top right) was broken at −90 °C (−130 °F). This temperature is close to the ductile-brittle transition temperature, as indicated by the presence of a small area of radial marks (between arrows B and arrows C) in the center of the otherwise fibrous fracture surface. Shear lips are present here also (at A's). The specimen in Fig. 443 (bottom left) was broken at −100 °C (−148 °F) and shows an area of radial marks (between arrows B and arrows C) much larger than that in Fig. 442, while retaining fibrous regions above arrows B (near the notch) and below arrows C. The shear lips here (at A's) are comparable to those in Fig. 441 and 442. The specimen in Fig. 444 (bottom right) was broken at −120 °C (−184 °F); its fracture surface contains a radial zone larger than that in Fig. 443, but still shows a small area of fibrous fracture below arrows C (opposite the notch). Also, the shear lips here (at A's) are appreciably smaller than those in Fig. 443. All at 10×

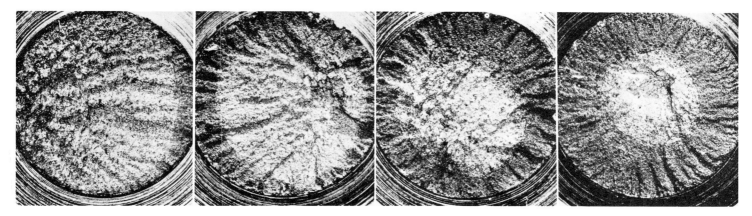

Fig. 445 to 448 Surfaces of four tension fractures, which illustrate the effect of decreasing stress on the fracture-surface characteristics of notched specimens of AISI 4340 steel broken at room temperature. Notch radius was 0.025 mm (0.001 in.). Microstructure of all four specimens was tempered martensite. The specimen in Fig. 445 (left) was broken by tension overload (notched tensile strength was 2006 MPa, or 291 ksi) and shows a relatively small fibrous zone at the right edge. The three other specimens were charged with hydrogen and then were subjected to sustained loading. The specimen in Fig. 446 (left center) broke in 1.65 h under a stress of 1379 MPa (200 ksi). The specimen in Fig. 447 (right center) broke after 5.35 h under a stress of 1034 MPa (150 ksi). The specimen in Fig. 448 (right) fractured after 5.5 h under a stress of 689 MPa (100 ksi). Note that the progressive decrease in fracture stress from Fig. 445 to 448 was related to a progressive increase in the size of the fibrous zone. All at ~8×

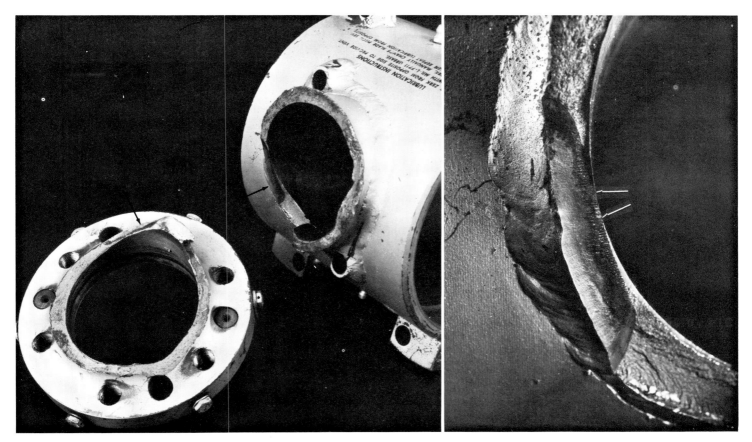

Fig. 449 Mating segments of an aircraft propeller hub of quenched and tempered AISI 4340 steel that broke by fatigue, showing both surfaces of the fracture, which occurred at the welded joint between the barrel of the hub and the fitting for attachment of the hub to the hydraulic cylinder. Arrows indicate on each fracture surface the location of the fatigue-crack origin—at the toe of the weld bead, where a slight weld overlap had formed. See also Fig. 450. 0.5×

Fig. 450 Fracture surface of the barrel portion of the broken hub in Fig. 449, showing the fatigue-crack origin. The fatigue crack was about 44 mm (1¾ in.) long, but had penetrated the hub wall completely for only about 3.2 mm (⅛ in.) (arrows). Note shear lip at inside of hub wall on either side of the fatigue crack. 2×

316 / AISI/SAE Alloy Steels

Fig. 451 Mating surfaces of a fatigue fracture in a blade arm of a forged aircraft propeller-hub spider of AISI 4340 steel and through the central pilot tube of the arm. The fatigue crack originated in the fillet between the bearing-retention flange and the blade arm. Hardness of the forging was 35 HRC. The origin is at arrow A (lower left), at an inclusion. The fatigue crack penetrated as far as the locations indicated by arrows B before final fast fracture occurred. Alignment of the bearing-race parting line (arrows C) was not as specified. Arrow D indicates the region of final fast fracture. See Fig. 452 for a higher magnification view of this region. 0.6×

Fig. 452 View of an area at arrow D in Fig. 451, in the region of final fast fracture. Fatigue patches are interspersed among areas of rapid crack advance, indicating that final fracture occurred in stages. 10×

Fig. 453, 454 Two views of the surface of a fatigue fracture in a 145-mm (5¾-in.) diam threaded piston rod of AISI 4340 steel heat treated to 341 HB. The full-face view in Fig. 453 (left) shows plainly the beach marks in the region of the primary fatigue crack, at top. Secondary fatigue cracks are visible on either side of it, as well as at the bottom. The oblique view in Fig. 454 (right) reveals that the fatigue cracks began at the roots of the threads, the origin of the primary crack being two threads to the right of the abutting secondary cracks. Although these reduced-size views do not permit detailed scrutiny, it seems evident that the thread roots were too sharp, causing stress concentrations locally. Beyond the fatigue-zone limits, failure was by radial fibrous fast fracture. Fig. 453: ~0.5×. Fig. 454: 0.67×

Fig. 455 Matching fracture surfaces of an AISI 4340 steel helicopter bolt that was broken by notch bending. The fine markings are fibrous tear ridges. Note the very large shear lips. 2.5×

Fig. 456, 457 Fig. 456 (top): View of a service fracture in a 125-mm (5-in.) diam suspension strut of a missile silo. AISI 4340 steel, heat treated to tensile strength of 1612.7 MPa (233.9 ksi), 11% elongation, and Charpy V-notch impact strength of 17.3 J (12.8 ft · lb). The fracture origin (arrow) is at a thread root, two full threads deep into the coupling. Actual size. Fig. 457 (bottom): View of the origin in the mating fracture surface (Fig. 458). Two wedge-shaped pieces had been broken from the strut perimeter at symmetrical locations on either side of the origin (arrows A); one remained in the coupling and is visible in Fig. 456. 1.5×

318 / AISI/SAE Alloy Steels

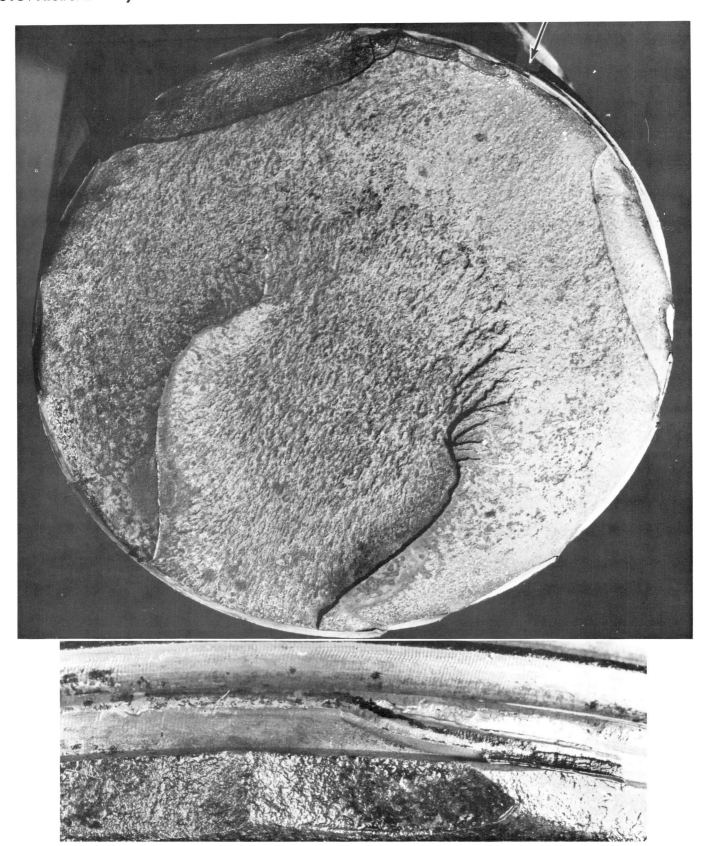

Fig. 458, 459 Figure 458 (top) is the mating fracture surface of the strut in Fig. 456. The fracture (origin at arrow) penetrated the strut at an angle, starting at the second thread within the coupling and emerging six threads lower on the opposite side. In both fracture surfaces, in the area opposite the origin, is a crudely circular crater with shear-lip borders. 1.3×. Figure 459 (bottom) is a view of the threads near the origin. Note that the thread behind the fracture surface is partly stripped and there are corrosion pits on the threads. Fracture was ascribed to thread defects, corrosion, and excessive strength with low toughness. 7×

Fig. 460 Surface of a tension-overload fracture in a notched specimen of AMS 6434 steel sheet that was broken at 27 °C (81 °F), showing small equiaxed dimples. The heavy curved lines are intersections of dimple surfaces. See also Fig. 461 and 462. TEM p-c replica, 2000×

Fig. 461 Surface of a tension-overload fracture in a notched specimen of the same AMS 6434 steel sheet as in Fig. 460, but broken at −73 °C (−99 °F). The dimples here are of the same size and nature as those in Fig. 460. See also Fig. 462. TEM p-c replica, 2000×

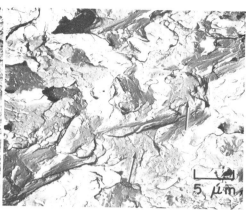

Fig. 462 Surface of a tension-overload fracture in a notched specimen of the same AMS 6434 steel sheet as in Fig. 460 and 461, but broken at −190 °C (−310 °F), showing a mixture of scattered dimples and (at arrows) cleavage facets. TEM p-c replica, 2000×

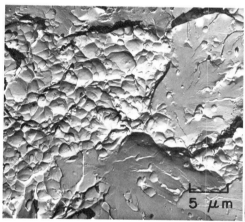
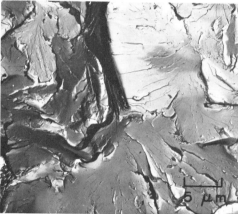
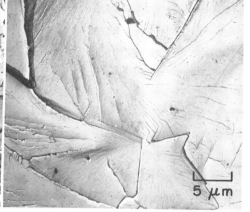

Fig. 463, 464, 465 Surfaces of fractures in three pressure vessels made of AMS 6434 steel. Figure 463 (left) shows equiaxed dimples, which resulted from tension overload, and quasi-cleavage facets. Figure 464 (center) shows quasi-cleavage facets plus (at bottom left) a small region of dimples. Figure 465 (right) shows cleavage facets in large ferrite grains; the facets contain many cleavage steps that originated at twist boundaries. TEM p-c replicas, all at 2000×

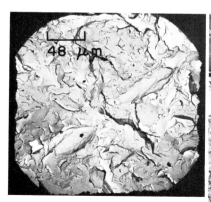

Fig. 466 Impact fracture in a specimen of AMS 6434 steel quenched and then tempered at 205 °C (400 °F) to a yield strength of about 1550 MPa (225 ksi) and broken at −196 °C (−321 °F), showing a woody type of cleavage-fracture surface. TEM p-c replica, 210×

Fig. 467 Impact fracture in a specimen of AMS 6434 steel (same properties as Fig. 466) broken at room temperature, showing fairly large equiaxed dimples, which are characteristic of tension-overload fractures. TEM p-c replica, 700×

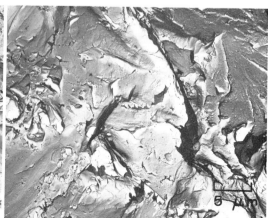

Fig. 468 Fracture surface of an air flask of AMS 6434 steel that was broken in service. This complicated surface contains several quasi-cleavage facets. TEM p-c replica, 2000×

320 / AISI/SAE Alloy Steels

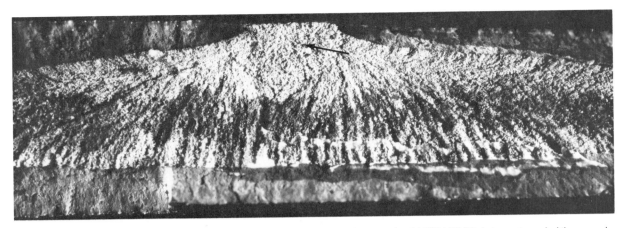

Fig. 469 Surface of a fracture in a wall of a flask made of AMS 6434 steel tempered at 205 °C (400 °F) that was stressed while exposed to tap water, producing stress-corrosion cracking. See Fig. 470 for a higher-magnification view of the region at arrow. 15×

Fig. 470 Higher-magnification view of fracture surface at arrow in Fig. 469, showing clearly the intergranular nature of the fracture, which resulted from stress-corrosion cracking. Grains generally are equiaxed, and the fracture has closely followed grain boundaries. TEM p-c replica, 3000×

Fig. 471, 472 Fracture of tie rod ball stud for automobile steering system due to high-cycle bending fatigue. The AISI 4615 part was carburized and tempered to 60 HRC on the surface. The stud was broken in the laboratory. Fig. 471 (left): Broken ball stud. Actual size. Fig. 472 (right): Fracture surface. 2.5× (General Motors Research Laboratories)

Fig. 473 Rotating bending fatigue fracture of an AISI 4817 shaft, carburized and hardened to 60 HRC on the surface. Fracture initiated in six fillet areas around the periphery, near the runouts of six grooves. Each fatigue area propagated separately, but uniformly, inward to final rupture at the center. Cause of fracture was traced to grinding damage in the fillet. The bright and shiny flat areas are regions of postfracture damage produced by the rubbing together of the broken portions of the shaft. (D.J. Wulpi, Consultant)

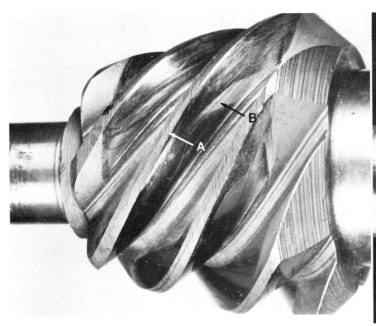

Fig. 474 View of a carburized spiral gear of AISI 4817 steel, showing fine subcase fatigue cracks in the gear-tooth tips (such as at arrow A) and flanks (such as at arrow B); these cracks are evidence of the first stages of spalling fatigue, which originated near the case-core interface because the load-carrying capacity of the gear was exceeded. Compare with Fig. 475. Actual size

Fig. 475 Companion gear to that in Fig. 474, (carburized AISI 4817 steel) showing subcase fatigue fractures initiated by fine cracks similar to those in Fig. 474. Large fragments have spalled away from the teeth. Fatigue beach marks can be seen in these complex fracture surfaces, especially at upper right. Actual size

Fig. 476 Surface of a fatigue fracture that began at six fillet areas around the periphery of a shaft, near the runouts of spiral grooves. The material is AISI 4817 steel; the shaft was carburized and heat treated to a surface hardness of 60 HRC. The fatigue cracks penetrated equally, indicating that stress concentrations at the six fillet areas were essentially equal. ~2.13×

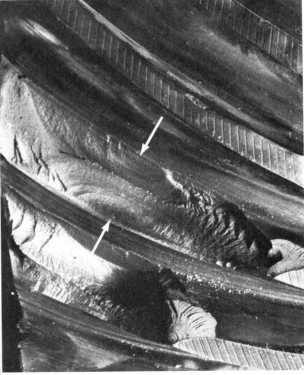

Fig. 477 Fatigue fractures in a spiral bevel pinion gear of AISI 4820H steel, case hardened to a depth of 1.1 to 2.0 mm (0.045 to 0.080 in.). Case hardness, 56 HRC; core hardness, 30 HRC. Fatigue cracks began in tooth roots (which were not shot peened), often forming on either side of a tooth (arrows) and meeting in the middle of the tooth. A deeper case and shot peening of the entire gear were recommended. Actual size

AISI/SAE Alloy Steels / 323

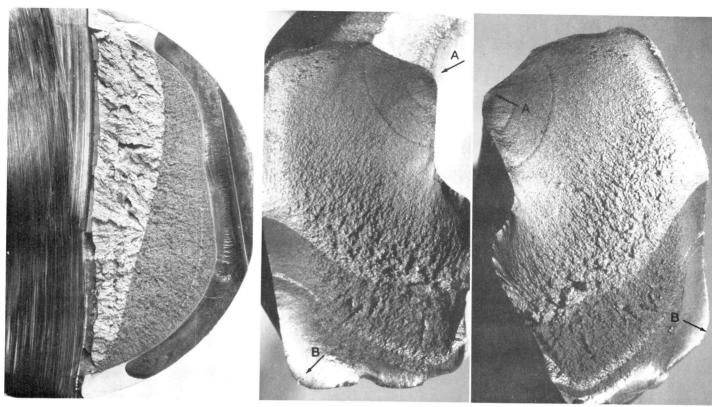

Fig. 478 Case-hardened pin of AISI 5046 steel with a hardness of 61 HRC in the case, 248 HB in the core, showing fatigue zone (right). The pin was cut (left), then broken (light area at center). Sectioning of the pin revealed subcase fatigue cracking. 2.25×

Fig. 479, 480 Views of the mating fracture surfaces of a suspension component that was fabricated of AISI 5132 steel and that broke in a fatigue test. Note the two fatigue-crack origins, one at arrows A (top) and the other at arrows B (bottom). The crack that was initiated at the origin at arrows B is the earlier crack, as indicated by the well-rubbed surface near that origin. Both at 3×

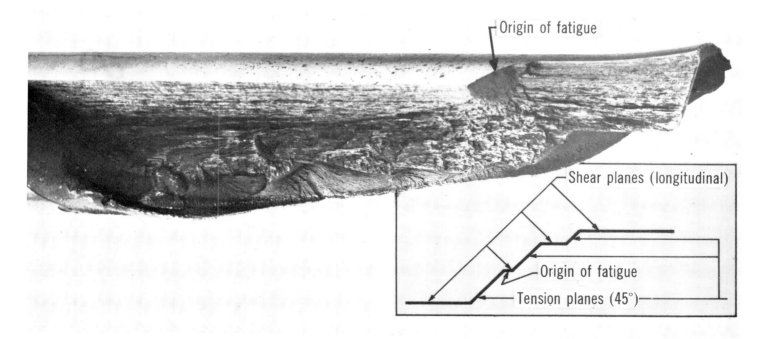

Fig. 481 Surface of a torsional-fatigue fracture in the cylindrical portion of a 35-mm (1 3/8-in.) diam torsion-bar spring of AISI 50B60 steel heat treated to a hardness of 53 HRC. The fatigue crack originated in a very small (0.38 mm, or 0.015 in. long) facet (labeled "Origin of fatigue") on a longitudinal shear plane and progressed alternately on 45° tension planes and longitudinal shear planes, as indicated in the diagram. Note that the semicircular fatigue patch surrounding the origin lies in the 45° plane. 1.25×

324 / AISI/SAE Alloy Steels

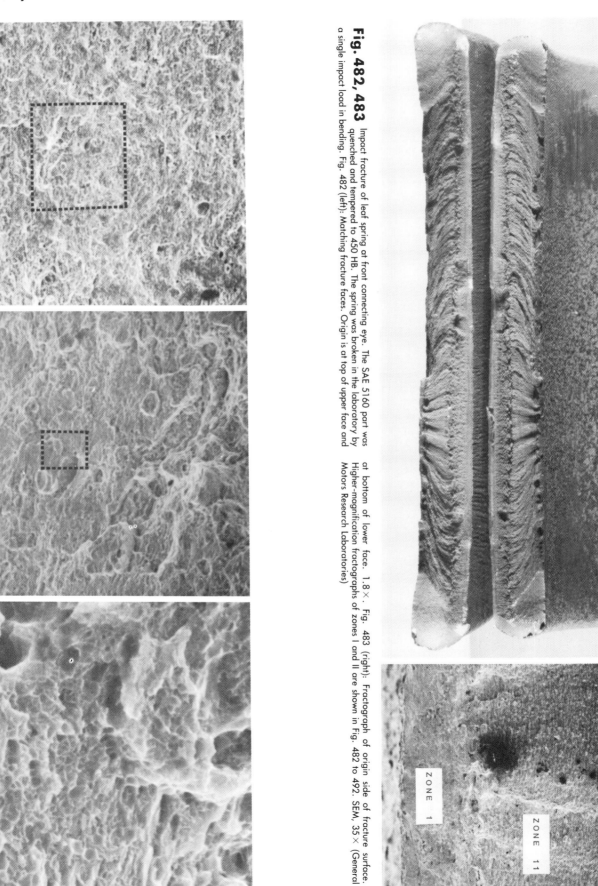

Fig. 482, 483 Impact fracture of leaf spring at front connecting eye. The SAE 5160 part was quenched and tempered to 450 HB. The spring was broken in the laboratory by a single impact load in bending. Fig. 482 (left): Matching fracture faces. Origin is at top of upper face and at bottom of lower face. 1.8×. Fig. 483 (right): Fractograph of origin side of fracture surface. Higher-magnification fractographs of zones I and II are shown in Fig. 482 to 492. SEM, 35× (General Motors Research Laboratories)

Fig. 484, 485, 486 Fracture features in zone I of fractograph of SAE 5160 leaf spring (Fig. 482 and 483). Fig. 484 (left): SEM, 300×. Fig. 485 (center): Enclosed area in Fig. 484. SEM, 1000×. Fig. 486 (right): Enclosed area in Fig. 485. Note numerous fine dimples. SEM, 10 000× (General Motors Research Laboratories)

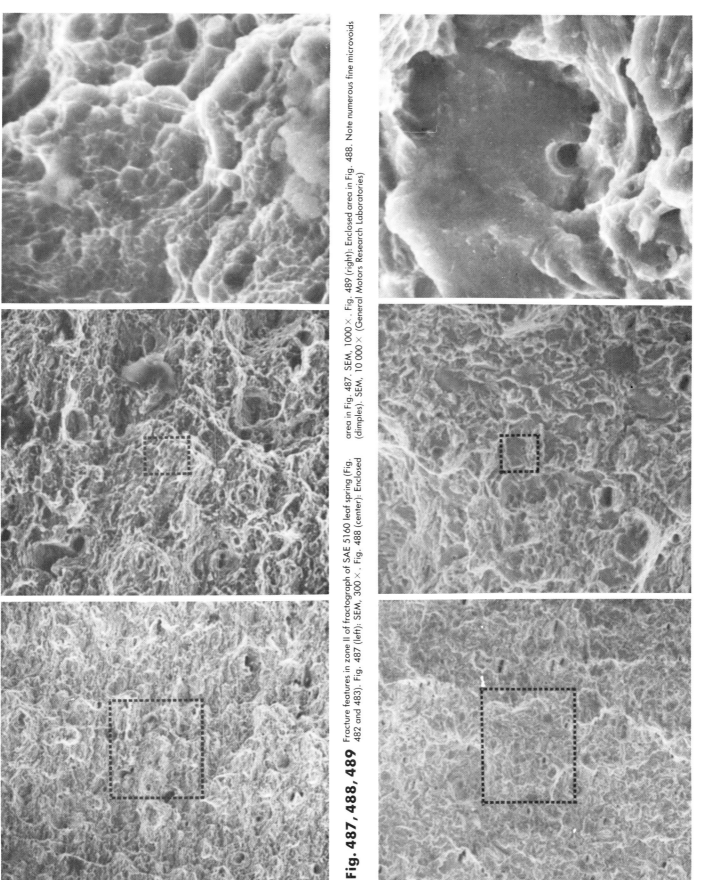

Fig. 487, 488, 489 Fracture features in zone II of fractograph of SAE 5160 leaf spring (Fig. 482 and 483). Fig. 487 (left): SEM, 300×. Fig. 488 (center): Enclosed area in Fig. 487. SEM, 1000×. Fig. 489 (right): Enclosed area in Fig. 488. Note numerous fine microvoids (dimples). SEM, 10 000×. (General Motors Research Laboratories)

Fig. 490, 491, 492 Fracture features in zone II of fractograph of SAE 5160 leaf spring (Fig. 482 and 483). Field differs from that in Fig. 487 to 489. Fig. 490 (left): SEM, 300×. Fig. 491 (center): Enclosed area in Fig. 490. SEM, 1000×. Fig. 492 (right): Enclosed area in Fig. 491. SEM, 10 000×. (General Motors Research Laboratories)

Fig. 493, 494 Fracture of brake spring due to seam defect in material. The AISI 5160H wire was drawn to 155 mm (0.531 in.) in diameter, hot wound, and quenched and tempered to 48 to 52 HRC. Fig. 493: Fractured brake spring. Note smooth, longitudinal split caused by seam. Actual size. Fig. 494: Close-up of one fracture face. 2× (P.W. Walling, Metcut Research Associates, Inc.)

Fig. 495, 496, 497 Fatigue failure of AISI 5160 initiated by ribbonlike inclusions. Fig. 495 (left): Secondary electron image of fracture surface. Ribbonlike inclusion is near the center. SEM, 40×. Fig. 496 (center): Backscattered electron image of field in Fig. 495 brings out details of fracture surface and enhances visibility of inclusion. SEM, 40×. Fig. 497 (right): Ribbonlike inclusions are revealed on the wall of this longitudinal crack associated with a fatigue fracture in AISI 5160. SEM, 75× (J.H. Maker, Associated Spring, Barnes Group Inc.)

Fig. 498 Fatigue crack origin on surface of spring was intentionally opened to reveal details of fracture initiation. Material: modified AISI 5160 (0.5% Cr, 0.1 to 0.2% V) wire. The spring was tested in the laboratory at stresses well in excess of those normally encountered in service. The surface-breaking portion of the fatigue origin is indicated on the fractograph. It initially appeared under the SEM as a faint line, transverse to the wire at the center, with each end at a slight angle. After intentional opening (see fractograph), the origin was shown to be roughly triangular in shape and defined by vertical ridges extending into the material. The balance of the fracture surface has a typical tensile failure pattern. SEM, 590× (J.H. Maker, Associated Spring, Barnes Group Inc.)

Fig. 499 Fatigue fracture in an AISI 6150 steel spring leg of an aircraft main landing gear, heat treated to a minimum tensile strength of 1550 MPa (225 ksi) and a hardness of 49 HRC. Visual inspection of the fracture surfaces showed that the fatigue crack originated at the location marked by arrows O. The appearance of the external surface of the spring leg indicated that the part had been shot peened, but that peening had been inadequate. The dimensions of the spring leg conformed to specifications. See also Fig. 500 to 504. 0.6×

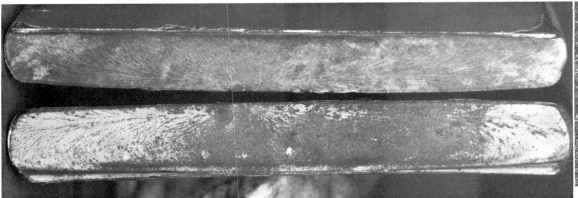

Fig. 500 Mating fracture surfaces of the broken main-landing-gear spring leg in Fig. 499. A small fatigue zone can be seen just to the left of center, at the edges shown adjacent here. This zone grew to a length of only about 5 mm (0.2 in.) and to a depth of slightly more than 2.5 mm (0.1 in.) before final fast fracture occurred. The remainder of each fracture surface shows features typical of rapid tearing. See Fig. 501 for a through-section of the spring leg. Actual size

Fig. 501 Profile of the spring leg in Fig. 499, showing the irregular fast-fracture surface, at right. 1% nital etch, 250×

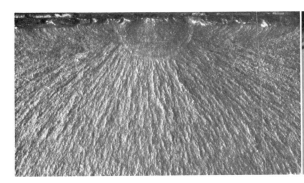

Fig. 502 A view of the fracture surface at bottom in Fig. 500, showing the fatigue zone at top center. Metallographic study of this region revealed partial decarburization to a depth of 0.71 mm (0.028 in.), which contributed to fracture. 4×

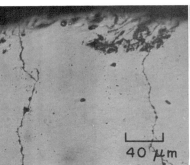

Fig. 503 Section through the spring leg in Fig. 499, near the fracture origin. Secondary cracks extend inward from the shot-peened surface (at top). 250×

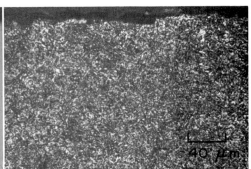

Fig. 504 Another section through the spring leg in Fig. 499, showing a profile of the fatigue-crack surface (at top). Compare the smoothness of this surface with the irregularity of the fast-fracture surface in Fig. 501. 1% nital etch, 250×

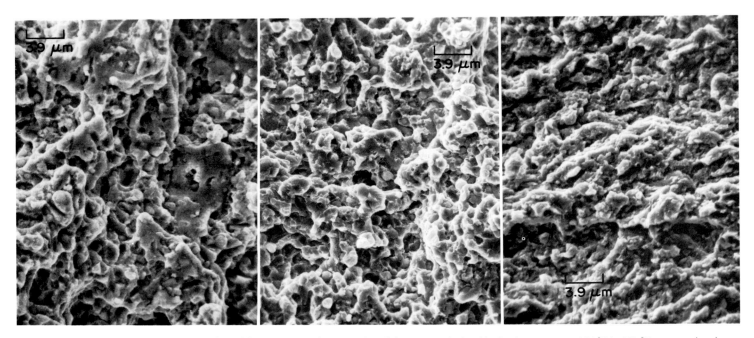

Fig. 505, 506, 507 Fracture surface of three specimens of AISI 52100 steel that were notched and broken by impact at −196 °C (−321 °F), presented to show the effect of rapid heating in austenitizing. The specimen in Fig. 505 (left) had been heat treated at 845 °C (1555 °F) for 20 min and oil quenched (a conventional heat treatment), then tempered at 175 °C (350 °F). The specimen in Fig. 506 (center) had been given two austenitizing cycles of rapid heating to 865 °C (1585 °F), holding for 1 min and oil quenching, and then had been tempered at 175 °C (350 °F). The specimen in Fig. 507 (right) had been heat treated at 1095 °C (2005 °F) for 10 min and quenched in warm oil, tempered for 3 h at 370 °C (700 °F), rapidly heated to 845 °C (1555 °F), held for 1 min and oil quenched, then tempered at 175 °C (350 °F). The successive decrease in size of cleavage facets and dimples in the three fractographs indicates successively decreasing size of acicular needles of martensite as a result of the different heat treatments. SEMs, all at 2250×

Fig. 508, 509 Fatigue fracture of truck front wheel spindle. The SAE 81B45 part was quenched and tempered to 30 HRC. Fig. 508 (left): Broken wheel spindle. 0.9×. Fig. 509 (right): Fracture surface. 2.5× (General Motors Research Laboratories)

Fig. 510 Bending-fatigue fracture in the heel of one tooth of a spiral bevel pinion of AISI 8617 steel, carburized and hardened to 57 HRC in the case. Fracture resulted from severe pitting. Note that pitting had begun in an adjoining tooth (near top). 0.75×

Fig. 511 Bending-fatigue fracture in two teeth of a reverse idler gear of AISI 8617 steel, carburized and hardened to 60 HRC in the case. Arrows point to the root fillets on both sides of each tooth, where fracture began due to excessive stress in these locations. ~2×

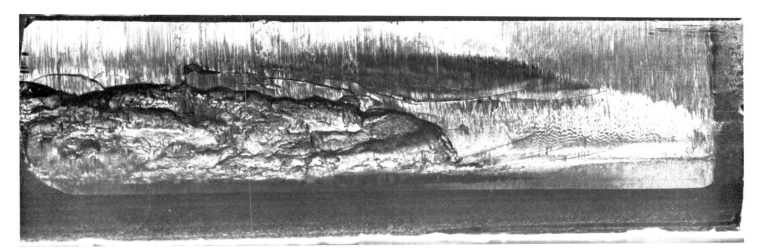

Fig. 512 Surface of a spalling-fatigue fracture in a single tooth of a heavily loaded final-drive pinion of AISI 8620 steel, carburized and hardened to 60 HRC in the case, showing vertical scratches, which indicate that appreciable abrasive wear took place also. The surface ripples at right suggest that a small amount of plastic flow occurred under the applied load. ~4×

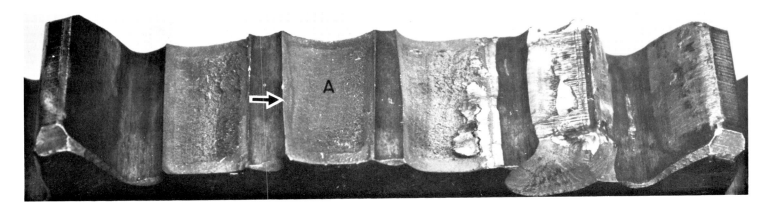

Fig. 513 Bending-fatigue fractures in several teeth of a spur gear of AISI 8620 steel, carburized and hardened to 60 HRC in the case. The tooth marked A apparently broke first, as the result of a fatigue crack that originated in the fillet to the left of the tooth (arrow). After this tooth broke off, fracturing of the teeth on each side of it was accelerated. 2×

330 / AISI/SAE Alloy Steels

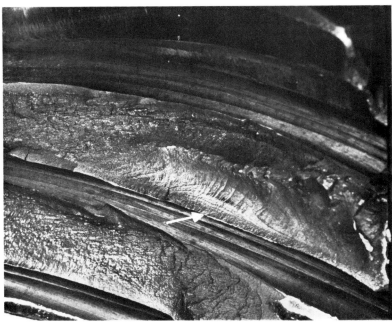

Fig. 514 Surface of a bending-fatigue fracture in a tooth (upper tooth in this view) of a large spiral bevel pinion of AISI 8620 steel carburized and hardened to 60 HRC at the surface. The arrow marks the fatigue-crack origin, in the root fillet. The absence of this tooth resulted in fracture of the tooth below by overload. 1.65×

Fig. 515 Surface of torsional-fatigue fracture in a splined shaft of AISI 8620 steel that was carburized and case hardened. Multiple fatigue cracks evidently formed at the roots of the splines and then joined to penetrate much of the case before final fast fracture occurred. 2×

Fig. 516, 517, 518 Figure 516 (left) is a view of a broken splined shaft of AISI 8620 steel, surface carburized to a case depth of 1.07 mm (0.042 in.) and hardened to 63 to 65 HRC; core hardness was 96 HRB. Specified case hardness was 57 to 62 HRC. The fracture resulted from torsional overload. Figure 517 (center) is a view of a surface of the fracture in Fig. 516. Brittle fracture of the splines occurred first; later, the soft, ductile core fractured in the transverse shear plane. Figure 518 (right) is a view of the O-ring region of the shaft in Fig. 516. A 45° torsion crack is visible (at A) at the edge of an oil hole; this crack is proof of the torsional nature of the loading that was imposed on the shaft. Fig. 516: 0.25×. Fig. 517 and 518: 1.5×

AISI/SAE Alloy Steels / 331

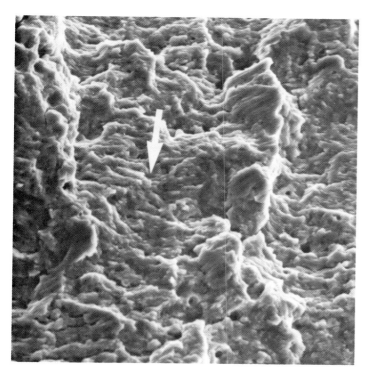

Fig. 519 Fatigue striations in AISI 8620. The roughly horizontal ridges reveal how the crack front advanced with each load application. The crack propagated in the direction of the arrow. SEM, 2100× (D.J. Wulpi, Consultant)

Fig. 520 Surface of a fracture in an AISI 8640 boom-point pin that broke during service. Steel was quenched and then tempered to a hardness of 269 HB. Sharp corners of mismatched transverse grease holes provided stress concentrations that generated two fatigue cracks (arrows). Actual size

Fig. 521 Surface of a fatigue fracture in a 200-mm (4-in.) diam axle of AISI 8640 steel with a hardness of approximately 30 HRC. Contained within a shrink-fitted collar (visible here), the nonrotating axle was subjected to bending stresses in three directions, which produced this unusual pattern of beach marks. 0.65×

Fig. 522 Surface of a fatigue fracture in a nonrotating axle of AISI 8640 steel with a hardness of approximately 30 HRC. Visible are beach marks, which indicate that the fatigue crack penetrated across more than 90% of the axle before final fast fracture occurred. The origin (arrow) was at a discontinuity in a cut thread. ~0.95×

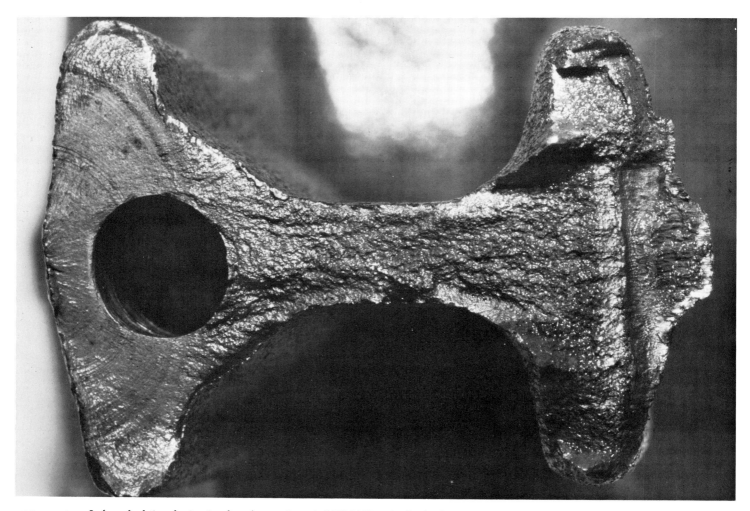

Fig. 523 Surface of a fatigue fracture in a forged connecting rod of AISI 8640 steel with a hardness of 26 to 27 HRC throughout. The rod broke after approximately 84 000 km (52 000 miles) of service. The fatigue-crack origin is at the left edge, at the flash line of the forging, but no unusual roughness of the flash trim there was discovered. No metallurgical cause of the fracture was detected. The fatigue crack progressed about halfway around the oil hole at left before final fast fracture occurred. Note the pronounced shear lip at the right edge. ~7×

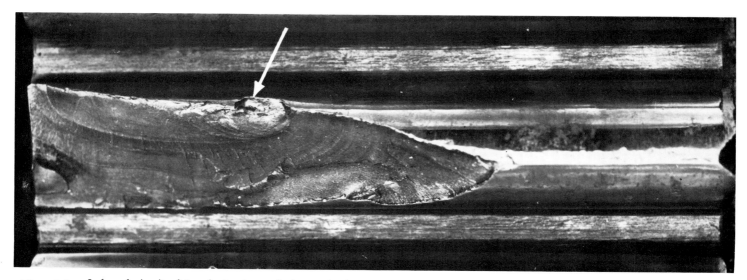

Fig. 524 Surface of a bending-fatigue fracture in a tooth of a sprocket-drive pinion of AISI 8650 steel induction hardened to 58 to 60 HRC. The fatigue-crack origin is at a deep pit (arrow) in the tooth surface. Severe end loading produced other, smaller pits in the surfaces of the teeth. Beach marks indicate that the fatigue crack penetrated a large portion of the tooth before a piece broke off. 1.6×

AISI/SAE Alloy Steels / 333

Fig. 525, 526 Figure 525 (left) is a fatigue fracture in the splined main drive shaft of a lift truck. The material is heat treated AISI 8645 steel. Hardness was 44 HRC throughout, except at the surfaces of some of the splines, where the hardness was 38 HRC. There was no apparent reason for the low hardness of these splines; decarburization, which was suspected initially, was not revealed by metallographic examination. Note the longitudinal cracks in the portion at left. Experimentally broken longitudinal sections of the shaft showed very "woody" fracture surfaces. Figure 526 (right) is a view of the fracture surface of the right portion of the shaft in Fig. 525. Sharpness of fillets is evident, such as at the arrows in both figures. See also Fig. 527 to 529. Both at 1.25×

Fig. 529 Longitudinal section through the same spline as in Fig. 528, displaying the size and even distribution of the inclusions in the steel. Some of the inclusions were found to be as long as 8 mm ($^5/_{16}$ in.). Etched in 1 part HCl, 1 part H_2O. 2×

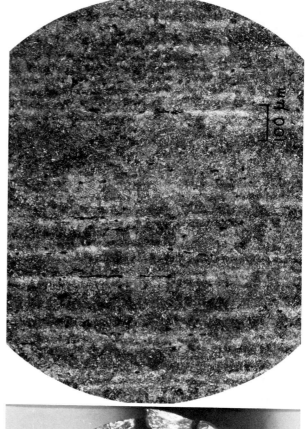

Fig. 528 Longitudinal section through a spline of the shaft in Fig. 525, showing inclusions, banding, and some free ferrite. The cause of the fracture was improper design; corners at bases of splines had adequate radii for most of the spline length, but these radii were reduced, forming sharp angles, at points near where the splines blended with the shaft. 2% nital etch, 100×

Fig. 527 Fracture surface of the left portion of the shaft in Fig. 525, showing the badly mangled splines separated by longitudinal cracks, which are evident also in Fig. 525. 1.25×

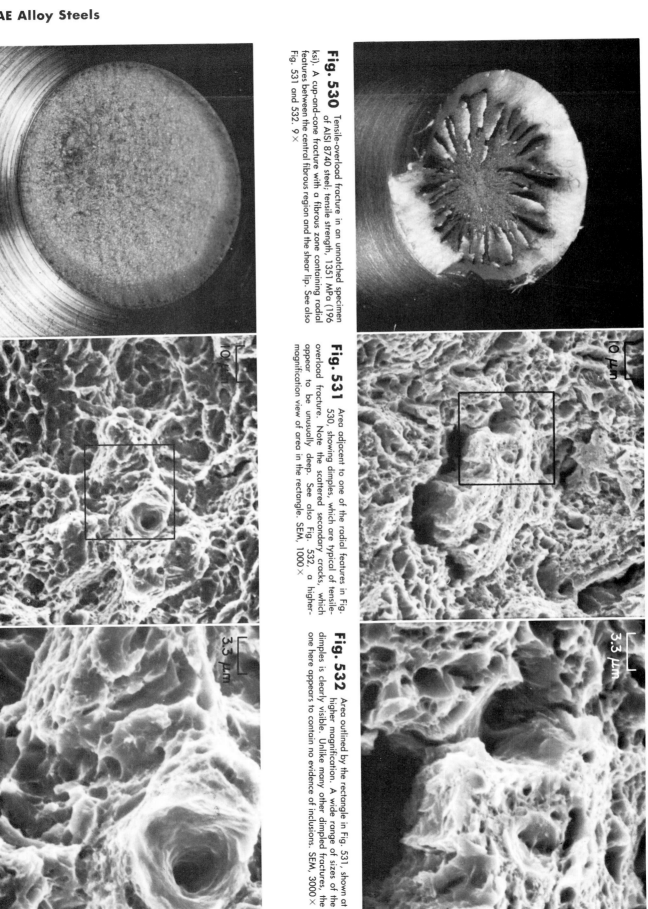

Fig. 530 Tensile-overload fracture in an unnotched specimen of AISI 8740 steel; tensile strength, 1351 MPa (196 ksi). A cup-and-cone fracture with a fibrous zone containing radial features between the central fibrous region and the shear lip. See also Fig. 531 and 532. 9×

Fig. 531 Area adjacent to one of the radial features in Fig. 530, showing dimples, which are typical of tensile-overload fracture. Note the scattered secondary cracks, which appear to be unusually deep. See also Fig. 532, a higher-magnification view of area in the rectangle. SEM, 1000×

Fig. 532 Area outlined by the rectangle in Fig. 531, shown at higher magnification. A wide range of sizes of the dimples is clearly visible. Unlike many other dimpled fractures, the one here appears to contain no evidence of inclusions. SEM, 3000×

Fig. 533 Surface of tensile-overload fracture in a notched specimen of AISI 8740 steel heat treated to a tensile strength of 2044 MPa (296.5 ksi). Fracture originated around the periphery of the specimen at the root of the notch. Radial features are evident. See also Fig. 534 and 535. 9×

Fig. 534 Area in the central final-fracture region in Fig. 533. The range of dimple sizes is similar to that in the unnotched fracture in Fig. 531, but only one secondary crack is evident, and it is of modest size. See Fig. 535 for a higher-magnification view of the area outlined by the rectangle. SEM, 1000×

Fig. 535 Higher-magnification view of the area in the rectangle in Fig. 534. Range in dimple size is more clearly evident here. Large dimple at upper right was presumably initiated by a large inclusion (not shown). SEM, 3000×

Fig. 536, 537 Grinding cracks in AISI 9310 helical pinion. Part was rough machined, carburized, hardened and tempered (56 to 58 HRC), and finish ground. Cracks, due to abusive final grinding, were discovered in a tooth root during nondestructive testing. Fig. 536 (left): Helical pinion. Arrows point to cracks. 2×. Fig. 537 (right): Fracture surface of opened crack (arrows) reveals a crack depth of 0.5 to 0.9 mm (0.020 to 0.035 in.). 5× (P.J. Walling, Metcut Research Associates, Inc.)

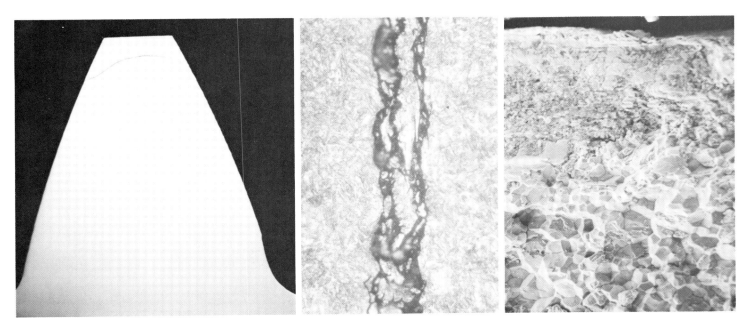

Fig. 538, 539 Quench crack in AISI 9310 spur gear. Part was rough machined, normalized and tempered, carburized, hardened and tempered (52 to 56 HRC), finish ground, and black oxide treated. A crescent-shaped surface crack was discovered 2.5 mm (0.1 in.) below the top land during nondestructive testing. Metallography revealed that the crack followed the case-core interface and was filled with metallic debris. Fig. 538 (left): Gear active profile shows crack at top land. 5×. Fig. 539 (right): Debris-filled quench crack. 2% nital, 750× (P.J. Walling, Metcut Research Associates, Inc.)

Fig. 540 Brittle intergranular fracture of AISI 9254 due to quench cracking. The crack initiated at a seam, 0.15 mm (0.006 in.) deep. The seam wall is the irregularly textured area at top in the fractograph. SEM, 200× (J.H. Maker, Associated Spring, Barnes Group Inc.)

Fig. 541 Mating fracture surfaces of commercial 75-mm (3-in.) diam grinding balls. Fracture was caused by severe ball-on-ball impacts in laboratory tests. Left top and bottom: Commercial forged and heat-treated low-alloy steel ball. Composition: 0.63% C, 0.90% Mn, 0.76% Si, 0.66% Cr, 0.018% P, and 0.011% S. Right top and bottom: Experimental manganese steel ball, cast and heat treated. Composition: 0.97% C, 5.3% Mn, 1.9% Cr, 1.0% Mo, 0.48% Si, 0.03% P, and 0.018% S. Outer rim of ball was under compression due to heat treatment and plastic deformation from impacts. (H.W. Leavenworth, Jr., U.S. Bureau of Mines)

Fig. 542 Catastrophic failure of 75-mm (3-in.) diam manganese steel grinding ball after 51 440 severe ball-on-ball impacts in a laboratory test. Composition: 1.21% C, 5.3% Mn, 1.8% Cr, 1.1% Mo, 0.46% Si, 0.1% Ni, and 0.1% Cu. Fracture initiated at an interior crack. (Note another such crack on the fracture surface at center.) The interior of the ball was under tension. The outer rim was under compression due to heat treatment and plastic deformation from impacts. (H.W. Leavenworth, Jr., U.S. Bureau of Mines)

AISI/SAE Alloy Steels / 337

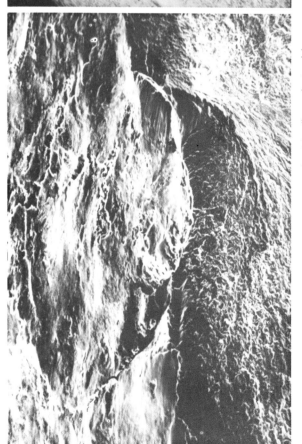

Fig. 543, 544 High-cycle fatigue fracture of valve spring for automobile engine. Material: Cr-V alloy steel (modified ASTM A232). Failure occurred during qualification testing at 17.56 million cycles. No material defect was found; stress-life combination was considered normal for this type of service. Fig. 543 (left): Fracture surface near origin. Note "factory roof" features usually associated with high-cycle fatigue. SEM, 50×. Fig. 544 (right): Same as Fig. 543, but at higher magnification. SEM, 200× (J.H. Maker, Associated Spring, Barnes Group Inc.)

Fig. 545, 546, 547 Fracture of compression spring after 4.3 million cycles in a laboratory test was caused by a large defect called a "scab." Failure origin was 45° away from the inner diameter of the spring, the point of maximum stress. Material: Oil-tempered and shot-peened modified Cr-V valve spring wire. Fig. 545 (left): Large scab that caused spring failure. SEM, 200×. Fig. 546 (center) and 547 (right): Scabs are also called "button" defects, because the depression looks like a buttonhole and the material left in it resembles a button. They are commonly caused by mechanical damage at some stage in wire production before final draw. Material: Armco 17-7PH (UNS S17700) wire. SEM, 40× and 200×, respectively (J.H. Maker, Associated Spring, Barnes Group Inc.)

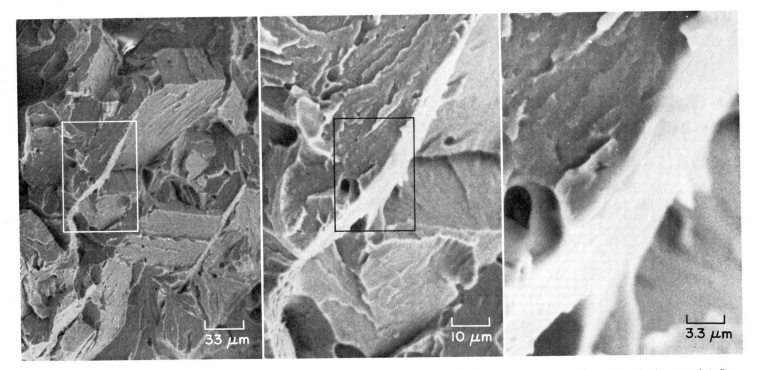

Fig. 548 Charpy impact fracture in 0.18C-3.85Mo steel after heat treatment in an inert atmosphere at 1200 °C (2190 °F) for 1 h, followed by an ice/10% brine quench. Many quasi-cleavage facets are visible. See Fig. 549 (an enlargement of the area in the rectangle). SEM, 300×

Fig. 549 Area outlined by the rectangle in Fig. 548, shown at higher magnification, which makes details of the quasi-cleavage more clearly evident. A few elongated dimples are present. Area in the rectangle here is shown at higher magnification in Fig. 550. SEM, 1000×

Fig. 550 Area outlined by the rectangle in Fig. 549, shown at higher magnification. The bright tilted facet may be a grain boundary or a tear ridge. An inclusion is faintly visible in the dimple about midway along the left edge. SEM, 3000×

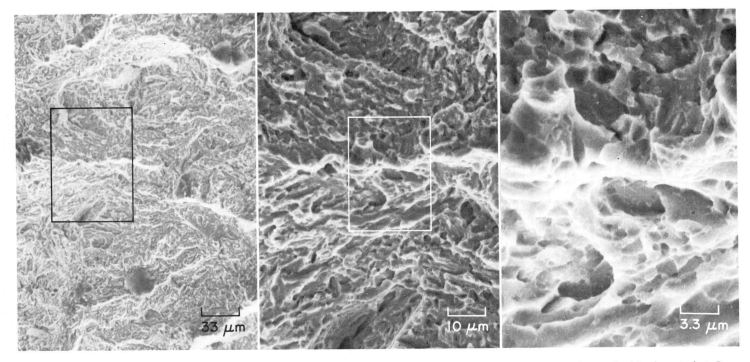

Fig. 551 Charpy impact fracture in 0.41C-4.2Mo steel after the same heat treatment as that for the specimen in Fig. 548. This fracture occurred by dimpled rupture, although the impact energy was only one-third that of Fig. 548. See also Fig. 552 and 553. SEM, 300×

Fig. 552 This view at higher magnification of the area outlined by the rectangle in Fig. 551 reveals that many of the dimples are highly elongated, exhibiting locally some features suggestive of cleavage. Area in rectangle here is shown enlarged in Fig. 553. SEM, 1000×

Fig. 553 Area outlined by the rectngle in Fig. 552, shown at triple the magnification. At this enlargement, many extremely fine dimples are visible, formed along the margins of larger voids. SEM, 3000×

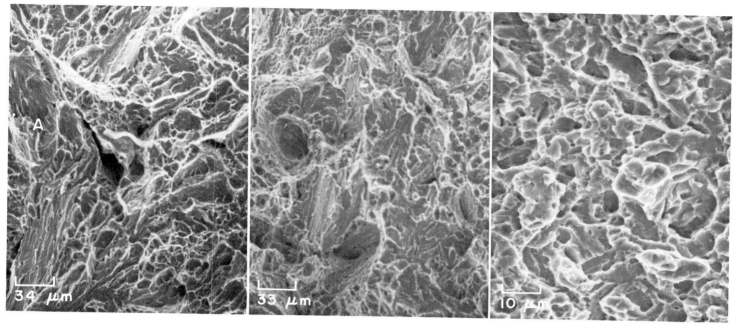

Fig. 554 This specimen was austenitized at 1205 °C (2200 °F). Yield strength was 1413 MPa (205 ksi). There appear to be large facets of quasi-cleavage, such as at A; elsewhere, the surface contains rather large dimples. Note the large void near the center. SEM, 290×

Fig. 555 This specimen was austenitized at 1095 °C (2000 °F). This fracture, like that in Fig. 554, displays quasi-cleavage facets; some of them are apparently rather deeply indented. There are also intervening regions showing clusters of equiaxed dimples. SEM, 300×

Fig. 556 This specimen, austenitized at 980 °C (1800 °F), shows none of the quasi-cleavage facets observed in Fig. 554 and 555. There appear to be many scattered dimples, but most of the surface consists of small facets, seemingly caused by cleavage. SEM, 1000×

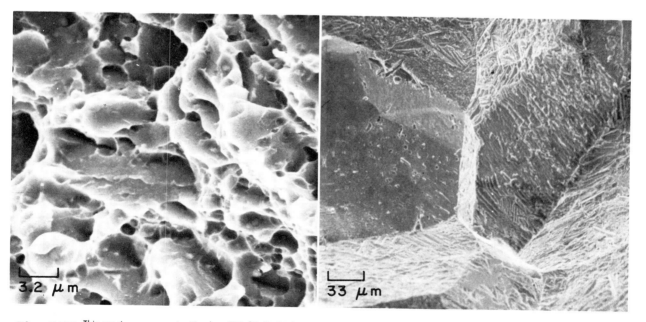

Fig. 557 This specimen was austenitized at 870 °C (1600 °F). Yield strength was 1324 MPa (192 ksi). The fracture surface exhibits a wide range of dimple sizes. The heat treatment resulted in a plane-strain fracture-toughness (K_{Ic}) value of 55 MPa$\sqrt{m}$ (50 ksi$\sqrt{in}$.); this contrasted with a value of 121 MPa$\sqrt{m}$ (110 ksi$\sqrt{in}$.) for the specimen in Fig. 554—probably because of incomplete solution of carbide particles at the lower austenitizing temperature for the specimen shown here. SEM, 3150×

Fig. 558 The heat treatment of this specimen was altered as follows: austenitize at 1205 °C (2200 °F) for 1 h, quench to 870 °C (1600 °F) and hold for 30 min, ice-brine quench, then refrigerate in liquid nitrogen. Fracture appears to have occurred by a combination of transgranular and intergranular cleavage. The K_{Ic} value of this specimen was even lower than that for the specimen in Fig. 557; the lower value is believed to be the result of a different distribution of carbide particles. SEM, 300×

Fig. 554 to 558 These fractographs show the effects of various austenitizing treatments on characteristics of fractures in an experimental secondary-hardening steel containing 0.3% C, 0.6% Mn, and 5.0% Mo. All specimens were austenitized for 1 h at the temperatures indicated, quenched in ice-brine, and then refrigerated for 30 min in liquid nitrogen. Plane-strain fracture-toughness specimens were then tested at room temperature as part of a program to develop high levels of toughness in steels that also maintain ultrahigh tensile strength.

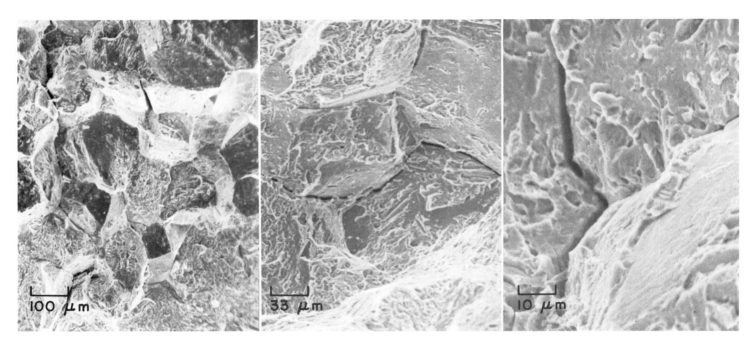

Fig. 559 An area of the fracture surface in which the fracture appears to be intergranular, with secondary cracks between the grains. The lower portion of the area may contain dimples. SEM, 100×

Fig. 560 Another area of the fracture surface, shown at higher magnification. This area has many characteristics of an intergranular fracture. Grain surfaces exhibit fine dimples, but no river patterns. SEM, 300×

Fig. 561 A third area of the fracture surface, at still higher magnification. This area shows a secondary crack penetrating deeply into the grain boundary. Small cleavage steps are evident. SEM, 1000×

Fig. 559, 560, 561 These fractographs are of a fracture-toughness specimen of 0.4C-0.6Mn-5Mo secondary-hardening steel that was tested as part of a program to develop fracture toughness in high-strength steel (the same program that yielded Fig. 554 to 558). This specimen was austenitized for 1 h at 1205 °C (2200 °F), ice-brine quenched, and refrigerated for 30 min in liquid nitrogen. Room-temperature yield strength was 1758 MPa (255 ksi). Plane-strain fracture-toughness (K_{Ic}) value was 76 MPa$\sqrt{m}$ (69 ksi$\sqrt{in.}$)—appreciably less than the value for the lower-carbon (0.3% C) steel in Fig. 554, which was given the same heat treatment.

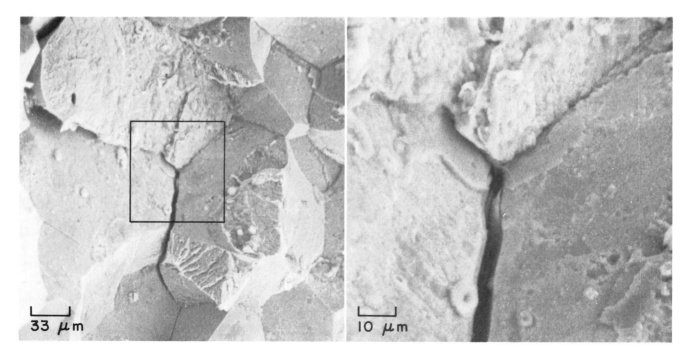

Fig. 562 A V-notch Charpy impact fracture in a specimen of 0.42C-2.1Mo-5.2Co alloy steel that was not as tough as had been expected. The fracture has been largely intergranular, although three facets appear to be transgranular cleavage facets. Note the secondary grain-boundary crack. Area in rectangle is shown enlarged in Fig. 563. SEM, 300×

Fig. 563 Area outlined by the rectangle in Fig. 562, shown at a higher magnification. The two facets in the lower portion are believed to be grain surfaces; facet in upper portion is believed to be a transgranular cleavage facet. The cause of the low toughness of this specimen was quench cracking. Small rings visible here may be artifacts. SEM, 1000×

AISI/SAE Alloy Steels / 341

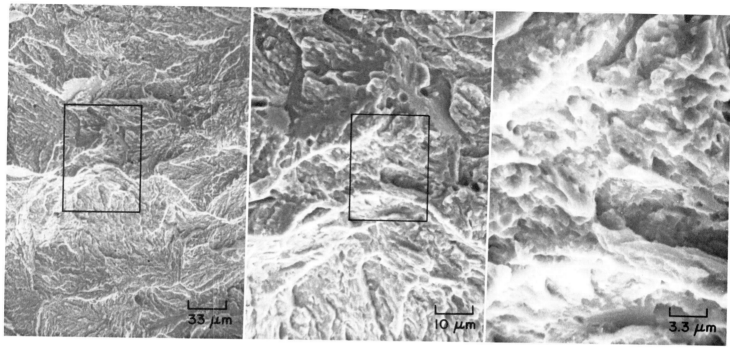

Fig. 564 Charpy V-notch impact fracture in 0.43C-3.85Mo-9Co alloy steel that was heat treated in an inert atmosphere at 1200 °C (2190 °F) for 1 h, then quenched in an agitated solution of ice and 10% brine. It appears that fracture was by a shear process. See also Fig. 565. SEM, 300×

Fig. 565 Higher-magnification view of the area within the rectangle in Fig. 564, showing more clearly the shear deformation and some isolated dimples. There appear to be small regions of local cleavage. See Fig. 566 for a higher-magnification view of the area in the rectangle here. SEM, 1000×

Fig. 566 Higher-magnification view of the area within the rectangle in Fig. 565, which reveals additional small dimples that are not readily apparent at the lower magnification. See Fig. 567 to 569 for the effects of tempering on the fracture surface. SEM, 300×

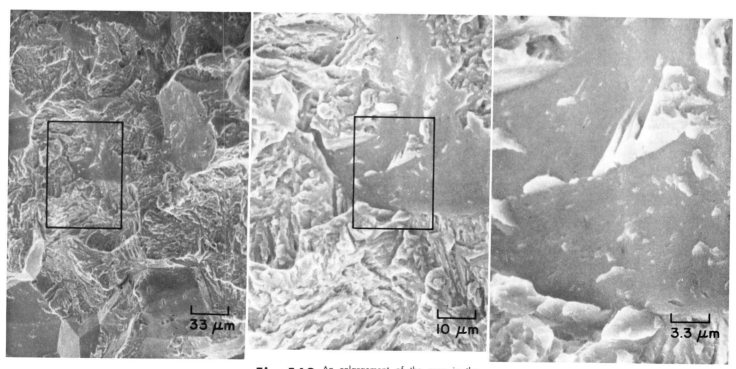

Fig. 567 Charpy impact fracture in a specimen of the same alloy steel and given the same austenitizing treatment as in Fig. 564, but then tempered for 1 h at 600 °C (1110 °F), which maximized secondary hardening, and air cooled. See also Fig. 568. SEM, 300×

Fig. 568 An enlargement of the area in the rectangle in Fig. 567. This specimen, which had the same yield strength as that of the as-quenched specimen in Fig. 564, fractured mainly by intergranular separation. See Fig. 569 for a higher-magnification view of the area in the rectangle here. SEM, 1000×

Fig. 569 Higher-magnification view of the area in the rectangle in Fig. 568. A tongue-like formation is visible (at left center) on the flat surface of a grain. Tempering has produced embrittling networks of carbide particles. SEM, 3000×

342 / AISI/SAE Alloy Steels

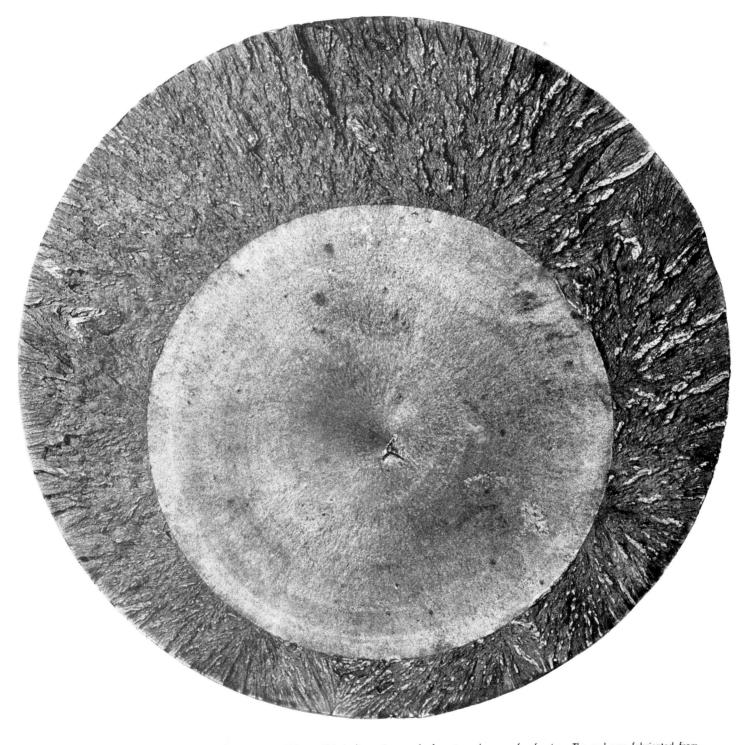

Fig. 570 Surface of a fatigue fracture in a 200-mm (8-in.) diam piston rod of a steam hammer for forging. The rod was fabricated from 0.26C-0.70Mn-0.87Ni-1Cr steel and heat treated to a hardness of 24 HRC at the surface and 17 HRC at the center of the section. This is an example of the fracture caused by pure tensile fatigue, in which surface stress concentrations are absent and a crack may start anywhere in the cross section. In this instance, the initial crack formed at a forging flake slightly below center, grew outward symmetrically, and ultimately produced a brittle fracture without warning. 0.93×

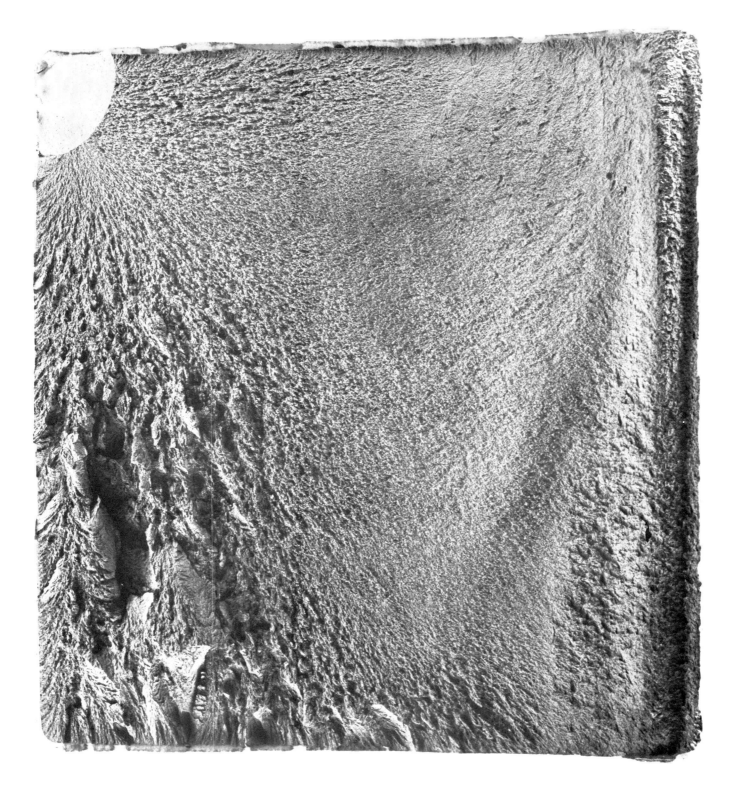

Fig. 571 Surface of a fracture in an equalizer bar, which supports the front end of a tractor, fabricated from D6B steel and heat treated to a hardness of 45 to 47 HRC. The bar was put into service for 200 h and then returned to the laboratory, where it was flexed in fatigue to fracture at 60×10^3 cycles. Note the very fine texture of the fatigue-crack zone at the upper left corner. It is concluded that because there are no beach marks beyond that small corner zone, the remainder of the fracture occurred in a catastrophic manner. Shear lips are evident along the top and right edges, and may also be present around the lower left corner. (D6B steel contains 0.45% C, 0.80% Mn, 1.00% Cr, 0.50% Ni, 1.00% Mo, and 0.10% V.) ~1.13×

Fig. 572 Four fracture surfaces of a hemispherical accumulator head of high-strength low-alloy steel (0.15% C, 0.8% Mn, 0.6% Cr, and 0.85% Ni) that broke into several pieces during pressure testing at room temperature. Hardness ranged from 28 to 39 HRC. Fracture originated at a thin spot in the wall, where banding was also present. At top are mating surfaces of a fracture near the origin, showing chevron marks and tongues, although tongues are more typical of low-temperature cleavage of iron-base alloys than of room-temperature cleavage. The fracture surface third from the top is very coarse-grained, whereas the surface at bottom is fine-grained and silky; the difference in appearance between these two surfaces may have resulted from a difference in rate of crack propagation. Actual size

ASTM/ASME Alloy Steels / 345

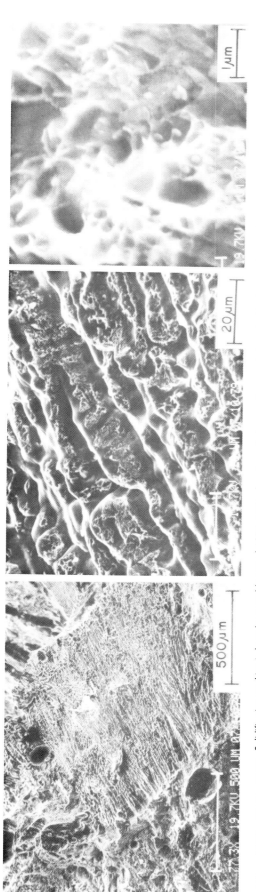

Fig. 573, 574, 575 Solidification cracking in laser-beam weldments of ASTM A372, class 6 (HY-80). Laser welds in 13-mm (0.5-in.) thick plate were made at a travel speed of 13 mm/s (30 in./min) using a beam power of 10.6 kW and a heat input of 0.83 kJ/mm (21.2 kJ/in.). Weldment toughness was measured at room temperature on a subsized 13-mm (0.5-in.) thick dynamic tear (DT) specimen notched in the weld. Solidification cracking of the hard and brittle fusion zone (untempered martensite) resulted in low toughness values typically little higher than 76 J (56 ft · lb). Toughness of the 3% Ni base metal (quenched and tempered martensite) generally exceeded 470 J (350 ft · lb). The fractographs of the DT specimen surface reveal considerable solidification cracking as well as some scattered porosity. Fig. 573 (left): SEM, 50×. Fig. 574 (center): SEM, 850×. Fig. 575 (right): SEM, 13 000×. (D.W. Moon and E.A. Metzbower, Naval Research Laboratory)

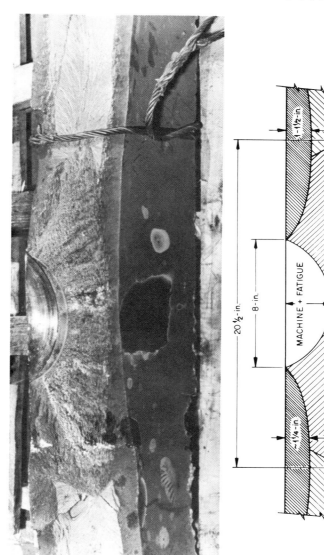

Fig. 576, 577 Intentional fracture by overpressurization of ASTM A508, class 2, pressure vessel. The intermediate test vessel (ITV) was fabricated as part of the Heavy Section Steel Technology (HSST) program at Oak Ridge National Laboratory, a government-sponsored effort aimed at gaging the toughness of nuclear pressure vessels. The ITV measured 1 m (39 in.) in outside diameter with a 150-mm (6-in.) thick wall and was designed for 67-MPa (9.7-ksi) internal pressurization. The preflawed vessel contained a semielliptical fatigue-cracked defect measuring 205 mm (8 in.) long and 65 mm (2⅝ in.) deep. Failure occurred at the 55 °C (130 °F) test temperature and an internal pressure of 198 MPa (28.7 ksi)—nearly three times the design pressure allowed by the ASME code and four times the probable operating pressure. Fig. 576 (top): Fracture surface. Note machined and fatigue precracked flaw at top center. Region of fatigue sharpened notch had a dimpled morphology; farther away from the flaw, a cleavage morphology. Fig. 577 (bottom): Schematic of fracture surface in Fig. 576. Initial failure was by ductile tearing. Flaw grew to approximately 500 mm (20 in.) long and 100 mm (4 in.) deep before onset of brittle propagation. (D.A. Canonico, C-E Power Systems, Combustion Engineering Inc.)

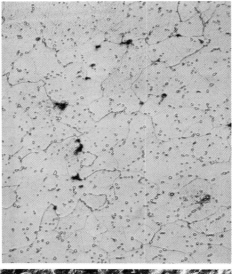

Fig. 578 Creep failure of steam boiler superheater tube. Material: normalized and tempered ASME SA213, grade T22 (2.25Cr-1Mo steel). Cracking occurred at a hot spot due to long-time exposure to tensile stresses induced by the internal pressure and service temperatures up to 705 °C (1300 °F). Spheroidization of carbides in the ferrite matrix was caused by exposure to temperatures in the 650 to 705 °C (1200 to 1300 °F) range. The black voids at grain boundaries are typical of creep damage. 3% nital, 500×. (J.R. Kattus, Associated Metallurgical Consultants Inc.)

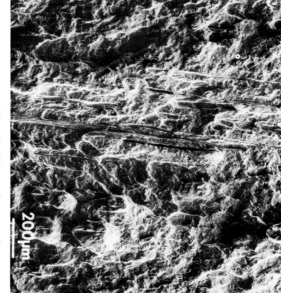

Fig. 579, 580 Effect of inclusions on fatigue crack propagation (FCP) in ASTM A514F. The conventionally melted electric furnace steel was water quenched from 900 °C (1650 °F), tempered at 620 °C (1150 °F), water quenched, stress relieved at 595 °C (1100 °F), and then air cooled. $\Delta K = 72$ MPa$\sqrt{m}$ (65 ksi$\sqrt{in.}$). Fig. 579 (left): Elongated features on T-L (standard transverse) orientation FCP test specimen. Direction of FCP is toward the top. This "streaking" in the FCP direction (and the major hot-rolling direction) is associated with inclusion formations present on the vertical faces of long protrusions or depressions, which also extend in the FCP direction. See also Fig. 281 to 283. SEM, 53×. Fig. 580 (right): Same as Fig. 579 but at higher magnification. See also Fig. 281 to 283. SEM, 1055×. (A.D. Wilson, Lukens Steel Company)

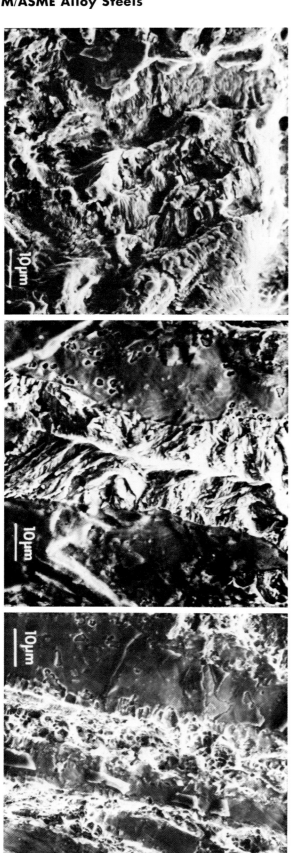

Fig. 581, 582, 583 Effect of inclusions on fatigue crack propagation (FCP) in ASTM A514F. Direction of FCP is toward the top. Fig. 581 (left): Area of fracture surface not near inclusion formations shows primarily ductile fatigue striations. Also note secondary cracking. Through-thickness (S-T orientation) FCP specimen at $\Delta K = 55$ MPa$\sqrt{m}$ (50 ksi$\sqrt{in.}$). SEM, 1020×. Fig. 582 (center) and 583 (right): Through-thickness (S-L orientation) fractures near inclusion formations at two levels of ΔK. At the low 22-MPa$\sqrt{m}$ (20-ksi$\sqrt{in.}$) level (Fig. 582), fatigue damage is by striation formation in the interinclusion material of an inclusion formation. At the high 55-MPa$\sqrt{m}$ (50-ksi$\sqrt{in.}$) level (Fig. 583), ductile fracture (microvoid coalescence) of the interinclusion material occurs. See also Fig. 579 and 580. SEM, both at 1020×. (A.D. Wilson, Lukens Steel Company)

ASTM/ASME Alloy Steels / 347

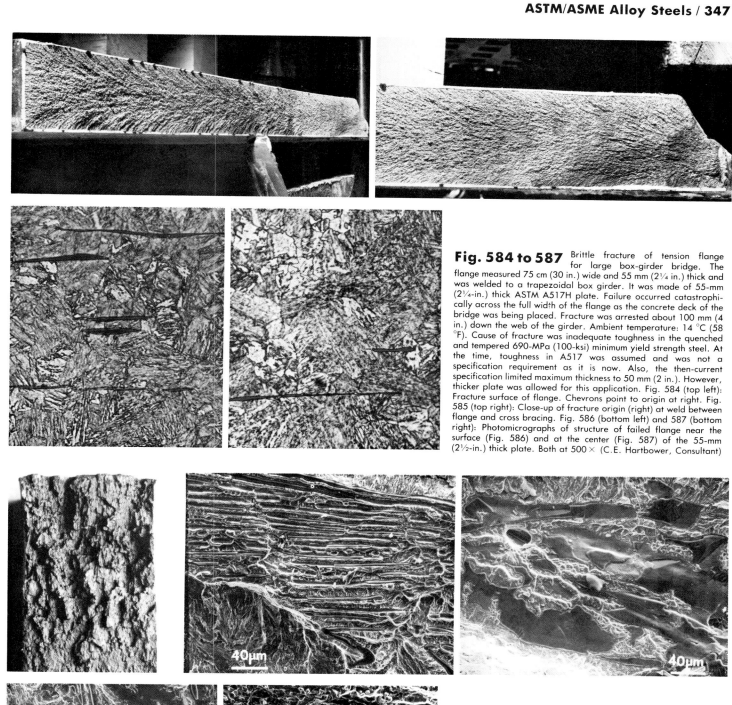

Fig. 584 to 587 Brittle fracture of tension flange for large box-girder bridge. The flange measured 75 cm (30 in.) wide and 55 mm (2¼ in.) thick and was welded to a trapezoidal box girder. It was made of 55-mm (2¼-in.) thick ASTM A517H plate. Failure occurred catastrophically across the full width of the flange as the concrete deck of the bridge was being placed. Fracture was arrested about 100 mm (4 in.) down the web of the girder. Ambient temperature: 14 °C (58 °F). Cause of fracture was inadequate toughness in the quenched and tempered 690-MPa (100-ksi) minimum yield strength steel. At the time, toughness in A517 was assumed and was not a specification requirement as it is now. Also, the then-current specification limited maximum thickness to 50 mm (2 in.). However, thicker plate was allowed for this application. Fig. 584 (top left): Fracture surface of flange. Chevrons point to origin at right. Fig. 585 (top right): Close-up of fracture origin (right) at weld between flange and cross bracing. Fig. 586 (bottom left) and 587 (bottom right): Photomicrographs of structure of failed flange near the surface (Fig. 586) and at the center (Fig. 587) of the 55-mm (2½-in.) thick plate. Both at 500× (C.E. Hartbower, Consultant)

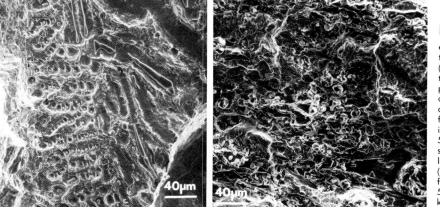

Fig. 588 to 592 Effect of inclusions on FCP in ASTM A533B. The conventionally melted electric furnace steel was water quenched from 900 °C (1650 °F), tempered at 670 °C (1240 °F), air cooled, stress relieved at 595 °C (1100 °F), and then furnace cooled. FCP specimen tested in the through-thickness (S-T, S-L) orientations. Direction of fatigue crack propagation is toward the top. Fig. 588 (top left): Fracture surface of S-L oriented FCP test specimen. Fracture surface appearance due to inclusion formation on depressions or plateaus on the fracture surface. Figures 589 to 591 show three kinds of MnS inclusions in type II MnS inclusion formations: highly elongated (Fig. 589, top center), flattened or pancaked (Fig. 590, top right), and small and round as a result of homogenization (Fig. 591, bottom left). SEM, Fig. 589 and 590: 230×, Fig. 591: 200×. Fig. 592 (bottom right): Alumina (Al_2O_3) galaxies on the through-thickness fracture surface of A533B. SEM, 200×. ΔK in Fig. 589, 590, and 592 = 25 MPa√m (23 ksi√in.); in Fig. 591, 58 MPa√m (53 ksi√in.). See also Fig. 593 to 598. (A.D. Wilson, Lukens Steel Company)

348 / ASTM/ASME Alloy Steels

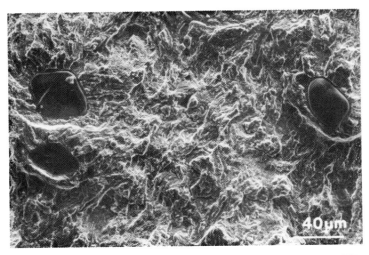

Fig. 593 Effect of inclusions on fatigue crack propagation (FCP) in ASTM A533B. Fractograph shows compact and well-dispersed type III MnS inclusions in calcium-treated electric furnace steel, through-thickness (S-L) orientation. Balance of inclusions were round, calcium-modified duplex types consisting of sulfide and aluminate phases. Decreasing the size and number of inclusions, and the number of inclusion formations, is responsible for the improved FCP behavior in the through-thickness orientations of calcium-treated steels. Direction of FCP is toward the top. $\Delta K = 26$ MPa$\sqrt{m}$ (24 ksi$\sqrt{in.}$). See also Fig. 588 to 592 and 594 to 598. SEM, 300× (A.D. Wilson, Lukens Steel Company)

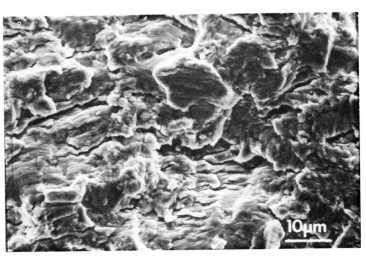

Fig. 594 Effect of inclusions on fatigue crack propagation (FCP) in ASTM A533B. Ductile fatigue striations and secondary cracking are present in this area remote from inclusion formations. Conventional electric furnace heat; L-T orientation; $\Delta K = 46$ MPa$\sqrt{m}$ (42 ksi$\sqrt{in.}$). Direction of FCP is toward the top. See also Fig. 588 to 593 and 595 to 598. SEM, 1200× (A.D. Wilson, Lukens Steel Company)

 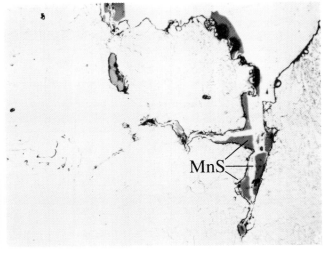

Fig. 595 to 598 Effects of inclusions on fatigue crack propagation (FCP) in ASTM A533B. In conventional electric furnace steel, protrusions perpendicular to the direction of FCP were noted on T-S and L-S fracture surfaces. The protrusions result from inclusion formations (type II MnS and Al_2O_3) acting as crack deflectors. Figures show T-S orientation features. FCP direction is toward the top. Fig. 595 (top left): Protrusions on fracture surface. Fig. 596 (top right): Metallographic cross section reveals MnS inclusions on a protrusion. 170×. Fig. 597 (bottom left): Protrusion. SEM, 15×. Fig. 598 (bottom right): Protrusion with Al_2O_3 galaxy. SEM, 28° tilt angle, 325×. Fig. 597 and 598 at ΔK of 40 MPa$\sqrt{m}$ (36 ksi$\sqrt{in.}$). See also Fig. 588 to 594. (A.D. Wilson, Lukens Steel Company)

 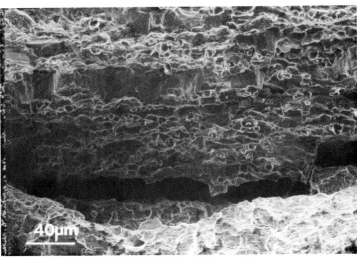

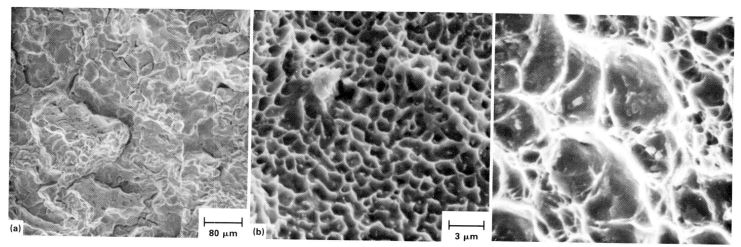

Fig. 599, 600 Cavitated intergranular fracture due to hydrogen attack of ASTM A533B. The specimen of pressure vessel steel—yield strength, 630 MPa (91.5 ksi)—was exposed for more than 1000 h to temperatures above 550 °C (1020 °F) and gaseous hydrogen at a pressure greater than 17 MPa (2.5 ksi). Under these conditions, hydrogen diffuses into the steel, reacts with thermodynamically less stable carbides, and forms bubbles of methane gas along grain boundaries. Mechanical properties plummet and, under impact, failure occurs by rapid coalescence of the methane bubbles, each nucleated at a submicron carbide or inclusion, along prior-austenite grain boundaries. These cavitated intergranular fractures are common in steels used for hydrogen service (hydrocracking or coal conversion processes, for example) and resemble both failures in "overheated" steels, where reprecipitation of fine grain boundary sulfides due to an overly high austenitization temperature provides the source of the voids, and high-temperature creep cavitation fractures, where voids form at carbides and grow by vacancy coalescence and matrix creep. Fig. 599 (left): Fracture surface of Charpy specimen. Note secondary cracking due to severe bubble coalescence along grain boundaries. SEM, 125×. Fig. 600 (right): Same as Fig. 599, but at higher magnification. Note nucleus at bottom of each cavity. SEM, 3330× (R.H. Dauskardt and R.O. Ritchie, University of California)

Fig. 601 Elevated-temperature fracture surface of Cr-Mo-V alloy steel specimen tensile tested at failure at 500 °C (930 °F) and an initial strain rate of 4.4×10^{-5}/s. Composition of the hot-rolled steel plate: 0.29% C, 1.01% Cr, 1.26% Mo, 0.25% V, 0.01% Mn, 0.59% Si, 0.037% P, 0.004% S, and 0.006% Sn. Heat treatment: two-step austenitize 1 h at 1300 °C (2370 °F), followed by 1 h at 950 °C (1740 °F), air cool; temper at 700 °C (1290 °F) to 28 HRC. Failure resulted from the nucleation, growth, and eventual coalescence of grain-boundary cavities. The cavities nucleated on complex sulfides. SEM, 2500× (D.P. Pope, University of Pennsylvania, and S.-H. Chen, Norton Christensen)

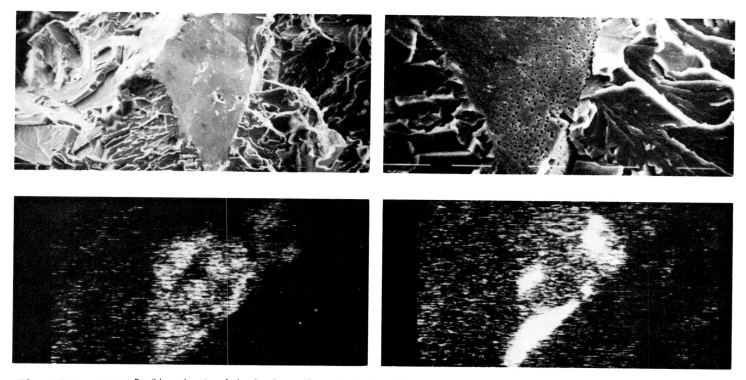

Fig. 602 to 605 Possible explanation of why phosphorus enhances the ductility of Cr-Mo-V steel. Composition of the hot-rolled steel plate: 0.27% C, 0.99% Cr, 1.25% Mo, 0.25% V, 0.01% Mn, 0.04% Si, 0.038% P, 0.004% S, and 0.005% Sn. Heat treatment: austenitize 1 h at 1100 °C (2010 °F), air cool, temper at 700 °C (1290 °F) to 28 HRC. Fig. 602 (top left) and 603 (top right): Fracture surface of specimen tensile tested at 500 °C (930 °F) and an initial strain rate of 4.4×10^{-5}/s. Test interrupted at yield and specimen subsequently broken by impact at −100 °C (−150 °F) in a scanning Auger microprobe. At elevated temperatures and low strain rates, this steel (like many other metals) fails intergranularly by grain-boundary cavitation. However, the grain-boundary facet at the center of Fig. 602 (SEM, 300×) is not uniformly cavitated, as revealed by the higher-magnification view in Fig. 603 (SEM, 850×). Fig. 604 (bottom left) and 605 (bottom right): Phosphorus and Sulfur maps, respectively, of the facet in Fig. 602 (both at 350×). When overlapped, these maps show that phosphorus-rich areas correspond to uncavitated regions of the grain-boundary facet, and sulfur-rich areas correspond to cavitated regions. Thus, when phosphorus is added to Cr-Mo-V steel, site competition between phosphorus and sulfur reduces sulfur segregation and, consequently, sulfide precipitation on grain boundaries. And, because cavities in this steel nucleate principally on sulfides, a reduction in sulfide population on grain boundaries leads to an improvement in ductility. (D.P. Pope, University of Pennsylvania, and S.-H. Chen, Norton Christensen)

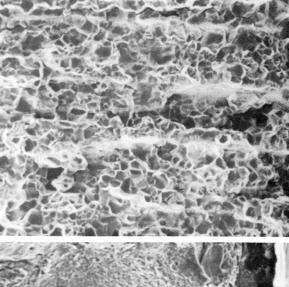
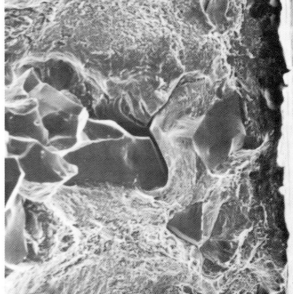
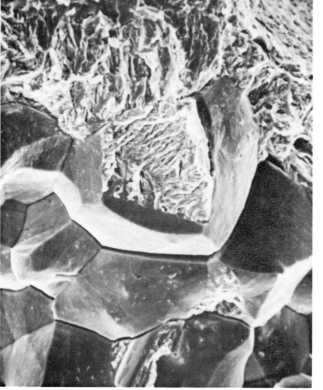
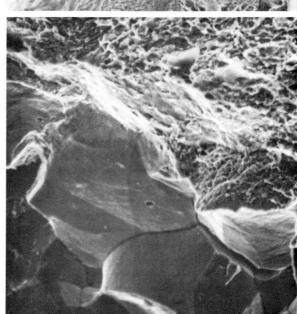

Fig. 606 to 610 Effect of hydrogen environment on fracture of an antimony-doped Ni-Cr steel. Composition of the vacuum-melted alloy: 0.30% C, 3.5% Ni, 1.7% Cr, and 0.06% Sb. Samples were cut from 13-mm (0.5-in.) thick hot-rolled plate perpendicular to the rolling direction and heat treated by austenitizing for 1 h at 1025 °C (1875 °F) and tempering for 1 h at 600 °C (1110 °F) to a grain size of ASTM 3 and a hardness of 30 HRC. Samples were then aged to 100 h at 480 °C (895 °F) to allow antimony to segregate to grain boundaries. Notched samples were fractured in air and in 170 kPa (1.7 atm) of hydrogen gas by slow loading in four-point bending. Fig. 606 (top left): A banded fracture surface, with locally different compositions, of antimony-doped sample fractured in air. A similarly banded structure was observed on samples fractured in hydrogen. SEM, 40×. Fig. 607 (top center) and 608 (top right): Higher-magnification views of fracture surface to sample broken in air. Failure was by stress-controlled intergranular fracture in solute-rich bands and by strain-controlled microvoid coalescence in solute-depleted regions. Auger electron spectroscopy confirmed that intergranular fracture was due to enhanced antimony segregation. SEM, 320× and 640×. Fig. 609 (bottom left) and 610 (bottom right): Fracture surface at two magnifications of sample broken in hydrogen. In this case, solute-depleted regions failed by quasi-cleavage. SEM, 320× and 640× (M.J. Morgan and C.J. McMahon, Jr., University of Pennsylvania)

Austenitic Stainless Steels / 351

Fig. 611 Surface of fatigue-crack fracture in a specimen of AISI type 301 stainless steel that was highly stressed, breaking in 2000 cycles. Crack growth was very irregular, with many pronounced offsets. At center, twin boundaries have affected crack propagation. Area in rectangle is shown at higher magnification in Fig. 612. SEM, 200×

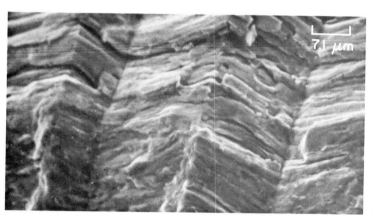

Fig. 612 A higher-magnification view of the area outlined by the rectangle in Fig. 611, which includes twin-boundary outlines. Several secondary cracks have formed at the roots of the fatigue striations. The striae, however, are not delineated well enough to permit a confident estimate of the spacing. SEM, 1400×

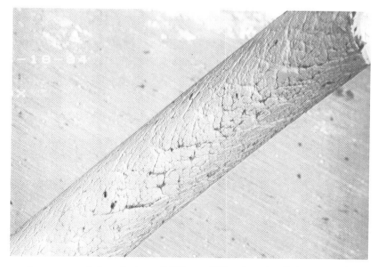

Fig. 613 Fatigue fracture of AISI type 302 spring wire. Failure initiated at grain-boundary damage called "alligatoring," a condition resulting from overetching during acid cleaning. Alligatoring is always detrimental to fatigue resistance and in extreme cases (such as this one) can lead to fracture of small index springs during coiling. SEM, 50× (J.H. Maker, Associated Spring, Barnes Group Inc.)

Fig. 614 Surface of a "rock candy" fracture in a bloom of AISI type 302 stainless steel. An abnormally high silicon content and inadvertent overheating at 1355 °C (2475 °F) for 45 min after the first blooming-mill pass caused excessive grain growth; the second pass then resulted in intergranular fracture. See also Fig. 615. 0.67×

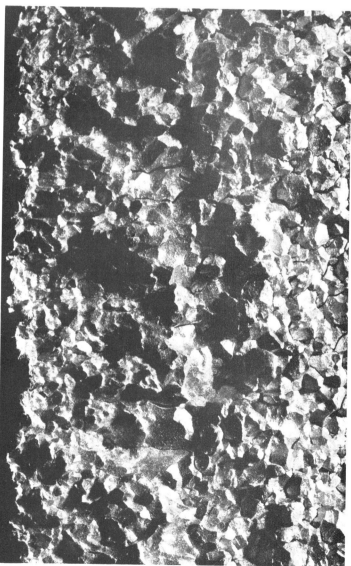

Fig. 615 Higher-magnification view of fracture surface in Fig. 614. This bloom fractured during rolling, splitting apart and wrapping itself around the roll. Note lack of deformation of grains. The fracture surface was cleaned with H_2SO_4 and was brightened with a solution containing HNO_3 and HF. 1.5×

352 / Austenitic Stainless Steels

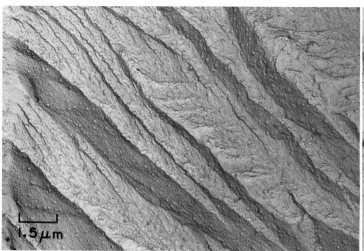

Fig. 616 TEM p-c replica of an area near A, close to the notch, in the fracture in Fig. 618, showing definite cleavage steps and essentially no evidence of fatigue striations. A thin oxide film is present. See also Fig. 617. 6500×

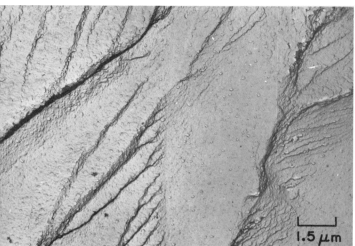

Fig. 617 TEM p-c replica of another area near A in Fig. 618, also showing evidence of cleavage steps. The fine marks at right, which are essentially parallel to the cleavage direction, probably are not fatigue striations. See also Fig. 619 to 626. 6500×

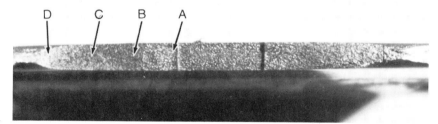

Fig. 618 Surface of a high-cycle fatigue fracture in a notched specimen of AISI type 302 stainless steel sheet subjected at 425 °C (800 °F) to tensile stress cycled from 24.8 to 124 MPa (3.6 to 18 ksi) at 1000 cycles/min. Fracture occurred after 1 625 450 cycles. The notch in the specimen was a center hole 0.25 mm (0.01 in.) in radius and 5 mm (0.20 in.) wide. Fatigue area became purple; fast-fracture area became straw colored, with shear and necking. See Fig. 616, 617, and 619 to 626 for views at A, B, C, and D. 5×

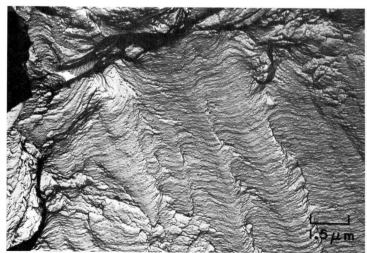

Fig. 619 TEM p-c replica of a third area near A in the fracture surface shown in Fig. 618. Here, fatigue striations are present in rather narrow and somewhat irregular patches. Some evidence of surface oxidation can be seen. 6500×

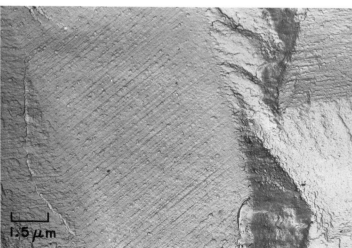

Fig. 620 TEM p-c replica of a fourth area near A in the fracture surface in Fig. 618, showing flat facets quite suggestive of intergranular cleavage. The parallel marks may not be fatigue striations, but rather features associated with the oxide film. 6500×

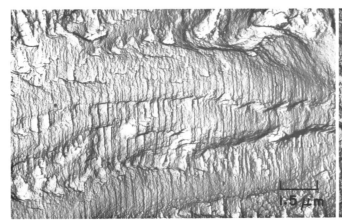

Fig. 621 TEM p-c replica of a fifth area near A in the fracture surface in Fig. 618. This area is characterized by narrow patches of extremely sharp and distinct fatigue striations. Again, the oxide film is apparent. 6500×

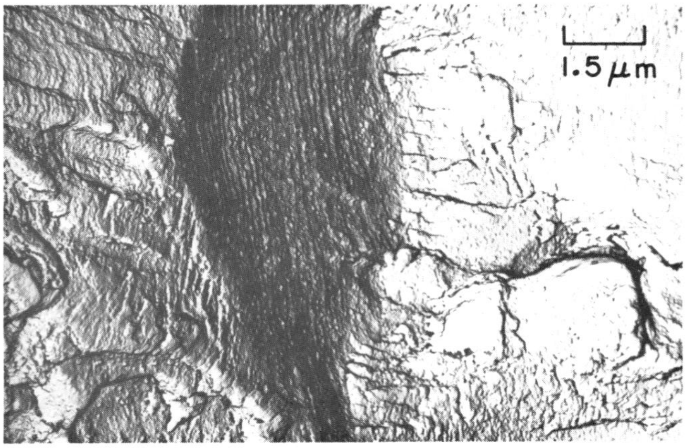

Fig. 622 TEM p-c replica of a sixth area near A in the fracture surface in Fig. 618, showing distinct fatigue striations. The markings to the right of the dark region are believed to be cleavage steps, not fatigue striations. 6500×

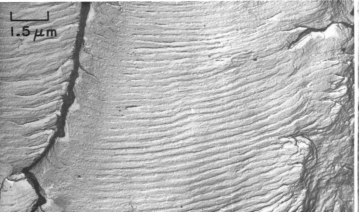

Fig. 623 TEM p-c replica of an area at B, approximately 2.3 mm (0.09 in.) from the notch, in the fracture surface shown in Fig. 618. The heavy, black mark at left is the result of a sharp difference in level at the junction of two crack fronts. The fatigue striations here are regular and well developed. 6500×

Fig. 624 TEM p-c replica of an area at C, about 4.4 mm (0.175 in.) from the notch, in Fig. 618. Note that the fatigue striations here are more widely spaced than those in Fig. 623, which indicates that the effective stress increased as the crack penetrated into the specimen. Partly formed dimples are visible above and right of center. 6500×

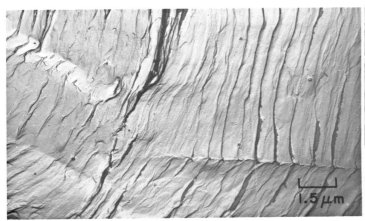
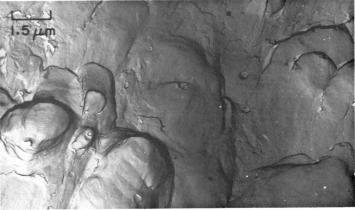

Fig. 625 TEM p-c replica of another area near C in Fig. 618, showing a further increase in fatigue-striation spacing. A junction of two crack fronts is visible at lower right. Secondary cracks appear to have formed at some striations. 6500×

Fig. 626 TEM p-c replica of an area at D, in the region of final fast fracture and about 11.7 mm (0.46 in.) from the notch, in Fig. 618. The elongated dimples confirm that final fracture was by shear. Fatigue striations ceased to form when this final stage of fracture began. 6500×

354 / Austenitic Stainless Steels

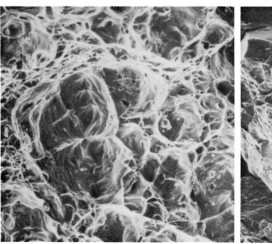 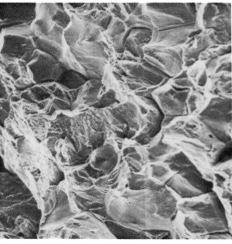 

Fig. 627, 628 Effect of strain rate in creep crack propagation on fracture mode in an AISI type 304 stainless steel tested at 540 °C (1000 °F). A ductile-to-brittle transition occurred at very slow strain rates. Fig. 627 (left): Microvoid coalescence (dimpled rupture) on fracture surface of sample tested at strain rate of 1.3 mm/min (0.05 in./min). SEM, 200×. Fig. 628 (right): Brittle, intergranular fracture surface of sample tested at 6 μm/min (240 μin./min). SEM 200× (J.E. Nolan, Westinghouse Hanford Company)

Fig. 629 Stress-corrosion cracking facets on the fracture surface of an AISI type 304 sample tested under constant load in $MgCl_2$ solution at 155 °C (310 °F). The steplike features are analogous to cleavage facets. See also Fig. 645. SEM, 3000× (R. Liu, N. Narita, C. Altstetter, H. Birnbaum, and N. Pugh, University of Illinois)

Fig. 630, 631, 632 Failure of AISI type 304 tube due to chloride stress-corrosion cracking (SCC) from the outside in. The oil feed tube was used in a paper mill, where relative humidity was 100%. Cracks and evidence of corrosion were found under the coating on the painted side of the tube. No damage was noted on the other, unpainted side. Torsional stresses in service and chlorides in the paint (at a level of 6600 ppm) provided the conditions necessary for chloride SCC. Fig. 630 (left): Crack on painted side of tube. Angle reveals influence of torsional stresses. Fig. 631 (center): Stress-corrosion cracks extending from the outside diameter of the tube (at top) toward its inner diameter. 10% oxalic acid electrolytic etch; 28×. Fig. 632 (right): Fractured tube. Branching of crack verifies that fracture occurred from the outside diameter (at top) in. 10% oxalic acid electrolytic etch, 28× (Z. Flanders, Packer Engineering Associates, Inc.)

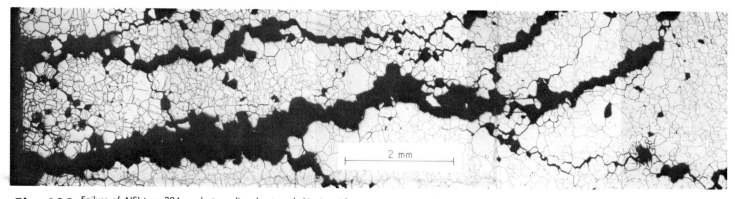

Fig. 633 Failure of AISI type 304 product gas line due to polythionic acid stress-corrosion cracking (SCC). The 205-mm (8-in.) diam Schedule 140 pipe was used in a coal-gasification pilot plant to transport product gas from the gasifier at a temperature of 430 °C (800 °F). The gas line failed by through-wall cracking around 270° of its circumference near a weld joining it to a 2.25Cr-0.5Mo steel tee. During plant shutdowns, the inside of the line was subjected to condensation of moisture and subsequent exposure to air. Chloride content of the condensate was 4.5 ppm. Also detected on the inside of the pipe were mixed iron and chromium sulfides, which had formed during elevated-temperature exposure to the sulfur-bearing product gas. The photomontage of the failed region of the stainless steel pipe reveals numerous intergranular, highly branched cracks originating at the inner surface (at left) of the pipe. Other metallographic work revealed that this region had been moderately sensitized due to the combined effects of welding and subsequent stress relief. Investigators concluded that polythionic acid ($H_2S_xO_6$, where x = 3 to 5) was produced when the condensate reacted in the presence of oxygen with the already existing sulfidic scale on the inner diameter of the pipe. The region near the dissimilar-metal weld provides both the required sensitized microstructure and either residual or differential thermal stresses. 10% oxalic acid electrolytic etch, 15× (D.R. Diercks, Argonne National Laboratory)

Fig. 634 to 638 Hydrogen-embrittlement failure of AISI type 304 bellows. The component was made by rolling sheet into tube, seam welding, and then forming. Continuous pressurization with gaseous hydrogen at 172 kPa (25 psi) induced hydrogen embrittlement at roots of bellows convolutes, regions where residual stresses from forming were at a maximum. Fig. 634 and 635 (top row): Bellows assembly and hydrogen-induced circumferential cracks at A and B. Fig. 636 (left center), 637 (bottom left), and 638 (bottom right): Microstructure of fifth convolute cross section, showing twins, transgranular and intergranular fracture modes, and secondary cracking. Oxalic acid electrolytic etch, 10×, 50×, and 200×, respectively (R.J. Schwinghamer, NASA Marshall Space Flight Center)

356 / Austenitic Stainless Steels

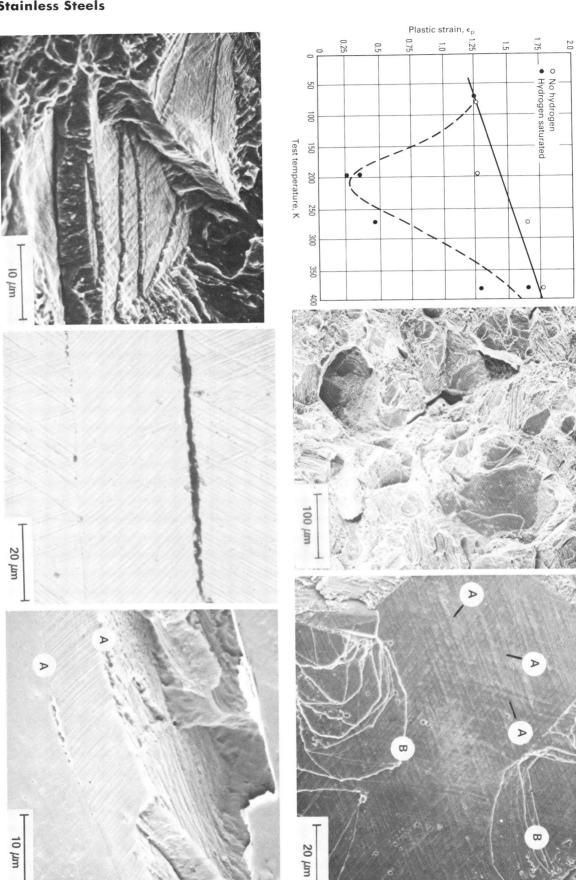

Fig. 639 to 644 Faceted fracture in hydrogen-damaged AISI type 304L. Hydrogen damage in type 304L supersaturated with hydrogen gas is most severe at −75 to −25 °C (−105 to −15 °F). The brittle fracture surfaces exhibit predominantly facets that form by parting along coherent twin boundaries, a process similar to cleavage. Although twin parting is unusual in face-centered cubic (fcc) alloys, similar facets have been observed in hydrogen-charged Tenelon (UNS S21400) and Nitronic 40 (UNS S21900) and in stress-corrosion cracking of type 304 in MgCl$_2$. Fig. 639 (top left): Ductility minimum in hydrogen-charged type 304L. Fig. 640 (top center): General view of fracture surface of sample tested at −75 °C (−105 °F). Note that facets are interconnected, are planar and flat, frequently exhibit steps 0.5 to 6.5 μm high, always have one to three sets of slip traces, and (unlike cleavage fractures) never show river patterns. Fracture plane of the facets as determined by x-ray diffraction to be {111} plane in the fcc austenite. SEM, 170×. Fig. 641 (top right): Slip traces (A) and steps (B) in type 304L sample tested at −75 °C (−105 °F). SEM, 950×. Fig. 642 (bottom left): View of facet at an angle. Note steps and slip traces. Traces arise by intersection of deformation bands with coherent twin boundaries on longitudinal section of fractured type 304L sample. Twin-boundary traces lie at an angle of 60° to 90° from the tensile axis (vertical in photomicrograph). 1000×. Fig. 643 (bottom center): Microcracking along coherent annealing twin boundaries on longitudinal section of fractured type 304L sample. Twin-boundary traces lie at an angle of 60° to 90° from the tensile axis (vertical in photomicrograph). 1000×. Fig. 644 (bottom right): Intentionally opened microcrack on polished longitudinal section. Note that features on crack face are identical to those on fracture surface facets (see Fig. 642), indicating that both facets and microcracks form by the same mechanism. SEM, 2000×. (G.R. Caskey, Jr., Atomic Energy Division, DuPont Company)

Austenitic Stainless Steels / 357

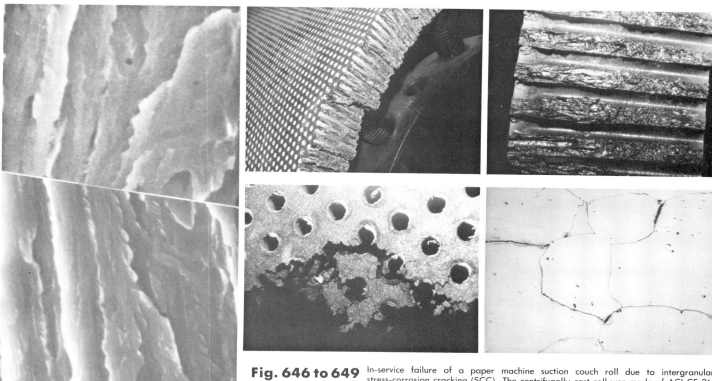

Fig. 645 Hydrogen embrittlement of AISI type 304 tested under constant load in pure hydrogen gas at 100 kPa (1 atm) and 25 °C (75 °F). The two-part fractograph compares matching fracture surfaces. Note the quasi-cleavage-type facets. They generally exhibit more ductility (have a rougher surface) than classic cleavage facets. See also Fig. 629. SEM, 6000× (R. Liu, N. Narita, C. Altstetter, H. Burnbaum, and N. Pugh, University of Illinois)

Fig. 646 to 649 In-service failure of a paper machine suction couch roll due to intergranular stress-corrosion cracking (SCC). The centrifugally cast roll was made of ACI CF-8M (~AISI type 316) and measured 8.8 m (29 ft) long and 1.2 m (49 in.) in diameter with a 65-mm (2.5-in.) wall thickness. Cracking initiated near the outside diameter of the roll, 2.7 m (9 ft) from the driven end, and extended roughly circumferentially 320° around the roll. An overall sensitized microstructure and no evidence of corrosive attack at the lightly stressed, undriven end of the roll indicated a stress-corrosion mode as opposed to plain intergranular corrosion. The specific corrosive agent was not identified. Fig. 646 (top left): Portion of crack surface (after mechanical separation). Cracks initiated at the drilled holes near, but not directly on, the outside diameter of the shell and then propagated from hole to hole and through the wall thickness. Fig. 647 (top right): Close-up of crack face reveals intergranular nature of failure. At the outside diameter (left) and 50 to 60% inward, the grain structure of the cast alloy is columnar. The remainder is predominantly equiaxed. Fig. 648 (bottom left): Evidence that cracking progressed from the outside to the inside diameter. Region of roll surface at center is completely separated, but is connected to the rest of the shell at the inside diameter. Fig. 649 (bottom right): Photomicrograph of structure near the inside diameter of the roll. Note intergranular cracking along sensitized, near-equiaxed grain boundaries. Ferric chloride etch, 20× (F.W. Tatar, Factory Mutual Research Corporation)

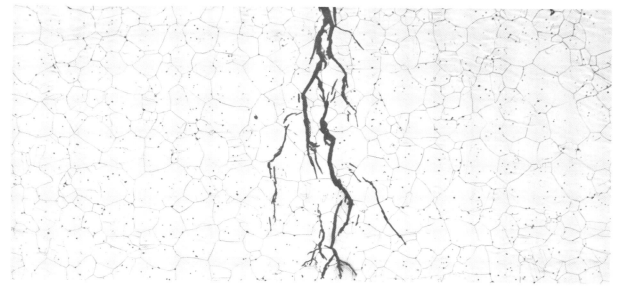

Fig. 650 Failure due to chloride stress-corrosion cracking (SCC) of an AISI type 316 pipe. The pipe served as a vent for the preheater-reactor slurry transfer line in a coal-liquefaction pilot plant. Although no material flowed through the vent line—a "dead leg"—the service temperature was low enough to permit formation of an aqueous condensate on its inside surface. Extensive cracking originated at the inside wall of the pipe. Examination of an etched transverse section of the crack tip (shown here) revealed transgranular cracking with a branching pattern characteristic of chloride SCC. 200× (J.R. Keiser and A.R. Olsen, Oak Ridge National Laboratory)

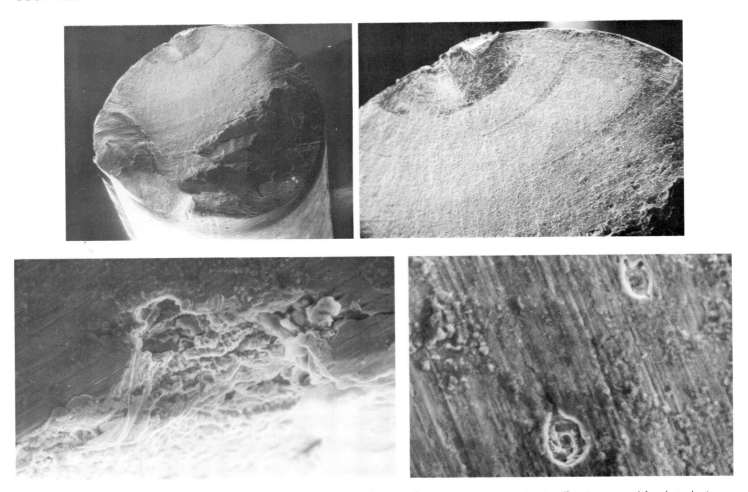

Fig. 651 to 654 Fatigue fracture of SIS 2343 (~AISI type 316L) wire due to cracking originating at corrosion pits. The wire was used for electrodes in an electrostatic precipitator that treated the gaseous effluent from a paper mill. Fig. 651 (top left): Fracture surface. Note beach marks. SEM, 20×. Fig. 652 (top right): Close-up of fracture origin (arrow) in Fig. 651. SEM, 50×. Fig. 653 (bottom left): High-magnification view of fracture origin and associated corrosion pit. Wire surface is at top, fracture surface at bottom. SEM, 620×. Fig. 654 (bottom right): Region on surface of wire near location of fracture, showing two corrosion pits. SEM, 1000× (C.R. Brooks, University of Tennessee)

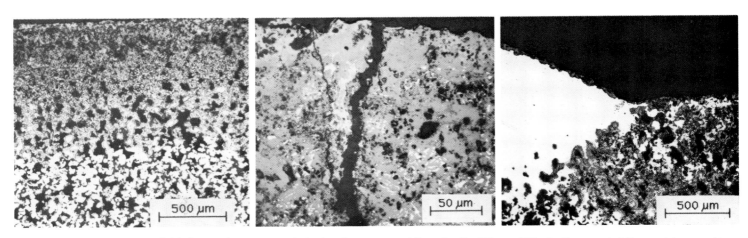

Fig. 655, 656, 657 Brittle fracture due to severe carburization of a porous metal filter tube. The P/M type 316L tube (5-μm nominal filter grade) was used in a coal-gasification pilot plant to remove entrained ash and char particles from the raw-product gas stream. Failure occurred after being in service for 8 h at 540 °C (1000 °F) in contact with a gas stream having the following composition (in vol%): 30% H_2O, 12.6% H_2, 0.43% Ar + O_2, 35.2% N_2, 8.4% CO, 11.7% CO_2, 1.0% CH_4, and 0.105% H_2S. Particulate matter in contact with the tube contained 32 to 34 wt% C. Carburization proceeded inward from the outside diameter of the tube in the direction of gas flow through the porous wall. Source of carbon was the char in the gas stream. Fig. 655 (left): Section through tube wall, outside diameter at top. Outer third of tube thickness is almost completely transformed to mixed $(Cr,Fe)_{23}C_6$ and $(Cr,Fe)_7C_3$ carbides. Balance of thickness has partially transformed, particularly in the vicinity of pores. Carbon content of the bulk material was 0.22 wt%, compared with the 0.03 wt% max specified for type 316L. Carbon content at the outside diameter was 1.40 wt%. 35×. Fig. 656 (center): Structure near through-wall crack at outside diameter of failed filter tube. Mixed carbides (gray phases) dominate. Only a very few small, light-colored islands of untransformed type 316L matrix remain. The volume increase associated with the formation of carbides has substantially reduced the amount of porosity (dark areas). 340×. Fig. 657 (right): Outside diameter of failed filter tube at region containing an axial seam weld (at left). Note that only the surface of the nonporous type 316L filler metal was transformed to carbides. 340× (D.R. Diercks, Argonne National Laboratory)

Austenitic Stainless Steels / 359

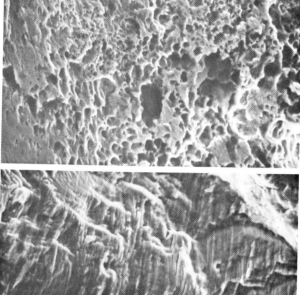

Fig. 658, 659, 660 Torsional overload fracture of cortical screw. The bone screw was machined from cold-worked AISI type 316L rod and then stress relieved. Its surface was electropolished and passivated. Fig. 658 (left): Fracture surface of screw. SEM, 28×. Fig. 659 (center): Close-up of edge of fracture surface. Note torsional shear bands and dimpled region typical of a ductile fracture. SEM, 117×. Fig. 660 (right): High-magnification view of dimpled-rupture portion of fracture surface reveals elongated dimples characteristic of a torsional (shear) failure mode. SEM, 1170×. (H.R. Shetty, Zimmer Inc.)

Fig. 661, 662, 663 Fatigue fracture of compression tube and plate (hip screw) orthopedic implant. The device was machined from an AISI type 316L hot forging. It was then mechanically polished, electropolished, and passivated. Fig. 661 (left): Fracture surface. Note beach marks. 5.2×. Fig. 662 (center): Fatigue striations on fracture surface. SEM, 1360×. Fig. 663 (right): Region of dimpled rupture in ductile overload zone of fracture surface in Fig. 661. Note burnishing. See also Fig. 670. SEM, 1040×. (H.R. Shetty, Zimmer Inc.)

360 / Austenitic Stainless Steels

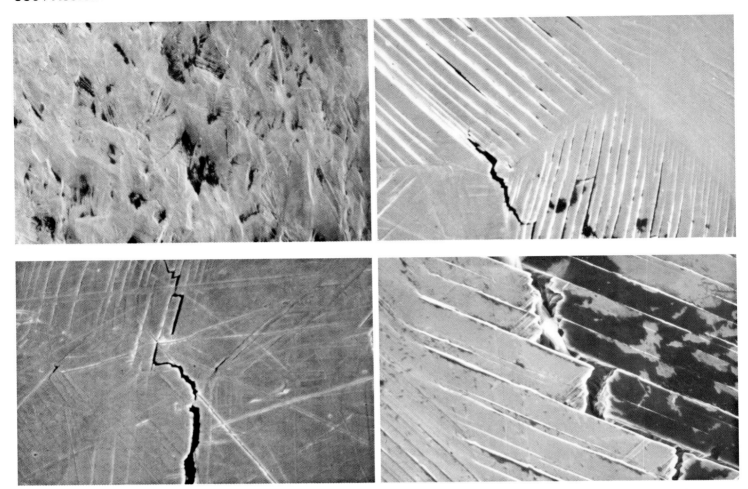

Fig. 664 to 667 Fatigue crack nucleation at slip bands in the plate of a hip screw (compression tube and plate) surgical implant. Material and processing same as in Fig. 661 to 663. Scanning electron photomicrographs are of the lateral (tension or screw-head-side) surface of plate subjected to fatigue loading. Fig. 664 (top left): Lateral surface of plate. Note slip-band activity and intrusions formed by slip bands. SEM, 187×. Fig. 665 (top right): Higher-magnification view of area on lateral surface of plate. Note microcracking at slip bands and the crack being formed at a grain-boundary triple point. SEM, 1160×. Fig. 666 (bottom left): Fatigue crack formed on lateral surface of plate. Growth is along slip planes. SEM, 983×. Fig. 667 (bottom right): Area on lateral plate surface near fracture origin. Features include slip-band cracking and crack growth along intersecting slip planes. SEM, 1400× (H.R. Shetty, Zimmer Inc.)

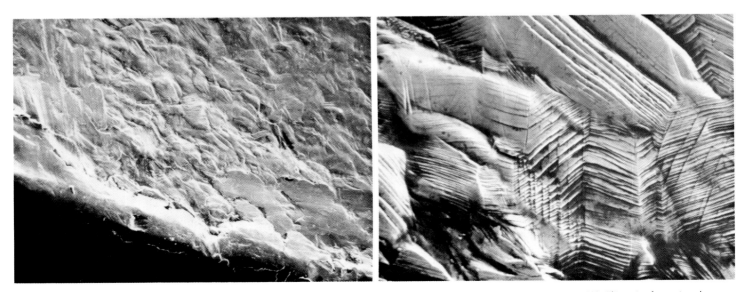

Fig. 668, 669 Fatigue crack nucleation in the plate of compression hip screw. Material and processing same as in Fig. 661 to 663. This pair of scanning electron photomicrographs shows the medial (compression or bone-side) surface of the failed plate. Fig. 668 (left): Medial surface near fracture origin. Ridges result from grain deformation. SEM, 86×. Fig. 669 (right): Higher-magnification view of representative area in Fig. 668. Note slip-band activity and intersecting slip bands. See also Fig. 664 to 667. SEM, 859× (H.R. Shetty, Zimmer Inc.)

Austenitic Stainless Steels / 361

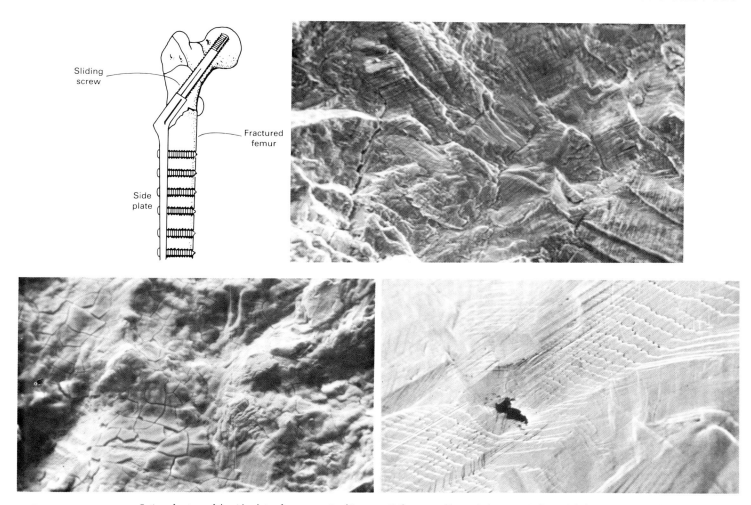

Fig. 670 to 673 Fatigue fracture of the side plate of a compression hip screw. Failure caused by a missing screw in the ninth hole. (Screws were placed in the eight other holes.) Premature crack initiation also may have resulted from local plastic deformation incurred when the plate was bent prior to insertion. Material: AISI type 316, surgical implant grade. Fig. 670 (top left): Drawing of typical compression hip screw. A Jewett nail (see Fig. 674 and 680) is similarly shaped, but the threaded shaft embedded in the head of the femur is replaced by a tri-finned "nail." Fig. 671 (top right): Region on fracture surface. Note fine striations and secondary cracks. Topography is typical of an austenitic stainless steel that fractured in the corrosive body environment. SEM, 450×. Fig. 672 (bottom left): One of many "mud crack" pattern areas on fracture surface. They are believed to be tenacious artifacts created by cracking of thin surface deposits of either corrosion products or dried body fluids. SEM, 450×. Fig. 673 (bottom right): Slip bands and crack on electropolished outside surface of plate near the fracture. Plastic deformation due to bending of the plate by the surgeon prior to insertion was apparently sufficient to initiate cracking. SEM, 1000× (C.R. Brooks and A. Choudhury, University of Tennessee)

Fig. 674, 675 Corrosive action of body fluids masks fracture surfaces of failed AISI type 316L Jewett nail (hip implant). Corrosion of fracture faces occurred because the patient remained mobile for an extended period after the implant failed. Fig. 674 (left): Fractured Jewett nail. Fig. 675 (right): Fracture surfaces after removal from femur. Corrosion and possibly some erosion due to rubbing together of the mating pieces have smoothed out the once-jagged fracture surface. Corrosion of the normally passivated stainless steel results when its protective oxide layer is ruptured by such processes as deformation exceeding the elastic limit of the material (component fracture) and the abrasive action between two metal pieces. 5× (R.J. Gray, Consultant)

362 / Austenitic Stainless Steels

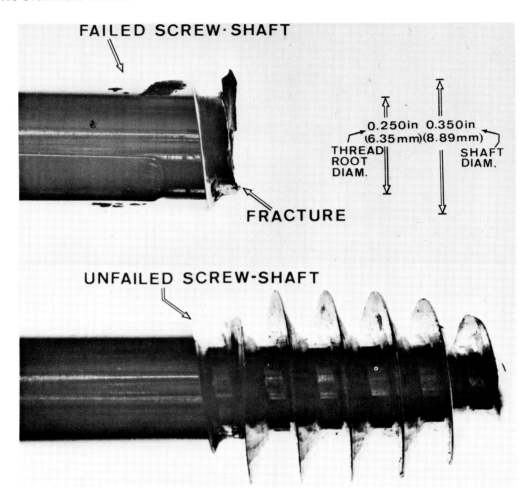

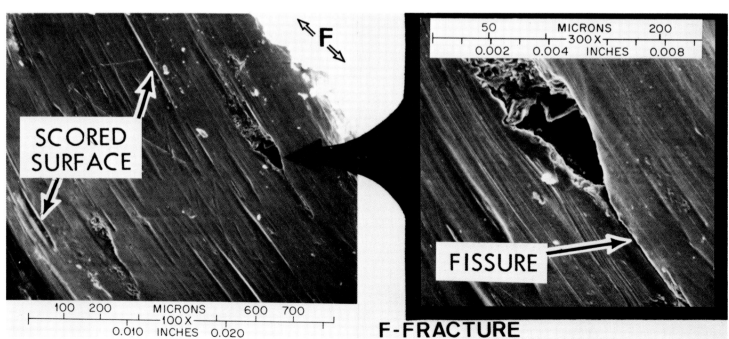

Fig. 676 to 679 Fatigue fracture of AISI type 316L sliding lag screw in compression hip screw (see Fig. 670). Failure occurred in forward ntoch of last of six thread roots. First five threads were embedded in the femoral head. Last root was in line with or close to the femoral neck fracture, where it would have been subjected to cyclic bending forces during patient mobility. Fig. 676 and 677 (top and center): Recovered portion of failed screw and an unused screw. Fig. 678 (bottom left): Scored surface of thread root near where fracture occurred. SEM, 78×. Fig. 679 (bottom right): Close-up of gouge on thread root surface in Fig. 678. Note fissure that developed due to low-cycle bending fatigue during patient mobility. SEM, 235× (R.J. Gray, Consultant)

Austenitic Stainless Steels / 363

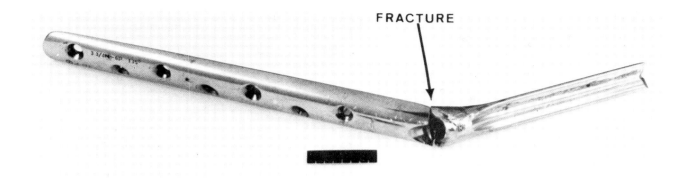

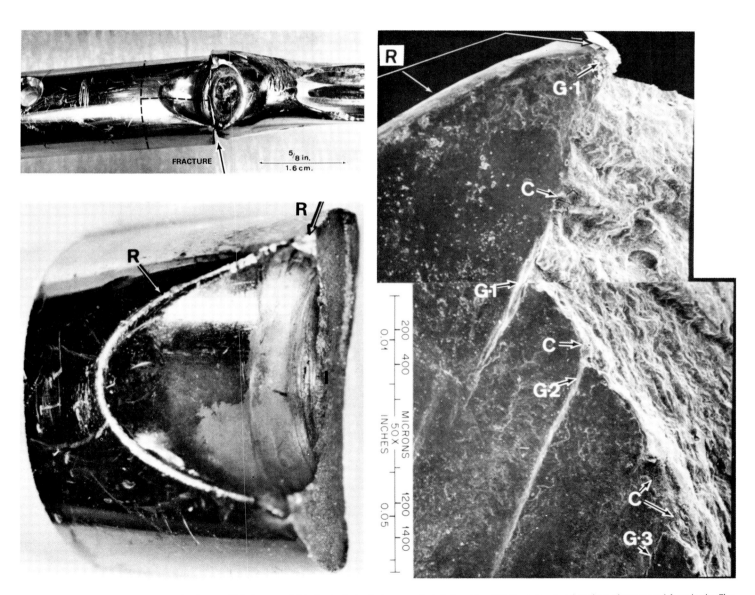

Fig. 680 to 683 Failure of AISI type 316L Jewett nail (hip implant) due to improper installation. Fig. 680 (top): Fractured implant after removal from body. The threaded tip of an impactor-extractor tool is inserted by the surgeon into a hole in the chamfered portion of the implant located where the side plate (left) meets the tri-finned nail (right). During installation, the threaded tip of the tool broke off. The jagged edge of the broken tool then gouged the rim of the chamfer. Even so, the operation was judged a qualified success. The implant failed 4 weeks after installation. Fig. 681 (left center): Close-up of fractured Jewett nail. Broken tip of impactor-extractor is in nail portion of implant at right. Dashed lines show cutting planes for subsequent microscopic examination. Fig. 682 (bottom left): Plate end of fractured impact. Fracture originates on ridge (R) of chamfer. The chamfer-fracture surface interface is indicated by I (see Fig. 684 to 687). Fig. 683 (bottom right): Close-up of area near fracture origin R in Fig. 682. Tensile stress was concentrated at groove G1, created by fractured edge of impactor-extractor tool. Fracture propagated along this groove. Also note second and third grooves (G2 and G3) and what appears to be corrosion (areas labeled C). SEM, 50× (R.J. Gray, Consultant)

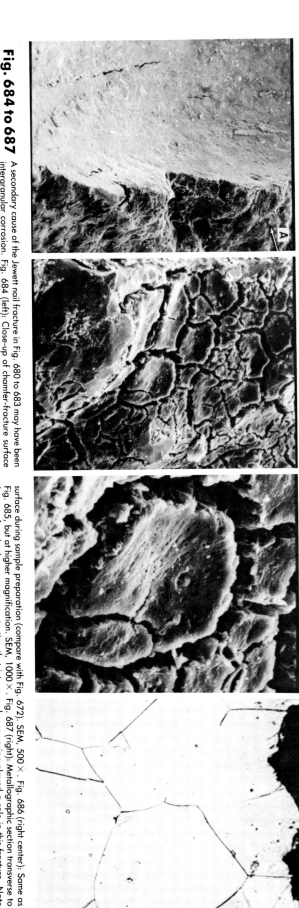

Fig. 684 to 687 A secondary cause of the Jewett nail fracture in Fig. 680 to 683 may have been intergranular corrosion. Fig. 684 (left): Close-up of chamfer-fracture surface interface (1 in Fig. 682). SEM, 50×. Fig. 685 (left center): Higher-magnification view of region A on fracture face in Fig. 684. Note evidence of intergranular corrosion. In some cases, however, these "mud crack" patterns are believed to be dried body fluids that were not completely removed from the fracture surface during sample preparation (compare with Fig. 672). SEM, 500×. Fig. 686 (right center): Same as Fig. 685, but at higher magnification. SEM, 1000×. Fig. 687 (right): Metallographic section transverse to fracture face lends support to contention that intergranular corrosion played a role in this fracture. Note how the "corrosion-etched grain size" in Fig. 686 is very similar to the etched grain size in this photomicrograph. 1000× (R.J. Gray, Consultant)

Fig. 688 to 691 Microvoid linking by creep in AISI type 316. Intergranular creep fracture depends on the nucleation, growth, and subsequent linking of voids on grain boundaries to form two types of cavities: wedge-type cavities (Fig. 688 to 690) and isolated, rounded-type cavities (Fig. 691). Fig. 688 (left): Wedge-type cavities in an AISI type 316 sample that failed by creep. 100×. Fig. 689 (left center): Wedge-type cavities (wedge cracks) form at triple points due to grain-boundary sliding and may be promoted by decohesion at interfaces between grain-boundary precipitates and the matrix, as shown here in a type 316 sample. 600×. Fig. 690 (right center): Creep failure due to wedge cracking at triple points produces a rough fracture surface with identifiable grain-boundary precipitates. For example, carbides and σ-phase particles are found on the fracture surface of this type 316 stainless steel. TEM replica, 10 000×. Fig. 691 (right): Rounded-type voids in type 316L weld metal. Sample failed in creep. These cavities are usually associated with cracking at triple points at high temperature. 100× (I. Le May, Metallurgical Consulting Services Ltd.)

Austenitic Stainless Steels / 365

Fig. 692, 693, 694 "Channel fracture" of irradiated AISI type 316. In this type of fracture, which is unique to highly irradiated materials, dislocation motion is confined to a few narrow zones or channels. The 20% cold-worked type 316 sample was irradiated at 385 °C (725 °F) to a neutron fluence of 1.28×10^{23} n/cm^2 ($E_n > 0.1$ MeV). Fig. 692 (left): Fracture surface of sample after tensile testing at 205 °C (400 °F). Fracture mode is shear. SEM, 43×. Fig. 693 (center): Higher-magnification view of fracture surface reveals the steplike topography of channel fracture. SEM, 1440×. Fig. 694 (right): Extreme close-up of a single channel on the fracture surface in Fig. 693. Note the numerous voids and the narrow zone at center, where dislocations have produced elongated voids by shear. TEM thin foil specimen, 57 000× (J.E. Nolan, Westinghouse Hanford Company)

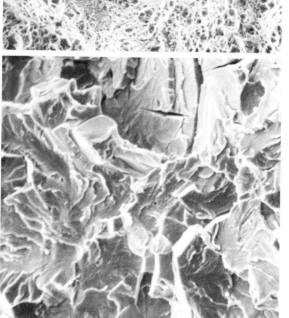

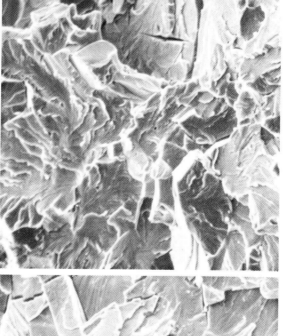

Fig. 695, 696 Ferritic iron-aluminum alloys have been proposed as substitutes for conventional chromium-bearing stainless steels, particularly the austenitic 300 series alloys with 15 to 28% Cr. Interest in these materials stems in large part from a desire to reduce U.S. dependence on imports of the strategic material chromium. One drawback, however, is the brittle nature of iron-aluminum alloys, especially those with higher aluminum contents. Shown here are fractographs of the flat fractures from tensile samples of an experimental iron-aluminum alloy (Fe-8Al-6Mo-0.8Zr-0.1C). Fig. 695 (left): Annealed 1 h at 825 °C (1515 °F), furnace cooled. Massive cleavage fracture (with 2% elongation) believed due to long-range order in the alloy. SEM, 750×. Fig. 696 (right): Annealed 1 h at 825 °C (1515 °F), water quenched. Fracture surface reveals a finer cleavage structure and evidence of increased ductility, probably the result of short-range order. Alloy had 19% elongation. SEM, 750× (H.W. Leavenworth, Jr., U.S. Bureau of Mines)

Fig. 697 Substitution of iron-aluminum alloys is one potential way to reduce dependence on conventional austenitic stainless steels with their high contents of the strategic element chromium (see Fig. 695 and 696). Another, perhaps more promising, alternative is to develop alloys that contain lesser amounts of chromium. An example is Fe-8Cr-12Ni-4.5Si-1Al-1Mn-0.05C. A ductile fracture characterized by dimple rupture is obtained in this material after annealing for 1 h at 1100 °C (2010 °F) and air cooling. Credit for the high ductility goes to a microstructure of low-carbon martensite. SEM, 250× (H.W. Leavenworth, Jr., U.S. Bureau of Mines)

366 / Martensitic Stainless Steels

Fig. 698, 699, 700 Surfaces of fractures in three AISI type 410 stainless steel links of a conveyor chain used in a food-processing operation. Heat treatment of the links had included box annealing at 995 °C (1825 °F) for 1 h, oil quenching, and tempering between 260 and 370 °C (500 and 700 °F) for 2 h. Measured hardness was 37 to 44 HRC. Service conditions included cleaning in water containing 2 ppm of chlorine at 70 °C (160 °F), cleaning in saturated steam at 120 °C (250 °F), cleaning in air at 120 °C (250 °F), and water cooling at 20 to 32 °C (68 to 90 °F). Figure 698 (left) shows the dark, oxidized surface of a heat-treat crack; the remainder of the fracture surface is brittle and shows no shear lip. Fig. 699 (center) shows a heat-treat crack (at top edge) and a shear lip (at bottom edge), with fast fracture between them. Figure 700 (right) shows a very small heat-treat crack (right side of top edge) and portions of a shear lip (at left edge and at top edge left of crack). See also Fig. 701 to 703. All at 8×

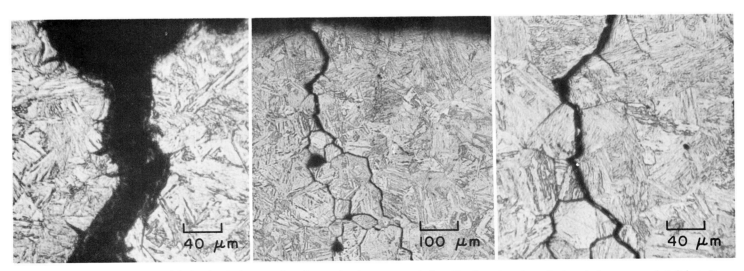

Fig. 701, 702, 703 Polished and etched sections through secondary heat-treat cracks branching from the primary fractures in the conveyor-chain links in Fig. 698 and 700, showing the intergranular nature of the cracks. Figure 701 (left) shows a crack in the link in Fig. 698. Note that the surfaces of this crack have oxide coatings; the short cracks branching from it were oxide induced. Figures 702 (center) and 703 (right) are views, at different magnifications, of a secondary crack in the link shown in Fig. 700. Again, oxide is present within the crack, as can be seen somewhat obscurely in Fig. 703. Note that this crack has branched to connect with nonmetallic inclusions. It is evident, from the presence of oxide on all crack surfaces, that both primary and secondary heat-treat cracks formed before the links were oil quenched from the annealing temperature. All etched in super picral. Fig. 701 and 703: 250×. Fig. 702: 100×

Martensitic Stainless Steels / 367

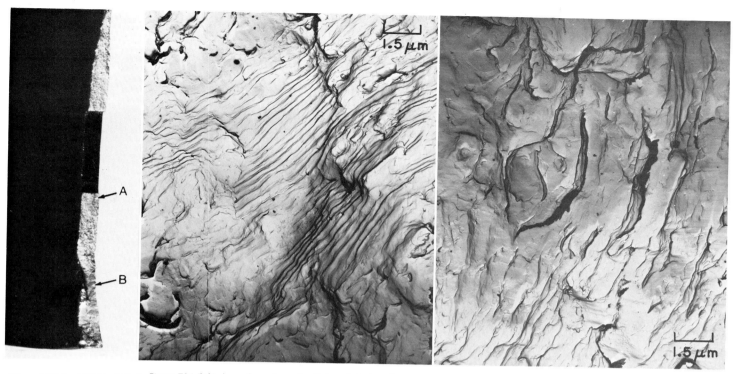

Fig. 704, 705, 706 Figure 704 (left) shows the surface of a high-cycle fatigue fracture in a 140 × 25 × 1.3 mm (5½ × 1 × 0.050 in.) notched specimen of AISI type 431 stainless steel oil quenched from 1040 °C (1900 °F) and tempered at 275 °C (525 °F) to a tensile strength of 1241 to 1379 MPa (180 to 200 ksi). The notch consisted of a center hole 0.25 mm (0.01 in.) in radius and 5 mm (0.2 in.) long. The specimen was subjected to tensile stress that was cycled from 154.4 to 772 MPa (22.4 to 112 ksi) at 1000 cycles/min, and fracture occurred after 5700 cycles. Figures 705 (center) and 706 (right) are TEM p-c replicas of two areas at A in Fig. 704, immediately below the notch. In Fig. 705, fatigue striations are well developed, and suggestions of partly formed dimples are present at a few locations. In Fig. 706, fatigue striations are much less regular than in Fig. 705; again, there are suggestions of partly formed dimples. See also Fig. 707 to 709. Fig. 704: 4.5×. Fig. 705 and 706: 6500×

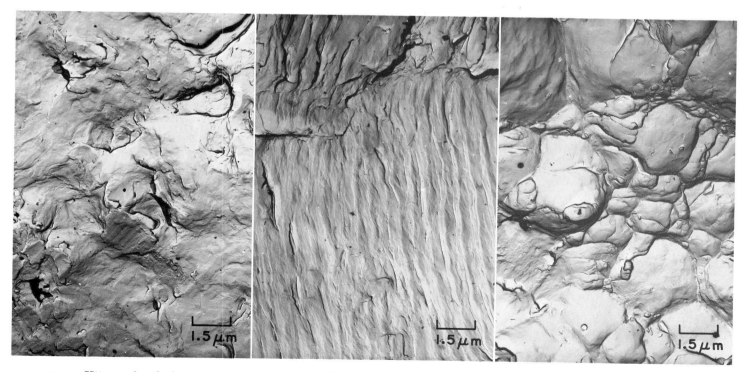

Fig. 707 TEM p-c replica of a third area at A in Fig. 704. This area exhibits faint but quite recognizable fatigue striations with spacings similar to those in Fig. 705. 6500×

Fig. 708 TEM p-c replica of a fourth area at A in Fig. 704. Regular, widely spaced fatigue striations resulted from increased stress due to reduced cross-sectional area. 6500×

Fig. 709 TEM p-c replica of an area at B in Fig. 704. The equiaxed dimples visible here resulted from tension overload during final fracture, but before final shear rupture. 6500×

368 / Martensitic Stainless Steels

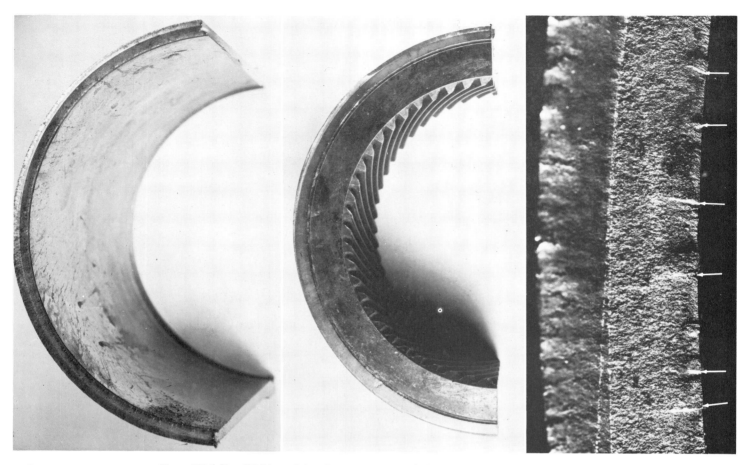

Fig. 710, 711, 712 Figures 710 (left) and 711 (center) show the mating segments of a fractured AISI type 501 stainless steel torque cylinder from the hub socket of an aircraft propeller. Cracking originated at an internal fillet at a shoulder between the smooth bore (Fig. 710) and the internal spline (Fig. 711). Figure 712 (right) shows clearly the inner fatigue zone and the outer zone of final tensile-shear fracture. The ratchet marks visible in Fig. 710 (at the bore) are shown more clearly in Fig. 712, a view that reveals many crack nuclei (at arrows). The fatigue cracks that propagated from these nuclei ultimately merged to form a common front. See Fig. 713 to 716 for TEM views of other regions of this fracture surface. Fig. 710 and 711: 0.8×. Fig. 712: 10×

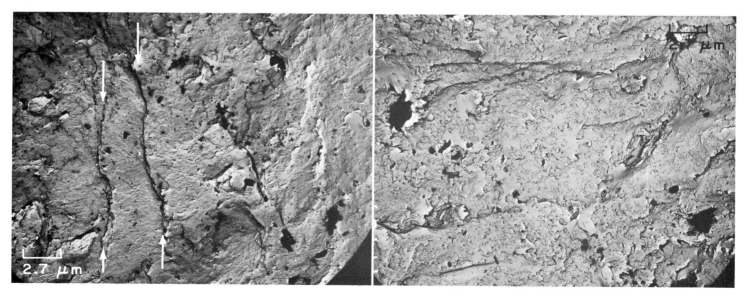

Fig. 713 TEM p-c replica of the fracture surface in Fig. 710, showing a region about 30 μm from a crack nucleus. Visible between the arrows are steps that separate fatigue patches. Fatigue striations are faintly visible within the patches. 3750×

Fig. 714 TEM p-c replica of the fracture surface in Fig. 711, showing a region about 30 μm from the inner edge. This view shows generally flat fatigue patches that are separated by steps, but no fatigue striations are visible within the patches. 3750×

Martensitic Stainless Steels / 369

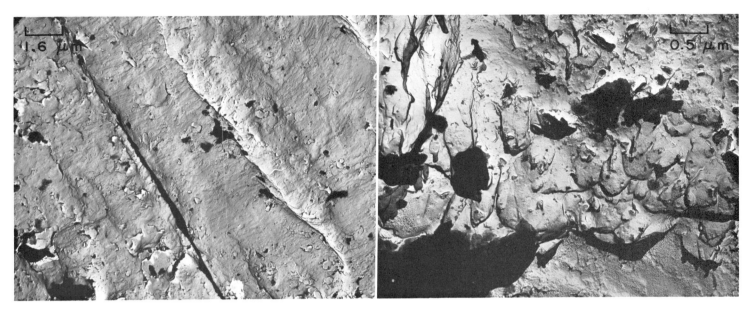

Fig. 715 TEM p-c replica of the fracture surface in Fig. 710, showing a region in the fatigue zone, but near the zone of final tensile-shear fracture. This replica reveals fatigue patches separated by steps and containing faint fatigue striations. 6125×

Fig. 716 TEM p-c replica of a portion of the shear lip in the fracture surface in Fig. 710, showing shear dimples, which resulted when microvoid coalescence occurred during final tensile-shear fracture. The dark areas are probably rust or "dirt." 18 750×

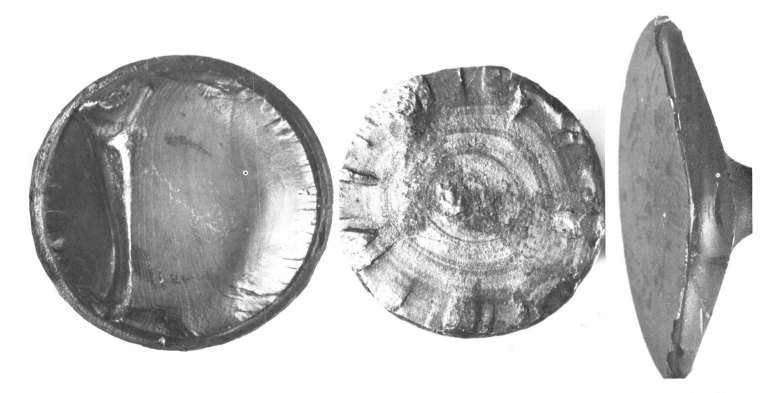

Fig. 717, 718, 719 Surfaces of fatigue fractures in three automotive intake valves forged of Silcrome-1 steel (0.45% C, 3.25% Si, and 8.5% Cr) and heat treated. Figure 717 (left) shows a transverse fracture in the stem of a valve with a hardness of 37 HRC. The stem rocked within the valve retainers, initiating many fatigue cracks around the periphery of the stem; these cracks later joined to form two main crack fronts. Very well-defined beach marks are visible in the fatigue zone at right. Figure 718 (center) shows a transverse fracture in the stem of a valve with a hardness of 38 HRC. The stem rotated in the valve retainers, which initiated many fatigue cracks around the periphery of the stem; these cracks penetrated nearly a third of the way to the center of the stem before uniting into a single circumferential crack front. Figure 719 (right) shows a fracture in the head of a valve with a hardness of 37 HRC. The fatigue-crack origin is at right, at the head-stem juncture. Note the peeled surface layer. Fig. 717 and 718: 10×. Fig. 719: 2×

370 / Precipitation-Hardening Stainless Steels

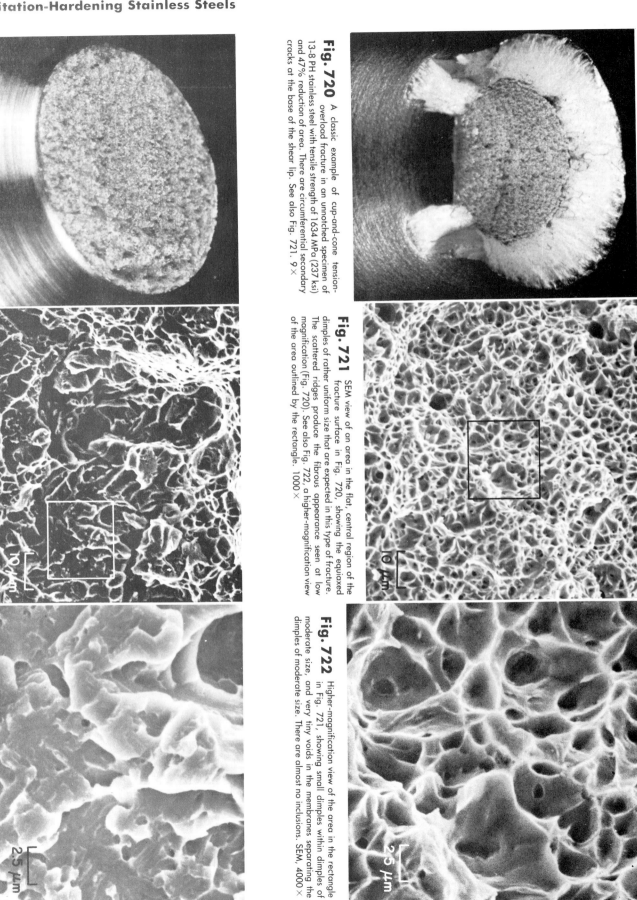

Fig. 720 A classic example of cup-and-cone tension-overload fracture in an unnotched specimen of 13-8 PH stainless steel with tensile strength of 1634 MPa (237 ksi) and 47% reduction of area. There are circumferential secondary cracks at the base of the shear lip. See also Fig. 721. 9×

Fig. 723 A typical tension-overload fracture in a notched specimen of 13-8 PH stainless steel with the same properties as those of the specimen in Fig. 720. The fracture surface appears uniformly fibrous, with a hint of a shear lip. See also Fig. 724. 9×

Fig. 721 SEM view of an area in the flat, central region of the fracture surface in Fig. 720, showing the equiaxed dimples of rather uniform size that are expected in this type of fracture. The scattered ridges produce the fibrous appearance seen at low magnification (Fig. 720). See also Fig. 722, a higher-magnification view of the area outlined by the rectangle. 1000×

Fig. 724 SEM view of an area of the fracture surface in Fig. 723, showing dimpled rupture in some regions (at left, for example) and, elsewhere, facets that suggest local quasi-cleavage without evidence of any conventional river patterns. See also Fig. 725, a higher-magnification view of the area in the rectangle. 1000×

Fig. 722 Higher-magnification view of the area in the rectangle in Fig. 721, showing small dimples within dimples of moderate size, and very tiny voids in the membranes separating the dimples of moderate size. There are almost no inclusions. SEM, 4000×

Fig. 725 Higher-magnification view of the area in the rectangle in Fig. 724. This SEM view shows that in the fine-scale structure there do appear to be some delicate cleavage steps, particularly at right. 4000×

Precipitation-Hardening Stainless Steels / 371

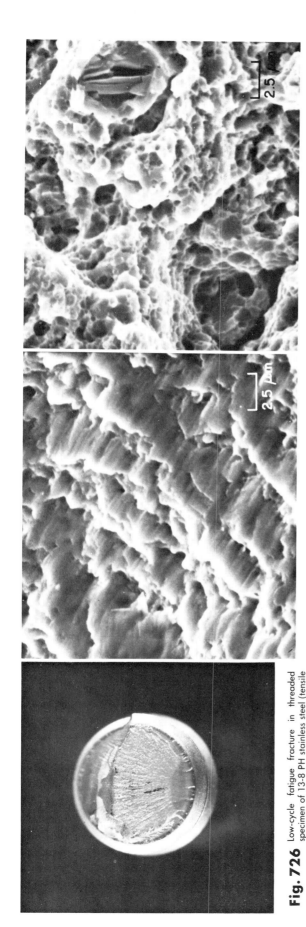

Fig. 726 Low-cycle fatigue fracture in threaded specimen of 13-8 PH stainless steel (tensile strength): 1634 MPa, or 237 ksi; 47% reduction of area) tested in tension-tension ($R = 0.1$) with maximum loading at 60% of tensile strength. Fracture, at 16 000 cycles, began at lower edge of crescent-shaped area at bottom. 6×

Fig. 727 SEM view of the fracture surface in Fig. 726, taken at the rounded lip at the bottom edge of the crescent-shaped area. Even this close to the crack origin, there are no fatigue striations. The irregular surface displays ridges formed by a shear process. 4000×

Fig. 728 Another SEM view of the fracture surface in Fig. 726, taken in the region of final fast fracture. Fatigue fracture has given way to dimpled rupture. Two pockets, one at the left and one at the right, contain features that resemble localized quasi-cleavage. 4000×

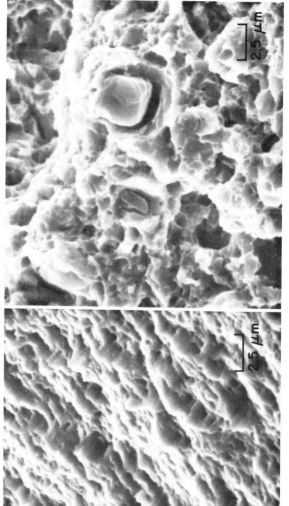

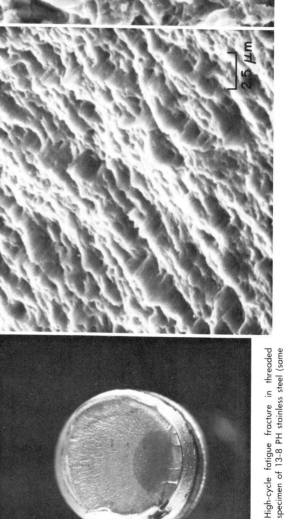

Fig. 729 High-cycle fatigue fracture in threaded specimen of 13-8 PH stainless steel (same properties as for specimen in Fig. 726) tested in tension-tension ($R = 0.1$) with maximum loading at 30% of tensile strength. Fracture occurred at 959 000 cycles. Note the crescent-shaped crack nucleus. See also Fig. 730 and 731. 6×

Fig. 730 SEM view of fracture in Fig. 729, taken as close as possible to the bottom edge of the origin region. This rough surface does not contain fatigue striations, but instead suggests a shear mechanism of fracture and bears considerable resemblance to Fig. 727. 4000×

Fig. 731 View of the fracture surface in the fast-fracture region in Fig. 729, showing basically equiaxed dimpled rupture, but also containing scattered features of quasi-cleavage. Some of the regions display heavy concentrations of extremely small dimples. SEM, 4000×

372 / Precipitation-Hardening Stainless Steels

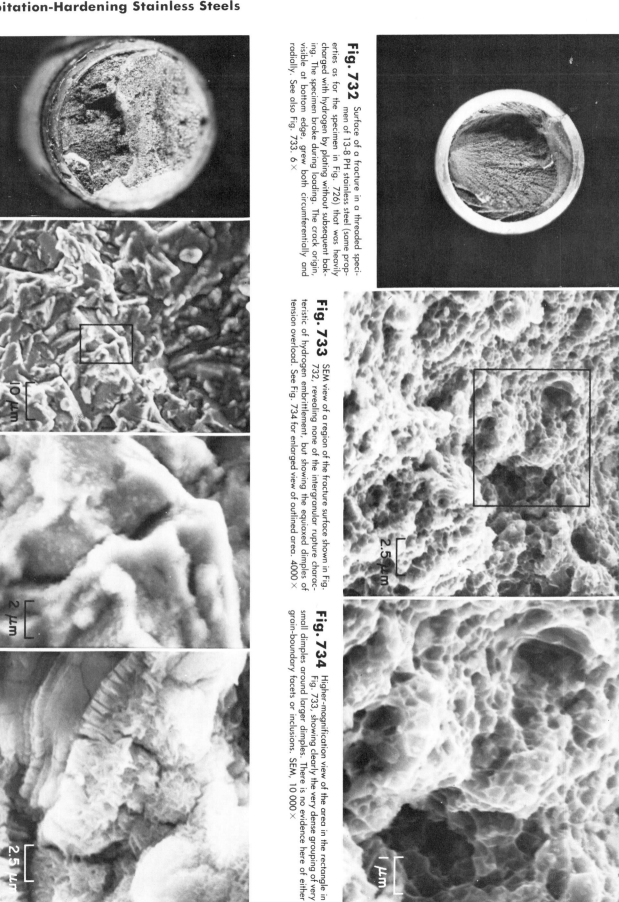

Fig. 732 Surface of a fracture in a threaded specimen of 13-8 PH stainless steel (same properties as for the specimen in Fig. 726) that was heavily charged with hydrogen by plating without subsequent baking. The specimen broke during loading. The crack origin, visible at bottom edge, grew both circumferentially and radially. See also Fig. 733. 6×

Fig. 733 SEM view of a region of the fracture surface shown in Fig. 732, revealing none of the intergranular rupture characteristic of hydrogen embrittlement, but showing the equiaxed dimples of tension overload. See Fig. 734 for enlarged view of outlined area. 4000×

Fig. 734 Higher-magnification view of the area in the rectangle in Fig. 733, showing clearly the very dense grouping of very small dimples around larger dimples. There is no evidence here of either grain-boundary facets or inclusions. SEM, 10 000×

Fig. 735 Stress-corrosion cracking fracture in threaded specimen of 13-8 PH stainless steel (same properties as in Fig. 726) loaded to 75% of tensile strength in a 3.5% NaCl solution. See also Fig. 736. 6×

Fig. 736 SEM view of crack origin area of fracture surface in Fig. 735, showing corrosion products that probably are oxides. Secondary cracks may or may not extend into the metal below. See also Fig. 737 and 738. 1000×

Fig. 737 Enlargement of the area in the rectangle in Fig. 736. The oxide coating appears to be quite continuous, except for the scattered secondary cracks. Compare with Fig. 738. SEM, 5000×

Fig. 738 Another higher-magnification view of the crack-origin area of the SCC fracture surface shown in Fig. 735, displaying a coating of hydroxide corrosion product containing many fine secondary cracks. Compare with the area shown in Fig. 737. SEM, 4000×

Precipitation-Hardening Stainless Steels / 373

Fig. 739 High-cycle fatigue fracture of the end cap for a high-pressure compressor used in a polyethylene process. The forging was made of Armco 15-5PH (UNS S15500), quenched and tempered to the H1075 condition. The end cap was located between the crosshead and plunger of the compressor, where it was subjected to repeated impacts by the tungsten carbide plunger. Failure initiated in a sharply radiused machining recess (see mating fracture surfaces). 1.55×. (E.V. Bravenec, Anderson & Associates, Inc.)

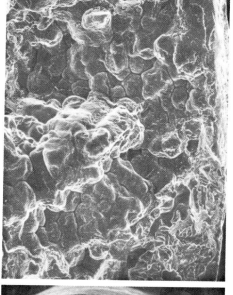

Fig. 740, 741, 742 Fracture of deflector yoke for aircraft main landing gear due to intergranular stress-corrosion cracking. Material: Armco 17-4PH (UNS S17400), 36 to 44 HRC. Seals were missing from both fractured and unbroken yokes. The inside of the broken part was filled with particulate matter, sand, and corrosion products, indicating that water had been present. Fig. 740 (left): Unbroken deflector yoke. Note pitting. 4×. Fig. 741 (center): Fracture surface. Box outlines area where evidence of intergranular attack was found. SEM, 12×. Fig. 742 (right): Boxed area in Fig. 741. Note intergranular nature of corrosion. SEM, 145×. (W.L. Jensen, Lockheed-Georgia Company)

374 / Precipitation-Hardening Stainless Steels

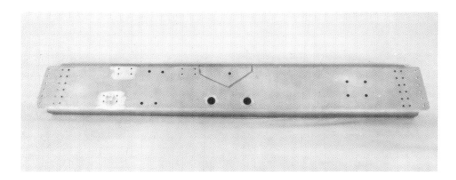

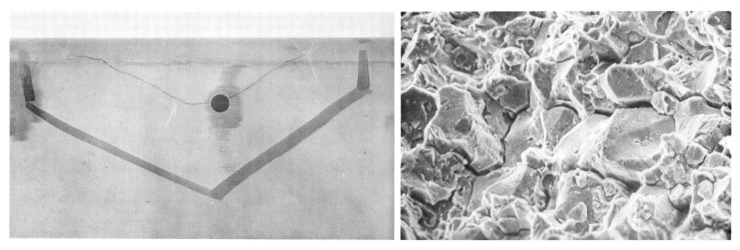

Fig. 743, 744, 745 Brittle intergranular fracture of Armco 17-7PH (UNS S17700) aircraft engine mount beam due to improper heat treatment. Hardness of the alloy in the specified TH1050 condition should have been 39 to 44 HRC, lower than the actual 46.5 HRC. Subsequent tensile testing of samples cut from the failed beam revealed identical tensile and yield strengths—1448 MPa (210 ksi)—and near-zero elongation. Other samples from the failed part were reheat treated according to TH1050 procedures and subsequently tensile tested. Properties of the reheat-treated material: tensile strength, 1407 MPa (204 ksi); yield strength, 1289 MPa (187 ksi); elongation in 50 mm (2 in.), 6.0%; hardness, 44 HRC. Fig. 743 (top): As-received engine mount beam. Area of cracking is outlined (top center). 0.17×. Fig. 744 (bottom left): Closer look at cracked area in Fig 743. 0.8×. Fig. 745 (bottom right): Fracture surface of opened crack. Note intergranular mode. SEM, 900× (W.L. Jensen, Lockheed-Georgia Company)

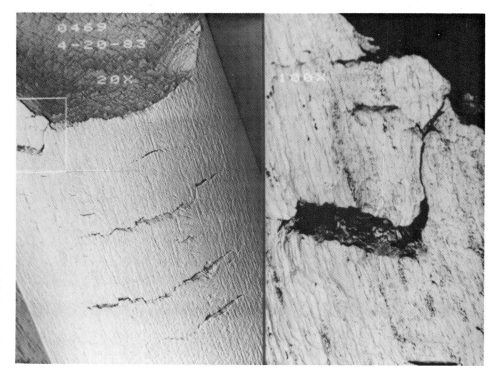

Fig. 746, 747 Fracture of Armco 17-7PH (UNS S17700) spring due to presence of "cross-check" defect. The wire broke during coiling. Cross-checks are probably caused by lubricant breakdown during wire drawing. Fig. 746 (left): Fracture surface (top) and surface of stainless steel wire. Origin is boxed. Note cross-check defects on wire. SEM, 20×. Fig. 747 (right): Higher-magnification view of fracture origin and associated cross-check. SEM, 100× (J.H. Maker, Associated Spring, Barnes Group Inc.)

Tool Steels / 375

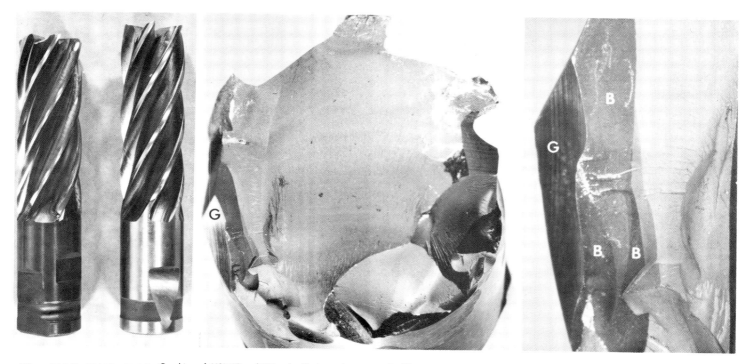

Fig. 748, 749, 750 Cracking of AISI M2 and M7 end mills due to improper rebuilding procedures. A robotic welding system was used to rebuild the worn cutting tools. Inadequate preheating and postweld heat treating resulted in a crack-prone weld-metal microstructure of brittle, untempered martensite. Fig. 748 (left): As-received end mills with cracked flutes. 0.38×. Fig. 749 (center): Fracture surface. G is the ground surface of the flute edge. 2.1×. Fig. 750 (right): Close-up of area G in Fig. 749. Areas labeled B are regions of brittle fracture. 5×. See also Fig. 751. (W.L. Jensen, Lockheed-Georgia Company)

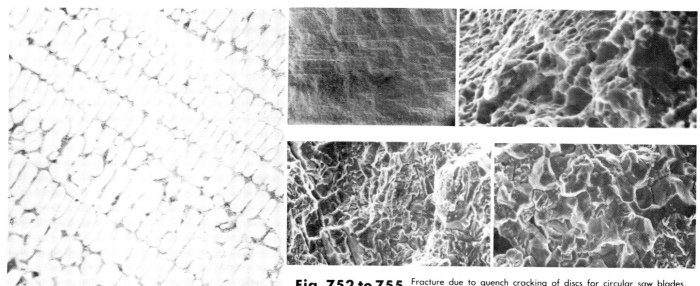

Fig. 751 Microstructure of as-deposited weld metal discussed in Fig. 748 to 750, which consists of untempered martensite dendrites and a carbide eutectic in interdendritic regions. This a brittle structure that is prone to cracking during welding. See the brittle fracture regions labeled B in Fig. 750. 5% nital, 500×. (W.L. Jensen, Lockheed-Georgia Company)

Fig. 752 to 755 Fracture due to quench cracking of discs for circular saw blades. Parts were made of a martempered low-alloy tool steel similar to type 234. Fig. 752 (top left): Woody fracture surface of saw disc that broke when dropped. 13×. Fig. 753 (top right): Fine and coarse dimples near the blade surface, in the decarburized zone. SEM, 2600×. Fig. 754 (bottom left): Fibrous zone at center of saw disc in Fig. 752. Note intergranular fracture along prior-austenite grains. The mechanism is probably related to phosphorus segregation at grain boundaries during austenitizing. SEM, 520×. Fig. 755 (bottom right): Quasi-cleavage topography of small and poorly defined cleavage facets, connected by shallow dimples, marks change from relatively slow intergranular cracking to more rapid, unstable crack propagation. SEM, 520× (C.R. Brooks, University of Tennessee, and D. Huang, Fuxin Mining Institute, People's Republic of China)

376 / Tool Steels

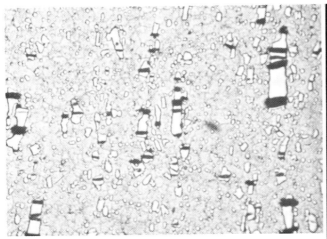

Fig. 756 Brittle in-service failure of diesel engine injector plunger. The part was made of AISI D2 bar. Heat treatment: air cool from 995 °C (1825 °F), cool to −75 °C (−100 °F), double temper at 175 °C (350 °F). Microstructure consists of carbides of various sizes dispersed in a martensitic matrix. The larger carbides are cracked, indicating that the material had been embrittled at some stage during its thermomechanical processing history. 3% nital, 280× (J.R. Kattus, Associated Metallurgical Consultants Inc.)

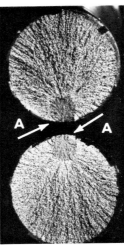

Fig. 757 Mating fracture surfaces of eccentrically loaded tensile specimen of A4N tool steel. Note fibrous zones of initial fracture (A). 2.25×

Fig. 758 A surface of a fatigue fracture that originated at the inside of a hollow drill of AISI W1 tool steel. Fracture initiation was attributed to distortion of the bore of the drill during the production of a hot upset joint. 1.8×

Fig. 759 Five fatigue-fracture surfaces that were produced in reversed bending in a pneumatic testing machine. The material is AISI W1 tool steel. The lighter areas of the fracture surfaces are the fatigue zones—all of which were initiated at the lower right corners. Note the lack of beach marks, which indicates that during each test the load amplitude and environment were uniform. The dark regions at the upper left corners are the regions of final fast fracture. 1.5×

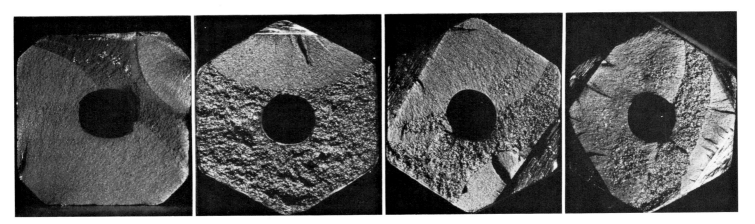

Fig. 760 to 763 Four fatigue-fracture surfaces that were produced in bending in the laboratory with the use of assorted notches to simulate service conditions; the material is AISI W1 tool steel. The specimen in Fig. 760 (left), which was tested without a notch and fractured after 432 000 cycles at 365 MPa (53 ksi), shows two crack origins at opposite corners where bending was maximum. The specimen in Fig. 761 (left center) had a single notch at top and broke after 20 500 cycles at 434 MPa (63 ksi); note the small shear lip around the region of fast fracture at bottom. The specimen in Fig. 762 (right center) was notched at two corners that were along an axis not in the plane of bending and broke after 50 000 cycles at 338 MPa (49 ksi); the two resulting fatigue cracks therefore had somewhat distorted shapes. The specimen in Fig. 763 (right) was notched at four sites—two in the bending plane and two in a plane at 60° from it—and broke after 10 260 cycles at 434 MPa (63 ksi). The four notches produced fatigue zones, which formed two major crack fronts that nearly penetrated the section. All at 2×

Tool Steels / 377

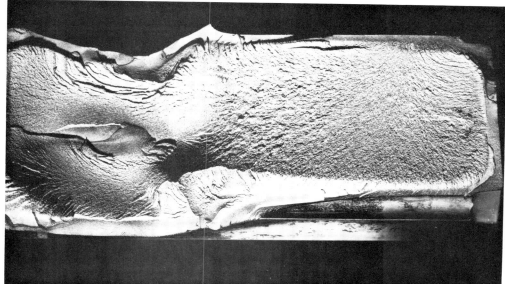

Fig. 764 Surface of a catastrophic fracture in an open-header die of AISI W2 tool steel that was case carburized to 0.95% C at the surface, heat treated in the range of 815 to 845 °C (1500 to 1555 °F), and water quenched. This die was used for cold forming and upsetting square heads. Upsetting stresses opened a small fatigue crack in the hardened case in the top edge near the upper right corner; it is to this area that the markings of radial fibrous fracture in the core of the die may be traced. The region of fast fracture encompasses nearly the entire fracture surface. Note the fine-grained, sharply defined case that frames the fibrous core. Actual size

Fig. 765, 766, 767 Light fractographs of fracture surfaces of three tensile-test specimens of tool steels. Figure 765 (top left): High-nickel AISI L6, showing a central fibrous area, a region of radial shear, and a shear lip. Figure 766 (bottom left) and Figure 767 (right): Low-carbon AISI L6 broken at 23 and 427 °C (73 and 800 °F), respectively, both showing cup-and-cone fractures. All at ~2.5×

Fig. 768 Surface of an impact fracture in a shear blade of AISI L6 tool steel heat treated to a hardness of 55 HRC. The impact generated the two crack fronts whose junction may be seen at the secondary fissure that enters the field of view at lower right; the radial markings on either side clearly meet this fissure at divergent angles. A third and separate crack front (at top left) met these two crack fronts at the wide secondary crack beneath the projecting shards. 3×

Fig. 769 Surface of a low-cycle fatigue fracture in the collar of a rivet-heading tool of AISI S1 tool steel heat treated to a hardness of 56 to 58 HRC. Following the development of the smooth fatigue zone at right, fracture occurred in a few hours of service, generating the unusual beach marks at center and left. 2×

378 / Tool Steels

Fig. 770 A "hammer-burst" fracture at the center of a 130-mm (5⅛-in.) diam, 38-mm (1½-in.) thick forged disk of AISI T1 tool steel. This longitudinal fracture shows a surface of either a lap or a cold shut; the appearance of such a fracture is often called "platy" or "laminated." 2.3×

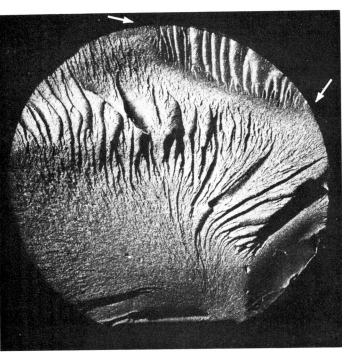

Fig. 771 An impact fracture in a 32-mm (1¼-in.) diam notched bar of annealed AISI T1 tool steel broken for inspection by two hammer blows. The crack from the the first blow started at the notch at bottom and halted at the arrows, penetrating the area at top right only with the second blow. See also Fig. 772 and 773. 2.6×

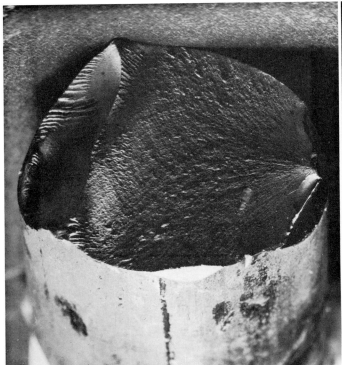

Fig. 772 An impact fracture in a 44-mm (1¾-in.) diam notched bar of AISI T1 tool steel that was broken for inspection same as the bar in Fig. 771. The crack resulting from the first hammer blow, which started at the notch at right, shows the network of intersecting fine markings called "Wallner lines." The crack that resulted from the second hammer blow is at left. 1.9×

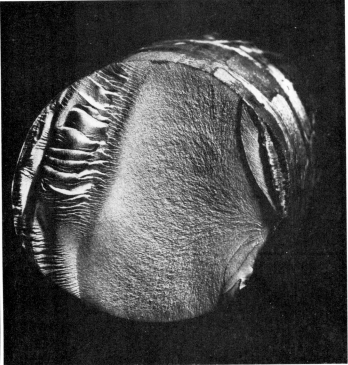

Fig. 773 An impact fracture in a notched bar of AISI M3 class 1 tool steel broken for inspection as in Fig. 771 and 772. The crack from the first hammer blow originated in the notch at right (origin is readily located by tracing back to the beginning of the fan-shaped pattern of radial marks); this crack penetrated the area at left with the second hammer blow. 1.7×

Tool Steels / 379

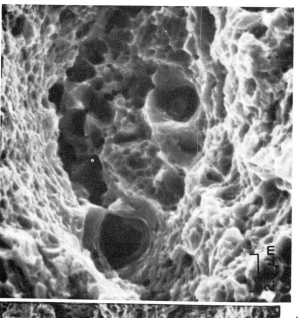

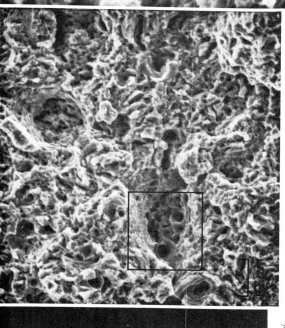

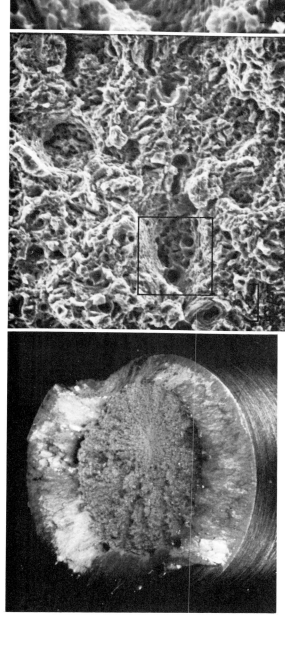

Fig. 774 Tension-overload fracture in an unnotched specimen of AISI H11 tool steel heat treated to a tensile strength of 2041 MPa (296 ksi) and 48% reduction of area. Note radial features between fibrous origin (just right of center) and shear lip. See also Fig. 775 and 776 for higher-magnification views of this fracture. 9×

Fig. 775 Another view of the fracture surface in Fig. 774, taken in the vicinity of the fracture origin, showing an unusually uneven array of dimpled facets with a scattering of large voids. See Fig. 776 for a higher-magnification view of the area in the rectangle. SEM, 1000×

Fig. 776 Higher-magnification view of the area in the rectangle in Fig. 775, showing one of the large voids and clearly revealing dimples of various sizes both at the bottom and along the side walls of the void. SEM, 4000×

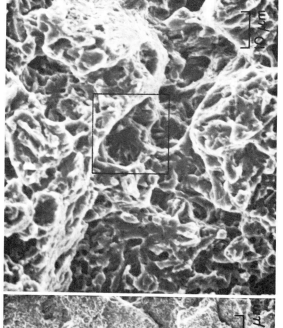

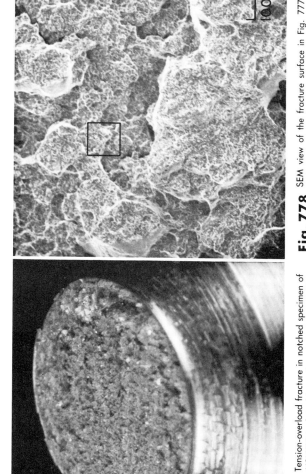

Fig. 777 Tension-overload fracture in notched specimen of AISI H11 tool steel (same heat treatment as in Fig. 774). Notched tensile strength ($K_t = 3.5$) is 2665 MPa (386.5 ksi). A flat surface, partly bounded by a hairline shear lip. See also Fig. 778 and 779. 9×

Fig. 778 SEM view of the fracture surface in Fig. 777. This surface seems to be generally dimpled, but at this relatively low magnification the features that appear to be dimples are not adequately resolved. See also Fig. 779, an enlarged view of area in rectangle. 100×

Fig. 779 Higher-magnification view of the area in the rectangle in Fig. 778, showing dimples that are associated with local quasi-cleavage. Very little evidence of secondary cracking is present. Area in rectangle is shown at higher magnification in Fig. 780. SEM, 1000×

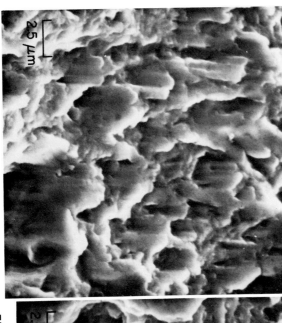

Fig. 780 Higher-magnification view of the area in the rectangle in Fig. 779. Small quasi-cleavage facets are visible, both within and adjacent to the larger dimples. SEM, 4000×

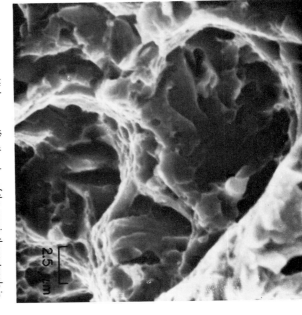

Fig. 783 View of the crescent in Fig. 781, showing a region farther from the outer edge than that in Fig. 782 and revealing ridges, formed during shear deformation, that have very irregular crests. SEM, 4000×

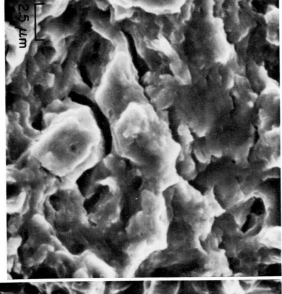

Fig. 781 Low-cycle fatigue fracture in threaded specimen of AISI H11 tool steel (some heat treatment and tensile strength as in Fig. 774) that broke after 21 000 cycles of tension-tension ($R = 0.1$); maximum load, 60% of tensile strength. Region A-A is fatigue portion of fracture. See also Fig. 782 to 785. 6×

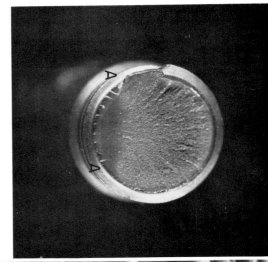

Fig. 784 Another view of the crescent in Fig. 781, showing a region still farther from the outer edge than that in Fig. 783. This region shows little evidence of conventional fatigue striations, but reveals small cleavage facets, many of which are separated by secondary cracks. SEM, 4000×

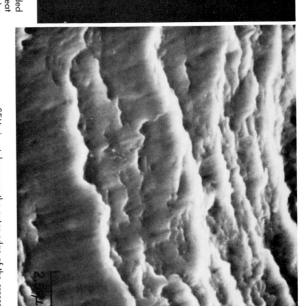

Fig. 782 SEM view taken near the outer edge of the crescent in Fig. 781. Evident are ridges that are attributed to mechanical damage during shear deformation. In this region, there appear to be no secondary cracks. 4000×

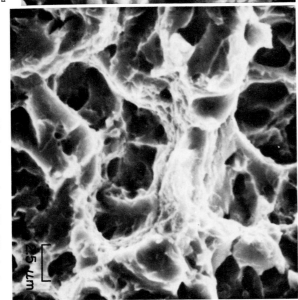

Fig. 785 View of the fracture surface in Fig. 781, showing a region in the zone of final fast fracture and revealing the equiaxed dimples expected in a tension-overload fracture. No secondary cracks are visible. SEM, 4000×

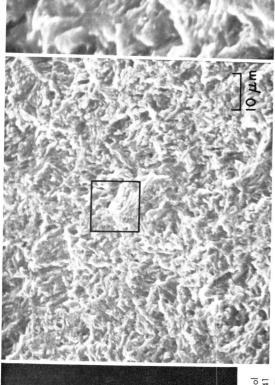

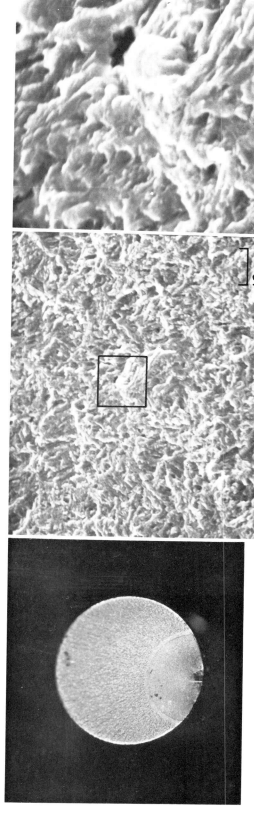

Fig. 786 High-cycle fatigue fracture in AISI H11 tool steel heat treated to a tensile strength of 2041 MPa (296 ksi) and 48% reduction of area, which broke after 775 000 cycles of tension-tension ($R = 0.1$) under a maximum load of 30% of tensile strength. The fatigue-crack origin is at the outer edge of the crescent-shaped region at bottom. See also Fig. 787 and 788 for higher-magnification views of this fracture. 6×

Fig. 787 An area within the crescent-shaped (fatigue) region of the fracture surface in Fig. 786. The surface contains many pockmarks, which resemble dimples. There are no suggestions of fatigue striations. See Fig. 788 for an enlargement of the area in the rectangle. SEM, 1000×

Fig. 788 Higher-magnification view of the area in the rectangle in Fig. 787. Some of the contours suggest striations, but they are very localized. Some features resemble cleavage facets, but do not have the degree of flatness that is normally expected of cleavage. SEM, 5000×

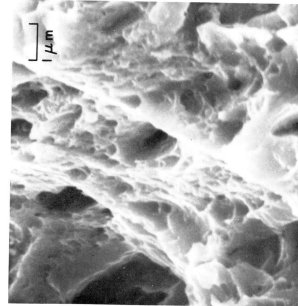

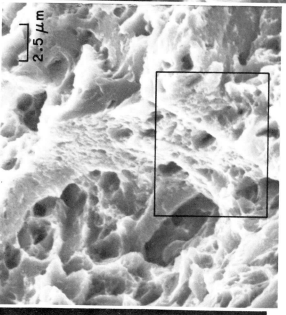

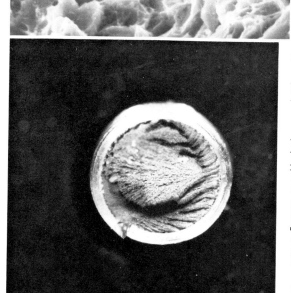

Fig. 789 Fracture caused by hydrogen embrittlement in threaded specimen of AISI H11 tool steel (same heat treatment and tensile strength as in Fig. 786). Hydrogen impregnation was by plating. Fracture occurred before the full sustained load could be applied and progressed around the circumference of the specimen. See also Fig. 790 and 791. 6×

Fig. 790 SEM view of the fast-fracture region of the fracture surface in Fig. 789, showing an unusual type of surface for a specimen heavily charged with hydrogen. Fracture occurred primarily by microvoid coalescence; there is no evidence of intergranular rupture. Area in rectangle is shown at higher magnification in Fig. 791. 4000×

Fig. 791 Higher-magnification view of the area in the rectangle in Fig. 790. Very fine dimples and large dimples are both present. A few features resemble cleavage facets, but the grain-boundary parting expected in hydrogen-embrittlement fractures is absent. SEM, 10 000×

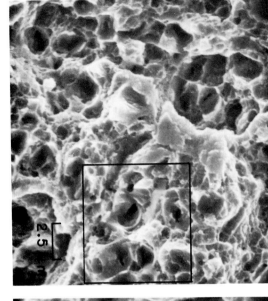

Fig. 792 Another view of the fracture surface in hydrogen-embrittled AISI H11 tool steel shown in Fig. 789. The surface consists almost entirely of dimpled rupture. A few facets are present and suggest possible local cleavage. See also Fig. 793. SEM, 4000×

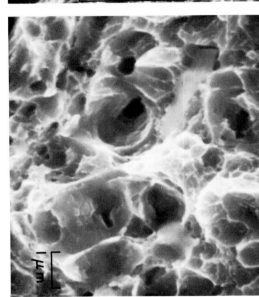

Fig. 793 Higher-magnification view of a region of which most is contained within the rectangle in Fig. 792. This region, like the one shown in Fig. 791, exhibits a considerable range in dimple size and no trace of grain-boundary parting. SEM, 10 000×

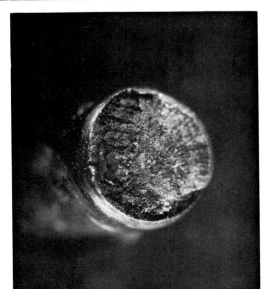

Fig. 794 Fracture caused by stress-corrosion cracking in a threaded specimen of AISI H11 tool steel (same heat treatment and tensile strength as in Fig. 786) at a sustained loading of 75% tensile strength in a 3.5% NaCl environment at room temperature. Appreciable corrosion is evident. See also Fig. 795 to 797. 6×

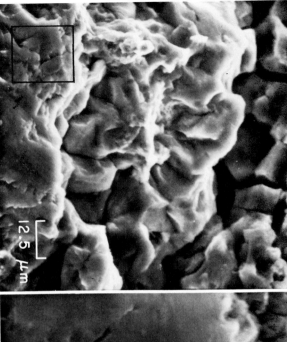

Fig. 795 SEM view of the fracture surface in fracture by grain-boundary cracking under NaCl attack. Some facets show no trace of corrosion. Deep secondary cracks are present. A few facets suggest quasi-cleavage. See also Fig. 796, 800×

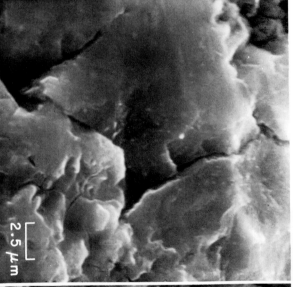

Fig. 796 Higher-magnification view of the area in the rectangle in Fig. 795, showing secondary cracking, which appears to follow some grain boundaries. The large, smooth facets resemble transgranular cleavage more than they resemble grain-boundary rupture. SEM, 4000×

Fig. 797 High-magnification view of a region of the fracture in Fig. 794 not shown in Fig. 795, displaying grain-boundary facets and some corrosion. Some of the deep secondary cracks may be transgranular. SEM, 4000×

Fig. 798 Surface of a fracture in a tensile-test specimen of 18% Ni, grade 300, maraging steel aged 3 h at 480 °C (900 °F), broken at room temperature. The crack-initiation zone (at center) is an example of extremely uneven fibrous fracture; there appear to be no radial marks leading to the base of the surrounding shear lip. The shear lip is of the pronounced type that is observed when a large reduction in cross-sectional area accompanies fracture. See also Fig. 799 to 802. 26×

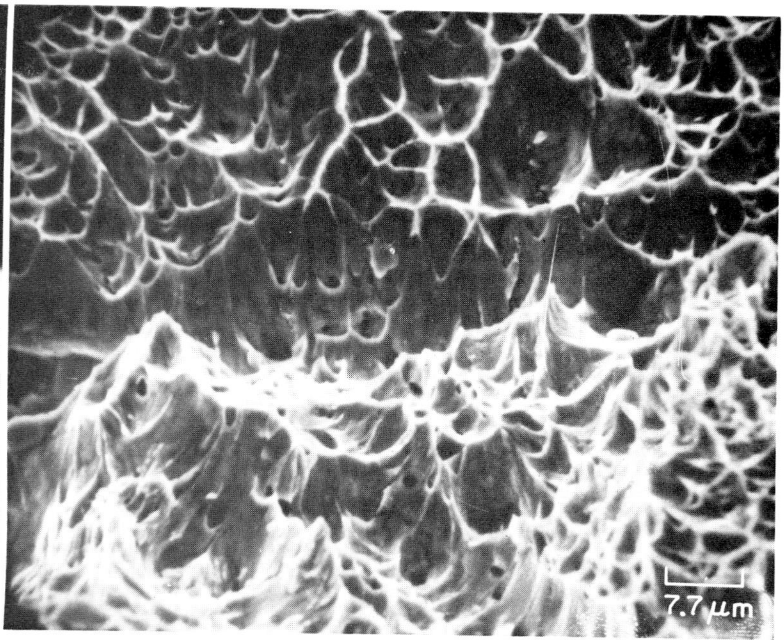

Fig. 799 SEM view of the fracture surface of the specimen of 18% Ni, grade 300, maraging steel in Fig. 798, showing a region in the area of transition from fibrous fracture in the central zone to fracture by shear in the surrounding lip. Visible in the lower portion of the fractograph are the equiaxed dimples (viewed obliquely) that are characteristic of fracture in pure tension; above are the elongated dimples of the shear lip. Note the horizontal boundary between these two types of fracture. See also Fig. 800 and 801. 1300×

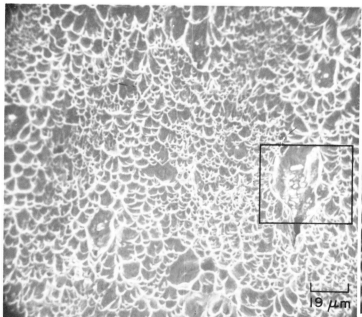

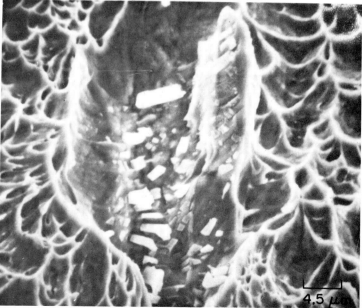

Fig. 800 View of the shear-lip region of the fracture surface in Fig. 798. Microvoid coalescence here has produced shear dimples with characteristic elongated shape. Note the large voids, which formed at inclusions of Ti(C,N). The matrix did not adhere to these inclusions, which allowed the voids to open early in the fracture process and to grow large. See Fig. 801 for a higher-magnification view of the area in the rectangle. SEM, 520×

Fig. 801 Higher-magnification view of the area in the rectangle in Fig. 800, showing one of the very large voids. Clearly visible is the array of Ti(C,N) inclusions that triggered the formation of this void. These may have been separate particles originally, or they may be segments of one or more larger particles that broke apart before the void opened. Except for a few tiny dimples visible here (lower right), the region in Fig. 800 was found to lack the dispersion of fine dimples normally present in surfaces of tensile-test fractures. SEM, 2200×

384 / Maraging Steels

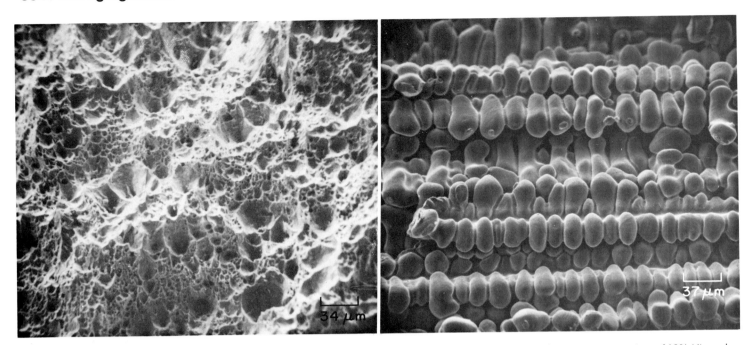

Fig. 802 An SEM view of the surface of the tensile-test fracture in 18% Ni, grade 300, maraging steel in Fig. 798, showing a portion of the central zone of the fracture, close to the origin. The surface here is composed of equiaxed dimples of two different sizes. The large dimples probably formed at Ti(C,N) particles; all the dimples were caused by particles of some sort. 290×

Fig. 803 Surface of tensile-test fracture in a cast specimen of 18% Ni, grade 300, maraging stgeel, showing a region where the fracture intersected a shrinkage cavity and exposed dendrites whose growth during solidification was arrested by a lack of molten metal within the cavity. In this view, the primary dendrite arms are oriented horizontally. See also Fig. 804 and 805. SEM, 270×

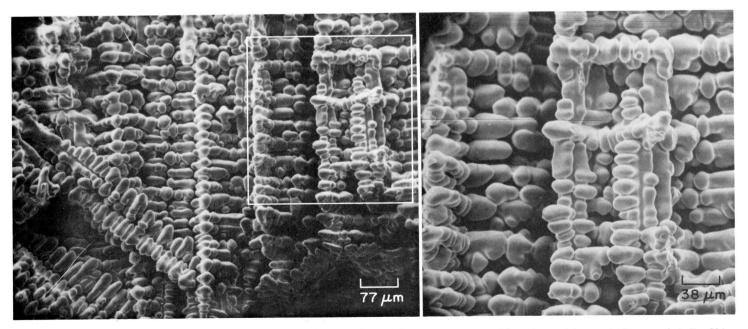

Fig. 804 Lower-magnification view of the exposed dendrites in the shrinkage cavity in the cast specimen of 18% Ni, grade 300, maraging steel in Fig. 803, showing dendrites that formed at different orientations. Had this casting been properly risered, the channels between the dendrite arms would have been filled, forming continuous metal. See also Fig. 805, which shows the area in the rectangle at higher magnification. SEM, 130×

Fig. 805 Enlarged view of the area in the rectangle in Fig. 804. Primary dendrite arms are oriented vertically here; secondary arms are at various orientations normal to the primary arms. Appreciable secondary-arm growth is evident, but the higher-order branching often found in castings has not occurred. SEM, 260×

Maraging Steels / 385

Fig. 806 Fracture surface of a fracture-toughness specimen of 18% Ni, grade 300, maraging steel; the heat treatment of this specimen was not reported. This view shows entry of the fatigue crack from bottom, and sharp onset of final fast fracture at center. See also Fig. 807 and 808; compare with Fig. 809. SEM, 26×

Fig. 807 SEM view, at higher magnification, of the fracture surface shown in Fig. 806, with the fatigue-crack surface at bottom and the fast-fracture surface at top. Note the secondary cracks and lack of clear fatigue striations in the fatigue area. See also Fig. 808. 300×

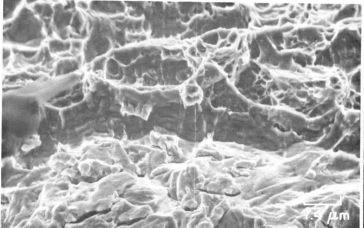
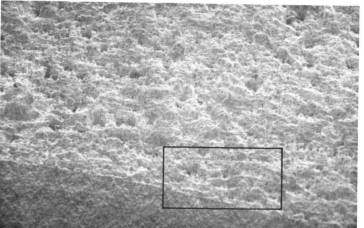

Fig. 808 Same fracture surface in 18% Ni, grade 300, maraging steel as in Fig. 806 and 807, shown here at still higher magnification. At bottom, fatigue has produced numerous secondary cracks. Note the stretched zone at center, at the transition from fatigue to final fast fracture. SEM, 1350×

Fig. 809 Fracture surface of a fracture-toughness specimen of 18% Ni, grade 300 maraging steel aged 3 h at 540 °C (1000 °F) and air cooled. Fatigue crack at bottom; ductile, tension-overload fracture at top. Note ridges, spaced at about 60 μm, in ductile region parallel to crack front. See also Fig. 806, 810, and 811. SEM, 47×

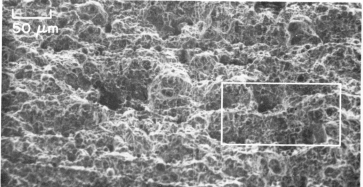
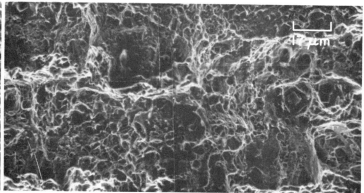

Fig. 810 Higher-magnification view of the area in the rectangle in Fig. 809. Tension-overload dimples are clearly shown, but no details of the fatigue zone (bottom) is evident. The discontinuous nature of the ridges in the tension-overload region is evident. See Fig. 811 for enlarged, composite view of area in rectangle here. SEM, 200×

Fig. 811 Composite of four SEM fractographs of the plane-strain region of the area of tension-overload fracture in the specimen shown in Fig. 809 and 810, taken from the area in the rectangle in Fig. 810. This shows more clearly the changes in surface orientation between the ridges. 600×

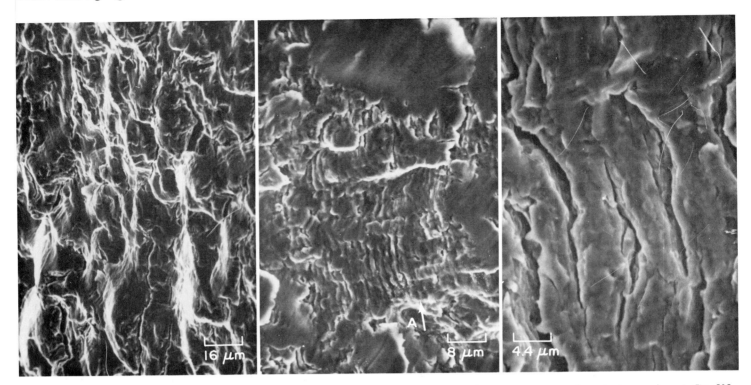

Fig. 812 A low-cycle fatigue fracture (at about 3000 cycles) of 18% Ni, grade 300, maraging steel that had been annealed 1 h at 815 °C (1500 °F) and air cooled. There is no evidence of clearly defined striations, but there is a system of more or less parallel secondary cracks. See also Fig. 813 and 814. SEM, 640×

Fig. 813 A view of a different area of the low cycle fatigue fracture in Fig. 812. Note that in this area quite regular striations are in evidence, many of them with fissures at their roots (as at A). Note also the several rubbed areas, particularly those at the top. See also Fig. 814. SEM, 1250×

Fig. 814 Same fracture as shown in Fig. 812 and 813, but an area at still another location, as seen at higher magnification. Here again is a system of repetitive fissures that appear to be quite deep and very irregular. There is little evidence of the conventional type of striations in this area. SEM, 2250×

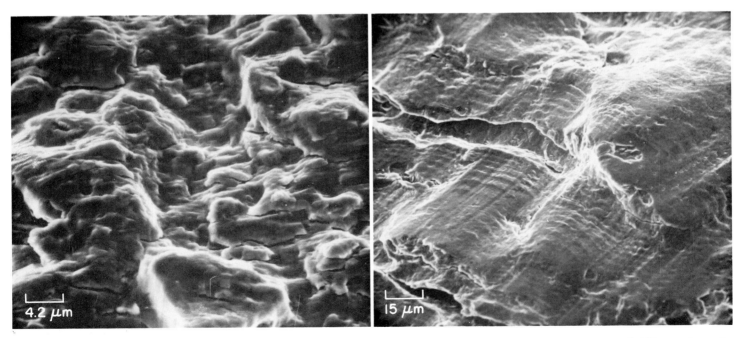

Fig. 815 Surface of a low-cycle fatigue fracture (at about 1000 cycles) in 18% Ni, grade 300, maraging steel that had been aged at 480 °C (900 °F) for 3 h and air cooled. This fracture surface displays a progression of rather fine but extremely irregular striations, separated here and there by secondary cracks. SEM, 2400×

Fig. 816 Low-cycle fatigue fracture of 18% Ni, grade 300, maraging steel (heat treatment not reported). This has relatively uniformly spaced fatigue striations with fewer secondary cracks than are seen in Fig. 812 to 815. The pattern of striations is similar to that produced by varied loading in service. SEM, 670×

Fig. 817, 818 Figure 817 (left) shows a slow-bend fracture of a weld specimen made by gas tungsten-arc butt welding two 19-mm (¾-in.) thick plates of 18% Ni maraging steel having a yield strength of 1724 MPa (250 ksi). A specimen 48 mm² × 180 mm (0.75 in.² × 7 in.) was cut across the weld. After the weld was ground flush, the specimen was notched and then fatigue-cracked in the center of the weld and parallel to the length of the weld. After being aged at 490 °C (915 °F), the specimen was broken at room temperature in slow, three-point bending. Fracture surface shows steps caused by splitting. A portion of one of these steps is shown at higher magnification in Fig. 818 (right); the surface here exhibits a network of high-reflectivity area, one example being identified by the arrows. The remaining areas of the surface were relatively dull. See also Fig. 819 to 822. Fig. 817: Actual size. Fig. 818: 8×

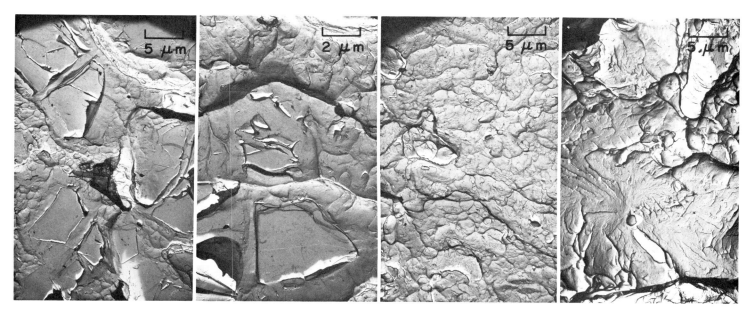

Fig. 819 to 822 Figures 819 and 820 (left and left center) are TEM views, at different magnifications, taken from one of the high-reflectivity areas in Fig. 818. The shiny aspect was due to flat platelets resulting from cleavage of particles of unidentified composition. Note that in both views there are tear dimples between the reflecting facets. Figure 821 (right center) reveals that the structure in the low-reflectivity area in this fracture surface consists of rather flat equiaxed dimples. Figure 822 (right) is from a companion 18% Ni maraging steel weld specimen to the one in Fig. 817 to 821; specimen here was hydrogen embrittled and then fractured by sustained loading at room temperature. This fracture surface shows dimples, together with quasi-cleavage facets that contain characteristic steps, river patterns, and tear ridges. Fig. 819, 821, and 822: TEM p-c replicas, all at 2000×. Fig. 820: TEM p-c replica, 5000×

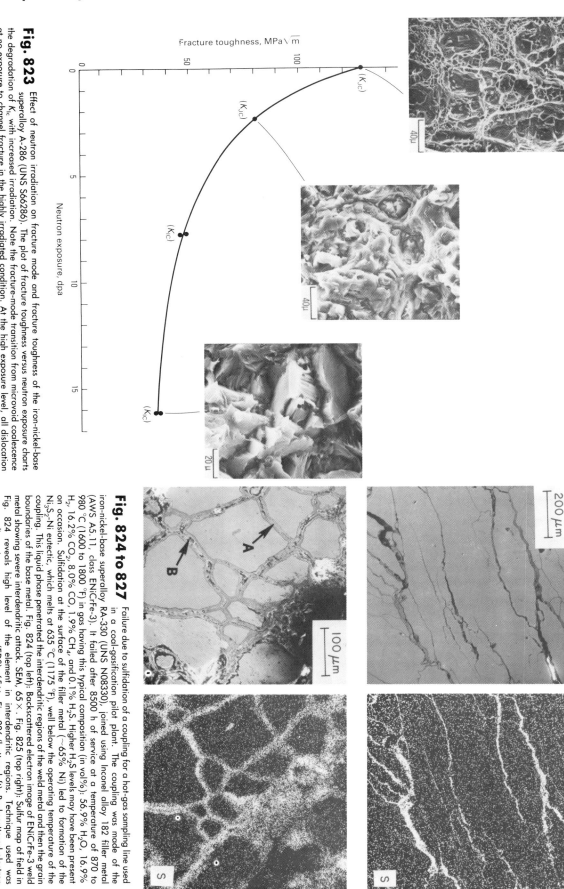

Fig. 823 Effect of neutron irradiation on fracture mode and fracture toughness of the iron-nickel-base superalloy A-286 (UNS S66286). The plot of fracture toughness versus neutron exposure charts the degradation of K_{Ic} with increased irradiation. Note the fracture-mode transition from microvoid coalescence at no exposure to channel fracture in the highly irradiated condition. At the high exposure level, all dislocation activity was localized in a planar slip band, and cracking eventually initiated and propagated along the resultant dislocation channels, causing fracture toughness to fall off. (J.E. Nolan, Westinghouse Hanford Company)

Fig. 824 to 827 Failure due to sulfidation of a coupling for a hot-gas sampling line used in a coal-gasification pilot plant. The coupling was made of the iron-nickel-base superalloy RA-330 (UNS N08330), joined using Inconel alloy 182 filler metal (AWS A5.11, class ENiCrFe-3). It failed after 8500 h of service at a temperature of 870 to 980 °C (1600 to 1800 °F) in gas having this typical composition (in vol%): 56.9% H_2O, 16.9% H_2, 16.2% CO_2, 8.0% CO, 1.9% CH_4, and 0.1% H_2S. Higher H_2S levels may have been present on occasion. Sulfidation at the surface of the filler metal (~65% Ni) led to formation of the Ni_3S_2-Ni eutectic, which melts at 635 °C (1175 °F), well below the operating temperature of the coupling. This liquid phase penetrated the interdendritic regions of the weld metal and then the grain boundaries of the base metal. Fig. 824 (top left): Backscattered electron image of ENiCrFe-3 weld metal showing severe interdendritic attack. SEM, 65×. Fig. 825 (top right): Sulfur map of field in Fig. 824 reveals high level of the element in interdendritic regions. Technique used was energy-dispersive x-ray spectroscopy (EDS), 65×. Fig. 826 (bottom left): Backscattered electron image of superalloy base metal adjacent to filler metal shows intergranular attack by Ni_3S_2-Ni (at B), most of which transformed to more stable chromium sulfide (at A). SEM, 130× (bottom right): Sulfur map of field in Fig. 826 reveals high levels of the element in grain boundaries. 130× (D.R. Diercks, Argonne National Laboratory)

Nickel-Base Superalloys / 389

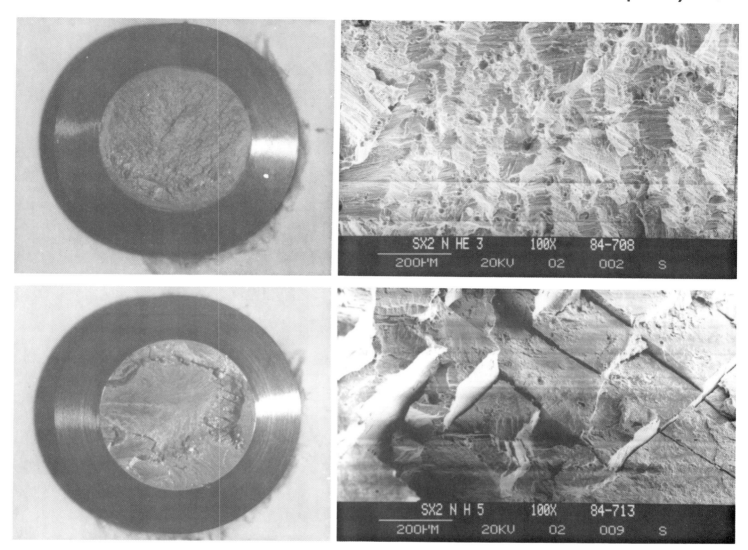

Fig. 828 to 831 Hydrogen-embrittlement fracture of a single-crystal nickel-base superalloy (CMSX-2) developed for gas turbine engine applications. Composition: 4.6% Co, 8.0% Cr, 0.6% Mo, 7.9% W, 5.6% Al, 0.9% Ti, 5.8% Ta, <0.1% Hf, 0.01% Zr, 0.005% C, remainder Ni. Heat-treated notched samples were stress relieved and then tensile tested in either helium (control samples) or hydrogen at 35 MPa (5 ksi) and room temperature. Cross-head speed, 2.12 μm/s; notched strength ratio H_2/He, 0.14. Fig. 828 (top left): Fracture surface of control sample tested in helium. 4.6×. Fig. 829 (top right): Higher-magnification view of fracture surface in Fig. 828. Note ductile failure mode. SEM, 95×. Fig. 830 (bottom left): Fracture surface of samples tested in hydrogen. 4.6×. Fig. 831 (bottom right): Higher-magnification view of fracture surface in Fig. 830. Note brittle crystallographic cracking. SEM, 95× (R.J. Schwinghamer, NASA Marshall Space Flight Center)

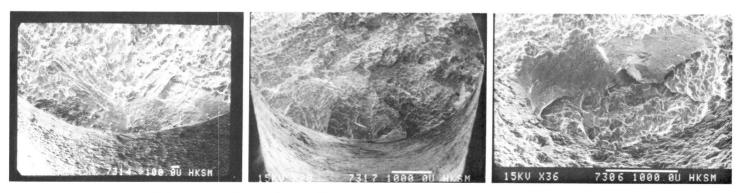

Fig. 832, 833, 834 Fatigue and creep fractures of case and hot isostatically pressed IN-738 tested in a salt environment (40% $MgSO_4$, 59% Na_2SO_4, 1% NaCl) at 705 °C (1300 °F). Samples heat treated 2 h at 1120 °C (2050 °F), air cool; 14 h at 845 °C (1550 °F), air cool. Fig. 832 (left): Fatigue fracture surface. Test conditions: 483 MPa (70 ksi) ± 207 MPa (30 ksi); 102 000 cycles at failure. Fractograph shows region of transgranular fracture initiation with characteristic "thumbprint" and river patterns pinpointing origin at surface (bottom). Failure was salt corrosion assisted. SEM, 29×. Fig. 833 (center): Fatigue fracture surface of salt-coated sample. Test conditions: 483 MPa (70 ksi) ± 345 MPa (50 ksi); 5000 cycles at failure. Fracture is transgranular with multiple surface origins. SEM, 13×. Fig. 834 (right): Fracture surface after stress-rupture testing at 896 MPa (130 ksi). Mechanism: salt corrosion-assisted intergranular fracture. Note grain-boundary decohesion. SEM, 23× (E.A. Schwarzkopf, J. Stefani, and J.K. Tien, Columbia University)

390 / Nickel-Base Superalloys

Fig. 835 Segment of a fractured second-stage gas-turbine wheel, cast from alloy 713C, that broke from fatigue in service. (About half of the disk portion of the wheel was never recovered.) The fracture origin (at arrow) was in a grinding-relief groove adjacent to the wheel-balancing pad. Additional circumferential cracks, not adjuncts of fracture, were found elsewhere in the grinding-relief groove. See Fig. 836 for a higher-magnification view of the rectangular area. 2×

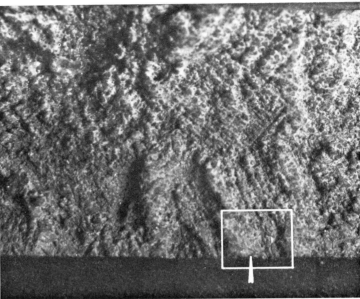

Fig. 836 Crack-origin region in the rectangle in Fig. 835, at higher magnification. Arrow marks the probable location of the crack nucleus. The surface of the grinding-relief groove is below the fracture surface. Measurements of the radius at the bottom of the groove gave values that were about 70% of the average desired and about 87% of the minimum. The excessive sharpness of the radius may have been conducive to cracking. See also Fig. 837. 10×

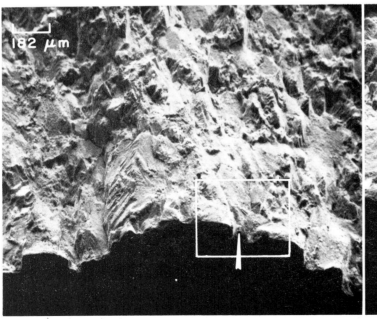

Fig. 837 SEM view taken of the outlined area in Fig. 836 after the fracture surface, which contained many white, gelatinous particles as received at the laboratory, was cleaned ultrasonically in 18% HCl solution for 1½ min and then washed in acetone and blown dry. The interface between the fracture and the grinding-relief groove was carefully examined in the origin area, and only one probable crack origin was discovered—that marked by the arrow. Area in rectangle is shown at higher magnification in Fig. 838. 55×

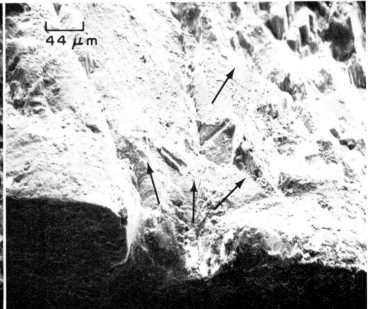

Fig. 838 A view at higher magnification of the outlined area (containing the probable origin of fracture) in Fig. 837. Numerous patches of fatigue striations are visible; the arrows indicate the directions of fatigue-crack propagation from what appears to be a surface defect within the grinding-relief groove. The fatigue facets were found to be transgranular, giving way, outside of the dark, crescent-shaped fatigue-crack area visible in Fig. 835, to ductile dimples. SEM, 225×

Nickel-Base Superalloys / 391

Fig. 839 A gas-producer turbine rotor cast of alloy 713LC that fractured after 440 h of service, as the result of hot corrosion fatigue. Fracture was abrupt, with three blades being thrown off. See Fig. 841 for a view of the area near the arrow. See also Fig. 840 and 842. 0.5×

Fig. 840 Fracture surfaces of the two broken turbine-rotor blades at the bottom in Fig. 839. Fatigue beach marks are faintly visible at right on the fracture surface of the lower blade. The region between the parallel black lines on each blade was examined by electron microscopy. 6×

Fig. 841 A macroetched view of a cracked turbine-rotor blade near the arrow in Fig. 839, displaying large columnar grains and a crack that initiated at the trailing edge (at left in this view) of the blade. This crack appears to have followed an intergranular path. Etched in 95 parts conc HCl, 5 parts 35% H_2O_2. 5.5×

Fig. 842 TEM p-c replica of a fracture surface of one of the turbine-rotor blades in Fig. 840. This view is typical of all the areas examined. A patch of fatigue striations is faintly visible at center between arrows. The fatigue crack is believed to have been initiated by hot corrosion, which was shown, by sectioning of blades, to exist at concentrations of lead. 2400×

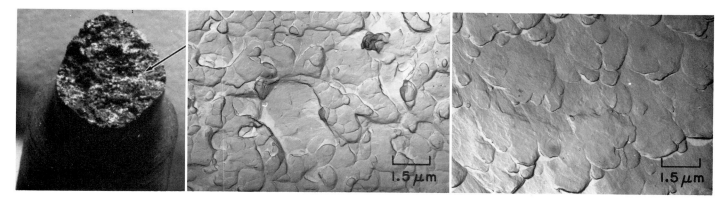

Fig. 843, 844, 845 Figure 843 (left) shows the surface of a dendritic stress-rupture fracture in a cast specimen of IN-100 nickel-base alloy that was annealed at 1175 °C (2150 °F) and loaded at 980 °C (1800 °F) to a tensile stress of 97 MPa (14 ksi). The specimen broke after 49 h of testing. Figure 844 (center) is a TEM p-c replica of an area at the arrow in Fig. 843, showing some evidence of intergranular separation and some facets of what appear to be cleaved particles (as at A) of intermetallic compounds. Figure 845 (right) is a TEM p-c replica of another area at the arrow in Fig. 843, showing evidence of glide-plane decohesion (at A's) as well as smooth areas of stretching (as at B). Some surface oxidation apparently occurred after fracture. Fig. 843: 8×. Fig. 844 and 845: 6500×

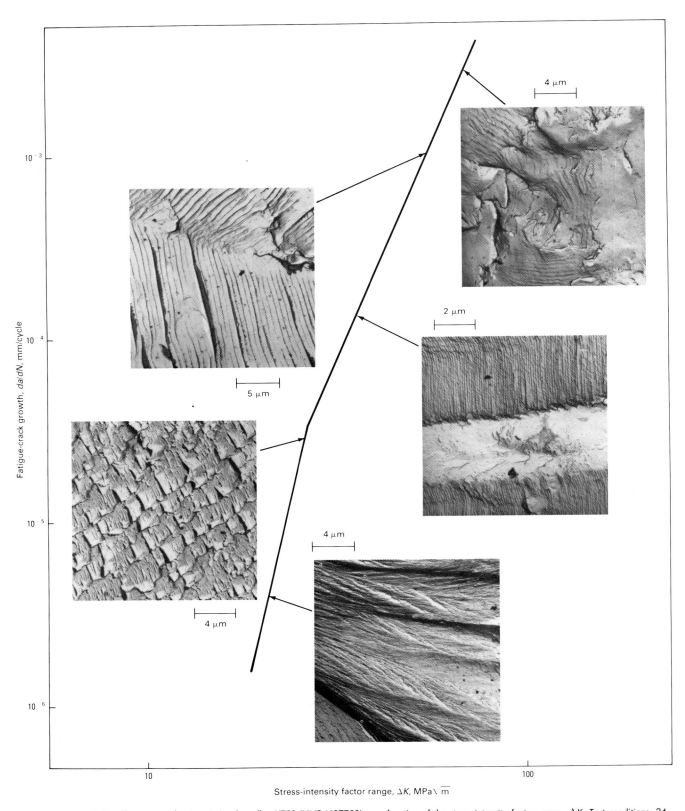

Fig. 846 Fatigue fracture mechanisms in Incoloy alloy X750 (UNS N07750) as a function of the stress-intensity factor range, ΔK. Test conditions: 24 °C (75 °F), 300 cycles/min, $R = 0.05$. The plot of fatigue crack growth rate, da/dN, versus ΔK shows that at high ΔK, the fatigue fracture surface exhibited well-defined striations and dimples. At progressively lower values of ΔK, a combination of fatigue fissures (associated with small secondary cracks) and striations was observed. At low ΔK, a highly faceted fracture surface resulted due to crystallographic fracture along intense slip bands. (J.E. Nolan, Westinghouse Hanford Company)

Nickel-Base Superalloys / 393

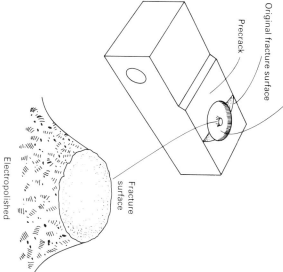

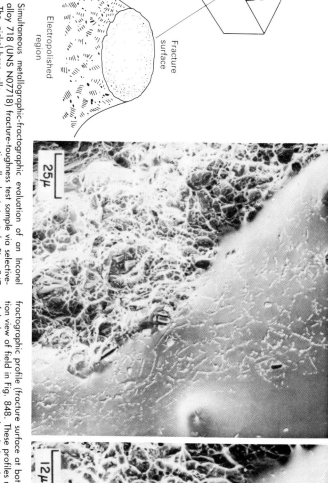

Fig. 847, 848, 849 Simultaneous metallographic-fractographic evaluation of an Inconel alloy 718 (UNS N07718) fracture-toughness test sample via selective-area electropolishing technique. The nickel-base alloy was conventionally heat treated. Fig. 847 (left): Schematic of selective-area electropolishing. Use of method enables simultaneous study of both the fracture surface topography and the underlying microstructure. Fig. 848 (center): Metallographic-fractographic profile (fracture surface at bottom left). SEM, 460×. Fig. 849 (right): Higher-magnification view of the large primary microvoids. These profiles reveal a duplex microvoid coalescence morphology. Growth of the large primary microvoids, which nucleated at ruptured (Ni,Ti)C inclusions, was preempted as many smaller microvoids nucleated at δ-phase precipitates. SEM, 950× (J.E. Nolan, Westinghouse Hanford Company)

Fig. 850, 851 Creep fracture in Inconel alloy MA754, a mechanically alloyed nickel-base superalloy stabilized by yttria. Fig. 850 (left): Fracture surface of longitudinal creep specimen tested at 760 °C (1400 °F) and 207 MPa (30 ksi) for 84 h. Fracture was intergranular. Note delamination along longitudinal and transverse grain boundaries (left). SEM, 190×. Fig. 851 (right): Fracture surface of long-transverse creep specimen tested at 980 °C (1800 °F) and 48 MPa (7 ksi) for 54 h. Fracture occurred along longitudinal grain boundaries (perpendicular to the applied stress). SEM, 50× (J.K. Tien, T.E. Howson, and J.E. Stulga, Columbia University)

Fig. 852 "Departure side pinning" during creep of Inconel alloy MA754. The mechanically alloyed nickel-base superalloy is stabilized by yttria (0.8% Y$_2$O$_3$). Samples were tested in creep to 2% strain at 760 °C (1400 °F) and 221 MPa (32 ksi) and then cooled on load. TEM examination revealed that dislocations were still associated with Y$_2$O$_3$ particles—the departure side pinning phenomenon. Etched in 60% ethanol, 30% n-butanol, 10% perchloric acid (70%). TEM thin foil, 76 800× (J.K. Tien, V.C. Nardonne, and D. Matejczyk, Columbia University)

394 / Nickel-Base Superalloys

Fig. 853 to 856 Fatigue fracture of Nimonic 115 at room temperature and 845 °C (1550 °F). Thermal treatment: solution treat 24 h at 1230 °C (2250 °F); furnace cool to 1000 °C (1830 °F) to nucleate precipitation; air cool; precipitation age for 96 h at 1000 °C (1830 °F); air cool. Test conditions: $R = -1$, constant strain rate ramps. Fig. 853 (left): Fracture surface of sample tested at room temperature. Failure is surface-initiated stage I with stage I propagation until overload. 5.7×. Fig. 854 (left center): Substructure of sample in Fig. 853 consists of dense, homogeneous dislocation tangles. Note well-defined slip bands that have sheared γ' precipitates. TEM thin foil, 12 400×. Fig. 855 (right center): Fracture surface of sample tested at 845 °C (1550 °F). Failure due to crack initiation in a stage II mode around most of the circumference. 5.8×. Fig. 856 (right): Substructure of sample in Fig. 855. Note similarity to substructure of sample tested at room temperature (Fig. 854). For example, γ' shearing due to coarse slip bands is evident. However, dislocation tangles are less dense. TEM thin foil, 11 400× (J.K. Tien and L. Fritzmeir, Columbia University)

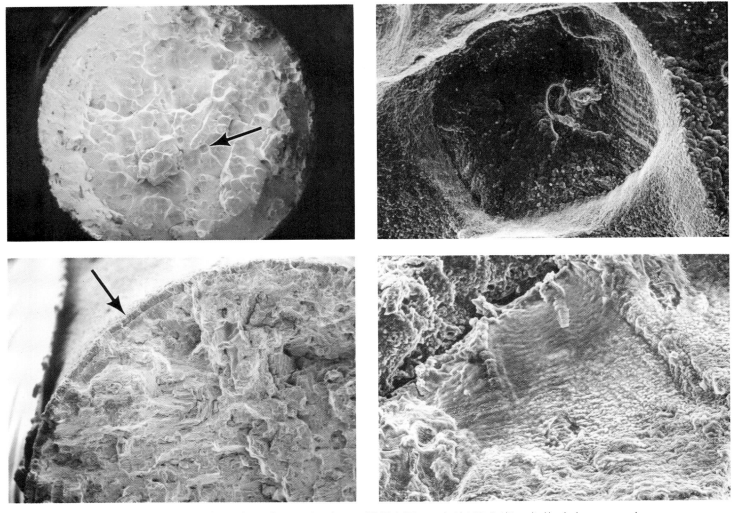

Fig. 857 to 860 Effect of thermal cycling on fatigue fracture of single-crystal PWA 1480 coated with NiCoCrAlY applied by the low-pressure plasma spray process. Fig. 857 (top left): Fracture surface of coated sample tested at 1050 °C (1920 °F) in isothermal low-cycle fatigue. The NiCoCrAlY coating was very ductile at the test temperature and did not crack. The well-protected superalloy failed via multiple internal cracking originating at microporosity (arrow). SEM, 12.5×. Fig. 858 (top right): High-magnification view of micropore at arrow in Fig. 857. SEM, 280×. Fig. 859 (bottom left): Fracture surface of coated sample cycled between 650 and 1050 °C (1200 and 1920 °F) in a thermomechanical fatigue test. The NiCoCrAlY coating was placed in tension during the low-temperature portion of the test, which caused it to crack in a few cycles. The nickel-base superalloy—now exposed to the atmosphere—failed at multiple surface locations (arrow) rather than at internal micropores (as in Fig. 857 and 858). SEM, 21×. Fig. 860 (bottom right): High-magnification view of fracture origin at arrow in Fig. 859. SEM, 260× (R.V. Miner and S.L. Draper, NASA Lewis Research Center)

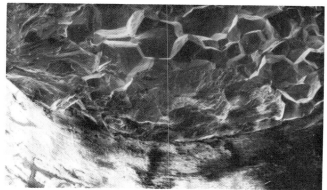

Fig. 861 Fatigue fracture of the nickel-base superalloy Udimet 720 tested in a salt environment at 705 °C (1300 °F). Fatigue conditions: 690 MPa (100 ksi) ± 207 MPa (30 ksi) 10 500 cycles to failure. Salt coating: 40% $MgSO_4$, 59% Na_2SO_4, 1% NaCl. Sample heat treatment: 4 h at 1170 °C (2140 °F), air cool; 4 h at 1080 °C (1975 °F), air cool; 12 h at 845 °C (1550 °F), air cool; 16 h at 760 °C (1400 °F), air cool. Fracture surface shows dual, transgranular fatigue origins. Initiation attributed to salt corrosion of grain boundaries located at surface and perpendicular to stress axis. Note salt coating on sample surface. SEM, 30× (E.A. Schwarzkopf, J. Stefani, and J.K. Tien, Columbia University)

Fig. 862 Fatigue fracture of Udimet 720 tested in air at 705 °C (1300 °F). Heat treatment and other test conditions same as in Fig. 861. Cycles to failure: 338 000, compared with 10 500 for the sample tested in a salt environment (Fig. 861). Fracture surface shows a transgranular "thumbprint," with river patterns pinpointing initiation sites. The major origin is where a grain boundary intersects the sample surface. SEM, 25× (E.A. Schwarzkopf, J. Stefani, and J.K. Tien, Columbia University)

Fig. 863, 864 Fatigue fracture of Udimet 720 tested in a salt environment at 705 °C (1300 °F). Sample heat treatment and salt composition same as in Fig. 861. Test conditions: 827 MPa ± 69 MPa (120 ksi ± 10 ksi), 5000 cycles to failure. Fig. 863 (left): Fracture surface shows intergranular origin at grain-boundary triple point (near bottom) with subsequent transgranular propagation. Because origin is near the surface, corrosion may have had an effect on fracture initiation. River patterns in transgranular portion of fracture reveal crack path. SEM, 23×. Fig. 864 (right): Higher-magnification view of fatigue fracture origin in Fig. 863. SEM, 82× (E.A. Schwarzkopf, J. Stefani, and J.K. Tien, Columbia University)

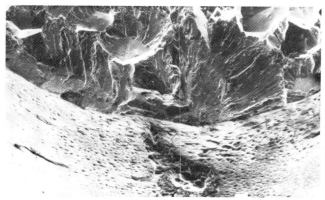

Fig. 865 Fatigue fracture of Udimet 720 tested in air at 705 °C (1300 °F). Heat treatment and other test conditions same as in Fig. 863 and 864. Cycles to failure: 66 000, compared with 5000 for the sample tested in a salt environment (Fig. 863 and 864). Fracture surface shows multiple transgranular fatigue origins near edge of sample. River patterns reveal initiation sites at specimen surface and grain boundaries perpendicular to the stress axis. SEM, 40× (E.A. Schwarzkopf, J. Stefani, and J.K. Tien, Columbia University)

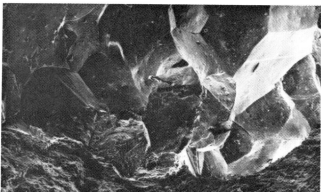

Fig. 866 Creep fracture of Udimet 720 tested in a salt environment at 705 °C (1300 °F). Sample heat treatment and salt composition same as in Fig. 861. Test stress: 896 MPa (130 ksi). Fractograph shows salt-coated sample surface and intergranular stress-rupture surface. Note salt corrosion of grain boundaries that intersect the sample surface. SEM, 40× (E.A. Schwarzkopf, J. Stefani, and J.K. Tien, Columbia University)

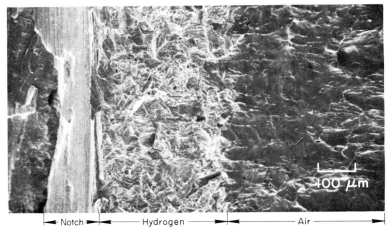

Fig. 867 Surface of a fracture in a bend-test specimen of Nickel 201 containing 0.02% C, 0.35% Mn, 0.01% S, 0.35% Si, 0.25% Cu, 0.04% Fe, remainder Ni. After the material had been annealed, a specimen 19 mm (¾ in.) wide by 3.17 mm (0.13 in.) thick was prepared, a notch was machined in the side of the specimen, and a fatigue crack was produced in the root of the notch. The specimen was then loaded in three-point bending (77-mm, or 3-in., support span) to give a fixed displacement rate of 0.0015 μm/s. The test was performed at room temperature in a stainless steel chamber equipped for testing in vacuum, in hydrogen, or in air. During the first portion of the test, the chamber was filled with hydrogen; partway through the test, the hydrogen was replaced with air. In this view, the notch is at the left, and just to the right of the root of the notch is a very shallow fatigue crack. The central region (light) is the portion of the fracture that was obtained in hydrogen, and the darker region at right is the portion that was subsequently produced in air. See Fig. 868 to 871 for higher-magnification views of the portions produced in hydrogen and in air, and of the transition between them. SEM, 100×

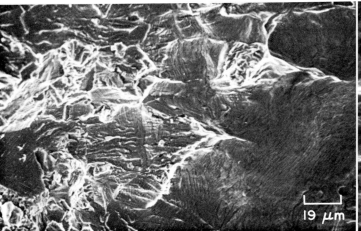

Fig. 868 View of an area of the fracture surface in Fig. 867 that shows the transition from fracture in hydrogen (left) to subsequent fracture in air (right). Much of the fracture produced in hydrogen shows transgranular cleavage. In the portion produced in air, flow occurred after the crack front has passed. SEM, 540×

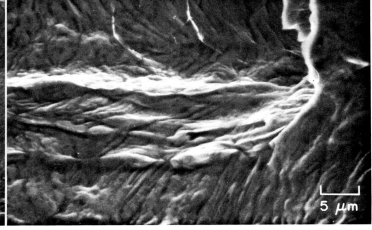

Fig. 869 View of the fracture surface in Fig. 867, showing a region from the portion of the fracture that was produced in air. Visible are suggestions of dimples, but these features have been largely obscured by the effects of the secondary flow that occurred in this very ductile metal following passage of the crack front. SEM, 2000×

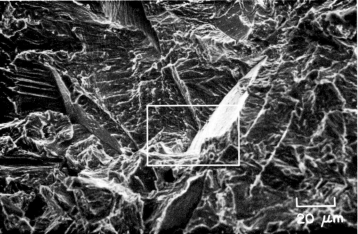

Fig. 870 View of the fracture surface in Fig. 867, showing a region from the portion of the fracture that was produced in hydrogen. As in the left portion of Fig. 868, the fracture here is mainly by transgranular cleavage. Scattered regions of intergranular separation were reported, but are not visible here. See also Fig. 871. SEM, 500×

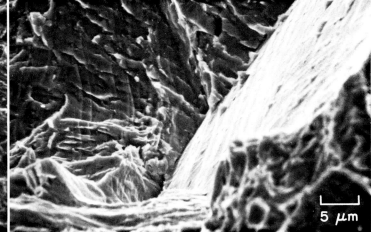

Fig. 871 Higher-magnification view of the area in the rectangle in Fig. 870, showing the ledgelike character of the cleavage facets. In contrast, a suggestion of dimpled rupture is visible a bottom right. Note the small tongues on the bright facet at right. SEM, 2000×

Nickel Alloys / 397

Fig. 872 Fracture surface of a specimen of Duranickel, vacuum solution treated and quenched, aged 1 h at 500 °C (930 °F) and cooled in air. The specimen, which was notched, precracked, and slow-bend fracture-toughness tested in hydrogen, broke by combined cleavage and intergranular fracture. See also Fig. 873. SEM, 160×

Fig. 873 A portion of the same fracture surface as that in Fig. 872, but at higher magnification, showing facets that contain cleavage river patterns (bottom right) and pyramid-shaped tongues (top left). SEM, 1000×

Fig. 874 Fracture in Duranickel that, except for being aged 4 h, was heat treated and tested same as Fig. 872. Fracture here is more completely intergranular, with less cleavage, than Fig. 872. Both fractures show secondary cracks. See also Fig. 875. SEM, 95×

Fig. 875 A view at higher magnification of the fracture in Fig. 874, showing it to be also a nearly wholly intergranular fracture, with deep secondary cracks at the grain boundaries. SEM, 200×

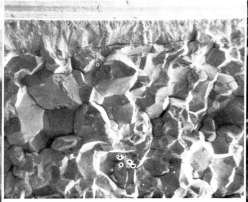

Fig. 876 Fracture in Duranickel that, except for being aged 8 h, was heat treated and tested same as Fig. 872 and 874. Notch and fatigue precrack are at top. An example of sustained-load intergranular fracture in hydrogen. See also Fig. 877 to 879. SEM, 20×

Fig. 877 A similar but larger view of the fracture in Fig. 876, showing notch at top but no fatigue crack. Fine grains at notch were presumably cold worked in machining and recrystallized in aging. SEM, 80×

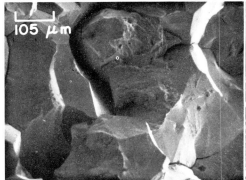

Fig. 878 Same fracture as that shown in Fig. 876 and 877, but an interior area. One or two cleavage facets are seen. Most of the fracture is intergranular, with large secondary cracks. SEM, 95×

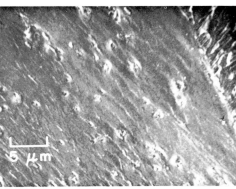

Fig. 879 A highly magnified view of the fracture in Fig. 876 to 878. Striations visible here show crack advances under constant bending deflection in hydrogen; compare with the fatigue striations in Fig. 880. SEM, 2000×

Fig. 880 High-magnification view of the fatigue precrack produced in air at base of notch in specimen in Fig. 876. Compare fatigue striations here with crack-advance features in Fig. 779. White marks at A's here are later fissures. SEM, 2000×

398 / Cobalt Alloys

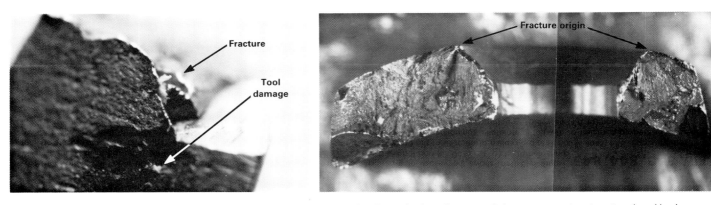

Fig. 881, 882 Fatigue fracture of a cast Vitallium (Co-30Cr-7Mo) surgical implant (side plate of Jewett nail) due to improper insertion. A tool used by the surgeon to contour the plate prior to its insertion into the patient also accidentally damaged the surface of the implant, introducing microcracks that led to fracture. Fig. 881 (left): Tool damage on surface of implant near fracture surface. 3×. Fig. 882 (right): The Jewett nail broke across a screw hole. Radial marks on the fracture surface (shown here) reveal the fatigue crack origin near the top of the hole. 10× (C.R. Brooks and A. Choudhury, University of Tennessee)

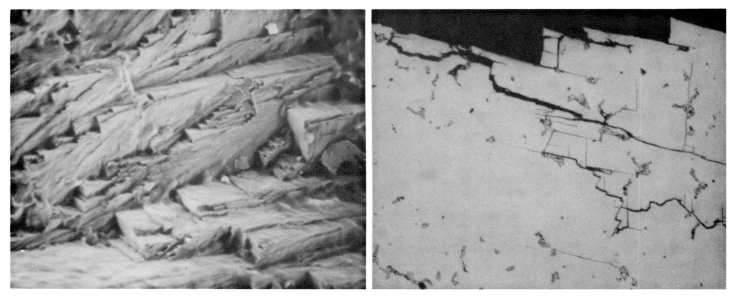

Fig. 883, 884 Fatigue fracture of cast ASTM F75 alloy (Co-28Cr-6Mo). Material was solution treated prior to testing on a cantilever-beam machine. Fig. 883 (left): "Stair step" fracture surface indicative of stage I fatigue. SEM, 515×. Fig. 884 (right): Photomicrograph reveals stage I fatigue cracking. $FeCl_3$ electrolytic etch, 200× (R. Abrams, Howmedica, Division of Pfizer Hospital Products Group, Inc.)

Fig. 885 Fatigue fracture of cast ASTM F75 alloy (Co-28Cr-6Mo). Material was hot isostatically pressed and solution treated prior to constant force amplitude, flexural fatigue testing. Fractograph shows region of stage I fatigue, characterized by a slip/cross-slip "stair step" morphology. Additional information on cobalt-base alloys used for surgical implants and their corresponding fracture morphologies can be found in the article "Failure Analysis of Metallic Orthopedic Implants" in Volume 11 of the 9th Edition of *Metals Handbook*. SEM, 532× (R. Abrams, Howmedica, Division of Pfizer Hospital Products Group, Inc.)

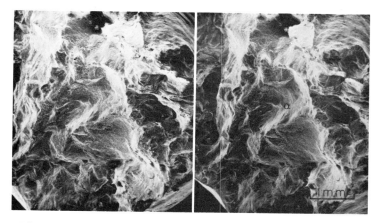

Fig. 886 Fracture at 425 °C (797 °F), 20 MPa (3 ksi). 11.5×

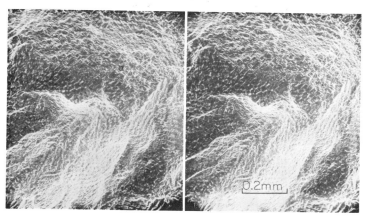

Fig. 887 Area at α in Fig. 886, at higher magnification. 60×

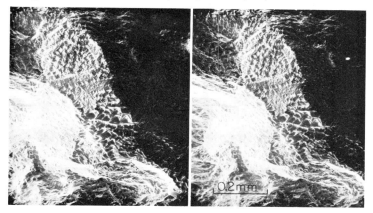

Fig. 888 A different area of the fracture in Fig. 886. 70×

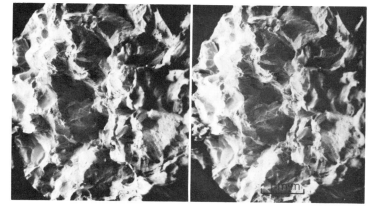

Fig. 889 Fracture at 440 °C (824 °F), 15.5 MPa (2.25 ksi). 11×

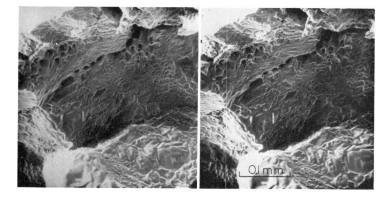

Fig. 890 A different area of the fracture in Fig. 889. 135×

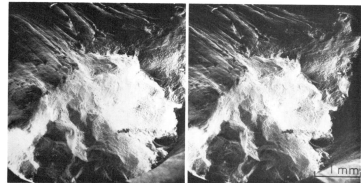

Fig. 891 Fracture at 525 °C (977 °F), 10.3 MPa (1.5 ksi). 12.5×

These stereo pairs of SEM fractographs, as well as those on the following page, are of fracture surfaces produced in specimens of high-purity copper (99.999% Cu) by creep testing to rupture at temperatures ranging from 425 to 642 °C (797 to 1188 °F) and at tensile stresses ranging from 7 to 20 MPa (1 to 3 ksi). The material was continuously cast into 19-mm (¾-in.) diam rod, swaged down to 14-mm (0.55-in.) diam, then machined into buttonhead specimens with gage sections 38 mm (1½ in.) long and 6.4 mm (¼ in.) in diameter. Before being tested, each specimen was heat treated in the creep-test furnace, first for 1 h at 400 °C (750 °F) in hydrogen and then for 16 h at 800 °C (1470 °F) in purified helium. The testing also was done in purified helium.

At low temperature or low stress, or both, the main mechanism of fracture was intergranular separation with accompanying microvoid coalescence, and the separated-grain facets were oriented at high angles relative to the tensile axis. At high temperature or high stress, or both, the fracturing was transgranular and highly plastic. At intermediate combinations of test temperature and stress, the main mechanism of fracture was the same as at low temperature or low stress (intergranular separation with accompanying microvoid coalescence), but the fracture surface was fairly smooth and had a fine texture.

The specimen in Fig. 886 to 888 was tested at 425 °C (797 °F) and 20 MPa (3 ksi) and attained a steady-state strain rate of 2.0×10^{-6}/min, fracturing in 76 h, 54 min. These views show a fine texture of dimples, which resulted from microvoid coalescence, on fairly smooth separated-grain facets. The area at α in the center of Fig. 886 is shown at higher magnification in Fig. 887; note the uniform size of the dimples. Figure 888, which is of an area of the fracture different from that in Fig. 886, shows dimples with shapes that vary depending on orientation of the grain facets relative to the tensile axis.

The specimen shown in Fig. 889 and 890 was tested at 440 °C (824 °F) and 15.5 MPa (2.25 ksi) and attained a steady-state strain rate of 7.0×10^{-7}/min, fracturing in 308 h, 30 min. These fractographs show two different areas of the fracture surface. The area shown in Fig. 889 displays separated-grain facets oriented at high angles relative to the tensile axis; the area shown in Fig. 890 (at higher magnification) clearly shows the small dimples on the grain facets.

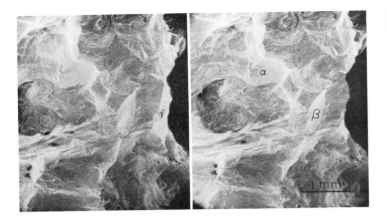

Fig. 892 Fracture at 540 °C (1004 °F), 10.3 MPa (1.5 ksi). 11.5×

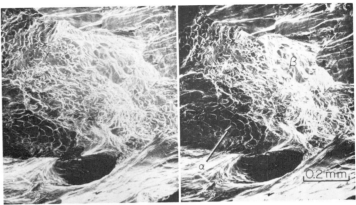

Fig. 893 Area at bottom of cavity at left center in Fig. 892. 60×

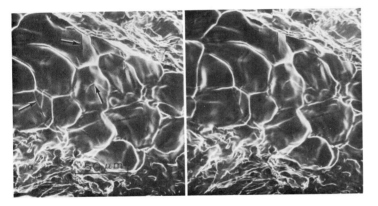

Fig. 894 Dimples indicated by arrow marked α in Fig. 893. 280×

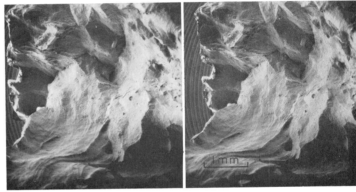

Fig. 895 Fracture at 602 °C (1116 °F), 10.3 MPa (1.5 ksi). 11.5×

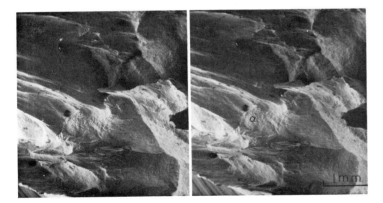

Fig. 896 Fracture at 642 °C (1188 °F), 7 MPa (1 ksi). 12×

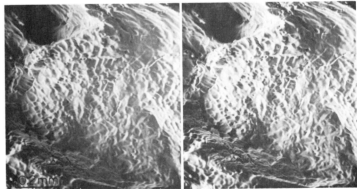

Fig. 897 Region at α in Fig. 896, at higher magnification. 60×

The specimen in Fig. 891 was tested at 525 °C (977 °F) and 10.3 MPa (1.5 ksi) and attained a steady-state strain rate of 5.0×10^{-7}/min, fracturing in 300 h, 6 min; its fracture surface has larger and smoother facets than those in Fig. 889 and 890. At top in this view is a region that fractured by transgranular rupture accompanied by plastic flow.

The specimen in Fig. 892 to 894 was tested at 540 °C (1004 °F) and 10.3 MPa (1.5 ksi) and attained a steady-state strain rate of 1.65×10^{-6}/min, fracturing in 67 h, 50 min. The fracture surface in Fig. 892 exhibits regions with a fine texture of dimples (such as at α and β), separated-grain facets (at top right) somewhat like those in Fig. 889, and highly plastically deformed features near the bottom edge of the large cavity at left center. Figure 893, from near the cavity in Fig. 892, shows different patterns of small dimples (at α and β) resulting from microvoid coalescence, and a region of extensive plastic flow (at bottom, adjacent to the large dimple). The dimples indicated by the arrow marked α are shown at still higher magnification in Fig. 894. The small arrows in Fig. 894 point to thermally etched networks of microstructural boundaries, which intersect the dimple boundaries and indicate that there is no relationship here of the dimple surfaces to crystallographic planes.

The specimen in Fig. 895 was tested at 602 °C (1116 °F) and 10.3 MPa (1.5 ksi) and attained a steady-state strain rate of 4.8×10^{-6}/min, fracturing in 40 h, 54 min; its fracture surface shows a region of intergranular fracture (above the center) similar to that in Fig. 889. Like the region at top in Fig. 891, the region in the lower half of Fig. 895 fractured by transgranular rupture accompanied by plastic flow.

The fracture surface of the specimen in Fig. 896, which was tested at 642 °C (1188 °F) and 7 MPa (1 ksi) attaining a steady-state strain rate of 2.0×10^{-6}/min and fracturing in 129 h, 24 min, displays one region with a fine texture of dimples (at α). Extreme plastic flow, particularly at the grain just to the right of center, is seen in stereo to project out like a knife edge. Figure 897 is a view at higher magnification of the region at α in Fig. 896, showing patterns of dimples of various sizes.

Copper Alloys / 401

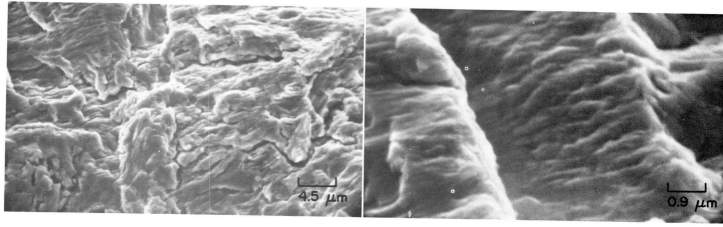

Fig. 898 Surface of a fatigue-test fracture in a notched specimen of OFHC copper that had undergone a 67% reduction in cross-sectional area by cold work before being tested in dry air. Region shown here is 3 mm (0.12 in.) from the notch, where slow crack growth caused irregular striations to form. Note the numerous secondary cracks. See also Fig. 899 to 902. SEM, 2200×

Fig. 899 Another region of the surface of the fatigue-test fracture in OFHC copper shown in Fig. 898, also 3 mm (0.12 in.) from the notch. At this higher magnification, the irregularity of the fatigue striations is clearly visible. There are no intergranular secondary cracks in this region, and there are no fissures between the fatigue striations. SEM, 11 000×

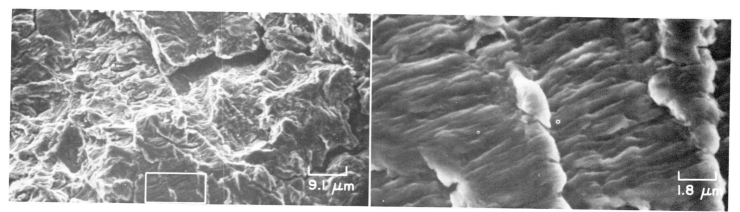

Fig. 900 A region of the surface of the fatigue-test fracture in OFHC copper shown in Fig. 898, 5 mm from the notch, where the rate of crack growth was greater than in the region in Fig. 898. Intergranular secondary cracks are clearly evident. Fatigue striations are not well resolved at this magnification. See Fig. 901 for a higher-magnification view of the area in the rectangle. SEM, 1100×

Fig. 901 Higher-magnification view of the area in the rectangle at the bottom of Fig. 900. Here, many of the fatigue striations have resulted in small fissures, but show the same irregularity displayed in Fig. 899; however, the steps visible here (at center and near right edge) appear to be much smaller than the one in Fig. 899. SEM, 5500×

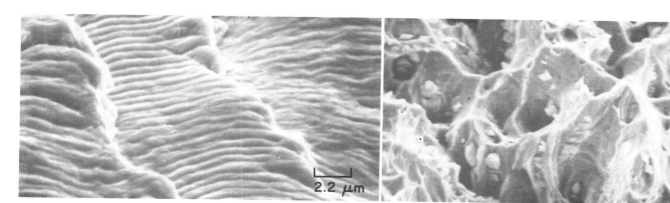

Fig. 902 Region of the fatigue-fracture surface in Fig. 898 to 901, 20 mm (0.8 in.) from the notch, where the rate of crack growth had increased considerably. The result of this high rate of crack growth is the array of very regular fatigue striations visible here, which are more typical of aluminum than of copper. SEM, 4600×

Fig. 903 The surface of a tensile-overload fracture in a specimen of free-machining copper, showing large dimples. Scattered throughout the structure are particles of Cu_2Te. The tellurium was added to the alloy to improve its machinability (the particles of telluride act as chip breakers). SEM, 1390×

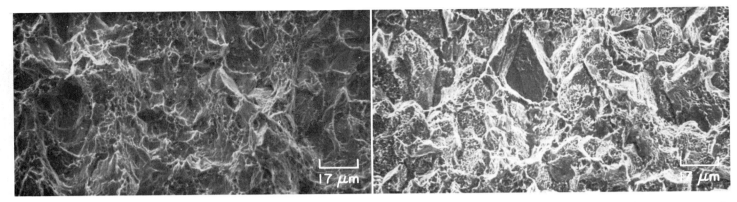

Fig. 904 Fracture surface of an underaged fracture-toughness test specimen of Cu-2.5Be alloy that had been aged for 1½ h at 260 °C (500 °F) prior to being tested in air. Tensile strength was 930 MPa (135 ksi). Fracture was transgranular and produced the wide variety of dimple sizes shown here. There are some large pores, but there is little evidence of secondary cracking. Compare with Fig. 905. SEM, 600×

Fig. 905 Fracture surface of a fully aged fracture-toughness test specimen of Cu-2.5Be similar to that in Fig. 904, but aged 3 h at 315 °C (600 °F) before being tested in air. Tensile strength was 1240 MPa (180 ksi). The dimples on the transgranular facets are much finer than in Fig. 904, and there is a network of secondary cracks that appear to be intergranular. One facet (at top center) appears to have formed by quasi-cleavage. SEM, 600×

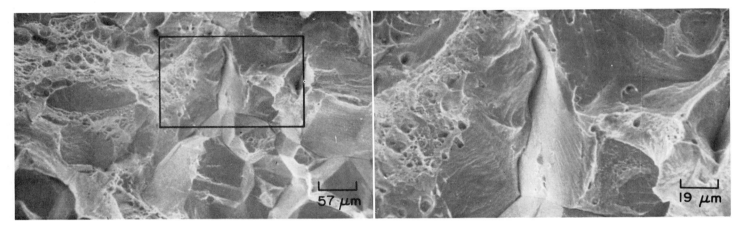

Fig. 906 Tensile-overload fracture in a fracture-toughness specimen of 64Cu-27Ni-9Fe alloy that underwent spinodal decomposition during heat treatment for 10 h at 775 °C (1425 °F). The surface contains many intergranular facets with intervening regions of dimpled transgranular facets. See Fig. 907 for a higher-magnification view of the area in the rectangle. SEM, 175×

Fig. 907 Higher-magnification view of the area in the rectangle in Fig. 906, revealing a fine structure of dimples, which resulted from microvoid coalescence, on the separated-grain facets. The regions of intergranular fracture here may have been caused by increased coarsening of the spinodal microstructure at the grain boundaries. SEM, 525×

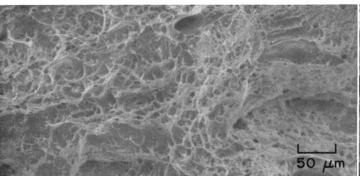
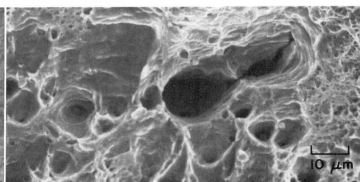

Fig. 908 Tensile-overload fracture in a fracture-toughness test specimen of the same 64Cu-27Ni-9Fe alloy as in Fig. 906, but here spinodal decomposition occurred during heat treatment at 775 °C (1425 °F) for 100 h. Only dimpled transgranular facets are visible (no intergranular facets); this may be due to a coarser spinodal structure resulting from the longer heat treatment. SEM, 200×

Fig. 909 Surface of the fracture in a fracture-toughness test specimen of the same 64Cu-27Ni-9Fe alloy as in Fig. 906 to 908, but which was heat treated at 775 °C (1427 °F) for 200 h. Very fine dimples can be seen among the larger ones. The large cavity at the center of this view is believed to have resulted from a localized separation at a grain boundary. SEM, 1000×

Copper Alloys / 403

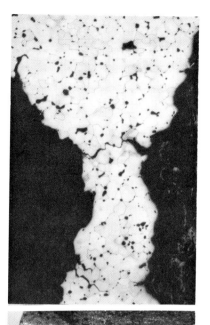

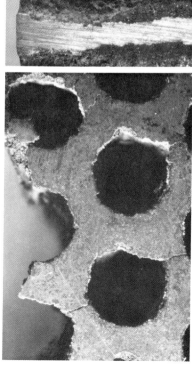

Fig. 910, 911, 912 Failure of a paper machine suction couch roll due to corrosion fatigue. The roll shell was made of centrifugally cast bronze (Cu-5Sn-5Pb-4Zn) similar to UNS C83600. Wall thickness at time of failure was approximately 45 mm (1.85 in.). Corrosion of base metal lining the drilled holes in the perforated shell decreased the widths of the ligaments between holes, which effectively increased the stress acting on the remaining ligament widths during normal roll operation. The stress increase combined with the corrosive service environment led to the initiation and propagation of corrosion-fatigue cracks. Fig. 910 (left): Outside diameter of bronze couch roll shell at location of primary crack (top). Note hole broadening due to corrosion and secondary cracking. 3.5×. Fig. 911 (center): Cross section through wall thickness. Note deposit buildups in drilled holes and severe under-deposit corrosion of the bronze ligaments directly below but not at the roll surface (top). 3.5×. Fig. 912 (right): Microstructure of cross section near and parallel to outside-diameter surface of couch roll (drilled holes at top and bottom). Note severe ligament corrosion and intergranular cracking, characteristics of corrosion fatigue in copper-base alloys. Potassium dichromate etch, 24.5×. (F.W. Tatar, Factory Mutual Research Corporation)

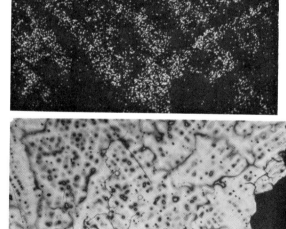

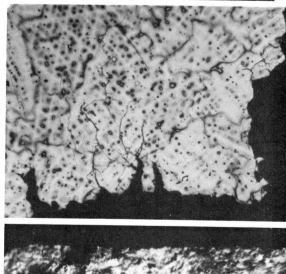

Fig. 913, 914 Failure of a paper machine suction press roll due to stress-corrosion cracking (SCC). The 90-cm (35 5/8-in.) diam perforated roll was made of a centrifugally cast tin bronze (90Cu-10Sn-1Pb) similar to UNS C90500. A circumferential crack was discovered in the shell and its rubber cover after more than 4 years of operation. The crack extended around 90% of the circumference. Significant pitting corrosion, cracks that initiated at the bottoms of the pits, and the transgranular mode of crack propagation all pointed to SCC as the cause of failure. Corrosion fatigue was ruled out because cracks initiated via this mechanism propagate intergranularly in copper-base alloys (see Fig. 910 to 912). Fig. 913 (left): Portion of crack face on a ligament between two drilled holes. Note transgranular fracture mode. 16×. Fig. 914 (right): Microstructure of cross section through crack face (bottom) of bronze shell. Note secondary cracking at bottoms of deep pits at drilled hole (left). Cracks are transgranular in the dendritic cast structure of the alloy and have many branches. Potassium dichromate etch, 33×. (F.W. Tatar, Factory Mutual Research Corporation)

Fig. 915, 916 Fracture of one of 1300 flange bolts that hold together sections of Philadelphia's William Penn statue. The 25-Mg (27-ton) statue is more than 100 years old and is located atop city hall, 150 m (500 ft) above ground. The bolt was made of a tin bronze (Cu-10Sn) similar to UNS C90500. It measured 19 mm (0.75 in.) in diameter and 75 mm (3 in.) long, and had been used in the as-cast condition. Date of fracture was not known. Initial examination of the fracture surface disclosed features resembling fatigue striations, giving rise to concern about the imminent failure of the remaining bolts. However, an in-depth analysis revealed that the cause of fracture was a coarse, unsound cast structure containing many voids. The "striations" that suggested a fatigue mechanism were actually the result of long-term preferential corrosion by the urban atmosphere of the exposed, heavily cored structure of the fracture face. Fig. 915 (left): Fracture surface of broken bronze bolt. Rippled appearance is not due to fatigue, but is instead the dendritic structure of the bronze as revealed via preferential corrosion (etching) by the atmosphere. SEM, 140×. Fig. 916 (right): Tin distribution map of field in Fig. 915 proves that "striations" really reflect the microstructure of the cast metal. 140× (J.E. Hanafee, Lawrence Livermore National Laboratory)

404 / Copper Alloys

Fig. 917 View of bottom of a Fourdrinier wire cloth of phosphor bronze C (C52100) that was removed from service after 14 days of operation. Heavy wear had occurred on both warp and shute (weft) wires. The fractures in the wires apparently occurred when the wires had worn almost completely through. The warp wires are shown horizontal in this view. 10×

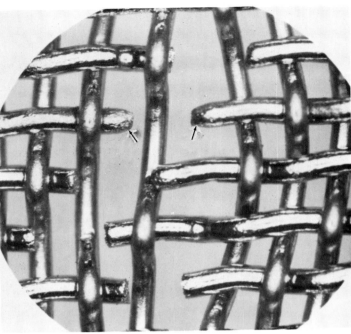

Fig. 918 Top view of a break in a Fourdrinier wire cloth of phosphor bronze C, showing fractures in shute (weft) wires. All but one of the fractures occurred by fatigue; the exception (at arrows) shows the necked-down profile of a tension fracture. ~50×

Fig. 919 Longitudinal section through a tension fracture in a warp wire of a Fourdrinier wire cloth of phosphor bronze C. Slip lines are visible throughout the microstructure. Plastic flow and reduction in area at the fracture (top) were attributed to excessive warp tension. $FeCl_3$ etch, 265×

Fig. 920 Longitudinal section through a fatigue fracture in a shute (weft) wire of Fourdrinier wire cloth of 80Cu-20Zn brass with 0.5% Sn added. The profile of the fracture is shown at top. Note corrosion pitting visible on the wire surface at left and at right. $FeCl_3$ etch, 265×

Cast Aluminum Alloys / 405

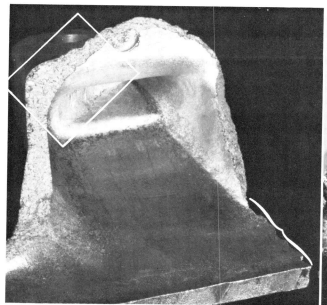

Fig. 921 A portion of a fractured carrier tray sand cast of an aluminum alloy intended to be 356.0-T6. Chemical analysis revealed that the copper and zinc contents were of an order of magnitude above the specified limits. The silicon content was approximately as dictated by the specification. The hardness measured 82 HB—appreciably above the 70 HB expected. Tensile strength was 179 to 193 MPa (26 to 28 ksi)—slightly below the specified minimum of 207 MPa (30 ksi). See Fig. 922 for the area in the rectangle and Fig. 923 for bracketed area at a magnification of 3×. Actual size

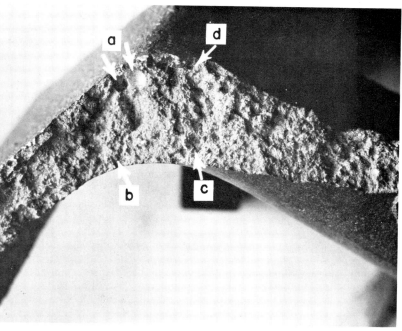

Fig. 922 Fracture surface of the shoulder of the sand-cast carrier tray shown in the rectangle in Fig. 921. The appearance of this surface suggests that fracture occurred in a brittle manner. The two arrows at "a" indicate two large gas cavities. The arrows at "b," "c," and "d" identify sites that were examined with a scanning electron microscope. (SEM fractographs of these sites are shown in Fig. 924, below, and Fig. 925, on the next page.) Note the absence of shear lips on the edges of this fracture surface, indicating a general lack of ductility. 3×

Fig. 923 This fracture surface is from the bracketed edge of the sand-cast carrier tray in Fig. 921. Like the surface shown in Fig. 922, the one here exhibits the characteristics of brittle fracture, there being no sign of a shear lip at any of these edges. Again, there is evidence of porosity; the arrows point to quite large gas cavities. The cavities could have been caused by a reaction between the liquid metal and organic binders in the sand mold, or could have resulted from entrapped air being carried into the mold by undue turbulence in pouring the casting. 3×

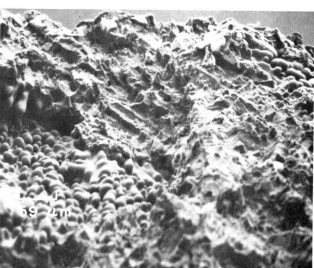

Fig. 924 Area at the lower edge of the fracture in Fig. 922, as indicated there by arrow at "b." Two regions of shrinkage porosity, at upper right and lower left, contain dendrite lobes, which existed as "free surfaces" within the cavities after the available liquid metal solidified. This behavior can be caused by an excessively high pouring temperature (which exaggerates total contraction in freezing) or by failure to provide proper risers. SEM, 170×

406 / Cast Aluminum Alloys

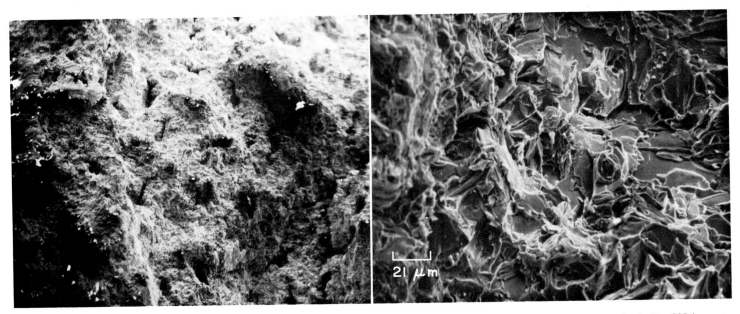

Fig. 925 Area at arrow "c" near the lower edge of the fracture in Fig. 922. At the low magnification here, it is not possible to identify details of the mechanisms of fracture. However, this view does show many cavities, some of them resembling dark slits, indicating quite generally distributed shrinkage porosity. Smeared region at upper left indicates mechanical damage—probably in handling of the carrier tray after fracture occurred. See also Fig. 926, at right. SEM, 18×

Fig. 926 A portion of the same fracture area as that in Fig. 925 (area at arrow "c" in Fig. 922), as seen at higher magnification. This view shows a complicated array of facets separated in many instances by indications of isolated dimples. The degree of flow in microvoid coalescence, however, did not contribute materially to ductility, and overall fracture was brittle in nature. This region does not contain any of the shrinkage porosity seen elsewhere in this fracture, but many contain nonmetallic inclusions. SEM, 475×

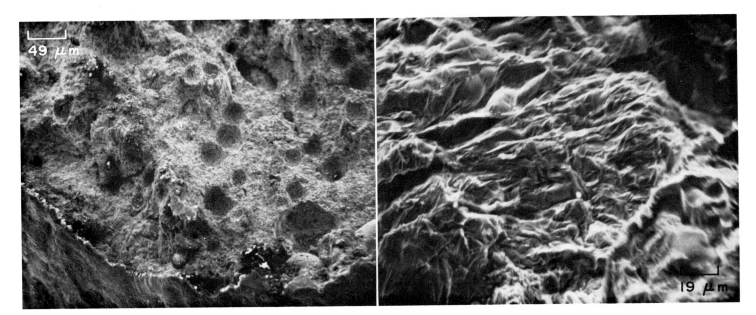

Fig. 927 This SEM view was taken in the area marked by arrow "d" in the fracture shown in Fig. 922. This area exhibits distribution of very small cavities, believed to be gas pockets, that are of a much different order of magnitude from those indicated by arrows at "a" in Fig. 922. The latter were about 1.5 mm (0.06 in.) in diameter, whereas the cavities here are about 25 μm across. At bottom and left, a small portion of the exterior surface of the casting is visible. See Fig. 928 for a view of this same fracture surface area at higher magnification. 205×

Fig. 928 Enlarged view of a part of the fracture surface area in Fig. 927. Some features vaguely resemble fatigue striations, but it is doubtful that they were caused by fatigue inasmuch as they do not seem to have been formed in an orderly sequence (and there are no beach marks in Fig. 922 and 923). Instead, the surface is very tumbled, with no suggestion of fatigue patches, and many regions bear whiskerlike features resembling some form of cleavage step. Lacking a significant amount of dimples, this surface can be rated as a brittle fracture. SEM, 520×

Cast Aluminum Alloys / 407

Fig. 929 A broken extension-housing yoke, part of a helicopter tail-rotor drive assembly, that fractured by fatigue in service. The fracture occurred through the left and right lugs at the points of attachment to the tail boom. The yoke was cast from aluminum alloy 356.0 and heat treated to the T6 temper. Chemical analysis showed that the composition of the yoke conformed to specifications, except for silicon, which was 0.05% below the required minimum. Hardness measurements ranged from 48 to 52 HRB (or about 81 to 85 HB), which was above the required minimum. Details of the fractures are shown in Fig. 930 to 938. Actual size

Fig. 930, 931 Both sides of each of the two lugs that were broken from the extension-housing yoke shown in Fig. 929. The inboard sides of the lugs are shown in Fig. 930 (left); the outboard sides, in Fig. 931 (right). Mechanical damage and loss of paint are evident. See also Fig. 932 and 933. Both at 2×

Fig. 932 Views of (at top) fracture surfaces of the upper lug in Fig. 930 and 931 and (at bottom) the mating surface of the yoke. Yoke surface is greatly deformed by repeated impacts, which occurred after fracture. 3×

408 / Cast Aluminum Alloys

Fig. 933 Views of (at top) fracture surfaces of the lower lug in Fig. 930 and 931 and of (at bottom) the mating surfaces of the yoke. Note discolored quadrant at top left corner of the bore of the lug (see also Fig. 934) and corresponding area in yoke fracture adjacent to bottom left corner of the bore. Edge of the discoloration is the beach mark of the fatigue crack that was the origin of final fracture. 3×

Fig. 934 Enlarged view of one fracture surface of the lug in Fig. 933, more clearly showing the fatigue-crack region at top left corner of the bore. Crack was nucleated at corner marked by arrow at "O." Fatigue area extended approximately to dashed line. See also Fig. 935 and 936. 8×

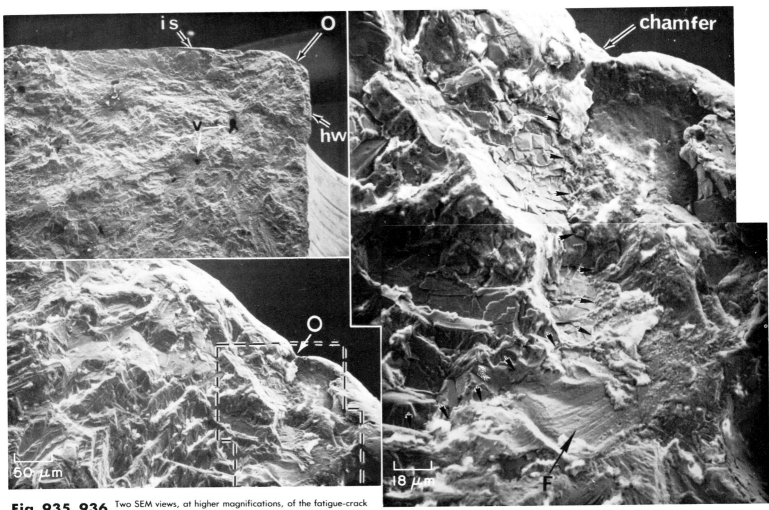

Fig. 935, 936 Two SEM views, at higher magnifications, of the fatigue-crack area of the lug fracture surface in Fig. 934. In the upper view (Fig. 935), arrow at "O" denotes the crack nucleus in the chamfer, arrow at "is" points to inboard side, arrow at "hw" indicates hole wall, and arrows at "v" point to voids. The lower view (Fig. 936) shows the area near the origin of the crack in Fig. 935 at higher magnification. In general, the crack propagated radially from point "O." See also Fig. 937. Fig. 935: 20×. Fig. 936: 200×

Fig. 937 This is the area within the dashed lines in Fig. 936, shown at higher magnification. The initial crack opened at the chamfer (upper right corner). The onset of fatigue fracture is indicated by the small arrows running from upper middle to lower left. The arrow at F points to an area of typical fatigue striations. The initial phase of fracture was brittle. SEM, 550×

Cast Aluminum Alloys / 409

Fig. 938 Fractured bell-crank fitting of cast aluminum alloy 356.0-T6. The fitting, which was from an aircraft rear horizontal elevator, fractured in a crash. No crack origin was found. See also Fig. 939 to 952. ~1.5×

Fig. 939 A more highly magnified view of the fracture in Fig. 938. The appearance here is typical of the entire fracture surface. There is little evidence of surface flow. See also Fig. 940. SEM, 60×

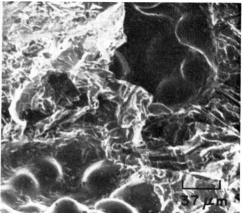

Fig. 940 Many of the features of the fracture surface in Fig. 939 are associated with shrinkage cavities of the type shown here. The round knobs are exposed secondary arms of dendrites. SEM, 270×

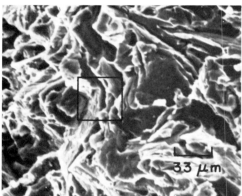

Fig. 941 A cellular region that was found to be characteristic of the fracture surface in Fig. 938. Although some features faintly suggest dimples, the higher-magnification view in Fig. 942 reveals no dimples. SEM, 300×

Fig. 942 Higher-magnification view of the area in the rectangle in Fig. 941. There are no suggestions of dimples visible here; instead, the surface exhibits platelets that have been bent and cracked. SEM, 1500×

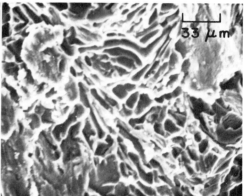

Fig. 943 View at another location in the fracture surface in Fig. 938. Cellular structure shown here and in Fig. 941 is attributed to the microstructure resulting from the freezing patterns and heat treatment of the casting. SEM, 300×

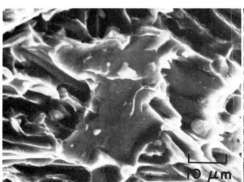

Fig. 944 Another view of the fracture surface shown in Fig. 938. Again, no evidence of true dimples, but cellular structure only. There are many small inclusions on the surface; most are not bonded to the matrix. SEM, 1000×

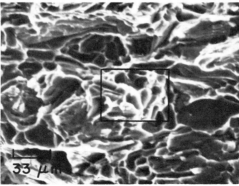

Fig. 945 An overload fracture in a miniature tensile-test specimen cut from the fitting in Fig. 938, displaying the same type of cellular surface structure produced by the service fracture. See also Fig. 946. SEM, 300×

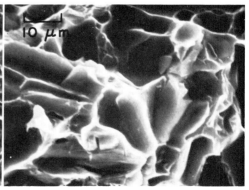

Fig. 946 Higher-magnification view of the area in the rectangle in Fig. 945. Although the features here strongly resemble those shown in Fig. 942, close scrutiny reveals the presence of tiny dimples in the edges of some cells. SEM, 1000×

410 / Cast Aluminum Alloys

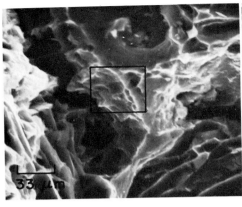

Fig. 947 View of the fracture surface shown in Fig. 945, but at another location. In this area, there are small, clearly visible dimples clustered in the central, outlined region, which is shown at higher magnification in Fig. 948. SEM, 300×

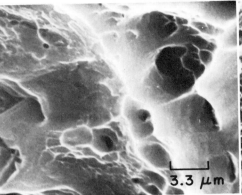

Fig. 948 Higher-magnification view of the area in the rectangle in Fig. 947, showing small dimples of several sizes and suggesting that the service fracture shown in Fig. 938 (with no dimples) was not the result of simple overload. SEM, 3000×

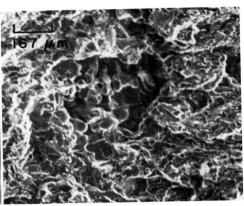

Fig. 949 Surface of a fracture in a miniature impact-test specimen cut from the broken aluminum alloy 356.0-T6 bell-crank fitting shown in Fig. 938. The appearance here is very similar to that of the service fracture in Fig. 939. SEM, 60×

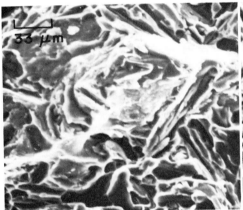

Fig. 950 View of the fractured test specimen in Fig. 949, but at another location. The appearance here is quite similar to that of the service fracture in Fig. 943. SEM, 300×

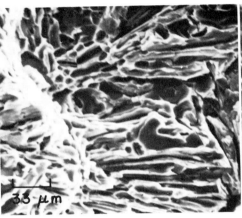

Fig. 951 Another view of the fracture surface shown in Fig. 949 and 950. Compare with the view of the service fracture in Fig. 941 and note the marked resemblance. SEM, 300×

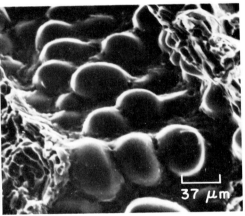

Fig. 952 View of fracture surface in Fig. 949 to 951, displaying a shrinkage cavity. Cavity is somewhat like those in Fig. 940, but with a much coarser dendrite cell structure. SEM, 270×

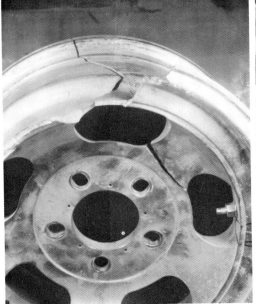
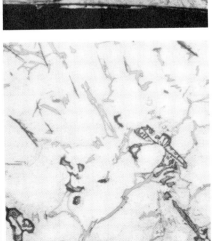

Fig. 953, 954, 955 Brittle fracture of die cast aluminum alloy 380.0 wheel. Failure initiated during normal operation at a cold shut defect in the as-cast part. Fig. 953 (left): Wheel after failure. Distance between fracture surfaces indicates that residual stresses were present. Fig. 954 (top right): Portion of fracture surface showing cold shut defect. Fig. 955 (bottom right): Presence of sharp-edged silicon (light gray) in the microstructure of the aluminum alloy is evidence that heat treating to increase toughness and relieve residual stresses was not performed. 0.5% HF etch, 400× (R.W. Bohl, University of Illinois)

Cast Aluminum Alloys / 411

Fig. 956, 957, 958 Fractures in a die cast aluminum alloy 518.0 housing of a chain-hoist hook, which is believed to have broken by overload in service. Figure 956 (left) is a view of the reassembled hook components, showing a transverse fracture (arrows) in the top surface of the housing. This fracture, which preceded and instigated final fracture of the housing, was above and parallel to a steel retaining pin (shown in Fig. 958, right), which secured the hook to the hoist chain. Figure 957 (center) is a side view of the housing, which consisted of two pieces held together by the retaining ring visible at top, showing the final fracture (roughly horizontal), the deformed retaining-pin hole (elongation of about 3% is indicated—a reasonable value for this alloy), and one end of the fracture in the top surface. See also Fig. 959 and 960. Fig. 956: ~1.1×. Fig. 957: 1.7×. Fig. 958: 2.4×

Fig. 959 Mating surfaces of the fracture in the side of the housing in Fig. 957. At right is the upper surface, with the retaining ring (shown at top in Fig. 957) still in place; visible in the bore of the retaining-pin hole is the fracture in the top surface of the housing (see Fig. 956). At left is the lower surface, with the retaining pin in place inside a collar. ~2.25×

412 / Cast Aluminum Alloys

Fig. 960 Enlarged view of the lower left portion of the top fracture surface (shown at right in Fig. 959) of the broken die cast housing in Fig. 956. Note gas cavities (at arrows), which are common interior features of satisfactory die castings. Bending of the retaining pin (see Fig. 958) is believed to have occurred after the transverse fracture of the top surface of the housing caused relocation of the load-bearing points on the pin. Low-cycle fatigue would have been a likely cause of fracture, but no fatigue beach marks could be found. 7.5×

Cast Aluminum Alloys / 413

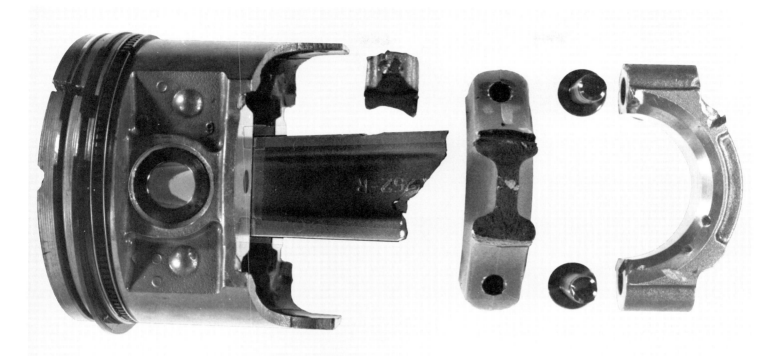

Fig. 961 Components of an experimental automotive connecting rod that fractured by overload during testing. (Shown also is the piston used in the test.) The cast rod had the following composition: 9.0% Si, 1.2% Fe, 3.7% Cu, 0.3% Mg, 0.5% Ni, 1.0% Zn, 0.2% Ti, remainder Al. Note that the shaft of the connecting rod suffered a compound fracture; the portion closest to the cap has a relatively straight, transverse fracture surface, and the portion connected to the piston shows a partially diagonal fracture, with the segment that broke away shown above the shaft. It is not known whether both fractures occurred in the shaft at the moment of initial fracture or whether the segment broke away later due to impact within the cylinder before the engine was shut off. The fracture surface visible here bears no features that in any way suggest a fatigue fracture. The 4340 steel bolts (left of cap at far right) also broke. See also Fig. 962 and 963. ~0.75×

Fig. 962 Fracture surface of the portion of the aluminum alloy connecting rod in Fig. 961 that was still attached to the piston after the test. It appears that each flange contains a chevron pattern, which suggests that there were two crack origins. This supports the theory that two successive fractures occurred in the rod. The mechanical damage in the right fracture surface indicates that the right origin probably led to the initial rupture. There are no indications of the beach marks that are characteristic of fatigue fractures. See also Fig. 963. 2×

Fig. 963 High-magnification view of one of the fracture surfaces of the fractured connecting rod shown in Fig. 961. It is evident that fracture was by dimpled rupture, which indicates that the fracture was caused by sudden overload. As in the views at about 0.75× in Fig. 961 and at 2× in Fig. 962, no indications of fatigue are visible here. The fracture appears quite "clean," with very little evidence of nonmetallic inclusions. SEM, 3000×

414 / Wrought Aluminum Alloys

Fig. 964, 965 Photograph and SEM fractograph showing the fracture surface of a small portable cylinder used for storage of helium gas under pressure that exploded while at rest in storage. The cylinder was approximately 70 mm (2¾ in.) in diameter by 250 mm (10 in.) long and had been formed from a 0.75-mm (0.03-in.) thick sheet of aluminum alloy 1100. Normal pressure in the cylinder, when full of helium, was 2 MPa (300 psi); maximum pressure, 4 MPa (600 psi). The cylinder broke when the bottom separated from the sidewall with explosive force, leaving a very even fracture surface; separation was probably at the very bottom of the sidewall. Figure 964 (left) shows the entire fracture surface of the sidewall at actual size. Figure 965 (right) is a magnified view (40×) of a portion of that fracture surface, showing several fracture levels, separated by offsets, containing crack arrests. Rupture began at the inside of the cylinder wall and terminated in a pronounced ductile shear lip (at top in Fig. 965). Note the secondary stress-corrosion cracks in the inside surface of the wall (at bottom in Fig. 965), which are parallel to the fracture surface. See also Fig. 966 and 967.

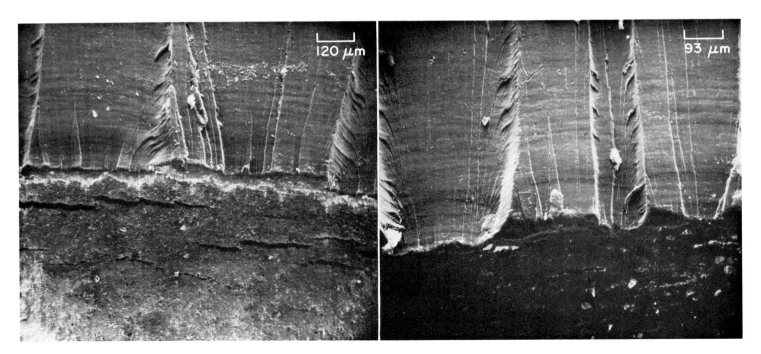

Fig. 966, 967 SEM views of two portions of the fracture surface of the cylinder in Fig. 964. These views, at higher magnification than Fig. 965, also show the crack arrests, the offsets between adjoining stress-corrosion-crack surfaces, and the numerous secondary cracks in the inside surface of the cylinder wall (at bottom) that are parallel to the fracture. In Fig. 967 (right), numerous corrosion pits are also visible. These pits are probably due to condensation of water vapor at the bottom of the cylinder that was carried by the helium gas. (Light-microscope examination revealed that cracks formed at bottoms of corrosion pits.) It is likely also that the forming operation created large stress concentrations at the junction between the sidewall and the bottom of the cylinder. Fig. 966: 83×. Fig. 967: 108×

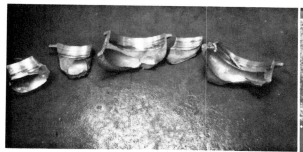

Fig. 968 Pieces of the hub of a forged aircraft main-landing-gear wheel half, which broke by fatigue. The material is aluminum alloy 2014-T6. Tensile specimens from elsewhere in the wheel had tensile strength of 493.7 MPa (71.6 ksi) and 8.9% elongation in the transverse direction, and tensile strength of 466.1 MPa (67.6 ksi) and 8.0% elongation in the longitudinal direction. See also Fig. 969. ~0.1×

Fig. 969 View of an area of one fracture surface of the broken hub in Fig. 968, showing the fatigue-crack origin. Visible are beach marks, which suggest that the origin is at the location marked by the arrow, presumably at one of several corrosion pits that were found on the surface of the hub. After having penetrated a short distance, the fatigue crack developed a step (dark facet); a second fatigue crack originated at one end of the step. 1.8×

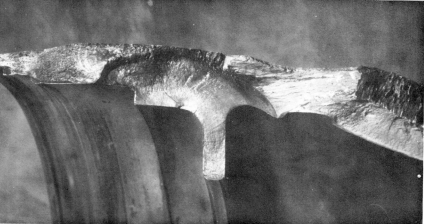

Fig. 970, 971 Figure 970 (left) shows the hub of a forged aluminum alloy 2014-T6 aircraft main-landing-gear wheel half, which broke in fatigue. A tensile specimen machined from the hub had tensile strength of 499.2 MPa (72.4 ksi), 12.1% elongation, and hardness of 143 to 150 HB, which are acceptable. The fatigue crack originated at the inside surface of the hub. Figure 971 (right) shows a fracture surface of the broken hub, showing the fatigue-crack origin. Clearly visible are beach marks, which indicate that the fracture began as a radial fatigue crack in a plane containing the axis of the wheel. Later, the crack turned to form a circumferential separation between the hub and web of the wheel half, as shown in Fig. 970. See also Fig. 972 and 973. Fig. 970: ~0.17×. Fig. 971: Actual size

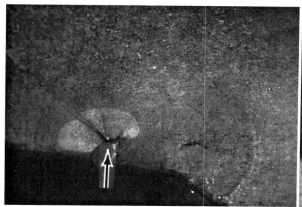

Fig. 972 Enlarged view of the fatigue-crack origin in Fig. 971, which plainly shows the region of initial penetration (light area). At arrow is a forging defect usually known as a bright flake. Note that grain flow is approximately parallel to the flake. ~5×

Fig. 973 Fracture surface of the broken hub in Fig. 970, showing an area of the circumferential separation between the hub and web. Bright flakes, similar to the defect that initiated the fatigue crack in Fig. 972, are visible at the arrows. These defects have been attributed to hydrogen damage. Actual size

416 / Wrought Aluminum Alloys

Fig. 974 Top surface of an extruded aluminum alloy 2014-T6 bottom cap of an aircraft wing spar, showing a fatigue fracture (center) that intersected one of the rivet holes indicated by the arrows. Hardness tests near the fracture gave an average value of 85 HRB, which is acceptable. See also Fig. 975 to 979. ~0.25×

Fig. 975 View of the fracture surface in Fig. 974. The rivet hole intersected by the fracture is abnormal, consisting of two overlapping holes (see Fig. 976). Beach marks, which are clearly visible, indicate that the fatigue crack began at the double-drilled rivet hole. 1.13×

Fig. 976 Higher-magnification view of the rivet-hole area of the fracture surface shown in Fig. 975. The double-drilled nature of the rivet hole is shown quite clearly here. Two fatigue cracks originated at this hole—one beginning at arrow A and growing to the left, and the other beginning at arrow B and growing to the right. 5×

Fig. 977 Polished and etched section through the cap in Fig. 974. At bottom is a fibrous interior structure typical of aluminum alloy extrusions. At top is one of the coarse-grained, recrystallized layers, 0.46 to 0.94 mm (0.018 to 0.037 in.) thick, at the top and bottom surfaces of the cap. Keller's reagent, 100×

Fig. 978, 979 Views of the top surface (Fig. 978, left) and bottom surface (Fig. 979, right) of the wing-spar cap in Fig. 974, with the mating fracture surfaces fitted together. The segment at right in each view was deformed after fracture, causing the gap. The double drilling of the rivet hole, noted in Fig. 975 and 976, is clearly shown (arrows). The edge distance, "e," was measured for each hole (as shown in Fig. 979 for the hole closer to the edge). Edge distance for the hole farther from the edge is about 8 mm (5/16 in.)—the minimum allowed for rivets of the size used here (4 mm, or 5/32 in. diam). For the closer hole, edge distance is about 75% of the minimum value, which increased the stresses in that section to excessive levels. Both at 5×

Wrought Aluminum Alloys / 417

Fig. 980 Fatigue fracture of an aluminum alloy 2014-T6 heat-treated forging. Details of the heat-treatment procedure were not available. Some machining was carried out on the forging prior to heat treatment. The aircraft structural component cracked in service. The horizontal lines on the fracture surface are grain boundaries. Fatigue striations are also visible, traversing the fracture face at roughly 60°. Note the absence of discontinuities at their intersections with the grain boundaries. SEM, 1800× (F. Neub, University of Toronto)

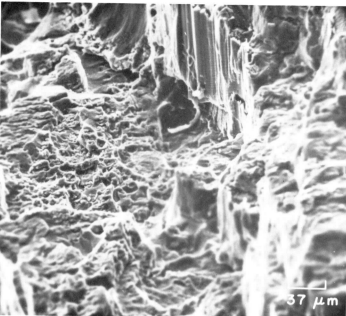

Fig. 981 Fracture surface of a fatigue-test specimen of aluminum alloy 2024-T3, showing a portion of the region of final fast fracture. Stress-intensity range (ΔK) was 21 MPa$\sqrt{m}$ (19 ksi$\sqrt{in.}$); the stress was applied in an argon atmosphere at room temperature at a frequency of 10 cps. The area has voids that may be moderate-size dimples. The vertical face is apparently a very large tear ridge or cleavage step joining two areas of dimpled rupture. See also Fig. 982 and 983. SEM, 270×

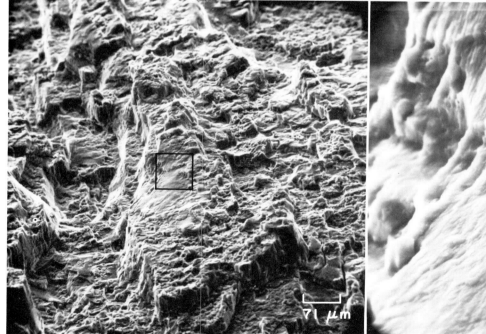

Fig. 982 A different area of the fracture surface shown in Fig. 981. This also exhibits vertical faces that are tear ridges. The surface is covered with voids, but at this low magnification it is not possible to decide whether or not they are dimples. The smooth central area outlined by the rectangle is shown enlarged in Fig. 983. SEM, 140×

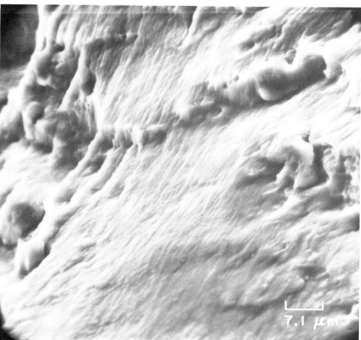

Fig. 983 A view of the rectangle-outlined area of the fracture surface in Fig. 982, at higher magnification. With this enlargement, it is evident that the area is not truly smooth, but rather that it bears a uniform array of extremely fine fatigue striations. Near the right edge is a small area of minute dimples. SEM, 1400×

418 / Wrought Aluminum Alloys

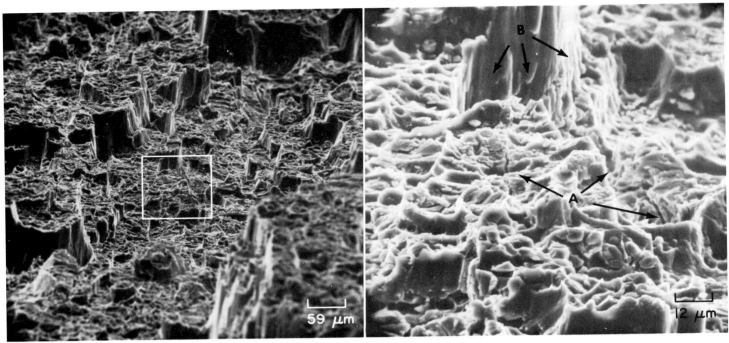

Fig. 984 Fracture surface of a fatigue-test specimen of aluminum alloy 2024-T3 tested at 23 °C (73 °F) in argon. The fatigue crack, similar in appearance to the one in Fig. 982, was produced by a stress-intensity range (ΔK) of 24.8 MPa$\sqrt{m}$ (22.6 ksi$\sqrt{in.}$) at a frequency of 10 cps. Much of the surface shows features resembling dimples, but the vertical "cliffs" are probably delaminations along grain boundaries. See Fig. 985 for a higher-magnification view of the area in the rectangle. SEM, 170×

Fig. 985 Area outlined by the rectangle in Fig. 984, as seen at higher magnification. This view provides a much clearer delineation of the fine details of the fracture surface and shows a combination of dimpled rupture and grain-boundary separation. Intergranular secondary fissures such as those marked by the arrows at A led to the formation of the vertical "cliffs" shown here (arrows at B) and in Fig. 984. Cracked and broken inclusions are visible at many locations. SEM, 850×

Fig. 986 Fracture surface of a fatigue-test specimen of aluminum alloy 2024-T3 that was tested in an environment of a 3.5% solution of NaCl in water. The stress-intensity range (ΔK) was 19.8 MPa$\sqrt{m}$ (18 ksi$\sqrt{in.}$) at 10 cps. The central region of this view contains patches of well-defined fatigue striations. In adjacent regions, there appear to be faintly defined striae that have been obscured by corrosion. In other regions, it is uncertain whether fatigue or cleavage was active. See Fig. 987 for area in rectangle. SEM, 260×

Fig. 987 View at higher magnification of the area in the rectangle in Fig. 986, showing the fatigue striations in finer detail. Note that superimposed on the fine striations at somewhat irregular intervals is a system of fissures, or perhaps more pronounced striations; the presence of these features may reflect either a repetitive variation in strain amplitude or stress, or periodic interruptions in the applied stress cycle (which allowed locally increased corrosion), or both. SEM, 1320×

Wrought Aluminum Alloys / 419

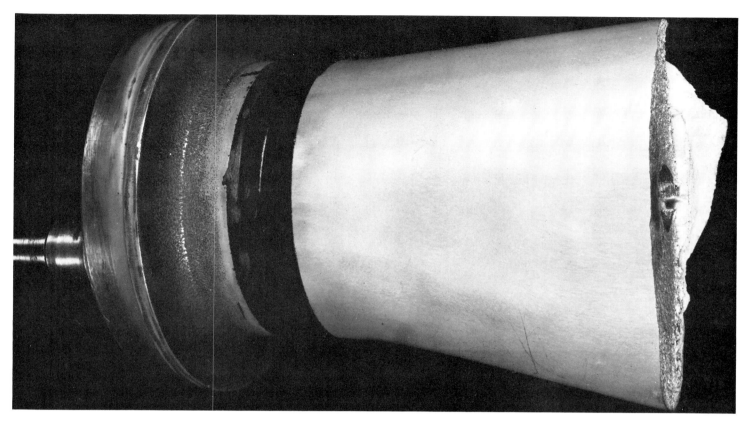

Fig. 988 View of the shank end of a fractured aircraft propeller blade fabricated of aluminum alloy 2025-T6. The blade broke by fatigue, which originated at an interior cavity that was provided to contain a balance weight comprised of compacted lead wool. Chemical analysis established that the blade was within specified composition limits. Hardness measurements (500-kg load) yielded an average value of 107 HB, which was above the required minimum of 100 HB. See also Fig. 989. Actual size

Fig. 989 Fracture surface of the shank end of the broken aircraft propeller blade in Fig. 988. The balance-weight cavity is visible at center, with the fatigue-crack origin at the upper edge (arrow). The fracture originated at the beginning of the radius that formed one end of the cavity. Examination of the cavity surface revealed severe roughness caused by tool marks and by corrosion pits. The combined effect of these tool marks and corrosion pits was considered to be the cause of crack initiation. 1.75×

420 / Wrought Aluminum Alloys

Fig. 990, 991, 992 Figure 990 (left) shows the surface of a fatigue fracture near the hub of an aluminum alloy 2025-T6 aircraft propeller blade. The fracture originated in a shot-peened fillet. Small fatigue cracks joined to form the main crack at A, which propagated to B-B and C-C before final fast fracture occurred. Figure 991 (center) shows a portion of the outside edge of the fracture surface in Fig. 990 between the arrows marked D, showing small, distinct fatigue cracks (at arrows) that had been present before final fast fracture. Figure 992 (right) is a view of the shot-peened fillet of a companion propeller blade, showing small fatigue cracks. Depth of the cold-worked layer produced by shot peening was nonuniform and averaged about 0.038 mm (0.0015 in.), instead of the stipulated 0.14 mm (0.0055 in.) minimum, which afforded inadequate surface fatigue strength. See also Fig. 993 to 995. Fig. 990: 0.75×. Fig. 991: 20×. Fig. 992: Actual size

Fig. 993 Surface of the shot-peened fillet of another companion propeller blade to that in Fig. 990, showing the same type of fatigue cracks as those in the propeller blade shown in Fig. 992. These cracks were present in large numbers in the fillet area. 20×

Fig. 994 A polished and etched section through the shot-peened fillet of the fractured blade in Fig. 990, showing two small fatigue cracks pulled open and blunted by plastic deformation. Note slip bands, which indicate that a slight amount of plastic flow occurred near cracks. Etched in Keller's reagent, 100×

Fig. 995 Another polished and etched section through the shot-peened fillet of the fractured blade in Fig. 990, showing a fatigue crack that grew from a surface flaw. The structure indicates that only very superficial shot peening occurred here. Etched in Keller's reagent, 500×

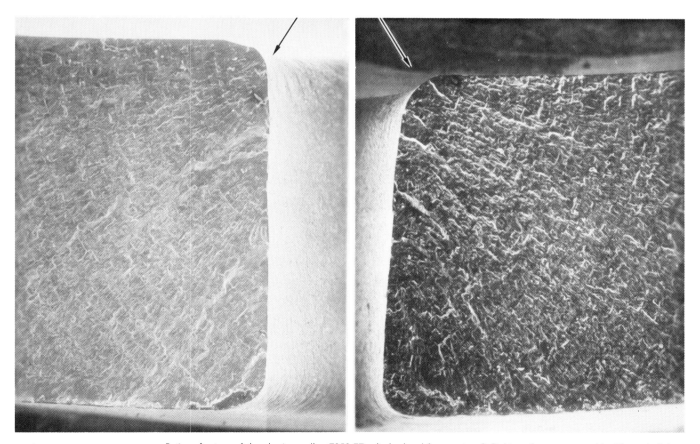

Fig. 996, 997, 998 Fatigue fracture of the aluminum alloy 7050-T7 cylinder head for an aircraft flight spoiler servo assembly. The end of the cylinder cracked after 1.25 million cycles in the endurance qualification test for the servo assembly. Five million cycles were required. Fracture initiated at the chamfers on the inside diameter of a drain hole. Striations were observed within 18 μm (0.0007 in.) of the origin, indicating that fracture was caused by excessively high cyclic stresses. Source of the high stress was traced to an error in the computer-aided design (CAD) software. Fig. 996 (top): Fracture surface with drain hole at bottom center. Note beach marks. "O" and corresponding arrows indicate fatigue origins. 2.5×. The as-received cylinder head is shown in the inset. Fig. 997 (bottom left) and 998 (bottom right): Close-ups of fatigue origins (arrows) in Fig. 996. SEM, 26× and 23× (W.L. Jensen, Lockheed-Georgia Company)

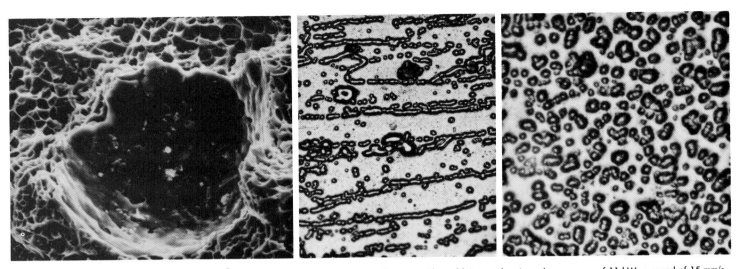

Fig. 999, 1000, 1001 Fractography of a laser beam weld in aluminum alloy 5456. The weld was made using a beam power of 11 kW, a speed of 15 mm/s (35 in./min), and a heat input of 0.74 kJ/mm (18.9 kJ/in.). Ductile fracture of the dynamic tear (DT) test sample occurred in the fusion zone due to softening of the zone and the presence of pores. Fig. 999 (left): Fracture surface of aluminum alloy 5456 DT sample. Large pore is surrounded by microvoids. Failure mode was microvoid coalescence. SEM, 850×. Fig. 1000 (center): Microstructure of alloy 5456 base metal. Elongated second-phase structures indicate rolling direction and strain-hardened nature of the material. Hardness: 87 HV. Etched in A-2 solution of Knuth system, 500×. Fig. 1001 (right): Microstructure of fusion zone where fracture occurred. No elongation of second phases is evident. Hardness: 80 HV. Etched in A-2 solution of the Knuth system, 1000× (E.A. Metzbower and D.W. Moon, Naval Research Laboratory)

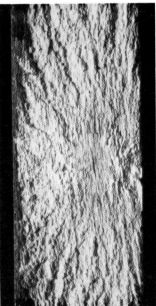

Fig. 1002, 1003 Fatigue fracture of an aluminum alloy 7175-T736 forging. The aircraft main landing gear component failed during a structural fatigue test. Cracking initiated at a dross inclusion at the surface of the part. Fig. 1002 (left): Portion of fracture surface of aluminum alloy 7075-T736 forging. Dross inclusion is at top. Note its spongy appearance. SEM, 300×. Fig. 1003 (right): High-magnification view of dross inclusion in Fig. 1002. It consists of octahedral spinel crystals ($MgAl_2O_4$) embedded in a spongy cluster of granulated oxides. SEM, 10 000× (F. Neub, University of Toronto)

Fig. 1004 An aluminum alloy 7075-T736 aircraft main landing gear forging, similar to that described in Fig. 1002 and 1003, which was shot peened on its inner-diameter surface to enhance fatigue resistance. The shot-peened part withstood cycles far beyond the number required for acceptance. One effect of peening was to drive the fracture-initiation site to a location well beneath the surface of the forging. The dross inclusion that was the origin of this fracture is the thin black line in the center of the nearly circular fatigue crack initiation area. 4.5× (C. Bryant, De Havilland Aircraft Company of Canada Ltd.)

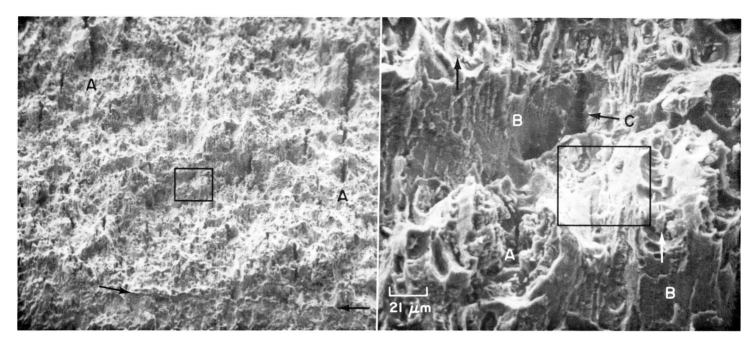

Fig. 1005 Fracture surface of a fracture-toughness test specimen of aluminum alloy 7075-T6, showing the zone of transition from the fatigue-precrack region (below arrows) to the tension-overload plane-strain fracture region (above arrows). Specimen was aged 24 h at 120 °C (250 °F); tensile strength was 593 MPa (86 ksi), and uniform elongation was 13%. Note the appreciable number of vertical delaminations (as at A's), which probably are grain-boundary separations, caused by the transverse tensile stress in the plane-strain region. Area in rectangle is shown enlarged in Fig. 1006. SEM (gold shadowed), 50×

Fig. 1006 Area outlined by the rectangle in Fig. 1005, as seen at ten times the magnification there. It is apparent that the matrix contained many second-phase particles that have undergone brittle fracture (arrows). The surface is quite complex, with some regions that appear to show clusters of minute dimples (A) and other regions that strongly resemble intergranular fracture (B's). At C is one of the numerous vertical fissures visible in Fig. 1005, presumably elongated during tension overload. See Fig. 1007 for a higher-magnification view of area in rectangle here. SEM (gold shadowed), 480×

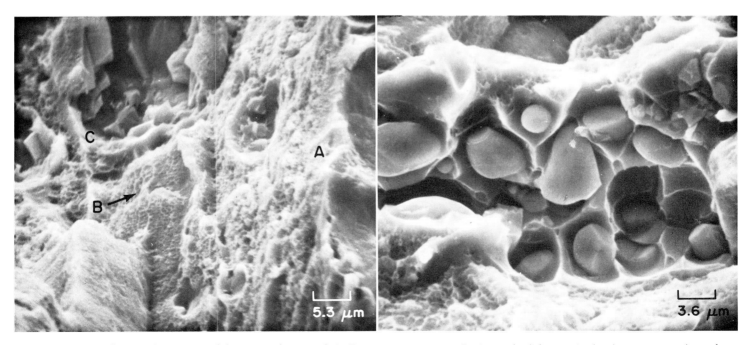

Fig. 1007 Higher-magnification view of the area in the rectangle in Fig. 1006. At A is a region of extremely fine dimples, which are slightly out of focus. At B, there appear to be grain-boundary facets bearing small, very shallow dimples. At C are fragments of a number of second-phase inclusions, many of which have separated from their original pockets in the matrix. SEM (gold shadowed), 1900×

Fig. 1008 Tension-overload fracture in the short-transverse plane of a specimen of aluminum alloy 7075-T6. At top and bottom are regions of quite small dimples. In the central portion of this view are large pockets in which the cleaved facets of intermetallic inclusions are visible. These inclusions, which are rich in iron and silicon, are only slightly bonded to the walls of the pockets. SEM, 2800×

424 / Wrought Aluminum Alloys

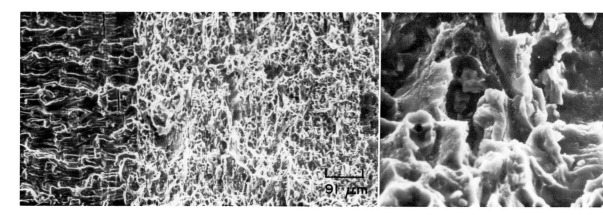

Fig. 1009 Specimen of aluminum alloy 7075-T6 broken in a slow-bend fracture-toughness test in air. Dark area, at left, is the fatigue-precrack surface, which shows no fatigue striae. At right is the tension-overload fracture surface; here, the surface appears dimpled, although this is uncertain at only 110×. See also Fig. 1010. SEM, 110×

Fig. 1010 A portion of the tension-overload fracture surface in Fig. 1009, as seen at ten times the magnification there, which reveals fine details. A number of medium-size dimples contain inclusions, some of which are broken. In contrast are the minute dimples visible at left and right in the upper portion of this view. SEM, 1100×

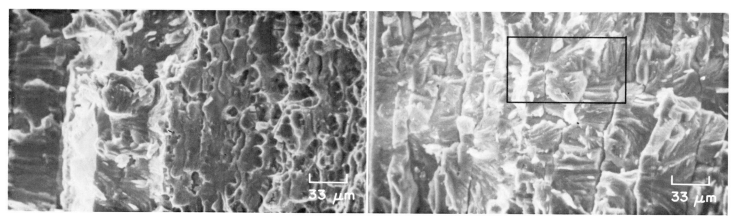

Fig. 1011 View of the transition between a slow-bend fracture induced in mercury vapor (left of center) and one occurring in air (right of center) in a specimen of aluminum alloy 7075-T6. The fracture induced in mercury vapor appears to have occurred by cleavage; the fracture in air exhibits dimples, which increase in frequency toward the right side of this fractograph. SEM, 300×

Fig. 1012 Fracture surface of an aluminum alloy 7075-T6 specimen broken in a slow-bend fracture-toughness test in mercury vapor. Obviously, this is a completely brittle fracture. Many facets show faint characteristics of quasi-cleavage. Note the large number of secondary cracks, possibly at grain boundaries, all of which are more or less parallel. Area in rectangle is shown enlarged in Fig. 1013. SEM, 300×

 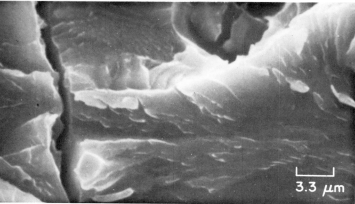

Fig. 1013 Higher-magnification view of the surface contours in the rectangle that is shown in Fig. 1012. This reveals the cleavage river patterns quite clearly; note the changes in fracture direction at right. This view also reveals the depth and continuity of the secondary cracks. See also Fig. 1014. SEM, 1000×

Fig. 1014 Another view, at still higher magnification, of the fracture surface in Fig. 1012 and 1013. At top is what appears to be a pocket holding a cracked inclusion. The remainder of the surface shows cleavage facets plus projections that are somewhat similar to the tongues seen in cold fractures of iron. SEM, 3000×

Wrought Aluminum Alloys / 425

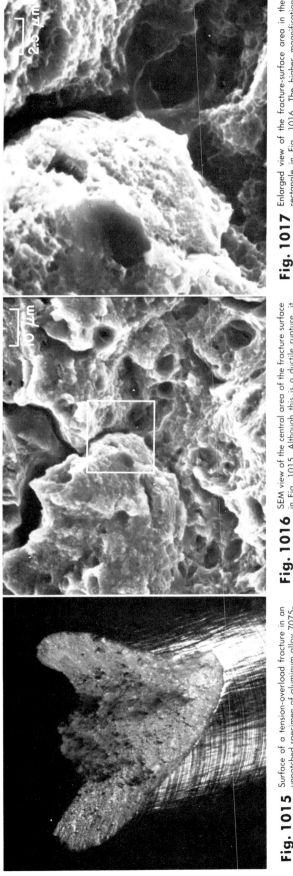

Fig. 1015 Surface of a tension-overload fracture in an unnotched specimen of aluminum alloy 7075-T6 having a tensile strength of 520 MPa (75 ksi), with 22% reduction of area. Surface is coarsely fibrous; shear lip has formed two opposing lobes. See also Fig. 1016 and 1017. 9×

Fig. 1016 SEM view of the central area of the fracture surface in Fig. 1015. Although this is a ductile rupture, it contains very deep secondary cracks. The major pores were sites of alloy second-phase particles that are no longer in place. See Fig. 1017 for an enlarged view of the area in the rectangle. 1000×

Fig. 1017 Enlarged view of the fracture-surface area in the rectangle in Fig. 1016. The higher magnification here makes it possible to observe the fine size of the dimples, which in general is less than half a micron. This is a remarkable uniformity of size in a fracture surface as rough as this one. SEM, 4000×

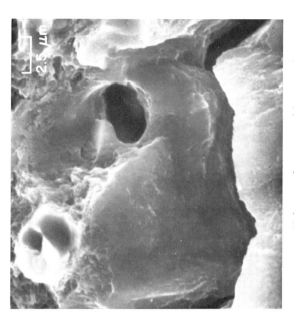

Fig. 1020 Higher-magnification view of the area in the rectangle in Fig. 1019. This fourfold enlargement makes visible the fine dimples surrounding the particle sites. Note that the surfaces of the secondary grain-boundary cracks are remarkably smooth. SEM, 4000×

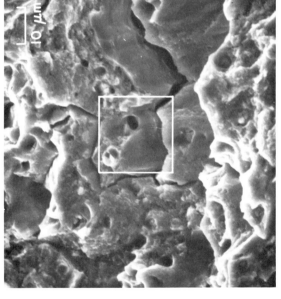

Fig. 1019 SEM view of the interior of the fracture surface in Fig. 1018. This shows that in general the secondary cracking is intergranular, although the primary rupture is not. Numerous pores contain alloy second-phase particles. See also Fig. 1020. 1000×

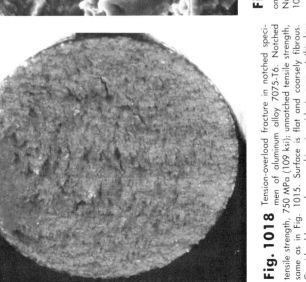

Fig. 1018 Tension-overload fracture in notched specimen of aluminum alloy 7075-T6. Notched tensile strength, 750 MPa (109 ksi); unnotched tensile strength, same as in Fig. 1015. Surface is flat and coarsely fibrous. Considerable secondary cracking is evident, even at this low magnification. See also Fig. 1019 and 1020. 9×

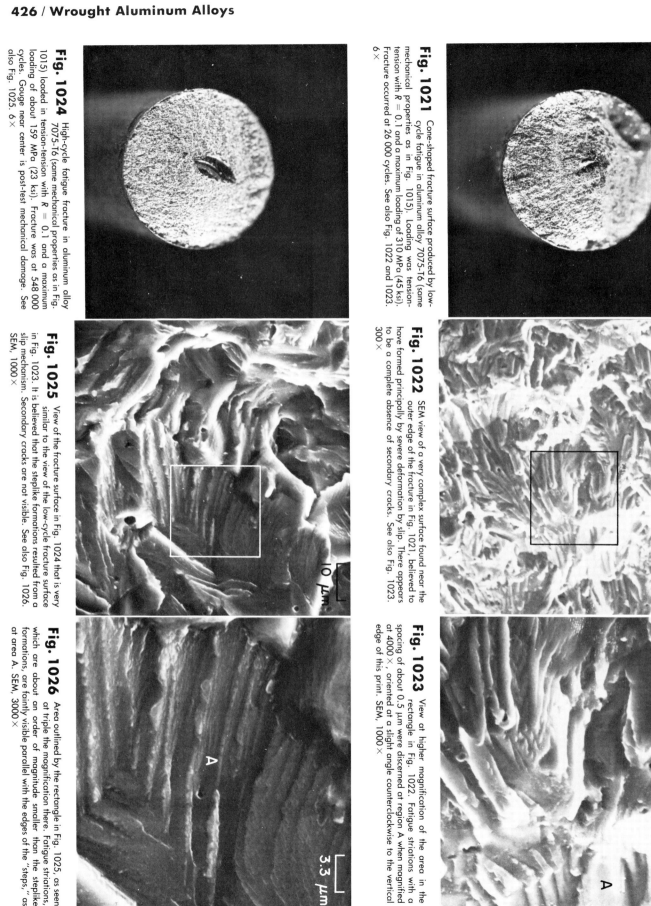

Fig. 1021 Cone-shaped fracture surface produced by low-cycle fatigue in aluminum alloy 7075-T6 (same mechanical properties as in Fig. 1015). Loading was tension-tension with $R = 0.1$ and a maximum loading of 310 MPa (45 ksi). Fracture occurred at 26 000 cycles. See also Fig. 1022 and 1023. 6×

Fig. 1022 SEM view of a very complex surface found near the outer edge of the fracture in Fig. 1021, believed to have formed principally by severe deformation by slip. There appears to be a complete absence of secondary cracks. See also Fig. 1023. 300×

Fig. 1023 View at higher magnification of the area in the rectangle in Fig. 1022. Fatigue striations with a spacing of about 0.5 μm were discerned at 4000×, oriented at a slight angle counterclockwise to the vertical edge of this print. SEM, 1000×

Fig. 1024 High-cycle fatigue fracture in aluminum alloy 7075-T6 (same mechanical properties as in Fig. 1015) loaded in tension-tension with $R = 0.1$ and a maximum loading of about 159 MPa (23 ksi). Fracture was at 548 000 cycles. Gouge near center is post-test mechanical damage. See also Fig. 1025. 6×

Fig. 1025 View of the fracture surface in Fig. 1024 that is very similar to the view of the low-cycle fracture surface in Fig. 1023. It is believed that the steplike formations resulted from a slip mechanism. Secondary cracks are not visible. See also Fig. 1026. SEM, 1000×

Fig. 1026 Area outlined by the rectangle in Fig. 1025, as seen at triple the magnification there. Fatigue striations, which are about an order of magnitude smaller than the steplike formations, are faintly visible parallel with the edges of the "steps," as at area A. SEM, 3000×

Wrought Aluminum Alloys / 427

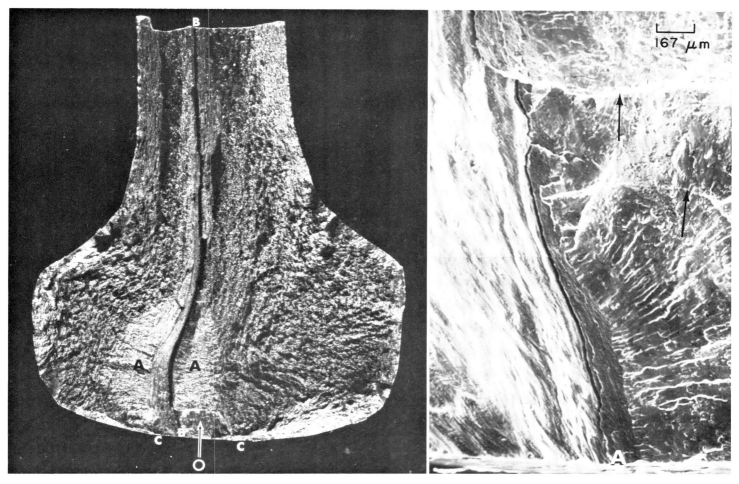

Fig. 1027 Surface of a crack in an aircraft wing-spar carry-through forging of aluminum alloy 7075-T6. The crack was discovered during inspection after 5269 h of service and was opened up. The external surface at edge C-C had been machined after forging. The regions marked A contain fatigue features that originated at a flaw (marked B-B) extending the full length of the segment. Area at O is shown enlarged in Fig. 1028. See also Fig. 1029 to 1034. ~3.5×

Fig. 1028 SEM view, at higher magnification, of the area at O in Fig. 1027. At left of the boundary (dark line running up from A) is the surface of the flaw (B-B in Fig. 1027); at right of it is a fatigue surface (A at right in Fig. 1027). Note the beach marks just above mid-height in the fatigue surface (at arrows). 60×

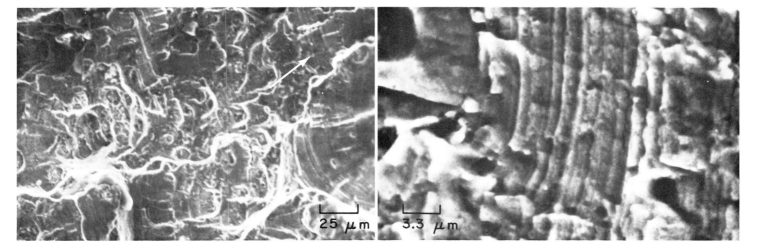

Fig. 1029 A typical region in an area A of Fig. 1027. Fatigue striations that are nearly vertical (arrow) are faintly visible. These were found to be parallel to the flaw (B-B in Fig. 1027) in both A areas in Fig. 1027. Scattered dimples are evident in locations adjacent to the fatigue striations. See also Fig. 1030. SEM, 400×

Fig. 1030 Highly magnified view of fatigue striations typical of those in areas A in Fig. 1027. In general, these striations were parallel with the flaw B-B. Dimples characterized the remainder of the fracture. Inclusions near the flaw suggested that it was the result of a pipe not cropped from the ingot. SEM, 3000×

428 / Wrought Aluminum Alloys

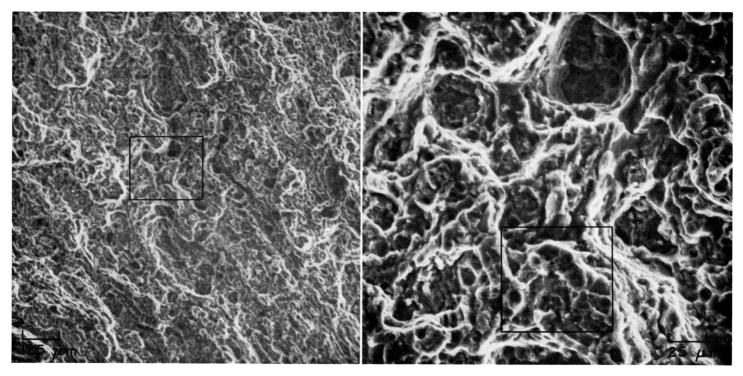

Fig. 1031 Region of the overload portion of the fracture in Fig. 1027. This region, exhibiting dimples characteristic of ductile rupture, is representative of the entire fracture, except for areas A and flaw B-B in Fig. 1027. Large dimples contain visible inclusions. See also Fig. 1032. SEM, 80×

Fig. 1032 Area in the rectangle in Fig. 1031, as seen at five times the magnification there. Large dimples show the inclusions that initiated them; fine dimples that are too small to be resolved at this magnification exist among the large dimples. See also Fig. 1033. SEM, 400×

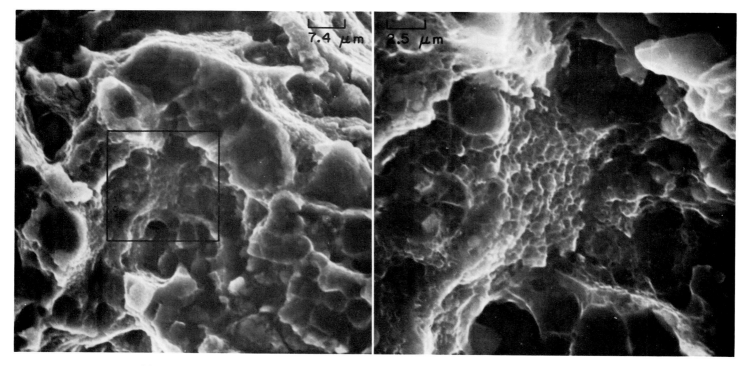

Fig. 1033 View at still higher magnification of the area in the rectangle in Fig. 1032. The fine dimples can just be distinguished on the surfaces adjacent to the large dimples. Many alloy second-phase particles are discernible, although some are so deep within the dimples as to be nearly invisible. See also Fig. 1034. SEM, 1340×

Fig. 1034 Greatly magnified view of the fracture region in Fig. 1031, showing the area in the rectangle in Fig. 1033. The region at center is made up of equiaxed dimples of quite uniform size. Note that several dimples are too deep for their bottoms to be lighted by an exposure proper for most of the dimples. SEM, 4000×

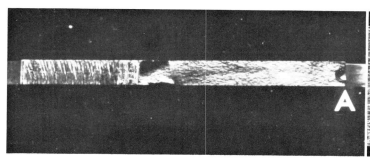

Fig. 1035 Fatigue-fracture surface of a bomb-rack side plate fabricated of aluminum alloy 7075-T6. Saw marks at left were made in opening up the fatigue crack for study. Crack origin was found to be at the edge of an attachment hole (marked A, at right). See Fig. 1036 for a view of the area near A at 5×. Actual size

Fig. 1036 Enlarged view of the area near A in Fig. 1035, showing corrosion (in dark area at right end) at the site of crack initiation. The side plate had been used as a fuel-tank support element in a 25-h ground-level qualification test of the tank in a corrosive environment under spectrum loading. See also Fig. 1037. 5×

Fig. 1037 Composite of three low-magnification SEM fractographs of the surface of the fatigue fracture at A in Fig. 1035 after ultrasonic cleaning. A small portion of the attachment hole is visible along the top of this view. The lack of beach marks here is perhaps due to the brevity of the test (25 h). Attempts to locate the exact point of crack initiation were not successful, because of the presence of tool marks at the edge of the attachment hole. It is believed, from the orientation of the fatigue striations at W, X, Y, and Z (see Fig. 1038 to 1041 for enlarged views of these areas), that the crack began at top right, near the edge of the attachment hole. 26×

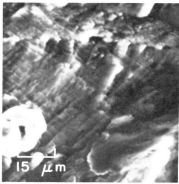

Fig. 1038 Area Z in Fig. 1037, which shows clear, well-defined fatigue striations. Normals to the striations point to the crack origin at the nearby corner of the attachment hole. SEM, 670×

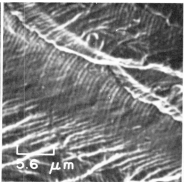

Fig. 1039 Area W in Fig. 1037, showing very regular fatigue striations. Like those in Fig. 1038, the striations here are oriented so that normals to them point to the corner near area Z. SEM, 1800×

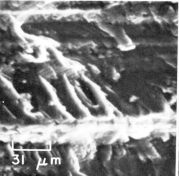

Fig. 1040 Area Y in Fig. 1037. This area, which is near the midthickness of the side plate, shows a herringbone pattern with ten times the spacing of the fatigue striations in Fig. 1039. SEM, 320×

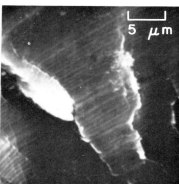

Fig. 1041 View of area X in Fig. 1037. This area, which is near the top left corner of the attachment hole, shows numerous fatigue facets that contain striations. SEM, 2000×

430 / Wrought Aluminum Alloys

Fig. 1042 Fatigue fracture in a notched plate specimen of aluminum alloy 7075-T6 subjected to cyclic stresses in air. (Notch is at bottom; sides of plate, at far left and right.) It is apparent that the fracture consists of at least three cracks that formed on various planes and temporarily grew independently. SEM, 26×

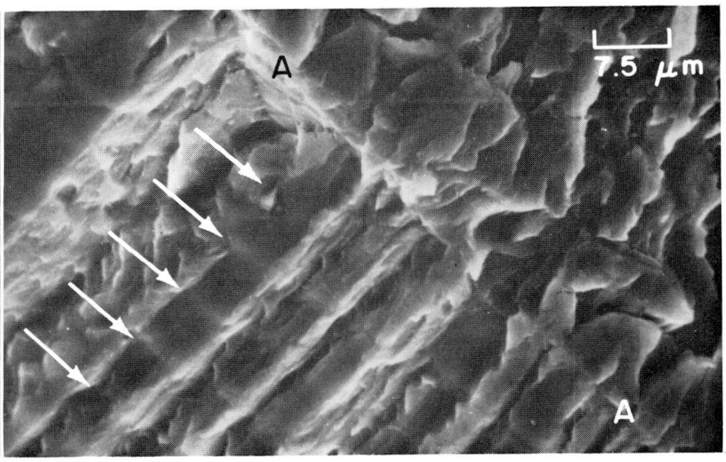

Fig. 1043 Fatigue fracture in a specimen of aluminum alloy 7075-T6 tested in air. The surface exhibits a pattern of brittle, widely spaced striations (at arrows) that are nearly parallel to a grain boundary (A-A). The tensile component (wide, flat region) of each striation was formed by a cleavage fracture. SEM, 1325×

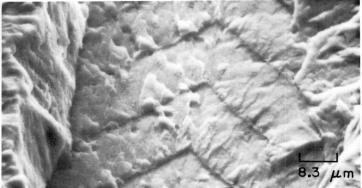

Fig. 1044 Corrosion-fatigue fracture in a specimen of aluminum alloy 7075-T6 tested in distilled water. Surface exhibits fatigue striations. The unusual sharp angle in the striations defines the location of a nearly vertical grain boundary in the central grain. Area in rectangle is shown at higher magnification in Fig. 1045. SEM, 600×

Fig. 1045 View of the area in the rectangle in Fig. 1044, at double the magnification there. Observe the wide, flat regions of the fatigue striations; these regions have progressed by cleavage. The unusual angled shape of the striations is the result of adherence of the crack fronts to crystallographic directions. SEM, 1200×

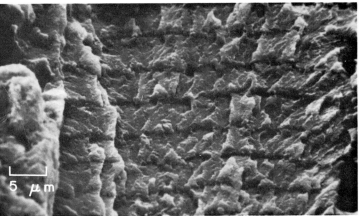

Fig. 1046 Corrosion-fatigue fracture in a specimen of aluminum alloy 7075-T6 that was tested in a 3.5% NaCl solution. Surface shows two types of striations: at left and at lower right are grains with ductile striations, and between them lies a grain with pronounced brittle striations. See also Fig. 1047. SEM, 500×

Fig. 1047 Brittle striations in a corrosion-fatigue fracture in a specimen of aluminum alloy 7075-T6 that, like the specimen in Fig. 1046, was tested in a 3.5% NaCl solution. Striations (horizontal here) have very uniform spacing. SEM, 2000×

Wrought Aluminum Alloys / 431

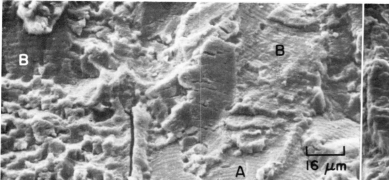

Fig. 1048 Corrosion-fatigue fracture in aluminum alloy 7075-T6 tested in a 3.5% NaCl solution. During testing, specimen was subjected to an applied electrical potential of −0.700 mV as measured against a standard calomel electrode. A mixture of ductile striations (at A) and brittle striations (at B) is evident, with some fissures. See also Fig. 1049. SEM, 625×

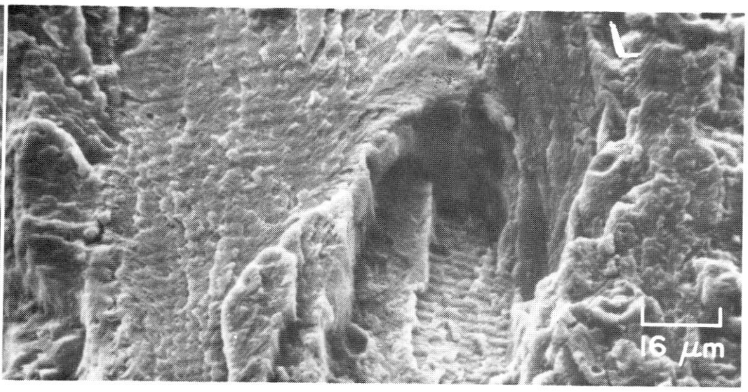

Fig. 1049 Corrosion-fatigue fracture in a specimen of aluminum alloy 7075-T6 tested in a 3.5% NaCl solution and, during testing, subjected to an applied electrical potential of −1.200 mV (versus −0.700 mV for the specimen in Fig. 1048). Surface shows brittle striations with typical topography; no evidence of striation fissures. See also Fig. 1050. SEM, 640×

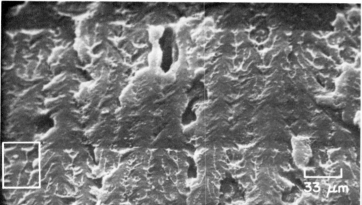
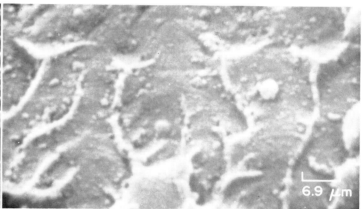

Fig. 1050 Corrosion-fatigue fracture in a specimen of aluminum alloy 7075-T6 that, like the specimen in Fig. 1049, was tested in a 3.5% NaCl solution while subjected to an applied electrical potential of −1.200 mV. Note the extremely large striation spacing; only two readily apparent adjacent brittle-striation peaks are included in this view. See also Fig. 1051. SEM, 300×

Fig. 1051 Higher-magnification view of the fracture in Fig. 1050, showing an area of which the right half is outlined by the rectangle there. The wide, flat portion of a single striation is visible, beginning at the peak near the top. The surface configuration in this region of cleavage separation suggests a set of river patterns, exaggerated by corrosion products. SEM, 1450×

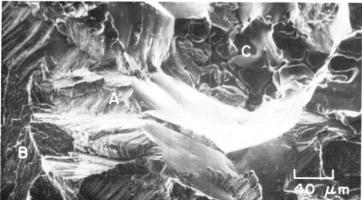
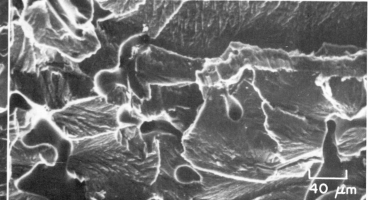

Fig. 1052 Fatigue-fracture surface of a test bar cast from aluminum alloy 7075, aged to the T6 temper and then tested in a 3.5% NaCl solution. A diagonal region of fatigue striations is visible at A; this region abuts areas of intergranular fracture (at B) and interdendritic porosity (at C). SEM, 250×

Fig. 1053 Fatigue-fracture in a cast single-crystal specimen of aluminum alloy 7075 that was aged to the T6 temper and then tested in a 3.5% NaCl solution. Most of the view is filled with an interdendritic network of solidification porosity, which did not halt or blunt the fatigue crack. SEM, 250×

432 / Wrought Aluminum Alloys

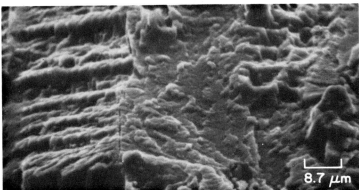

Fig. 1054 Surface of a corrosion-fatigue fracture in a specimen of aluminum alloy 7075-T6 that was tested in a 3.5% NaCl solution. Shown here are portions of two grains that exhibit brittle striations; the difference in appearance of their striations is caused by the difference in grain orientation. The vertical dividing line is the boundary between the grains. SEM, 1150×

Fig. 1055 An area of the corrosion-fatigue fracture surface of an aluminum alloy 7075-T651 that was tested in a 3.5% NaCl solution under a cyclic stress-intensity range (ΔK) of 6.6 MPa$\sqrt{m}$ (6 ksi$\sqrt{in.}$) at 10 cps. Ductile and brittle striations would be expected in such a fracture, but cannot be resolved at this magnification. See also Fig. 1056. SEM, 120×

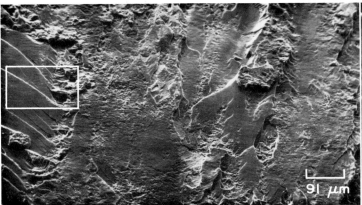
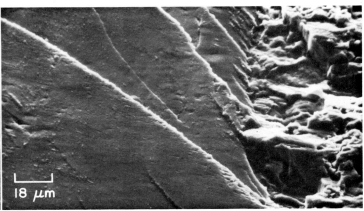

Fig. 1056 Another area of the fracture surface of which a portion is shown in Fig. 1055. This possesses similar features, but the surface is flatter, lacking the blufflike cleavage steps present in Fig. 1055. Again, fatigue striations are unresolved. See also Fig. 1057. SEM, 110×

Fig. 1057 Area in the rectangle in Fig. 1056, at higher magnification, showing the edge of the smooth area believed to possess patches of ductile striations. Compare these features with those of the 7475-T6 specimen (also tested in 3.5% NaCl) in Fig. 1058 and 1059. SEM, 560×

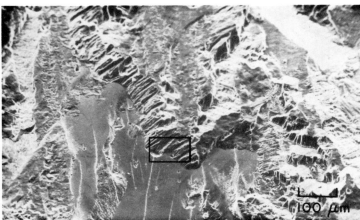
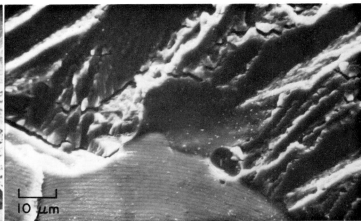

Fig. 1058 Corrosion-fatigue fracture in aluminum alloy 7475-T6 tested in a 3.5% NaCl solution under a cyclic stress-intensity range (ΔK) of 13 MPa$\sqrt{m}$ (12 ksi$\sqrt{in.}$) at 10 cps, showing contrasting features: ductile striations (unresolved) in smooth regions; brittle striations in adjacent grains. See also Fig. 1059. SEM, 100×

Fig. 1059 The area outlined by the rectangle in Fig. 1058, as seen at ten times the magnification there, which makes the fine fatigue striations in the smooth region clearly visible. Above are brittle striations forming, at a steep incline, what is in effect a cleavage step up to a second fracture level. SEM, 1000×

Fig. 1060 Surface of a fracture in an aircraft lower-bulkhead cap fabricated from an extrusion of aluminum alloy 7075-T6. The cap was cut open to expose the fracture for inspection, and evidence of corrosion was found along the edge at which fracture was initiated (bottom edge in this view). Secondary cracks had produced "leaves" (at arrows) in the fracture surface. See also Fig. 1061 to 1069. 8×

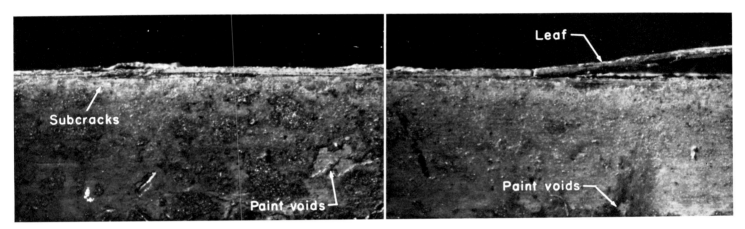

Fig. 1061 A portion of the painted external surface of the fractured aircraft lower-bulkhead cap in Fig. 1060, showing a profile of the fracture surface (at top) and voids in the paint film. Subcracks are evident immediately below and parallel with the fracture surface. Compare with Fig. 1062. 8×

Fig. 1062 Another portion of the painted external surface of the cap in Fig. 1060. Fracture surface, shown in profile at top, contains a "leaf"; compare this view of the "leaf" with that in Fig. 1060, which is normal to the fracture surface. As in Fig. 1061, paint voids are evident. 8×

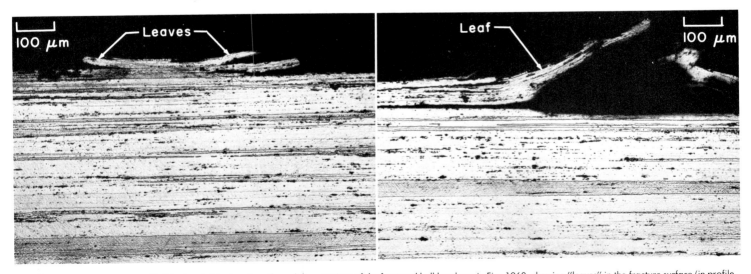

Fig. 1063, 1064 Polished and etched sections through two portions of the fractured bulkhead cap in Fig. 1060, showing "leaves" in the fracture surface (in profile, at top). Note the elongated grains, which are parallel with the fracture surface. Note also that both the primary and secondary cracks are intergranular. Intergranular corrosion has nearly separated several of the "leaves" from the fracture surface. Stress-corrosion cracking was favored by imposition of tensile stresses in installation and by exposure to a marine atmosphere in service. Both etched in Keller's reagent, 100×

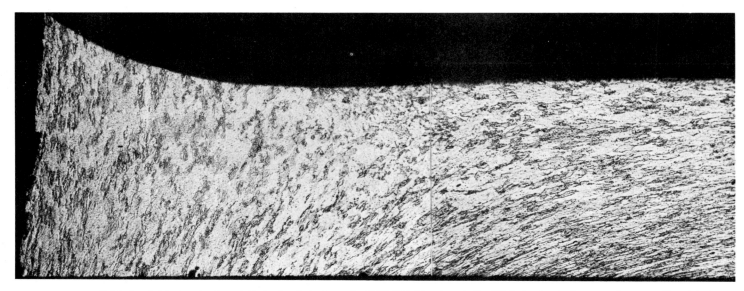

Fig. 1065 Polished and etched section through the fractured aluminum alloy 7075-T6 bulkhead cap in Fig. 1060, showing the fracture surface in profile at left. Note change in grain flow produced during extrusion, from a longitudinal orientation at right to transverse at left; transverse orientation results in grains being parallel to the fracture surface, a condition that aggravates stress-corrosion cracking. Etched in Keller's reagent, 20×

Fig. 1066 TEM p-c replica of an area of the fracture surface in Fig. 1060 near the edge at which fracture was initiated, showing corrosion debris on a very flat intergranular surface. The original separated-grain facets have been corroded away. 3000×

Fig. 1067 TEM p-c replica taken from another area of the fracture surface in Fig. 1060 near the edge at which fracture was initiated. This area, like the one in Fig. 1066, shows that cracking was intergranular and was accompanied by severe corrosion. 3000×

Fig. 1068 TEM p-c replica of an area of the fracture surface in Fig. 1060 deeper into the fracture than Fig. 1066 and 1067; here, corrosion of the separated-grain facets is evident, although not as severe as in areas nearer the edge at which fracture was initiated. 3000×

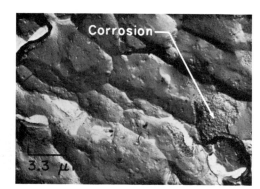

Fig. 1069 Area of the fracture surface in Fig. 1060 still farther from the initiation edge than Fig. 1068. This surface, although still intergranular, is relatively free of corrosion products. In a portion of the fracture formed later, evidence of fatigue was found. Final fracture was by shear. TEM p-c replica, 3000×

Fig. 1070 TEM p-c replica of the surface of an intergranular fracture in an aircraft landing-gear component of forged aluminum alloy 7075-T6 that fractured as the result of stress-corrosion cracking. Note the secondary cracks between the grains. 1000×

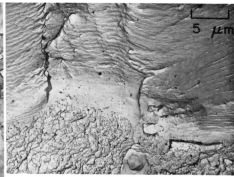

Fig. 1071 TEM p-c replica of the surface of a fracture in a specimen of aluminum alloy 7075-T6 that first was cracked by fatigue and then was broken by impact. Stretching and serpentine glide are evident at about mid-height, at the root of the fatigue crack; below are tear dimples formed on impact. 2000×

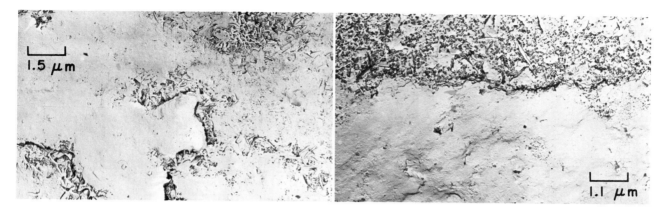

Fig. 1072, 1073 TEM p-c replicas of two areas of the surface of a fracture in a forged aluminum alloy 7075-T6 support for a cargo-door cylinder. The fracture, which originated at a cold shut, formed at an attachment hole that extended above and below the cold shut. These TEM replicas reveal that the fracture surface is flat and encrusted with oxide particles. Fig. 1072: 6500×. Fig. 1073: 9300×

Fig. 1074 Side view of a piece of fractured aluminum alloy 7079-T6 bogie beam of an aircraft main landing gear. The fracture surface is at top, in profile. The very straight edge at lower right is a saw cut that was made to separate the fracture area from the remainder of the beam to facilitate inspection. Hardness of the beam was 84 HRB. See also Fig. 1075 and Fig. 1076. 0.25×

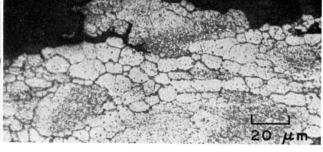

Fig. 1075 Polished and etched section through the light region at arrow "d" in Fig. 1076. At top in this view is the profile of the fracture surface. Note that the fracture usually followed grain boundaries. Cracks branching from the fracture surface are also evident. Etched in Keller's reagent, 500×

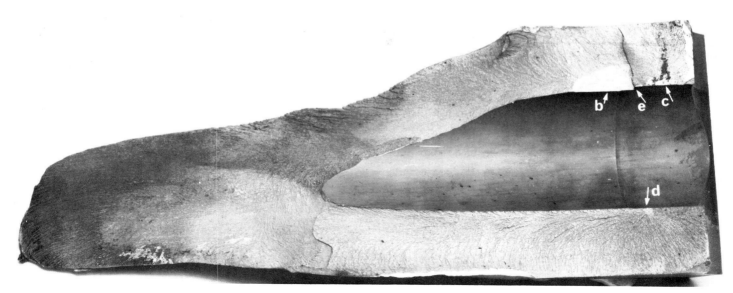

Fig. 1076 Surface of the fracture in the bogie beam shown in Fig. 1074. Three stress-corrosion cracks caused the final fast fracture, which was a tearing type of rupture. The intergranular attack characteristic of stress corrosion is shown in Fig. 1075. The three stress-corrosion cracks are the light areas marked by arrows "b," "c," and "d." The crack surfaces appear flat, but contain many small steps. Some of these steps were found to have been pulled away from the fracture surface in a leaf fashion, indicating overlapping or branching cracks. At arrow "e" is a pronounced step, which hides a portion of area "b." ~0.4×

436 / Wrought Aluminum Alloys

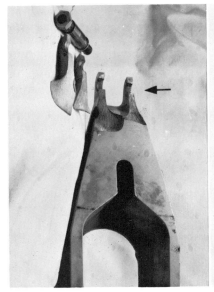

Fig. 1077 Portion of a broken aircraft landing-gear actuator beam of Al alloy 7079-T6, showing stress-corrosion fractures (at arrow). 0.17×

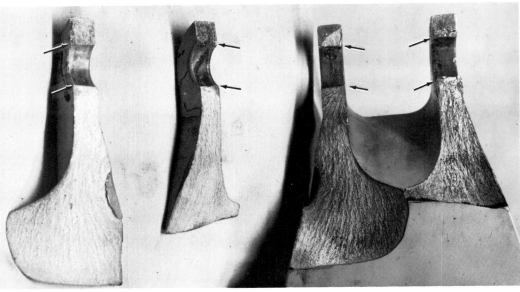

Fig. 1078 Higher-magnification view of the surfaces of the stress-corrosion fractures shown in Fig. 1077. These fractures originated at the interior edges of boltholes in clevis-attachment lugs, at the locations designated by the arrows. An appreciable amount of evidence of both corrosion and secondary cracking was found at these origins. Visible are chevron marks, which also confirm these locations as the origins of fracture. 0.63×

Fig. 1079 Fracture surface outboard of the bolthole in the clevis-attachment lug at right in Fig. 1078. Entire surface is typical of stress-corrosion cracking. Origin is at arrow (lower right corner of surface). 4×

Fig. 1080 Fracture surface inboard of the bolthole in the clevis-attachment lug in Fig. 1079. The primary stress-corrosion crack, which originated at arrow, is within the area enclosed by the dashed line (lower left). The remainder of this surface shows chevron marks typical of final fast fracture. 4×

Fig. 1081 Fracture surface outboard of the bolthole in the lug at left in Fig. 1078. Here, as in Fig. 1079, the entire surface is typical of stress-corrosion cracking. Origin is at arrow (upper right corner of surface). 4×

Fig. 1082 Fracture surface inboard of the bolthole in the clevis-attachment lug in Fig. 1081. The primary stress-corrosion crack, which originated at the arrow, and which is enclosed by the dashed line, is larger than that in Fig. 1080. This actuator beam had hardness normal for 7079-T6: 152-157 HV (10-kg load). 4×

Wrought Aluminum Alloys / 437

Fig. 1083 Tensile-overload fracture in a specimen of a superplastic eutectic alloy containing 67% Al and 33% Cu. The material was cast, and the as-cast ingot was extruded at 430 °C (805 °F). Testing was performed at 0.025 mm/s (0.001 in./s) and at a controlled temperature of 450 °C (840 °F). The surface shows coarse dimples and an irregular contour. There is some suggestion of a shear lip, but it is not sharply defined. See also Fig. 1084. SEM, 70×

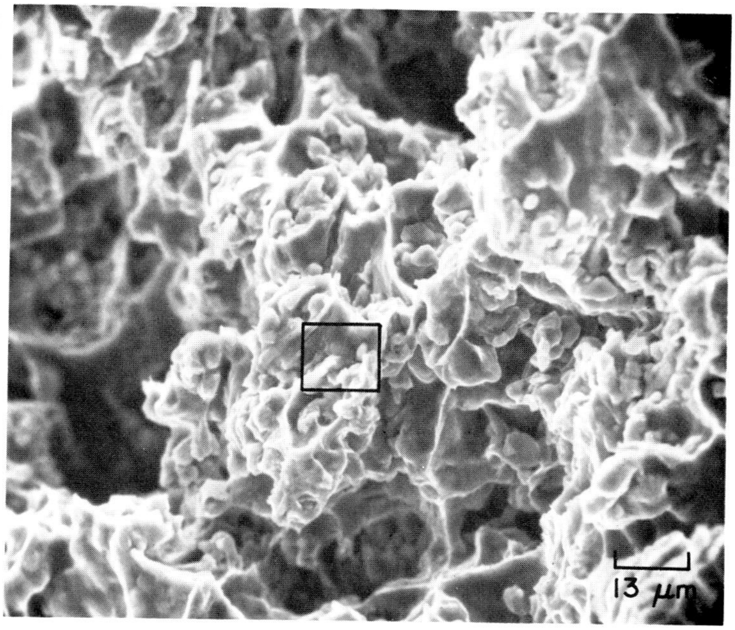

Fig. 1084 Higher-magnification view of the area in the rectangle in Fig. 1083, showing more clearly the details of the fracture surface. It should be noted that the grain size of this specimen is approximately 7 μm. Note also that a majority of the dimples contain one or more particles—probably particles of the $CuAl_2$ phase of the eutectic. See Fig. 1085 for a higher-magnification view of area in rectangle here. SEM, 770×

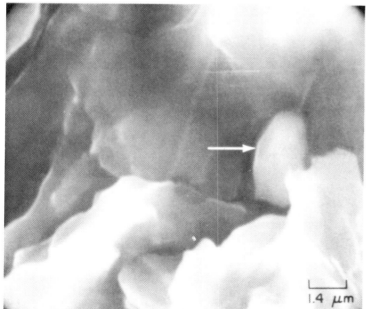

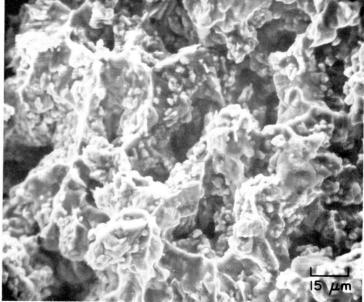

Fig. 1085 Higher-magnification view of the area outlined in Fig. 1084. The contours of the facets in the right foreground show that the Al solid-solution matrix phase of this eutectic alloy was highly ductile. Arrow points to a spheroidal particle that should be of the $CuAl_2$ phase; most of the surface of this particle appears to be detached from the matrix. SEM, 7000×

Fig. 1086 Tensile-overload fracture in a specimen of 67Al-33Cu alloy that was cast, extruded, and tested the same as the specimen in Fig. 1083. In this specimen, the grain size is only 2 μm. It appears that both the large dimples and the $CuAl_2$ particles here are smaller than those in Fig. 1084. SEM, 670×

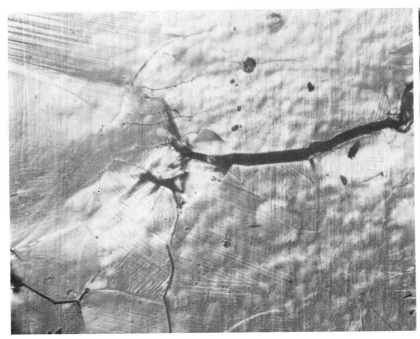

Fig. 1087 Deformation at the tip of a fatigue crack in Al-5.6Zn-1.9Mg. The experimental alloy was heavily cold worked, solution treated for 1.5 h at 465 °C (870 °F) in flowing argon, aged for 24 h at 120 °C (250 °F) to develop peak hardness, and then tested in fatigue. Sample electropolished in a mixture of perchloric acid, ethanol, butyl cellusolve, and water and observed under phase-contrast illumination. SEM, 200× (R.E. Ricker, University of Notre Dame, and D.J. Duquette, Rensselaer Polytechnic Institute)

Fig. 1088 Transgranular corrosion-fatigue crack propagation in a solution-treated and peak-aged Al-5.6Zn-1.9Mg sample tested in humid nitrogen gas. Compare with Fig. 1091 and 1092. SEM, 5000× (R.E. Ricker, University of Notre Dame, and D.J. Duquette, Rensselaer Polytechnic Institute)

Fig. 1089 The external surface (top) and the corrosion-fatigue fracture surface (bottom) of a solution-treated and peak-aged Al-5.6Zn-1.9Mg sample tested in high-purity deaerated water. SEM, 100× (R.E. Ricker, University of Notre Dame, and D.J. Duquette, Rensselaer Polytechnic Institute)

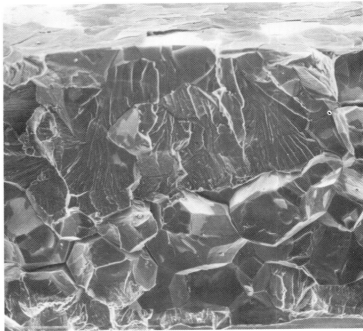

Fig. 1090 The external surface (top) and the corrosion-fatigue fracture surface (bottom) of a solution-treated and peak-aged Al-5.6Zn-1.9Mg sample tested in deaerated 0.5 mol NaCl at a cathodic potential of −1.6 V (SCE). Compare with Fig. 1089, 1093, and 1095. SEM, 100× (R.E. Ricker, University of Notre Dame, and D.J. Duquette, Rensselaer Polytechnic Institute)

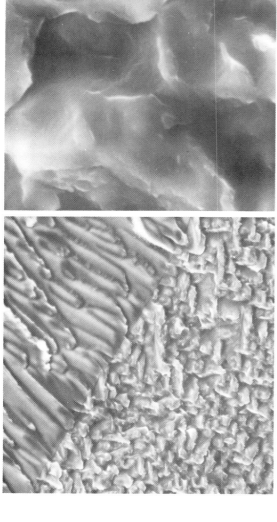

Fig. 1091, 1092, 1093 Fatigue crack propagation in a solution-treated and peak-aged Al-5.6Zn-1.9Mg sample tested in an inert environment (dry nitrogen gas). Compare with the examples in Fig. 1088 to 1090, 1094, and 1095 of corrosion fatigue in this experimental alloy. Fig. 1091 (left): Region of transgranular fatigue crack propagation. SEM, 500×. Fig. 1092 (center): High-magnification view of fine fatigue striations on a transgranular portion of the fracture surface. SEM, 10 000×. Fig. 1093 (right): Typical region of intergranular fracture in an Al-5.6Zn-1.9Mg alloy tested in an inert environment. SEM, 500× (R.E. Ricker, University of Notre Dame, and D.J. Duquette, Rensselaer Polytechnic Institute)

Fig. 1094, 1095 Corrosion-fatigue crack propagation in solution-treated and peak-aged Al-5.6Zn-1.9Mg tested in deaerated 0.5 mol Na_2SO_4. Fig. 1094 (left): Region of transgranular crack propagation. Direction of crack growth is toward the top left. Compare with Fig. 1088, 1091, and 1092. SEM, 2000×. Fig. 1095 (right): Region of intergranular crack propagation. Crack growth direction is toward the top. Compare with Fig. 1093. SEM, 1000× (R.E. Ricker, University of Notre Dame, and D.J. Duquette, Rensselaer Polytechnic Institute)

440 / P/M Aluminum Alloys

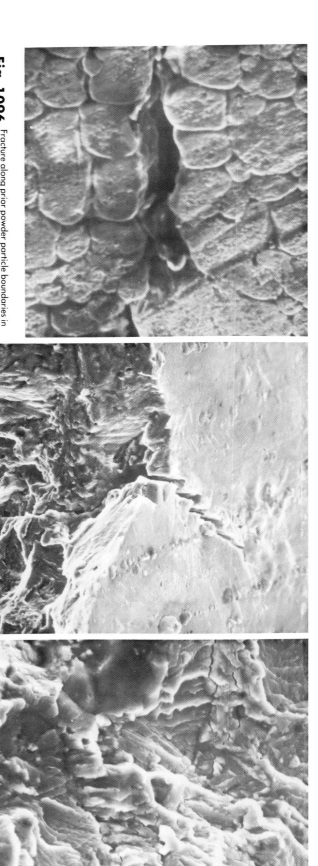

Fig. 1096 Fracture along prior powder particle boundaries in cold-rolled Al-4.2Mg-2.1Li. The experimental P/M alloy was solution treated for 1 h at 510 °C (950 °F), water quenched, naturally aged, and then cold rolled to a 4% thickness reduction in one pass. The cracks at particle boundaries reflect the low ductility typical of aluminum-lithium alloys. SEM, 375×. (R.E. Ricker, University of Notre Dame, and D.J. Duquette, Rensselaer Polytechnic Institute)

Fig. 1097, 1098 Corrosion-fatigue crack initiation and propagation in a solution-treated and peak-aged Al-4.2Mg-2.1Li P/M alloy tested in deaerated high-purity water. Fig. 1097 (left): View of external surface (top) and fracture surface (bottom). SEM, 500×. Fig. 1098 (right): Higher-magnification view of fracture surface. SEM, 1000×. (R.E. Ricker, University of Notre Dame, and D.J. Duquette, Rensselaer Polytechnic Institute)

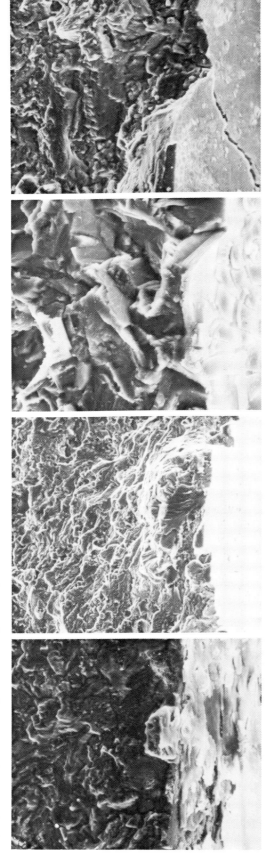

Fig. 1099, 1100 Corrosion-fatigue crack propagation in a solution-treated and peak-aged Al-4.2Mg-2.1Li P/M alloy tested in deaerated 0.5 mol Na₂SO₄. Fig. 1099 (left): View of external surface (top) and fracture surface (bottom). SEM, 400×. Fig. 1100 (right): Corrosion products on the fracture surface. SEM, 800×. (R.E. Ricker, University of Notre Dame, and D.J. Duquette, Rensselaer Polytechnic Institute)

Fig. 1101, 1102 Corrosion-fatigue crack propagation and initiation in a solution-treated and peak-aged Al-4.2Mg-2.1Li P/M alloy tested in deaerated 0.5 mol NaCl at a cathodic potential of −1.6 V (SCE). Fig. 1101 (left): View of external surface (top) and fracture surface (bottom). SEM, 200×. Fig. 1102 (right): Corrosion products on both the external surface and the fracture surface. SEM, 200×. (R.E. Ricker, University of Notre Dame, and D.J. Duquette, Rensselaer Polytechnic Institute)

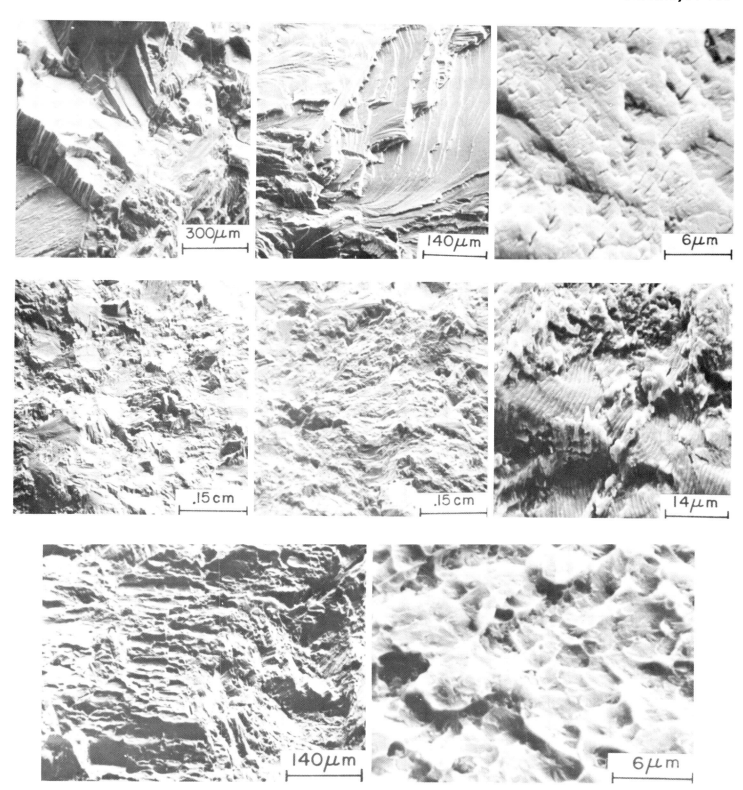

Fig. 1103 to 1110 Fatigue crack growth fracture topography in a Ti-6Al-2Sn-4Zr-2Mo-0.1Si (Ti-6242, UNS R54620) forging, $\alpha + \beta$ processed prior to β heat treatment and aging. Compact tension specimen tested in air at 25 °C (75 °F). As crack growth rate (da/dN) and stress-intensity factor (ΔK) increased, the fracture surface changed from one characterized by large transgranular facets to one exhibiting intergranular facets and dimples. Crack growth direction is from left to right. Fig. 1103, 1104, and 1105 (top row): Fracture surface at low ΔK (<25 MPa$\sqrt{}$m, or 23 ksi$\sqrt{}$in.) and da/dN of 0.005 μm/cycle. Note large transgranular facets in Fig. 1103 and 1104. Intermixed with the facets were regions of pronounced secondary cracking (Fig. 1105). SEM; 60×, 125×, and 2900×. Fig. 1106 (second row, left): At intermediate ΔK (25 to 38 MPa$\sqrt{}$m, or 23 to 35 ksi$\sqrt{}$in.) and a da/dN of 0.1 μm/cycle, the amount of faceting decreased, being replaced by more ductile modes of fracture. SEM, 12×. Fig. 1107 and 1108 (second row, center and right): At high ΔK (>38 MPa$\sqrt{}$m, or 35 ksi$\sqrt{}$in.) and da/dN of 0.5 μm/cycle, the fracture surface was characterized by striations, microvoids, and dimples. SEM, 12× and 1250×. Fig. 1109 and 1110 (bottom row): Final separation occurred along prior-β grain boundaries (Fig. 1109, da/dN = 20 μm/cycle), with the intergranular fracture surface exhibiting dimples (Fig. 1110, fast fracture region). SEM, 150× and 3500× (J.A. Ruppen and A.J. McEvily, University of Connecticut)

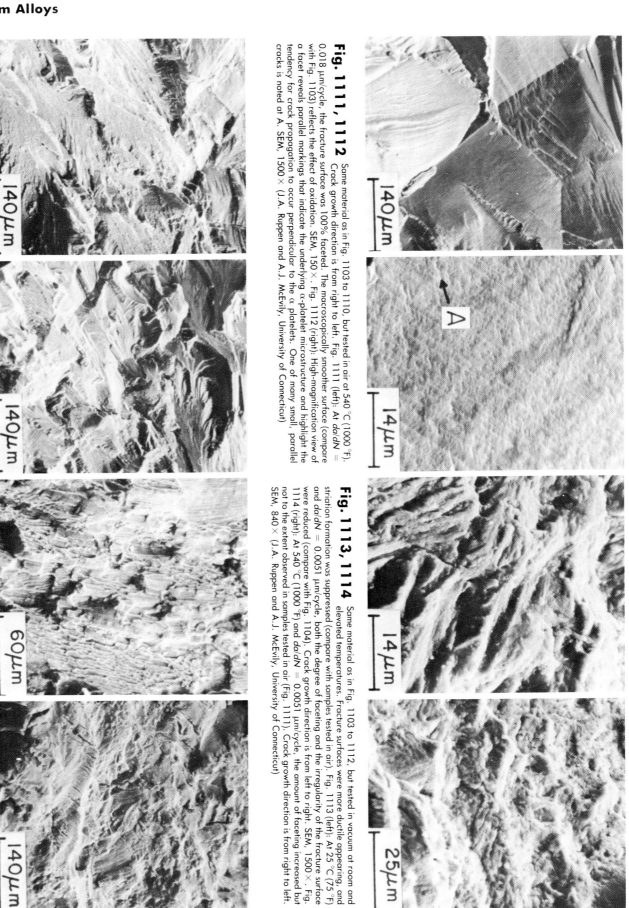

Fig. 1111, 1112 Same material as in Fig. 1103 to 1110, but tested in air at 540 °C (1000 °F). Crack growth direction is from right to left. Fig. 1111 (left): At da/dN = 0.018 μm/cycle, the fracture surface was 100% faceted. The macroscopically smoother surface with Fig. 1103) reflects the effect of oxidation. SEM, 150×. Fig. 1112 (right): High-magnification view of a facet reveals parallel markings that indicate the underlying microstructure and highlight the tendency for crack propagation to occur perpendicular to the α platelets. One of many small, parallel cracks is noted at A. SEM, 1500× (J.A. Ruppen and A.J. McEvily, University of Connecticut)

Fig. 1113, 1114 Same material as in Fig. 1103 to 1112, but tested in vacuum at room and elevated temperatures. Fracture surfaces were more ductile appearing, and striation formation was suppressed (compare with samples tested in air). Fig. 1113 (left): At 25 °C (75 °F) and da/dN = 0.0051 μm/cycle, both the degree of faceting and the irregularity of the fracture surface were reduced (compare with Fig. 1104). Crack growth direction is from left to right. SEM, 1500×. Fig. 1114 (right): At 540 °C (1000 °F) and da/dN = 0.0051 μm/cycle, the amount of faceting increased but not to the extent observed in samples tested in air (Fig. 1111). Crack growth direction is from right to left. SEM, 840× (J.A. Ruppen and A.J. McEvily, University of Connecticut)

Fig. 1115 to 1118 Fatigue crack growth fracture topography in a Ti-6Al-2Sn-4Zr-2Mo-0.1Si (Ti-6242, UNS R53620) forging, β processed prior to α + β heat treatment and aging. (Contrast with the α + β forging in Fig. 1103 to 1114.) Compact tension specimen tested in air at room and elevated temperatures. Crack growth direction is from left to right in room-temperature fractographs, and from right to left in elevated-temperature fractographs. Fig. 1115 and 1116 (left and left center): At low da/dN (0.025 μm/cycle) and at 25 and 540 °C (75 and 1000 °F), respectively, the β forging exhibited a less irregular fracture surface consisting of smaller facets (compare with Fig. 1103 and 1111). Features in this microstructure-sensitive regime are related to the prior-β grain size of the β forging and the colony size of the α + β forging. SEM, 150×. Fig. 1117 and 1118 (right center and right): At higher da/dN (2.5 μm/cycle) and at 25 and 540 °C (75 and 1000 °F), respectively, fracture modes were the same in both β and α + β forged material, but the β forging exhibited elongated dimples corresponding to the underlying α/β platelet microstructure (Fig. 1117) and more secondary cracking at elevated temperature (Fig. 1118). SEM, 350× and 150× (J.A. Ruppen and A.J. McEvily, University of Connecticut)

Fig. 1119, 1120 Same material as in Fig. 1115 to 1118, but tested in vacuum at room temperature. The β forging exhibited a sharp change in fracture appearance corresponding to a da/dN of 0.1 μm/cycle. Crack growth direction is from right to left. Fig. 1119 (left): Fracture surface at $da/dN = 0.0076$ μm/cycle. Crack growth is structure sensitive. SEM, 700×. Fig. 1120 (right): Fracture surface at $da/dN = 0.13$ μm/cycle. Crack growth is now structure insensitive, characterized by a rougher surface with dimples and voids. SEM, 700× (J.A. Ruppen and A.J. McEvily, University of Connecticut)

Fig. 1121 Ductile overload fracture of a tensile specimen of Ti-6Al-4V ELI (ASTM F136, UNS R56401). The wrought alloy was annealed for 1 h at 760 °C (1400 °F) and air cooled prior to testing. The fracture surface is characterized by essentially 100% dimpled rupture. SEM, 1000× (R. Abrams, Howmedica, Pfizer Hospital Products Group Inc.)

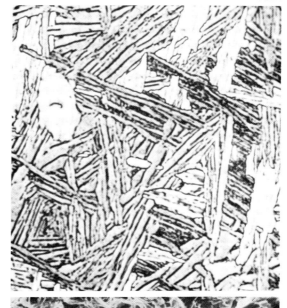

Fig. 1122, 1123, 1124 Ductile fracture of laser beam welded Ti-6Al-2Nb-1Ta-1Mo (Ti-6211, UNS R56210). A plate measuring 13 mm (0.5 in.) thick was welded using beam power of 8 kW, speed of 28 mm/s (65 in./min), and heat input of 0.29 kJ/mm (7.4 kJ/in.). Fig. 1122 (left): Fracture surface of the dynamic tear test specimen indicates 100% ductile rupture via microvoid coalescence. Mechanical properties of the weld were excellent, even though some porosity is evident. SEM, 500×. Fig. 1123 (center): Microstructure of the titanium alloy base plate features the basketweave appearance of Widmanstätten α. Primary α and β phase are also present. Hardness: 32 to 40 HRC. Keller's etch, 1000×. Fig. 1124 (right): Microstructure of the fusion zone features needlelike martensitic α surrounded by some β phase, and the boundaries of the elongated β grains present prior to cooling. Hardness: 32 to 38 HRC. Keller's etch, 500× (E.A. Metzbower and D.W. Moon, Naval Research Laboratory)

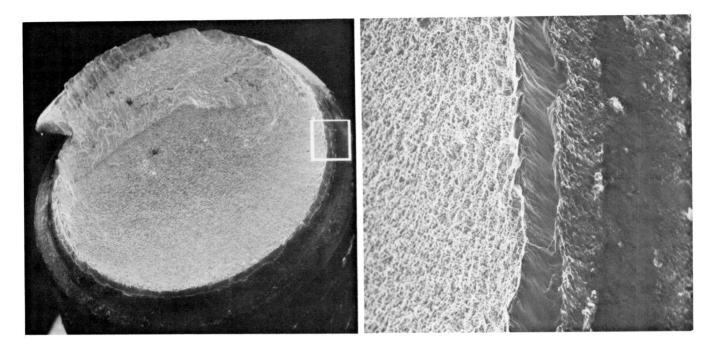

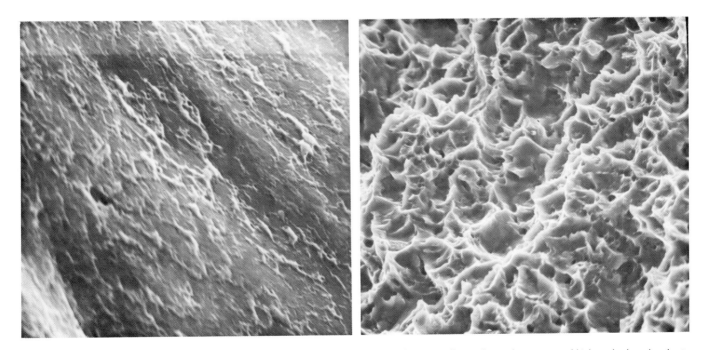

Fig. 1125 to 1128 Fracture of Ti-6Al-4V (UNS R56400) threaded fasteners during installation due to the presence of high-angle shear bands at thread roots. The nut (collar) of the aerospace fastener is designed to break in two when a specific installation torque is reached. The joint is then created by the remaining, tightened half of the collar and its mating threaded pin. In this case, however, the titanium alloy pins from a single manufacturer had been breaking before their collars "torqued off." Comparisons were made with pins produced by other manufacturers that did not fail during installation. The only significant differences observed were in the extent of shear-band formation and the orientation of shear bands near the roots of the rolled threads, both of which can be controlled by altering the tool-workpiece geometry. Shear-band formation was more extensive in the failed pins, and the bands are oriented at a higher angle to the pin axis. (More details are given in Fig. 1129 to 1132.) Fig. 1125 (top left): Typical fracture surface of failed pin. Fracture was flat and propagated across the pin through a single thread root. SEM, 10×. Fig. 1126 (top right): Boxed area in Fig. 1125. Note the relatively featureless shear band on the circumference of the fracture surface. It is at a 45° angle to the pin axis. SEM, 100×. Fig. 1127 (bottom left): Small, partially formed dimples on the shear-band portion of the fracture surface. SEM, 5000×. Fig. 1128 (bottom right): Balance of fracture surface away from the shear band was typical of dimpled rupture in Ti-6Al-4V. The equiaxed shape of the dimples indicate that a tensile stress was operative. SEM, 1000× (G. Hopple, Lockheed Missiles & Space Company, Inc.)

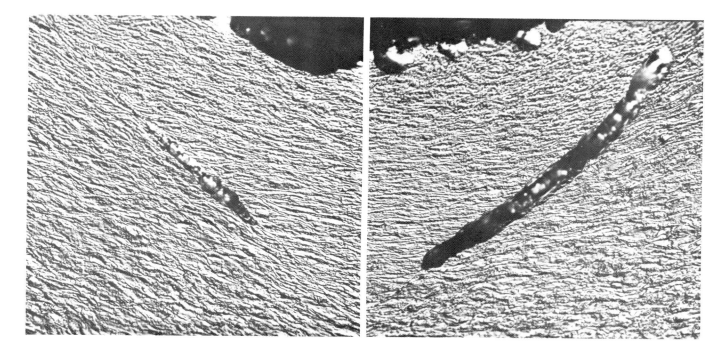

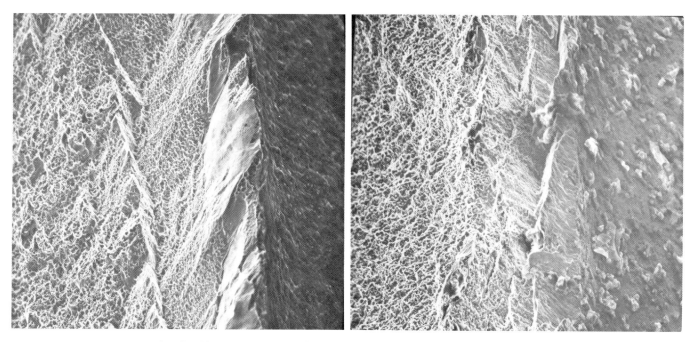

Fig. 1129 to 1132 Shear-band formation in Ti-6Al-4V threaded fasteners that fractured during installation (see Fig. 1125 to 1128). Shear bands form in thread roots of fasteners during the cold thread rolling process. In Ti-6Al-4V, they readily form during the highly localized plastic deformation that accompanies mechanical working at temperatures below 705 °C (1300 °F). The role of shear-band orientation in fracture resistance is significant: A tensile stress is needed to initiate shear band fracture, and the higher the angle of the shear band to the pin axis, the greater the tensile component of the uniaxial stress applied to the fastener during installation. A suitably high tensile stress is required regardless of the extent of shear-band formation. Fig. 1129 (top left) and 1130 (top right): Void initiation and coalescence, respectively, in a shear band at a thread root adjacent to the fracture. Rounded voids lend support to contention that tensile stress initiated fracture. Duplex etch, Kroll's solution and H_2O_2-KOH. Differential interference contrast microscopy, 500×. Fig. 1131 (bottom left) and 1132 (bottom right): Comparison of shear-band orientation in tensile-tested Ti-6Al-4V fasteners made by producer of failed pins (Fig. 1131) and a producer of pins that did not fail during installation (Fig. 1132). Shear-band angles are higher—40° to 50° versus 20° to 30°—in the former. SEM, 200× (G. Hopple, Lockheed Missiles & Space Company, Inc.)

446 / Titanium Alloys

Fig. 1133, 1134 Figure 1133 (left) shows two portions of a fractured titanium alloy Ti-6Al-4V second-stage compressor disk from a jet engine. Of the three radial fractures, the one at upper right, which passes through the bolthole marked "a," was found to be the primary fracture, exhibiting several fatigue-crack origins adjacent to the hole. The two other fractures showed no sign of fatigue. Figure 1134 (right) shows a surface of the fracture through the bolthole at arrow "a" in Fig. 1133. Several fatigue cracks originated at locations within brackets "c" and "d." Deep tool marks in the bore of the hole are visible between arrows "e" and "f"; a section through the hole marked "b" in Fig. 1133 showed no such deep tool marks. Average tensile properties of two specimens cut from the disk were 1007 MPa (146 ksi) tensile strength, 17.5% elongation, and 39% reduction of area. See also Fig. 1135 to 1140. Fig. 1133: 0.2×. Fig. 1134: 0.9×

Fig. 1135 Enlarged view of the fatigue-crack regions (indicated by brackets) at the edges of the bolthole in Fig. 1134. The cracks grew in both directions from the hole. 3×

Fig. 1136 Separate, higher-magnification views of the bracketed regions of the edges of the compressor-disk bolthole in Fig. 1134 and 1135, shown correctly aligned opposite each other. Visible are beach marks in several fatigue-crack zones (at arrows). These separate cracks penetrated only a short distance before joining to form common fronts, which then advanced toward both the rim and the bore of the disk. 8×

Fig. 1137 View of the bore of the bolthole at arrow "a" in the fractured titanium alloy Ti-6Al-4V compressor disk in Fig. 1133, showing more clearly the deep tool marks faintly visible in Fig. 1134 to 1136. The arrows mark locations of longitudinal secondary cracks in the bore of the hole. 15×

Fig. 1138 View of the edge of the bolthole through which passed one of the secondary radial fractures (at left or at bottom) in the compressor disk in Fig. 1133. Note the irregular shear-overload cracks at the edge of the hole, which show no characteristics of fatigue, and the irregular secondary cracks in the bore. 4×

Fig. 1139, 1140 Figure 1139 (left) shows a section through one of the fatigue-crack zones at one of the bolthole edges in Fig. 1136, taken transversely to the hole. The fracture surface is shown in profile at right, and the bore of the hole is shown in profile at top. Note the layer of cold-worked metal above the dashed line. Figure 1140 (right) shows a section through the same general area as that in Fig. 1139, but taken longitudinally to the hole. Here, as in Fig. 1139, a cold-worked layer can be seen (at top); severe surface flow in the bore of the hole is also visible. This damaged condition of the bore surface, and the deep tool marks shown in Fig. 1137, probably caused the fatigue fracture. Both etched in HF plus HNO_3, in H_2O. 100×

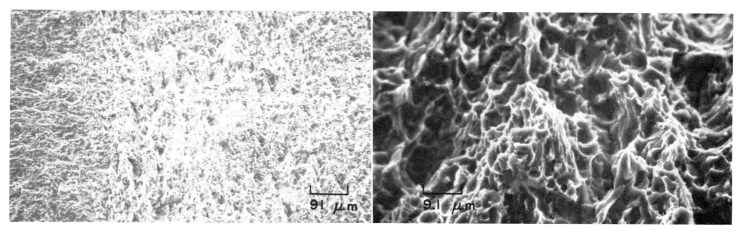

Fig. 1141 Fracture surface of a ductile fracture-toughness specimen of titanium alloy Ti-6Al-4V that was solution treated for 40 min at 830 °C (1525 °F), water quenched, aged at 510 °C (950 °F), then loaded in three-point bending (in air). See also Fig. 1142 to 1145. SEM, 110×

Fig. 1142 Higher-magnification view of the fracture surface in Fig. 1141, typical of fractures produced in this alloy in air after various solution treatments. Entire surface shows dimpled fracture of fine, equiaxed α phase that ruptured in ductile shear. SEM, 1100×

448 / Titanium Alloys

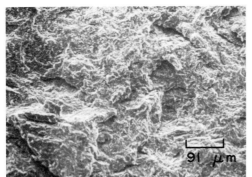

Fig. 1143 Fracture surface of a specimen of Ti-6Al-4V alloy similar to the specimen shown in Fig. 1141 and having the same history, but tested in a hydrogen atmosphere. The principal features here are numerous secondary cracks. See also Fig. 1144. SEM, 110×

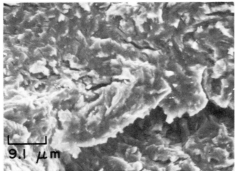

Fig. 1144 Higher-magnification view of the fracture surface shown in Fig. 1143. The embrittlement of this specimen by hydrogen was slight (about 12%). Fracture was by a mixture of shear rupture and transgranular cleavage. See also Fig. 1145. SEM, 1100×

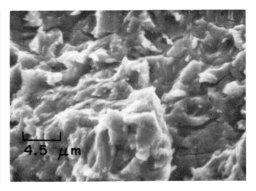

Fig. 1145 Same fracture surface as in Fig. 1143 and 1144, but shown at even greater magnification. Many small secondary cracks are present among moderate-sized dimples. SEM, 2200×

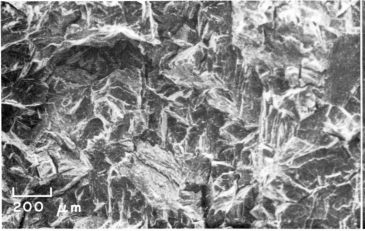

Fig. 1146 A Ti-6Al-4V fracture-toughness specimen identical to that in Fig. 1155 (next page), except broken in hydrogen at 25 °C (77 °F). This fracture surface appears to be very brittle, showing intergranular secondary cracks that follow exceedingly angular paths. Note the larger cleavage facets. See also Fig. 1147. SEM, 50×

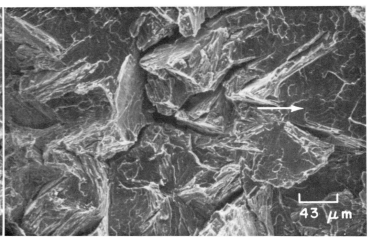

Fig. 1147 Higher-magnification view of the fracture surface in Fig. 1146, showing quasi-cleavage facets and detailed river patterns. Note particularly the unusual concentric pattern of steps at arrow and compare it with the similar pattern of steps that is shown at higher magnification in Fig. 1148. SEM, 230×

Fig. 1148 Higher-magnification view of the fracture surface in Fig. 1146, showing an area different from that in Fig. 1147. As in Fig. 1146, the very angular secondary cracking is noteworthy and is believed to have been influenced by interfaces between β phase and acicular α phase. Note the fine parallel steps on the angular facets at bottom right. SEM, 230×

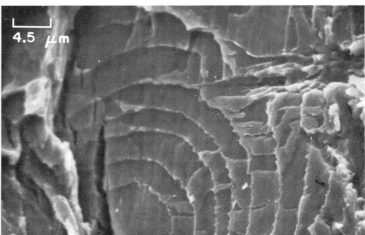

Fig. 1149 Another view of the fracture surface in Fig. 1146 to 1148, at even higher magnification, showing a region containing a very unusual "terraced" facet that is similar to the area shown and discussed in Fig. 1147. It is possible that this relatively flat area was a prior-β grain. Also note that the "terraces" have cleavage steps of their own. SEM, 2200×

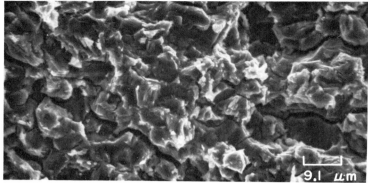

Fig. 1150 Fracture surface of a fracture-toughness specimen of titanium alloy Ti-6Al-4V that was heat treated for 40 min at 955 °C (1750 °F), stabilized* and then tested at 25 °C (77 °F) in hydrogen. The deep secondary cracks are interpreted as following the boundaries between the grains of primary α and the β matrix. See also Fig. 1151. SEM, 1100×

Fig. 1151 View of the fracture surface in Fig. 1150, seen at higher magnification, which shows in greater detail the fine secondary cracks; these are transgranular, in contrast to the network of major secondary cracks. The fracture is brittle, but the facets are small and irregular and have no detectable river patterns. SEM, 5000×

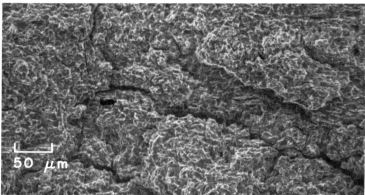
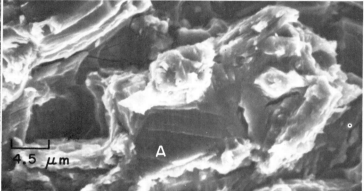

Fig. 1152 Fracture surface of a fracture-toughness specimen same as in Fig. 1150, except this specimen received a 24-h soak at 955 °C (1750 °F) before undergoing the stabilizing treatment.* A gross network of secondary intergranular cracks is evident, providing steps that separate different levels of the main crack advance. See also Fig. 1153. SEM, 200×

Fig. 1153 View of the fracture surface in Fig. 1152, at 11 times the magnification there. Note the intricate rupture patterns. Some of the main cracks are believed to follow the interfaces between the continuous β-phase matrix and the dispersed acicular α phase. Observe the "terraced" area at A; similar areas are visible in Fig. 1149. SEM, 2200×

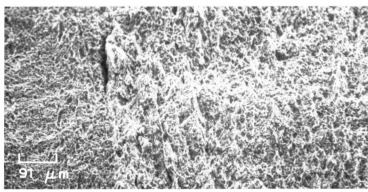
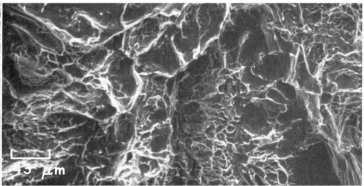

Fig. 1154 Fracture surface of a fracture-toughness specimen of titanium alloy Ti-6Al-4V heat treated 40 min at 955 °C (1750 °F) and water quenched, aged at 510 °C (950 °F), and tested in hydrogen. The fatigue-precrack region is at left. The tensile-overload region, at right, closely resembles that of the specimen in Fig. 1158. SEM, 110×

Fig. 1155 Fracture surface of a fracture-toughness specimen of titanium alloy Ti-6Al-4V heat treated 40 min at 1040 °C (1900 °F), stabilized,* and then tested in air at 25 °C (77 °F). Note the tremendous range in size of the dimples. See also Fig. 1146 to 1149. SEM, 230×

NOTE: All structures shown on this page are continuous β phase with dispersed α phase.

*Stabilizing consisted of furnace cooling to 704 °C (1300 °F) from the heat-treating temperature, holding 1 h, furnace cooling to 593 °C (1100 °F), holding 1 h and air cooling. This treatment was used to secure a large grain size to increase the susceptibility of the alloy to slow-strain-rate embrittlement in hydrogen.

450 / Titanium Alloys

Fig. 1156 Fracture surface of a specimen of titanium alloy Ti-6Al-4V mill-annealed sheet that was fracture-toughness tested in air. At left (dark) is fatigue-precrack region; at right is region of tensile-overload fracture, which is dimpled and has aligned elongated cavities suggesting localized slip. See also Fig. 1157. SEM, 110×

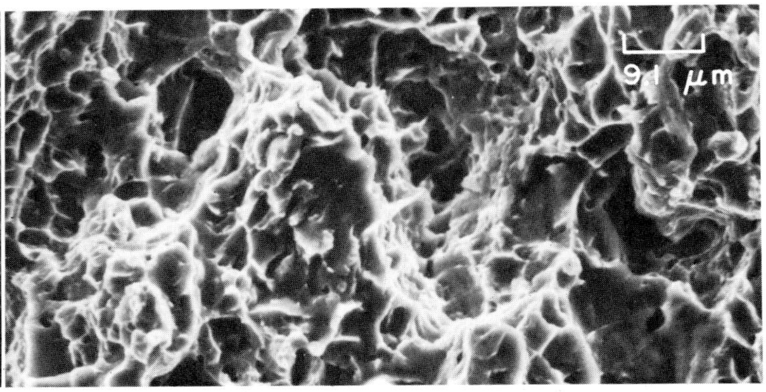

Fig. 1157 Portion of the tensile-overload fracture region of the fracture surface in Fig. 1156, as seen at ten times the magnification there. This displays a rather uniform size of equiaxed dimples, shows essentially no inclusions, and contains no indication of secondary cracking; in summary, this is a thoroughly ductile fracture. SEM, 1100×

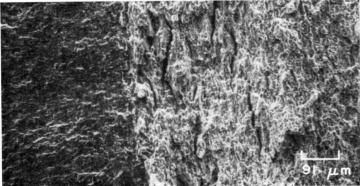

Fig. 1158 Fracture surface of a specimen of the same material and condition as in Fig. 1156 (mill-annealed Ti-6Al-4V sheet), but fracture-toughness tested in hydrogen at 1 atm. The fatigue-precrack region (dark area at left), produced in air, is identical with that in Fig. 1156, but note the deep cracks among the dimples in the tensile-overload region at right. See also Fig. 1159. SEM, 110×

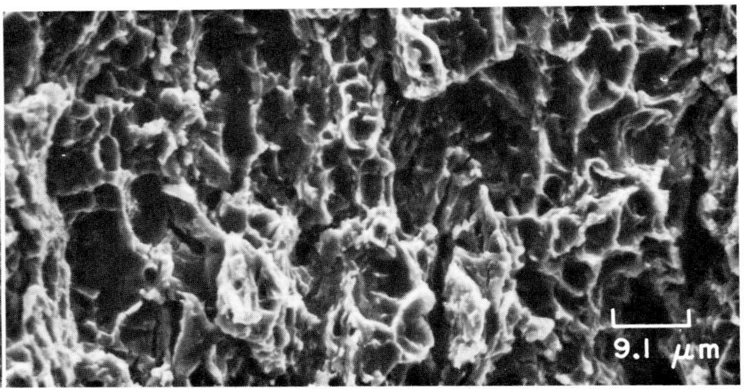

Fig. 1159 View of the region of tensile-overload fracture in Fig. 1158, at ten times the magnification there. In spite of the effects of hydrogen, this region is quite similar in appearance to the tensile-overload region of the air-tested specimen in Fig. 1157. However, note that here there are many fine, parallel secondary cracks scattered throughout the dimpled surface. SEM, 1100×

Fig. 1160 Same as Fig. 1158, except that before being fracture-toughness tested in hydrogen, this specimen was heat treated at 705 °C (1300 °F) for 2 h and air cooled. In contrast to the fracture surface shown in Fig. 1158, no dimples are evident here. The surface exhibits giant steps. See also Fig. 1161. SEM, 110×

Fig. 1161 Area outlined by the rectangle in Fig. 1160, as seen at ten times the magnification there. The rim of the central step in Fig. 1160 is at top right, and secondary fissures extend diagonally across the view. This surface contains no large cleavage facets and has been characterized as "feathery."

NOTE: All structures shown on this page are continuous α phase with dispersed β phase.

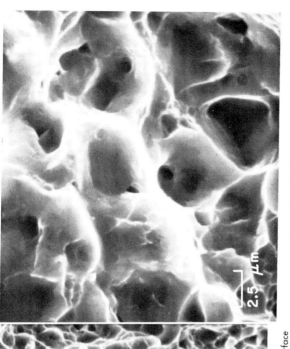

Fig. 1164 Area outlined by the rectangle in Fig. 1163, as seen at four times the magnification there. Note the numerous small dimples that are situated within the larger dimples. This ductile fracture exhibits no sign of secondary cracking. SEM, 4000×

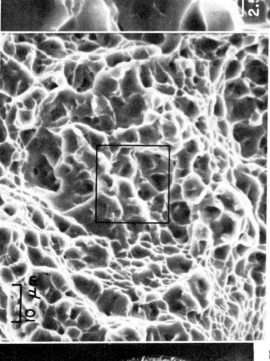

Fig. 1163 View of the fibrous central zone of the fracture surface in Fig. 1162, showing the equiaxed dimples expected in this region. The dimples are relatively uniform in size, with none being extremely large. A few inclusions are evident. See Fig. 1164 for an enlarged view of the area in the rectangle. SEM, 1000×

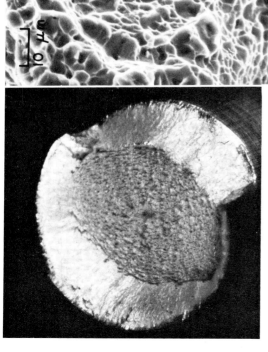

Fig. 1162 Tensile-overload fracture surface of an unnotched specimen of titanium alloy Ti-6Al-4V heat treated to tensile strength of 1158.8 MPa (168.5 ksi) and 47% reduction of area. A classic example of cup-and-cone fracture having a flat, fibrous central zone. See also Fig. 1163. 9×

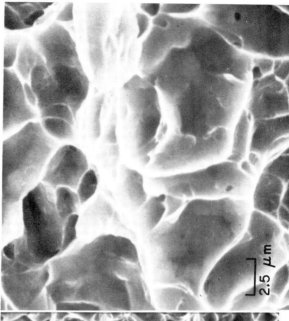

Fig. 1167 View a higher magnification of area in the rectangle in Fig. 1168. Note the small dimples that are present both at the edges of and within the larger dimples. SEM, 4000×

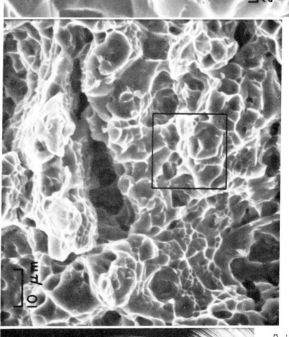

Fig. 1166 Center of fracture surface in Fig. 1165. In contrast to the fracture surface of the unnotched specimen in Fig. 1163, dimples of various sizes are displayed here, and some secondary cracking is detectable. See also Fig. 1167. SEM, 1000×

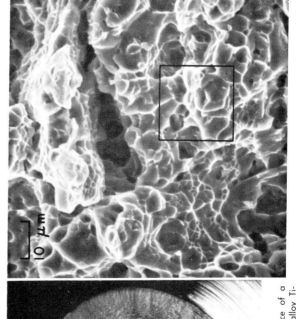

Fig. 1165 Tensile-overload fracture surface of a notched specimen of titanium alloy Ti-6Al-4V heat treated to the same mechanical properties as specimen in Fig. 1162 and with a notched tensile strength of 1696 MPa (246 ksi). Crack nucleus is below center, at right. See also Fig. 1166. 9×

452 / Titanium Alloys

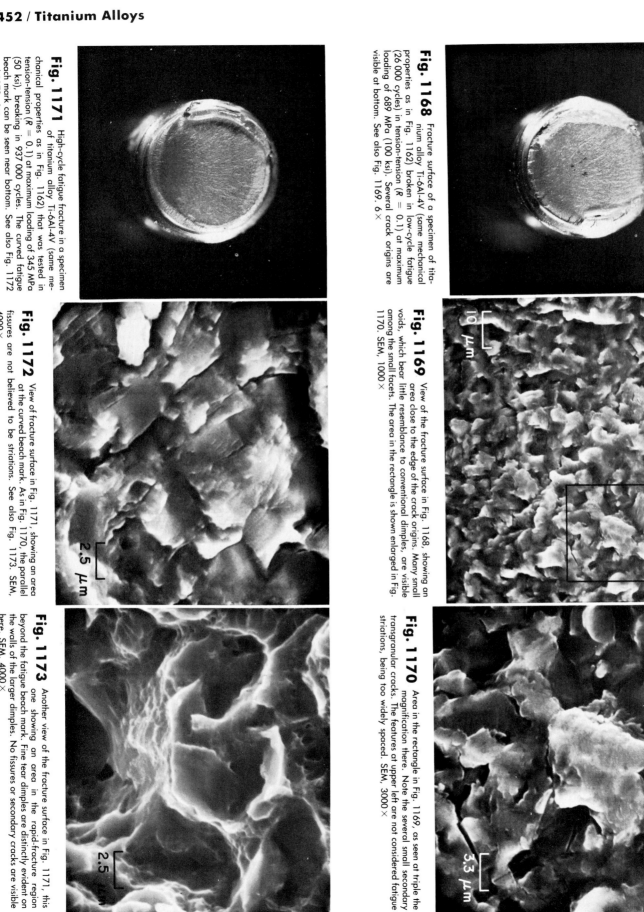

Fig. 1168 Fracture surface of a specimen of titanium alloy Ti-6Al-4V (same mechanical properties as in Fig. 1162) broken in low-cycle fatigue (26 000 cycles) in tension-tension ($R = 0.1$) at maximum loading of 689 MPa (100 ksi). Several crack origins are visible at bottom. See also Fig. 1169. 6×

Fig. 1169 View of the fracture surface in Fig. 1168, showing an area close to the edge of the crack origins. Many small voids, which bear little resemblance to conventional dimples, are visible among the small facets. The area in the rectangle is shown enlarged in Fig. 1170. SEM, 1000×

Fig. 1170 Area in the rectangle in Fig. 1169, as seen at triple the magnification. Note the several small secondary transgranular cracks. The features at upper left are not considered fatigue striations, being too widely spaced. SEM, 3000×

Fig. 1171 High-cycle fatigue fracture in a specimen of titanium alloy Ti-6Al-4V (same mechanical properties as in Fig. 1162) that was tested in tension-tension ($R = 0.1$) at maximum loading of 345 MPa (50 ksi), breaking in 937 000 cycles. The curved fatigue beach mark can be seen near bottom. See also Fig. 1172 and 1173. 6×

Fig. 1172 View of fracture surface in Fig. 1171, showing an area at the curved beach mark. As in Fig. 1170, the parallel fissures are not believed to be striations. See also Fig. 1173. SEM, 4000×

Fig. 1173 Another view of the fracture surface in Fig. 1171, this one showing an area in the rapid-fracture region beyond the curved beach mark. Fine tear dimples are distinctly evident on the walls of the larger dimples. No fissures or secondary cracks are visible here. SEM, 4000×

Titanium Alloys / 453

Fig. 1174 Fracture resulting from stress-corrosion cracking in a specimen of titanium alloy Ti-6Al-4V that was stressed in methanol. The crack path is transgranular (by cleavage). See also Fig. 1175. TEM p-c replica, 2000×

Fig. 1175 A companion fractograph to that in Fig. 1174, from the same specimen of titanium alloy Ti-6Al-4V, but showing an area farther from the fracture origin. Visible are what appear to be a few small dimples. TEM p-c replica, 2000×

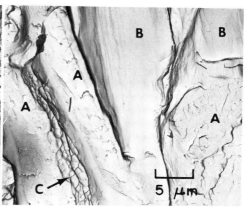

Fig. 1176 Bending fracture in a specimen of titanium alloy Ti-6Al-4V that broke by stress-corrosion cracking in 3.5% NaCl (areas A), but that also exhibits regions of stretching (areas B) and a step containing what appear to be dimples (arrow at C). TEM p-c replica, 2000×

Fig. 1177 Surface of a fracture in a specimen of titanium alloy Ti-7Al-1Mo-1V that was broken in a drop-weight tear test, consisting mostly of dimples. A feature that apparently is a tear ridge (vertical) is visible at far right. TEM p-c replica, 2000×

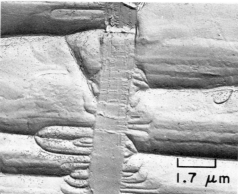

Fig. 1178 Fracture produced by tension overload in a specimen of titanium alloy Ti-7Al-2Nb-1Ta, showing staining and very large dimples. Vertical band at center is where crack initiated at a second-phase region and propagated to left and right. TEM p-c replica, 6000×

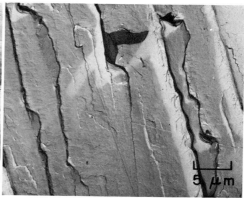

Fig. 1179 Fracture resulting from stress-corrosion cracking in a chill cast specimen of titanium alloy Ti-7Al-2Nb-1Ta that was stressed in distilled water, exhibiting facets that were produced by essentially pure cleavage. TEM p-c replica, 2000×

Fig. 1180 Cleavage fracture in a specimen of titanium alloy Ti-7Al-2Nb-1Ta, produced by stress-corrosion cracking in methanol. Note the regular array of fissures, which are parallel to the crack front. See also Fig. 1181. TEM p-c replica, 2000×

Fig. 1181 Companion fractograph to that in Fig. 1180, from the same specimen, but taken at the stress-corrosion crack front. In general, the facets here show features of cleavage; in contrast, the facets at top center show what apparently are ripples. TEM p-c replica, 2000×

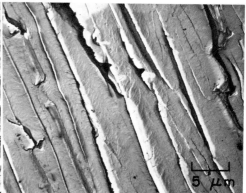

Fig. 1182 Cleavage fracture in another specimen of titanium alloy Ti-7Al-2Nb-1Ta, also produced by stress-corrosion cracking in methanol. Note the similarity of the cleavage facets here to those in Fig. 1179, which were produced in distilled water. TEM p-c replica, 2000×

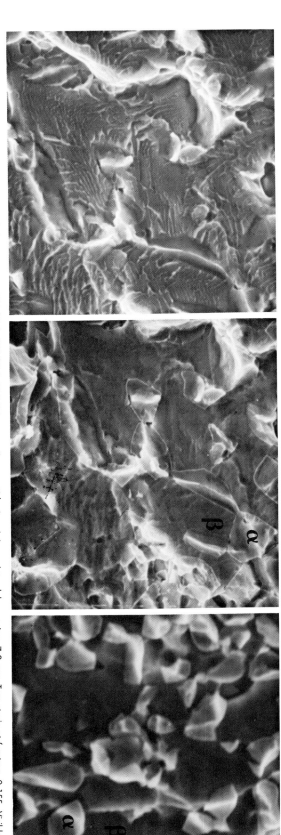

Fig. 1183, 1184, 1185 Tensile fractures at −196 °C (−320 °F) in Ti-5.8Mn, an α-β alloy. Volume fraction β: 0.64. In all cases, 9-kg (20-lb) ingots were β forged to 75-mm (3-in.) billets, followed by β extrusion to 23 mm (0.92 in.) in diameter and α-β swaging to 16 mm (0.62 in.) in diameter. Etchant (when used): 17 mL benzalkonium chloride, 35 mL ethanol, 40 mL glycerine, 25 mL HF. Fig. 1183 (left) and 1184 (center): Some region unetched and etched, respectively, on fracture surface of alloy annealed 200 h at 700 °C (1290 °F) and water quenched. Alpha particle size: 4.4 μm. Interalpha spacing: 7.8 μm. True strain at fracture: 0.155. Visible are regions of ductile α and brittle β matrix. Note deformation markings in β and voids in α. SEM, 2100×. Fig. 1185 (right): Etched fracture surface of alloy annealed 24 h at 700 °C (1290 °F) and water quenched. Alpha particle size: 2.35 μm. Interalpha spacing: 4.17 μm. True strain at fracture: 0.10. Fracture is brittle except for some local α ductility. SEM, 2000×. (H. Margolin and R.V. Vijayaraghavan, Polytechnic Institute of New York)

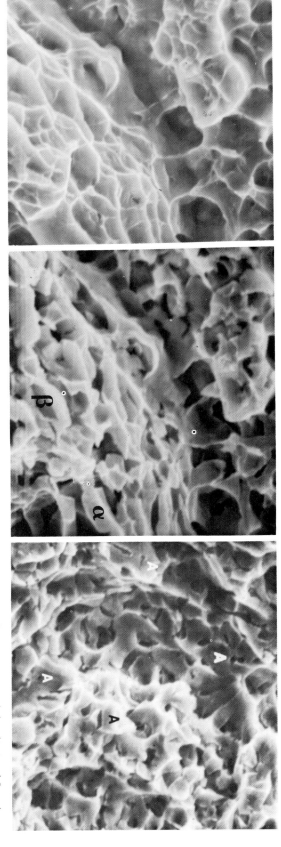

Fig. 1186, 1187, 1188 Tensile fractures at −196 °C (−320 °F) in Ti-3.9Mn alloy. Volume fraction β: 0.38. Beta particle spacing: 2.1 μm. Interbeta spacing: 4.5 μm. Processing and heat treatment same as in Fig. 1189 and 1190. Etchant (when used) same as in Fig. 1183 to 1185. Fig. 1186 (left) and 1187 (center): Unetched and etched, respectively, fracture surface of tensile specimen with prior strain at room temperature of 0.140. True strain at subzero fracture: 0.253. Material machined using coolant during final stages of specimen preparation. Compared with Fig. 1184 and 1185, average dimple size is smaller and there is comparatively less thinning of β at shear regions. SEM, 3000×. Fig. 1188 (right): Etched fracture surface of specimen machined without coolant. True strain at fracture: 0.144. Cooling during machining substantially increases fracture strain. Fractograph is of region near origin, which was close to specimen surface. Note wavy nature of fracture propagation and large number of β particles at dimple boundaries (A). Dark regions are β particles. SEM, 1600×. (H. Margolin and R.V. Vijayaraghavan, Polytechnic Institute of New York)

Titanium Alloys / 455

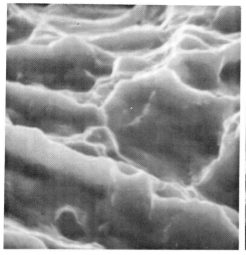 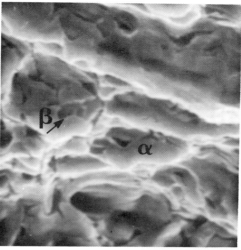 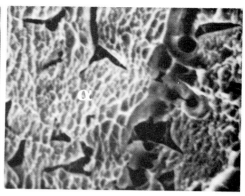

Fig. 1189, 1190 Tensile fracture at room temperature in Ti-3.9Mn, an α-β alloy with a β volume fraction of 0.38, β particle size of 2.1 μm, and interbeta spacing of 44.25 μm. Processing same as in Fig. 1183 to 1185. Heat treatment: anneal 24 h at 700 °C (1290 °F), water quench. True strain at fracture was 0.68. Note particles inside voids. Fig. 1189 (left): Unetched fracture surface. SEM, 3800×. Fig. 1190 (right): Etched fracture surface. Etchant same as in Fig. 1183 to 1185. SEM, 3000× (H. Margolin and R.V. Vijayaraghavan, Polytechnic Institute of New York)

Fig. 1191 Tensile fracture at −196 °C (−320 °F) in Ti-1.8Mn, an α-β alloy with a β volume fraction of 0.17. Beta particle size: 6.30 μm. Interbeta spacing: 30.8 μm. Material processed as in Fig. 1183 to 1185. Heat treatment: 30 days at 780 °C (1435 °F); furnace cool over 3 days to 700 °C (1290 °F); 24 h at 700 °C (1290 °F); water quench. Fracture surface etched as in Fig. 1183 to 1185. Note shear dimples in α and β (finer dimples in α). True strain to fracture: 0.90 (0.93 at room temperature). SEM, 1250× (H. Margolin and R.V. Vijayaraghavan, Polytechnic Institute of New York)

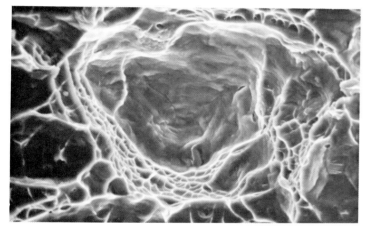 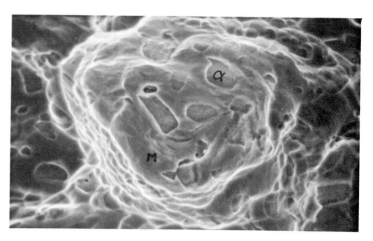

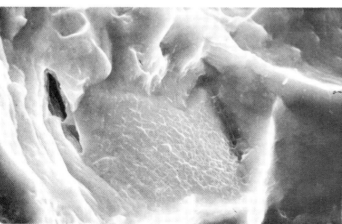 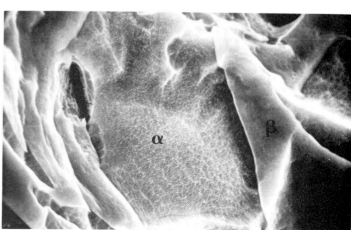

Fig. 1192 to 1195 Tensile fractures at room temperature in Ti-5Mo-4.5Al-1.5Cr. The α-β alloy was processed as in Fig. 1183 to 1185. Etchant (when used) same as in Fig. 1183 to 1185. Fig. 1192 (top left) and 1193 (top right): Unetched and etched, respectively, fracture surfaces of alloy having α particles in a martensite matrix. Volume fraction α: 0.30. Alpha particle size: 9.0 μm. Interalpha spacing: 20.0 μm. Fracture was ductile. Note shear at the boundary of the large void and particles inside the void. SEM, 570× and 620×. Fig. 1194 (bottom left) and 1195 (bottom right): Unetched and etched, respectively, fracture surfaces of alloy having α particles in a retained β matrix. (Tilt slightly different in etched Fig. 1195.) Alpha particle size: 11.6 μm. Interalpha spacing: 8.6 μm. Note large void at α-β interface close to shear. SEM, 3250× (H. Margolin and R.V. Vijayaraghavan, Polytechnic Institute of New York)

456 / Miscellaneous Metals and Alloys

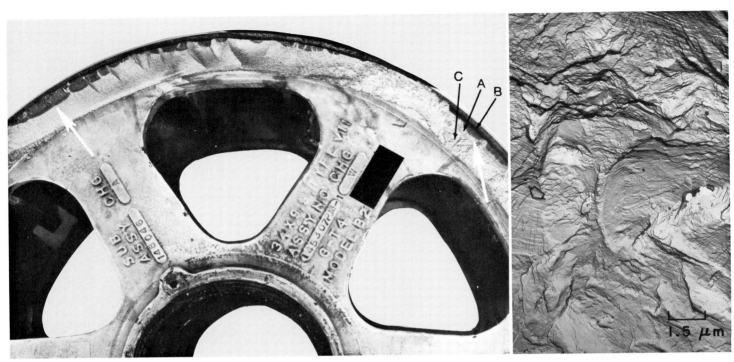

Fig. 1196 Portion of an aircraft landing wheel of sand-cast magnesium alloy AZ81A-T4, showing a corrosion-fatigue fracture that occurred in service at the locking-ring channel. Nominal tensile strength was 276 MPa (40 ksi). The fracture extended halfway around the perimeter of the wheel. Many discolored, crescent-shaped fatigue-crack nuclei were found; two are indicated by white arrows. See Fig. 1197 to 1200 for TEM views of areas at A, B, and C. 0.3×

Fig. 1197 TEM p-c replica of an area at the fatigue-crack nucleus at A in the fracture surface in Fig. 1196, showing "brittle" fatigue striations (such as at the arrow at left center). 6500×

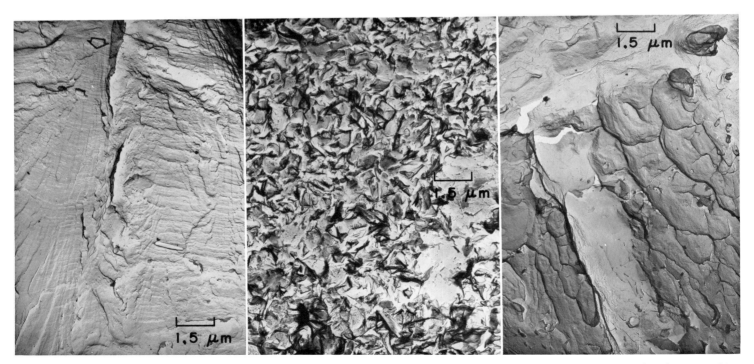

Fig. 1198 TEM p-c replica of another area at A in Fig. 1196. Fine, roughly horizontal striations are visible in a fatigue patch at the arrow at top center; the marks at the right include widely spaced horizontal fissures. 6500×

Fig. 1199 TEM p-c replica of an area at B in Fig. 1196. In this area, the fracture surface is largely obscured by corrosion products. The few facets that are visible have essentially no distinguishable features. 6500×

Fig. 1200 TEM p-c replica of an area at C in Fig. 1196, in the region of final fast fracture. This area exhibits rather flat, elongated dimples. The region at left center in this view appears to have been stretched. 6500×

Miscellaneous Metals and Alloys / 457

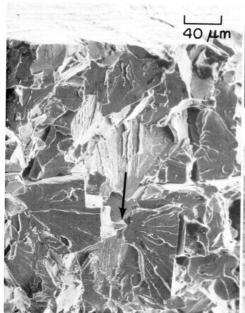

Fig. 1201 Surface of a slow-bending fracture produced at −160 °C (−256 °F) in Fe-3.9Ni alloy deoxidized with aluminum and titanium, which was forged at 1260 °C (2300 °F), hot rolled at 1040 °C (1905 °F), annealed at about 1000 °C (1830 °F), and air cooled. Root of specimen notch is at top. Fracture was by cleavage that began at cracked Ti(C,N) particles (as at arrow). See also Fig. 1202. SEM, 250×

Fig. 1202 Higher-magnification view of the slow-bending fracture in Fig. 1201. During testing, specimen temperature was controlled by an isopentane bath cooled with liquid nitrogen. Ferrite grain size, 30 μm. The angle of the notch was 45°, and the nominal radius of the root of the notch was 0.25 mm (0.01 in.). The fracture originated at cracked Ti(C,N) particles (as at arrow). SEM, 1200×

Fig. 1203 Surface of a slow-bending fracture that was produced at −130 °C (−200 °F) in an Fe-0.3Ni alloy that had been melted, deoxidized, forged, rolled, and heat treated exactly the same as the higher-nickel alloy shown in Fig. 1201 and 1202. See also Fig. 1204 to 1206. SEM, 550×

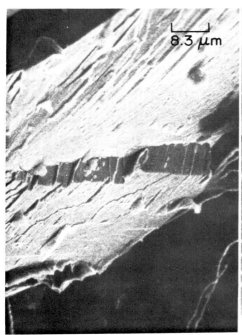

Fig. 1204 Another view of the slow-bending fracture in Fig. 1201, showing a different region of the fracture surface at higher magnification. Here, as in the specimen shown in Fig. 1201 and 1202, fracture was by quasi-cleavage. Also shown is the intersection of a twin boundary with the cleavage plane. See also Fig. 1205. SEM, 1200×

Fig. 1205 Surface of a bending fracture in a specimen of the same alloy and the same history as that in Fig. 1203, showing assorted river patterns. Note twist boundary at center, which has generated several cleavage steps. Although the higher nickel content of the alloy in Fig. 1201 gave greater strength, the aspects of subzero impact rupture are about the same for both alloys. See also Fig. 1206. SEM, 750×

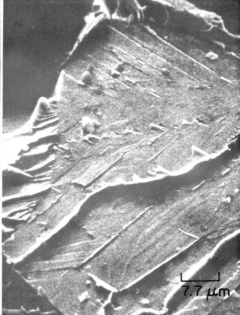

Fig. 1206 Another view of the bending fracture in Fig. 1205, showing a different region at higher magnification. This region best demonstrates the difference between this Fe-0.3Ni alloy and the Fe-3.9Ni alloy in Fig. 1201 and 1202 by displaying many small "tongues," or microtwins, which are charcteristic of Fe-0.3Ni and also of unalloyed (Armco) iron, but not of Fe-3.9Ni. SEM, 1300×

458 / Miscellaneous Metals and Alloys

Fig. 1207 Both fracture surfaces of a specimen that was not aged after austenitizing. See also Fig. 1208. 4.8×

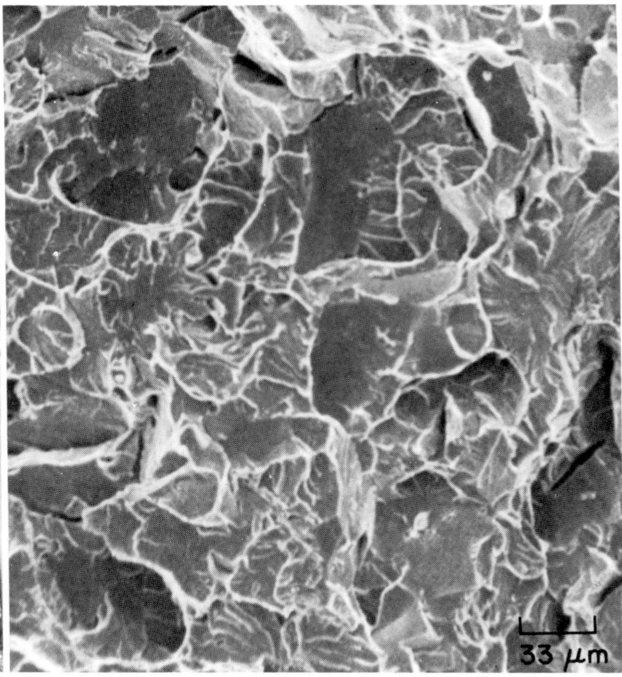

Fig. 1208 This SEM fractograph of the unaged specimen in Fig. 1207 shows quasi-cleavage facets. The impact energy was less than 27 J (20 ft · lb). 300×

Fig. 1209 Both fracture surfaces of a specimen aged at 650 °C (1200 °F) 2 h and air cooled. See also Fig. 1210. 4.8×

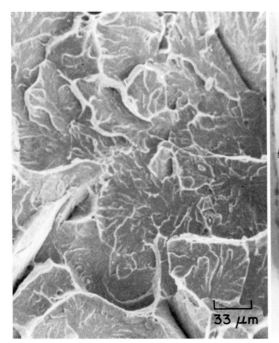

Fig. 1210 This SEM view of the specimen in Fig. 1209 shows that aging at 650 °C (1200 °F) produced almost no change in fracture-surface characteristics from those of unaged specimen in Fig. 1207. Impact energy was less than 27 J (20 ft · lb). 300×

Fig. 1211 Both fracture surfaces of a specimen aged at 700 °C (1290 °F) 2 h and air cooled. Impact energy was about 203 J (150 ft · lb). Much deformation accompanied fracture. See also Fig. 1212. 4.8×

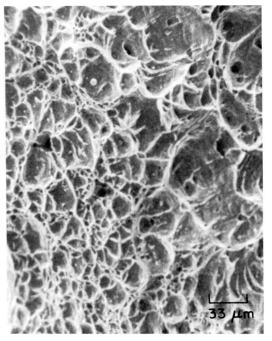

Fig. 1212 This SEM fractograph of the specimen in Fig. 1211, which was aged at 700 °C (1290 °F), shows dimpled rupture instead of the quasi-cleavage facets observed in the specimens in Fig. 1208 and 1210. 300×

NOTE: Fig. 1207 to 1218 present six related pairs of light and SEM fractographs of specimens of an Fe-12Ni-0.5Ti alloy intended for use at cryogenic temperatures. The series shows the effect on Charpy impact energy at −196 °C (−321 °F), and on fracture-surface characteristics, of the temperature of aging after austenitizing at 900 °C (1650 °F) for 2 h and air cooling.

Fig. 1213 Both fracture surfaces of a specimen aged at 750 °C (1380 °F) 2 h and air cooled. Impact energy was about 203 J (150 ft · lb); deformation is similar to that in Fig. 1211. See also Fig. 1214. 4.8×

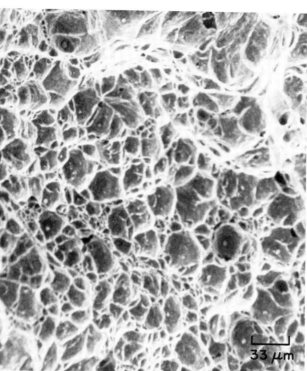

Fig. 1214 SEM fractograph of the specimen in Fig. 1213. This fracture, like that of the specimen in Fig. 1212, shows dimpled rupture. 300×

Fig. 1215 Specimen aged at 800 °C (1470 °F) 2 h and air cooled. Unlike Fig. 1213, this fracture shows little deformation, and impact energy was less than 27 J (20 ft · lb). See also Fig. 1216. 4.8×

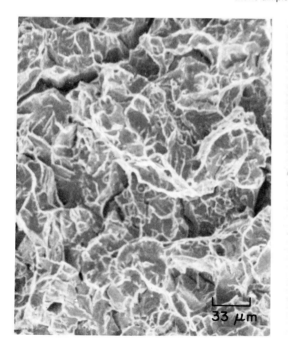

Fig. 1216 SEM view of specimen in Fig. 1215. Fracture appears to have propagated by a mixture of intergranular separation and cleavage, yet shows indications of some dimples. Note the deep secondary intergranular cracks. 300×

Fig. 1217 Specimen aged at 850 °C (1560 °F) 2 h and air cooled. The fracture is much coarser than that of the specimen in Fig. 1215, but is similar to it in other respects. See also Fig. 1218. 4.8×

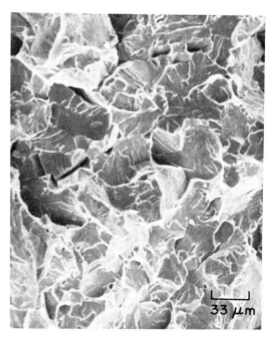

Fig. 1218 This SEM fractograph of the specimen in Fig. 1217 shows that the fracture, like that in Fig. 1216, exhibits cleavage facets and short, deep intergranular cracks. This specimen also had impact energy of less than 27 J (20 ft · lb).

NOTE: Fig. 1207 to 1218 present six related pairs of light and SEM fractographs of specimens of an Fe-12Ni-0.5Ti alloy intended for use at cryogenic temperatures. The series shows the effect on Charpy impact energy at −196 °C (−321 °F), and on fracture-surface characteristics, of the temperature of aging after austenitizing at 900 °C (1650 °F) for 2 h and air cooling.

Fig. 1219 Fracture surface in an iron-base alloy containing 7 at.% (6.4 wt%) Cr and 1 at.% (3.2 wt%) Ta, pulled in tension at room temperature after solution treatment for 1 h at 1320 °C (2410 °F) and water quenching, then aging at 700 °C (1290 °F) for 40 min and air cooling. Fracture was by transgranular cleavage believed to have originated at a grain-boundary network of Laves-phase Fe_2Ta precipitate. See also Fig. 1220 to 1222. SEM, 100×

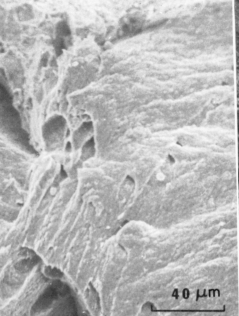

Fig. 1220 Fracture surface in a specimen of the same Fe-Cr-Ta alloy, and with the same heat treatment, as in Fig. 1219, but tensile tested at 200 °C (390 °F). In addition to cleavage, this fracture shows features of quasi-cleavage, but the higher test temperature provided greater ductility and a number of elongated dimples made their appearance. See also Fig. 1221 and 1222. SEM, 500×

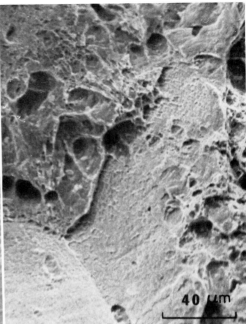

Fig. 1221 Same Fe-Cr-Ta alloy, and with the same heat treatment, as in Fig. 1219 and 1220, but fractured by tension overload at 400 °C (752 °F). Its appearance suggests rupture along two grain boundaries (at bottom center), as well as an increased quantity of dimples compared with the number present in Fig. 1220. The dimples here are also much more equiaxed than those in Fig. 1220. See also Fig. 1222. SEM, 500×

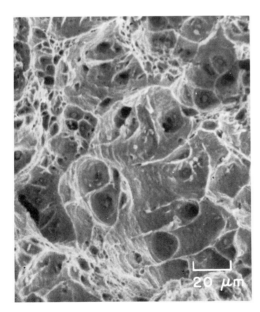

Fig. 1222 Fracture surface in a specimen of the same Fe-Cr-Ta alloy, and with the same heat treatment, as in Fig. 1219 to 1221, but which was broken in tension at 600 °C (1112 °F). At this test temperature, the ductility was sufficient to eliminate completely both cleavage and intergranular mechanisms of failure, and fracture occurred entirely by microvoid coalescence. SEM, 500×

Fig. 1223 Room-temperature tensile fracture by cleavage of an Fe-6.2Ta alloy solution treated 1 h at 1400 °C (2550 °F), water quenched, aged 1 h at 700 °C (1290 °F) and air cooled. Laves-phase Fe_2Ta precipitate, uniformly distributed in the matrix and believed present at grain boundaries, caused brittleness (<1% tensile elongation). See also Fig. 1224. SEM, 100×

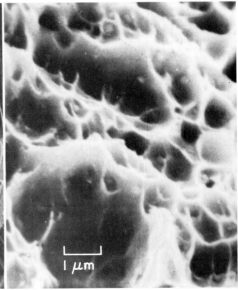

Fig. 1224 Room-temperature tensile fracture in a specimen of the same alloy as in Fig. 1223, given the same heat treatment plus a spheroidizing treatment of 10 min at 1100 °C (2010 °F) and air cooling. Spheroidization of precipitates at grain boundaries and within grains, plus a finer grain size than in Fig. 1223, resulted in dimpled fracture and 22% tensile elongation. SEM, 10 000×

Miscellaneous Metals and Alloys / 461

Fig. 1225 Surface of a "rock candy" fracture in a notched as-cast specimen of Alnico alloy broken by impact at room temperature, showing large, shiny facets. The brittle nature of this specimen is reflected in its unnotched tensile strength of only 34 MPa (5 ksi). See also Fig. 1226 to 1231. 5.5×

Fig. 1226 TEM p-c replica of an area of the facet at the arrow in Fig. 1225, showing features resulting from transgranular cleavage. Pronounced cleavage steps in central region indicate that fracture progressed from bottom to top. The margins of this central region are believed to have formed at grain boundaries; the cause of the marked irregularity of the margin at top is not known. 6500×

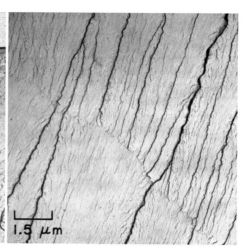

Fig. 1227 TEM p-c replica of a second area of the facet at the arrow in Fig. 1225, showing features resulting from transgranular cleavage and cleavage steps that cross a tilt boundary extending from lower right to upper left. The approximate direction of crack propagation was from lower left to upper right. 6500×

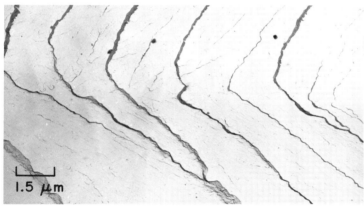

Fig. 1228 TEM p-c replica of a third area of the facet at the arrow in the fracture surface shown in Fig. 1225. Although no sharp boundary is visible, it appears certain that the cleavage steps, which entered from the top in this view, cross a tilt boundary at mid-height, where they change direction. 6500×

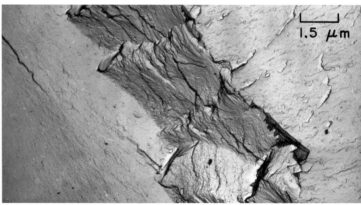

Fig. 1229 TEM p-c replica of a fourth area of the facet at the arrow in Fig. 1225, showing evidence that here fracture occurred in two different directions and on two different levels. The appearance of the fracture region that connects these two levels (diagonal dark band) has been markedly affected by local microstructure. 6500×

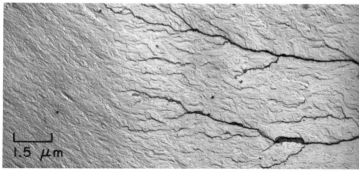

Fig. 1230 TEM p-c replica of a fifth area of the facet at the arrow in Fig. 1225. The fracture entered this area from the left and, at a tilt boundary, began the formation of cleavage steps, creating a river pattern "flowing" toward the right. 6500×

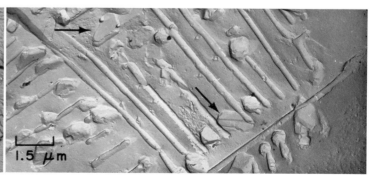

Fig. 1231 TEM p-c replica of a sixth area of the facet at the arrow in Fig. 1225. This area contains features that could be cleaved second-phase particles or imprints of pulled-out particles. Also visible is a geometric array of rodlike second-phase particles. 6500×

462 / Miscellaneous Metals and Alloys

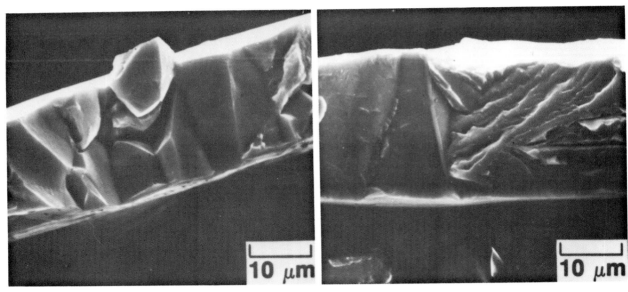

Fig. 1232, 1233 Fracture-mode transition in rapidly solidified ribbons of Fe-40Al (at.%). The material was made by chill block melt spinning. Samples were tested in tension, with the tensile force applied parallel to the ribbon length. Fig. 1232 (left): Fracture surface of as-spun ribbon. Fracture mode is primarily intergranular. SEM, 1600×. Fig. 1233 (right): Fracture surface of ribbon heat treated in helium for 1 h at 1000 °C (1830 °F). Fracture mode is primarily transgranular cleavage. SEM, 1600× (D. Gaydosh, NASA Lewis Research Center)

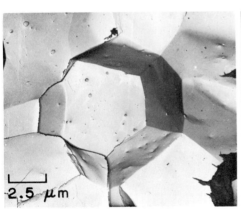

Fig. 1234 A very clean intergranular fracture in sintered tungsten that was broken by impact. Note the several very flat separated-grain facets; small sintering pores can be seen in several of the facets. TEM p-c replica, 4000×

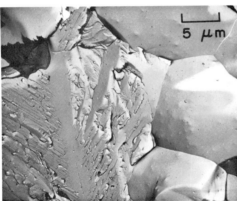

Fig. 1235 Impact fracture in sintered tungsten that had been recrystallized, showing separated-grain facets (at right) and cleavage facets (at left); the cleavage facets exhibit twins in the {112} plane. TEM p-c replica, 2000×

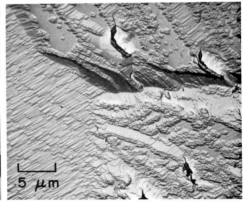

Fig. 1236 Surface of an impact fracture in a single crystal of tungsten that was etched with H_2O_2 after fracture. Cleavage along both the {100} and {113} planes is evident. TEM p-c replica, 2000×

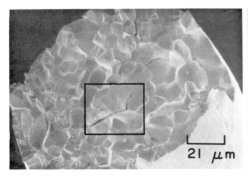

Fig. 1237 Surface of a brittle, intergranular fracture, produced by bending, in a polycrystalline iridium wire (0.127-mm, or 0.005-in., diam) that had been annealed in vacuum for 2 h at 1200 °C (2190 °F). See Fig. 1238 for an enlarged view of the area in the rectangle. SEM, 470×

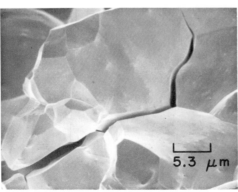

Fig. 1238 Higher-magnification view of the area in the rectangle in Fig. 1237. This entire area exhibits separated-grain facets and shows no evidence of plastic flow. SEM, 1900×

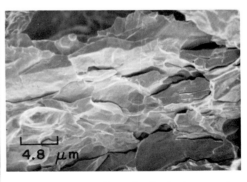

Fig. 1239 Surface of a brittle, intergranular fracture, produced by bending, in an iridium sheet (rolled to a thickness of 0.076 mm, or 0.0003 in.) that had been annealed for 2 h at 1200 °C (2190 °F) in vacuum. Note the deep secondary cracks between the elongated grains. 2100×

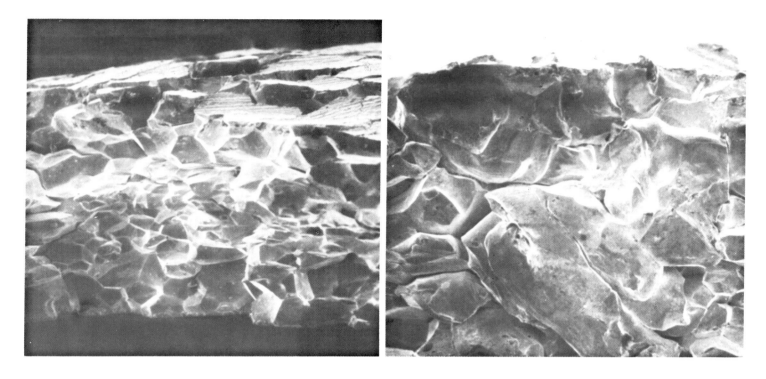

Fig. 1240 to 1243 Intergranular fracture at elevated temperatures in DOP-26, a thorium-doped Ir-0.3W alloy. A high-rate biaxial punch test was used to fracture a disk specimen measuring 52 mm (2 in.) in diameter and 0.635 mm (0.025 in.) thick. The intrinsic brittleness of iridium makes it unique among face-centered cubic metals. This ordinarily application-limiting property has been taken advantage of here to provide striking examples of exclusively intergranular fracture. Fig. 1240 (top left): Fracture surface of disk aged 18 h at 1500 °C (2730 °F) and then biaxially tested at 1440 °C (2625 °F). SEM, 115×. Fig. 1241 (top right): Another section of the fracture in Fig. 1240. SEM, 185×. Fig. 1242 (bottom left): Surface cracking in an iridium alloy disk annealed for 1 h at 1500 °C (2730 °F) and then biaxially tested at 1000 °C (1830 °F). SEM, 60×. Fig. 1243 (bottom right): High-magnification view of large crack near center of disk in Fig. 1242. SEM, 1000× (T.G. George, Los Alamos National Laboratory)

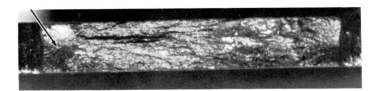

Fig. 1244 Surface of a tension-overload fracture in a notched specimen of stress-relieved TZM alloy (Mo-0.5Ti-1.0Zr) sheet. Unnotched tensile strength was 972 MPa (141 ksi). The specimen was notched on each edge to a depth of 2.5 mm (0.1 in.) with a jeweler's saw and broken at room temperature; notched tensile strength was 1007 MPa (146 ksi). Fracture produced distinct flat regions at the origins of each notch. See Fig. 1245 and 1246 (below) for TEM replicas of areas at the arrow. 9×

Fig. 1247 Surface of a high-cycle fatigue fracture in a stress-relieved specimen of 90Ta-10W alloy sheet containing a central transverse notch 5 mm (0.2 in.) wide. Tensile strength was 552 MPa (80 ksi). The specimen was loaded in tension at room temperature to a stress cycled from 80 to 400 MPa (11.6 to 58 ksi) at 1000 cycles/min and broke after 15 900 cycles. Considerable necking accompanied fast fracture. See also Fig. 1248 and 1249 (below) for TEM replicas of areas at the arrow. 6×

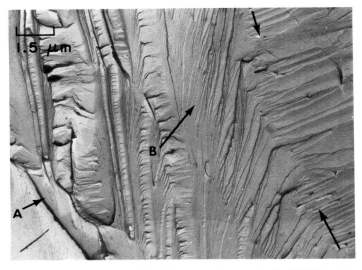

Fig. 1245 TEM p-c replica of an area at the arrow in Fig. 1244 (above). Here, the fracture surface shows transgranular cleavage. Tilt boundary at A; elsewhere, twist boundaries (between arrows) and suggestions of cleavage feather marks (at B). 6500×

Fig. 1248 TEM p-c replica of an area at arrow in Fig. 1247 (above). A few fatigue striations are visible at top and at right. Remainder of surface here is unusual, consisting of what appear to be very flat-bottomed, shallow dimples. 6500×

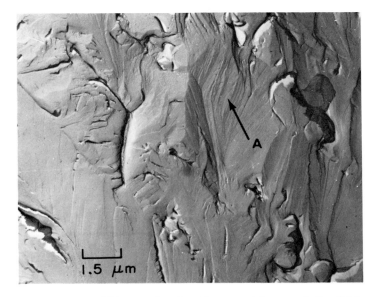

Fig. 1246 TEM p-c replica of another area at the arrow in Fig. 1244 (above). The features visible here are cleavage facets with involved steps but few clear river patterns. The features at A bear some resemblance to feather marks. 6500×

Fig. 1249 TEM p-c replica of another area at the arrow in Fig. 1247 (above), showing distinct fatigue striations at center and at upper left. Elsewhere, the surface bears scattered and indistinct marks that resemble fatigue striations. 6500×

Metal-Matrix Composites / 465

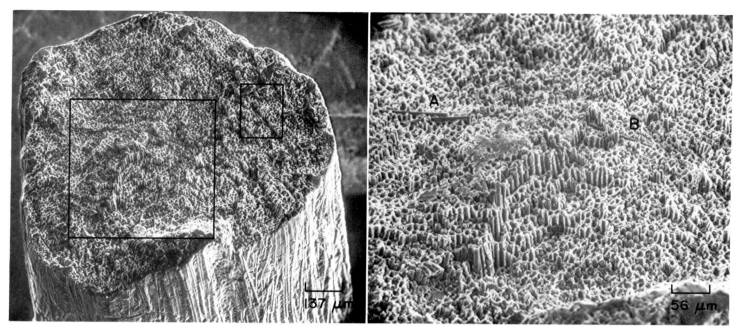

Fig. 1250 Surface of a fracture in a longitudinal tensile-test specimen of a carbon (graphite)-magnesium composite having a tensile strength of 640 MPa (93 ksi). The approximate makeup of the composite was 40% graphite fibers and 60% magnesium matrix (99.5% pure), by volume. See Fig. 1251 and 1252 for higher-magnification views of the areas outlined by the large and small rectangles, respectively. See also Fig. 1254. SEM, 75×

Fig. 1251 Higher-magnification view of the area in the larger rectangle in Fig. 1250. At A are two transversely oriented features that appear to be graphite fibers that either are lying loose on the surface or were inadvertently misaligned during manufacture. The feature at B is probably a ridge. Much of the surface exhibits broken fibers, the ends of which are roughly in the same plane as the fracture in the surrounding matrix, but at center are exposed fibers with the appearance of "palisades." SEM, 180×

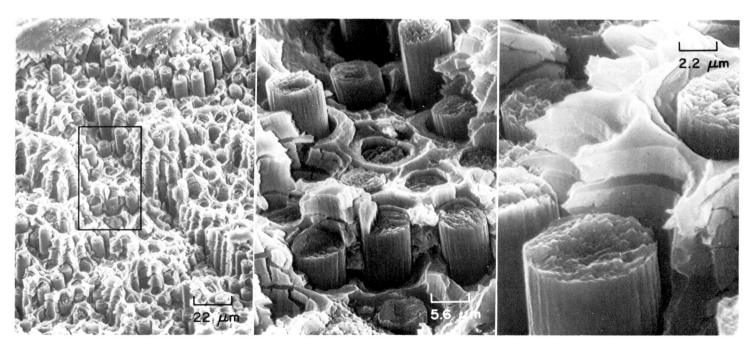

Fig. 1252 Higher-magnification view of the area in the smaller rectangle in Fig. 1250, showing "palisades," some of which are the reverse of those in Fig. 1251 in that the fibers have broken at the bottom of the "palisades," exposing the sides of the holes in the matrix. See Fig. 1253 for an enlarged view of the area in the rectangle. SEM, 450×

Fig. 1253 Enlargement of the area in the rectangle in Fig. 1252, showing that the magnesium matrix fractured in a brittle manner; note the many transverse fissures in the sides of the holes. Note also that the surfaces of the holes in the matrix reflect faithfully the fluted contours of the graphite fibers. SEM, 1800×

Fig. 1254 View at higher magnification of an area of the fracture surface shown in Fig. 1250. The rough surfaces of the broken graphite fibers are clearly visible here. Also compare these rough surfaces with the smooth cleavage facets on the broken tungsten fibers shown in Fig. 1257. SEM, 4500×

466 / Metal-Matrix Composites

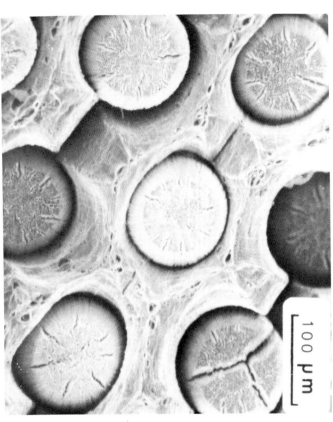

Fig. 1255 Tensile fracture at room temperature in a unidirectionally reinforced composite consisting of 45 vol% NS-55 tungsten fibers in an aluminum alloy 6061 matrix. Fiber diameter: 110 µm. Composite was fabricated by diffusion bonding alternating layers of metal foils and aligned fiber arrays for 30 min at 69 MPa (10 ksi) and 480 °C (895 °F). Tensile strength of the metal-matrix composite was in close agreement with rule-of-mixture values. (Compare with Fig. 1256 and 1257.) Fracture surface reveals interlaminar failure of the aluminum matrix (features of this type are usually obliterated in conventional metallography). Inadequate bonding between matrix foils may have been caused by thin surface oxides that did not dissolve during bonding. Also note separations at fiber-matrix interfaces. The absence of a reaction layer indicates that the strength of the composite derives primarily from mechanical locking. Both fiber-matrix and matrix-matrix interface separations were due to transverse tensile stresses resulting from necking of both fibers and matrix. Radial cracks propagating along fiber grain boundaries were caused by the circumferential stresses generated during reduction of the cross-sectional area (~50%) during necking. Higher-magnification views revealed a fibrous, ductile fracture surface. SEM, 250×. (C. Kim, Naval Research Laboratory)

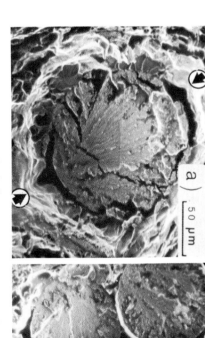

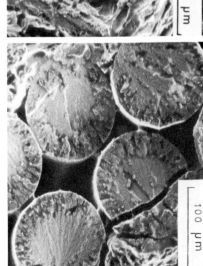

Fig. 1256, 1257 Tensile fracture at room temperature in a unidirectionally solidified composite consisting of 45 vol% NS-55 tungsten fibers in an AISI 1010 carbon steel matrix. Fiber diameter: 110 µm. Composite was fabricated by diffusion bonding alternating layers of metal foils and aligned fiber arrays for 30 min at 69 MPa and 760 °C (1400 °F). Tensile strength was considerably less than that predicted by the rule of mixtures. Compare with Fig. 1255, Fig. 1256 (left): Typical fracture surface reveals adequate fiber-matrix diffusion bonding with the ductile matrix deformed around the fiber. The unbonded regions (arrows) are at points where the macroscopic pressure direction during bonding was tangent to the fiber circumference. The fiber failed in a brittle transgranular mode. However, each fiber had a 15- to 20-µm-thick annular feature at its periphery consisting of numerous secondary fractures propagating along the boundaries of the elongated grains. The river pattern in the core of the fiber indicates that these secondary cracks originated inside the peripheral ring. Subsequent analyses revealed that the annular fracture surface feature corresponded exactly with an iron diffusion front. Cause of the lower-than-expected tensile strength of the composite was fiber embrittlement due to contamination by elemental iron. SEM, 300×. Fig. 1257 (right): Lower-magnification view of fracture surface in Fig. 1256 better shows annular rings on fiber surfaces. Major (transverse) fiber fractures originated at the circumference of the smooth interior core region. Note the inadvertent omission of a foil layer between two fiber layers. SEM, 150×. (C. Kim, Naval Research Laboratory)

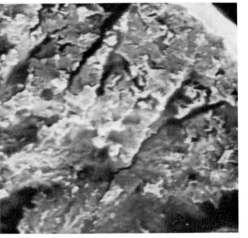

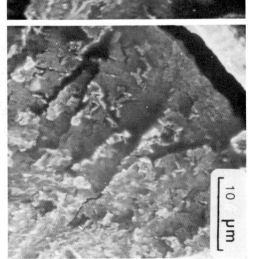

Fig. 1258, 1259 Longitudinal cracking in the annular portion of the tungsten fiber fracture surface in Fig. 1256 and 1257 is revealed in these mating-surface fractographs (one is an inverted image). The transverse fracture proceeded radially inward. The good matchups between the upper left of the fiber, and the intergranular longitudinal cracks propagated toward the upper left of the fiber, and the intergranular longitudinal cracks propagated radially inward. The good matchups between the traces of the longitudinal cracks on the mating fracture surfaces indicate they nucleated independent of and prior to the transverse fracture. SEM, 2025×. (C. Kim, Naval Research Laboratory)

Fig. 1260 Fracture due to fiber splitting in tungsten fiber reinforced metal-matrix composites. When the tungsten/aluminum and tungsten/steel composites in Fig. 1255 to 1259 fractured under transverse tension loading, the predominant failure mode was fiber splitting, as shown here for the 45 vol% tungsten/1010 steel composite. More than 80% of the fibers split. In the 45 vol% 6061 aluminum alloy matrix material, 70% of the fibers split. SEM, 250× (C. Kim, Naval Research Laboratory)

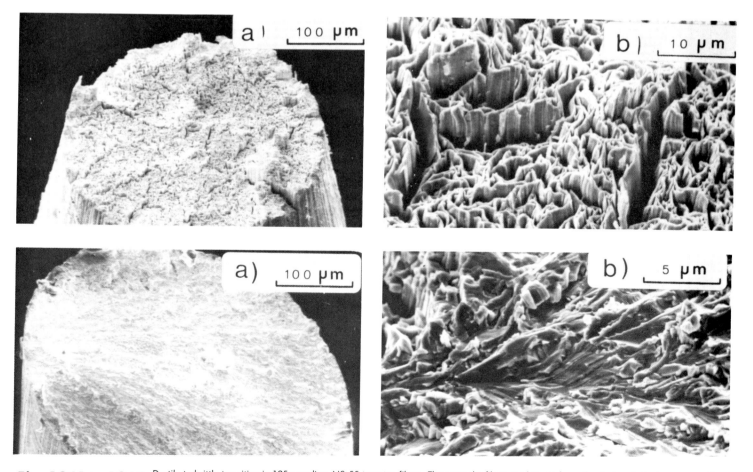

Fig. 1261 to 1264 Ductile to brittle transition in 125-μm-diam NS-55 tungsten fibers. These are the fibers used to reinforce the steel and aluminum composites in Fig. 1255 to 1260. The transition is thought to be brought about by strain-induced grain-boundary migration and polygonization. Polygonization promotes brittle behavior by increasing the effective stress concentration at grain boundaries. Samples were isothermally annealed for 30 min and then tensile tested at room temperature. Fig. 1261 (top left) and 1262 (top right): Tungsten fiber annealed at 800 °C (1475 °F), which is below the transition temperature. Note considerable necking and the fibrous, ductile fracture surface. SEM, 210× and 2100×. Fig. 1263 (bottom left) and 1264 (bottom right): Tungsten fiber annealed at 950 °C (1740 °F), which is above the transition temperature. Failure was brittle (no necking) and fracture surface is flat, exhibiting transgranular cleavage. SEM, 210× and 4150× (C. Kim, Naval Research Laboratory)

468 / Metal-Matrix Composites

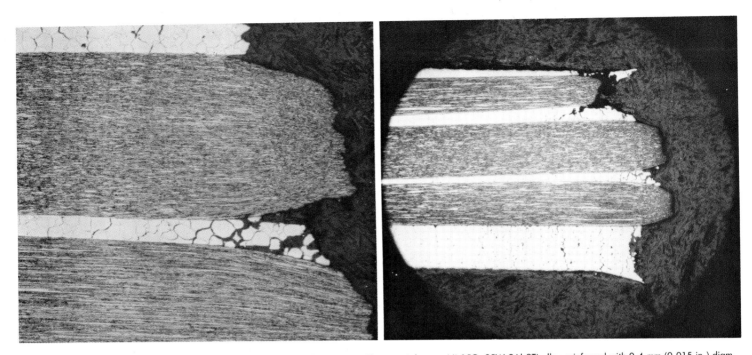

Fig. 1265, 1266 Fracture of metal-matrix composite tensile specimen. The material was a Ni-15Cr-25W-2Al-2Ti alloy reinforced with 0.4-mm (0.015-in.) diam tungsten fibers. The composite was fabricated by slip casting matrix alloy powder around the fibers and then hot isostatically pressing the preform at 1095 °C (2000 °F). The fiber failed in a ductile manner, while the matrix failed in a brittle manner because of poor bonding between individual powder particles. Murakami's reagent, 25× and 100× (D.W. Petrasek, NASA Lewis Research Center)

Fig. 1267 Fracture of metal-matrix composite stress-rupture specimen. Matrix alloy and processing same as in Fig. 1265 and 1266. Reinforcing fibers, however, were W-1ThO$_2$, 0.5 mm (0.02 in.) in diameter. Specimen failed after 17.5 h at 1095 °C (2000 °F) and 241 MPa (35 ksi). Necking indicates that fibers failed in a ductile manner. Compare with Fig. 1268. Murakami's reagent, 100× (D.W. Petrasek, NASA Lewis Research Center)

Fig. 1268 Fracture of metal-matrix composite stress-rupture specimen. Material and processing same as in Fig. 1265 and 1266. The composite failed after 3 h at 1205 °C (2200 °F) and 103 MPa (15 ksi). Fibers failed in a brittle mode. Compare with Fig. 1267. Murakami's reagent, 100× (D.W. Petrasek, NASA Lewis Research Center)

Fig. 1269, 1270 Fracture of metal-matrix composite low-cycle fatigue specimen. Material: Fe-24Cr-4Al-1Y matrix reinforced with W-1ThO$_2$ fibers, 0.2 mm (8 mil) in diameter. The composite was fabricated by applying a matrix alloy powder "cloth" to a fiber mat via use of a plasticizer. A monotape was then made by hot pressing at 1095 °C (2000 °F). Several monotapes were hot pressed together to form a panel, from which fatigue specimens were machined. The composite failed after 200 cycles at 760 °C (1400 °F). Cup-and-cone fiber fractures indicate a ductile failure mode. SEM, 100× and 250× (D.W. Petrasek, NASA Lewis Research Center)

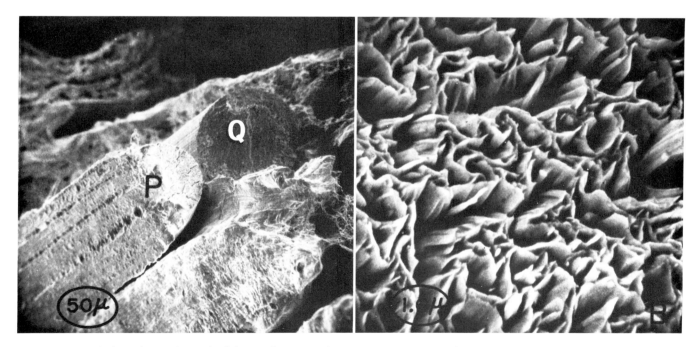

Fig. 1271 Surface of a tension-overload fracture in a composite consisting of tungsten fibers in a silver matrix. The tungsten fiber marked P has undergone a ductile fracture with necking, whereas fiber Q has suffered a sharp, flat, transverse cleavage fracture. See also Fig. 1272. SEM, 320×

Fig. 1272 Higher-magnification view of an area of tungsten fiber P in Fig. 1271. Deformation during fracture has thinned the walls of the dimples to sharp edges, which shows the highly ductile nature of the fracture. Note the uniformity in height of the walls. Compare with Fig. 1262. SEM, 10 000×

470 / Cemented Carbides

Fig. 1273 Eta phase on the fracture surface of a 94WC-6Co alloy tested in four-point bending. The large, relatively smooth region in the center is η phase, a hard and brittle cubic compound—$(Co_3W_3)C$—that adversely affects fracture toughness of the cemented carbide. Some embedded carbide grains protrude from the phase; others appear to have been removed, exposing the sockets that contained them. In both cases, the fracture path followed η-WC interfaces. Microporosity associated with η-phase formation during sintering may account for the apparent role of the phase as a site of fracture initiation. TEM cellulose acetate replica, 4300×. (S.B. Luyckx, University of the Witwatersrand, South Africa)

Fig. 1274, 1275, 1276 Brittle fractures of 97WC-3Co alloys tested in four-point bending. Fig. 1274 (left): Transgranular (cleavage) fracture of a large, triangular WC grain. Note river patterns covering most of the grain surface. TEM Formvar replica, 11 200×. Fig. 1275 (center): Partly intergranular (smooth grains) and partly transgranular (rougher-appearing grains) fracture. Fracture surface was first etched in 5% HCl to remove the cobalt binder and reveal the role of WC in the fracture process. TEM Formvar replica, 4000×. Fig. 1276 (right): Partly intergranular (smooth grains) and partly transgranular fracture. The trapezoidal WC grain at center (transgranular fracture) exhibits Wallner lines (indicated by arrow), which result from the interaction between the advancing crack front and a simultaneously propagating elastic wave. Fracture surface was etched in 5% HCl to remove the cobalt. TEM Formvar replica, 12 000×. (S.B. Luyckx, University of the Witwatersrand, South Africa)

Fig. 1277 Fracture in four-point bending of a 94WC-6Co alloy. Analysis of mating fracture surfaces (shown here) was needed to accurately gage the fractions of intergranular and transgranular cleavage. Crack propagation was primarily along WC-Co interfaces. Although fracture was brittle on a macroscopic scale, areas of ductile dimpled rupture are visible in the cobalt matrix. SEM, 2000×. (S.B. Luyckx, University of the Witwatersrand)

Ceramics / 471

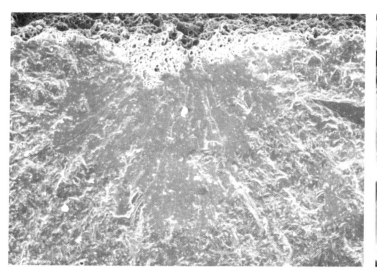 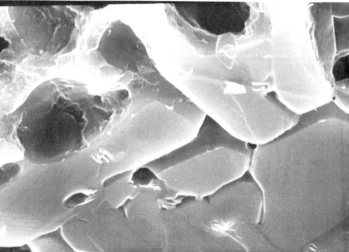

Fig. 1278, 1279 Fracture of α-SiC initiated at corrosion pits. The boron- and carbon-doped SiC was injection molded and pressureless sintered. The sample was then coated with 2 mg/cm^2 (0.07 oz/ft^2) Na$_2$SO$_4$ and exposed for 48 h at 1000 °C (1830 °F) in 0.1% SO$_2$/O$_2$ gas. Corrosion products were removed with a solution of 10% HF in water and the sample was broken in four-point bending at a strain rate of 0.5 mm/min (0.02 in./min). Fig. 1278 (left): Fracture surface shows radial crack lines emanating from the origin, which was a corrosion pit (top). SEM (30° tilt), 115×. Fig. 1279 (right): Close-up of pit-ceramic interface reveals preferential grain-boundary attack in advance of the pit. SEM (30° tilt), 4300× (J.L. Smialek and N.S. Jacobson, NASA Lewis Research Center)

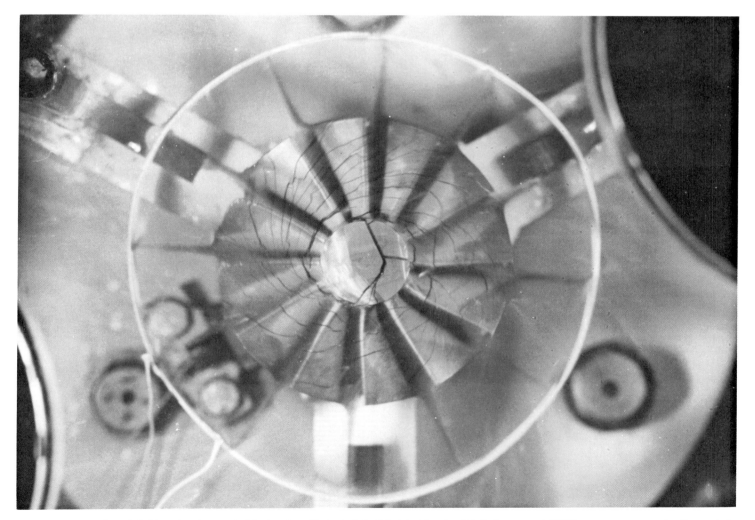

Fig. 1280 Ceramic turbine rotor at moment of fracture in ambient spin test. Material: sintered α-SiC. Test was run at 97 000 rpm—an average rotor tip speed of 560 m/s (1840 ft/s). The rotor was an 86 000 rpm design built as part of the Department of Energy's Advance Gas Turbine program. (Sohio Engineered Materials Company and Allison Gas Turbine Division, General Motors Corporation)

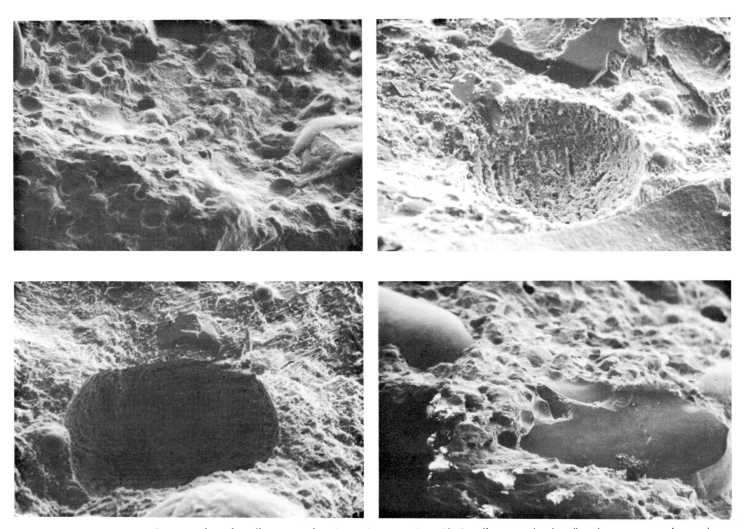

Fig. 1281 to 1284 Fracture surface of a sulfur concrete featuring noninterconnecting voids. Its sulfur cement bonds well to the aggregate to form a dense concrete. (Compare with Fig. 1287 to 1289.) Fig. 1281 (top left): Backscattered electron (BE) display of sulfur concrete with noninterconnecting voids. 17×. Fig. 1282 (top right): Higher-magnification view of portion of field in Fig. 1281 has a noninterconnecting void at its center. BE image, 170×. Fig. 1283 (bottom left): The structure within a void region. BE image, 165×. Fig. 1284 (bottom right): Fractograph shows aggregate embedded in the sulfur cement matrix. BE image, 17× (H.W. Leavenworth, Jr., U.S. Bureau of Mines)

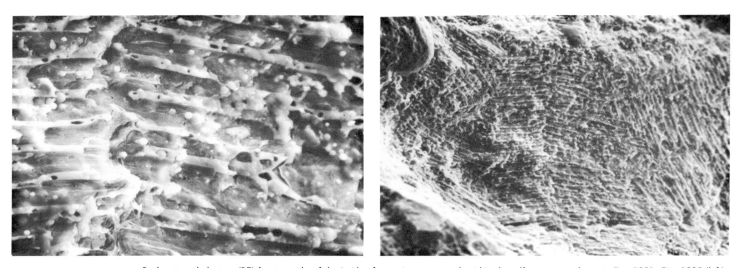

Fig. 1285, 1286 Backscattered electron (BE) fractographs of the inside of a noninterconnected void in the sulfur concrete shown in Fig. 1281. Fig. 1285 (left): Note two-phase structure. BE image, 170×. Fig. 1286 (right): Higher-magnification view of field in Fig. 1285. The two phases consist of polymeric polysulfides and crystalline sulfur. BE image, 860× (H.W. Leavenworth, Jr., U.S. Bureau of Mines)

Concrete and Asphalt / 473

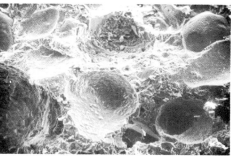

Fig. 1287, 1288, 1289 Fracture surface of a sulfur concrete having a high density of interconnecting voids, resulting in a porous, weak material. Aggregate used in this concrete had a polished-appearing surface that did not bond well to the sulfur cement. (Compare with Fig. 1281 to 1284.) Fig. 1287 (left): Smooth surfaces of interconnected voids that resulted from use of a polished aggregate. SEM, 120×. Fig. 1288 (center): Fractograph of sulfur concrete with high void density. SEM, 55×. Fig. 1289 (right): Backscattered electron image of field in Fig. 1288 reveals sulfur crystals in void regions. 55× (H.W. Leavenworth, Jr., U.S. Bureau of Mines)

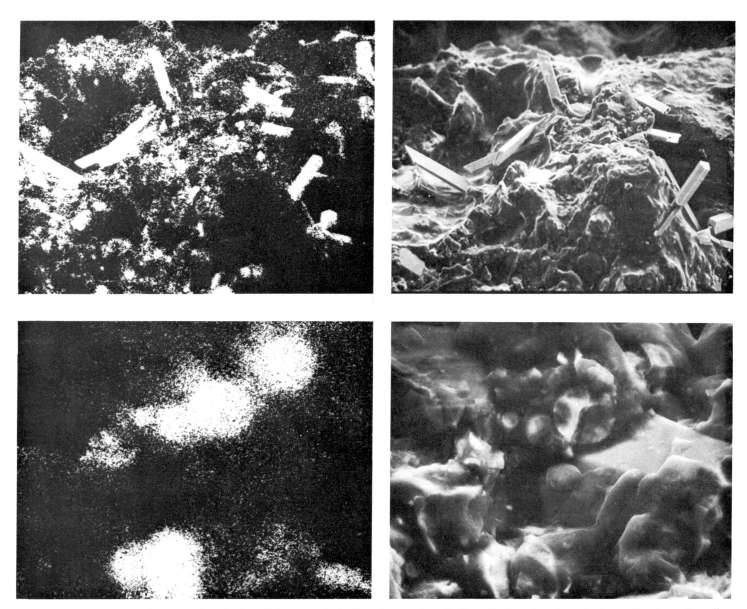

Fig. 1290 to 1293 Fracture surface of sulfur asphalt showing distribution and morphology of sulfur within the asphalt phase. Faceted needles of crystalline sulfur form via unconfined growth into asphalt microvoid regions. Spherical sulfur particles are due to confined growth in the asphalt phase. Fig. 1290 (top left): K_α x-ray map shows sulfur distribution in asphalt. 100×. Fig. 1291 (top right): Backscattered electron (BE) image of field in Fig. 1290 reveals faceted, needlelike sulfur crystals. 100×. Fig. 1292 (bottom right): Higher-magnification sulfur x-ray map. 1000×. Fig. 1293 (bottom right): BE image of field in Fig. 1292 reveals spherical sulfur particles. 1000× (H.W. Leavenworth, Jr., U.S. Bureau of Mines)

474 / Resin-Matrix Composites

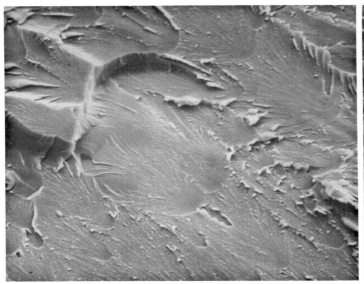

Fig. 1294 Fracture surface of a Fiberite 934 epoxy resin specimen that failed in tension. The specimen was tested at 25 °C (77 °F) in the dry condition (see note below). Fracture, which initiated at an internal defect, progressed from lower right to upper left. The parabolic markings were produced by cavity nucleation and growth ahead of the advancing crack front. The beginning of the final region where tearing occurs is shown at upper left. SEM, 30° specimen tilt, specimen sputter coated with gold, 1000× (L. Clements, San Jose State University, and J.C. Liu, Cornell University)

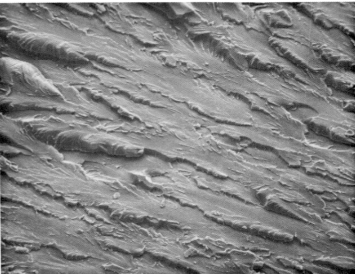

Fig. 1295 Fracture surface of a Fiberite 93 epoxy resin specimen that failed in tension. The specimen was tested at 96 °C (205 °F) in the dry condition (see note below). As in Fig. 1294, fracture initiated at an internal defect and progressed from lower right to upper left. Shown is the final failure region where tearing occurred. This epoxy specimen is more ductile than the one tested at 25 °C (Fig. 1294). SEM (same conditions as in Fig. 1294), 200× (L. Clements, San Jose State University, and J.C. Liu, Cornell University)

NOTE: The epoxy specimens shown above in Fig. 1294 and 1295 were of Fiberite's 934 epoxy resin, which is one of the standard "state-of-the-art" 175 °C (350 °F)-cure epoxy resin systems used in high-performance aerospace composites. The epoxy was gravity cast at room temperature between glass plates and cured at 175 °C (350 °F) according to the manufacturer's recommendations. Tensile specimens were then cut from the cast sheet. Both of the specimens shown above were tested "dry," which means that they were dried before testing in a vacuum oven at 80 °C (175 °F) until no further weight loss was detected. They were tested in a dry oven at the temperatures noted.

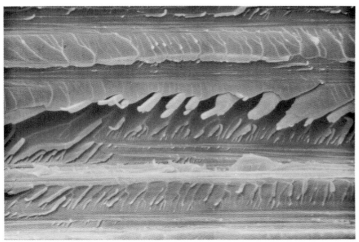

Fig. 1296 Fracture surface of a transverse (90°) carbon/epoxy specimen* that failed in tension. The specimen was tested in the dry condition at 25 °C (77 °F). The surface shows several common features of transverse tensile failure surfaces: (1) branching patterns left on resin when a single branched crack moved across resin, (2, 4, 7) areas where fiber pulled away from resin upon failure, (3) tongues of resin left when cracks moving through resin in two directions on two distinct levels came together, (5) impression of grooves in fiber that pulled away, and (6) branches emanating from debonding of fiber in area 5. SEM (same conditions as in Fig. 1294), 2000× (L. Clements, San Jose State University, and J.C. Liu, Cornell University)

Fig. 1297 Fracture surface of transverse (90°) carbon/epoxy specimen* that failed in tension. The "room moisture" specimen was tested at 25 °C (77 °F) at a moisture content of ~0.6 wt%. A few fiber breaks are shown in this fractograph. These are not common in transverse tensile failure surfaces (unless the fibers are somewhat misoriented). Note that the fibers fail at angles to the fracture surface that range from almost perpendicular to very oblique. Note that all fibers retain a great deal of resin on their surfaces, which indicates that the interfacial bond was strong and that the primary failure mode was resin failure. SEM (same conditions as in Fig. 1294), 1000× (L. Clements, San Jose State University, and J.C. Liu, Cornell University)

*See the comments on failure characteristics of the transverse composite specimens described in Fig. 1296 to 1299 on the following page. Additional information on the processing of the composites described in Fig. 1296 to 1310 as well as the associated test conditions can be found on page 476.

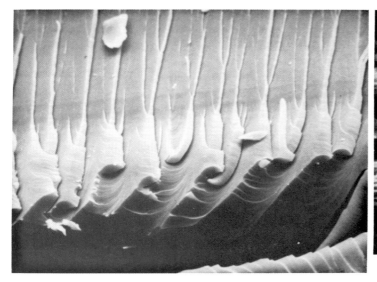

Fig. 1298 Fracture surface of a transverse (90°) carbon/epoxy specimen* that failed in tension. The "wet" specimen was tested at 25 °C (77 °F) at a moisture content of 1.5 wt%. It was cycled five times between 74 and 25 °C (165 and 77 °F) at a relative humidity of 100% before testing. Fractograph shows one area of an epoxy-rich interlaminar region, between layers of fiber/resin tape. SEM (same conditions as in Fig. 1294), 2000× (L. Clements, San Jose State University, and J.C. Liu, Cornell University)

Fig. 1299 Fracture surface of a transverse (90°) carbon/epoxy specimen* that failed in tension. The "wet" specimen was tested at 25 °C (77 °F) at a moisture content of ~1.5 wt%. Same pretest cycling as in Fig. 1298. Fractograph shows at (1) and (5) hackles formed when secondary cracks in the resin propagated as a crack front moved from left to right; at (2) and (4) bare fibers where interfacial debonding occurred upon failure; and at (3) an impression left in the resin by a debonded fiber (right side) and an area where two or more cracks intersected (left side). SEM (same conditions as in Fig. 1294), 2000× (L. Clements, San Jose State University, and J.C. Liu, Cornell University)

NOTE: In transverse (90°) specimens such as those shown in Fig. 1296 to 1299, the fibers run perpendicular to the loading direction, and thus parallel to the failure surface. Unlike homogeneous, isotropic materials, composite failure surfaces are quite complex and will show a variety of different features in different locations. Transverse specimens, however, can usually be found to have a clearly predominant failure type. If the fiber/matrix interfacial bond is weak, such specimens will tend to fail by debonding (interface separation). If the interface is strong, however, matrix failure will be the predominant failure mode. Most current fiber/epoxy systems have a strong interfacial bond and fail by this mode. (The exception is for aramid fibers, where the fiber itself is transversely weak and may fail by splitting.) In such failures, some fibers will also be torn or broken, but these are few. There also will be some areas where the interface is weak, so some debonding may be seen, but it will not predominate.

Fig. 1300 Fracture surface of longitudinal (0°) carbon/epoxy specimen* failed in flexure. The "wet" specimen was tested at 96 °C (205 °F) at a moisture content of ~2 wt%. This fractograph shows a typical three-point-bend flexural failure surface. It illustrates the dramatic difference between the tensile and compressive sides of the failure surface. Note the long, protruding fibers on the tensile side, and the crushed, crumpled fibers on the compressive side. SEM (same conditions as in Fig. 1294), 20× (L. Clements, San Jose State University, and J.C. Liu, Cornell University)

Fig. 1301 A different view and higher magnification of the carbon/epoxy flexural specimen* shown in Fig. 1300. Once again, the difference between the compressive (left) and the tensile (right) sides of the fracture surface is dramatic. SEM (same conditions as in Fig. 1294), 50× (L. Clements, San Jose State University, and J.C. Liu, Cornell University)

*Additional information on the processing of the composites described in Fig. 1296 to 1310 as well as the associated test conditions can be found on the following **page**.

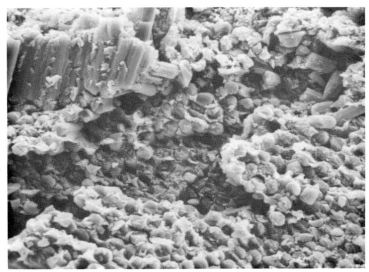

Fig. 1302 Another view of the carbon/epoxy flexural specimen shown in Fig. 1300 and 1301. This fractograph shows a typical area in the compressive failure region at a higher magnification. Note the crushed and crumpled fibers. See also Fig. 1303. SEM (same conditions as in Fig. 1294), 500× (L. Clements, San Jose State University, and J.C. Liu, Cornell University)

Fig. 1303 A typical area in the tensile failure region of the carbon/epoxy flexural specimen shown in Fig. 1300 and 1301. Note the long, protruding fibers with some attached epoxy remaining. This area resembles typical regions in a specimen failed in pure tension. See also Fig. 1302. SEM (same conditions as in Fig. 1294), 500× (L. Clements, San Jose State University, and J.C. Liu, Cornell University)

Fig. 1304 Fracture surface of a longitudinal (0°) carbon/epoxy specimen failed in tension. The specimen was tested in the dry condition (see below for explanation) at 25 °C (77 °F). The fibers are broken at many different levels; the fiber breaks are perpendicular to the fiber (and load) direction. There is substantial epoxy adhering to the fiber surfaces, indicating that the interfacial bond strength is fairly high. SEM (same conditions as in Fig. 1294), 500× (L. Clements, San Jose State University, and P.R. Lee, United Technologies)

Fig. 1305 Same carbon/epoxy specimen as shown in Fig. 1304, but showing a somewhat less typical fracture area. Although the fibers still protrude at varying levels, the overall appearance is more flat than in Fig. 1304. There are also some holes where fairly short fibers have pulled out. SEM (same conditions as in Fig. 1294), 500× (L. Clements, San Jose State University, and P.R. Lee, United Technologies)

NOTE: The composite specimens shown in Fig. 1296 to 1310 all consisted of *circa* 1978 Thornel 300 fibers in Narmco 5208 epoxy resin. The composites were all fabricated by lay-up of prepreg tape, and were autoclave cured 175 °C (350 °F) according to the manufacturer's recommendations. Specimens were then end-tabbed and cut in accordance with the ASTM D3039. The 0° and 90° tensile specimens were tested in accordance with ASTM D3039, and the 0° flexural three-point-bend specimens were tested in accordance with ASTM D790. The composites were dried before testing in the same manner as the epoxy specimens described in Fig. 1294 and 1295. "Dry" specimens were tested as-dried and had essentially 0 wt% moisture. "Room" moisture specimens had about 0.6 wt% moisture, having been placed after drying in a chamber at 60 °C (140 °F) and 60% relative humidity for at least 2 months and then placed in a drawer in the laboratory for several months before testing. "Wet" specimens had about 1.5 to 2.0 wt% moisture, having been placed for several months in an environmental chamber at 60 °C (140 °F) and 100% relative humidity. The composites were tested at moisture conditions designed not to alter their conditioning, and at the temperatures noted.

Resin-Matrix Composites / 477

Fig. 1306 Fracture surface of a longitudinal (0°) carbon/epoxy specimen* failed in tension. The specimen was tested at 25 °C (77 °F) in the dry condition. This fractograph shows the surface of some broken fibers and the associated hackles formed in the epoxy upon separation of the bundle from an adjacent one. The amount of epoxy remaining on the fibers indicated a fairly strong interfacial bond. SEM (same conditions as in Fig. 1294), 2000× (L. Clements, San Jose State University, and P.R. Lee, United Technologies)

Fig. 1307 Fracture surface of a carbon/epoxy "wet" specimen* tension tested at 96 °C (205 °F) at a moisture content of ~2 wt%. Fractograph shows a bundle of fractured fibers, impressions of carbon fibers in remaining epoxy, and pieces of epoxy that have broken loose but not fallen away from the bundle. This fracture morphology is somewhat unusual in that the broken fiber surfaces are not exactly perpendicular to the fiber axis, which indicates that the failure was not 0° tensile in nature. SEM (same conditions as in Fig. 1294), 2000× (L. Clements, San Jose State University, and P.R. Lee, United Technologies)

Fig. 1308 Fracture surface of a longitudinal (0°) carbon/epoxy specimen* that failed in tension. Specimen was tested at 25 °C (77 °F) in the dry condition. This fractograph shows typical tensile failure of the epoxy resin in a local resin-rich region. This results when fibers break and produce overload tensile failure in the resin. SEM (same conditions as in Fig. 1294), 2000× (L. Clements, San Jose State University, and P.R. Lee, United Technologies)

Fig. 1309 Fracture surface of a carbon/epoxy "wet" specimen* tested at 25 °C (77 °F) at a moisture content of 2.0 wt%. Fractograph shows typical appearance of a local failure of two fibers where the fracture originated at the point of contact between the two fibers. SEM (same conditions as in Fig. 1294), 2000× (L. Clements, San Jose State University, and P.R. Lee, United Technologies)

Fig. 1310 Fracture surface of a longitudinal (0°) carbon/epoxy specimen* failed in tension. Specimen was tested in the dry condition at 96 °C (205 °F). The region surrounding the broken fiber at the center of the fractograph was extremely resin-rich. Note how the resin failure radiated from the fiber break and progressed to adjacent areas. This is an unusual fracture morphology in that fractured fibers rarely produce such a flat failure. SEM (same conditions as in Fig. 1294), 2000× (L. Clements, San Jose State University, and P.R. Lee, United Technologies)

*Additional information on the processing of the composites described in Fig. 1296 to 1310 as well as the associated test conditions can be found on the previous page.

Fig. 1311 to 1316 Fractures in short beam shear, compression, and flexure (three-point bending) of woven carbon fabric/phenolic prepreg composite panels. The flat laminate test specimens represent the ablative materials used on the Space Shuttle's solid rocket motor nozzle. Flexure and compression laminates were fabricated from four prepreg plies; shear specimens, from nine. Prepreg layups were vacuum bagged and cured for 5 h at 155 °C (310 °F) and 7 MPa (100 psi). Each pair of photos shows specimens before and after testing. The orientation is edge-on (parallel to the plane of the laminate). Surfaces were milled. Fig. 1311 and 1312 (top row): Carbon/phenolic shear specimen before and after testing. White lines highlight the interlaminar shear failure in Fig. 1312. 25× and 10×. Fig. 1313 and 1314 (center row): Carbon/phenolic compression specimen before and after testing. Load was applied parallel to the plane of the laminate. Note ply buckling in Fig. 1314. 20×. Fig. 1315 and 1316 (bottom row): Carbon/phenolic flexure specimen before and after testing. 15× and 12×. (R.J. Schwinghamer, NASA Marshall Space Flight Center)

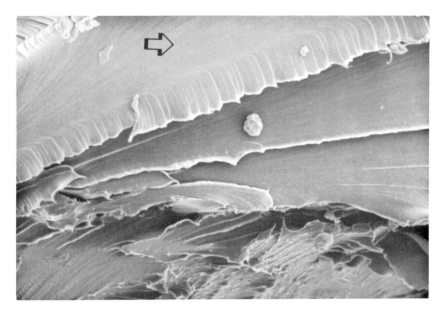

Fig. 1317 Quasi-brittle fatigue crack propagation in 3.2-mm (0.13-in.) thick polycarbonate sheet. Arrow indicates direction of crack growth. At this thickness, polycarbonate shows features characteristic of both brittle (microcracking) and ductile (thinning and fibrillation) fracture. Note "river patterns" on the plane at top that formed during tearing. Irregular features similar to "tire tracks" at bottom left are formed during the fatigue process when randomly oriented microcracks are created on different planes in front of the notch. When the main crack propagates, it continuously switches from one microcrack plane to the other. SEM, 300× (A.S. Moet, Case Western Reserve University)

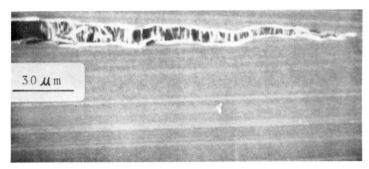 

Fig. 1318, 1319 Crack-growth mechanisms in three-point bending of notched 0.4-mm (16-mil) thick specimens of linear polyethylene (density = 0.964 g/cm^3). The notch, made by a razor blade, was held open during observation in the scanning electron microscope. Fig. 1318 (left): Notched three-point-bend specimen after 215 min at 42 °C (110 °F) and constant load. Stress intensity = 0.30 MPa$\sqrt{m}$ (0.27 ksi$\sqrt{in.}$). Note the craze-like damage zone emanating from the notch (far left). Damage zone propagated at 0.16 μm/min prior to crack growth. SEM, 540×. Fig. 1319 (right): Specimen in Fig. 1318 after 367 min at 42 °C (110 °F) and constant load. Stress intensity = 0.36 MPa$\sqrt{m}$ (0.33 ksi$\sqrt{in.}$). Damage mechanism is microcrazing in advance of the propagating crack. SEM, 315× (N. Brown and X. Lu, University of Pennsylvania)

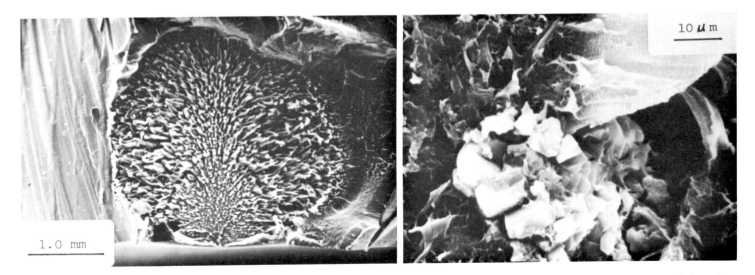

Fig. 1320, 1321 Failure of polyethylene copolymer pipe due to slow crack growth. The 60-mm (2.4-in.) diam, 6-mm (0.25-in.) wall thickness, extruded gas pipe (density = 0.937 g/cm^3) failed after 211 h in a laboratory test at 80 °C (175 °F) and an internal water pressure of 17 kPa (120 psi). Failure initiated near the inner surface of the pipe. Fig. 1320 (left): Fracture surface near origin (exposed by sectioning). SEM, 15×. Fig. 1321 (right): High-magnification view of failure origin reveals foreign particles. Spectroscopy subsequently determined that aluminum was present. Contaminants may have been aluminum oxide. SEM, 1300× (N. Brown, University of Pennsylvania)

480 / Polymers

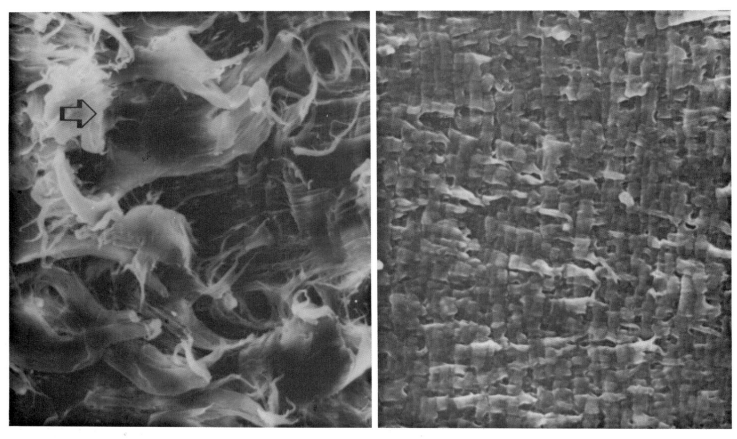

Fig. 1322 Tearing and fibrillation in medium-density polyethylene tested in creep at room temperature. Arrow indicates direction of crack propagation. Fracture stress $(\sigma_f) = 15.5$ MPa (2.25 ksi). Applied stress = 0.37 σ_f. SEM, 200× (A.S. Moet, Case Western Reserve University)

Fig. 1323 Fatigue striations in medium-density polyethylene. Crack growth direction is left to right. Loading conditions: $\sigma_{min}/\sigma_{max} = R = 0.5$, where $\sigma_{max} = 0.3\ \sigma_{yield}$ and $\sigma_{yield} = 22.6$ MPa (3.278 ksi); frequency = 0.5 Hz. SEM, 200× (A.S. Moet, Case Western Reserve University)

Fig. 1324, 1325 Chevronlike features on fracture surface of polyimide recorder belt point to origin of failure. The 75-μm (3-mil) thick belt was used on Spacelab I. Cause of failure was traced to a design deficiency. SEM, 120× and 240× (R.J. Schwinghamer, NASA Marshall Space Flight Center)

Electronic Materials / 481

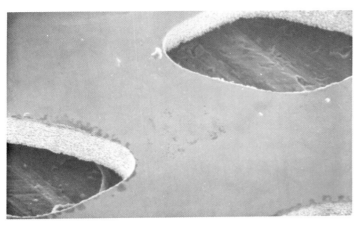
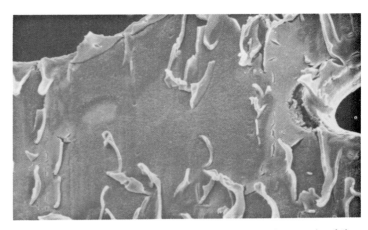

Fig. 1326, 1327 Effect of exposure to the atomic oxygen environment found in low earth orbit (LEO) on a silver solar cell interconnect. The pure silver foil, 40 μm (1.5 mil) thick, was exposed to the LEO environment for 40 h at 100 °C (210 °F) in the spacecraft velocity direction. The sample was flown on Space Shuttle flight 8. The original surface of the interconnect (Fig. 1326, left) was converted to an oxide of silver (Fig. 1327, right). Both reflectivity and electrical conductivity decreased as a result. SEM, 3800×. (R.J. Schwinghamer, NASA Marshall Space Flight Center)

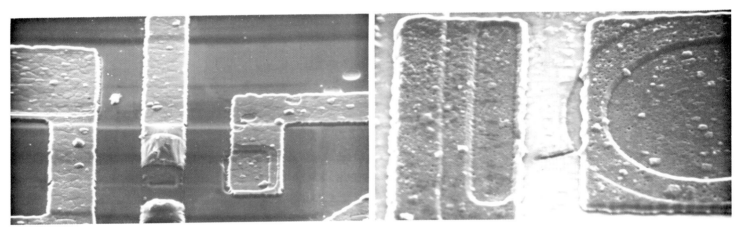

Fig. 1328, 1329 Integrated circuit (IC) defects. Material: pure aluminum vacuum deposited on Si/SiO$_2$. Fig. 1328 (left): Oxide flaws under the metallized layer on this IC produced a thin oxide layer that promoted voltage "punchthrough" and failure (center). SEM, 800×. Fig. 1329 (right): Shown here is a base emitter for input transistors. The cracklike defect at center was caused by electrical discharge across the junction. SEM, 3360×. (R.J. Schwinghamer, NASA Marshall Space Flight Center)

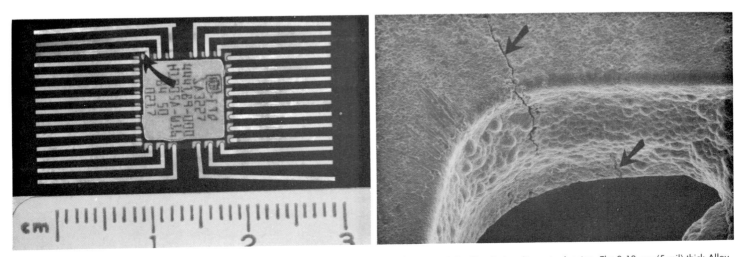

Fig. 1330, 1331 Fatigue failure of L-shaped electronic flat pack leads due to cyclic, eccentric loading during ultrasonic cleaning. The 0.13-mm (5-mil) thick Alloy 42 (42Ni-58Fe) leads were nickel electroplated, soldered (Cu-Ag alloy) to tungsten, and then gold plated. Fig. 1330 (left): A 28-lead electronic flat pack. Cracking was predominantly observed in the smallest L-shaped leads (arrow). 2.5×. Fig. 1331 (right): Cracks in L-shaped leads were found in the solder and near the edge of the solder in both top and bottom fillets (arrows). See also Fig. 1332 to 1334. SEM, 300×. (T. O'Donnell, P. Tung, A. Shumka, R. Ruiz, and J. Okuno, California Institute of Technology)

Fig. 1332, 1333, 1334 SEM fractographs of a failed electronic flat pack lead. Fig. 1332 (top): Cracks on both the top and bottom lead surfaces. The solder surface and post-cracking solder zone corrosion products are noted in the foreground. The extent of corrosion indicates solder thickness on the lead solder side. This lead was mechanically overloaded to reveal the fracture surface. SEM, 400×. Fig. 1333 (bottom left): Low-angle close-up of the lead fracture across the thickness of the lead (arrow in Fig. 1332). Postfracture corrosion in foreground; center to top, fatigue. SEM, 2000×. Fig. 1334 (bottom right): High-magnification view of the mixed fracture mode showing evidence of fine fatigue striations. SEM, 800× (T. O'Donnell, P. Tung, A. Shumka, R. Ruiz, and J. Okuno, California Institute of Technology)

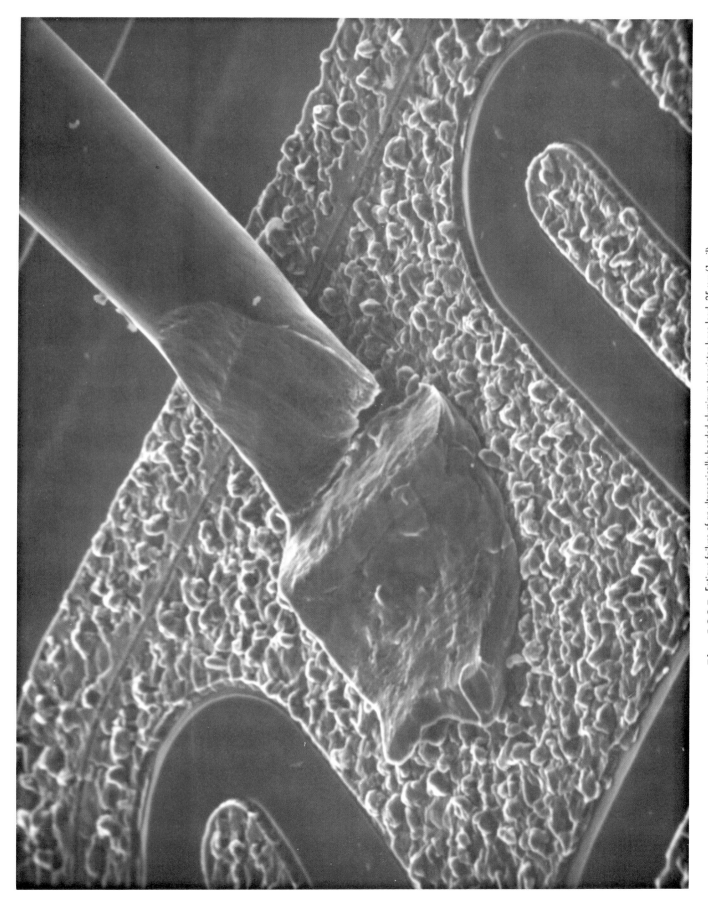

Fig. 1335 Fatigue failure of an ultrasonically bonded aluminum transistor base lead, 25 μm (1 mil) in diameter. The transitor was power cycled until bond failure occurred. SEM, 1500× (R.J. Schwinghamer, NASA Marshall Space Flight Center)

484 / Electronic Materials

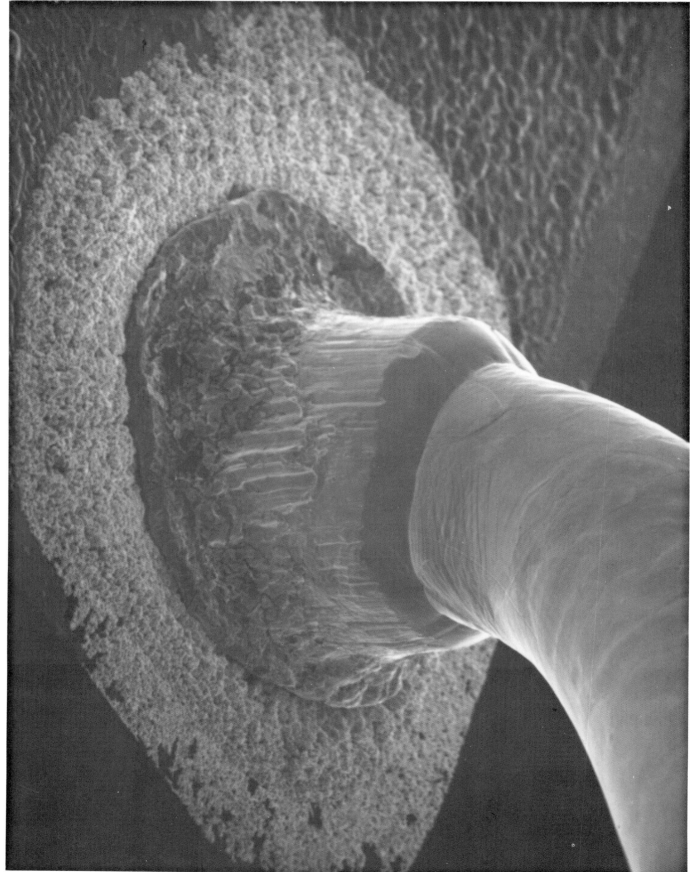

Fig. 1336 Thermocompression ball bond defect between a gold wire, 25 μm (1 mil) in diameter, and vacuum-deposited aluminum. Note the relatively thick intermetallic layer that formed during power cycling. SEM, 2750× (R.J. Schwinghamer, NASA Marshall Space Flight Center)

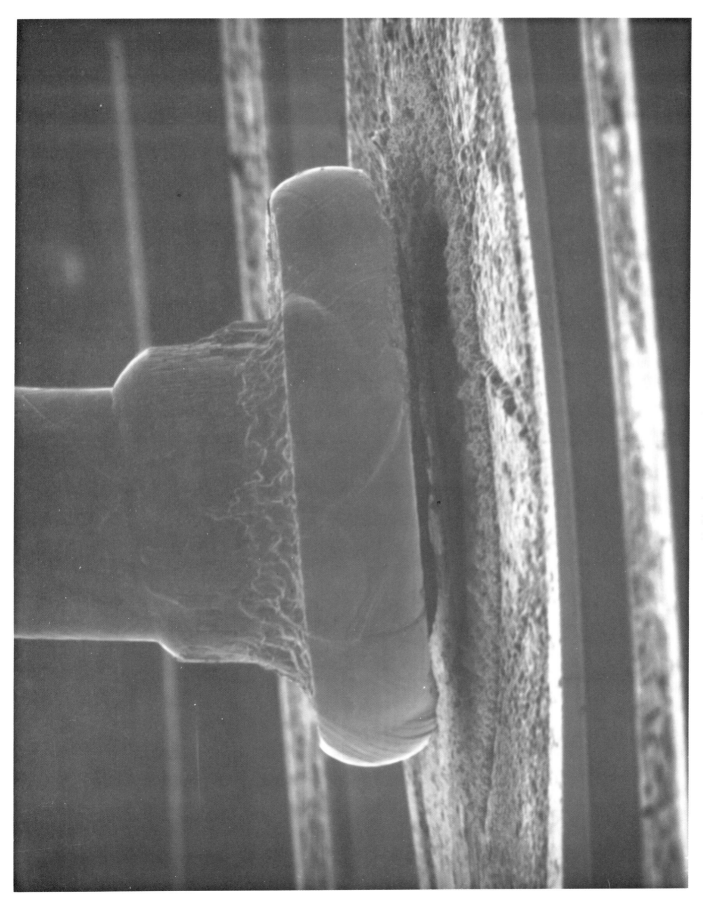

Fig. 1337 Same type of ball bond as in Fig. 1336. In this case, intermetallic formation weakened the bond, causing it to lift. SEM, 2700× (R.J. Schwinghamer, NASA Marshall Space Flight Center)

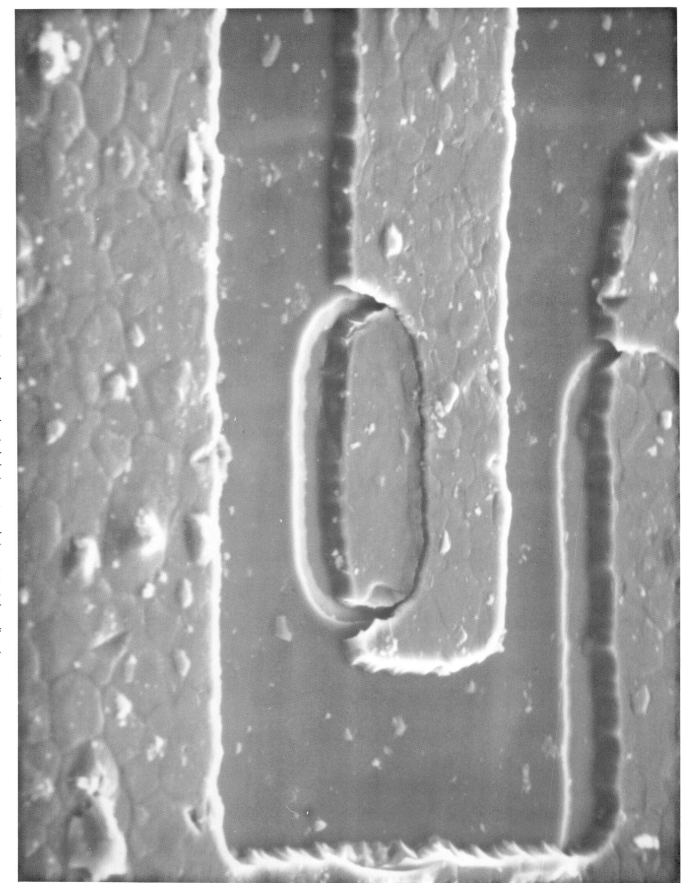

Fig. 1338 Misregistration of vacuum-deposited aluminum at an ohmic contact window allowed etchant to tunnel around the perimeter of the window, resulting in poor step coverage. The metallized layer was deposited on Si/SiO$_2$. SEM, 5850× (R.J. Schwinghamer, NASA Marshall Space Flight Center)

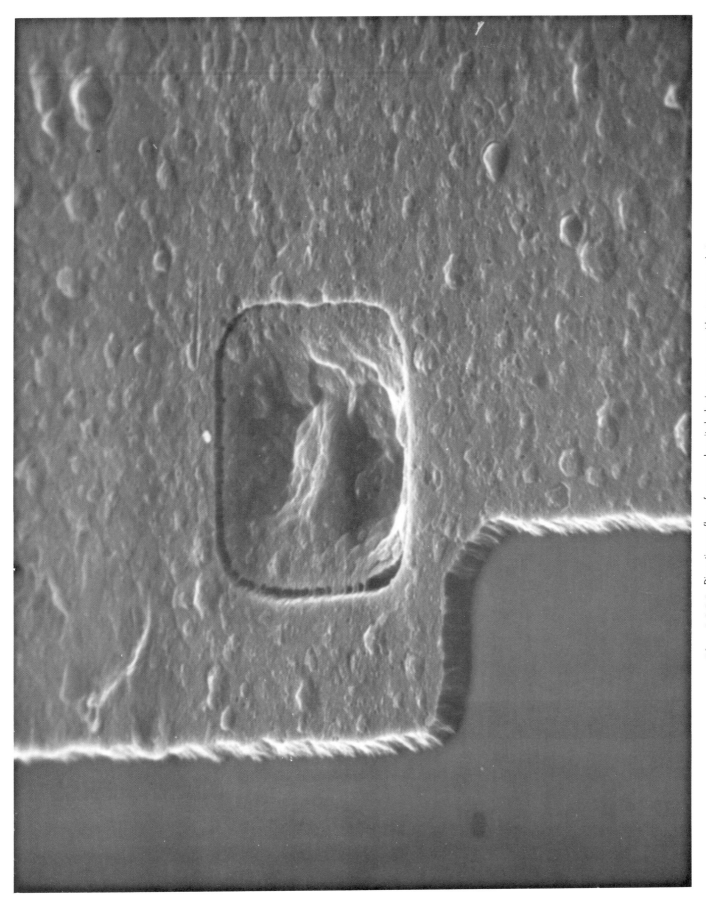

Fig. 1339 Discontinuous flow of vacuum-deposited aluminum over an oxide step at an ohmic contact window. The metallization was applied to Si/SiO$_2$. SEM, 6000×. (R.J. Schwinghamer, NASA Marshall Space Flight Center)

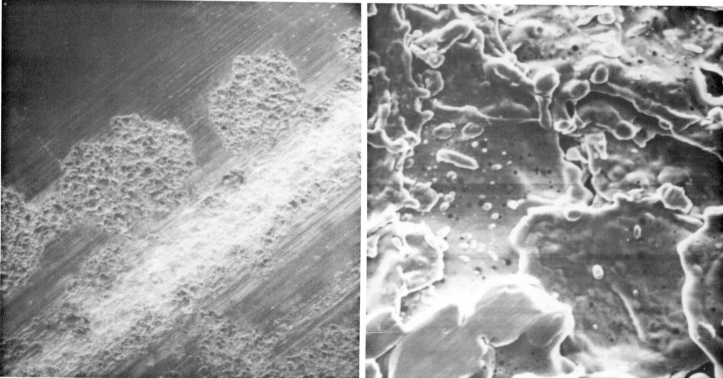

Fig. 1340 to 1343 Failure of brush/slip ring assembly due to arcing. This failure mechanism is always associated with interfacial melting and a high rate of metal transfer in both directions across the interface. Brushes were made of 0.13-mm (5-mil) diam high-purity silver wires; the slip ring, of oxygen-free copper (UNS C10100). Sliding was performed in humidified CO_2, which acted as a lubricant. Failure occurred at a high current of 50 A. Fig. 1340 (top left): Silver brush ends alloyed with copper as a result of interfacial melting. SEM, 215×. Fig. 1341 (top right): Same as Fig. 1340, but at higher magnification. SEM, 370×. Fig. 1342 (bottom left): Pits on copper slip ring caused by arcing. SEM, 1800×. Fig. 1343 (bottom right): Same as Fig. 1342, but at higher magnification. SEM, 3600× (M. Garshasb and R.W. Vook, Syracuse University)

Metric Conversion Guide

This Section is intended as a guide for expressing weights and measures in the Système International d'Unités (SI). The purpose of SI units, developed and maintained by the General Conference of Weights and Measures, is to provide a basis for world-wide standardization of units and measure. For more information on metric conversions, the reader should consult the following references:

- "Standard for Metric Practice," E 380, *Annual Book of ASTM Standards,* Vol 14.02, 1986, American Society for Testing and Materials, 1916 Race Street, Philadelphia, PA 19103
- "Metric Practice," ANSI/IEEE 268–1982, American National Standards Institute, 1430 Broadway, New York, NY 10018
- *Metric Practice Guide—Units and Conversion Factors for the Steel Industry,* 1978, American Iron and Steel Institute, 1000 16th Street NW, Washington, DC 20036
- *The International System of Units,* SP 330, 1986, National Bureau of Standards. Order from Superintendent of Documents, U.S. Government Printing Office, Washington, DC 20402-9325
- *Metric Editorial Guide,* 4th ed. (revised), 1985, American National Metric Council, 1010 Vermont Avenue NW, Suite 320, Washington, DC 20005-4960
- *ASME Orientation and Guide for Use of SI (Metric) Units,* ASME Guide SI 1, 9th ed., 1982, The American Society of Mechanical Engineers, 345 East 47th Street, New York, NY 10017

Base, supplementary, and derived SI units

Measure	Unit	Symbol
Base units		
Amount of substance	mole	mol
Electric current	ampere	A
Length	meter	m
Luminous intensity	candela	cd
Mass	kilogram	kg
Thermodynamic temperature	kelvin	K
Time	second	s
Supplementary units		
Plane angle	radian	rad
Solid angle	steradian	sr
Derived units		
Absorbed dose	gray	Gy
Acceleration	meter per second squared	m/s^2
Activity (of radionuclides)	becquerel	Bq
Angular acceleration	radian per second squared	rad/s^2
Angular velocity	radian per second	rad/s
Area	square meter	m^2
Capacitance	farad	F
Concentration (of amount of substance)	mole per cubic meter	mol/m^3
Conductance	siemens	S
Current density	ampere per square meter	A/m^2
Density, mass	kilogram per cubic meter	kg/m^3
Electric charge density	coulomb per cubic meter	C/m^3
Electric field strength	volt per meter	V/m
Electric flux density	coulomb per square meter	C/m^2
Electric potential, potential difference, electromotive force	volt	V
Electric resistance	ohm	Ω
Energy, work, quantity of heat	joule	J
Energy density	joule per cubic meter	J/m^3
Entropy	joule per kelvin	J/K
Force	newton	N
Frequency	hertz	Hz
Heat capacity	joule per kelvin	J/K
Heat flux density	watt per square meter	W/m^2
Illuminance	lux	lx
Inductance	henry	H
Irradiance	watt per square meter	W/m^2
Luminance	candela per square meter	cd/m^2
Luminous flux	lumen	lm
Magnetic field strength	ampere per meter	A/m
Magnetic flux	weber	Wb
Magnetic flux density	tesla	T
Molar energy	joule per mole	J/mol
Molar entropy	joule per mole kelvin	$J/mol \cdot K$
Molar heat capacity	joule per mole kelvin	$J/mol \cdot K$
Moment of force	newton meter	$N \cdot m$
Permeability	henry per meter	H/m
Permittivity	farad per meter	F/m
Power, radiant flux	watt	W
Pressure, stress	pascal	Pa
Quantity of electricity, electric charge	coulomb	C
Radiance	watt per square meter steradian	$W/m^2 \cdot sr$
Radiant intensity	watt per steradian	W/sr
Specific heat capacity	joule per kilogram kelvin	$J/kg \cdot K$
Specific energy	joule per kilogram	J/kg
Specific entropy	joule per kilogram kelvin	$J/kg \cdot K$
Specific volume	cubic meter per kilogram	m^3/kg
Surface tension	newton per meter	N/m
Thermal conductivity	watt per meter kelvin	$W/m \cdot K$
Velocity	meter per second	m/s
Viscosity, dynamic	pascal second	$Pa \cdot s$
Viscosity, kinematic	square meter per second	m^2/s
Volume	cubic meter	m^3
Wavenumber	1 per meter	1/m

Conversion factors

To convert from	to	multiply by
Angle		
degree	rad	1.745 329 E − 02
Area		
in.^{2}	mm^2	6.451 600 E + 02
in.^{2}	cm^2	6.451 600 E + 00
in.^{2}	m^2	6.451 600 E − 04
ft^2	m^2	9.290 304 E − 02
Bending moment or torque		
lbf · in.	N · m	1.129 848 E − 01
lbf · ft	N · m	1.355 818 E + 00
kgf · m	N · m	9.806 650 E + 00
ozf · in.	N · m	7.061 552 E − 03
Bending moment or torque per unit length		
lbf · in./in.	N · m/m	4.448 222 E + 00
lbf · ft/in.	N · m/m	5.337 866 E + 01
Current density		
A/in.^{2}	A/cm^2	1.550 003 E − 01
A/in.^{2}	A/mm^2	1.550 003 E − 03
A/ft^2	A/m^2	1.076 400 E + 01
Electricity and magnetism		
gauss	T	1.000 000 E − 04
maxwell	μWb	1.000 000 E − 02
mho	S	1.000 000 E + 00
Oersted	A/m	7.957 700 E + 01
Ω · cm	Ω · m	1.000 000 E − 02
Ω circular-mil/ft	μΩ · m	1.662 426 E − 03
Energy (impact, other)		
ft · lbf	J	1.355 818 E + 00
Btu (thermochemical)	J	1.054 350 E + 03
cal (thermochemical)	J	4.184 000 E + 00
kW · h	J	3.600 000 E + 06
W · h	J	3.600 000 E + 03
Flow rate		
ft^3/h	L/min	4.719 475 E − 01
ft^3/min	L/min	2.831 000 E + 01
gal/h	L/min	6.309 020 E − 02
gal/min	L/min	3.785 412 E + 00
Force		
lbf	N	4.448 222 E + 00
kip (1000 lbf)	N	4.448 222 E + 03
tonf	kN	8.896 443 E + 00
kgf	N	9.806 650 E + 00
Force per unit length		
lbf/ft	N/m	1.459 390 E + 01
lbf/in.	N/m	1.751 268 E + 02
Fracture toughness		
ksi $\sqrt{\text{in.}}$	MPa$\sqrt{m}$	1.098 800 E + 00
Heat content		
Btu/lb	kJ/kg	2.326 000 E + 00
cal/g	kJ/kg	4.186 800 E + 00

To convert from	to	multiply by
Heat input		
J/in.	J/m	3.937 008 E + 01
kJ/in.	kJ/m	3.937 008 E + 01
Length		
Å	nm	1.000 000 E − 01
μin.	μm	2.540 000 E − 02
mil	μm	2.540 000 E + 01
in.	mm	2.540 000 E + 01
in.	cm	2.540 000 E + 00
ft	m	3.048 000 E − 01
yd	m	9.144 000 E − 01
mile	km	1.609 300 E + 00
Mass		
oz	kg	2.834 952 E − 02
lb	kg	4.535 924 E − 01
ton (short, 2000 lb)	kg	9.071 847 E + 02
ton (short, 2000 lb)	kg × 10^3(a)	9.071 847 E − 01
ton (long, 2240 lb)	kg	1.016 047 E + 03
Mass per unit area		
oz/in.^{2}	kg/m^2	4.395 000 E + 01
oz/ft^2	kg/m^2	3.051 517 E − 01
oz/yd^2	kg/m^2	3.390 575 E − 02
lb/ft^2	kg/m^2	4.882 428 E + 00
Mass per unit length		
lb/ft	kg/m	1.488 164 E + 00
lb/in.	kg/m	1.785 797 E + 01
Mass per unit time		
lb/h	kg/s	1.259 979 E − 04
lb/min	kg/s	7.559 873 E − 03
lb/s	kg/s	4.535 924 E − 01
Mass per unit volume (includes density)		
g/cm^3	kg/m^3	1.000 000 E + 03
lb/ft^3	g/cm^3	1.601 846 E − 02
lb/ft^3	kg/m^3	1.601 846 E + 01
lb/in.^{3}	g/cm^3	2.767 990 E + 01
lb/in.^{3}	kg/m^3	2.767 990 E + 04
Power		
Btu/s	kW	1.055 056 E + 00
Btu/min	kW	1.758 426 E − 02
Btu/h	W	2.928 751 E − 01
erg/s	W	1.000 000 E − 07
ft · lbf/s	W	1.355 818 E + 00
ft · lbf/min	W	2.259 697 E − 02
ft · lbf/h	W	3.766 161 E − 04
hp (550 ft · lbf/s)	kW	7.456 999 E − 01
hp (electric)	kW	7.460 000 E − 01
Power density		
W/in.^{2}	W/m^2	1.550 003 E + 03
Pressure (fluid)		
atm (standard)	Pa	1.013 250 E + 05
bar	Pa	1.000 000 E + 05
in. Hg (32 °F)	Pa	3.386 380 E + 03
in. Hg (60 °F)	Pa	3.376 850 E + 03
lbf/in.^{2} (psi)	Pa	6.894 757 E + 03
torr (mm Hg, 0 °C)	Pa	1.333 220 E + 02

To convert from	to	multiply by
Specific heat		
Btu/lb · °F	J/kg · K	4.186 800 E + 03
cal/g · °C	J/kg · K	4.186 800 E + 03
Stress (force per unit area)		
tonf/in.^{2} (tsi)	MPa	1.378 951 E + 01
kgf/mm^2	MPa	9.806 650 E + 00
ksi	MPa	6.894 757 E + 00
lbf/in.^{2} (psi)	MPa	6.894 757 E − 03
MN/m^2	MPa	1.000 000 E + 00
Temperature		
°F	°C	5/9 · (°F − 32)
°R	°K	5/9
Temperature interval		
°F	°C	5/9
Thermal conductivity		
Btu · in./s · ft^2 · °F	W/m · K	5.192 204 E + 02
Btu/ft · h · °F	W/m · K	1.730 735 E + 00
Btu · in./h · ft^2 · °F	W/m · K	1.442 279 E − 01
cal/cm · s · °C	W/m · K	4.184 000 E + 02
Thermal expansion		
in./in. · °C	m/m · K	1.000 000 E + 00
in./in. · °F	m/m · K	1.800 000 E + 00
Velocity		
ft/h	m/s	8.466 667 E − 05
ft/min	m/s	5.080 000 E − 03
ft/s	m/s	3.048 000 E − 01
in./s	m/s	2.540 000 E − 02
km/h	m/s	2.777 778 E − 01
mph	km/h	1.609 344 E + 00
Velocity of rotation		
rev/min (rpm)	rad/s	1.047 164 E − 01
rev/s	rad/s	6.283 185 E + 00
Viscosity		
poise	Pa · s	1.000 000 E − 01
stokes	m^2/s	1.000 000 E − 04
ft^2/s	m^2/s	9.290 304 E − 02
in.^{2}/s	mm^2/s	6.451 600 E + 02
Volume		
in.^{3}	m^3	1.638 706 E − 05
ft^3	m^3	2.831 685 E − 02
fluid oz	m^3	2.957 353 E − 05
gal (U.S. liquid)	m^3	3.785 412 E − 03
Volume per unit time		
ft^3/min	m^3/s	4.719 474 E − 04
ft^3/s	m^3/s	2.831 685 E − 02
in.^{3}/min	m^3/s	2.731 177 E − 07
Wavelength		
Å	nm	1.000 000 E − 01

(a) kg × 10^3 = 1 metric ton

SI prefixes—names and symbols

Exponential expression	Multiplication factor	Prefix	Symbol
10^{18}	1 000 000 000 000 000 000	exa	E
10^{15}	1 000 000 000 000 000	peta	P
10^{12}	1 000 000 000 000	tera	T
10^{9}	1 000 000 000	giga	G
10^{6}	1 000 000	mega	M
10^{3}	1 000	kilo	k
10^{2}	100	hecto(a)	h
10^{1}	10	deka(a)	da
10^{0}	1	BASE UNIT	
10^{-1}	0.1	deci(a)	d
10^{-2}	0.01	centi(a)	c
10^{-3}	0.001	milli	m
10^{-6}	0.000 001	micro	μ
10^{-9}	0.000 000 001	nano	n
10^{-12}	0.000 000 000 001	pico	p
10^{-15}	0.000 000 000 000 001	femto	f
10^{-18}	0.000 000 000 000 000 001	atto	a

(a) Nonpreferred. Prefixes should be selected in steps of 10^3 so that the resultant number before the prefix is between 0.1 and 1000. These prefixes should not be used for units of linear measurement, but may be used for higher order units. For example, the linear measurement, decimeter, is nonpreferred, but square decimeter is acceptable.

Abbreviations and Symbols*

a crack length; crystal lattice length along the a axis

A ampere

A area; ratio of the alternating stress amplitude to the mean stress

Å angstrom

ac alternating current

AES Auger electron spectroscopy

AIME American Institute of Mining, Metallurgical and Petroleum Engineers

AISI American Iron and Steel Institute

AMS Aerospace Material Specification (of SAE)

ANSI American National Standards Institute

API American Petroleum Institute

ASME American Society of Mechanical Engineers

ASTM American Society for Testing and Materials

at.% atomic percent

atm atmosphere (pressure)

b crystal lattice length along the b axis

B thickness

bal balance or remainder

bcc body-centered cubic

Btu British thermal unit

c crystal lattice length along the c axis

CL confidence limits

cm centimeter

cpm cycles per minute

cps cycles per second

CRT cathode ray tube

CSFS computer-simulated fracture surface

CVN Charpy V-notch (impact test or specimen)

d day

d used in mathematical expressions involving a derivative (denotes rate of change); depth; diameter

$\mathcal{D}$ fractal dimension

da/dN fatigue crack growth rate

DBTT ductile-brittle transition temperature

dc direct current

diam diameter

DIC differential interference contrast

dm decimeter

DPH diamond pyramid hardness (Vickers hardness)

DT dynamic tear (test)

DWTT drop-weight tear test

e natural log base, 2.71828; electron

E energy; modulus of elasticity; embrittlement index; electrochemical potential

EDS energy-dispersive spectroscopy

Eq equation

et al. and others

ETP electrolytic tough pitch (copper)

eV electron volt

f frequency; focal length

F force

fcc face-centered cubic

FF fast-fast (wave form)

Fig. figure

ft foot

FS fast-slow (wave form)

FSS fatigue striation spacing

g gram

gal gallon

GPa gigapascal

h hour

H height

HAZ heat-affected zone

HB Brinell hardness

hcp hexagonal close-packed

HK Knoop hardness

HR Rockwell hardness (requires scale designation, such as HRC for Rockwell C hardness)

HSLA high-strength low-alloy (steel)

HV Vickers hardness (diamond pyramid hardness)

Hz hertz

J joule

J crack growth energy release rate (fracture mechanics)

J_c creep J-integral

J_f cyclic J-integral

K Kelvin

K stress-intensity factor

ΔK stress-intensity factor range

K_c plane-stress fracture toughness

K_{Ic} plane-strain fracture toughness

K_{Iscc} threshold stress intensity for stress-corrosion cracking

K_{th} threshold crack tip stress-intensity factor

keV kiloelectron volt

kg kilogram

km kilometer

kPa kilopascal

ksi kips (100-lb) per square inch

kV kilovolt

L length

L liter

lb pound

LME liquid-metal embrittlement

ln natural logarithm (base e)

m meter

M molar solution; magnification

M_c camera magnification

M_e enlarging magnification

M_f temperature at which martensite formation finishes during cooling

M_s temperature at which martensite starts to form from austenite on cooling

M_t total magnification

mg milligram

*Additional abbreviations and symbols and their respective definitions can be found in Tables 1, 2, and 4 in the article "Quantitative Fractography" in this Volume.

Mg megagram
min minimum; minute
MJ megajoule
mL milliliter
mm millimeter
mol% mole percent
MPa megapascal
mph miles per hour
N newton
N fatigue life (number of cycles); normal solution
N_f number of cycles to failure
NA numerical aperture
NACE National Association of Corrosion Engineers
NASA National Aeronautics and Space Administration
NBS National Bureau of Standards
ND normal direction (of a sheet)
NDE nondestructive evaluation
NDT nil-ductility transition; nondestructive testing
NDTT nil-ductility transition temperature
nm nanometer
No. number
OD outside diameter
oz ounce
p page
P applied load
$\mathcal{P}$ profile parameter
Pa pascal
p-c plastic-carbon (replica)
PDF probability density function
pH negative logarithm of hydrogen-ion activity
PH precipitation hardenable
P/M powder metallurgy
ppb parts per billion
ppm parts per million
psi pounds per square inch
R ratio of the minimum stress to the maximum stress
RA reduction of area

RD rolling direction (of a sheet)
Ref reference
rpm revolutions per minute
RSC reversed sigmoidal curve
s second
SAE Society of Automotive Engineers
SCC stress-corrosion cracking
SCE saturated calomel electrode
SF slow-fast (wave form)
SHE standard hydrogen electrode
SI Système International d'Unités
S_i image distance
SLR single-lens-reflex (camera)
SME solid-metal embrittlement
S_o object distance
SS slow-slow (wave form)
STA solution treated and aged
t thickness; time
T temperature
T_m, T_M melting temperature
tcp topologically close-packed
TD transverse direction (of a sheet)
TEM transmission electron microscopy
TME tempered martensite embrittlement
TTS tearing topography surface
TW-C truncated wave with compressive dwell
TW-T truncated wave with tensile dwell
TW-TC truncated wave with tensile and compressive dwell (hold)
UNS Unified Numbering System (ASTM-SAE)
UTS ultimate tensile strength
V volt
vol volume
vol% volume percent
W watt
W width
W_a atomic weight
WDS wavelength-dispersive spectroscopy
wt% weight percent
yr year

Z atomic number
° angular measure; degree
°C degree Celsius (centigrade)
°F degree Fahrenheit
⇌ direction of reaction
÷ divided by
= equals
≈ approximately equals
≠ not equal to
≡ identical with
> greater than
⩾ much greater than
≥ greater than or equal to
∞ infinity
∝ is proportional to; varies as
∫ integral of
< less than
≪ much less than
≤ less than or equal to
± maximum deviation
− minus; negative ion charge
× diameters (magnification); multiplied by
· multiplied by
Ω ohm
/ per
% percent
+ plus; positive ion charge
√ square root of
~ approximately; similar to
α angle
Δ change in quantity; an increment; a range
ε strain
$\dot{\varepsilon}$ strain rate
μin. microinch
μm micron (micrometer)
ν Poisson's ratio
ξ topographic index
π pi (3.141592)
ρ density
σ tensile stress
τ shear stress

Greek Alphabet

A, α alpha	I, ι iota	P, ρ rho
B, β beta	K, κ kappa	Σ, σ sigma
Γ, γ gamma	Λ, λ lambda	T, τ tau
Δ, δ delta	M, μ mu	Υ, υ upsilon
E, ε epsilon	N, ν nu	Φ, φ phi
Z, ζ zeta	Ξ, ξ xi	X, χ chi
H, η eta	O, o omicron	Ψ, ψ psi
Θ, θ theta	Π, π pi	Ω, ω omega

Index

A

Abbe's criterion, for microscope magnification 80
Abbreviations, and symbols 492-494
Aberration, spherical, in SEM imaging .. 167-168
Abrasion
    as fatigue crack origin 263
    high-carbon steels 285
Abrasive cutoff wheel cutting, specimen ... 76, 92
Absorption, of hydrogen 124
Abusive final grinding, cracking from 335
Accelerating voltage, SEM imaging 167
Acetic acid, as ferrous cleaning agent 75
Acicular needles, martensite 328
Acid cleaning
    alligatoring from 351
    chemical etching 75-76
    hydrogen embrittlement by 22
Acidic solutions, as corrosive environment 24
Acrylic lacquers, as preservative 73
Adiabatic shear
    bands, in titanium alloy 43
    defined 31
    strain rates for 31-33, 43
Adsorption, of hydrogen 124
Aging
    double 34, 47
    effect on embrittlement 34
    effect on iron fracture surfaces 458-459
    nickel alloys 397
    quench, as embrittlement 129-130
    strain, as embrittlement 129-130
    TEM for 129
Air
    dry, for fracture preservation 73
    humid, fracture effects 72
    nickel alloy cracking in 396
    and vacuum, fatigue fractures in .... .48, 55
Air blast cleaning, of fractures 74
AISI/SAE alloy steels. See also *AISI/SAE alloy steels, specific types; Steel(s)*.
    austenitization effects 339
    bolts, spontaneous rupture 299
    Charpy impact fracture 338
    drill pipe, corrosion fatigue fracture 291
    fractal analysis 212-214
    fractographs 291-344
    fracture/failure causes illustrated 216
    fracture surface, pure tensile fatigue 342
    high-strength low-alloy, fracture surfaces 344
    hydrogen flaking 125
    quasi-cleavage facets, dimples, and voids ... 330
    spontaneous sulfide-SCC fracture 299
    temper embrittlement 134
AISI/SAE alloy steels, specific types. See also *AISI/SAE alloy steels; Steel(s)*.
    AISI 304 (SUS 304), effect of frequency and wave form effect on fatigue properties 62
    AISI 508 B60, torsional fatigue fracture 323
    AISI 1040 bolts, quench cracks 149
    AISI 1040 bolts, SCC failure 151
    AISI 1070, transverse fracture 142, 163
    AISI 1085, cathodic cleaning 75
    AISI 1085, ultrasonic cleaning 74
    AISI 1340, corrosion fatigue fracture 291
    AISI 4130, brittle fracture 291
    AISI 4130, effect of stress intensity factor range on fatigue crack growth rate 57
    AISI 4130, fatigue fracture surface 293
    AISI 4130, frequency and wave form effects on fatigue properties 59
    AISI 4130, hydrogen-embrittled 31
    AISI 4130, metallographic study, hitchpost failure 292
    AISI 4140, ductile fracture 298
    AISI 4140, embrittlement by liquid cadmium 30, 39
    AISI 4140, fatigue fracture surface 295
    AISI 4140, fracture surface, near weld toe... 294
    AISI 4140, improper heat treatment 298
    AISI 4140, microstructures, with temper embrittlement 153
    AISI 4140, service fracture 297
    AISI 4140, tire tracks 23
    AISI 4142, splitting 106
    AISI 4146, improper angular hardening 300
    AISI 4150, star and beach marks 301
    AISI 4315, hydrogen embrittlement 301
    AISI 4315, tension overload fracture 301
    AISI 4320, fatigue failure 111, 120
    AISI 4340, Charpy impact fractures 314
    AISI 4340, dimples, SEM fractograph 207
    AISI 4340, dimples with inclusions 65, 67
    AISI 4340, effect of frequency and wave form on fatigue properties 59
    AISI 4340, effect of lead on fracture morphology 30, 38
    AISI 4340, effect of stress intensity factor range on fatigue crack growth rate 57
    AISI 4340, effects of decreasing stress 315
    AISI 4340, embrittlement 214
    AISI 4340, fatigue fracture surface 303
    AISI 4340, fractal analyses 212-215
    AISI 4340, fractal dimensions 213
    AISI 4340, fractographic analysis 302
    AISI 4340, fracture appearance, impact energy vs test temperature 109
    AISI 4340, fretting wear 308
    AISI 4340, hammer blow mechanical failure 305
    AISI 4340, hydrogen damage 302
    AISI 4340, hydrogen embrittlement 306
    AISI 4340, improper heat treatment 309
    AISI 4340, low-cycle fatigue fracture 308
    AISI 4340, mating fracture surface 310
    AISI 4340, mating segments, fatigue fracture 315-316
    AISI 4340, pre-existing crack as fracture origin 65
    AISI 4340, profile angular distributions 203
    AISI 4340, quasi-cleavage in hydrogen-embrittled 31
    AISI 4340, radial fracture 312
    AISI 4340, roughness parameters 213
    AISI 4340, service failure 317
    AISI 4340, stringers on fracture surface 67
    AISI 4340, tensile fracture 103
    AISI 4340, tension overload fracture 304, 311-313
    AISI 4340, true profile length values 200
    AISI 4340, unfavorable grain flow 67-68
    AISI 4615, high-cycle bending fatigue fracture 321
    AISI 4817, fatigue fracture surface 322
    AISI 4817, rotating bending fatigue fracture 321
    AISI 4817, subcase fatigue cracking 322
    AISI 5046, fatigue zone, subcase fatigue cracking 323
    AISI 5132, mating fracture surface 323
    AISI 5140H, effect of strain rate on fracture appearance 31, 41
    AISI 5160H, fracture from seam 326
    AISI 5160, ribbonlike inclusions 326
    AISI 5160 wire spring, fracture from seam 63, 64
    AISI 6150, fatigue fracture surface 327
    AISI 8617, bending fatigue fracture 329
    AISI 8620, bending fatigue fracture 330
    AISI 8620, fatigue striations 331
    AISI 8620, spalling fatigue fracture 329
    AISI 8620, torsional overload fracture 330
    AISI 8640, beach marks and final fast fracture 331
    AISI 8640, service fracture surface 331
    AISI 8645, fatigue fracture surface 333
    AISI 8740, decohesive rupture 24
    AISI 8740, tensile-overload fracture 334
    AISI 9254, brittle intergranular fracture 335
    AISI 9310, fatigue fracture, reversed cyclic bending 120
    AISI 9310, inclusion in service fracture surface 66
    AISI 52100, effect of rapid heating in austenitizing 328
    AMS 6434, impact fracture 319
    AMS 6434 steel sheet, tension overload fracture 319
    AMS 6434, stress-corrosion cracking 320
    AMS 6434, tension overload fracture 319
    Cr-V alloy, high-cycle fatigue fracture 337
    D6B, fracture surface 343
    SAE 21-4N (EV 8) steel, photo-illumination effects 87
    SAE 51 B60 railroad spring, torsion failure .. 121
    SAE 81 B45, fatigue fracture surface 328
    SAE 4150, overtempering 301
    SAE 4150, reversed torsional fatigue fracture 301
    SAE 5160, impact fracture, with mating surface 324-325
Alcohols, as organic cleaning solvents 74
Alconox, as ferrous and aluminum detergent 74-75
Alkaline solutions, in chemical etching cleaning 75
Alligatoring, austenitic stainless steels 351
Alloy steels. See also *AISI/SAE alloy steels; ASTM/ASME alloy steels*.
    miscellaneous, figure numbers for 216
    miscellaneous, fracture causes illustrated 216
    properties, historical study 2
Alnico alloy, fracture topography 461
Alternating stress, effect on fatigue cracking ... 15
Aluminum. See also *Aluminum alloys; Aluminum alloys, specific types; P/M aluminum alloys*.
    hydrogen embrittlement 23-24, 124
    Monte Carlo electron trajectories in 167
    as polymer contaminant 479
    pure, integrated circuit defects 481
    SCC of 28
    transistor base lead, fatigue failure 483
    vacuum-deposited, electronic defects 484, 486-487
Aluminum alloys. See also *Aluminum*.
    embrittlement by low-melting alloys 29

Aluminum alloys (continued)
  fatigue striations in 176
  forging, SCC fracture by decohesion 18, 25
  high-strength, grain-boundary separation 174
  hydrogen-embrittled, types 23-24
  intergranular corrosion 126
  in metal-matrix composites 466
  SCC of 18, 25, 28, 133
  vacuum effects 46
Aluminum alloys, specific types
  2014-T6, aircraft component fatigue failure .. 175
  2014-T6, knobbly structure 33
  2024, nondestructive dimple profiles 199
  2024-T3, fatigue striations 19, 20
  2124-UT, hydrogen-embrittled 33
  2219-T851, water vapor effect on crack
    propagation rate 40, 52
  6061, halide-flux inclusion 65, 67
  6061-T6, striations on fatigue crack fronts 23
  7050 low copper, hydrogen-embrittled 33
  7075-T6, effect of corrosion on fatigue
    strength 43, 54
  7075-T6, fatigue fracture, resistance spot
    weld 66, 67
  7075-T6, SEM and TEM fractographs,
    compared 187-188
  7075-T6, stress-corrosion fracture 28, 35
  X7091, true profile length values 200
  7475, effect of stress intensity factor range on
    fatigue crack growth rate 57
  7475-T7651, fatigue striation spacing 22
  Al-4Cu, fractured, SEM projected facets 195
  Al-4Cu, fractured, serial sectioning profile .. 198
  Al-4Cu, fracture surface
    measurements 205-206
  Al-4Cu, SEM fractograph, calculation of
    features 206-207
  Al-4Cu, true profile length values 200
Aluminum-copper alloys, facets, SEM
  projection 195
Aluminum-lithium alloys, ductility 440
Aluminum stub substrates, SEM specimens .. 172
Ammonia
  as corrosive environment 24
  undissociated, effect in high-carbon steels ... 282
Ammonium citrate, as ferrous cleaning agent .. 75
Ammonium oxalate solutions, as ferrous
    cleaning agents 75
Angular distribution
  fractal analysis 212
  profile, for fracture surface area 202-204
Annular fracture surface, metal-matrix
    composites 466
Anodic dissolution
  effect on cleavage 42
  as SCC mechanism 25
Antihalation films, camera lens 84
Antimony
  cast polycrystalline, early TEM 6
  -doped ASTM/ASME alloy steels 350
  fractured ingots, history 1
  segregation to grain boundaries 350
Aperture
  camera lens, selection 80-81
  number to f/number conversions for
    Macro-Nikkor lenses 81
  optimum, problem of 80-81
  size effects in SEM imaging 168
Applied stress
  effect, embrittlement 29
  effect, striation spacing 120
  examining 92
  periodic interruptions, wrought aluminum
    alloys 418
  pulsation from 304
Arcing, brush/slip ring assembly failure from .. 488
Arc strikes
  fatigue crack initiation at 261
  in fracture surfaces, medium-carbon steels ... 255
Area
  as basic convex figure quantity 194
  of closed figure 195
  fractal analysis 214-215

fracture surface 201-205
  of fracture surface, importance 193
  of irregular fracture surfaces 211-215
  parametric relationships 204-205
  profile angular distributions for 202-204
  ratios, for partially oriented surfaces 201
  stereological relationships 196
  and surface parameters 200
  total facet surface 202
  triangular elements for 201-202
  true, and true length 204
  true fracture surface, importance 211
  true mean facet 208
  true, vertical sectioning for ... 198-199, 211-212
Artifacts, in replicas 184-185
Asphalt
  fracture/failure causes illustrated 217
  sulfur, fracture surface 473
Assumption of randomness, in projected
    images 194-195
Astigmatism, in SEM illuminating/imaging
    system 167
ASTM/ASME alloy steels. See also *ASTM/ASME
    alloy steels, specific types; Steels(s)*.
  fractographs 345-350
  fracture/failure causes illustrated 216
  hydrogen flaking 125
  temper embrittlement 134
ASTM/ASME alloy steels, specific types. See
    also *ASTM/ASME alloy steels; Steel(s)*.
  ASME SA213, creep failure 346
  ASTM A325 bolt, SCC in 133
  ASTM A372, solidification cracking,
    laser-beam welds 345
  ASTM A490, bolt specimens 103, 104
  ASTM A508 class II, overheating 146
  ASTM A508, fracture by
    overpressurization 345
  ASTM A514F, effect of inclusions on fatigue
    crack propagation 346
  ASTM A517H, brittle fracture 347
  ASTM A533B, cavitated intergranular fracture,
    hydrogen attack 349
  ASTM 533B pressure vessel, hydrogen effects on
    fracture appearance 37, 51
  ASTM A533B, effect of inclusions on fatigue
    crack propagation 347, 348
  Cr-Mo-V, elevated-temperature fracture
    surface 349
  Cr-Mo-V, phosphorus effect on cavities 349
  Cr-Mo-V, phosphorus effect on ductility 349
  Ni-Cr antimony-doped, hydrogen effects 350
Astroloy. See *Nickel-base superalloys, specific
    types*.
Asymptotic curvature, fractal 211-212
Atomic number imaging, electrons and uses
    for 168
Atomic oxygen, effect on silver solar cell
    interconnect 481
Austenitic stainless steels. See also *Austenitic
    stainless steels, specific types; Steel(s)*.
  fatigue striations 21
  ferritic iron-aluminum, brittleness of 365
  fractographs 351-365
  fracture/failure causes illustrated 217
  hydrogen effects 39
  intergranular corrosion 126, 142
  irradiation embrittlement 127
  SCC failures of 133
  sigma-phase embrittlement 132
  temperature effects 50-52
Austenitic stainless steels, specific types. See also
    Austenitic stainless steels.
  AISI 301, fatigue fracture surface 351
  AISI 301, hydrogen-embrittled 31, 39, 52
  AISI 302, alligatoring 351
  AISI 302, high-cycle fatigue fracture 352
  AISI 302, hydrogen effect 39, 52
  AISI 302, rock-candy fracture 351
  AISI 304, chloride SCC 354
  AISI 304, effect of strain rate on creep crack
    propagation 354
  AISI 304, hydrogen embrittlement 355, 357

AISI 304L, hydrogen damaged 356
  AISI 304, polythionic acid SCC 354
  AISI 304, SCC facets 354
  AISI 316, channel fracture 365
  AISI 316, chloride SCC 357
  AISI 316, fatigue fracture appearance 51, 56
  AISI 316, intergranular SCC 357
  AISI 316L, fatigue fracture, orthopedic
    implant 359-364
  P/M 316L, brittle fracture 358
  SIS 2343, corrosion pit cracking 358
Austenitization
  aging after, effect in iron alloy 458
  AISI/SAE alloy steels 298
  effect on fracture characteristics 339
  effect on fracture toughness 340
  incomplete, AISI/SAE alloy steels 291
  phosphorus segregation during, tool steels ... 375
  rapid heating effects in 328
Axis of symmetry, assumed 202-203
Axle grease, as fracture preservative 73

B

Background papers, fractographic 78
Backscattered electrons
  fractographs, sulfur concrete fracture
    surfaces 472
  and secondary electrons, compared 168
  SEM illuminating/imaging system 167-168
Bacteria, thiobacillus, as corrosive 245
Baking, postplate, high-carbon steels 284
Ball-on-ball impact fracture(s), AISI/SAE alloy
    steels 336
Bands
  AISI/SAE alloy steels 333
  deformation, butterflies as 115, 134
  shear 32, 42
  transformed 32
Band saw cutting, for macroscopic
    examination 92
Basic solutions, as corrosive environment 24
Beach marks
  AISI/SAE alloy steels 301, 322, 331, 332
  austenitic stainless steels 358, 359
  cast aluminum alloys 408
  circular 273
  and circular spall 114, 125-126
  ductile iron crankshaft 228
  in fatigue fracture 111-112
  as fatigue striations 175
  high-carbon steels 281, 285
  martensitic stainless steels 369
  in medium-carbon steels 260, 267, 273-276
  oval 273
  superalloys 391
  titanium alloys 446, 452
  tool steels 377
  wrought aluminum alloys 415, 416, 421, 427
Beam, electron, in SEM imaging 167-168
Bearing cap bolts, sulfide SCC failure 299
Bearings
  butterflies in 115, 134
  spalling fatigue in 114-115
Bearing steels, rolling-contact fatigue 115, 134
Bell jar, carbon 173
Bellows length, photomacrography 79
Bending
  high-cycle fatigue, alloy steels 296
  historical studies 3
Bending fatigue fracture(s)
  alloy steel gear teeth 329-330
  AISI/SAE alloy steels ... 296, 321, 329-330, 332
  iron 220
  rotating, alloy steels 321
  surfaces, tool steels 376
Bending impact fracture, low-carbon steel 242
Bending overload fracture, classic, medium-
    carbon steels 272
Biaxial stress, effect on dimple rupture 31, 39
Bifurcated fiber optic light source,
    fractographic 79

Blanking fracture, effect in high-carbon
  steels..................................285
Blistering, as hydrogen damage 124, 125, 142
Blowholes, in iron........................221
Blunting, at crack tip....................15, 21
Blur circles, photographic..................84
Body-centered-cubic metals
  embrittlement.............................123
  hydrogen embrittlement....................22
  intergranular fracture in..................123
  iron, slip lines...........................219
  in low temperatures, effect on fracture mode..33
  state of stress effects on.................31
  strain rate effect on......................31
Body fluids, dried, as corrosive............361
Bohr model of atom..........................168
Bolts
  AISI/SAE alloy steels, spontaneous rupture..299
  bearing cap, sulfide SCC in................299
  bronze, preferential corrosion.............403
  chevrons from fracture origin, SEM
    fractographs.............................170
  failures, fractographic study..............248
  fractured, SEM fractographs of radial marks..169
  quench cracks..............................131
Bonding
  in metal-matrix composites.................466
  sulfur-cement..............................472
  ultrasonic, aluminum transistor base lead..483
Brasses
  corrosion pitting fracture.................404
  SCC of.................................28, 36
Breaking, final, of specimens................77
Bright-field illumination
  coarse-grain iron alloy..................93-94
  and dark-field, SEM images, compared........92
  for slip..................................121
Bright flake, wrought aluminum alloys.......415
Brittle, defined.............................173
Brittle fracture(s). See also *Brittle; Brittle intergranular fracture(s); Brittleness.*
  AISI/SAE alloy steels..................291, 335
  ASTM/ASME alloy steels.....................347
  austenitic stainless steels......354, 356, 358
  cast aluminum alloys..........405-408, 410
  cemented carbides..........................470
  ductile irons.........227, 231-232, 235-237
  effect of grain size...................106-107
  granular, macrograph.......................103
  historical study............................5
  in-service, tool steels....................376
  intergranular...............174-175, 335, 354
  interpretation of......................105-111
  low-carbon steels......................243, 249
  macroscopic characteristics................107
  magnesium matrix, metal-matrix composites..464
  malleable iron.............................238
  materials illustrated in...................217
  matrix, in composites......................468
  medium-carbon steels.........258, 270, 273
  metal-matrix composites...............466-468
  microscopic characteristics................109
  polymer...................................479
  by pure tensile fatigue....................342
  with river patterns, ductile iron..........230
  SEM characterized.....................174-175
  splines, alloy steels......................330
  titanium alloys........................449, 454
  tool steels................................375
  wrought aluminum alloys....................424
Brittle intergranular fracture(s). See also *Brittle fracture(s); Brittleness.*
  austenitic stainless steels................354
  low-carbon steel..........................245
  precipitation-hardening stainless steels...374
  steel alloy............................30, 38
Brittleness. See also *Brittle fracture(s); Brittle intergranular fracture(s).*
  examining..................................92
  of ferritic iron-aluminum alloys...........365
  in iron-base alloys........................460
  as temperature-dependent...................106
  weld metals................................375

Brittle striations, formation................35
Bronzes. See also *Copper; Copper alloys.*
  couch roll shell, primary crack............403
  shell, SCC microstructure..................403
Brush cleaning, of fractures................74
Burning
  alloy steels...............................300
  for macroscopic examination................92
  in medium-carbon steels....................259
  in steels..................................127
Burnishing, macroexamination of.............72
Butterflies, in bearings...............115, 134
Button defects. See *Scabs.*

C

Cadmium
  liquid, low-alloy steel embrittlement by...30, 39
  as low-melting embrittler...................29
Calcium, effects on inclusions and cracking,
  low-carbon steels..........................247
Cameras
  35-mm single-lens-reflex................78-79
  magnification...............................80
  view.....................................78-79
Carbides
  embrittling effect, AISI/SAE alloy steels..341
  spheroidization, ASTM/ASME alloy steels..346
Carbon
  -FeO reaction in iron......................221
  historical studies...........................3
  thin films.................................173
Carbon planchets, for SEM specimens.........172
Carbon replicas
  techniques, TEM.............................7
  two-stage..................................182
Carbon steels. See also *Carbon steels, specific types; High-carbon steels; Low-carbon steels; Medium-carbon steels; Steel(s).*
  in metal-matrix composites.................466
  SCC failures..............................133
Carbon steels, specific types. See also *Carbon steels.*
  AISI 1042, temperature effect on fracture
    mode...................................33, 45
  AISI 1060, knobbly structure................33
  AISI 1080, temperature effect on fracture
    mode...................................33, 44
  AISI 1085, cathodic cleaning............75-76
Carbontetrachloride, as carcinogenic cleaning
  agent.......................................74
Carburization
  -caused brittle fracture, austenitic stainless
    steels...................................358
  historical study..........................1-2
Carcinogenic organic solvents, cleaning......74
Care, of fractures............................72
Carpet plot. See also *Fracture surface map.*
Carpet plots
  fractured titanium alloy...................172
  stereo imaging............................171
  titanium alloy, by stereophotogrammetry....198
Case hardening, effects, visual examination..72
Cast aluminum alloys. See also *Aluminum; Cast aluminum alloys, specific types.*
  chemical analysis..........................405
  effects of freezing and heat treatment.....409
  experimental, transverse fracture surface..413
  fractographs..........................405-413
  fracture/failure causes illustrated........217
  shrinkage cavities.........................409
Cast aluminum alloys, specific types. See also *Cast aluminum alloys.*
  356.0-T6, brittle fracture............405-406
  356.0-T6, fatigue fracture.............407-408
  356.0-T6, service fracture.............409-410
  A357 blade, porosity in.................66, 67
  A357-T6 gear housing, shrinkage void..66, 67
  A357-T6, inclusion in fracture surface..65, 66
  A357-T6, shrinkage void....................67
  380.0, brittle fracture....................410
  518.0, overload fracture in service....411-412

Castings, quality control tests.........141-142
Cast irons
  ductile, fracture modes....................230
  granular brittle fractures.................103
Catastrophic failure, alloy steels............336
Catchlights, on fracture surfaces...........83-84
Cathode ray tube
  as synchronized with SEM imaging
    system...............................169-171
  in x-ray analyses..........................168
Cathodic cleaning, of fractures..............75
Cavitation
  creep, as decohesive rupture................20
  formation by methane gas bubbles.......37, 51
  grain-boundary..............19, 26, 219, 349
  intergranular creep rupture by..........19, 26
  in iron....................................219
  and slip lines, iron.......................219
Cavities. See also *Shrinkage cavities.*
  in alloy steel.............................349
  and dimples, compared..................20, 220
  elongated, titanium alloys.................450
  fatigue fracture from.....................419
  gas, cast aluminum alloys............405-406
  grain-boundary.........................219, 349
  in intergranular iron fracture.............219
  in iron................................219-220
  nodule-bearing, in ductile iron............229
  nucleation................................122
  r-type................................122, 140
  separation, copper alloys..................402
  shrinkage..............................140, 160
  types, austenitic stainless steels.........364
Cellulose acetate replica(s)
  with acetone..............................180
  for SEM imaging.......................171-172
  tape, as fracture preservative..............73
  tape, for light microscopy.............94-95, 99
Cellulose nitrate, with amyl-, ethyl-, or methyl
  acetate, for replicas.....................180
Cemented carbides. See also *Cemented carbides, specific types.*
  fractographs..............................470
  fracture/failure causes illustrated........217
  fracture toughness........................470
Cemented carbides, specific types
  94WC-6Co, eta-phase, fracture surface.....470
  94WC-6Co, mating fracture analysis........470
  97WC-3Co, brittle fractures...............470
Centerline cracks, low-carbon steel..........244
Central fibrous region, alloy steels.........334
Ceramics. See also *Ceramics, specific types.*
  fractographs..............................471
  at fracture...............................471
  fracture/failure causes illustrated........217
Ceramics, specific types
  Al_2O_{32} + 3 glass, true profile length..200
  Al_2O_{32}-glass, area/length parametric relation..204
  alpha-SiC, corrosion pitting fracture.....471
Channel fracture, austenitic stainless steels..365
Characteristic x-rays, defined...............168
Charpy V-notch impact test,
  fractures.....................106, 108-110, 341
Chemical analysis, cast aluminum
  alloys................................405, 407
Chemical etching, as fracture cleaning
  technique................................75-76
Chevron marks
  AISI/SAE alloy steels..................306, 307
  ASTM/ASME alloy steels....................347
  as brittle.............107-108, 111, 173, 258
  cast aluminum alloys......................413
  causes................................107-108
  on cleavage fracture surface................13
  fracture, brittleness and ductility of....108
  from fracture origin, SEM fractographs....170
  fractures, crack front in..................107
  medium-carbon steels....255-258, 260-261, 269
  polymers..................................480
  visual examination.................107, 111-113
  wrought aluminum alloys...............429, 436
Chill
  casting, irons............................219

498 / Index

Chill (continued)
 inverse, ductile iron fracture from 227
 tests, applications 141
Chisel-point fractures, fcc metals 100
Chisel steels, early fractographs 5
Chlorides
 effect in alloy steels 291
 solution, as corrosive environment 24
 stress-corrosion cracking 357
Chromic acid, as cleaning agent 75
Chromium
 depletion, austenitic stainless steels 51
 embrittlement sources 123
 plating, effect in AISI/SAE alloy steel
  fracture 297
 steel, feather markings 18
Chromium-molybdenum steels, creep
  embrittlement 124
Circular fluorescent-light tubes,
  photographic 83
Circular spall, in steels 113-115, 123-128
Clamshell markings. See *Beach marks; Fatigue
  striations.*
Cleaning techniques
 air blast 74
 brush 74
 cathodic 75
 chemical etching 75-76
 of fracture surfaces 73-77, 179-183
 organic solvents 74
 replica-stripping 74
 water-base detergent 74
Clear acrylic lacquers, as fracture
  preservatives 73
Cleavage. See also *Cleavage facets; Cleavage
  fractures; Cleavage steps; Intergranular
  fractures; Transgranular cleavage fractures.*
 alloy steels 338
 Alnico alloy 461
 austenitic stainless steels 352
 crack path, low-carbon steel 117
 defined 13-14
 effect of anodic dissolution 42
 historical study 4
 intergranular and transgranular, compared ... 290
 iron alloys 223, 459
 local, tool steels 382
 low-carbon steel 174
 materials illustrated in 217
 and mechanical twinning 4
 metal-matrix composites 467
 planes 13
 and ripples, compared 453
 river patterns on 252, 424
 steps 223, 352
 surface features 13-18
 tools steels 382
 transcrystalline, iron 222
 transgranular 175, 461, 467
 as transgranular fracture mode, SEM
  defined 175
 wrought aluminum alloys 418
Cleavage facets
 AISI/SAE alloy steels 319
 and dimples, sizes of 328
 formation 339
 light fractographs 93-95
 molybdenum alloy 464
 nickel alloys 396-397
 titanium alloys 453
Cleavage fractures
 AISI/SAE alloy steels 302-303, 319
 in Armco iron, shear step 224
 brittle, ductile iron 227
 of corrosion products 29
 defined 13
 ductile irons 232, 236, 237
 by early TEM study 6
 effect of subgrain and grain boundaries ... 17
 etch pits on 101
 feather pattern, steps 18
 flat, high-purity iron 219
 formation 13, 17

 in hydrogen-embrittled stainless steel ... 31
 iron-aluminum alloys 365
 iron-base alloys 224, 365, 457, 460
 light fractographs 94, 99
 low-carbon steels 249, 252
 low-melting metals 30, 38
 by mercury vapor embrittlement 30, 38
 nickel alloys 297
 in polycrystalline metals 252
 with river patterns, tongues, grain
  boundary 224
 by SCC, titanium alloys 453
 surfaces 13, 17-18
 at three different magnifications 175
 titanium alloys 30, 38, 453
 transgranular 302, 460
 transition to dimple rupture 33, 45
 woody, alloy steels 319
 wrought aluminum alloys 424, 430
Cleavage steps 13, 263
 AISI/SAE alloy steels 301
 Alnico alloy 461
 in Armco iron 18
 austenitic stainless steels 352-353
 giant, titanium alloys 450
 in iron 17, 457
 medium-carbon steels 263
 precipitation-hardening stainless steels ... 370
 terraced facets with 448
 titanium alloys 448, 450
 wrought aluminum alloys 417, 432
Cliffs, wrought aluminum alloys 418
Closing crack, in fatigue. See also *Cracks.* ... 15
Coarsening, alloy steels 292
Coated lenses 84
Coatings
 sputter 173
 surface 72-73, 83
 thermal evaporation 173
Cobalt alloys. See also *Cobalt alloys, specific
  types.*
 fractographs 398
 fracture/failure causes illustrated 217
 in metal-matrix composites, ductile dimpled
  rupture 470
Cobalt alloys, specific types
 ASTM F75 cast, fatigue fracture 398
 ASTM F75, stage I fatigue fracture
  appearance 16
 Vitallium, fatigue fracture 398
Cold cracking, examination/interpretation . 137-138
Cold shuts
 as discontinuities, defined 64-65
 tool steels 378
 wrought aluminum alloys 435
Collimated electron beam, SEM imaging ... 167
Collisions, elastic and inelastic 168
Colors
 fatigue area, austenitic stainless steels ... 352
 of fracture, crack growth measurement by ... 120
 temper, of cracks 65
Columnar fractures 2
Commercially pure titanium, vacuum effects on
  fatigue 48-49
Composites. See *Metal-matrix composites; Resin-
  matrix composites.*
Compression, effects in resin-matrix
  composites 478
Computer-aided design 421
Concrete
 fracture/failure causes illustrated 217
 sulfur, fracture surfaces 472-473
Conductivity, specimen, for SEM imaging ... 171
Configuration parameter, surface roughness ... 200
Connectivity, as roughness parameter 201
Contamination line, nondestructive profiles ... 199
Contour plots
 fractured titanium alloy 172
 stereo imaging 171
Convergent magnetic lenses, SEM 167
Conversion oxide film replicas, formation ... 181
Convex figure, basic quantities for 194
Cooling, titanium alloys 454

Copper
 contamination, in low-carbon steel 249
 dimple formation 173
 free machining 401
 grain-boundary embrittlement, low-carbon
  steel 246
 high-purity 399-400
 oxygen-free high-conductivity 401
 serpentine glide formation 17
Copper alloys. See also *Copper; Copper alloys,
  specific types; High-purity copper; Oxygen-
  free high-conductivity copper.*
Copper alloys
 fractographs 399-404
 fracture/failure causes illustrated 217
 oxygen-free high-conductivity 401
 SCC failures 133
Copper alloys, specific types
 64Cu-27Ni-9Fe, spinodal decomposition ... 402
 64Cu-27Ni-9Fe, tensile overload fracture ... 402
 80Cu-20Zn brass, corrosion pitting fracture ... 404
 90Cu-10Sn-1Pb, stress-corrosion cracking ... 403
 Cu-10Sn (tin-bronze), preferential corrosion ... 403
 Cu-2.5Be, fully aged, strength 402
 Cu-2.5Be, underaged fracture-toughness
  test 402
 Cu-5Sn-5Pb-4Zn, corrosion fatigue failure ... 403
 phosphor bronze C (C52100) wire cloth, fatigue
  fracture from warp tension 404
Copper-zinc alloys, fracture history 1
Corrosion. See also *Corrosion fatigue; Corrosion
  fatigue fracture(s); Corrosion pitting;
  Corrosion products; Corrosive environments;
  Pitting corrosion.*
 in AISI/SAE alloy steels 318
 in bolts 248
 copper alloys 403
 electronic materials 482
 fatigue, and SCC, unified theory 42
 fracture, cleaning of 73-76
 general 41
 hot 391
 -induced, hydrogen-assisted fracture, low-carbon
  steels 248
 intergranular, austenitic stainless steels ... 364
 kinetics 41
 in low-carbon steel 243, 248
 passive 41-42
 preferential, from urban atmosphere 403
 prevention, surface coatings for 73
 sites, from dot maps 168
 strain amplitude and stress effects on ... 418
 under-deposit, copper alloys 403
Corrosion fatigue
 crack growth rate 41
 defined 36
 in low-carbon steel 245, 250
Corrosion fatigue fracture(s)
 copper alloys 403
 crack propagation, P/M aluminum alloys ... 440
 low-carbon steel 250
 transgranular 438
 wrought aluminum alloys 432, 438
Corrosion pitting. See also *Pitting; Pitting
  corrosion.*
 AISI/SAE alloy steels 291, 318
 austenitic stainless steels 358
 in ceramic 471
 copper alloys 404
 wrought aluminum alloys ... 414, 415, 419
Corrosion products
 AISI/SAE alloy steels 299, 306
 austenitic stainless steels 37, 361
 defined 29
 effect on crack propagation 133
 effect on dimple rupture 29
 electronic materials 482
 on hip implant 37
 hydroxide 372
 on intergranular fracture surface 37
 in low-carbon steel 245
 magnesium alloy 456
 niobium alloy 37

Index / 499

P/M aluminum alloys 440
precipitation-hardening stainless steels 372
in SCC fractures 27
Corrosive environment. See also *Stress-corrosion cracking.*
effect on dimple rupture 24-29
liquid, effects and types 36, 41-46
types 24
Crack. See *Crack arrest; Crack arrest marks; Crack closure; Crack front; Crack growth; Cracking; Crack initiation; Crack length; Crack opening; Crack origin; Crack path; Crack propagation; Cracks; Crack tip; Fatigue crack growth; Fatigue crack growth rate; Fatigue crack propagation; Fatigue striations; Fatigue striation spacings.*
Crack arrest, mechanism of 15
Crack arrest marks. See also *Fatigue striations.*
in hydrogen-embrittled titanium alloys 23, 32
wrought aluminum alloys 414
Crack closure 15
decarburization softening-induced enhancement 38
as mechanical damage 72
by partial slip reversal 21
Crack deflection, model 206
Crack front
advancing, mechanical damage 72
defined 176
direction, wrought aluminum alloys 430
located by beach marks 112
primary, ductile irons 233
striations on 23
Crack growth
direction, titanium alloys 441-442
effect of vibration 109
fatigue 14-18, 420
and fatigue striation spacings 205
modes, medium-carbon steels 273
OFHC copper 401
in polyethylene 480
stable, mechanical damage in 72
structure sensitivity, titanium alloys 442-443
from surface flaw 420
surface roughness from 276
Cracking
cold 137-138
hot 138
pre- 75, 236-237, 397
stress-relief 139-140
surface corrosion from 72
weld 137-140
Crack initiation
AISI/SAE alloy steels 302
fatigue 112
as fatigue stage I 175
malleable iron 238
P/M aluminum alloys 440
by tool marks and corrosion pits 419
Crack length
and crack path preference 201
mean, calculated 207
measuring 77
true, calculated 207
Crack opening
secondary 77
by slip 21
Crack origin. See also *Fracture origin.*
at abraded surface area 263
determined by beach marks 112
fractured medum-carbon steel shell 257
lap as 64, 65
medium-carbon steels 261, 272
multiple 272
superalloys 390
wrought aluminum alloys 415
Crack path. See also *Fracture path.*
ductile and brittle 102
in low-carbon steel 245
in SCC 133
tortuosity 38, 206
transgranular, SCC caused 133-134, 152
in white iron 239

Crack propagation. See also *Fatigue crack propagation.*
by alternate slip 15, 21
direction, and fatigue striation spacings 205
ductile iron 228
effect of twin boundaries, austenitic stainless steels 351
environmental effects 35
as fatigue stage II 175
historical studies 3
in iron 222-223
in low-carbon iron 222
malleable iron 238, 239
in monotonic fracture 229
rate, effect of grain-boundary cavitation on ... 38
SEM studies 169
Cracks
in broken/unbroken aluminum alloy, compared 121, 137, 138
corrosion of 72
high-speed, effects in low-carbon steel 252
as hydrogen damage 124
nucleation, austenitic stainless steels 360
part-through 72
pre-existing, as fracture origin 65
primary, opening 77
secondary, opening 77
separations, measuring 77
Crack tip
alternate slip 15, 21
blunting 15, 21
dislocation nucleation 30
plastic zone 15, 16
resharpening 21
slip at 15, 21
Creep
crack propagation, austenitic stainless steels .. 354
damage, as indicator of fracture mode 62-63
effect on fracture 59
embrittlement, of chromium-molybdenum steels 124
failure, ASTM/ASME alloy steels 346
failure, by wedge cracking 364
fatigue life prediction by 123
matrix 349
microvoid linking by 364
rupture, defined 18-20
-rupture embrittlement 123-124
stages 19, 25
tests, medium-density polyethylene 480
tests, for high-temperature effects 121
Creep embrittlement 123-124
Creep rupture(s)
defined 18-20
embrittlement, interpreting 123-124
high-purity copper 399
intergranular, austenitic stainless steels 364
strain rates for 31
superalloys 389, 393, 395
Critical resolution point. See also *Fractal plot.* 211
Cross-check defect, precipitation-hardening stainless steels 374
CRT. See *Cathode ray tube.*
Cryogenic temperatures, effect on fatigue .. 52-53
Crystalline fractures 2
Cup-and-cone fracture(s)
in alloy steels 302
examining 98
in fcc metals 100
medium-carbon steels 253
precipitation-hardening stainless steels 370
studies 3
tensile 3, 98, 100, 102, 253
titanium alloys 451
tool steels 377
zones 102
Curling, in TEM replicas 181
Cutting
abrasive wheel 92
band saw 92
specimen 76-77
by ultramicrotome 199

Cyclic frequency, effect on fatigue cracking. See also *Frequency.* 15
Cyclic loading, effect on fatigue. See also *Loading.* 53-54, 111
Cyclic stresses, wrought aluminum alloys, See also *Stress.* 421, 430

D

Dark-field illumination fractographs
coarse-grain iron alloy 93-94
light-field, and SEM, compared 92-100
Data signals, SEM, origin and detection 168
Dealloying, as SCC mechanism 25
Debris, from field fractures 92
Decarburization. See also *White zone.*
alloy steels 300, 333
effect of hydrogen 37-38, 51
high-carbon steels 288
partial, alloy steels 327
Decohesion. See also *Decohesive rupture(s); Intergranular decohesion fracture(s).*
austenitic stainless steels 364
glide-plane 391
along grain boundaries, elongated grains .. 23
along grain boundaries, equiaxed grains .. 23
intergranular, band formation by hydrogen 37, 51
intergranular, effect of strain rate 41
intergranular, SCC fracture by 27
nodule, ductile irons 236
tearing, in welds 139
through weak grain-boundary phase 23
Decohesive rupture(s). See also *Decohesion; Grain-boundary separation; Intergranular brittle fracture.*
creep rupture 18-20
defined 18-20
intergranular, steels 31
materials illustrated in 217
mechanisms 18
in precipitation-hardenable stainless steel ..18, 24
by SEM imaging 174-175
Deep-field microscopy, of fractures 96
Deflection model, crack path tortuosity by 206
Deformation
bands, butterflies as 115, 134
flow, medium-carbon steels 258
with fracture, iron alloy 458-459
kinematic analysis 166
Delaminations. See also *Splitting.*
causes for 105
defined 104
superalloys 393
wrought aluminum alloys 418
De La Pirotechnia (V. Biringuccio) 1
Delayed fracture, martensitic low-carbon steel .. 248
Dendrite arms, maraging steels 384
Departure side pinning, superalloys 393
Deposition, film, by shadowing 172
Depth-of-field
fractographic effects 78, 80-81, 87-88
optimum aperture for 80
SEM 166
Depth, three-dimensional measurement 207
Dessicator, for fracture preservation 73
Desulfurization, in low-carbon steel 247
Detail fracture(s)
high-carbon steels 288
as rail fracture mode 117, 136
rail head 289
Dezincification, definition and mechanics of ... 26
Differential interference contrast microscopy
for as-polished samples 95, 100
titanium alloys 445
Diffusional flow, effect of slip lines, iron 219
Diffusion bonding, metal-matrix composites ... 466
Dimple rupture(s). See also *Dimple(s); Dimple shape; Dimple size; Ductile fracture(s); Microvoids.*
as aging effect, iron alloy 458
alloy steels 291, 334

Dimple rupture(s) (continued)
  biaxial tension effects...................31, 39
  cast aluminum alloys......................413
  cemented carbides........................470
  corrosive environmental effects..........24-29
  defined..................................12-13
  ductile iron............................235-237
  environmental effects....................22-35
  exposure to low-melting metals..........29-30
  equiaxed, precipitation-hardening stainless
    steels..................................371
  hydrogen effects.........................22-24
  intergranular, in steel......................14
  iron-aluminum alloys.......................365
  iron-base alloys....................365, 459-460
  materials illustrated in....................217
  and microvoid coalescence................12-13
  as partially oriented surface..............203
  precipitation-hardening stainless steels..370-371
  shear band formation..................444-445
  stereo pair, titanium......................171
  and strain rate..........................31-33
  stress state effects.....................30-31
  surface, profile angular distribution......203
  surface contour cavities, iron.............223
  temperature effects......................33-35
  titanium alloys...............171, 443-445
  tool steels................................382
  transition to cleavage fracture..........33, 44
  true area value, steel.....................203
  wrought aluminum alloys..............417, 418
Dimple(s). See also *Dimple ruptures; Dimple shape; Dimple size*.
  alloy steels...... 292, 293, 303-304, 308, 319,
                                   324-325, 338
  austenitic stainless steels............353, 359
  cast aluminum alloys..................409-410
  as cavities, in iron.......................220
  clusters, wrought aluminum alloys..........423
  complex, precipitation-hardening stainless
    steels..................................372
  copper/copper alloys...........399-400, 402
  within dimples, titanium alloys............451
  as ductile fracture mechanism................4
  elongated...12-16, 173-174, 304, 338, 353, 442
  and fatigue striations.....................177
  fine, on void margins......................338
  flat shear.................................304
  formation in copper........................173
  in fractured steel, SEM fractograph........207
  and grain-boundary cavities, compared......20
  grain-boundary, in low-carbon steel........240
  height and width measurement...............207
  high-purity copper.................... 399-400
  with inclusions..................65, 67, 219
  intergranular facet transition to..........306
  intergranular fracture with, titanium alloys...441
  in irons...............................219, 220
  maraging steels........................385, 387
  martensitic stainless steels...............367
  by nondestructive profiling................199
  oval-shaped, formation......................13
  precipitation-hardening stainless steels..371-372
  pure titanium...............................16
  SEM for.....................................96
  on shear fractures..............12, 15-16
  superalloys................................392
  on tear fracture....... 12, 15-16, 387, 452
  tear dimples on............................452
  tension overload...........................385
  three-dimensional, quantitative
    measurement.............................208
  titanium/titanium alloys........16, 441-442, 451
  tool steels............................375, 379
  true mean area of..........................207
  true mean intercept length.................207
  two-dimensional, quantitative
    measurement.........................206-207
  wrought aluminum alloys.....417, 423-424, 427
Dimple shape
  determined by loading......................173
  effect of direction, principal stress.......30
  effect of stress state......................12

Dimple size
  acicular needles, effect of................328
  AISI/SAE alloy steels......................334
  and asymmetrical strain.................12, 16
  average, titanium alloys...................454
  defined....................................206
  magnification effect.......................425
  microvoid effects...........................12
  and particle size..........................101
  range, titanium alloys.....................449
  temperature effects.....................34, 46
  titanium alloys.......................449, 454
  wrought aluminum alloys....................424
Diode sputter coater, for SEM specimens....173
Direct carbon method, TEM replication........7
Direct carbon replicas, formation...........181
Directionality
  crack propagation, FSS and.................205
  in parametric relationships, partially-oriented
    surfaces................................201
Discoloration, cast aluminum alloys. See also
  Colors...................................408
Discontinuities
  AISI/SAE alloy steels......................331
  cold shuts...............................64-65
  as fracture initiation sites................64
  inclusions, defined.........................65
  laps.....................................64-65
  leading to fracture......................63-68
  porosity.................................65, 67
  seams....................................64-65
  segregation.................................67
  unfavorable grain flow...................67-68
Dislocation
  as fracture mechanism.......................12
  irradiated materials.......................365
  nucleation, crack tip.......................30
  pile-ups, microvoid coalescence at..........12
  substructure................................52
Display system, SEM illuminating/imaging....169
Divacancies, corrosion-generated..........42-43
Dot maps, from x-ray analysis...............168
Double-aging, effect on embrittlement....34, 47
Double-cup fractures, fcc metals............100
Dross inclusions
  composition................................422
  wrought aluminum alloys....................422
Dry air, for fracture preservation........73-74
Dry cutting, of specimens....................76
Ductile, defined. See also *Dimple rupture(s); Ductile fracture(s); Ductile rupture(s); Ductile tearing; Ductility*.
Ductile fracture(s). See also *Dimple rupture(s); Ductile rupture(s)*.
  AISI/SAE alloy steels......................298
  ASTM/ASME alloy steels.....................346
  dimples in.................................220
  of fibers, metal-matrix composites.........468
  interpretation of.........................96-98
  irons................................220, 223
  light fractographs......................94, 99
  by liquid lead embrittlement................38
  low-carbon steel bolts.....................248
  maraging steels............................385
  materials illustrated in...................217
  metal-matrix composites................467-468
  micromechanism...............................4
  microscopic features........................97
  polymer....................................479
  radial marks, SEM fractographs.............169
  superalloys................................389
  surface information, SEM imaging.....173-174
  tensile, appearance........................173
  titanium alloys............443, 450, 451, 455
  wrought aluminum alloys...............422, 425
Ductile iron. See also *Ductile iron, specific types*.
  brittle fracture........227, 231-232, 235-237
  commercial pearlitic, fatigue crack
    propagation.........................228-229
  ductile tearing........................230-236
  ductile-to-brittle transition...........231, 236
  ferritic, high load fatigue fracture surface....229

  fractographs..........................227-237
  fracture/failure causes illustrated........216
  fracture, slow monotonic loading...........230
  high-silicon ferritic......................229
  pearlitic and ferritic, fatigue fracture
    surfaces................................229
  spur gear, brittle cleavage................227
Ductile iron, specific types
  80-60-03 induction hardened, fatigue-crack
    origin..................................228
  80-60-03 induction hardened, fatigue-test
    fracture................................228
  ASTM A536 grade 100-70-03, brittle
    cleavage................................227
Ductile rupture(s). See also *Ductile fracture(s)*.
  shear, titanium alloys.....................447
  titanium alloys.......................443, 447
  wrought aluminum alloys...............425, 428
Ductile tearing
  ASTM/ASME alloy steels.....................345
  in ductile irons.......................230-236
Ductile-to-brittle transition, ductile
  irons.................................231, 236
Ductile-to-brittle transition temperature
  AISI/SAE alloy steels......................314
  defined.....................................34
  in ductile iron............................231
  effect of neutron irradiation..............127
  factors influencing....................105-106
Ductility
  aluminum-lithium alloys....................440
  ASTM/ASME alloy steels, phosphorus
    enhancement.............................349
  effect of temperature, titanium alloy...35, 48
  examining...............................72, 92
  graphite effect, in gray irons.............226
  increased, in ASTM/ASME alloy steels.......349
  in low-carbon steels.......................240
  phosphorus-enhanced........................349
Dwell time
  defined.....................................59
  effect on fatigue crack growth rate (*da/dN*)...60
  effect on mean stress.......................63
  effect on striation spacing.............60, 61

E

Effective stress intensity, and fatigue striation
  spacing...................................205
Elastic collisions, as electron signals.....168
Electrical conductivity, atomic oxygen effect in
  silver....................................481
Electroless nickel plating, for edge
  retention..............................95, 100
Electrolytes, for cleaning of ferrous fractures...75
Electron(s). See also *Secondary electrons*.
  backscattered..........................167-168
  beam, in SEM imaging..................167-168
  gun, in SEM imaging system.................167
  signals, SEM...............................168
  trajectories, tungsten and aluminum........167
Electron fractography
  defined......................................1
  history....................................4-8
Electronic materials
  aluminum transistor base lead, fatigue
    failure.................................483
  ball bond, defects.....................484-485
  brush/slip ring assembly, failure due to
    arcing..................................488
  fractographs...........................481-488
  fracture/failure causes illustrated........217
  integrated circuits, electrical discharge
    defect..................................481
  L-shaped flat pack leads, loading
    failure..............................481-482
  ohmic contact window, defects in vacuum-
    deposited aluminum..................486-487
  silver solar cell interconnect, effect of atomic
    oxygen environment......................481
Electronic pencil, for SEM
  micrographs...........................194, 207

Elevated temperature. See also *High temperature; Temperature.*
air, effect on overload fracture
 surfaces 35, 49-50
effect on dimple rupture 34-35
effect on environments 49
fracture surface, ASTM/ASME alloy steels .. 349
titanium alloys 442
Elevation, of nonplanar fracture surfaces 211
Elongated dimples, defined. See also
 Dimple(s). 173-174
Embrittlement. See also *Grain-boundary embrittlement; Hydrogen embrittlement; Stress-corrosion cracking.*
885 °F (475 °C) 136-137
AISI 4340 steels 214
aluminum alloy, by mercury 30, 38
causes 123
creep-rupture 123-124
effect of temperature 29
fiber, metal-matrix composites 466
graphitization 124
historical study 2
hydrogen 124-126
intergranular corrosion 126
from intergranular fractures 173
liquid-metal 29, 126-127
by low-melting metals 29
neutron irradiation 127
overheating 127-129
oxygen, in iron 222
quench aging 129-130
quench cracking 130-132
sigma-phase 132-133
solid-metal, defined 29-30
strain aging 129-130
stress-corrosion cracking 133-134
temper 134-135
tempered martensite 135-136
thermal 136
titanium alloys 448
Energy-dispersive x-ray spectroscopy 2, 168
Engineered materials
brittle fracture 109
effect of oxidation 35
fracture/failure causes illustrated 217
fracture modes 12-22
types illustrated 216
Environment
atomic oxygen, in low earth orbit 481
defined 22
effect on dimple rupture 22-35
effect on fatigue 35-63
effects of 22-63
elevated temperature effects 49
gaseous, effects on fatigue 36-41
inert, creep effects in 59
liquid, effects on fatigue 36, 41-46
loading, effect on fatigue 36, 53-54
for SCC of aluminum/aluminum alloys 28
for SCC of brass 28
for SCC of steels 27
for SCC of titanium alloys 28
temperature, effects on fatigue 36, 49-53
urban, tin-bronze preferential corrosion by .. 403
vacuum, effects on fatigue 36, 46-49
Equiaxed dimples. See also *Dimple(s).*
AISI/SAE alloy steels 302, 304, 319, 339
conical 14
in copper 173
defined 173
formation 12-14
and hemispheroidal dimples 173
in low-carbon iron 223
maraging steels 383, 387
martensitic stainless steels 367
precipitation-hardening stainless steels .. 370, 371
shape 12-14
with spheroidal particles 127
titanium alloys 444-445
tool steels 380
triaxial stress effect 31, 40
wrought aluminum alloys 428

Equicohesive temperature, defined 121
Errors
magnification, as distortion 196
perspective, as distortion 196
in roughness parameters 193
Etched sections, photolighting of 84, 87, 88
Etching
for etch pits 96, 101
extraction replicas 183
fractures 96
grain boundaries 96
over-, alligatoring as 351
for temper embrittlement 134
Evaporation, thermal 172-173
**Everhart-Thornley electron
 detector** 93-94, 168
Examination. See also *Visual examination.*
macroscopic 91-93
preliminary visual, fractures 72-73
visual 91-165
Excitation volume, effects in SEM imaging ... 167
Experimental techniques, quantitative
 fractography 194-199
Exposure(s)
guide for Polaroid photos 89
light meters and 85-86
for macro Luminar lenses 81
test 86-87
Extraction replicas
formation 182-183
of grain boundary particles, maraging steel .. 183
single-stage, procedure for 183

F

Face-centered-cubic metals
effect of low temperature on dimples in 33
effects, state of stress 31
embrittlement 123
hydrogen embrittlement 22
tensile fractures in 100
Facets
acicular needle effect on 328
of alloy steel fractures 308, 319
aluminum-copper alloy, SEM projection 195
areas measured 208
of Armco iron cleavage fracture 224
cleavage 224, 319, 328, 380, 397, 424
cleavage, and tongues, compared 424
dimpled, tool steels 379
fatigue fracture 14, 19
fracture appearance 128
grain, in oxygen-embrittled iron 222
grain-separated, high-purity copper 399-400
intergranular, AISI/SAE alloy steels 308
intergranular, transition to dimples 306
of nickel alloy fracture 397
overheating 146, 147
prototype, true profile length values 200
quasi-cleavage 319, 338, 339
terraced, titanium alloys 448
of tool steel fractures 379-380
total, surface area of 202
transgranular to intergranular, titanium
 alloys 441
Factory roof, AISI/SAE alloy steels 336
Failure analysis
historical study 2
macrofractography for 91
replication procedures for 94-95
types illustrated 217
Faraday cage, in SEM 168
Fast fractures
AISI/SAE alloy steels 295, 311, 327
high-carbon steels 280
magnesium alloy 456
mechanical damage in 72
medium-carbon steels 260-262, 264, 267
precipitation-hardening stainless steels 371
tool steels 376
unstable, as fatigue stage III 175
woody 281

Fatigue. See also *Fatigue crack growth; Fatigue crack growth rate; Fatigue crack propagation; Fatigue crack propagation rate; Fatigue fracture(s); Fatigue striations; Fatigue striation spacings; Fracture(s); Fracture surface(s).*
damage, and fatigue life prediction 207
defined 14-18, 111, 175
effect of frequency and wave form 58-63
environmental effects 35-63
as fracture mode 35-63
high-cycle, AISI/SAE alloy steels 296
historical studies 3
lamellar structure and 4
life prediction 207
materials illustrated in 217
mechanical properties affecting 49
patches, martensitic stainless steels 368-369
precracking 75, 236-237, 397, 479
quasi-brittle, polycarbonate sheet 479
as slip process 35
stages of 175
superalloys 389
as transgranular fracture, SEM defined .. 175-176
wrought aluminum alloys 418
zone, terminating edge 267
Fatigue crack growth. See also *Crack growth; Cracking; Fatigue.*
effect of vacuum 46
fracture topography, titanium alloys 441-443
from surface flaw 420
Fatigue crack growth rate (da/dN)
cryogenic temperature effects 52-53
cyclic loading effects 53-54, 62
dwell time effects 60
gases, effects on 40, 52
and fatigue striation spacing 41
fracture mode changes with 441
high, microvoid coalescence formation .. 119-120
measurement 120-121
second-phase particles/inclusions, effect on ... 16
stress intensity factor range, effect on 56-58
Fatigue crack initiation, SEM analysis 169
Fatigue crack propagation. See also *Fatigue.*
cast aluminum alloys 408
corrosion, P/M aluminum alloys 440
corrosion, wrought aluminum alloys 439
ductile iron 228
effect of inclusions, ASTM/ASME alloy
 steels 346-348
high-carbon steels 289, 290
hydrogen-assisted 302
in low-carbon steel, effect of calcium 247
mechanism of 15, 21
medium-carbon steels 275
radial, wrought aluminum alloys 415
rate, effect of embrittling or corrosive
 environments 35
transgranular, wrought aluminum alloys .. 438-439
Fatigue fracture(s). See also *Fatigue; Fracture surfaces; High-cycle fatigue fracture; Low-cycle fatigue fracture(s).*
bending-plus-torsional, medium-carbon
 steels 260
of bolt 112, 120
cast aluminum alloys 407-408
cobalt alloys 398
copper 401
in coupling pins 113, 120
defined 14-18
in drive shaft 112, 120
ductile iron 228-229
environments, types affecting 35-63
facetlike 120
frequency and wave form effects 58-63
furrow-type 44, 54
in gaseous environments 36-41
high-carbon steels 277, 279-280, 283, 288
high-cycle, martensitic stainless steels 367
hydrogen effect on surface appearance ... 37, 51
illumination techniques for 85
from improper heat treatment, AISI/SAE alloy
 steels 309

502 / Index

Fatigue fracture(s) (continued)
  interpretation of 111-121
  in liquid environments 36, 41-46
  loading effect on 36, 53-54
  low-carbon steel 243, 251
  low-cycle, high-carbon steels 283
  macroscopic characteristics 111-121
  markings, interpretation 112, 118, 119
  materials illustrated in 217
  mechanisms 4-5
  medium-carbon
    steels 258, 259, 260, 263, 269, 273
  and monotonic fracture surfaces, compared .. 229
  oxygen-free high-conductivity copper 401
  profile 15, 22
  SEM fractograph of aluminum alloy 175
  sequence to 111
  stages 14-21
  and stress intensity factor range 54-58
  with striations 16-18
  superalloys 394
  surface, high-carbon steels 279
  temperature, effect on 36, 49-53
  as transgranular, with slip-plane fracture 117
  in vacuum 36, 46-49
  wrought aluminum alloys 415-421
Fatigue fracture surface. See *Fatigue fracture(s); Fracture surface(s).*
Fatigue life
  effect of strain rate, stainless steels 59
  effect of wave form 62, 63
  improper heat treatment effects 309
  predicting, by creep 123
Fatigue precracking
  ductile irons 236, 237
  nickel alloys 397
  ultrasonic cleaning of 75
Fatigue striations. See also *Beach marks; Crack arrest; Crack arrest marks; Fatigue striation spacings; Quasi-striations.*
  AISI/SAE alloy steels 331
  aluminum alloy 19, 20, 176
  angle, as grain boundary locator 430
  ASTM/ASME alloy steels 346, 348
  austenitic stainless steels .. 39, 52, 352-353, 359
  beach marks as 175
  brittle 119, 430, 432, 456
  cast aluminum alloys 406
  copper alloys 403
  and dimples 177
  ductile 119, 247, 346, 431-432
  in ductile iron 229
  early Zapffe TEM 6
  electronic materials 482
  and fissures, compared 294
  formation, interinclusion 247
  formation, titanium alloys 442
  fracture characteristics with 16-18
  as fracture mechanism 4-5
  high-carbon steels 290
  on joining crack fronts 23
  and lamellar spacing, compared 290
  light fractographs 94, 96-97
  low-carbon iron 220, 222
  in low-carbon steel 177, 247
  maraging steels 385, 386
  martensitic stainless steels 367
  measurement 121
  in medium-density polyethylene 480
  as microscopic feature in
    fatigue 118-121, 137-138
  nickel alloys 205-206, 397
  in nickel, check on precision
    matching 205-206
  OFHC copper 401
  on plateaus, schematic 23
  roots, austenitic stainless steels 351
  shadowing technique for 172
  with slip traces 15
  from sulfur-containing atmospheres 41, 53
  superalloys 390, 391, 392
  tantalum alloys 464
  titanium alloys 441

  wrought aluminum alloys 417-418,
                                 426-429, 431-432, 439
Fatigue striation spacings. See also *Fatigue striations.*
  austenitic stainless steels 353
  brittle 35
  da/dN as 16
  defined 15
  determined, example case 205
  dwell time effect 60, 61
  and fatigue crack propagation rate 41
  as fatigue mechanism 4-5
  as a function of applied stress 120
  loading conditions, effect on 15, 22
  local variations 19
  martensitic stainless steels 367
  prediction of 48
  and projected images 205
  temperature, effect on 49-50
  variations, aluminum alloy 22
  wrought aluminum alloys 431
Feather markings
  on chromium steel 18
  on cleavage fracture surface 13
  defined 13
  titanium alloys 450
Ferritic ductile irons, fracture modes 228-237
Ferritic iron-aluminum alloys, brittleness 365
Ferritic rings, ductile iron 230
Ferrous alloys
  fatigue striations in 176
  fracture/failure causes illustrated 217
Fiber optic light sources
  bifurcated 79
  for lens flare and ghost images 84-87
  for visual examination 78
Fibers. See also *Fibrous fractures; Fibrous fracture surface; Tungsten fibers.*
  fracture, transverse 466
  graphite 465
  resin-matrix composites 474-478
  resin-matrix composites, fracture sequence ... 474
  splitting, as fracture mode 467
  tungsten 466
Fibrillation, in polymers 479, 480
Fibrous fractures 2, 313-314
Fibrous fracture surfaces. See also *Fracture surface(s).*
  AISI/SAE alloy steels 311, 313-316
  ductile, metal-matrix composites 466
  maraging steels 383
  precipitation-hardening stainless steels 370
  titanium alloys 451
  tool steels 376-377
  wrought aluminum alloys 425
Field-emission guns, as electron source 167
Field fractures, debris from 92
Film, fractographic 85, 169
Films
  antihalation 84
  deposition, by shadowing 172
  intergranular embrittlement by 110
  oxide, austenitic stainless steels 352-353
  passive rupture, in SCC and corrosion
    fatigue 42
Final breaking, of specimens. See also *Fast fractures.* 77
Fisheyes. See *Flakes.*
Fissures
  AISI/SAE alloy steels 294
  intergranular secondary, wrought aluminum
    alloys 418
  magnesium alloy 456
  OFHC copper 401
  titanium alloys 452, 453
Flakes. See also *Hydrogen flaking.*
  AISI/SAE alloy steels 310
  bright 415
  forging, as crack initiation site 342
  graphite, gray iron fracture at 225
  high-carbon steels 285
Flaking. See *Flakes; Hydrogen flaking.*
Flame cutting, of specimens 76

Flame hardening, medium-carbon steels .. 265, 266
Flash lines
  fatigue-crack origin at 332
  forging, medium-carbon steels 258
Flat cleavage fracture, irons 219
Flaws
  internal, fatigue fracture from 279
  surface, fatigue crack growth from 420
  surface, fatigue fracture from 279
Flexure, effects in resin-matrix
  composites 475-478
Flow
  grain. See *Grain flow.*
  plastic, tear ridges from 224
Fluorescent-light tubes, circular 83
Flutes
  defined, as fracture 20-21
  in titanium alloys 27, 28
  tool steels 375
Flux inclusion, overload fracture by 65, 67
Focusing
  camera 79-80
  in SEM illuminating/imaging system 167
Forging defect, wrought aluminum alloys 415
Forward sigmoidal curve, fractal analysis 213
Fractal analysis. See also *Fractal dimensions; Fractal plot.*
  of AISI 4340 steels 213-214
  experimental background 211-212
  experimental procedure 212-213
  fractal equation for irregular surfaces ... 213-215
  of fracture surfaces 211-215
  linearizaton of RSC fractal curves 213
  mathematical concept 211
  modified fractal dimensions 213
  profile parameters 212
  summary 214-215
  surface roughness parameters 212
Fractal dimensions
  applicability 211, 214-215
  Mandelbrot 213
  modified 213
  and roughness parameters, 4340 steel 213
Fractal plot
  with linearized RSCs 214
  RSC behavior 212-213
  theoretical linear, applicability 211
Fractographs
  bolt failure study by 248
  content and materials in 216
  dark-field illumination, light-field illumination,
    SEM, compared 92-100
  light, materials types 1, 216
  replica, compared 95, 100
  SEM, calculations of features in 206-207
  SEM, material types 216
  single SEM, in quantitative fractography 194
  TEM and SEM, compared 185-192
  TEM, material types 216
Fractography. See also *Photography.*
  advantages as SEM application 8
  defined 1
  depth-of-field effects 78, 80-81, 87-88
  electron 1, 4-8
  fracture origin location by 91
  history of 1-11
  lighting for 81-85
  macrofractography, defined 1
  microfractography, defined 1
  optical, defined 1
  purpose of 1
  quantitative 8, 193-210
  SEM 8, 173-176
  stereomicroscope for 87-88
  TEM applied, history 6
Fractology. See *Fractography.*
Fracture appearance. See also *Fracture(s).*
  facets 128
  frequency, effect on 58, 60
  hydrogen effects on 124
  and impact energy vs temperature 109
  surface, effect of heat treatment on hydrogen-
    embrittled aluminum alloy 33

Index / 503

surface, effect of microvoid nucleation ... 12
surface, titanium alloy, heat treatment, and microstructure effects ... 32
temperature effects on ... 49-53
Fractured parts. See also *Fracture(s); Part(s).*
  lighting of highly reflective ... 84, 88
  photographic setups for ... 78
  photography of ... 78-90
Fracture initiation ... 64, 103
Fracture interpretation. See also *Fracture(s)* ... 72
  brittle fracture ... 105-111
  ductile fracture ... 96-98
  fatigue fracture ... 111-121
  high-temperature fractures ... 121-123
  tensile-test fractures ... 98-105
Fracture lines. See *Radial marks.*
Fracture maps, types of ... 33, 44
Fracture modes. See also *Fracture(s); specific fracture types.* ... 12-22
  cleavage ... 13-14
  creep damage and ... 62
  decohesive rupture ... 18-20
  dimple rupture ... 12-13
  effect of fatigue crack growth rate ... 441
  effect of low temperatures ... 33
  effect of stress state ... 30-31
  fatigue ... 14-18
  flutes ... 20-21
  miscellaneous, materials illustrated in ... 217
  mixed, materials illustrated in ... 217
  quasi-cleavage ... 20
  slow, monotonic loading, ductile irons ... 231
  tearing topography surface ... 21-22
  transition ... 33, 44
  types illustrated ... 217
  unique ... 20-22
Fracture origin. See also *Crack origin; Fracture(s).* ... 91
  chevrons emanating from ... 170
  and fracture cause ... 72
  location, fractography for ... 91
  visual examination ... 72
Fracture path. See also *Crack path; Fracture(s).*
  effect of electrochemical potential ... 35
  fractal analysis ... 212
  interdendritic, low-carbon steels ... 249
  medium-carbon steels ... 263
  and microstructure, correlated ... 195-196
  preference index ... 201
  sinusoidal ... 109, 116
  tortuosity, determined ... 206
  types, engineering alloys ... 12
Fracture path preference index, linear ... 201
Fracture profiles. See also *Profiles.*
  examining ... 95, 100
  by light microscope ... 94, 99
  sections ... 95-96
Fracture(s). See also *Fast fractures; Fractographs; Fractography; Fracture appearance; Fractured parts; Fracture interpretation; Fracture modes; Fracture origin; Fracture path; Fracture profiles; Fracture specimens; Fracture studies; Fracture surfaces; Fracture toughness.*
  care and handling ... 72
  cleaning techniques ... 73-76
  columnar ... 2
  crystalline ... 2
  discontinuities leading to ... 63-68
  effects of environment ... 22-63
  etching ... 96
  fibrous ... 2
  granular ... 2
  initiation ... 64, 103
  markings, types ... 91
  materials illustrated in ... 217
  and microstructure, correlating ... 201
  mixed modes ... 176
  modes of ... 12-22, 176
  path preference ... 201
  patterns, historical study ... 2
  photography of ... 78-90
  profile sections ... 95-96

propagation ... 91
sectioning ... 76-77
sequence ... 91
silky ... 2
sources illustrated ... 217
studies, history of ... 1-8
test, as quality control application ... 140-143
tests, historical ... 3-4
texture, photographing ... 82-85
topography, titanium alloys ... 441-443
Type I through Type VII (historical) ... 1
types of causes illustrated ... 216
visual examination ... 91-93
vitreous ... 2
woody, historical ... 1-3
Fracture specimen(s). See also *Specimen(s); Test specimen(s).*
  care and handling ... 72
  fracture-cleaning techniques ... 73-76
  nondestructive inspection ... 77
  opening secondary cracks ... 77
  preliminary visual examination ... 72-73
  preparation ... 72-77
  preservation techniques ... 73
  sectioning ... 76-77
Fracture studies
  nineteenth century ... 2-3
  sixteenth to eighteenth centuries ... 1-2
  twentieth century ... 3-8
Fracture surface map. See also *Carpet plot.*
  titanium alloy, by stereophotogrammetry ... 198
Fracture surface(s). See also *Cleaning techniques; Cleavage; Fatigue fracture(s); Fibrous fracture surfaces; Nondestructive inspection; Quantitative fractography; Sectioning; Surface(s).*
  arbitrary test volume enclosing ... 199
  area, importance of ... 193
  cause of failure from ... 12
  chemical etching, and ultrasonic cleaning ... 75
  cleaning, for TEM ... 179-183
  cleavage ... 13, 17-18
  computer-simulated, true profile length ... 200
  effect of microvoid coalescence ... 12, 13
  effect of photo-illumination ... 82-89
  effect of section thickness ... 105
  elevated temperature, grain-boundary cavities ... 349
  equations, basic ... 194-196
  fatigue, ductile iron ... 228
  fatigue, interpreting markings ... 112, 118, 119
  fatigue, tire tracks ... 23
  flat and shiny, AISI/SAE alloy steels ... 304
  fractal analysis of ... 211-215
  high-carbon steels ... 279, 280, 288
  history ... 1-3
  mapping, photogrammetry for ... 197
  markings, types ... 91
  mating, effect of dimples ... 13
  measurements, aluminum alloy example ... 205-206
  mechanical damage to ... 92-93
  medium-carbon steels ... 260, 263, 273
  morphologies, ASTM/ASME alloy steels ... 345
  morphologies, ductile iron ... 235
  nondestructive inspection effects ... 77
  photography of ... 78-90
  plane of polish, projection plane and, correlated ... 196
  plane-strain ... 308
  P/M aluminum alloys ... 440
  preparation/preservation ... 72-77
  profile, AISI/SAE alloy steels ... 327
  profile, medium-carbon steels ... 255
  and projected images, parametric relationships ... 202
  prototype faceted, profile angular distribution ... 203
  roughness ... 199-205
  service fracture, AISI/SAE alloy steels ... 331
  Silcrome-1 martensitic stainless steel ... 369
  steel spring wire, screw marks ... 280
  tool steels ... 376

with triangular elements ... 202
true area, importance ... 211
Fracture toughness. See also *Toughness.*
  AISI/SAE alloy steels ... 340
  cemented carbides ... 470
  copper alloys ... 402
  effect of neutron irradiation ... 388
  specimen, ultrasonic cleaning of ... 74
Freon TF, as organic cleaning solvent ... 74
Frequency, effect on fatigue and wave form ... 58-63
Fretting
  medium-carbon steels ... 262
  wear, AISI/SAE alloy steels ... 308
FSS. See *Fatigue striation spacings.*
Furrow-type fatigue fracture, titanium alloy ... 44, 54
Fusion, incomplete, as discontinuity ... 65
Fusion zone
  softening, wrought aluminum alloys ... 422
  titanium alloys ... 443

G

Galling
  AISI/SAE alloy steels ... 295
  high-carbon steels ... 285
Gallium, as low melting embrittler ... 29
Gamma modulation, in SEM display systems ... 169
Gas cavities. See also *Cavities; Gases.*
  cast aluminum alloys ... 405-406, 412
Gaseous environments, effect on fatigue ... 36-41
Gases
  effect on fatigue ... 36-41
  helium, under pressure ... 414
  hydrogen, titanium embrittlement by ... 23, 32
  methane, along grain boundary ... 349
  reactive, as fatigue environment ... 35
Gas pockets, cast aluminum alloys ... 406
Gear teeth
  AISI/SAE alloy steels, fractured ... 298
  bending-fatigue fractures ... 329-330
  high-carbon steels, fractured ... 277
  subcase fatigue cracking ... 322
General corrosion
  defined ... 41
  surface, low-carbon steel ... 250
Geometry
  of image formation, SEM ... 196
  methods, in quantitative fractography ... 198
  of specimen/instrument, SEM effects ... 168
Ghost images, photographic ... 84
Gleeble test, low-carbon steels ... 240
Glide-plane decohesion, superalloys ... 391
Globular particles
  FeO, in iron ... 223
  oxide inclusions, in iron ... 220
Gold
  use in thermal evaporation ... 172
  wire, and vacuum-deposited aluminum, ball bond ... 484
Gouge marks, alloy steels ... 295
Grain. See also *Grain boundary; Grain-boundary embrittlement; Grain-boundary separation; Grain flow; Grain size.*
  dropping ... 126
  equiaxed, in SCC fracture ... 320
  facets, dimples on ... 399-400
  facets, oxygen-embrittled iron ... 222
  ferrite, cleavage facets in ... 319
  flow ... 84, 88, 434
  large columnar, superalloys ... 391
  surfaces, AISI/SAE alloy steels ... 340
Grain boundary. See also *Grain; Grain-boundary cavitation; Grain-boundary separation; Grain size.*
  alligatoring ... 351
  as anodic in SCC process ... 25-26
  attack, preferential, in ceramic ... 471
  black voids at ... 346
  cementite films, embrittling effect ... 123
  cleavage fracture, iron ... 224

Grain boundary (continued)
    copper film 157
    cracking, tool steels 382
    decohesion 18, 23
    delaminations, wrought aluminum alloys 418
    effect in hydrogen embrittlement 23
    effect of microvoid nucleation at 12
    effect on cleavage fracture formation 13, 17
    embrittlement, low-carbon steel............ 246
    etching 96
    facets, dimples in 240
    fatigue fracture, AISI/SAE alloy steels 296
    ferrite, high-carbon steels 278
    melting, AISI/SAE alloy steels 300
    melting point constituents 18
    methane gas along 349
    microvoid coalescence at 12
    migration, metal-matrix composites 467
    nitrides (or carbonitrides) in 282
    profile angular distributions for 202
    SCC fracture along 320
    secondary cracking 222, 340, 397
    sliding 19, 25, 121-123, 140-141, 364
    slip lines, irons 219
    strengthening, by embrittlement 111
    sulfide precipitation 349
    surface, iron impact fracture 222
    titanium alloys 443
Grain-boundary cavitation
    ASTM/ASME alloy steels 349
    effect on crack propagation rate 38
    intergranular creep rupture by 19, 26
    irons 219
    nucleation, growth, coalescence 349
Grain-boundary separation. See also *Decohesive rupture(s); Intergranular brittle fracture.*
    copper alloys 402
    final, along titanium alloys 441
    high-strength aluminum alloy 174
    phosphorus, embrittlement by 29
    phosphorus, tool steels 375
    wrought aluminum alloys 418, 423
Grain-corner cracks. See *Triple-point cracking; Wedge cracks.*
Grain dropping, defined 126
Grain flow
    change, wrought aluminum alloys.......... 434
    fatigue fracture through 67-68
    lighting of deeply etched specimens for ... 84, 88
    unfavorable, as discontinuity 67-68
Grain size
    corrosion-etched 364
    effect on brittle fracture 106-107
    effect on embrittlement 29
    fine, AISI/SAE alloy steels 305
    historical study 1, 3
    mean intercept length 207
    wrought aluminum alloys 437
Granular fractures. See also *Intergranular fracture(s); Transgranular fracture(s).*
Graphite
    effect in malleable iron 238
    morphology, effect in gray irons 226
    nodules, ductile irons 229-237
Graphite fibers. See also *Fibers.*
    metal-matrix composites, misaligned 465
Graphite flakes, gray iron fracture at 225
Graphitization
    as embrittlement, interpreting 124
    morphology, effect on gray iron fracture 226
    in tool steels, test disks 141, 162
Grating replicas, for SEM internal calibration 167
Gray irons
    brittleness of 123
    fractographs 225-226
    fracture/failure causes illustrated 216
    hypereutectic, graphite effects 226
    sand cast, fracture at flake 225
Gray levels, in SEM imaging display 169
Grinding
    abusive final, AISI/SAE alloy steel failure... 335
    damage, AISI/SAE alloy steels 321
    relief groove, superalloys................. 390

H

Halide-flux inclusion, overload fracture by .. 65, 67
Hall-Petch equation 106
Hammer-burst fracture, tool steels 378
Handling, of fractures 72
Hardening, secondary, austenitizing effects... 341
Hardness
    of adiabatic shear bands 32
    effect of lattice 32
    phase transformation 32-33
Hard vacuum, defined 46
Haynes 556, effect of temperature 34, 47
Heat-affected zone
    cold cracks in 137, 155-156
    cracking, medium-carbon steels............ 254
    decohesive rupture 18
    graphite formation in 124
    stress-relief cracking................... 139
    striations 21
Heat tinting, for replicas................. 181
Heat-treat cracks
    martensitic stainless steels 366
    secondary, as intergranular 366
    types 65
Heat treatment
    cracks, martensitic stainless steels 366
    effect, AISI/SAE alloy steels 298
    effect, hydrogen-embrittled aluminum alloy ... 33
    effect on fracture appearance, low-copper aluminum alloy 33
    effect on fracture appearance, titanium alloy .. 32
    fracture effect, medium-carbon steels 272
    history 1-3
    improper, AISI/SAE alloy steels 309
    nitrided/nitrogenized layer from 282
    after overheating 127
    postweld, microstructural effects........... 375
    rapid, effects in AISI/SAE alloy steels 328
    wrought aluminum alloys 417
Height
    of dimples, measurement in fractured steel .. 207
    of fracture surface roughness 200
    three-dimensional measurement............. 207
Helium gas, exploded under pressure 414
Hemispheroidal dimples, defined 173
Herringbone pattern. See *Chevron marks.*
Hexagonal-close-packed metals
    effect of low temperature on dimples in 33
    effects, state of stress 31
    hydrogen embrittlement 22
Hexamethylenetetramine, as chemical etchant 75-76
High-carbon steels. See also *High-carbon steels, specific types.*
    fractographs 277-290
    fracture/failure causes illustrated 216
    hydrogen flaking........................ 125
    shell crack and detail fracture 288
High-carbon steels, specific types
    AISI 10 B62, oxidation effects 280
    AISI 10 B62, steel wire, fatigue fracture surface.............................. 280
    AISI 1053, fatigue fracture surface, fracture origin............................... 277
    AISI 1055, fatigue fracture surface 279
    AISI 1060, fatigue fracture from seam 281
    AISI 1060, fatigue fracture from surface flaw 279, 281
    AISI 1060, torsional overload fracture 278
    AISI 1070, complex fatigue fracture 281
    AISI 1070, heat-treating failure 282
    AISI 1070, inadequate removal of blanking fracture 285
    AISI 1070, torsional fatigue fracture 282
    AISI 1074, galling failure 285
    AISI 1074, hydrogen embrittlement 284
    AISI 1074, low-cycle fatigue fracture 283
    AISI 1074, mechanical or lubrication failure 283
    AISI 1074, scab failure 284
    AISI 1075, embrittlement failure by hydrogen-assisted flakes....................... 285
    AISI 1095, fatigue fracture surface 288
    ASTM A228, hydrogen embrittlement failures........................... 286
    ASTM A230, atypical fracture surface 287
    ASTM A230, fatigue failure from seam 287
High-cycle fatigue fracture(s). See also *Fatigue; Fracture(s).*
    AISI/SAE alloy steels.................... 337
    martensitic stainless steels 367
    precipitation-hardening stainless steels ..371, 373
    titanium alloys 452
    tool steels 381
    wrought aluminum alloys................. 426
High-purity copper
    effects of stress and temperature 399
    fracture modes 399-400
    fracture surfaces and mechanisms 399-400
    intergranular separation 399
High-purity deaerated water, wrought aluminum alloys............................... 438
High-purity iron, flat cleavage fracture....... 219
High-temperature fractures
    effect of temperature 52
    interpretation of 121-123
High-temperature materials, specific types
    Incoloy 800, effect of stress intensity factor range on fatigue crack growth rate 58
    Incoloy 800, effects of temperature........ 52
    Incoloy 800, ridges and striations in sulfidizing atmosphere 41, 53
    Incoloy 800, wedge cracking 26
    Inconel 600, effect of frequency and wave form on fatigue properties................... 60
    Inconel 625, wedge cracking 26
    Inconel 718, effects of stress intensity factor range on fatigue crack growth rate 58
    Inconel X-750, effect of frequency and wave form on fatigue properties 59-60
    Inconel X-750, effect of stress intensity factor range on fatigue crack growth rate 57
    Inconel X-750, effect of temperature 52
    Inconel X-750, effect of temperature on double-aged............................... 47
    Inconel X-750, effect of vacuum 48
High-temperature oxidation, surface effects 35, 72
High-vacuum diffusion pump, SEM......... 171
Hilger-Watts stereoscope 171
Hip screws, austenitic stainless steels 359-364
Histograms
    of facet areas......................... 208
    of profile angular distributions 203
History, of fractography 1-11
    electron fractography 4-8
    microfractography, development of 3-4
    quantitative fractography 8
    before twentieth century 1-3
Hold times. See *Dwell time.*
Homogenization, inclusion formation 347
Hot corrosion, superalloys 391
Hot cracking
    defined 18
    electron beam weld..................... 157
    examination/interpretation 138
    in stainless steels 123
Hot shortness 2, 126-127
Humid air, fracture effects.................. 72
Hydride formation, embrittlement by 124
Hydrochloric acid, in chemical etching cleaning 75-76
Hydrogen. See also *Embrittlement; Hydrogen damage; Hydrogen embrittlement.*
    atmosphere, AISI/SAE alloy steels in 292
    attack, in ASTM/ASME alloy steels.... 349-350
    austenitic stainless steels 39, 356
    corrosion, bolts....................... 248
    effect on antimony-doped Ni-Cr alloy steels 350
    effect on dimple rupture 22-24
    effect on fatigue fracture appearance.. 30, 37, 51
    flaking............................... 125

gaseous, titanium embrittlement by......23, 32
nickel alloy cracking in 396
Hydrogen damage
  AISI/SAE alloy steels....................302
  bright flakes in 415
Hydrogen embrittlement
  AISI/SAE alloy steels........293, 301, 305-307
  of aluminum 23-24
  austenitic stainless steels 355
  causes 22-23
  in decohesive rupture 18, 24
  effects on dimple rupture 22-24
  examination and interpretation 124-126
  high-carbon steels...................284, 285
  low-carbon steels 248
  maraging steels 387
  precipitation-hardening stainless steels 372
  premature spring failures from...........286
  as SCC mechanism, steels...............23-25
  of stainless steels 30, 355, 372
  superalloys 389
  of titanium 23, 448
  tool steels 381-382
Hydrogen flaking, defined 125
Hydrogen sulfides, effect on alloy steels......299

I

Illumination
  effect in fracture surface, medium-carbon
    steels 257
  phase-contrast 438
  photographic, effects on fracture surfaces.. 82-89
  in SEM imaging 167-168
  shift, photo effects 86
  ultraviolet 84-85
Illuminators
  Nicholas 81
  stereomicroscope........................78
Image(s)
  formation, geometry of 196
  projected, quantitative fractography 194-196
  stereo, fractographic 87-88
  width, photomacrographic 80
Imaging. See also *Stereo imaging.*
  SEM, clarity 168
  stereo, SEM display systems 171
  stereoscopic, in quantitative
    fractography 196-197
  system, SEM........................167-168
  thermal-wave 169
Impact fractures
  AISI/SAE alloy steels....... 314, 319, 324, 328
  iron 222
  tool steels 377, 378
Implants, orthopedic, austenitic stainless
  steels 359-364
IN-738. See *Nickel-base superalloys, specific
  types.*
Incident electrons, Monte Carlo projections ...167
Incident light
  for discontinuities.......................63
  meter, effects 85
Incipient melting. See *Burning.*
Inclusions
  AISI/SAE alloy steels....................333
  in aluminum alloy 19
  copper, effect in low-carbon steel 249
  in dimples 65, 67, 174
  as discontinuities, defined 65
  dross 422
  effect in SCC 26
  effect on fatigue crack propagation 346, 347
  effect on striation 16
  as fracture origin, AISI/SAE alloy steels 303
  globular oxide, in iron 220
  high-carbon steels......................277
  illuminated and photographed 86
  intermetallic, wrought aluminum alloys....423
  in iron, dimples from 219
  in low-carbon steels, effect of calcium 247
  manganese oxide/sulfide, in iron 221
  microvoid coalescence at 12
  from pipe.............................427
  ribbonlike, AISI/SAE alloy steels 326
  second-phase, wrought aluminum alloys....423
  service fracture, steel forging 65, 66
  size and distribution 333
  sorting by dot maps 168
  spheroidal oxide, in iron 220
  stringers 140-141, 161
  sulfide, in white iron 239
  titanium, in maraging steels 383
  ultrasonic cleaning effects 75
  wrought aluminum alloys................418
Incoloy alloys. See *High-temperature materials,
  specific types; Nickel-base superalloys,
  specific types.*
Incomplete fusion, as discontinuity 65
Inconel alloys. See *High-temperature materials,
  specific types; Nickel-base superalloys,
  specific types.*
Index of surface roughness, defined.........201
Indium, as low-melting embrittler 29
Induction hardening, improper, effect in alloy
  steel 300
Induction melting, vacuum, in irons 219
Inelastic collisions, as electron signals 168
Information system (SEM imaging)
  electron signals 168
  in situ studies 169
  specimen/instrument geometry effects.....168
  thermal-wave imaging 169
  x-ray signals 168
Ingot steels, early fractographs...............5
Inhibited sulfuric acid, in cathodic cleaning ...75
Initiation. See also *Crack initiation; Fatigue
  fracture initiation.*
  fatigue cracks 112
  of fracture 103
In situ **studies,** SEM imaging 169
Instrumentation
  geometry effects, in SEM 168
  SEM 166-171
Integrated circuit defects, electronic materials..481
Interface
  melting, electronic materials 488
  twin-matrix, iron.......................224
Intergranular cavities, iron 219
Intergranular corrosion
  austenitic stainless steels 364
  as embrittlement, interpretation and
    examination 126
Intergranular creep fractures
  austenitic stainless steels 364
  by grain-boundary cavitation 19, 26
  by triple-point cracking 19, 25
  by wedge cracking 26
Intergranular decohesion fractures 24, 31
  aluminum alloys 35
  austenitic stainless steels 27, 34
  bands, fatigue fracture in hydrogen......37, 51
  of brass 28, 36
  effect of corrosive or embrittling environment 35
  embrittling effect, low-melting metals 29
  and hydrogen embrittlement, steel 31
  by SCC 24, 27-28, 36
  of steels 27
Intergranular dimple rupture
  effect of microvoid nucleation 12
  in steel 14
Intergranular embrittlement, by films or
  segregation 110
Intergranular fracture(s). See also *Cleavage
  fracture(s); Intergranular corrosion; Creep
  fractures; Intergranular decohesion fractures;
  Intergranular dimple rupture.*
  AISI/SAE alloy steels....... 293, 299-300, 305,
                                        307-308, 335
  ASTM/ASME alloy steels 349-350
  austenitic stainless steels 352-353, 355
  austenitizing effect 339
  in bcc metals 123
  of bolts 299
  brittle..................30, 38, 109, 174-175
  cleavage, high-carbon steels 290
  copper alloys..........................402
  from creep rupture 18-19, 25
  decohesive............................24
  in engineering alloys 12
  by grain-boundary cavitation 349
  by grain-boundary sliding 121-123, 140-141
  high-carbon steels......................284
  high-purity copper 400
  hydrogen-assisted, low-carbon steel 248
  iridium and iridium alloy 462-463
  irons..................................219
  light fractographs 94, 98
  by liquid cadmium embrittlement 30, 39
  low-carbon steels 240
  medium-carbon steels...................271
  nickel alloys 397
  oxygen-embrittled Armco iron...........222
  from quench cracking 131, 148-149
  rock-candy appearance 110-111
  salt corrosion assisted, superalloys 389
  SCC, in precipitation-hardening stainless
    steels 373
  separation, iron alloy 459
  sintered tungsten 462
  superalloys 391, 393
  titanium alloys 441
  tool steels 375
  and transcrystalline cleavage, iron........222
  types 110, 121-123
  wrought aluminum alloys........ 431, 434, 439
Intergranular microvoid coalescence,
  defined 128
Intergranular secondary cracking, OFHC
  copper 401
Intergranular separation
  alloy steels 341
  high-purity copper 399
  iron alloy 459
Interlamella spacing, high-carbon steels......290
Interlaminar failure, in composites......466, 478
Intermetallic inclusions, wrought aluminum
  alloys................................423
Internal defects, resin-matrix composite failure
  from 474
Interpretation, of fractures..............96-123
Inverse chill, ductile iron fracture from 227
Iridium
  fractured sheet, secondary cracks 462
  intergranular fracture 463
  polycrystalline, brittle intergranular fracture..462
Iron-aluminum alloys, experimental 365
Iron-base superalloys, fracture/failure causes
  illustrated.............................217
Iron-chromium alloys, sigma-phase
  embrittlement 132
Iron-chromium-aluminum alloy, solidification
  in 140, 160
Irons
  bending fracture, oxide inclusions 220
  blowholes 221
  bright-field, dark-field, and SEM images,
    compared 92
  cast, granular brittle fracture 103
  cleavage fractures, twist boundary 17
  ductile, figure numbers for 216
  embrittlement sources..................123
  fractographs 219-224
  fracture/failure causes illustrated 216
  fracture types, historical....................1
  grain-boundary cavitation 219
  gray, figure numbers for 216
  high-purity, flat cleavage fracture 219
  intergranular fracture and transcrystalline
    cleavage............................222
  light fractography images, compared ...93-94
  low-carbon, dimpled ductile rupture 220
  low-carbon, high oxygen, tensile-test
    fracture 223
  low-carbon, oxide inclusion 222
  malleable, figure numbers for 216
  slip lines 219
  transgranular cleavage fracture 460

Irons (continued)
  white, figure numbers for 216
  woody fractures 1-3
  wrought, impact fracture 224
Irons, specific types
  Armco, cleavage fracture 224
  Armco, oxygen-embrittled 222
  Armco, shear step 224
  Armco, slip-band cracks 224
  Armco, slip steps 224
  Armco, tear ridges 224
  Armco, tilt boundary, cleavage steps, river
    patterns 18
  Fe-0.3C-0.6Mn-5.0Mo, quasi-cleavage
    fracture 26
  Fe-0.3Ni, fracture surface 457
  Fe-0.6Mn-5.0Mo, quasi-cleavage fracture .. 26
  Fe-3.9Ni, cleavage fracture 457
  Fe-4Al, fracture mode transition 462
  Fe-8Ni-2Mn-0.1Ti, facet areas measured .. 208
  Fe-Cr-Ta, cleavage and quasi-cleavage ... 460
  Fe-Cr-Ta, tension overload fracture 460
Iron sulfide, particles 219
Irradiated channel fracture, austenitic stainless
  steels 365

J

Jamming, gear tooth fracture by 277
Jernkontoret fracture tests (Arpi) 3
Jewett nail
  austenitic stainless steels 361, 363-364
  cobalt alloy 398
J-integral, defined 15

K

Ketones, as organic cleaning solvent 74
Keyway
  fatigue fracture origin at 260
  fretting wear, AISI/SAE alloy steels 308
  in tapered shaft, peeling fracture at ... 253
Kinematic SEM analysis, of deformation 166
Kinetics
  corrosion 41
  repassivation 42
Knobbly structure
  in alloy steel fracture surface 33, 43
  aluminum alloy fracture surface 33
  cast aluminum alloys 409
  formation 33
K shell, defined 168

L

Lacquers, clear acrylic, as fracture
  preservatives 73
Lamellar structure
  ductile iron 229
  as fatigue mechanism 4
  fine, irons 221
  fractured pearlite, and striations, compared .. 119
  pearlite, high-carbon steels 290
Lamellar tearing
  examination/interpretation 138-139
  formation 158
Laminates, resin-matrix composites 478
Lanthanum hexaboride, as electron source ... 167
Laps
  as crack origin, steel 64, 65
  as discontinuities, defined 64-65
  low-carbon steel fracture from 252
  rolling, fracture from 64
  tool steels 378
Laser beam weld
  solidification cracking 345
  wrought aluminum alloys 422
Lattice hardening 32
Lead
  effect on fracture morphology, steel ... 30, 38
  fractured ingots, history 1
  as low-melting embrittler 29
Leaves, aluminum alloys 433
Length. See also *Crack length.*
  intercept 194
  of linear feature 195
  mean intercept 195
  mean, of discrete linear features 195
  mean perimeter, of closed figures 195
  perimeter 194, 195
  projected, fractal analysis 212
  ratios, for partially oriented surfaces . 201
  stereological relationships 196
  true, and true area, parametric relationships .. 204
  true, defined 199-200
  true, fractal analysis 212
  true, values for dimpled and prototyped faceted
    4340 steel 203
  true, values for various materials 200
  true profile, defined 199-200
Lens aperture, selection 80-81
Lenses
  coated 84
  convergent magnetic, in SEM imaging 167
  flare problem 84-87
  macro luminar 81
  Macro-Nikkor 81
  photographic 79
  selection of apertures 80-81
  stigmators 167
Light. See also *Illumination; Lighting.*
  fractography 93-96
  incident, for discontinuities 63
  meters, fractographic 85-86
  sources, fractographic 81-82
Light-field illumination fractographs
  dark-field and SEM, compared 92-100
  material types in 216
  texture in 83
Light fractography
  deep-field microscopy 96
  defined 93
  etching fractures 96
  fracture profile sections 95-96
  replicas for light microscopy 94-95
  taper sections 96
Lighting, photographic
  basic, illustrated 82
  direct and oblique 78
  for etched sections 84
  for highly reflective parts 84
  parallel 83
  ring 83
  techniques 82-85
  tent 88
  with ultraviolet illumination 84-85
Light microscopy
  dark-field fractograph, and SEM image,
    compared 92
  embrittlement phenomena 123-137
  historical study 4
  interpretation of fractures 96-123
  light fractography 93-96
  quality control applications 140-143
  replicas for 94-95
  and visual examination 91-165
  weld cracking 137-140
Light photomacrography, scanning 81
Lineal roughness parameter, R_L 199-200
Linearization, of reversed sigmoidal
  curves 212-215
Linear polyethylene, crack-growth
  mechanisms 479
Line spall, in steels 113-115, 129-134
Liquid cadmium, low-alloy steel embrittlement
  by 30, 39
Liquid lead, embrittling effect, alloy steel .. 30, 38
Liquid mercury, decohesive fracture from .. 18, 25
Liquid-metal embrittlement
  defined 29-30
  examination and interpretation 126-127
  four forms 126, 143
Liquid nitrogen. See *Nitrogen.*
Liquid(s)
  corrosive 35, 77
  environments, effect on fatigue 36, 41-46
  as penetrants, effects 77
Lithium, as low-melting embrittler 29
Loading
  conditions 4-5, 12-15, 72
  effect on crack propagation and striation .. 15
  effect on fatigue 15, 36, 53-54
  end 332
  examining 72, 92
  failure, electronic materials 481-482
  as fatigue mechanism 4-5
  fatigue parameters based on 54
  fracture, modes 14
  imbalance, effect in medium-carbon steels .. 273
  mean stress, effect on fatigue cracking .. 15
  slow, in four-point bending 350
  stress intensity factor range, effect on .. 54
  torsional, AISI/SAE alloy steels 330
  uniform and nonuniform, crack
    effects 175-176
  visual examination, effect on 72
Loading conditions
  as fatigue mechanism 4-5
  mechanical damage from 72
  types of 12-15
Low-carbon iron
  ductile rupture 220
  oxide inclusion 222
Low-carbon steels. See also *Low-carbon steels,
  specific types; Steel(s).*
  brittle fracture 249
  cleaning 76
  cleavage, fractograph 174
  cleavage crack path 117
  copper contamination 249
  corrosion fatigue fracture surface 250
  fractographs 240-252
  fracture/failure causes illustrated ... 216
  intergranular fracture 240
  shear deformation and shear lips 244
  temperature effect on fracture modes .. 33, 45
  tension fractures 240
  ultra-, effect of temperature on fracture
    mode 34, 46
Low-carbon steels, specific types
  AISI 15 B22, delayed fracture 248
  AISI 1019 shaft, fatigue fracture 243
  AISI 1020, centerline cracks 244
  AISI 1020 shaft, brittle fracture 243
  AISI 1025, brittle intergranular fracture .. 245
  AISI C-1080, stress-corrosion fracture .. 27, 35
  ASME SA178, internal corrosion fatigue
    cracking 245
  ASTM A178, grain-boundary embrittlement
    failure 246
  ASTM A516-70, calcium effects on inclusions
    and fatigue crack propagation 247
  ASTM A517-70, fatigue crack propagation .. 247
  SAE 1010 tie rod, bending impact fracture .. 242
  SAE 1010 tie rod, in-service fatigue
    fracture 241
Low-cycle fatigue fracture(s). See also *Fatigue
  fracture(s).*
  AISI/SAE alloy steels 308
  bending, austenitic stainless steels .. 362
  high-carbon steels 283
  maraging steels 386
  metal-matrix composites 469
  precipitation-hardening stainless steels .. 371
  titanium alloys 452
  tool steels 377, 380
  wrought aluminum alloys 426
Low earth orbit, atomic oxygen environment
  of 481
Low-melting metals, effect on dimple
  rupture 29-30
L shell, defined 168
Lubrication breakdown, high-carbon steels .. 283
Lüders bands, from strain aging 129, 148
Luminar lenses, and optimum aperture
  concept 81

M

Machinability, of copper ... 401
Machining marks
  AISI/SAE alloy steels ... 296
  coarse, low-carbon steel ... 251
  leading to fatigue fracture ... 251
Macrocrack propagation, in fatigue ... 117-118
Macroexamination, visual ... 72-73
Macrofractography ... 1, 3, 91
Macro-Nikkor lenses, aperture selection ... 80-81
Macroscopes
  illustrated ... 79
  mounted ... 79
  techniques ... 91-93
Macroshear stress, in medium-carbon steels ... 253
Magnesium alloys, corrosion fatigue fracture ... 456
Magnetic electron microscopes ... 6
Magnetic-particle inspection, surface effects ... 77
Magnification
  camera, determining ... 80
  defined, stereoscopic methods ... 196
  error ... 196
  and focusing, photomacrographic ... 79-80
  in SEM illuminating/imaging system ... 167
Malleable irons
  fractographs ... 238-239
  fracture/failure causes illustrated ... 216
Malleable irons, specific types
  ASTM A47 grade 32510, fracture sequence ... 238
  ASTM A220 grade 50005, microcracking ... 239
Mandelbrot fractal dimension ... 213
Manganese oxide, inclusions in iron ... 221
Manganese oxysulfide, effect in low-carbon steel ... 249
Manganese sulfide, inclusions ... 221, 263
Maps
  phosphorus ... 349
  photogrammetry for ... 197
  sulfur ... 349
  x-ray ... 167, 473
Maraging steels. See also *Maraging steels, specific types.*
  cobalt-free, grain-boundary precipitates ... 183
  cobalt-free high-titanium, thermal embrittled ... 136, 154
  fractographs ... 383-387
  fracture/failure causes illustrated ... 217
  shallow dimples ... 14
  thermal embrittlement ... 136, 154
Maraging steels, specific types
  18% Ni grade 300, fibrous fracture ... 383
  18% Ni grade 300, fracture toughness ... 385
  18% Ni grade 300, low-cycle fatigue fracture ... 386
  18% Ni grade 300, slow-bend fracture ... 387
  18% Ni grade 300, tensile-test fracture ... 384
Martensite
  acicular needles of ... 328
  tempered, in ductile iron, crack growth ... 228
  transformed ... 26, 32, 42
Martensitic stainless steels
  fractographs ... 366-369
  fracture/failure causes illustrated ... 217
Martensitic stainless steels, specific types
  AISI 410, fracture surfaces ... 366
  AISI 431, high-cycle fatigue fracture ... 367
  AISI 501, mating segments, fatigue fracture ... 368-369
  AISI 4340, light fractographs ... 83
  Silcrome-1, fatigue fracture surfaces ... 369
Martensitic transformation ... 26, 32, 42
Mating fracture surface(s)
  AISI/SAE alloy steels ... 294, 318, 323
  automotive bolt ... 274
  cast aluminum alloys ... 408, 411-412
  crack origins ... 13, 323
  drive shaft ... 258
  effect of dimples ... 13
  elongated manganese sulfide inclusions ... 26
  fatigue, low-carbon steel ... 251
  fracture origin ... 13, 323
  martensitic stainless steels ... 368
  matching dimples on ... 12
  medium-carbon steels ... 255, 258, 268, 274
  overload ... 411-412
  sectioned ... 268
  sudden overload failure ... 294
  torsional overload fracture, high-carbon steels ... 278
  wrought aluminum alloys ... 416
Matrix
  creep, alloy steels ... 349
  separation, effect of crack tip plastic zone ... 16
Maximum stress, direction effect on dimple shape ... 13
Mean curvature, convex figures ... 195
Mean free distance, stereological relationships ... 195
Mean stress
  effect of dwell time ... 63
  as loading condition, effect on fatigue cracking ... 15
  and ramp rates ... 63
Mean tangent diameter, convex figures ... 195
Measurement
  quantitative, of dimples ... 206-207
  statistical vs individual ... 207-208
  unit, in fractal analysis ... 211-215
Mechanical damage
  AISI/SAE alloy steels ... 305, 337
  cast aluminum alloys ... 406, 407, 413
  common types ... 72
  defined ... 72
  high-carbon steels ... 283
  scab, high-carbon steels ... 284
  during shear deformation, tool steels ... 380
  wrought aluminum alloys ... 426
Mechanical twinning. See *Microtwins; Twinning.*
Medium-carbon steels. See also *Medium-carbon steels, specific types.*
  automotive bolt, fatigue failure ... 274
  fatigue fracture ... 258
  fractographs ... 253-276
  fracture/failure causes illustrated ... 216
  I-beam, fatigue fracture surface ... 263-264
  plate, shear bands ... 42
  single-overload torsional fracture ... 275
  steel axle housing, fatigue fracture ... 276
  weld, HAZ cracking ... 254
Medium-carbon steels, specific types
  AISI 1030 tapered shaft, torsional fatigue or "peeling" fracture ... 253
  AISI 1033, effect of temperature on fracture ... 254
  AISI 1035, brittle fracture ... 258
  AISI 1035, cup-and-cone tensile fracture ... 253
  AISI 1038 modified, fatigue fracture ... 259
  AISI 1039 shaft, fatigue fracture surface ... 262
  AISI 1040, fatigue fracture surface ... 260
  AISI 1041, fatigue fracture surface ... 260, 261
  AISI 1041, fracture by reverse stressing ... 262
  AISI 1041, fretting in keyed spindle ... 262
  AISI 1041, torsional fatigue fracture surface ... 261
  AISI 1045 crane gear, effects of flame hardening ... 265
  AISI 1045, fracture surfaces ... 266-268
  AISI 1046, fatigue fracture surface ... 274
  AISI 1046, reversed bending fatigue ... 269
  AISI 1050, bending overload fracture ... 272
  AISI 1050, fatigue fractures ... 269, 273
  AISI 1050, fatigue fracture surface ... 273
  AISI 1050, rotating bending failure ... 273
  AISI 1144, fatigue fracture surface ... 263
  ASTM A515 grade 70, crack mating surfaces ... 255
  ASTM A515 grade 70, fractured shell ... 256-257
  SAE 1050 modified, brittle fracture ... 270-271
Mercury
  liquid, decohesive fracture from ... 18, 25
  as low-melting embrittler ... 29
  vapor, as embrittler ... 30, 38, 424
Metallographic sectioning methods, for profiles ... 198-199
Metal-matrix composites
  fractographs ... 465-469
  fracture/failure causes illustrated ... 217
  stress rupture ... 468
Metal-matrix composites, specific types
  carbon (graphite)-magnesium, tensile fracture ... 465
  Fe-24Cr-4Al-1Y with W-1ThO fibers, low-cycle fatigue fracture ... 469
  Ni-15Cr-25W-2Al-2Ti with tungsten fibers, ductile fracture ... 468
  NS-55 tungsten-Al 6061 matrix, tensile fracture ... 466
  NS-55 tungsten fibers-AISI 1010 carbon steel, tensile fracture ... 466
  tungsten fibers with silver matrix, ductile and transverse cleavage fractures ... 469
Metals
  low-melting, embrittling by ... 29-30
  partly converted, fractures ... 1-3
  types illustrated ... 216
Metal shadowing. See also *Shadowing.* ... 7
Meters, light, for photography ... 85-86
Methane gas bubbles
  cavitation ... 37, 51
  rapid coalescence of ... 349
Metric conversions, guide and references ... 489-491
Microcracking
  austenitic stainless steels ... 356
  brittle, polycarbonate sheet ... 479
  ductile irons ... 235
  inititation, ductile iron ... 233, 234
  malleable iron ... 239
  at slip bands ... 360
  through sulfide inclusions, malleable iron ... 239
Microcrazing, in linear polyethylene ... 479
Microelectronic devices, thermal-wave imaging of ... 169
Microfractography ... 1, 3-4
Microinclusions, deep shell crack from ... 288
Micromechanisms, of fracture ... 179
Micron bar, on micrographs ... 167
Microplastic deformation, ductile irons ... 236
Microporosity
  in brittle cleavage fracture, ductile iron ... 227
  cemented carbides ... 470
Microscopes. See also *Macroscopes; Scanning electron microscope(s); Scanning electron microscope(s); Stereomicroscopes; Transmission electron microscopes.*
Microscopy
  deep-field ... 96
  light ... 91-165
Microstructure
  AISI/SAE alloy steels, austenitization effects ... 292
  Alnico alloy ... 461
  austempered ductile iron, strength ... 232
  brittle cleavage fracture, ductile iron ... 227
  bronze shell ... 403
  effect in fatigue ... 14
  effect of high temperatures ... 121
  effect on tensile ductility ... 101
  fractal analysis ... 212-215
  fracture appearance, hydrogen-embrittled titanium alloy ... 32
  and fracture path, correlated ... 195-196, 201
  high-carbon steel, torsional overload fracture ... 278
  high-purity copper ... 399-400
  hypereutectic gray iron ... 226
  inadequate heat treatment effects, cast aluminum alloys ... 410
  intergranular, AISI/SAE alloy steels ... 300
  medium-carbon steels, shell fracture ... 256-257
  in planar sections ... 198
  spinoidal, coarsening in copper alloys ... 402
  and surface topography ... 393
  titanium alloy base plate ... 443
  transverse, high-carbon steels ... 278

Microstructure (continued)
  upper-bainite, stereo-pair photographs 89
  weld-metal, from inadequate pre- and postweld heat treatment 375
  wrought aluminum alloys, fusion zone softening and pores 422
Microtearing. See also *Tearing.*
  in ferritic ductile iron 234
Microtongues. See also *Tongues.*
  in ductile irons 232
Microtwins, iron alloy 457
Microvoid coalescence. See also *Dimple(s); Dimple rupture(s).*
  AISI/SAE alloy steels 293
  ductile iron 230-231
  as ductile mechanism 96
  by ductile rupture, titanium alloys 443
  effect on dimple size 12
  as failure mechanism 12
  in fatigue fracture 119-120
  growth, effect on fracture surface 13
  high-purity copper 399-400
  intergranular 14, 128
  intergranular dimple rupture, steel 14
  iron-base alloy fracture by 220, 460
  linking by creep, austenitic stainless steels ... 364
  in low-carbon steel bolts 248
  maraging steels 383
  martensitic stainless steels 369
  as micromechanism of ductile fracture 4
  schematic under Modes I, II 16
  strain-controlled, ASTM/ASME alloy steels 350
  superalloys 388, 393
  titanium alloys 441, 445
  tool steel fracture by 381
Microvoid nucleation, effect on fracture surface appearance 12
Microvoids. See *Microvoid coalescence.*
Mirrors, photographic effects 83
Mixed fracture modes, SEM 176
Mode I loading condition, tear effect on dimple shape 12-14
Mode II loading condition, shear effect on dimple shape 12-14
Mode III loading conditions, shear effect on dimple shape 12-14
Modes, of fracture 12-71
Moist air, as corrosive environment 24
Molten salts, as corrosive environment 24
Molybdenum
  alloy, tension-overload fracture 464
  early TEM 6
  embrittlement sources 123
  vacuum-arc-cast high oxygen 6
Monel
  alloy 400, expansion joint, cleaning 76
  failed in liquid mercury 25
Monotonic fracture
  and fatigue fracture, compared 229
  slow, pearlitic ductile irons 230
Monte Carlo projection, electron trajectories 167
M shell, defined 168
Mud cracks, austenitic stainless steels 361

N

Naphtha, as organic cleaning solvent 74
Necking, in tensile-test fractures 98-105
Neutron irradiation
  effect on fracture mode/toughness, superalloys 388
  as embrittlement, examination and interpretation 127
Nicholas illuminator, fractographic 81
Nickel alloys. See also *Nickel alloys, specific types; Nickel-base superalloys; Nickel-base superalloys, specific types.*
  cast, stage I fatigue fracture appearance 19
  fatigue striations, precision matching 205-206

fractographs 396-397
fracture/failure causes illustrated 217
hydrogen embrittlement 124
Nickel alloys, specific types
  201, bend-test fracture surface 396
  Duranickel, fracture surface 397
Nickel-base superalloys
  effects of reactive atmospheres 40
  embrittlement in 123
  fracture/failure causes illustrated 217
Nickel-base superalloys, specific types
  Astroloy, effect of stress intensity factor range on fatigue crack growth rate 57-58
  Astroloy, effects of vacuum on fatigue 48
  Astroloy, fatigue fracture in air/vacuum 48, 54
  CMSX-2, single-crystal, hydrogen embrittlement 389
  IN-100 nickel-base, dendritic stress-rupture fracture 391
  IN-718, effect of frequency on fracture appearance 58, 60
  IN-738, effect of frequency and wave form on fatigue properties 63
  IN-738, fatigue and creep fractures 389
  Incoloy X750, fatigue fracture mechanisms 392
  Incoloy 800, effect of stress intensity factor range on fatigue crack growth rate 58
  Incoloy 800, effects of temperature 52
  Incoloy 800, ridges and striations in sulfidizing atmosphere 41, 53
  Incoloy 800, wedge cracking 26
  Inconel 600, corrosion fatigue 45-46
  Inconel 600, effect of frequency and wave form on fatigue properties 60
  Inconel 625, wedge cracking 26
  Inconel 718, effects of stress intensity factor range on fatigue crack growth rate 58
  Inconel 718 (UNS N07718), evaluation 393
  Inconel X-750, effect of frequency and wave form on fatigue properties 59-60
  Inconel X-750, effect of stress intensity factor range on fatigue crack growth rate 57
  Inconel X-750, effect of temperature 52
  Inconel X-750, effect of temperature on double-aged 47
  Inconel X-750, effect of vacuum 48
  Inconel X-750, fatigue fracture appearance ... 58
  Inconel X-750, fatigue fractures in air/vacuum 48, 55
  Inconel X-750, light/SEM fractographs, compared 94, 96
  Udimet 720, creep fracture 395
  Udimet 720, fatigue fracture 395
  Udimet 720, multiple transgranular fatigue origins 395
Nickel-chromium-molybdenum-vanadium steel rotor, effect of stress intensity range range factor on fatigue crack growth rate 56-57
Nickel-copper alloys, low-melting metal embrittlement 29
Nil-ductility casting, gray irons 226
Niobium alloys, corrosion products on intergranular fracture surface 37
Nitric acid, as titanium cleaning agent 75
Nitrides
  in grain boundaries, high-carbon steels 282
  solution, as corrosive environment 24
Nitrogen, liquid, iron cooling by 219
Nondestructive inspection
  effects 77
  profile generation 199
Nonferrous alloys, fracture/failure causes 217
Nonmetallic stringers, formation 65
Nonuniform loading, effects on cracking 175-176
Nucleation
  crack, AISI/SAE alloy steels 307
  dislocation, crack tip 30
  fatigue crack, austenitic stainless steel implants 360

grain-boundary cavity 20
microvoid, effects on fracture surface 12

O

Oblique illumination
  effects on fatigue fracture 85
  for medium-carbon steels 275
  photographic 83
  replica fractograph 100
  and vertical lighting, compared 87
Oblique photography, by view camera 78
Observation. See also *Examination; Visual examination.*
  direct, of thin-foil specimens 179
Ohmic contact window, vacuum-deposited aluminum defects 486, 487
Oil, fresh, as fracture preservative 73
One-step replicas, TEM 7
One-step temper embrittlement. See *Tempered martensite embrittlement.*
On-site photomacrography, setups for 78
Opened crack
  austenitic stainless steels 356
  fracture surface, AISI/SAE alloy steels 335
  precipitation-hardening stainless steels 374
  primary 77
  secondary 77
Optical fractography, defined 1
Optimum-aperture concept 80-81
Organic-fiber brush, for fracture cleaning 74
Organic solutions, as corrosive environment 24
Organic solvents
  chlorinated, carcinogenic 74
  for fracture cleaning 74
Orientation effects, SEM image recording 169
Orthopedic implants, austenitic stainless steels 359-364
Orthophosphoric acid, for oxide coating removal 75
Oval-shaped dimples, formation 13
Overheating
  as embrittlement, examination/interpretation 127-129
  facets, steel alloy 146
  fracture, alloy steel 144
  fracture, vanadium-niobium plate steel 145
  historical study 2
  in low-carbon steel 246
  transverse fracture 142, 163
Overload fracture 12
  AISI/SAE alloy steels 294, 298-299
  cast aluminum alloys 409, 411-413
  dimple rupture, effect of elevated temperature 35, 49-50
  ductile 101, 299, 443
  at flux inclusion 65, 67
  high-carbon steels 278
  macrograph 101
  medium-carbon steels 258
  sudden 294, 413
  titanium alloys 443
  torsional 258, 278
Overloading, spall formation by 114
Overpressurization, fracture by 345
Overtempering, alloy steels 301
Oxidation
  AISI/SAE alloy steels 305
  of clean fracture surface, for replicas 181
  as effect of high temperature 35
  effect on dimple rupture 35
  effect on slip reversal 15
  fracture, cleaning of 73-76
  high-carbon steels 280
  high-temperature 72
  intergranular, AISI/SAE alloy steels 300
  in low-carbon steel 246
  titanium alloys 442
Oxide coatings, removal of 75
Oxide dispersion-strengthened materials, dimple size 12

Oxide(s)
effect in hydrogen damage............ 125-126
effect on fatigue crack growth rate.........41
film, austenitic stainless steels......... 352-353
flaws, in integrated circuits...............481
as inclusions.......................65, 220
martensitic stainless steels................366
spheroidal.......................220, 300
surface, effects, metal-matrix composites....466
temperature, effect on.....................35
wrought aluminum alloys..................435

Oxygen
atomic, in low earth orbit................481
effect of temperature.....................35
-embrittled iron........................222
as embrittler of fcc metals................123

Oxygen-free high-conductivity copper, fatigue test fracture surfaces...................401

P

Palisades, broken graphite fibers as..........465
Palladium-shadowed plastic-carbon replica.................................185
Papers, background, for photomacrography....78
Parallax
determination, by stereo imaging..........197
excessive, in SEM imaging...............167
Parallel-axes photography method..........88
Parallel lighting, photomacrographic..........83
Parameters, profile and surface roughness, fractal analysis........................212
Parametric relationships
fracture surface and projected images.......202
profile and surface roughness, equation.....212
true area/length, plotted.................204
Partially oriented surfaces
dimpled fracture as....................203
parametric methods for.................204
Particle fracture model, fatigue failure.......207
Particles
boundaries, P/M aluminum alloys..........440
cleavage fracture from..................457
density and spacing, effect on tensile fracture........................ 100-101
effect on fatigue striation..................16
opaque, AISI/SAE alloy steels............308
second-phase.................. 16, 219, 423
size............................101, 207
spheroidal, iron.......................223
sulfide, effect on ridge formation.......41, 53
sulfur, in asphalt......................473
surface defects, iron....................220
Part(s). See also *Fractured parts.*
fractured, photography of............. 78-90
fracture sources illustrated...............217
photolighting of highly reflective........84, 88
Passive corrosion, defined................ 41-42
Passive film rupture, in SCC and corrosion fatigue................................42
Pearlite, spheroidized, high-carbon steel microstructure..........................278
Pearlitic ductile irons, fracture modes................ 228-230, 235-236
Peeling fracture, medium-carbon steel........253
Peening. See *Shot peening.*
Perfectly oriented surface, true area and length...............................204
Perimeter
length, as basic figure quantity........ 194-195
mean, of closed figures.................195
Periodic marks, by SCC of brass.............28
Persistent slip bands, defined..............117
Perspective
distortion, effects of stereo imaging........171
effect in SEM imaging.............169, 171
error, as distortion....................196
Petroleum-base compounds, solvent-cutback..............................73
P-F test, Shepherd, for tool steels......141, 162
Phase transformation, and hardness........ 32-33
Phosphoric acid, as ferrous cleaning agent......75

Phosphorus
for ductility enhancement, ASTM/ASME alloy steels..........................349
as embrittler...........................29
maps................................349
segregation, tool steels..................375
Photogrammetry. See also *Stereophotogrammetry.*
methods, quantitative fractography..... 197-198
stereo-, contour map and profiles by........197
Photographs, materials illustrated...........216
Photography. See also *Fractography.*
35-mm single-lens-reflex cameras....... 78-79
auxiliary equipment.....................89
depth-of-field effects........ 78, 80-81, 87-88
exposures, test...................... 86-87
film..................................85
focusing........................... 79-80
of fractured parts/surfaces............. 78-90
lens aperture selection................ 80-81
lens conversion tables...................81
lenses................................79
lighting techniques................... 82-83
light meters........................ 85-86
light sources....................... 81-82
magnification, determining................80
microscope systems.....................79
scanning light photomacrography..........81
setups for fractured parts.................78
stereo images...................... 87-88
test exposures...................... 86-87
view cameras...................... 78-79
visual examination, preliminary............78
Photomacrographs, preparation of........ 78-90
Photomacrography. See also *Fractography; Photography; Photomacrographs.*
auxiliary equipment.....................89
central optical path.....................79
scanning light..........................81
view camera systems for.................79
Photomicrographs, SEM.................194
Photomontage, austenitic stainless steels......354
Picture element. See *Pixel.*
Pipe, wrought aluminum alloy...............427
Pitting. See also *Corrosion; Corrosion pitting; Pitting corrosion.*
AISI/SAE alloy steels...............329, 332
by arcing............................488
austenitic stainless steels................358
ceramic..............................471
copper alloys.........................403
corrosion................... 41, 43, 243, 358
in medium-carbon steels................259
precipitation-hardening stainless steels......373
in SCC............................25, 27
Pitting corrosion. See also *Corrosion; Corrosion pitting; Pitting.*
copper alloys.........................403
defined...............................41
low-carbon steel......................243
martensitic stainless steels................43
Pixel size, effect on nonfractal behavior......211
Plain carbon steels, fracture map.............44
Planar sections, for profile generation........198
Plane of focus. See *Focusing.*
Plane-strain fractures
AISI/SAE alloy steels..................308
tension-overload, maraging steels..........385
wrought aluminum alloys................423
Plastic coatings, for fracture surfaces..........73
Plastic deformation. See also *Deformation; Microplastic deformation.*
of ductile/brittle fractures................173
in ductile irons.................... 227-237
effect on fatigue cracks, wrought aluminum alloys..............................420
high-purity copper................ 399-400
impact..............................336
in medium-carbon steels................258
shear bands, titanium alloys..............445
Plastic flow
AISI/SAE alloy steels..................329
high-purity copper....................400

microscopic, tear ridges from.............224
wrought aluminum alloys................420
Plateaus
ASTM/ASME alloy steels...............347
in ductile irons.......................231
fatigue striations on.....................23
multiple, crack propagation on............16
Platelets
cast aluminum alloys...................409
microstructure, titanium alloys............442
Plate steel, hydrogen flaking...............141
Plating
chromium, effect in AISI/SAE alloy steel fracture............................297
electroless nickel, for edge retention.....95, 100
hydrogen-charging, precipitation-hardening stainless steels......................372
hydrogen embrittlement by............22, 30
Platinum-carbon alloys, for thermal evaporation...................... 172-173
Plots
carpet..............................172
contour.............................172
fractal......................... 211-214
Ply buckling, resin-matrix composites........478
P/M aluminum alloys. See also *Aluminum; P/M aluminum alloys, specific types.*
experimental.........................440
fractographs.........................440
fracture/failure causes illustrated..........217
P/M aluminum alloys, specific types
Al-4.2Mg-2.1Li, fracture along powder particle boundaries..................440
Al-4.2Mg-2.1Li, corrosion-fatigue cracking...........................440
Pockmarks, tool steels....................381
Point location, as stereoscopic method... 196-198
Polaroid films, fractographic............85, 169
Polaroid Instant 35-mm slide camera system............................ 78-79
Polaroid photographs, exposure guide........89
Polycarbonate sheet, quasi-brittle fatigue crack propagation.........................479
Polycrystalline metals, cleavage fractures.....252
Polyethylene
copolymer pipe, slow crack growth failure.............................479
linear, crack growth mechanisms..........479
medium-density, fatigue striations..........480
medium-density, tearing and fibrillation.....480
Polygonization, tungsten fibers.............467
Polyimide, fracture surface................480
Polymeric polysulfides, in concrete..........472
Polymers
fractographs..................... 479-480
fracture/failure causes illustrated..........217
linear polyethylene, crack growth mechanisms........................479
polycarbonate sheet, quasi-brittle fatigue crack propagation.........................479
polyimide, failure origin................480
Polystyrene, with benzene, for replicas.......180
Polythionic acid stress-corrosion cracking, austenitic stainless steels...............354
Polyvinyl formal, with ethylene dichloride, for replicas.............................180
Porosity
cast aluminum alloys............ 67, 405-406
as discontinuities, defined.............65, 67
ductile iron cleavage fracture from.........227
as hydrogen damage...................124
interdendritic, wrought aluminum alloys....431
random............................66, 67
shrinkage........................ 405-406
solidification, wrought aluminum alloys....431
wrought aluminum alloys...........422, 425
Postweld heat treatment cracking, examination...................... 139-140
Power cycling, effect on thermocompression ball bond..............................484
Precipitation
within dimples, x-ray analysis............174
effect in SCC..........................26

Precipitation-hardening stainless steels
fractographs.................................370-374
fracture/failure causes illustrated..........217
Precipitation-hardening stainless steels, specific types
 13-8 PH, cup-and-cone tension overload fracture....................................370
 13-8 PH, high-cycle fatigue fracture.........371
 13-8 PH, low-cycle fatigue fracture..........371
 13-8 PH, SCC fracture........................372
 Armco 15-5 PH, high-cycle fatigue fracture....................................373
 Armco 17-7 PH, brittle intergranular fracture....................................374
 Armco 17-7 PH, fracture by cross-check defect......................................374
Precision matching study, fatigue striations in nickel.......................205-206
Precracking, fatigue...........75, 236-237, 397
Preferential corrosion, tin-bronze alloy......403
Preferential grain-boundary attack, ceramic fracture...471
Preliminary visual examination
 for photography..............................78
 specimen....................................72-73
Preparation
 and preservation, specimen...................72-77
 of SEM specimens...........................171-173
Preservation
 and preparation, specimen...................72-77
 techniques..................................73
Primary cracks, opening........................77
Primary creep, defined.........................19
Probability parameter, as roughness parameter.....................................201
Profile angular distributions, for fracture surface area.................................202-204
Profile generation
 metallographic sectioning methods...........198-199
 nondestructive..............................199
 from replicas...............................199
Profile parameters, fracture surface roughness...........................199-200, 212-215
Profile roughness parameters............212-215
Profiles
 alloy steels............................214, 327
 examining.................................95, 100
 fatigue fracture...........................15, 22
 for fractal analysis.......................212
 fractal properties........................211-215
 fracture................15, 22, 95-96, 212, 214
 of fracture surface roughness................199-205, 212-215
 generation, quantitative fractography...198-199
 by light microscope.........................94, 99
 matching, fatigue striations in nickel...205-206
 modified fractal curve.....................212
 nondestructive..............................199
 from replicas...............................199
 roughness, parameters......................199-200
 sections...................................95-96
 and surface roughness parameters, relationship............................212
Projected images
 assumption of randomness..................194-195
 and fatigue striation spacing, correlated....205
 fracture path and microstructure, correlated...............................195-196
 and fracture surface, parametric relationships...............................202
 quantitative fractography..................194-196
 single SEM fractograph.....................194
 and spatial features, stereological relationships...............................196
Projection plane, basic quantities and relations....................................194-196
Projection stereology, imaging by............194
Prototype facets, true area values...........203
Pulsation, AISI/SAE alloy steels..............304
Pumps, vacuum, for SEM systems................171
Pure titanium
 commercial, effects of vacuum on fatigue..48-49
 commercial, fatigue striations...............20

dimples, SEM stereo pair....................171
effect of stress intensity factor range on fatigue crack growth rate.........................57

Q

Quality control
 examination technique.................140-143
 tensile testing for........................101
Quantitative fractography. See also *Fracture surface(s); Photography; Roughness parameters; Surface(s).*..........193-210
 analytical procedures....................199-205
 angular distributions, profile............201-204
 applications, example cases...............205-208
 area, fracture surface, estimating.......201-205
 defined...................................193
 development..............................193-194
 experimental techniques..................194-199
 geometrical methods........................198
 goal and history............................8
 metallographic sectioning methods......198-199
 nondestructive profiles....................199
 partially oriented surfaces................201
 photogrammetric methods..................197-198
 profile generation.......................198-199
 profile parameters......................199-200
 profiles from replicas....................199
 projected images.........................194-196
 research, summary of......................211
 roughness parameters....................199-205
 statistical vs individual measurement...207-208
 stereoscopic methods.......................198
 surface parameters......................200-201
 triangular elements.....................201-202
Quantitative stereoscopy, defined...........171
Quasi-brittle fatigue, polycarbonate sheet....479
Quasi-cleavage fracture
 AISI/SAE alloy steels................303, 305, 319
 austenitic stainless steels.................357
 in austenitized iron alloy..................26
 defined......................................20
 ductile irons..........................232, 235
 facets, titanium alloys....................448
 facets, unaged iron alloy...............458-459
 by hydrogen embrittlement..................31
 iron alloy.............................457-459
 malleable iron.............................238
 precipitation-hardening stainless steels..370-371
 rosette tensile fractures with.............104
 solute-depleted ASTM/ASME alloy steels...350
 stress state effect on appearance.......31, 40
 tool steels......................26, 375, 379, 380
Quasi-striations, fatigue fracture..........15, 22
Quench aging
 defined...................................129
 as embrittlement, examination/interpretation.........................129-130
Quench cracks
 AISI/SAE alloy steels......................305
 brittle intergranular fracture by..........335
 as embrittlement, examination/interpretation.........................130-132
 low-carbon steel...........................251
 service fracture by.........................65
 tool steels................................375

R

Radial fracture. See also *Radial marks.*
 AISI/SAE alloy steels......................312
 zone.......................................311
Radial marks
 AISI/SAE alloy steels....294, 302, 312-314, 334
 and circular spall....................113, 123
 cobalt alloy..............................398
 curved, star or rosette fracture as....103-104
 fine/coarse...........................102-103
 and line spall........................114, 129-130
 SEM fractographs..........................169
 studies.....................................3

 tool steels...........................377-378
Radiation, embrittlement effects..............127
Rail steels, fracture modes......115, 117, 135-137
Ramp
 defined.....................................59
 rates, and mean stress.....................63
Randomly oriented surface, true area and length.......................................204
Randomness, in projected images..........194-195
Random porosity, cast aluminum fracture surface..................................66, 67
Rapid coalescence, intergranular fracture by...349
Raster pattern, SEM..........................167
Ratchet marks
 defined...................................112
 martensitic stainless steels...............368
 medium-carbon steels............260, 269, 272
Rebounding, effects in white iron.............239
Rebuilding, improper, tool steel cracking....375
Reciprocity failure, defined..................86
Red rust. See *Corrosion products.*
Reflected-light meter, for photomacrography...85
Reflectivity
 of fracture surfaces, and photolighting...84-88
 of silver, atomic oxygen effects...........481
Reheat cracking, examination.............139-140
Repassivation
 effect on SCC...............................25
 kinetics....................................42
Replicas
 artifacts in...........................184-185
 cellulose acetate, SEM....................171
 extraction.............................182-183
 fractographic, compared................95, 100
 grating...................................167
 for light microscopy......................94-95
 profiles from.............................199
 sectioning................................199
 shadowing........7, 95, 100, 172, 183-184
 single-stage............................179-182
 two-stage.................................182
Replica-stripping cleaning, of fractures......74
Replication
 extraction replicas....................182-183
 procedures, for light microscopy.........94-95
 single-stage replicas..........7, 179-182
 tape, cellulose acetate....................73
 techniques..................................7-8
 two-stage replicas.....................7, 182
Replication techniques
 one-step replicas..............7-8, 179-182
 shadowing...........7-8, 95, 100, 172, 183-184
 TEM..7
 two-step replicas......................7, 182
Resin-matrix composites
 carbon epoxy wet specimen, local failure...477
 carbon epoxy wet specimen, tension tested..477
 carbon phenolic shear specimen............478
 fractographs...........................474-478
 fracture/failure causes illustrated........217
 fracture morphology.......................477
 longitudinal carbon epoxy specimen, failed in flexure............................475-476
 longitudinal carbon epoxy specimen, failed in tension............................476-477
 overload tensile resin failure............477
 ply buckling..............................478
 resin failure.............................474
 transverse carbon epoxy, fracture sequence....................................474
 transverse carbon epoxy room moisture specimen, failed in tension..............474
 transverse carbon epoxy wet specimen, failed in tension..................................475
 woven carbon fabric phenolic prepreg panels, short beam shear, compression, and flexure fractures...................................478
Resin-matrix composites, specific types
 Fiberite 934 epoxy resin, fracture from internal defect.....................................474
 Fiberite 93 epoxy resin, fracture from internal defect.....................................474
Resolution, photographic...................80-81

Reversed bending fatigue fracture
AISI/SAE alloy steels 298
medium-carbon steels 269
Reversed sigmoidal curves, linearization
of 213-214
Reversed stressing, medium-carbon steel 262
Reversed torsional fatigue fracture, alloy
steels 301
Reverse slip. See also *Slip.*
AISI/SAE alloy steels 298
Ridges. See also *Tearing; Tear ridges.*
AISI/SAE alloy steels 304, 331
austenitic stainless steels 360
high-carbon steels 289
maraging steels 385
precipitation-hardening stainless
steels 370-371
by shear 371
from sulfur-containing environments 41, 53
tear, AISI/SAE alloy steels 293
tear, wrought aluminum alloys 417
tool steels 380
Ring assembly, arcing failure of 488
Ring illumination, circular fluorescent-light
tube 83
Ripples
AISI/SAE alloy steels 329
and cleavage, compared 453
mechanism 13
from slip step formation 16
River patterns
AISI/SAE alloy steels 291
Alnico alloy 461
in Armco iron 18
in cleavage
fractures 13, 172, 175, 252, 397, 424
corrosion fatigue, aluminum alloy 43, 54
defined 13
in ductile irons 230-231
on fracture surfaces 13, 252
in iron 18, 219, 222-224
low-carbon steel 252
metal-matrix composites 466
nickel alloys 397
polymers 479
precipitation-hardening stainless steels 370
superalloys 395
and thermal evaporation 172
titanium alloys 448
wrought aluminum alloys 424
R_L. See *Roughness parameters.*
Robinson detector 168
Rock-candy fracture
Alnico alloy 461
austenitic stainless steels 351
as intergranular 110-111
low-carbon steel 249
Rolling-contact fatigue, in bearings 115, 134
Rolling lap, fracture from 64
Rosette (star) fractures, as radial
marks 103-104
Rotary bending fatigue fracture, medium-carbon
steel 272
Rotary pumps, for SEM vacuum system 171
Roughness
fracture surface 199-205, 276
index of surface 201
parameters 4340 steel 213
parameters, quantitative fractography ... 199-205
surface, fractal analysis 211-215
Roughness parameters. See also *Surface roughness parameters.*
average peak-to-trough height 200
fractal dimension, irregular planar curve 200
fracture path preference index 201
probability parameter 201
profile configuration 200
profile parameters 199-200
quantitative fractography 199-205
surface 200-201
true length 199-200
vertical 200
R_P. See *Roughness parameters.*

R_S. See *Surface roughness parameters.*
RSCs. See *Reversed sigmoidal curves.*
Rub marks, fatigue 118-119
Ruled surface, true area and length 204
Rupture(s). See also *Decohesive rupture(s); Dimple rupture(s); Fracture(s).*
intergranular, and transcrystalline
cleavage 222
low-carbon steels 240
materials illustrated in 217
spontaneous, AISI/SAE alloy steels 299
titanium alloys 449

S

SAE alloy steels. See *AISI/SAE alloy steels.*
Salts, molten 24
Sampling
for fractal analysis 212
plan 212
point 196-198
Sand-cast gray iron, flaking fracture 225
Saw cutting
of specimens 76
wrought aluminum alloys 429, 435
Scabs
alloy steel fracture by 337
as mechanical damage 284
Scanning coil, in SEM imaging 167
Scanning electron microscope(s)
applied to fractography 8
for dimples 96
fractographs, compared with dark-field
illumination and light-field illumination
fractographs 92-100
geometry of image formation 196
history 7-8
literature 8
schematic cross section 166
vacuum system 171
Scanning electron microscopy 166-178
capabilities, defined 166
dark-field, and light-field illumination
fractographs, compared 92-100
display system 169-171
of ductile fractures 173-174
fractographs 92-100, 173-176, 185-192, 216
fractographs, types of 216
fractography 173-176
illuminating/imaging system 167-168
image and dark-field light microscope
fractograph, compared 92
information system 168-169
instrumentation 166-171
of intergranular brittle fracture 174-175
preparation, of fractographs 185, 192
preparation, of specimens 171-173
replication procedures 94-95
specimen demagnetization for 77
stereo, tilt method 171
stereo pair, flat cleavage fracture, irons 219
for stereo viewing 192
and TEM fractographs, compared 185-192
of transgranular fracture modes 175-176
vacuum system 171
with x-ray analysis, advantage 168
Scanning light photomacrography 81-82
Scraping artifacts, TEM replicas 185
Screws, hip, austenitic stainless steels 359-364
Seams
AISI/SAE alloy steels 326, 335
in bolts 131, 149
brittle intergranular fracture from 335
as discontinuities, defined 64-65
high-carbon steels 281, 287
quench cracking from 131, 149
walls, alloy steels 335
wire springs fracture from 63
Season cracking, SCC of brass as 28
Seawater. See also *Water.*
corrosion, low-carbon steel 250
as corrosive environment 24

Secondary cracking
AISI/SAE alloy steels ... 293-294, 299, 301, 306,
308, 327, 334
angular, titanium alloys 448
ASTM/ASME alloy steels 346, 348
austenitic stainless steels 353
circumferential, precipitation-hardening
stainless steels 370
copper alloys 403
of corrosion products 29
deep intergranular, iron alloy 459
in dimpled surface, titanium alloys 450
ductile irons 232
at elevated temperatures, titanium alloys ... 442
and fatigue striations 20
intergranular 47, 401, 425, 448
iridium 462
low-carbon steel 244
in low-melting metal embrittlement 30, 38
at machining marks 251
nickel alloys 397
OFHC copper 401
opening 77
in oxygen-embrittled iron 222
parallel, maraging steels 386
precipitation-hardening stainless steels 372
as splitting 68
stress-corrosion 27
titanium alloys 441, 448-450, 452
tool steels 377, 380, 382
transgranular 449, 452
transverse, AISI/SAE alloy steels 301
wrought aluminum alloys 414, 424, 425, 434
Secondary cracks, visual macroanalysis 72
Secondary creep, defined 19
Secondary electrons. See also *Electrons.*
and backscattered electrons, compared ... 168
image, coarse-grain iron alloy 93-94
in SEM imaging 168
Secondary flow, nickel alloys 396
Secondary hardening, austenitizing effects ... 341
Secondary-phase structures, for rolling
direction 422
Second-phase particles
brittle fracture, wrought aluminum alloys ... 423
cleaved, Alnico alloy 451
iron sulfide 219
large, effect on striation 16
microvoid coalescence at 12
wrought aluminum alloys 428
Sectioning
for macroscopic examination 92
medium-carbon steels 268
metallographic, methods 198-199
planar 198-199
profile parameters from 194
replica 199
sampling plan 212
serial, profile from 198
transverse 179
vertical, for true fracture surface
area 198-199, 211-212
Sections
fracture profile 95-96
taper 96
thickness, effect on fracture surface 105
Segregation
antimony 350
in decohesive rupture 18
as discontinuity, defined 67
grain-boundary, phosphorus 29
intergranular embrittlement by 110
phosphorus, tool steel 375
sulfur 349
transverse fracture from 142, 163
Self-similitude, in rough planar curves 211
Semifractal plot. See also *Fractal plot.* 211
Separation
cavities, copper alloys 402
circumferential, wrought aluminum alloys ... 415
final, titanium alloys 441
intergranular, high-purity copper 399
matrix, effect of crack tip 16

Serial sectioning. See also *Sectioning.*
   for fractal analysis 212-213
   for fracture surface area 194
   profile, fractured aluminum-copper alloy 198
Serpentine glide
   defined................................... 13
   formation, in copper...................... 17
   from slip step 16
   wrought aluminum alloys.................. 434
Service fractures
   AISI/SAE alloy steels............ 297, 317, 319
   cast aluminum alloys 409-412
   fatigue, low-carbon steel 241, 251
   macroscopic examination 91-93
   mating fracture 251
   rotary bending fatigue fracture, medium-carbon steels 272
   superalloys 390, 391
   tool steels 376
   wrought aluminum alloys.................. 417
Setups, fractographic....................... 78
Shadow cast. See *Shadowing.*
Shadowing
   as film deposition 172
   methods.............................. 183-184
   of replicas 95, 100, 183-184
   single-stage plastic replicas 184
   in TEM replication 7
   two-stage plastic-carbon replicas 184
Shear. See also *Shear bands; Shear deformation; Shear dimples; Shear fracture(s); Shear lips.*
   adiabatic 31-33, 42-43
   elongated dimple formation 13
   elongated voids, irradiated stainless steels ... 365
   as fatigue modes......................... 12
   final fast fracture, austenitic stainless steels .. 353
   ridges, precipitation-hardening stainless steels 371
   on secondary slip systems, furrow-type fracture 44, 54
   short beam, resin-matrix composites........ 478
   step, in iron............................. 224
Shear bands
   adiabatic 31-33, 42-43
   formation and fracture, titanium alloys 444-445
   orientation 445
   along slip planes 32
   torsional, austenitic stainless steels 359
Shear deformation
   low-carbon steel 244
   mechanical damage during, tool steels 380
   ridge formation during, tool steels 380
Shear dimples
   maraging steels 383
   martensitic stainless steels 369
   titanium alloys 455
Shear fractures
   AISI/SAE alloy steels..................... 341
   effect of striation 16
   elongated dimples.................... 12, 15-16
   maraging steels 383
   martensitic stainless steels 367
   precipitation-hardening stainless steels 371
Shear lips
   alloy steels ...294, 302, 306-307, 311, 314-315, 318, 332-334
   on crater 318
   ductile, wrought aluminum alloys 414
   low-carbon steel 244
   maraging steels 383
   martensitic stainless steels 366, 369
   medium-carbon steels 253
   precipitation-hardening stainless steels 370
   tool steels 379
   wrought aluminum alloys.............. 414, 425
Shell cracks, high-carbon steel.............. 288
Shelling. See *Spalling.*
Shepherd P-F test, for tool steels 141, 162
Short beam shear, resin-matrix composites............................... 478
Short-time tensile tests, for high-temperature properties measurement 121, 139

Shot peening
   effect on dross inclusion, wrought aluminum alloys..................................422
   inadequate, alloy steel failure by...........327
   superficial, effect in wrought aluminum alloys..................................420
Shrinkage. See also *Shrinkage cavities.*
   area, iron................................223
   porosity.............................405-406
   void, cast aluminum alloy 67
Shrinkage cavities. See also *Cavities; Dimple(s).*
   cast aluminum alloys 409
   iron.....................................220
   low-carbon steel 249
   maraging steels 384
   solidification in......................140, 160
   and white spot, iron......................220
Shrinkage porosity, cast aluminum alloys 405-406
Sigma phase
   defined..................................132
   embrittlement 132-133, 150
Sigmoidal curves, reversed.............212-215
Silicates, as inclusions..................... 65
Silky fractures 2
Silver, pure, effect of atomic oxygen 481
Single bending impact load, medium-carbon steel fracture 270
Single-lens-reflex cameras, 35-mm 78-79
Single stage direct carbon replicas, formation...............................181
Single-stage replicas
   conversion oxide film 181-182
   direct carbon 181
   production methods 179
   technique, schematic 180
   thick plastic........................ 180-181
   thin plastic......................... 179-180
Sintered tungsten. See *Tungsten.*
Slag stringers, as woody fracture pattern 224
Sliding, grain-boundary.................19, 25
Slip. See also *Slip planes; Slip reversal; Slip steps; Slip traces.*
   at crack tip 15, 21
   cross-slip, cobalt alloy 398
   distance, in particle fracture model 207
   environmental effects 35
   fatigue as process of...................... 35
   light microscopy for 106
   lines, in irons 219
   localized, titanium alloys 450
   at low temperatures 33
   progression, by DIC illumination 121
   severe deformation, wrought aluminum alloys..................................426
Slip bands
   AISI/SAE alloy steels.....................298
   in Armco iron 224
   austenitic stainless steels 360
   cracks 224, 298
   formation, in fatigue.....................117
   persistent 117
   superalloy fracture surface along 392
   wrought aluminum alloys.................420
Slip/cross slip, cobalt alloys.................398
Slip planes
   cracking................................. 14
   displacement, on dimples 13
   fracture, in fatigue 117
   shear bands along......................32, 32
Slip reversal
   effect of oxidation 15
   effect on crack propagation 35
   partial 15, 21
   in vacuum 46
Slip steps
   formation, and serpentine glide, ripples from 16
   on grain-boundary cavities................219
   in iron..................................224
   wrought aluminum alloys.................426
Slip traces
   with fracture striations 15

and striations, compared 119
Slit-island technique, for surface roughness ... 200
Slow-bend fracture
   iron alloy................................457
   maraging steels 387
   toughness test, LME 30, 38
Slow monotonic fracture, ductile irons... 230-232
SLR cameras. See *Cameras.*
Small-bubble artifact, in replicas 184
Sodium carbonate, in cathodic cleaning 75
Sodium cyanide, in cathodic cleaning 75
Sodium hydroxide
   as ferrous cleaning agent 75
   solutions, in cathodic cleaning............. 75
Solder, cracked, in electronic materials............................ 481-482
Solidification
   cracking, ASTM/ASME alloy steels.........345
   in gray iron 226
   historical study 2
   in iron-chromium-aluminum alloy......140, 160
   maraging steels 384
   porosity, wrought aluminum alloys.........431
   in shrinkage cavity 140, 160
Solid-metal embrittlement, defined....... 29-30
Solvent-cutback petroleum-base compounds, as fracture preservatives 73
Spalling
   AISI/SAE alloy steels.................322, 329
   as fatigue failure 113-115, 329
   in forged hardened steel rolls..............113
   medium-carbon steels 275
   in white iron 239
Spatial features
   and projected images, stereological relationships 196
   stereometry/profile analysis for 194
Specimen(s). See also *Fracture specimens; Test specimen(s).*
   conductivity, for SEM imaging 171
   fracture sources illustrated 217
   fracture toughness, ultrasonic cleaning of..... 74
   geometry, effects in SEM 168
   preparation/ preservation......... 6-7, 72-77, 171-173, 179
   sectioning and cutting................. 76-77
   for SEM 171-173
   for TEM 6-7, 179
   thin-foil, direct observation of 179
   tilt, effects in SEM imaging 168
Spherical aberration, SEM illuminating/imaging 167-168
Spinodal decomposition, copper alloys.......402
Spiral gouges, high-carbon steels............280
Splines, mangled, alloy steels 333
Splitting
   fiber, metal-matrix composites 467
   maraging steels 387
   in tensile specimens 104-107
   in unfavorable grain flow 67-68
Spontaneous rupture, AISI/SAE alloy steels 299
Spot metering, photographic 79, 85
Spot size, effects in SEM imaging 167
Spot-type meter, for photomacrography 85
Sputter coating, SEM specimens.............173
Stage I, fatigue fracture..................14, 19
Stage II, fatigue fracture 14, 19-21
Stage III, fatigue fracture 14, 16
Stainless steels. See also *Austenitic stainless steels; Martensitic stainless steels; Precipitation hardening stainless steels; Stainless steels, specific types; Steels.*
   duplex, by 885 °F embrittlement...........155
   ferritic, 885 °F (475 °C) embrittlement 136, 155
   hot cracking in 123
Stainless steels, specific types
   13-8 PH, hydrogen embrittlement........... 30
   17-4 PH, intergranular SCC-caused decohesive fracture 24
   25CR-12Ni cast, sigma-phase embrittlement 150

301, effect of stress intensity factor range on fatigue crack growth rate 57
302, effect of stress intensity factor range on fatigue crack growth rate 57
304, effect of frequency and wave form on fatigue properties 59, 60
304 wire, SCC failure 133, 152
312 weld metal, sigma-phase embrittlement 151
316, effect of frequency and wave form on fatigue properties 59, 60
316, effect of stress intensity factor range on fatigue crack growth rate 57
316, effect of vacuum on fatigue 48
316, fatigue fracture in air/vacuum 48, 55
321, effect of frequency and wave form on fatigue properties 59, 60
347, effect of frequency and wave form on fatigue properties 59, 60
AISI 4340, hydrogen embrittlement 31
SUS 304, effect of cyclic load on fatigue crack rate....................................... 62
SUS 304, effect of frequency and wave form on fatigue properties 62
Stair step fracture, cobalt alloy............. 398
Star marks, alloy steels...................... 301
Star (rosette) fractures, as radial marks .. 103-104
Static fatigue, hydrogen embrittlement as 124
Statistical measurement, and individual measurement, compared......... 207-208
Steady-state creep, defined 19
Steady-state strain rate, high-purity copper 399-400
Steel(s). See also *AISI/SAE alloy steels; ASTM/ASME alloy steels; Austenitic stainless steels; High-carbon steels; Low-carbon steels; Maraging steels; Martensitic stainless steels; Medium-carbon steels; Precipitation-hardening stainless steels; Stainless steels; Tool steels.*
  chromium, feather markings 18
  coating, exposing and cleaning 73
  effect of temperature on dimple size 34, 46
  embrittlement by low-melting alloys 29
  fracture types 1-3
  high-carbon, figure numbers for 216
  high-strength, effect of high temperature on overload fracture................... 35, 50
  hot shortness 126-127
  hydrogen embrittlement 23, 25, 31, 124
  hydrogen flaking......................... 125
  intergranular dimple rupture.............. 14
  low-carbon, figure numbers for 216
  medium-carbon, figure numbers for 216
  overheating 127-129
  pearlitic eutectoid, TTS fracture 28
  plate, hydrogen flaking 141
  SCC of 27
  sulfide inclusions 14
Steel(s), specific types
  15 B28 induction-hardened, replica fractographs of fatigue fracture, compared........95, 100
  15 B28, photo-illumination effects 85
  1008 strip, photo effects of lighting 87
  1039 fatigue specimen, photolighting effects .. 84
  1040, cleavage fractures................... 175
  1085, stereo-pair photographs 89
  4340, light fractographs 83
  8620, photo-illumination effects 86
  API-5LX grade X42 carbon-manganese pipeline, effect of stress intensity factor range on fatigue crack growth rate 56
  API grade X-60 line pipe, sinusoidal fracture path 116
  Grade X42 pipeline, effect of hydrogen on fatigue fracture appearance 37, 51
  HY-130 bainitic, TTS fracture...........21, 28
  HY-180, stress-corrosion fracture 27, 34
  RQC-90 steel plate, cold cracks 156
Stereogrammetry. See *Stereophotogrammetry.*
Stereo imaging. See also *Imaging.*
  fractographic............................ 87-88
  parallax determination 197
  SEM display systems 171

Stereology, in quantitative fractography... 193-196
Stereometry, for three-dimensional spatial analysis 194
Stereomicroscopes, See also *Microscopes.*... 78-79
Stereo-pair photographs 87-89
Stereo-pair viewer, effects 87
Stereophotogrammetry. See also *Photogrammetry.*
  carpet plot, titanium alloy 198
  contour map and profiles by 197
  SEM 171
Stereoscan, development of 8
Stereoscopes, Hilger-Watts................ 171
Stereoscopic imaging
  fracture examination...................... 92
  methods............................. 196-198
  in quantitative fractography 196-198
Stereo viewing, TEM and SEM fractographs .. 192
Stigmators, as SEM lens................... 167
Strain. See also *Strain aging; Strain rate.*
  asymmetrical, effect on dimple size 12, 16
  -hardening relief, mechanism of 42
  increasing to fracture, malleable iron 238
  -induced martensitic transformation, and SCC 26
  localization, zone of 42
Strain aging
  as embrittlement, examination/ interpretation....................... 129-130
  Lüders bands from 129, 148
  stretcher-strain formation by 129, 148
Strain rate
  effect on creep crack propagation, austenitic stainless steels....................... 354
  effect on dimple rupture............... 31-33
  effect on embrittlement 29
  effect on fatigue life, stainless steels........ 59
  effect on fracture appearance, steel alloy ..31, 41
  effect on intergranular creep rupture 19
  effect on strength 121
  moderately high, effects 31
  steady-state, effect in high-purity copper 400
  very high, effects 31-33
  very low, effects 31
Strain wave form. See *Wave form.*
Streaking, alloy steels..................... 346
Strength
  effect of temperature 121
  excessive, effect in AISI/SAE alloy steels ... 318
  fatigue surface, wrought aluminum alloys ... 420
Stress. See also *Applied stress; Stress concentration.*
Stress
  applied, effect on embrittlement 29
  applied, examining 92
  biaxial, effect on dimple rupture 31, 39
  decreasing, effects in AISI/SAE alloy steels .. 315
  effect in dimple rupture 30-31
  effect in high-purity copper 399-400
  effect in visual examination 72
  macroshear, effect in medium-carbon steels ... 253
  principal, direction effects on dimple shape ... 30
  -relief cracking 139-140
  reverse, medium-carbon steel fracture by 262
  state, effect on dimple shape 12
  state, tensile specimen 104
Stress alloying. See *Liquid-metal embrittlement; Solid-metal embrittlement.*
Stress concentration
  crack formation at....................... 111
  effect on fatigue cracking................. 322
  by forming operations, wrought aluminum alloys 414
  from visual examination 72
Stress-corrosion cracking. See also *Corrosion; Corrosive environment; Stress-corrosion fracture(s).*
  AISI/SAE alloy steels......27, 299, 305, 320
  of aluminum 28
  of brass 28
  chloride, austenitic stainless steels......... 354
  cleavage fracture, titanium alloys 453
  compressive-stress 25
  copper alloys 403

and corrosion fatigue, unified theory 42
decohesion fracture, aluminum alloy forging................................18, 25
decohesive rupture from.................18, 24
defined.................................... 24
effect on dimple rupture................ 24-29
examination/interpretation 133-134
facets, austenitic stainless steels 354, 357
flutes and cleavage from 28
fracture mode change by 26-27
grain-boundary separation by 174
intergranular 373, 357
mechanisms of 25-27
precipitation-hardening stainless steels .. 372-373
of steels.................................. 27
sulfide, alloy steels...................... 299
titanium alloys 28-29, 453
wrought aluminum alloys..........414, 433-436
Stress-corrosion fracture(s). See also *Stress-corrosion cracking.*
  austenitic stainless steel 34
  in brass 36
  path, effect of electrochemical potential ...27, 35
  in steel.................................. 34
Stress intensity, effective, and FSS 205
Stress intensity factor (K)
  defined.........................15-16, 54, 56
  effect on fracture modes................. 441
Stress intensity factor range
  defined.............................. 54, 56
  effect on fatigue 54-58
  effect on fatigue crack growth rate 56-58
  wrought aluminum alloys.............417, 418
Stress-relief cracking
  examination/interpretation 139-140
  metallographic example 159
Stress rupture. See also *Stress-relief cracking; Stress rupture fractures.*
  cracking.............................. 139-140
  tests, for high-temperature effects 121
Stress rupture fracture(s)
  examination 139-140
  metal-matrix composites................. 468
  superalloys 391
Stretcher-strain formation, by strain aging 129, 148
Stretching
  titanium alloys 453
  wrought aluminum alloys................ 434
Striations. See *Fatigue striations.*
Striation spacing. See *Fatigue striation spacings.*
Stringers
  deep shell crack from 288
  defined.................................. 65
  as inclusion 65, 67
  nonmetallic, high-carbon steels 285
  of slag, as woody fracture 224
  sulfide, AISI/SAE alloy steels 294
Stripping, of carbon replicas 181
Structural alloys, overload failure mechanism 12
Studies, fracture, history of................ 1-8
Subcase fatigue fracture, alloy steel..... 322-323
Subgrain boundary, effect on cleavage fracture 17
Substrates, for SEM specimens 172
Subsurface cracking
  in cleavage, low-carbon steel............. 117
  high-carbon steels 280
  medium-carbon steels 261
Subsurface(s)
  cracking 117, 261, 280
  fracture, ductile irons 236
  thermal-wave imaging of 166, 169
Sulfidation, superalloys failure by 388
Sulfides
  in alloy steels 14, 291, 294, 299
  effect in hydrogen embrittlement 125
  as inclusions 14, 65, 239
  intergranular cracking by 123
  in malleable iron 239
  in mating fracture, medium-carbon steels.... 263
  particles, effect on ridge formation 41, 53

514 / Index

Sulfides (continued)
precipitation, on grain boundaries . . . 349
SCC by, alloy steels . . . 299
stringers, AISI/SAE alloy steels . . . 294
Sulfur
asphalt, fracture surface . . . 473
attack, effect on fatigue crack growth rate . . . 41
bacterial effects on . . . 245
cement, bonding . . . 472
concrete, fracture surfaces . . . 472-473
-containing atmospheres, ridges and striations . . . 41, 53
crystalline, in concrete . . . 472
distribution and morphology, in asphalt . . . 473
as embrittler of fcc metals . . . 123
enrichment, crack tip region . . . 41
maps . . . 349, 388
segregation . . . 349
Sulfuric acid, inhibited, in cathodic cleaning . . . 75
Superalloys
fractographs . . . 388-395
fracture/failure causes illustrated . . . 217
Superalloys, specific types. See also *Nickel-base superalloys; Superalloys.*
713C, service failure . . . 390
713LC, service fracture . . . 391
A-286 iron-nickel-base (UNS S66286), effect of neutron irradiation on fracture mode/toughness . . . 388
MA754, creep fracture . . . 393
MA754, departure side pinning . . . 393
Nimonic 115, fatigue fracture . . . 394
RA-330 iron-nickel-base (UNS N08330), sulfidation failure . . . 388
single-crystal PWA 1480, effects of thermal cycling . . . 394
Surface coatings
antireflective, for photography . . . 83
effects, visual examination of . . . 72
for fracture preservation . . . 73
and sectioning . . . 77
sputter and thermal evaporation . . . 173
Surface roughness parameters . . . 212-215
defined . . . 212
for fractal analysis . . . 212
and profile roughness parameter, calculated . . . 212
surface roughness . . . 200-201
surface volume . . . 201
topographic index . . . 201
Surface(s). See also *Fracture surface(s); Subsurface cracking; Subsurface(s); Surface coatings; Surface roughness parameters.*
debris, cleaning techniques for . . . 73-76
defects, on inclusions . . . 220
films, passive, as SCC mechanism . . . 25
fractal analysis of . . . 211-215
fractured, photography of . . . 78-90
irregular, fractal plots . . . 211-215
irregular, RSC profiles . . . 213
nondestructive inspection effects . . . 77
partially oriented, parametric relationships for . . . 201
perfectly oriented, true area and length . . . 204
prototype faceted, true profile length values . . 200
randomly oriented, true area and length . . . 204
ruled, true area and length . . . 204
topography, and microstructure . . . 393
volume, as surface roughness parameter . . . 201
Swaging, in irons . . . 219
Symbols, and abbreviations . . . 492-494
Système International d'Unités, guide for . . . 489-491

T

Tangent diameter, as basic figure quantity . . . 194
Tantalum
alloy, high-cycle fatigue fracture . . . 464
grains, striations . . . 20-21
heat-exchanger, fatigued . . . 20
Taper sections, of fractures . . . 96
Tap water, as corrosive . . . 24, 320
Tear. See *Tear dimples; Tear fracture(s); Tearing.*
Tear dimples . . . 12, 15
alloy steels . . . 304
maraging steels . . . 387
titanium alloys . . . 452
Tear fractures
effect of striations . . . 16
elongated dimples . . . 12, 15
surface (Mode I), dimple shape . . . 12, 14
Tearing. See also *Microtearing; Ridges; Tear dimples; Tear ridges.*
complex . . . 28
by decohesion, in welds . . . 139
in ductile fracture . . . 173
in ductile irons . . . 230-236
historical study . . . 2
lamellar . . . 138-139, 158
by machining, AISI/SAE alloy steels . . . 296
polymers . . . 479, 480
rapid, AISI/SAE alloy steels . . . 327
resin-matrix composites . . . 474
transgranular . . . 305
Tearing topography surface fracture
appearance . . . 28-29
defined . . . 21-22
Tear ridges. See also *Ridges; Tearing.*
AISI/SAE alloy steels . . . 304
in aluminum alloy . . . 19, 20
in Armco iron . . . 224
and fluting . . . 20-21
in iron . . . 26
maraging steels . . . 387
titanium alloys . . . 453
in TTS fractures . . . 21, 28
wrought aluminum alloys . . . 417
Tectyl 506 surface coating . . . 73
Tellurium, effect in copper . . . 401
TEM. See *Transmission electron microscopy; Transmission electron microscope(s).*
TEM direct carbon replicas, iron . . . 224
TEM p-c replicas, iron . . . 224
Temper
effects, AISI/SAE alloy steels . . . 341
embrittlement . . . 134-135
over-, AISI/SAE alloy steels . . . 301
Temperature. See also *Elevated temperature; High temperature.*
cryogenic, effect on fatigue . . . 52-53
ductile-to-brittle transition, defined . . . 34
effect, austenitic stainless steel . . . 50-52
effect, dimple rupture . . . 33-35
effect, dimple size . . . 34, 46
effect, ductile iron fracture . . . 231
effect, embrittlement rate . . . 29
effect, high-purity copper . . . 399-400
effect, intergranular creep rupture . . . 19
effect, medium-carbon steels . . . 254
effect, slip . . . 33
effect, strength . . . 121
equicohesive . . . 121
as fatigue environment . . . 35
high, effect on dimple rupture . . . 34-35
low, effect on dimple rupture . . . 33-34
and oxidation . . . 35
Tempered martensitic embrittlement . . . 135-136
Tensile fatigue, pure . . . 342
Tensile fractures. See also *Tensile-test fracture.*
cup-and-cone, studies . . . 3
shear, elongated dimple formation . . . 13
shear, martensitic stainless steels . . . 368-369
Tensile shearing . . . 13, 368-369
Tensile-test fracture. See also *Tensile fracture.*
carbon steel casting . . . 105
ductile and brittle . . . 102
interpretation of . . . 98-105
iron ingot . . . 219
low-carbon, high oxygen iron . . . 223
maraging steels . . . 384
titanium alloys . . . 454
types, fcc metals . . . 100
Tension . . . 3, 13
Tension overload fractures
AISI/SAE alloy steels . . . 304, 311-315, 334
copper alloys . . . 402
equiaxed dimples, precipitation-hardening stainless steels . . . 372
free-machining copper . . . 401
iron-base alloy . . . 460
martensitic stainless steels . . . 367
precipitation-hardening stainless steels . . . 370, 372
titanium alloys . . . 449
tool steels . . . 379
wrought aluminum alloys . . . 423, 425, 437
Tent lighting, for reflective parts . . . 88
Terminal creep, defined . . . 19
Terrace cracks, by ductile shearing . . . 139
Tertiary creep, defined . . . 19
Test exposures, fractographic . . . 86-87
Test specimen(s). See also *Fracture specimens; Specimens.*
fracture sources illustrated . . . 217
preparation/preservation . . . 72-77
Texture
dimples as . . . 399-400
fracture, photographing . . . 82-85
high-purity copper, temperature and stress effects . . . 399
Thermal cycling, superalloys . . . 394
Thermal embrittlement, examination/interpretation . . . 136
Thermal evaporation films, SEM specimens . . . 172-173
Thermal molding, for replicas . . . 181
Thermal stress, alloy steels . . . 291
Thermal-wave imaging, SEM . . . 169
Thermocompression ball bonds
defect between gold wire/vacuum-deposited aluminum . . . 484
intermetallic formation effects . . . 485
Thick single-stage plastic replicas, formation . . . 180-181
Thin films, carbon . . . 173
Thin-foil specimens, direct observation . . . 179
Thinning, in polymers . . . 479
Thin single-stage plastic replicas, formation . . . 179-180
Thiobacillus bacteria, low-carbon steel fracture from . . . 245
Thread defects, AISI/SAE alloy steels . . . 318
Three-point bending, nickel alloys . . . 396
Threshold crack tip stress intensity factor, effect on SCC . . . 24-25
Through-wall cracking, austenitic stainless steels . . . 354
Tilt. See also *Tilt boundary; Tilt-twist boundary.*
in area measurement . . . 208
method, stereo SEM . . . 171
specimen, effect in SEM imaging . . . 168
Tilt boundary
in Armco iron . . . 18
defined . . . 13
schematic . . . 17
Tilting stages, photographic . . . 88
Tilt method, stereo SEM . . . 171
Tilt-twist boundary
defined . . . 13
in iron . . . 17
Tin, as low-melting embrittler . . . 29
Tin-bronze alloys, preferential corrosion . . . 403
Tire tracks
AISI/SAE alloy steels . . . 298
defined . . . 18
on fatigue fracture surface . . . 23
polymer . . . 479
on quenched-and-tempered steel . . . 23
and striations, compared . . . 119
Titanium. See also *Titanium alloys; Titanium alloys, specific types.*
commercially pure . . . 16, 20, 48-49, 57, 65, 171
effect of stress intensity factor range on fatigue crack growth rate . . . 57
effect of vacuum on fatigue . . . 48-49
elongated dimples on shear fracture surface . . . 16

fatigue striations 20
hydrogen embrittlement 23
pure, furrow-type fatigue fracture 44, 54
stereo pair dimples 171
weld fracture from incomplete fusion 65
Titanium alloys
effect of vacuum 47-48
embrittlement by low-melting alloys 29
fatigue striations in 176
fractographs 441-455
fracture/failure causes illustrated 217
fracture topography 441-443
hydrogen embrittlement 23, 32, 124
nitric acid cleaning 75
SCC of 28-29
Titanium alloys, specific types
7075-T6, low-melting metal embrittlement
 fracture 30, 38
IMI 155 (British commercially pure), effect of
 stress intensity factor range on fatigue crack
 growth rate 57
IMI-685, effect of frequency and wave
 form on fatigue properties 60-62
Ti-0.35O, fluting and cleavage 27
Ti-10V-2Fe-3Al, carpet plot by
 stereophotogrammetry 198
Ti-10V-2Fe-3Al, fractured, SEM stereo pair,
 carpet plot, and contour plot 172
Ti-24V, true profile length values 200
Ti-28V, true profile length values 200
Ti-3.9Mn, tensile fracture 454
Ti-3.9Mn, voids with particles 455
Ti-5Al-2.5Sn, tensile brittle fracture 454
Ti-5Al-2.5Sn, effect of frequency and wave
 form on fatigue properties 60-62
Ti-5Al-2.5Sn, hydrogen-embrittled 23, 32
Ti-5Mo-4.5Al-1.5Cr, tensile ductile
 fracture 455
Ti-6A-4V, effect of heat treatment and
 microstructure on fracture appearance 32
Ti-6A-4V, hydrogen embrittled 32
Ti-6Al-2Nb-1Ta-0.8Mo, effect of temperature
 on ductility 35, 48
Ti-6Al-2Nb-1Ta-1Mo laser beam weld, ductile
 fracture 443
Ti-6Al-2Sn-4Zr-2Mo-0.1Si, fatigue crack
 growth fracture topography 441-443
Ti-6Al-2Sn-4Zr-6Mo, effect of frequency and
 wave form on fatigue properties 62
Ti-6Al-2Sn-4Zr-6Mo, high-temperature effect,
 overload fracture 35, 49
Ti-6Al-4V, brittle fracture surface 448
Ti-6Al-4V, cup-and-cone fracture 451
Ti-6Al-4V, dimple size and secondary
 cracking 451
Ti-6Al-4V, effect of biaxial tension on dimple
 rupture 31, 39
Ti-6Al-4V, effect of frequency and wave form
 on fatigue properties 62
Ti-6Al-4V, effect of stress intensity factor range
 on fatigue crack growth rate 57
Ti-6Al-4V ELI, ductile overload fracture 443
Ti-6Al-4V, feathery fracture surface 450
Ti-6Al-4V, grain-boundary cracking 449
Ti-6Al-4V, high-cycle fatigue fracture 452
Ti-6Al-4V, low-cycle fatigue fracture 452
Ti-6Al-4V, radial fractures 446
Ti-6Al-4V, secondary cracks 448-449
Ti-6Al-4V, shear band fracture 444-445
Ti-6Al-4V STA alloy rocket motor, adiabatic
 shear bands 43
Ti-6Al-4V, stress-corrosion cracking 452
Ti-6Al-4V, tensile overload fracture ... 449, 450
Ti-6Al-4V, tool mark fracture 447
Ti-6Al-4V, TTS fracture 29
Ti-6Al-5Zr-0.5Mo-0.25Si, striation
 spacing 60, 61
Ti-6Al-6V-2Sn, effect of elevated temperature
 on overload fracture 35, 50
Ti-6Al-6V-2Sn, effect of stress intensity factor
 range on fatigue crack growth rate 57
Ti-7Al-1Mo-1V, fracture surface with tear
 ridge 453

Ti-7Al-2Nb-1Ta, cleavage fracture 453
Ti-7Al-2Nb-1Ta, staining and large
 dimples 453
Ti-7Al-2Nb-1Ta, stress-corrosion
 cracking 453
Ti-8Al-1Mo-1V, fluting 27, 28
Ti-8Al-1Mo-1V, hydrogen-embrittled 23, 32
Ti-8Al-1Mo-1V, SEM and TEM fractographs,
 compared 189-191
Ti-8Al-1Mo, flutes and cleavage 27
Toluene, as organic cleaning solvent 74
Tongues
AISI/SAE alloy steels 341
and cleavage facets, compared 424
on cleavage fracture surface 13, 252
formation 13
iron 219, 224, 457
low-carbon steels 252
medium-carbon steels 267
micro-, in ductile irons 232
pyramid-shaped, nickel alloys 397
on steel weld metal 17
Tool marks
AISI/SAE alloy steels 296
cobalt alloy 398
as fracture origin, medium-carbon steels 258
titanium alloys 446-447
wrought aluminum alloys 419
Tool steels. See also *Tool steels, specific types.*
early fractographs 5
fractographs 375-382
fracture/failure causes illustrated 217
fracture tests for 141, 162
high-carbon 141, 162
low-alloy, quench cracking fracture 375
quench cracking in 130
Shepherd P-F test for 141, 162
Tool steels, specific types
234, quasi-cleavage fracture 26
234 saw disk, dimple rupture 14
AISI A4N mating fracture surface, fibrous
 zones 376
AISI D2, brittle in-service failure 376
AISI H11, high-cycle fatigue fracture 381
AISI H11, hydrogen embrittlement 381-382
AISI H11, low-cycle fatigue fracture 380
AISI H11, tension overload fracture 379
AISI L6 high-nickel, fracture surfaces 377
AISI L6, impact fracture 377
AISI L6 low-carbon, cup-and-cone
 fracture 377
AISI M2, cracking from improper
 rebuilding 375
AISI M3, impact fracture 378
AISI M7, cracking from improper
 rebuilding 375
AISI S1, low-cycle fatigue fracture 377
AISI T1, hammer-burst fracture 378
AISI T1, impact fracture 378
AISI T2 high-speed, TEM and SEM
 fractographs, compared 186
AISI W1, fatigue fracture surfaces 376
AISI W1, simulated service fractures 376
AISI W2, catastrophic failure surface 377
Topographic index, as surface roughness
 parameter 201
Torsion
deformation, alloy steels 304
high-cycle fatigue, AISI/SAE alloy
 steels 296
historical studies 3
Torsion fractures
brittle, medium-carbon steels 273
elongated dimples 16
fatigue, alloy steels 323
fatigue, high-carbon steels 282
fatigue, medium-carbon steels 253
low-carbon steel 250
medium-carbon steels 253, 268
overload, alloy steels 330
overload, austenitic stainless steels 359
overload, high-carbon steels 278
Tortuosity, crack path 38, 206

Toughness. See also *Fracture toughness.*
inadequate, ASTM/ASME alloy steels 347
maraging steels 385
Transcrystalline cleavage, and intergranular
 rupture, iron 222
Transformation stress cracks, service fracture
 by 65
Transgranular cleavage fracture(s). See also
 *Cleavage fracture(s); Transgranular
 fracture(s).* 175
AISI/SAE alloy steels 301
cemented carbides 470
high-carbon steel 290
low-carbon steel 240, 243
molybdenum alloy 464
nickel alloys 396
titanium alloys 453
Transgranular fracture(s)
austenitic stainless steels 355
austenitizing effects 339
chloride SCC in austenitic stainless steels ... 357
copper alloys 403
engineering alloys 12
fatigue 175-176
high-purity copper 399-400
at low temperatures 121
radial marks, SEM fractographs 169
stress-corrosion, cleavage steps 18
superalloys 389
thumbprint, superalloys 395
Transgranular stress corrosion, cleavage
 fracture by 18
Transition temperature. See also *Temperature.*
ductile-to-brittle 34
Transmission electron microscope(s)
history 5-6
literature 7
replication techniques 7
specimens for 6-7
Transmission electron microscopy 179-192
artifacts in replicas 184-185
cleaning fracture surfaces 179-183
fractographs, and SEM fractographs,
 compared 185-192
fractographs, for stereo viewing 192
fractographs, method of preparation ... 185, 192
material types 216
replication, fracture surface
 preparation 179-183
replication procedures 94-95
and SEM fractographs, compared 185-192
shadowing of replicas 183-184
specimen preparation 179
Transverse fractures
cast aluminum alloys 411-413
martensitic stainless steels 369
shear, high-strength steel, fractograph 174
Transverse sectioning, for TEM 179
Triangular elements, for fracture surface
 area 201-202
Triaxial stress
effect on fracture mode 30-31
in stainless steel 31, 40
Trichloroethylene, as carcinogenic cleaning
 agent 74
Triple-point cracking. See also *Wedge
 cracks.*
austenitic stainless steels 364
in fatigue 119
intergranular creep rupture by 19, 25
Tripod, photographic 89
True area. See also *Area.*
of fracture surface, importance 211
True length. See also *Length.*
defined 199-200
fractal analysis 212
TTS fractures. See *Tearing topography surface
 fracture(s).*
Tungsten. See also *Tungsten fibers.*
-aluminum metal-matrix composites, splitting
 fracture 467
crystal, cleavage planes 462
embrittlement sources 123

Tungsten (continued)
  Monte Carlo electron trajectories in 167
  sintered, intergranular fracture 462
  -steel metal-matrix composites, splitting
    fractures 467
Tungsten fibers. See also *Fibers; Tungsten.*
  brittle transgranular failure 466
  ductile-to-brittle transition effects 467
  embrittlement 466
  metal-matrix composites 466
  splitting fracture 467
Turbomolecular pumps, for SEM vacuum
    system 171
Twin boundaries, austenitic stainless
    steels 351, 356
Twin-matrix interface, iron cleavage fracture
    along 224
Twinning
  austenitic stainless steels 355
  formation, low-carbon steel 252
  light microscopy for 106
  mechanical, and cleavage 4
  parting, austenitic stainless steels 356
Twist boundary
  cleavage steps from 319
  defined 13
  low-carbon steel 252
  molybdenum alloy 464
  schematic 17
Two-stage replicas, technique 7, 182
Two-step plastic-carbon technique, TEM
    replication 7

U

Ultralow-carbon steel, temperature effect on
    fracture mode 34, 46
Ultramicrotome, replica cutting by 199
Ultramicrotomy, diamond knife 179
Ultrasonic agitation, organic solvents 74
Ultrasonic cleaning
  of corroded surface 74
  extraction replica 183
  of fatigue precrack region 75
  organic solvents 74
  for surface coating removal 73
Ultraviolet illumination, effects 84-85
Underbead crack, weld metal 155
Unfavorable grain flow. See *Grain flow.*
Uniaxial stress, effect on fracture mode 30-31
Uniform loading, effect on cracking 176
Unstable fast fracture, as fatigue stage 175
Upsetting stresses, tool steel fracture by 377
Urban environment, effect on tin-bronze
    alloys 403

V

Vacancy coalescence, alloy steels 349
Vacuum
  and air, fatigue fractures compared 48, 55
  effect, aluminum alloys 46
  effect, Astroloy 48
  effect, fatigue 36, 46-49, 120
  effect, Inconel X-750 48
  effect, titanium alloys 47-48
  effect, stainless steel 48
  as fatigue environment 35
  induction melting, irons 219
  nickel alloy cracking in 396
  system, SEM imaging 171
  vapor deposition, for replica shadowing 180
Vacuum bell jar, for carbon thin films 173
Vacuum-deposited aluminum, as electronic
    material defects 484, 486-487
Vacuum deposition 180, 484-487
Valve springs, failure from seam 64
Vanadium-niobium alloy, overheating 145
Vapor
  bubbles 180-181
  deposition, aluminum 484-487

water 40, 52
Vertical lighting
  and oblique lighting, compared 87
  photomacrographic 83
Vertical roughness parameter, defined 200
Vertical sectioning, for true fracture surface
    area 198-199, 211-212
Vibration
  effect, low-carbon steel 245
  effect, crack growth 109
View camera
  fractographic 78-79, 85
  with scanning light photomacrography
    system 82
Visual examination
  embrittlement phenomena 123-137
  fractographic 78
  interpretation of fractures 96-123
  and light microscopy 91-165
  preliminary, of specimen 72-73
  quality control applications 140-143
  sequence for fractured components 92
  techniques 91-93
  weld cracking 137-140
Vitreous fractures, studies 2
Voids
  alloy steels 339
  black, at grain boundaries 346
  cast aluminum alloys 408
  dimple formation along 338
  initiation, titanium alloys 445
  irradiated materials 365
  maraging steels 383
  nodule-nucleated, ductile irons 236, 237
  noninterconnecting, sulfur concrete 472
  -nucleating sites, sulfide inclusions as 14
  with particles, titanium alloys 455
  precipitation-hardening stainless steels 370
  rounded 445
  r-type cavities as 122, 140
  shrinkage 67
  titanium alloys 445, 452, 455
  tool steels 379
Voltage, accelerating, in SEM 167
Volume, surface, as roughness parameter 201

W

Wallner lines
  cemented carbides 470
  formation 13-14
  and striations, compared 119
  tool steels 378
  on WC-Co fracture surface 18
Water
  corrosion, precipitation-hardening stainless
    steels 373
  distilled, as corrosive 24, 438
  high-purity deaerated 438
  tap, as corrosive 24, 320
  vapor, effect on fatigue fracture
    appearance 40, 52
Water-base detergent cleaning, of
    fractures 74
Wave form
  defined 59
  effect on fatigue 58-63
  and frequency 58-63
  strain, in fatigue testing 62
**Wavelength-dispersive x-ray
    spectroscopy** 168
Wax technique, for TEM replicas 185
Wedge cracks. See also *Triple-point
    cracking.*
  austenitic stainless steels 364
  formation, in decohesive rupture 19, 25-26
  as intergranular 121-123
Wedge tests, application 141, 161
Weld cracking. See also *Weld(s).*
  cold cracking 137-138
  examination/interpretation 72, 137-140
  hot cracking 138

lamellar tearing 138-139
  stress-relief cracking 139-140
Weld(s)
  defect, effect in medium-carbon steel 258
  deposit, as crack origin 254, 268
  flange, fractured medium-carbon steel 256
  fracture, titanium, from incomplete fusion ... 65
  HAZ, striations 21
  laser beam, ductile fracture 422
  penetration, inadequate, as discontinuity 65
  photo effects of lighting on 87-88
  resistance spot, fatigue fracture 66, 67
  robotic 375
  toe, hardness, fractured medium-carbon
    steel 256
  underbead crack 155
White irons
  fractographs 238-239
  fracture/failure causes illustrated 216
  high-chromium, shear/tensile stress
    cracking 239
White spots, and shrinkage cavity, iron 220
White zone, high-carbon steel 288
Widmanstätten structures, titanium
    alloys 23, 443
Width, of dimples, measurement 207
Wollaston prism, in fatigue study 121
Woody fractures
  alloy steels 319, 333
  appearance, in tensile fractures 104
  high-carbon steels 281
  studies 1-3
  in iron 224
  surface, tool steels 375
  in wrought iron 224
Wrought aluminum alloys
  cliffs 418
  extrusions, fibrous interior structure 416
  fractographs 414-439
  fracture/failure causes illustrated 217
  superelastic eutectic, tensile-overload
    fracture 437
Wrought aluminum alloys, specific types. See
    also *Wrought aluminum alloys.*
  1100, fractured by gas explosion 414
  2014-T6, fatigue fracture 415-416
  2014-T6 heat-treated forging, fatigue
    fracture 417
  2024-T3, dimple rupture 417
  2024-T3, fatigue fracture 418
  2025-T6, corrosion pit and tool mark
    fracture 419
  2025-T6, fatigue failure by inadequate shot
    peening 420
  5456, laser beam weld, ductile fracture 422
  7050-T7, fatigue fracture from cyclic
    stress 421
  7075-T6, brittle fracture 424
  7075-T6, cleavage fracture 424, 430
  7075-T6, cold shut fracture 435
  7075-T6, cone-shaped fracture surface 426
  7075-T6, corrosion and leaves 433
  7075-T6, corrosion-fatigue fracture 430-433
  7075-T6, corrosion fatigue fracture, ductile
    striations 431, 432
  7075-T6, corrosion fatigue with brittle
    striations 430, 432
  7075-T6, fatigue fracture 427, 429
  7075-T6, fatigue fracture by cyclic
    stress 430
  7075-T6, high-cycle fatigue fracture 426
  7075-T6, intergranular and stress-corrosion
    cracking 434
  7075-T6, intergranular fracture 431
  7075-T6, solidification porosity 431
  7075-T6, stress-corrosion cracking 433-436
  7075-T6, stretching and serpentine
    glide 434
  7075-T6, tension-overload fracture 423-425
  7075-T6, tension-overload plane-strain
    fracture 423
  7075-T736 forging, effect of peening on dross
    inclusion 422

7175-T736 forging, fatigue fracture by dross inclusion ... 422
7475-T6, corrosion fatigue fracture ... 432
67Al-33Cu, tensile overload fracture ... 437
Al-5.6Zn-1.9Mg, corrosion-fatigue fracture ... 438
Al-5.6Zn-1.9Mg, fatigue crack propagation ... 439
Al-5.6Zn-1.9Mg, fatigue crack tip deformation ... 438
Al-5.6Zn-1.9Mg, transgranular corrosion fatigue crack propagation ... 438
Wrought iron, woody impact fracture ... 224
Wrought products, quality control tests ... 141-142

W-type cracks. See *Triple-point cracking; Wedge cracks.*

X

X-ray diffraction, studies ... 4
X-ray fluorescent analysis, low-carbon steel ... 249
X-rays
  characteristic, defined ... 168
  maps ... 168, 473
  origin, in Bohr atom model ... 168
  signals, SEM ... 168
Xylene, as organic cleaning solvent ... 74

Y

Yellow (K2) photographic lens filter, with ultraviolet illumination ... 84

Z

Zone
  of strain localization, medium-carbon steel ... 42
  transformed, medium-carbon steel ... 42